HIGHWAY DESIGN
AND TRAFFIC SAFETY
ENGINEERING
HANDBOOK

HIGHWAY DESIGN AND TRAFFIC SAFETY ENGINEERING HANDBOOK

Ruediger Lamm

University of Karlsruhe
Karlsruhe, Germany

Basil Psarianos

National Technical University of Athens
Athens, Greece

Theodor Mailaender

Mailaender Ingenieur Consult
Karlsruhe, Germany

McGRAW-HILL

New York San Francisco Washington, D.C. Auckland Bogotá
Caracas Lisbon London Madrid Mexico City Milan
Montreal New Delhi San Juan Singapore
Sydney Tokyo Toronto

Library of Congress Cataloging-in-Publication Data

Lamm, Ruediger.
    Highway design and traffic safety engineering handbook / Ruediger
Lamm, Basil Psarianos, Theodor Mailaender.
        p. cm.
    Includes bibliographical references and index.
    ISBN 0-07-038295-6
      1. Roads—Design and construction.   2. Roads—Safety measures.
3. Traffic safety.   I. Psarianos, Basil.   II. Mailaender, Theodor.
III. Title.
TE175.L36   1999
625.7′25—dc21   98-41928
                                             CIP

McGraw-Hill

A Division of The McGraw·Hill Companies

2 3 4 5 6 7 8 9 0   DOC/DOC   9 0 3 2 1 0 9

ISBN 0-07-038295-6

*The sponsoring editor for this book was Larry S. Hager, the editing supervisor
was Paul R. Sobel, and the production supervisor was Clare B. Stanley. It was
set in Times Roman by Renee Lipton of McGraw-Hill's Professional Book
Group composition unit.*

Printed and bound by R. R. Donnelley & Sons Company.

McGraw-Hill books are available at special quantity discounts to use as premi-
ums and sales promotions, or for use in corporate training programs. For more
information, please write to the Director of Special Sales, McGraw-Hill, 11 West
19th Street, New York, NY 10011. Or contact your local bookstore.

This book is printed on acid-free paper.

*The main author dedicates this book
to the memory of his father Fritz Lamm*

and to

*Professor Emeritus
Dr.-Ing. Dr.h.c. Dr.-Ing.E.h. Dr.-Ing.E.h.D.sc. (hon),
Bundesminister a.D. Hans Leussink*

Associate Authors:

Elias M. Choueiri*
Ministry of Transport, Beirut, Lebanon

Ralf Heger
Dresden University of Technology, Dresden, Germany

Rico Steyer
Dresden University of Technology, Dresden, Germany

Language Editors:

John C. Hayward
Michael Baker, Jr., Inc., Pittsburgh, PA, U.S.A.

Elias M. Choueiri*
Ministry of Transport, Beirut, Lebanon

Jeffrey A. Quay
Michael Baker, Jr., Inc., Pittsburgh, PA, U.S.A.

---

*He is Director-General in the Ministry of Transport, Beirut, Lebanon. (He has held several positions at U.S. and Lebanese government agencies, and at public and private universities, colleges, and research organizations. He has published numerous articles, and presented many papers at international conferences.)

CONTENTS*

Part 1   Network (NW)

Part 2   Alignment of Nonbuilt-Up Roads (AL)

*READERS: Please note in the "Contents" only the 2-digit and 3-digit section numbers are listed, whereas the book contains a subdivision into 4-digit numbers.

Chapter 6. Overview   6.1

Chapter 7. Basic Procedure in Road Planning and Design with Special Emphasis on Environmental Protection Issues   7.1

Chapter 8. Relevant Speeds   8.1

Chapter 9. Safety Criteria I and II   9.1

Chapter 10. Driving Dynamics and Safety Criterion III   **10.1**

Chapter 11. General Alignment Issues with Respect to Safety   **11.1**

Chapter 12. Horizontal Alignment   **12.1**

Chapter 13. Vertical Alignment   **13.1**

Chapter 19. Human Factors 19.1

Chapter 20. Road Safety Worldwide 20.1

Chapter 21. Summary of Part 2 "Alignment" 21.1

Part 3 Cross Sections of Nonbuilt-Up Roads (CS)

Chapter 22. Methodical Procedure 22.3

Chapter 23. Overview 23.1

Chapter 24. Fundamentals for the Dimensions of the Cross-Sectional Design Elements 24.1

Chapter 25. Cross Section Design   **25.1**

Chapter 26. Summary of Part 3 "Cross Sections"   **26.1**

General Conclusion of Parts 1 to 3

PREFACE

The aim of this book is to provide a comprehensive presentation about the interrelationships between highway design, driving behavior, driving dynamics, and traffic safety. To do this, recent knowledge and practical experience had to be collected, classified, and arranged to permit practical application in all relevant fields of highway design and traffic safety.

Roadway construction has not, and will not, match the pace of traffic growth worldwide. Highway engineering under these conditions requires explicit considerations of traffic safety, as well as environmental and spatial compatibility. Thus, special emphasis has been given in this book to the discussion, analysis and evaluation of qualitative safety characteristics, and to the development of quantitative safety criteria, in order to achieve

- design consistency,
- operating speed consistency, and
- driving dynamic consistency

in highway design.

The book is organized around three important highway engineering parts: Part 1: Network (NW), Part 2: Alignment (AL), and Part 3: Cross Sections (CS).

Most of the main chapters in Part 2 and Part 3 are subdivided into two subchapters:

1. Recommendations for Practical Design Tasks; and
2. General Considerations, Research Evaluations, Guideline Comparisons and New Developments.

Subchapters 1 and 2 have to be regarded as a unit.

This enables the reader to focus on the "how" in Subchapter 1 and to understand the "why" in Subchapter 2. Both subchapters are organized so that they mostly cover identical subjects.

Recommendations for Practical Design Tasks in Subchapter 1 are valid in many countries, based on the research of the authors, as elaborated in Subchapter 2.

Subchapter 2, "General Considerations, Research Evaluations, Guideline Comparisons and New Developments," consists of discussions of background research, practical acknowledged experiences, evaluations of numerous existing highway geometric design guidelines, compilations and comparisons of limiting and standard values for relevant design elements in different countries, as well as new developments. The examination, revision and elaboration of proposals for rules in highway geometric engineering also draws upon an in-depth literature review. This review covered national and international journals, proceedings, and technical research reports.

Generally speaking, modern Highway Geometric Design Guidelines from many countries (for example: Australia[37,507], Austria[35,36], Canada[608], United Kingdom[139], France[700], Greece[453,454,456], South Africa[663], Sweden[686], and Switzerland[687-693] have influenced the development of the book. In particular the American Policy on Geometric Design of Highways and Streets[5] and the German Guidelines for the Design of Roads[243,244,246,247] were drawn upon extensively.

The book is based, to a great extent, on research—examining highway engineering with special emphasis on geometric design and traffic safety—conducted over the last two decades (particularly by the principal author) which has led to the development of three Safety Criteria for distinguishing good, fair, and poor design practices on newly designed or existing roadway sections and a Safety Module for evaluating road networks. It takes into account numerous research studies from international colleagues and organizations, such as, for example, References: Brannolte et al.[68,69], Brilon/Weiser[77], Durth/Lippold[159], Hall et al.[257], Harwood et al.[268], HUK[30], Krammes et al.[359], Krammes/Garnham[361], Lay[466], Leutzbach/Zoellmer[476], Lunenfeld/Alexander[489], McLean/Morrall[508], Messer[512], O'Cinnéide[553], Steierwald/Buck[674], TRB[710], U.S. DOT[728,729], Wegman[746], Zegeer et al.[774–783].

Taken together, Subchapters 1 and 2 present specific recommendations and research under-pinnings for all relevant geometric design fields and provide the most current and extensive infor-mation source available treating highway design and traffic safety engineering.

One important aim of the book is that it is not solely directed to new designs—the basic con-cept of most international guidelines—but that it is also strongly related to the safety evaluation of existing (old) alignments, in order to give the responsible authorities qualitative and quantita-tive information about appropriate countermeasures. Note that the existing alignments, it applies to, are estimated to make up at least 70 to 80 percent of the rural road networks worldwide.

Development of both Subchapters 1 and 2 followed the same general logic:

- Organization by subject of the investigations corresponding to the individual chapters of the two main Parts "Alignment" and "Cross Sections."
- Arrangement of essential research results based on the literature review as they applied to the individual part.
- Discussion and comparison of recommendations.
- Evaluation of different rules and contradictions with other new design concepts.
- Comparison of design fundamentals as well as alignment and cross-sectional elements of different guidelines with those of the developed parts.
- Understanding and explanation of the entire design process with respect to how the design elements and element sequences or element superimpositions actually affect driving behav-ior and accident characteristics.
- Suggestions for improvement and possible changes regarding the design rules of today.
- Elaboration of quantitative safety evaluation processes, in order to fill the gaps which exist between the desired safety level according to the design guidelines of the countries under study and the actual physical safety which prevails on the road.

Important references, which support the content of the individual chapter or subchapter are listed under the respective headline or are introduced at the beginning and at the end of the cited material from other authors, guidelines or research reports.

Special chapters dealing with traffic safety issues worldwide, human factors, and vehicular safety are included in Part 2 "Alignment."

The book is written for a national and international audience, consisting of:

Academia

- The book is valuable to all educators and students in graduate-level courses on highway geometric design, traffic safety related to highway design, or road surveying. It is also usable for undergraduate study as a senior-level elective course. The book analyzes and gives a complete and detailed insight to all aspects of highway geometric design and the corresponding relationships with traffic safety.
- Colleagues who might benefit from using it as an extensive desk-top text, supplementary and reference book in their classrooms, or as a background tool for research purposes.
- Students who would need it for practical application and as background information to clarify facts in the subject area.

University Research and Research Institutions

- The content of the book can be directly implemented in all post-graduate studies and research with respect to highway design and safety-related programs.
- Scientists who would use it as a basis for their research work because it provides a world-wide view of current research and standards.
- Post-graduate students who would use it as a unique source for evaluating the current state of the art in order to accomplish future research developments.

National and International Consultings

- Consultants who would use the book to estimate the safety level of highway geometric design tasks, to develop cost-effective alternatives, and to design environmentally friendly roads.
- Consultants that work internationally should be particularly interested, since the recommended design rules were elaborated on an international basis. The same is true for internationally interested researchers and academics.

Highway Agencies at Federal, State, or Local Level

- Highway agencies that would use the book to evaluate the overall road network or specific roadway sections with respect to quantitative and qualitative safety issues, to determine whether or not certain measures for new designs, redesigns or rehabilitation strategies are safety effective and thus to use it as the basis which would aid them in obtaining government funds.

Highway Engineering Professionals in General

- All highway engineers concerned with the design of roads or acting as safety auditors for a third party would use the book as a profound reference for implementing sound design decisions and for conducting safety controls.

Policy Makers and Researchers Focusing on Updating Geometric Guidelines

- The book is valuable for the authorities of those countries all over the world which lack up-to-date guidelines for the design of roads, since the book proposes concrete solutions for all individual design problems, based on recent research and experience worldwide, considering at the same time important safety issues.
- This also applies even to the authorities of those countries, which have a tradition in developing guidelines or policies for the design of highways. They can benefit from the content of the book, since information from all over the world will be at their disposal. This will help them to examine, compare, revise, and update existing guidelines, if necessary, and learn from others' experience, judgment, and research.

In general, the book is an invaluable source of information for scientists, consultants, highway agencies, educators, and students in the field of highway design and traffic safety engineering.

Although this handbook is believed to be correct at the time of its printing, it does not accept responsibility for any consequences arising from the use of the information contained in it. People using the information contained in the book should apply, and rely upon their own skill and judgment to the particular issue which they are considering.

Ruediger Lamm

ACKNOWLEDGMENTS

Special thanks are extended to the following organizations and individuals which/who supported, directly or indirectly, the development of this book:

Institute for Highway and Railroad Engineering, University of Karlsruhe, Karlsruhe, Germany. Directors of the Institute: Professor Emeritus Dr.-Ing. H.G. Krebs (1966–1981)†, Professor Emeritus Dr.-Ing. Dr.h.c. E.-U. Hiersche (1982–1996), Univ.-Prof. Dr.-Ing. R. Roos (1996–Present).

Institute for Transport Studies, University of Karlsruhe, Karlsruhe, Germany. Director of the Institute: Professor Emeritus Dr.-Ing., Dr.-Ing. E.h. W. Leutzbach (1962–1991).

Committee: Guidelines for the Design of Roads (RAS), Part: "Alignment (RAS-L)" of the German Road and Transportation Research Association (Membership "Lamm": 1964–1984). Chairmen: Dr.-Ing. Dr.-Ing. E.h. G. Koeppel (1964–1978), Professor Emeritus Dr.-Ing. W. Durth (1978–1997).

Committee: Guidelines for the Design of Roads (RAS), Part: "Cross Sections (RAS-Q)" of the German Road and Transportation Research Association. Chairmen: Chief Director H. Schliesing (1978–1988), Univ.-Prof. Dr.-Ing. W. Brilon (1989–Present).

Committee on Geometric Design (A2A02) of the National Research Council, Transportation Research Board, U.S.A., (Membership "Lamm": 1983–1997). Chairmen: Mr. G.M. Nairn, Jr. (1979–1985), Dr. J.C. Glennon (1985–1991), Prof. Dr. J.M. Mason, Jr. (1991–1997), Prof. Dr. D.B. Fambro (1997–Present).

Consulting Agencies:
Mailaender Ingenieur Consult GmbH, Karlsruhe, Germany, President: Dipl.-Ing. T. Mailaender, Vice President: Dipl.-Ing. K. Brust.

AKG Software Consulting GmbH, Ballrechten-Dottingen, Germany, President: Dipl.-Ing. A.K. Guenther.

CTI Engineering Co., Ltd., Tokyo, Japan, President: Y. Ishii, Manager: S. Kakido.

NAMA, Consulting Engineers and Planners SA, Athens, Greece, Sector Director for Transportation: G. Soilemezoglou.

Michael Baker Jr. Inc., Pittsburgh, PA, U.S.A., President: Dr. J.C. Hayward.

German Railroad Inc., Division "Network," Project Center: Southwest, Karlsruhe, Germany, Director: Dipl.-Ing. A. Samaras.

Foundations, Universities, and Ministries:
International Road Federation, Washington, D.C., U.S.A.; National Science Foundation, Washington, D.C., U.S.A.; State University of New York Research Foundation, Albany, NY, U.S.A.; National Technical University of Athens, Athens, Greece; Texas Transportation Institute, The Texas A&M University System, College Station, TX, U.S.A.; Ministry for Environment, Regional Planning, and Public Works, Athens, Greece; Regional Offices of the New York State Department of Transportation: Albany, Rochester, Syracuse, and Watertown, NY, U.S.A.; Ohio Department of Transportation, Columbus, Ohio, U.S.A.; Ministry for Economy, Traffic and Regional Development, Wiesbaden, State of Hessen, Germany. Regional Offices of the Ministry of Environment and Transportation, State of Baden-Wuerttemberg, Germany: Freiburg, Karlsruhe, Stuttgart, Tuebingen, and the following Highway Departments: Bad Mergentheim, Bad Saeckingen, Besigheim, Calw, Donaueschingen, Ehingen, Ellwangen, Freiburg, Heidelberg, Heilbronn, Kaiserslautern (State of Rhineland-Palatinate), Karlsruhe, Kirchheim, Munich (State

of Bavaria), Offenburg, Reutlingen, Riedlingen, Schorndorf, Schwaebisch Hall, and Weilheim (State of Bavaria).

Special thanks are extended: Margrita Wiedemann for her excellent support in organizing and typing the manuscript; to Martin R. Lamm for his continuous support in translating and reviewing the manuscript; to Dipl.-Ing. Olaf Eberhard for his scientific and organizational support; to Dipl.-Ing. Hans Messmer for data processing; and to the technical drawers: Gudrun Broeker, Peter Hack, Guiseppe Biundo.

Finally, I thank the following Associations and Colleagues for granting Copyright permission, from which important contributions were used: "From *A Policy on Geometric Design of Highways and Streets,* Copyright 1990 by the American Association of State Highway and Transportation Officials, Washington, D.C. Used by permission." Austroads Incorporated "*Rural Road Design Guide, 1989,*" Haymarket NSW, Australia. Research Association for Traffic- and Road Engineering "*Guidelines for the Alignment (RVS 3.23) and for Construction (RVS 3.31),*" Vienna, Austria. Road and Transportation Research Association "*Guidelines for the Design of Roads (RAS),*" Cologne, Germany. Swiss Association of Road Specialists (VSS) "*Various Swiss Norms (SNV 640...),*" Zuerich, Switzerland. D.W. Harwood, U.S.A., Portions of Sections 15.2.2 and 15.2.3, Part 2: Alignment. R.A. Krammes, U.S.A. and M. Garnham, England, Section 11.2.2, Part 2: Alignment. M.G. Lay, Australia, Portions of Section 5.3, Part 2: Alignment. J.A. Reagan and W.A. Stimpson, U.S.A., Section 10.2.1.3, Part 2: Alignment. F.C.M. Wegman, The Netherlands, Section 20.1, Part 2: Alignment.

SI* (MODERN METRIC) CONVERSION FACTORS

Conversion from SI Units	Conversion to SI Units
Lengths	**Lengths**
1 cm = 0.3937 in	1 in = 2.54 cm
1 m = 3.2808 ft	1 ft = 0.3048 m
1 km = 0.6214 mi	1 mi = 1.6093 km
Areas	**Areas**
1 cm^2 = 0.1550 in^2	1 in^2 = 6.4516 cm^2
1 m^2 = 10.7639 ft^2	1 ft^2 = 0.0929 m^2
1 km^2 = 0.3861 mi^2	1 mi^2 = 2.590 km^2
Volumes	**Volumes**
1 l = 0.2642 gal	1 gal = 3.7854 l
1 l = 0.035313 ft^3	1 ft^3 = 28.3169 l
1 m^3 = 35.3133 ft^3	1 ft^3 = 0.02832 m^3
Velocities	**Velocities**
1 km/h = 0.6214 mi/h	1 mi/h = 1.6093 km/h
Mass	**Mass**
1 kg = 2.2046 lb	1 lb = 0.4536 kg
Force	**Force**
1 N = 0.2248 lb	1 lb = 4.4482 N
Pressure or Stress	**Pressure or Stress**
1 N/m^2 = 0.02088 lb/ft^2	1 lb/ft^2 = 47.880 N/m^2

*SI is the Symbol for the International System of Measurement

LIST OF ACRONYMS

a	acceleration/deceleration [m/sec^2]
a, b, c, d, e, f	cross section groups
A I to A VI	road categories
A	parameter of the clothoid [m]
A_{max}	maximum parameter of the clothoid [m]
A_{min}	minimum parameter of the clothoid [m]
AADT	average annual daily traffic [veh. per day]
A, B, C, D, E	road category groups

or

A, B, C, D, E, F	level of consistency (LOC)
ACD	accident cost density
	[monetary unit per km per year of the country under study]
ACR	accident cost rate [monetary unit per 100 veh.-km]
$ACR_{R,W,M}$	relative weighted mean accident cost rate
	[monetary unit per 100 veh.-km]
$ACR_{W,M}$	weighted mean accident cost rate
	[monetary unit per 100 veh.-km]
AD	accident density [acc. per km per year]

or

AD	algebraic difference in grade [%]
A_E	parameter of the egg-shaped clothoid [m]
AG	at-grade
APF	abdomenal peak force [kN]
A_R	common clothoid parameter of the symmetrical reversing clothoid [m]
AR	accident rate [acc. per 10^6 veh.-km]
$AR_{R,W,M}$	relative weighted mean accident rate
	[acc. per 10^6 veh.-km]
$AR_{W,M}$	weighted mean accident rate
	[acc. per 10^6 veh.-km]
b	(half) vehicle width [m]
B II to B IV	road categories
C III, C IV	road categories
C	carryover factor [-]

or

C	center points of circular curves

	or
C	nominal clearance between vehicles [m]
CCR	curvature change rate of a roadway section with similar road characteristics [gon/km]
CCR_S	curvature change rate of the single circular curve with transition curves [gon/km]
$\overline{CCR_S}$	average curvature change rate of the single curve for the observed roadway section without considering tangents [gon/km]
CF	cumulative number of traffic fatalities
$C_{F/SI}$	combined average personal damage amount for fatally and seriously injured persons [monetary unit for the country under study]
C_{LI}	average personal damage amount for slightly injured persons [monetary unit]
C_{PD}	property damage costs [monetary unit]
CS	circular curve to spiral (circular curve to clothoid)
C_W	aerodynamic drag coefficient [-]
d	distance between two circular curves with respect to clothoids [m]
	or
d	distance from the edge of the traveled way to the rotation axis with respect to the superelevation runoff [m]
D IV, D V	road categories
D	wheelbase [m]
	or
D	viewing distance [m]
DC	degree of curve [deg./100ft] or [deg./100 m]
DT	distance traveled [m]
e	superelevation rate [%]
e_{max}	maximum superelevation rate [%]
e_{min}	minimum superelevation rate [%]
$(-e)$	negative superelevation rate [%]
e_b	superelevation rate at the beginning of the superelevation runoff [%]
e_e	superelevation rate at the end of the superelevation runoff [%]
e_D	superelevation deficiency to be upgraded [%]
E V, E VI	road categories
E	feature expectation factor [-]
	or
E	distance between the center of the outside rails and the edge of the pavement of the adjacent road [m]
EMVR	exclusive motor vehicle road
f	tangent offset [m]
	or
f	friction factor [-]
f_R	friction factor in radial (side) direction [-]

f_{Rmax}	maximum friction factor in radial (side) direction [-]
f_{Rperm}	maximum permissible side friction factor [-]
f_{RA}	side friction assumed [-]
f_{RD}	side friction demanded [-]
f_T	friction factor in tangential direction [-]
f_{Tmax}	maximum friction factor in tangential direction [-]
f_{Tperm}	maximum permissible tangential friction factor [-]
F	centrifugal force [N]
	or
F	force in general [N]
	or
F/F_{max}	(maximum) friction force [N]
FA	projected frontal area of passenger cars [m^2]
F_A	front overhang [m]
F_L	aerodynamic drag force [N]
F_R	radial (side) force [N]
FRK	fatality rate [fat. per 10^9 veh.-km]
F_T	tangential (longitudinal) force [N]
FTC	full trailer combination
g	acceleration of gravity [m/sec^2]
G	longitudinal grade [%]
G_{min}	minimum grade in distortion section [%]
G_{max}	maximum longitudinal grade [%]
GS	grade separated
h	height of center of gravity from road surface [m]
	or
h	height of sideslope [m]
h_1	height of eye above the road surface [m]
h_2	height of object above the road surface [m]
h_3	height of vehicle headlights above the roadway surface [m]
H	distance between the top of the rail and the upper edge of the road surface [m]
HIC	head injury criterion [-]
HL	plan view
HPC	head protection criterion [-]
ICS	intermediate cross section
IT	independent tangent
K	K-factor or rate of vertical curvature [m/%]
l	length of the front overhang [m]
L	length of the arrestor bed [m]
	or
L	length of curve or section [m] or [km]

	or
L	length of the vehicle wheelbase [m]
	or
L	length of vertical curve [m]
	or
L/L_A	length of the clothoid [m]
L_{min}	minimum lengths of circular curves [m]
	or
L_{min}	minimum tangent length between curves in the same direction of curvature [m]
L_{max}	maximum tangent length [m]
L_{cl}	length of clothoid [m]
L_{cr}	length of circular curve [m]
Ld	diverge taper length (lane addition), [m]
$L_e/L_{e\,min}$	(minimum) length of the superelevation runoff [m]
Li	merge taper length (lane drop), [m]
Lr	lateral displacement taper length [m]
LW	lane width [m]
L_Z	length of widening attainment [m]
m	vehicle mass [kg]
	or
m	median
n	number of lanes [-]
	or
n	utilization ratio of side friction [%]
NIT	non-independent tangent
OSD	opposing sight distance [m]
PSD	passing sight distance [m]
PSPF	public symphysis peak force [kN]
Q	vehicle weight force (vehicle weight), [N]
r	rolling radius [m]
	or
r	rolling resistance [%/100]
	or
r	rate of rotation of the traveled way [% per sec]
R	radius of circular curve [m]
R_{min}	minimum radius of curve [m]
ΔR	tangent offset (shift) of the clothoid [m]
R^2	coefficient of determination [-]
R'	radius of the right rear wheel trajectory [m]
R_f	workload potential rating for average conditions [-]

R_i	radius of the trajectory of the inside rear wheel [m]
R_o	radius of the trajectory of the outer edge of the front overhang [m]
RDC	rib deflection criterion [mm]
RGT	road for the general traffic
R_R/R_E	surrogate radii with respect to clothoids [m]
RR	curve radii ratio [-]
RRR	restoration, rehabilitation, resurfacing strategies
RS	recommended speed sign [km/h or mph]
R_V	radius of vertical curve [m]
R_{VC}/R_{VCmin}	(minimum) radius of crest vertical curve [m]
R_{VS}/R_{VSmin}	(minimum) radius of sag vertical curve [m]
s	paved shoulder
Δs	rate of change or relative grade of the superelevation runoff [%]
$max\Delta s$	maximum relative grade of the superelevation runoff [%]
$min\Delta s$	minimum relative grade of the superelevation runoff [%]
S_1	perception-reaction distance [m]
S_2	braking distance [m]
S	sight distance [m]
	or
S	sight distance factor [-]
	or
S	center of gravity [-]
	or
S	sum of all property and personal damages in the time period observed [monetary unit of the country under study]
SC	spiral to circular curve (clothoid to circular curve)
SCR_T	driver workload expressed by the transformed skin conductance reaction [-]
SCS	standard cross section
S_l	lateral safety space [m]
SS	spiral to spiral (clothoid to clothoid)
SSD/SSD_{min}	(minimum) stopping sight distance [m]
ST	spiral to tangent (clothoid to tangent)
S_v	vertical safety space [m]
SW	shoulder width [m]
t_R	driver perception reaction time [sec]
T	length of the investigated time period [years]
	or
T	tangent length with respect to vertical curves [m]
TL	tangent (transition) length between two successive curves [m]
TL_{max}	necessary acceleration/deceleration length to reach $V85_{Tmax}$ between curves 1 and 2 [m]

TL_{min}	necessary acceleration/deceleration length between curves 1 and 2 [m], also see TL_C
T_L	long tangent length of the clothoid [m]
TL_C	acceleration or deceleration distance between curve 1 and curve 2 [m], also see TL_{min}
TL_L	long tangent length [m]
TL_S	short tangent length [m]
TS	tangent to spiral (tangent to clothoid)
T_S	short tangent length of the clothoid [m]
TTI	thoracic trauma index [m/sec^2]
U	driver unfamiliarity factor [-]
	or
U	vehicle track width on curve [m]
v	speed [m/sec]
V	speed [km/h]
	or
V	entering velocity [km/h]
VC	viscous criterion [m/sec]
V_d	design speed [km/h]
V_f	desired travel speed [km/h or mph]
VL	profile
V_p	project speed [km/h]
V_{perm}	permissible speed or speed limit [km/h]
V_0	speed at the beginning of the braking maneuver [km/h]
V_1	speed at the end of the braking maneuver [km/h]
V15 (TR)	15th-percentile speed of trucks [km/h]
V85	85th-percentile speed of passenger cars under free flow conditions on clean, wet road surfaces [km/h]
	or
V85 (PC)	85th-percentile speed of passenger cars [km/h]
$V85_T$	85th-percentile speed in tangent [km/h]
$V85_{Tmax}$	maximum 85th-percentile speed in tangent [km/h]
$\overline{V85}$	average 85th-percentile speed for the observed roadway section without considering tangents. $\overline{V85}$ represents a good estimate for the design speed (V_d) of existing (old) alignments [km/h]
	or
$\overline{V85}$	average value of $V85_i$ and $V85_{i+1}$ [km/h]
$\Delta V85$	difference between 85th-percentile speeds of successive curves [km/h]
$\Delta V85_T$	difference between 85th-percentile speeds in tangents and curves
w	pavement width [m]
	or
w	width of sideslopes [m]

w_c	pavement width on curve [m]
w_t	pavement width on tangent [m]
Δw	pavement widening [m]
WL_l	driver workload value of the previous feature [-]
WL_n	driver workload value of the actual feature [-]
x	horizontal length in plan with respect to vertical curves [m]
x_v	horizontal length to vertex [m]
$X, Y; X_m, Y_m$	coordinates with respect to clothoids
y'	tangent offset (vertical) with respect to vertical curves
$y(x)$	height difference from the beginning of the vertical curve [m]
Z	additional clearance to compensate for difficulty of driving on curves [m]
α	angle of superelevation rate [gon]
β	deflection angle [degrees, basis 360° or 400 gon]
β_n	constants
$\gamma_i \tau_i, \alpha_i$	angular changes [gon]
γ	air density [kg/m^3]
	or
γ	central angle of curve [degrees, basis 360° or 400 gon]
ϑ	right front wheel angle [rad]
ϑ_c	initial wheel angle at the beginning of the circular curve [rad]
ζ	wheel slip [%]
κ'	curvature of the right rear wheel trajectory [m^{-1}]
μ_G	skidding value [-]
μ_p	peak friction factor [-]
τ	deflection angle of the clothoid [gon]
ϑ	rotational angle [rad]
ϕ	upper divergence angle of headlight beam [°]
ω	tire angular speed [rad/sec]

P · A · R · T · 1

NETWORK (NW)

CHAPTER 1
THE CONCEPT OF FUNCTIONAL CLASSIFICATION*

The American Association of State Highway and Transportation Officials (AASHTO)[5] reports the following on highway systems, classification, and functional relationships. The classification of highways into different operational systems, functional classes, or geometric types is necessary for communication among engineers, administrators, and the general public. Different classification schemes for road networks have been applied for different purposes in rural and urban regions in various countries. For example, see Refs. 5, 244, 304, 456, 663, 675, 686, and 700.

A complete functional design system provides a series of distinct travel movements. The six recognizable stages in most trips include main movement, transition, distribution, collection, access, and termination. Figure 1.1 shows a hypothetical highway trip using a freeway, where the main movement of vehicles is uninterrupted, high-speed flow. When vehicles are approaching their destinations using the freeway, vehicles reduce their speeds on freeway AASHTO ramps, which act as transition roadways. The vehicles then enter moderate-speed arterials (distributor facilities) that bring them nearer to the vicinity of their destination neighborhoods. Next, the vehicles enter collector roads that penetrate neighborhoods. The vehicles finally enter local access roads that provide direct approaches to individual residences or other termination points.[5] These six stages will be categorized in the following into the relevant functional levels: connector, collector, access, and local.

Functional classification is used to group streets and highways according to the type of service they are intended to provide. A schematic illustration of this basic idea, developed by AASHTO,[5] is shown in Fig. 1.2.

In Fig. 1.2a, the desired lines of travel are straight lines connecting trip origins and destinations (circles). The widths of the lines indicate the relative amount of travel desired on that line of travel. The sizes of the circles indicate the relative trip-generating and attracting power of the places shown. Because it is impractical to provide direct-line connections for every desired line, trips normally are channelized on a rural road network in the manner shown in Fig. 1.2b. Heavy travel movement is served directly or nearly directly; lighter travel movement is channeled into somewhat indirect paths. The facilities in Fig. 1.2 are labeled local access, collector, and arterial (connector), which are terms that describe their functional relationship. In this scheme, the functional hierarchy is also seen to be related to the hierarchy of trip distances, travel time, and highway geometric design levels served by the network.[5]

Associated with the idea of traffic categorization is the dual role that the highway and street network plays in providing: (1) access to property and (2) travel mobility. *Access* is a fixed

---

*Elaborated based on Ref. 5.

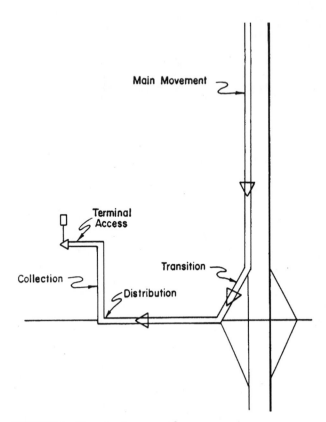

FIGURE 1.1   Hierarchy of movement.[5]

requirement of the defined area while *mobility* can be provided at varying levels of service. Mobility can incorporate several qualitative elements, such as riding comfort and absence of speed changes, but the most basic factor is operating speed or trip travel time. Local rural facilities emphasize the land access function. Arterials for main movement (connector function) emphasize the high level of mobility for through movement. Collectors approximately offer balanced service for both functions. This scheme is illustrated conceptually in Fig. 1.3.[5]

Thus, the two major considerations in classifying highway and street network functionally are access and mobility. The conflict between serving through movement and providing access to a dispersed pattern of trip origins and destinations necessitates the differences and gradations in the various functional types. Regulated limitation of access is necessary on arterials to enhance their primary function of mobility. Conversely, the primary function of local roads and streets is to provide access (implementation of which causes a limitation of mobility). The extent and degree of access control is thus a significant factor in defining the functional road category,[5] as will be shown in the following sections.

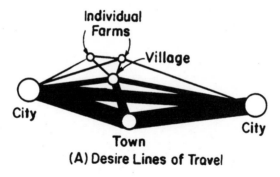

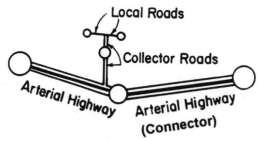

FIGURE 1.2   Channelization of trips in a typical rural road network.[5]

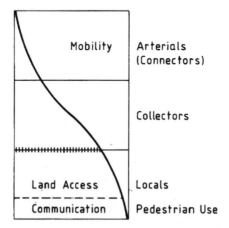

FIGURE 1.3   Relationship of functionally classified systems in service traffic mobility and land access. (Elaborated and modified based on Ref. 5.)

CHAPTER 2
PRINCIPLES FOR ROAD NETWORK DESIGN*

2.1 INTRODUCTION

The interrelationships between the different types of living spaces, such as housing, working, education, service, and recreation, as well as the various economic interdependencies among them, require different transportation modes, which have to be tuned to the respective traffic tasks. Transportation modes are intended to make life easier for people. Therefore, depending on traffic demands, transportation modes should be planned and designed in such a way that they are safe, effective in terms of utility and cost, and environmentally friendly.

It is the task of traffic planning to determine the mode and extent of the traffic demands, in order to create decision criteria for design, construction, and operation of the individual transportation modes in cooperation with the parties involved and concerned.

Part 1, "Network (NW)," should be valid for all roads that are devoted to general traffic, as well as for roads exclusively for motor vehicles. The road network, as a carrier of motorized and nonmotorized individual traffic and road-bound public transportation, is part of the overall traffic infrastructure, which also includes the railroad network, the network of waterways, the aviation network, the pipeline network, and communications systems. The road network is the *essential element* in developing rural and urban areas, which means that road planners and designers should pay adequate attention to connecting roads to other existing transportation modes and to making arrangements for individual mobility and transit operations. Accommodation for different traffic systems should take into account a number of criteria, including regional and urban planning, the country's economy, environmental protection, and social development.

Within the road system itself, the road-bound public transportation, as well as individual traffic modes (like pedestrian, bicycle, and motor vehicle), has to be regarded as a combined traffic system.[244] Individual traffic modes have to be sponsored where they offer the best advantage from the viewpoint of regional, economical, ecological, and social development.

Part 1, "Network," as it is seen here, should lead first of all to an improvement in, and an adaption of, the existing road infrastructure.

Road network design essentially influences the spatial development as well as the regional and local space structure. Therefore, it is necessary to fine-tune road network design to regional and urban planning and to the planning tasks in other specific fields. In this connection, it may also become essential to minimize avoidable increases in traffic, wherever possible, and to plan for the unavoidable traffic, in all its modes, in such a way that the living and environmental conditions can be improved for the citizens.[671]

The task of road network design should be to arrange and design the individual road sections according to their respective functions within the scope of transportation and regional planning.

---

*Elaborated based on Refs. 244, 260, and 433.

In this connection, the network has to be seen as an entirety, independent of the site (inside or outside built-up areas).

Part 1, "Network," represents the basis for a classification system of road categories. The functionally classified design of individual network elements will be provided in such a way that a direct connection with Parts 2 and 3, "Alignment (AL)" and "Cross Sections (CS)," is ensured.

When there are conflicting goals, the use of "NW" always requires careful concerted action, for example, between the demands of traffic, regional planning, preservation of natural beauty and wildlife, and protection of landscape. A deviation from the rules and quantifiable assessments developed in "NW" makes sense only if a sound evaluation process leads to a better solution with respect to the conflicting goals.[244]

2.2 IMPORTANT DEMANDS

In the framework of road network design, one can distinguish between traffic-related functions (mobility/connection and access) and nontraffic-related functions (local or pedestrian use). Traffic-related and nontraffic-related functions can be superimposed on each other on individual roadway sections in various ways. The task of road network and road space design is to solve conflicts between these functions by considering traffic safety, environmental compatibility, and costs. Furthermore, the distribution of tasks among other transportation modes has to be considered.

Another important principle for the creation of sound networks is the desire to achieve, as far as possible, equivalent living conditions in the observed region or for the whole country. In this connection, the road network serves to ensure accessibility to all important living spaces.

With regard to the negative effects of motorized road traffic on the environment, the interactions between roadways and roadside surroundings are of special significance in the design of road networks. In this connection, it is necessary to improve traffic safety and to minimize land use, interferences with natural resources, damages to the landscape and municipal cultural goods, and exhaust and noise emissions.

Traffic-related and nontraffic-related functions are combined for every network section with different demands. For mobility (connector) functions the network design depends on the significance of the connection. However, the traffic-related demand has to be adjusted to the location of the road (inside or outside built-up areas), the utilization of the areas adjacent to the road (with or without buildings), access demands, and the intensity of local or pedestrian use functions.

Therefore, for the design of an individual roadway section, superimposing different functions can lead to conflicts. Such conflicts are more difficult to solve, the more intensive mobility (connector), access, and local (for example, pedestrian use) functions have to be resolved at the same time. This means that the highway engineer has to decide what possibilities exist to reduce functional conflicts and which functions have priority, if necessary.[244] Important network demands are listed in Step 1 of Fig. 3.1.

In the following sections, a system of functional classification of roads will be presented in order to develop a sound hierarchy of road categories. Those road categories are needed as a basis for Parts 2 and 3, "Alignment (AL)" and "Cross Sections (CS)."

CHAPTER 3
CLASSIFICATION OF ROADS*

3.1 ROAD FUNCTIONS

The functional classification of the road network is determined essentially by regional, urban planning, and environmental protection considerations.

To design the road network, a classification of the network parts is necessary based on their numerous functions for connector, collector, and local demands. The superimpositions of these functions are expressed by category groups, and the connector functions are further subdivided into connector functional levels corresponding to their traffic-relevant significance. The combination of the category group and the connector functional level results in the determination of a road category (see the flowchart in Fig. 3.1).

The road network design determines and assesses the relevant road categories for the planning, design, and operation of roads. The scope of road network design is differentiated into traffic-related functions (connector and collector) and nontraffic-related functions (local and pedestrian use).

3.1.1 Mobility (Connector) Functions (Step 2, Fig. 3.1)

Roads outside built-up areas primarily serve mobility (connector) functions. The goal of road network planning is to design the connector functions in such a way that sound traffic flow and good traffic quality can be guaranteed for individual road sections. However, in assessing quality levels, the goals of saving travel time and transportation costs and guaranteeing sufficient traffic safety have to be compared carefully with the goal of environmental protection. Therefore, the rating and design of roads or roadway sections—depending on the respective connector function—is based on adequate travel speeds. Thus, for connector functions, the sound assessment of "speeds" is an essential criterion.

3.1.2 Access (Collector) Functions (Step 2, Fig. 3.1)

Roads inside built-up areas mainly serve access (collector) functions. Furthermore, accessibility has to be provided to areas adjacent to the road. Within such a road section, the collector function causes terminating and originating traffic. For these traffic purposes, only low requirements exist on "speed." Collector functions may be partially superimposed on connector functions.

---

*Elaborated based on Refs. 244, 260, and 433.

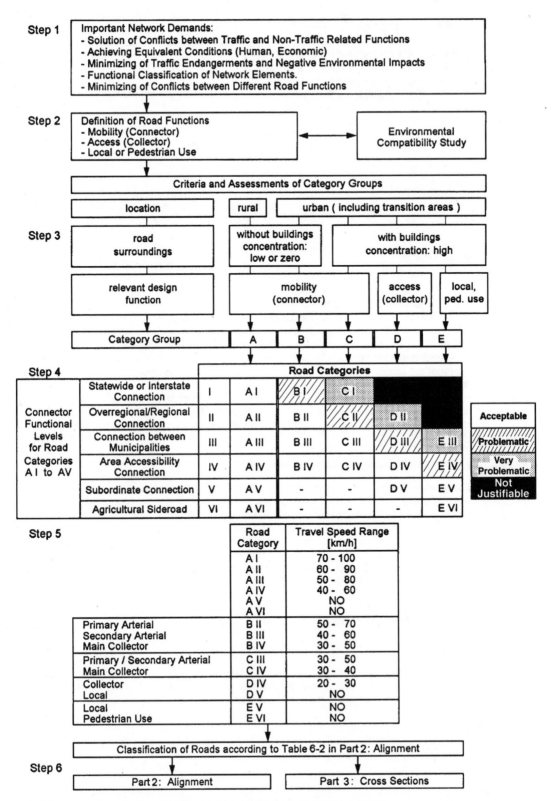

FIGURE 3.1 Flowchart for Part 1, "Network." (Elaborated based on Refs. 244, and 260.)

3.1.3 Local or Pedestrian Use Functions (Step 2, Fig. 3.1)

The local function (pedestrian use or communications function) is typical for roads in built-up areas. This function results from residential activities, like children playing, walking, shopping, etc. Therefore, "speed" is of no interest for this function.

3.2 ROAD CATEGORY GROUPS (STEP 3, FIG. 3.1)

Conflicts, which have to be solved by road network and road space design, normally exist between local, collector, and connector functions.

Accordingly, roads are classified sectionwise by the following criteria:

- *Location:* inside (urban) or outside (rural) built-up areas.
- *Degree of concentration of buildings on the road:* low or high.
- *Relevant design function:* mobility (connector), access (collector), and local or pedestrian use.

A rational combination of the three criteria results in five possible groups of roadways with common functional and road user demand characteristics, that is, five road category groups—A, B, C, D, and E (see Fig. 3.1, Step 3).

Roads of Category Group A.   Rural principal arterial systems (interstates, freeways) used for substantial statewide or interstate travel and movement outside built-up areas and rural minor arterial systems for linkage of cities, larger towns, and other traffic generators, such as recreational or production centers and integrated intercounty services. The functional characteristic of highways which belong to category group A is mobility (connector functions). Access and pedestrian traffic are not important (see Fig. 3.1, Step 3).

Roads of Category Group B.   Mainly primary arterials used in suburban areas with a low (or zero) concentration of buildings on the road. The functional characteristic of roads which correspond to this group is mobility (Fig. 3.1, Step 3). Limited access is not excluded for this category group. Roads that belong to this group are found in areas where land availability is limited and separation of motorized traffic from nonmotorized traffic is not as pronounced as that of category group A. The design standards of this group are generally lower than those of category group A. For pedestrian and bicycle traffic, roads of category group B can have, according to volume and safety aspects, separate sidewalks or bicycle lanes along the shoulder.

Roads of Category Group C.   Mainly arterials used in a network of collectors in urban areas with high concentrations of buildings on the road. Their main function is mobility (Fig. 3.1, Step 3), although they provide access functions to the buildings directly adjacent to the street. Commonly, there are sidewalks on both sides, usually behind the curb, and it is desirable to have bicycle lanes on one side and parking lanes on both sides. Therefore, conflicts may arise between connector, collector, and parking functions. Even though the cross section of a category group C road may have a higher standard than the cross section of a road of category groups A or B, its design and operating speed is considerably lower. Furthermore, it is very important to integrate roads of this category group into the urban environment and development in order to decrease the impact of traffic on the environment, as far as possible, and provide high-quality living standards for the inhabitants. Thus, careful consideration of the different functional demands is necessary.

Roads of Category Group D.   All urban roads for which the main purpose is to provide access possibilities to areas adjacent to the road (Fig. 3.1, Step 3). Within such a roadway section, terminating and originating traffic is present due to the collector function. Through traffic may also be present and, at certain times during the day, those roads may even take over connector functions. As a result, a number of varying road-user demands have to be taken into consideration here and have to be balanced. In addition, many safety problems arise from the operation of roads that correspond to this category group because of heavy pedestrian and bicycle use which conflicts with vehicles seeking access routes and parking areas. For this reason, the separation of the different road-user groups must always be sought in order to lessen dangerous traffic situations.

Roads of Category Group E.   All local urban roads on which pedestrian use plays a key role (Fig. 3.1, Step 3). Through traffic normally does not use these roads and the access (collector) function is only partially allowed. Volumes up to 250 vehicles/h do not generally disturb the pedestrian traffic on these roads. Road space has to be designed for the mixed application by different road-user groups.[244] They serve mainly residential, pedestrian, and communications functions.

3.3   *CONNECTOR FUNCTIONAL LEVELS (STEP 4, FIG. 3.1)*

While the previously described road category groups provide a functional classification of roads, they do not provide the designer with adequate quantitative measures that would enable selection of the design elements of the highway which are appropriate for individual road categories within the network. Thus, a further categorization is needed in order to distinguish, on the basis of mobility, between the significance of different connection demands.

The meaning of mobility depends on the importance of the centers it connects. This statement holds true for both rural and urban environments. For instance, the importance of a major road which connects two states (known as an interstate in the United States) is higher than that of a road which connects two adjacent villages within a county. The importance of every center that is connected by a road depends on statewide, regional, and urban planning. According to the German Guidelines for the Design of Roads, Part: "Guide for the Functional Classification of the Road Network,"[244] an urban area is associated with four levels of urban operations and can be characterized as a high, medium, or low level center or as not being a center at all:

- A high-level center denotes an urban area which includes major administrative services, cultural and various economic activities, and regional services.
- A medium-level center is an urban area which takes care of the daily and special needs of the inhabitants and where specific industrial and commercial activities and services are met.
- A low-level center is an urban area where mainly the daily needs of the inhabitants are met.
- An urban area that does not meet any of the previously mentioned three levels is regarded as not being a center at all.

The preceding scheme of centers forms the basis for describing rural mobility and for defining the importance of a connection, expressed by the so called connector functional level. Similar considerations have been conducted for urban mobility, access, and pedestrian use; however, since this book deals mainly with rural roads, they will not be discussed here in detail.

In addition to the categorization of urban centers presented here, the following aspects also are of concern for classifying sound connector functional levels:

1. Assignment of a city or town, or a part thereof, to a certain mobility level should be carried out using existing or planned transportation modes. In this connection, harmony between existing transportation modes is necessary in order to define the significance of a connection for a highway route. Ignoring other transportation modes when classifying connector functional levels could lead to a wrong level definition.

2. Similar considerations also apply to those nodes of a road which cannot be assigned directly to an urban area. Such nodes are recreational areas and specific traffic generating sites that do not represent urban areas, that is, airports, sports fields, university campuses, large industrial parks, etc. Special care has to be taken when assigning a connector functional level to those areas or locations.

3. The last case concerns all roads that connect two countries or have importance within a continent. These road connections normally must be of a special type because of the importance they play with respect to international road transport.

For the following parts, "Alignment" and "Cross Sections," it is very important to assign sound design qualities to individual roadway sections. Therefore, based on German[244] and South African[663] experiences, a stepwise transition from the highest to the lowest connector functional level is introduced for rural networks in the left part of Fig. 3.1, Step 4. Overall, the hierarchy of the connector functional system consists of six levels according to the type of service they provide, denoted by the Roman numbers I through VI, with I being the highest and VI being the lowest.

A combination of the German and South African concepts[244,663] leads to the following categorization scheme:

1. *Connector functional level I:* Connection between the national road system and those of the neighboring countries. Designated in Step 4 of Fig. 3.1 as *statewide or interstate connection.*

2. *Connector functional level II:* Linkage between state or provincial capitals, main centers of population, and production centers. Designated in Step 4 of Fig. 3.1 as *overregional or regional connection.*

3. *Connector functional level III:* Connection between local centers of population. Linking districts, local centers of population, and developed areas with the principal arterial system of levels I and II. Designated in Step 4 of Fig. 3.1 as *connection between municipalities.*

4. *Connector functional level IV:* Linking the locally important traffic generators with their rural hinterlands. Designated in Step 4 of Fig. 3.1 as *(large) area accessibility connection.*

5. *Connector functional level V:* Providing service to the smaller communities. Designated in Step 4 of Fig. 3.1 as *subordinate connection.*

6. *Connector functional level VI:* Provides access to farmland or forests. Designated in Step 4 of Fig. 3.1 as *agricultural sideroad.*

Of course, certain overlappings between successive connector functional levels are not only possible but have to be expected.

Levels I to III should provide a high degree of mobility for longer trip lengths. Therefore, they need to provide a high level of service with high design speeds; however, a certain geometrical progression is needed with respect to the different design levels.

Levels IV and V serve a dual function by accommodating the shorter trips and feeding the arterials. They must provide some degree of mobility. An intermediate design speed and level of service is required.

Level VI has relatively short trip lengths, with property access as its main function. A design speed does not play any role.[663]

The proposed categorization scheme for connector functional levels is considered appropriate for most nations.

3.4   DESIGNATION OF ROAD CATEGORIES (STEP 4, FIG. 3.1)

The combination of a road category group (Step 3, Fig. 3.1) and an appropriate connector functional level, as determined from regional or urban planning processes,[244] will result in a road category (Step 4, Fig. 3.1). Because of the different high-conflict potential between the demands of the road surroundings (expressed by the category group) and the relevant connector functional level of a roadway section, not all possible combinations of road categories are desirable from the planning point of view. The road categories designated by shaded areas in Step 4 of Fig. 3.1 denote those cases for which the mixture of different road functions is not fully compatible with the respective road category group; indeed, the darker the shaded area is, the more incompatible the mixture of different road functions is with a specific category group. This incompatibility makes it very difficult for such a road category to satisfy both the traffic- and nontraffic-related road-user demands. When such a case arises, especially in urban areas, which are already developed, efforts should be directed toward separating the road-user demands with respect to mobility, access, and pedestrian use functions. If such an approach is not totally feasible, then care should be given to avoid compromising safety.

Finally, Step 4 in Fig. 3.1 presents 15 possible road categories that result from the sound combination of a certain road category group and an appropriate road function. According to the approach described earlier, the highest road category represents a freeway of category A I, whereas the lowest road category is a local residential road with mainly pedestrian use functions of category E VI.

Overall, the following road categories are justified (compare Steps 4 and 5 of Fig. 3.1):

A I through A VI

B II through B IV

C III and C IV

D IV and D V

E V and E VI

Using a complex determination process based on travel distances and travel times, appropriate ranges for travel speeds on weekdays were established with respect to the individual road categories in Step 5 of Fig. 3.1.[244]

These travel speeds represent the basic assumptions for the arrangement of recommended speed limits and for the design speeds in Part 2, Chap. 6 (see Table 6.2), which are also valid for Part 3, "Cross Sections." The established travel speeds for the individual road categories may be considered appropriate for most nations.

In Part 1, "Network," the functional classification of roads stood in the forefront. One of the main aspects of this book "safety" is only indirectly considered. Therefore, procedures to recognize, analyze, and evaluate different endangerment levels of roads in networks are discussed in Secs. 18.1 and 18.2 of Chap. 18 in Part 2, "Alignment."

CHAPTER 4
TRANSITION TO THE FOLLOWING DESIGN PARTS

Consideration about design quality for horizontal and vertical alignments, as well as for cross sections, are mainly based on the developed road categories. Table 6.2 of Part 2, "Alignment (AL)," reveals the importance of the road category for the design process. As can be seen from Table 6.2, important design and operational characteristics, like decisions about the kind of traffic, speed limit, cross section, intersection access, or design speed depend on the road category. This statement is also confirmed by Table 6.3 of "AL," which reveals that different design principles and safety evaluation processes have to be applied, based on the significance of the road category in the network.

The present book deals primarily with roads in rural and suburban networks, which means that mobility (connector) functions outside built-up areas with low or zero concentration of buildings become relevant. Therefore, the road categories A I through A V, as well as B II and B III, form the basis for the following design Parts 2 and 3, "Alignment (AL)" and "Cross Sections (CS)." In this connection, it should be mentioned that road category A VI will not be considered, since this category is unimportant in considering mobility functions.

P · A · R · T · 2

ALIGNMENT OF
NONBUILT-UP ROADS (AL)

CHAPTER 5
INTRODUCTORY CONSIDERATIONS

"What has to be considered in establishing modern highway geometric design recommendations?" remains an exciting, thought-provoking question in the field of highway engineering. While several important goals in highway geometric design, such as function, traffic quality (capacity), economy, and aesthetics, are reasonably well understood today, deficiencies still exist in the proper analysis and evaluation of the impact of highway geometric design on traffic safety and environmental protection, as well as the interrelationships between human factors and vehicular involvements.

5.1 SAFETY AND ROAD DESIGN

Unfortunately, most people are unaware of how large a problem unsafe traffic operations represent on a worldwide basis. The tragic consequence of traffic accidents puts unsafe traffic operations on a par with war or drug use, as an example of irresponsible social behavior which must change. People need to be made aware of, and assume responsibility for, the possible effects of their driving behavior on themselves and on others. Most drivers, however, have little understanding of traffic risks.[702]

This lack of awareness and responsibility may be an important reason why more than 500,000 people are killed—or about one life every minute—and over 15 million suffer injuries as a result of road accidents every year worldwide. Of the millions who are injured, tens of thousands are maimed for life. The financial cost is many thousands of millions of dollars annually.[311,702] The most recent estimate by the Environmental and Prognosis Institute in Heidelberg, Germany, indicates that 50 million people will die and 1.1 billion will be injured from road traffic accidents worldwide between 1995 and the year 2030, if the development of the motor vehicle traffic remains unchanged in the future.[51,724] Put in the context of the 1995 population of nations, this represents the death of about 90 percent of the population of France and injury to every man, woman, and child in China.

Related to the fatality situation in western Europe, for example, statistics show that traffic deaths alone correspond to between three and four crashes of big jumbo jets every week. We would never tolerate such unsafe performance by airlines, but why do we then tolerate a similar number of deaths from road traffic accidents?[702]

It is estimated that over 50 percent of highway fatalities occur on two-lane rural roads outside built-up areas. Half of these fatalities occur on curved roadway sections.[425] Generally speaking, curved roadway sections and the corresponding transition sections present a great opportunity for reducing accident frequency and severity.

Multilane, median-separated highways, on the other hand, are much safer. For example, the U.S. Interstate system and the comparable German autobahn system represent the safest road classes, recording only about 10 percent of total traffic fatalities, despite the fact that about 25 percent of the vehicle kilometers driven are normally done on these roads.[106] Multilane highways are normally designed more generously with sound curvilinear alignments included in the design of these roads, particularly in western Europe.

Even though the human factor may be identified as a major cause of accidents, it is virtually impossible to control and difficult to design for the driver's frame of mind and physical condition. The highway engineer cannot influence alcohol abuse or seat-belt usage and has little capability to improve driver judgment at intersections. However, good geometric design should help to control traffic operating speeds and to reduce accidents brought on by excessive speeds inconsistent with conditions or geometry.

Many of these speed errors may be related to inconsistencies in the horizontal alignment, which causes the driver to be surprised by sudden changes in the road's characteristics and exceed the critical speed of a curve and lose control of the vehicle. These inconsistencies can and should be controlled by the engineer when a roadway section is designed or improved.[470]

Two-lane rural roads exhibit higher accident rates and severity than multilane highways. Therefore, special emphasis should be placed on this portion of the road network when designing, redesigning, or conducting restoration, rehabilitation, or resurfacing projects.

To improve the highway engineer's ability to analyze rural roads and to provide safer designs, three quantitative safety criteria have been developed, which, when properly applied, are intended to provide rural two-lane highways with:

- Design consistency
- Operating speed consistency
- Driving dynamic consistency

These criteria are the *main focus* of the traffic safety portions of this book. In addition, other quantitative and qualitative safety measures are presented with respect to single design elements, as well as with respect to the combination and superimposition of element sequences in horizontal, vertical, and three-dimensional alignment.

Furthermore, an attempt was made to include the influence of "human factors," based on sound psychological and physiological assumptions, in order to be able to evaluate their impact on geometric design and safety. The same is true for the motor vehicle itself and vehicular components in connection with safety considerations.

5.2 *ENVIRONMENTAL PROTECTION*

Without doubt, modern highway engineering should respond to the mobility requirements of the citizens as well as to the highly developed economy. Providing traffic service to the commercial needs of users is imperative, and is certainly a part of the quality of life we are used to today.[26] In the meantime, the increasing impairment of the human living space is of concern to all of us.

Therefore, the preservation and conservation of natural resources is one of the most important tasks for the future, and it requires balance and care by the highway design professional. The goals of protecting nature and landscape, soil and water, and avoiding the disturbance of the ecological balance should be some of the engineer's main priorities. However, the phrase "environmental protection" in western Europe is very often used as an argument to abandon any further expansion of the highway network. This is unfortunate since with today's knowledge base and practice, it is possible to provide *both* improved mobility *and* a protected environment. To do so, the design professional must understand that:[26,209]

1. European governments have serious policies aimed at protecting nature and the environment. However, by abandoning future completion of traffic routes, many other social problems cannot be solved, economic growth and employment cannot be ensured, and traffic safety cannot be improved.

2. Environmental protection is a necessary, non-negotiable requirement for the highway engineer. In the planning, design, and/or construction phases, the relevant consequences of a project with respect to the environment have to be investigated, evaluated, and balanced against other public and private interests.

3. In examining the environmental compatibility, all planning stages and the resulting adjustments can become difficult, extensive, time consuming, and expensive—but necessary.

4. Environmental protection is expensive. In the majority of cases, it is estimated that between 5 and 20 percent of project costs are spent on environmental protection in western Europe. However the public, through their acceptance of government policy requiring these enhancements, supports the high value placed on a protected environment.[26,209]

The governments of the European Union have enacted Laws about Environmental Compatibility Examinations (ECE) for traffic projects. They require engineers to conduct environmental compatibility studies (ECS) on the potential environmental impact of planned highways.

These studies must be done *before* starting the design and construction phases of a highway construction project and must consider all existing natural, ecological, and cultural resources which are important to the integrity of any region.

The ECS normally consists of two parts:[305,465]

- A space-related sensitivity investigation
- A comparison of alternatives (see also Chap. 7)

Incorporating a new design component ECS into the guideline framework is a significant step forward for examining and evaluating environmental protection issues in the highway geometric design process.

5.3   HISTORICAL DEVELOPMENT OF HIGHWAY GEOMETRIC DESIGN AND CONSTRUCTION*

Today's complex and widely spread traffic systems, like the U.S interstate or the German autobahn, express the needs of highly industrialized nations. But these systems do have predecessors which can be traced back to ancient civilizations. The following sections will go as far back as necessary in order to provide a better review of the interesting, complex development processes related to highway design and construction.

The development of civilizations on this globe has always been connected with different means of transportation. Without the early, simple "transportation systems," the development of different cultures and civilizations could not have been easily realized. One of the most important requirements for the settlement of hunters and gatherers was the construction of very simple traffic routes, which enabled the exchange of goods, plants, animals, or even knowledge between different tribes.[373] However, such exchanges were not only a bilateral business but also a multilateral barter.

Consequently, the first marketplaces were conveniently located alongside large rivers or natural ports. Oriental caravansaries or North American forts are examples of these marketplaces.

In addition to trade and military roads, religion also played a key role in ancient days in promoting well-engineered road networks. All known religions have had places of pilgrimage and worship, as well as holy places which often attracted a great number of people. Examples include the Oracle of Delphi and the Pyramides of the Mayan people.

As the religious and/or political powers of tribes grew, the traffic to holy places experienced a fast increase. Consequently, one important demand was to construct long-lasting paved routes in order to reach these places or destinations with relative safety and comfort. Because the materials for these "roads," in relation to ancient monuments, had to be moved over long distances, spe-

---

*Elaborated based on Refs. 373, 386, 466, and 680.

cial transport roads consisting of very thick and strong pavement were built. Unfortunately, most of these special transport roads were destroyed over the years. Here lies the secret: How were the people of ancient days able to build those fantastic and difficult civil engineering masterpieces?[373]

The very beginning of road construction is well described by Loewe[484] as follows:

> As soon as it was no longer a question of simple transportation by human carrier or by pack and ride animals and the use of vehicles became relevant, narrow and steep footpaths and mule tracks no longer could be sufficient, but wider routes with moderate grades became necessary, which at the same time provided a sufficient strength in order to bear the loaded vehicle wheels.

5.3.1 Paths and Roads in Ancient Civilizations

Before the widespread use of wheeled vehicles, some form of road construction had to be developed for sleds which were used, for example, to transport Assyrian and Egyptian squared stones and columns (Fig. 5.1). Both the loads and the distances moved were enormous.

The oldest known, naturally paved road, is the Faiyumhollow, which is located 70 km south of Cairo. It was built in 4600 B.C. This 13-km-long road connected a basalt quarry with waterways in order to transport the black squared basalt stones to the river Nile.[581] Nowadays it is common knowledge that waterways were the most important transportation mode at that time. By means of sleds, the Egyptians were able to tow squared stones from the banks of the river Nile to places adjacent to the pyramids. From there they were hauled along piled-up ramps to the exact construction spot. During the construction of the pyramids, those ramps were raised step by step and afterwards removed. In this connection, it should be noted that the Egyptians realized the advantage of the wheel through the conquest of Egypt by the Hyksos tribe.[373]

FIGURE 5.1   Assyrian transport of a godly image.[635]

Around 2000 B.C., the Minos tribe built a paved "stone road" with a length of nearly 50 km that ran from the capital of Crete, Knossos, via the mountains of Crete and the city of Gortyra (Gortys), to the ports on the south shore of the Greek Islands. That road can be called a "technical monument." It was equipped with ditches on both sides of the cross section, and consisted of a 20-cm-thick layer of sandstone covered with clay mortar and topped with basalt slabs. Available information indicates that, as early as 500 B.C., gravel was used for road construction purposes in Crete.[581]

Between 2000 and 3000 B.C., the ring-shaped Stonehenge was constructed. The stones, up to 20 m long, were transported over waterways and land for a distance of 180 km. From the banks of the Avon River, the stones were transported to the construction site on a so-called avenue. This avenue, similar to a slipway, was removed after the construction was over. The aim of the ring-shaped construction remains a secret. Supposedly, it served as a site for some form of religious rituals. It could also have been used as an astronomical observatory.

Another form of early paved roads is the so-called plank path, which consisted mainly of a series of split oak or pine trunks (see Fig. 5.2). Many different plank paths have been discovered in the swamp areas of northern Germany. They can be traced back to the bronze age (1800 to 1200

FIGURE 5.2   Plank path near Oldenburg, Germany.[373]

B.C.). Researchers assume that the plank paths were located in the course of the four amber roads. The amber roads connected the Mediterranean Sea with the North and Baltic seas of Germany. Trade relations with the area where amber was found increased quickly because amber was very popular for use in women's jewelry. The main marketplace for amber at that time was Hallstadt, Austria, where amber was exchanged for bronze and salt. In addition to the amber roads, there existed other famous roads like the Incense Road, which stretched from the south of Arabia to the Mediterranean Sea, or the Silk Roads between China and Mesopotamia and the ports of Syria.

The Phoenicians used a certain technique for road construction, on which the Greeks later expanded. The first holy Greek roads, which led mostly to worship places, were track road types, where the wheels moved along tracks chiseled into the rocks. Special provisions were developed for moving wagons from one track to another[366,635] (see Fig. 5.3*).

It should be noted that the Greeks established a law for road construction, which set the width of the vehicle track at 4 ft, 4 in (about 1.40 m). The latter was later adopted by the Romans as a wheel base measurement. Another example relevant to the art of building roads in Greece refers to the more than 10,000 steps carved out of rocks on the way to Parnassus; these steps are still visible today. (Similar roads can be found in the road network of the Inkas. The road of the Kings was about 5700 km long and passed through the Andes, at around 3600 m above the sea level.[581]

Meanwhile, technically designed and paved roads, which were by far shorter than the previously discussed "natural roads," were built by the Egyptians. The first paved roads were the so-called procession roads. Nearly all temple surroundings or holy places consisted of a system of procession roads which, for the most part, were lined up with columns, sphinxes, or statues of

---

*The main author regrets that the sources of some of the photographs referenced to here are unknown. Prof. Lamm collected these photographs over the past 25 years of lecturing and cannot recall from where or from whom he got these photographs.

FIGURE 5.3   Track-grooved road with a certain kind of switch. (*Source:* Unknown.)

FIGURE 5.4   Procession road to the Temple of Ammun near Luxor, Egypt, during the reign of Ramses II (1290 to 1223 B.C.).[373]

dynasties. An example would be the procession road to the Temple of Ammun near Luxor, Egypt (see Fig. 5.4).

For instance, from the lightness and elegance of Tutenkhamen's chariot, which is shown in Fig. 5.5, one can draw the conclusion that the roads inside cities or within temple surroundings had to be relatively flat. The flatness of roads also was quite important for transporting the godly figures of Assyria and Babylonia during religious ceremonies. For this reason, the roads in Assur were equipped with tracks which allowed the movement of wagons (see Fig. 5.6). Because the tracks were filled with elastic material prior to starting the procession, the godly figures could thus be moved without any dangerous vibrations.[373]

The use of bitumen was known at that time for filling gaps between bricks. During the reign of Nebuchadnezzar II of Babylon (605 to 562 B.C.), the 16-m-wide procession road Aibur Schabu

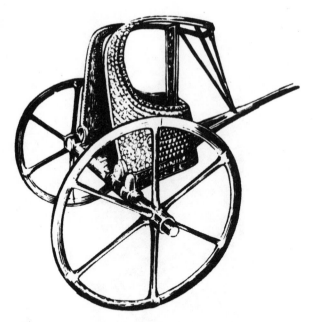

FIGURE 5.5   The chariot of the Pharaoh Tutenkhamen, Egypt, about 1350 B.C.[373]

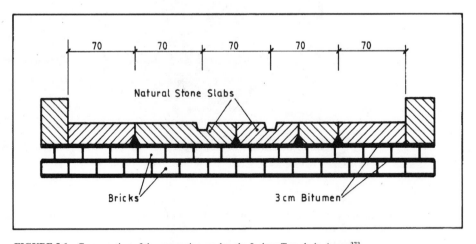

FIGURE 5.6   Cross section of the procession road to the Ischtar Temple in Assur.[373]

was built (Fig. 5.7). This road led from the King's castle, via the Ischtar Gate and the Hanging Gardens of Semiramis, to the temple complex within the city. The top layer of the road consisted of 35-cm-thick limestone slabs. Every slab had the inscription: "I am Nebuchadnezzar, King of Babylon, son of Nabopolassars, King of Babylon. I paved the road of Babel for the procession of the great Chief Marduk with Schadi limestone slabs. Marduk, Chief give us eternal life."[373]

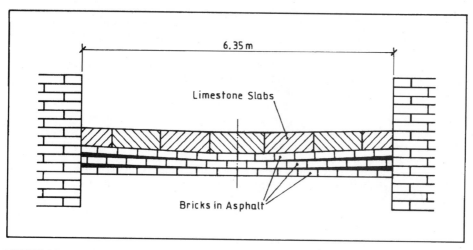

FIGURE 5.7 Cross section of the procession road Aibur Schabu in Babylon, about 600 B.C.[373]

In addition to religious and trade reasons, another reason for the extensive road construction during those time periods was to administer the huge empires. For this reason, the roads had to serve the following:

1. Extensive, fast relocation of armies to quell rebellions

2. Easily spreading important news throughout the empire

3. Safe and effective transport of tributes from conquered nations.

The knowledge base for road construction dates back to the old Persian empire. A high level of roadway engineering had to be developed in order to administer this huge empire. Good examples are the King's roads of Darius (550–486 B.C.). The most important King's road was the road between Sardes in Lydia and Susa in west Iran. It should be noted that the King's roads did not directly connect cities. Instead, they bypassed the cities at certain distances in order to provide an uninterrupted transport of goods and news, such as during times of rebellion. Thus, the national roads passed through inhabited and safe regions of the empire. These roads were equipped in such a way as to provide comfort and safety to users, by using road facilities like rest houses, service areas, and military camps. The road enabled the creation of the first postal service during the fifth century B.C.[484]

The King's roads had a uniform roadway width which enabled, with the exception of rough areas, the traffic in opposite directions to pass easily. The King's road network was only sparsely paved and provided security for the Royal Army. Only royal officials or messengers were allowed to use these roads. These officials traveled in horse-drawn wagons from one safe camp to another. Messengers, however, had to ride during day and night times. They were provided with special royal horses and required 10 days to travel the 2500-km-long distance between Sardes and Susa or 250 km/day.[373]

In this connection, it should be noted that, so far, much reliance has been placed on the major roads that survived for today's archaeologists. However, it should not be forgotten that most early roads did not survive, so the presented samples may be biased.

There is also evidence of very early road construction activities in Peru, India, and China.[635] For instance, several important roads connected the administrative and trade centers in the area of the Persian Gulf with China. In this connection, the Silk Road, which was a collection of caravan routes since 300 B.C., should be mentioned. The most famous traveler on this system was probably Marco Polo on his way from Venice to Beijing between 1271 and 1275. In the fourteenth century the Silk Road was still very important.

According to Lay,[466] the old China maintained a large network of roads which were paved with bricks and included large bridges. The Chinese network was established during the western Zhou dynasty period or around 900 B.C. The main design element at that time was the tangent. Trees were planted along the road, in order to give the road an avenue character. In the former capital, Xian, there existed nine streets (9 m wide), which were built in a chessboard pattern. These roads were connected by 7-m-wide circular roads and 5-m-wide secondary streets. At that time, special officials were in charge of maintaining the roads.

The most powerful development in road construction occurred during the Ch'in and Han dynasties or around 200 B.C. The emperor Shi, who also built the Great Chinese Wall, gave the order to establish postal roads across all of the Chinese empire. These roads reached even the farthest locations of the empire and were about 15 m wide with excellent shoulders. Every 10 m a tree was planted along the road.

For the emperor himself, as well as for officials in the administration and in the military, an extra lane was reserved on the main roads. During the Tang dynasty, 700 yr after the Emperor Shi, the Chinese road network was about 40,000 km. It is interesting to note that the first road engineers were well-known in China, and their names were Tu Mao, Yu Xu, and Shan, the first female road engineer.[466]

Methods for building roads can also be traced back to around 2000 B.C. in India. A considerable number of early roads were paved with bricks, with bitumen serving as mortar between stones, and were equipped with subdrainage systems. The "Rigweda," probably the oldest preserved written report (around 1500 B.C.), tells us quite a bit about great roads. Around 300 B.C. the Indians connected the Persian King's road from Susa near the town of Rawalpindi with a well-equipped 3000-km-long road, which led, via Delhi and Allahabad, to the capital Patna and to the mouth of the Ganges River. A special ministry was in charge of administering this road.[466]

Roman Roads.   Roman road construction and design were highly developed and enabled the Romans to spread an amazing network of military roads and trade routes, which led from Rome to most parts of the expanding empire. In the year 312 B.C., the well-known Via Appia or Appian Way was built between Rome and Kapua (Fig. 5.8). About 100 yr later, the famous northbound Via Flaminia was constructed. By the end of the Roman empire, the length of the major military roads was 8000 to 10,000 geographical mi (1 geographical mi = 7 km). In addition, many minor roads interconnected the larger main network. In total, Rome at its height of power had about 400,000 km of stabilized roads, of which more than 80,000 km were excellent long-distance roads. Sixteen main roads led into the capital Rome at that time, and pavement widths up to 24 m were common.

This road network served the Romans for the administration of their empire from Scotland to Ethiopia and from Spain to Mesopotamia. No comparable road network was developed until the twentieth century.

Roman roads were predominantly straight and, if possible, elevated to dominate the surroundings (Fig. 5.9).

Furthermore, the Roman engineers avoided bottlenecks with their designs. The aerial photograph in Fig. 5.10 reveals the typical alignment of a Roman military road. The road, which is no longer visible at the ground level, still stands out clearly as a bright straight ribbon from the surroundings. A modern road can be seen in the background.

Longitudinal grades of up to 10 percent appear to have been present in the Roman network, with steeper grades used only in exceptional cases. In the outskirts of Rome the archaeologist Rondelet discovered a separation of the roadway width used by Roman engineering. In the center was a cambered surface for drainage purposes (Fig. 5.11), which was accomplished normally at each site by a narrower strip or sidewalk. These strips were separated from the pavement through curb-like elevations (Fig. 5.12).

Roman Pavement Structure.   The minimum thickness of Roman roads was between 80 and 100 cm. At that time, a road ditched into the soil was called *Via Ruta* and a path carved out of the rock *Via Rupta*. These words were preserved in the French and English languages as rue and *route*. Similarly, the word *Via Strata*, which means the strewed way, survived in the German language

FIGURE 5.8   Via Appia Antica near Rome.[635]

as *strasse* and in the English language as *street*. Thus, we can credit the Romans with beginning not only the road design technology but also the language necessary to describe it!

These *strewed ways* were understood to be minor loam or water-bonded roads. Besides, there existed roads with chalk or different kinds of mortar-bonded surfaces and the important ones were paved. The surface consisted of natural stones—found mostly in rural areas (Fig. 5.8)—or rectangular plates—found mainly in urban areas (Fig. 5.13).

In mountainous terrain, wagons often moved along tracks which were chiseled into the rocks (Fig. 5.3). This saved the Romans from leveling off the whole pavement, and reduced the danger of carts sliding laterally.[366] To make this design effective, the Greeks and the Romans had to have had some form of vehicle standard administration, which was able to enforce uniform wheelbases for the wagons on those roads. Figure 5.14 shows the effects of such standards by the uniform grooves in a Roman pavement.

Roman roads were properly constructed using a more or less standard bearing body out of crushed stones with a decreasing gradation of stone material from the bottom to the top and a decreasing layer thickness (Fig. 5.15). They were water resistant and were designed and con-

FIGURE 5.9 Original straight and elevated Roman road still in use today, Mesopotamia. (*Source:* Lamm.)

structed according to uniform rules, standardized throughout the empire, and effectively administered by the organizational talent of the Romans.[366]

From a road construction point of view, the pavement structure of many Roman main roads was unmatched. For instance, for a thickness of 1 m (or 1.25 m in exceptional cases), the pavement structure was composed of three to five layers (Fig. 5.15), which were partially built by applying mortar. The bottom layer consisted of larger stones, set as hand-pitched stone subbase, while a mortar layer was used for the surface course, which was completed for main roads by a pavement set in mortar. The interim layers were constructed with more coarse-grained or fine-grained stone material.[635]

To pass through swamps, the Romans built regular plank roadways based on fascine construction and different plank layers. Examples of this construction were found in the surroundings of Oldenburg, Germany[484] (see Fig. 5.16).

The Romans also bring us early examples of highly developed postal services (Figs. 5.17 to 5.19) and milestone marking systems (Fig. 5.20).

It is interesting to note that Roman engineers continue to have a direct impact on highway safety—even in 1995. On March 6 of that year, the *Badenia Newest News,* ran the following headline and story:

Are the Romans Guilty of Traffic-Related Fatalities?

London: The in a straight-line constructed roads by the old Romans through the Cotswolds, West of London, cause a rapidly increasing number of fatalities in the idyllic forest area. On these road sections 23 persons died in 1994, 12 more than 1993. Jackie Harris, the traffic safety engineer for the county Clousestershire, reported: "Through the Cotswolds pass several roads, which were constructed during the Roman occupation. They are especially endangered since their straight alignments tempt motorists to speed."

Thus, some 20 centuries after its design and construction, this first "modern" highway system continues to affect the traveling public.

FIGURE 5.10   Aerial photograph of a Roman military road, Great Britain.[373]

FIGURE 5.11   Cambered Roman road surface, Middle East. (*Source:* Lamm.)

5.3.2 The Development of Roads from the Eighth to the Nineteenth Century

Eighth Century to the Seventeenth Century.   The split of the Roman Empire into East and West empires began around the year 300. In 476, the West empire was destroyed following an offensive by the Teutons under the leadership of Odoaker. Since Karl the Great (800), a considerable portion of the empire had been reunited. Unlike the development of the West Empire, the East Empire existed until the fall of Constantinople in 1453, after which the Turks and later the Russians took control of this empire until 1917. During that time, no further development of the Roman road network took place.[373]

With respect to the West Empire, Karl the Great used the old Roman strategy, which meant that road engineers used military campaigns for building new roads to and within occupied regions. Thus, it can be concluded that Karl the Great initiated the building of new roads and improving existing ones. He used the road network to demonstrate his imperial power and to administer justice. The reign of Karl the Great, however, was too short for further successful development of Roman heritage.

Afterward, the traffic routes in the European countries fell into extremely poor conditions as a result of a lack of systematic and professional maintenance (Fig. 5.21 on page 5.23). Consequently, the efficient Roman road network fell into disrepair.

According to Lay,[466] several important road activities which have taken place in Europe should be noted, as will be shown in the following sections.

In the ninth century, the Caliph Abd-ar-Rahman II gave the order to pave the streets of Cordoba with bricks. At that time, Cordoba was the most developed city of Spain as well as the largest city in Europe. He didn't, however, consider connector roads between the different caliphates because a major portion of the freight traffic was transported by sea. Thus, his efforts remained limited.

FIGURE 5.12   Roman road layouts.[373, 630]

FIGURE 5.13 Roman road, paved with rectangular stone plates, Middle East. (*Source:* Lamm.)

FIGURE 5.14 Grooves in a Roman road pavement. (*Source:* Unknown.)

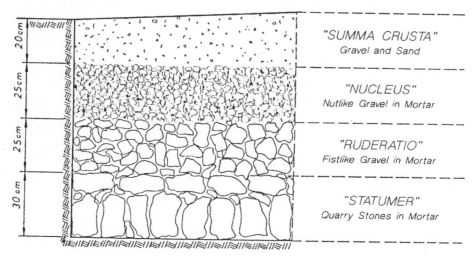

FIGURE 5.15   Roman road structure.[635]

FIGURE 5.16   Excavation of a plank path in northern Germany.[373]

FIGURE 5.17   Roman mail coach, relief from the second or third century, found near Maria Saal, Austria.[635]

During the same time period, one of the most progressive powers, the Normans, built the Waräger Road, which led from the Baltic Sea to Constantinople, in order to provide a link with the Far East. The most important cities on that route were Kiev and Novgorod. The big advantage of that road was that it provided the Europeans with a way to avoid the Mediterranean Sea, which was controlled then by the Moorish people for quite a long time. The journey took place more on rivers than on roads. An interesting portion of that road can still be found in the city of Novgorod, Russia.

At the beginning of the second millennium, travel activity increased as a result of the gradual increase in the power of a centralized Europe and the increase in religious demands, such as building new churches and cathedrals all over the continent. Another reason was the development of important trade fairs in order to exchange knowledge and experiences. The connection between towns led slowly to a simple road network, without, of course, following a coordinated planning concept.

For instance, one result of the lack of a coordinated planning concept is the fact that the two main roads in England, the Fosse and the Icknield Way, didn't extend to the capital London. In 1102, King Heinrich I of England gave the order to build a road from London via Wenlock Edge to Wales. The reason for building this road was to have better control over the Welsh people.[466]

In 1285, England established a law with respect to widening roads to accommodate two vehicles. To protect travelers from attacks by highway robbers, another law paved the way for clearing a corridor of 60 m from all trees and for filling up ditches. In order to improve the lifespan of roads, iron wheels and spikes in wheels were prohibited.[373]

Furthermore, Lay reported the following: "During the 11th century the basic principle was that every house and work location inside a city should have a direct access to a road or at least to a path. This fact led to increased activities with respect to roads and ways inside built-up areas. At the end of the 13th century, most of the cities which were rapidly developing had some paved streets. The paving craft was born at that time."

In addition, an interesting change took place with respect to the words "street, way, lane, or path" which were replaced by the new word "road" in the English language. This linguistic alter-

FIGURE 5.18   One-axle Roman travel cart in front of a post station, when changing horses, Emperor time (drawing, nineteenth century).[635]

ation proved the increasing importance of roads. Even though several activities occurred in the Middle Ages to build new roads or to maintain existing ones, the efforts, however, were limited to just a few regions. Generally, it can be concluded that roads after the Middle Ages and in subsequent centuries were in poor condition. Many reports have documented the fact that humans, vehicles, and horses had almost disappeared into potholes and roadside ditches. Animals, especially pigs, caused extensive damage to roads. Very often, people destroyed roads on purpose in order to use the road materials for their own properties. The roads were used to store garbage, sand, and coal or were used as marketplaces. In addition, roads offered possibilities for expanding personal estates. The first major efforts which were undertaken to improve roadway conditions in Europe occurred in the seventeenth century, especially in France and in Great Britain.[466]

Similar developments took place in Japan, as reported in Ref. 303. The Edo Shogunate, for instance, set the width of major roads to at least 11 m, and of secondary roads to at least 5.5 m. The road pavement consisted of gravel and small stones. It was 3 cm deep, compacted, and covered with sand. Planting Japanese cedars along the Nikko Highway started in 1625. The first British minister who came to Japan, Sir Rutherford Alcock, described his impressions of a Japanese road as follows:

> Their highway, the Tokaido, the imperial road throughout the kingdom, may challenge comparison with the finest in Europe. Broad, level, carefully kept and well constructed, with magnificent avenues of timber to give shade from the scorching heat of sun, it is difficult to exaggerate their merit.

People traveled in those days in Japan on roads which were equipped with various road facilities and services, built along the roadway at intervals of about 16 km. A milepost system was

FIGURE 5.19 Two-axle Roman travel cart, relief, Middle East. (*Source:* Lamm.)

FIGURE 5.20 Roman milestone, Via Maris, Middle East. (*Source:* Lamm.)

FIGURE 5.21   Typical poor roadway conditions in the Middle Ages. [Jan Brueghel's (1568–1625), *Busy Road through a Forest*.][373]

established in the second half of the sixteenth century, and postal service by horses were common. The German Engelbert Kaempfer, who came to Japan for a Dutch trading house reported:

> An unbelievable number of people travel the highways of this country every day. The reason is the high population of this country but another is, unlike other nations, the Japanese travel extremely often.

At the end of the Edo Shogunate, Ernest Satow, a British diplomat, expressed his surprise about the 1680 paved Hakone Road as follows: "The pass, which climbs the range of mountains by an excellent road, was paved with huge stones after the manner of the Via Appia."[303]

It should be noted that Japanese land routes were made exclusively for people and horses; horse-drawn carriages were rare events up to the beginning of the nineteenth century. This was the main reason for always keeping roads in good condition, because maintaining the roads was relatively easy. What was really surprising was the fact that cleaning roads and conducting regular maintenance work were performed not by the shogunate or the government of the feudal clans but by the residents alongside the roads. The reason for this may be the common belief of the Japanese people, which is that roads are not the exclusive property of the feudal class, but they are a "public property" and thus belong to everyone or to the society as a whole.[303]

Eighteenth and Nineteenth Centuries.   The sharp population increases in the eighteenth and nineteenth centuries led to corresponding traffic increases in urban and rural areas. For instance, the population of London, which nearly doubled between 1780 and 1820, caused road improvements to become very important for the exchange of goods. The industrial revolution of the eigh-

FIGURE 5.22 Luxury road of the epoch of Louis XIV.[373]

teenth century increased military demands, and fewer social restraints led to a considerable increase in traffic and to calls for roads with adequate pavement. Accordingly, since the mid-eighteenth century, the technique for road design and construction has improved considerably. The leading nation at that time in the art of road design and construction was France (see Fig. 5.22).

In the year 1622, Nicolas Bergier, a lawyer from Reims, published *The History of the Big Roads of the Roman Empire*. This book was widely read in France, and was even used by German road administrators. This book led to a renaissance of Roman road construction without providing, however, any major technical improvements. Hubert Gautier proposed another technique for road construction in his book *Traite de la Construction des Chemins,* which was published in 1693. According to Gautier, the road subbase should consist of a thick layer of large stones, topped with stone pavement material, and confined between two parallel curbs.[466]

Ecole Polytechnique. In 1747, the first school on bridge and road engineering was established in Paris, France. Since 1795 this school has been known as Ecole Polytechnique under the director Perronet. The development of the science of road engineering at this school gave France the advantage over other European countries in the technical matters of road design and construction (Fig. 5.23). The military expeditions of Napoleon, for which good roads were needed, enhanced the importance of developing new and better methods of road construction in France. During Napoleon's reign, the main roads in France were as wide as 17 m and used a 5-m-wide superelevated pavement.[386]

Based on the scientific principles developed at Ecole Polytechnique, Napoleon was able to reconstruct and complete a road network of state routes in less than 15 yr (Fig. 5.24).

Napoleon also initiated the construction of the first modern Alps route across the Simplon Pass between 1803 and 1805, because the latter was impassable and prevented the French Army from engaging in the battle of Marengo[373] (see Fig. 5.25).

Furthermore, it is interesting to note that during the same time period, the highway administration chief of the duchy Sachsen-Weimar in Germany, the privy councilor Johann Wolfgang· Goethe (who would think of him in connection with road construction?), was also concerned with

FIGURE 5.23 Perronet, director of the Ecole des Ponts et des Chaussees supervising road construction in a painting by Claude Vernet (1775).[373]

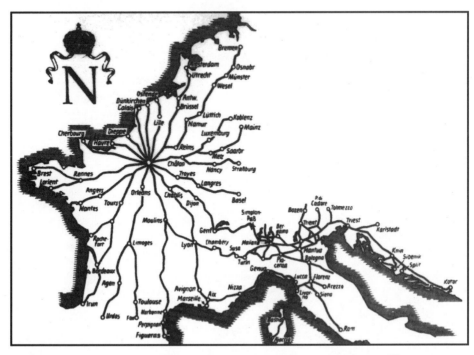

FIGURE 5.24 The state routes in the French Empire at the time of Napoleon I around 1810.[373]

FIGURE 5.25   French soldiers crossing the Simplon Pass in the year 1800.[373]

financing road construction, especially with respect to road maintenance. Similar to today's fight for assigning motor vehicle and mineral oil taxes, he attempted to get the road and bridge tolls and the gate and pavement penny used for road construction purposes. This escort money was a relic from the troubled times of the Middle Ages, when the sovereigns demanded such money for armed protection when passing their land. Based on prescriptive laws, this money was still paid in the eighteenth century, and Goethe demanded that the funds be directed solely toward road construction and maintenance. Furthermore, it is interesting to note that he requested that any beginning signs of road damages be taken care of at once using suitable materials and that a proper drainage of the structure of the road always be provided.[366]

One of the first educational successors of Ecole Polytechnique was the Polytechnic School at Karlsruhe—known today as the University of Karlsruhe (TH), Germany, near the French border.

This school was founded in 1825. In the study program of the year 1835, lectures on road construction and trackage engineering were already a part of the curriculum. The school catalogue of 1855 lists lectures, between 10 to 12 h/week, on:[386]

- Theory and construction of vehicles
- Construction of roads and their appurtenances
- Construction of tracks, railroads, and their appurtenances
- Construction of bridges
- Comparison of the natural and artificial shipping with the traffic on roads and railroads

The following interesting developments are reported by Lay:[466]

> The first instructions to build broken stone roads were published in 1831, based on the experience and knowledge of the Frenchman Pierre-Marie Jérome Trésaguet. These can be regarded as the first published standard for road construction. The new construction technique is based on the concepts of Gautier. The so-called Trésaguet-pavement consisted of broken stones of up to 20 cm in size, which were placed on a gently sloped subgrade. The gaps between the stones were filled with smaller broken stones (some kind of a hand-pitched stone subbase). The subbase was topped with a layer of gravel, in order to achieve a smooth rolling road surface. The advantage of the new construction, as compared to the older ones, was an easy maintenance of the pavement. Another view of Trésaguet was to place large drainage ditches on both sides of the road in order to effectively drain the road pavement. In addition, it should be noted that Trésaguet created the official position known as "Road Master." The road master was in charge of maintaining the road and its facilities. It should not be forgotten that the road master position, which was founded in France during the 19th century, still applies today.

The British stonemason Thomas Telford built his first important bridge in 1787; this followed a considerable number of bridges, ports, channels, and roads. He was one of the leading British civil engineers, and was the founder of the Institution of Civil Engineers in 1820. With respect to roads, he was appointed by the British Government as the main person to conduct the first transportation survey in order to improve the infrastructure of the Scottish Highlands. Under his direct supervision, around 1500 km of roads and nearly 1000 bridges were built until 1822. They are still regarded today as key elements of the road network of northern Scotland. Furthermore, Telford recognized the damaging effect of water on the body of a road. Consequently, he paid special attention to drainage demands. Thus, he used, for the first time, partly cut cubic stones as the subbase of a level subgrade. For drainage reasons, the subbase was gently sloped, and the gaps between the stones were filled with gravel. On top of the subbase, a second 15-cm-thick base course consisting of smaller stones (<6 cm in diameter) was used. For the wearing surface, he used a mixture of different sizes of gravel. This construction technique provided stability for 10 kg/1-mm tire width and was nearly impermeable. This technique increased the life span of the road considerably. Additionally, he attempted, as often as possible, to raise the road surface above the surroundings in order to achieve a more effective drainage of the body of the road. If this could not be achieved, the roadside was drained in order to avoid any potholes near the traveled road.[466]

In 1820 the Scotsman John Loudon McAdam (1756–1836) introduced the Macadam road construction method, which consisted of different layers of single-sized stones ranging from 19 to 13 mm, in order to satisfy a steady increase of traffic demands by means of relatively fast-moving horse-drawn coaches on small iron-covered wheels. In 1816, McAdam became the chairman of the local department for toll roads in Bristol. This position allowed him to practically realize his new ideas, which he presented on several occasions in front of the English Parliament in the years 1810, 1819, and 1823. At the same time, he published his first book, entitled *Remarks about the Present System of Roadmaking.* His second book, *Practical Essay on the Scientific Repair and Preservation of Public Roads,* was published in 1819.[466]

The basic principle of the Macadam road construction method is, according to Lay, that it used small, broken, and sharp-edged gravel. McAdam observed that a 25-cm-thick subbase of such material had the same strength and stability as the more expensive hand-pitched stone subbase. One assumption for this new construction method was a dry and well-drained subbase. Another important element of McAdam's mixture was the size of gravel used. For a 20-cm-thick base course, he used crushed gravel with a maximum size of 7.5 cm, whereas for a 5-cm-thick wearing surface, he used gravel with a maximum size of 2.0 cm. The surface course was partially mixed with tar. Such a Macadam road construction method was able to satisfy a tire width of 18 kg/1 mm. The principle and secret of this construction method lie in the structural interconnections between the compacted sharp-edged crushed stones. Summarizing, it can be stated that the name McAdam will always be connected with the invention of the Macadam or crushed gravel road.[466]

Paving roads with cobblestones, laid on a special base course, dates back to the year 1860. A further breakthrough in road construction occurred with the invention of the steamroller by the French engineer Lemoine[478] (see Fig. 5.26).

FIGURE 5.26   Steamroller at the beginning of road construction. (*Source:* Unknown.)

Development in Germany.   At the beginning of the nineteenth century paved roads in Germany existed, to some extent, within towns or as routes to get to the summer residences of the aristocrats. Road construction activities were overwhelmingly unsuccessful because of a lack of financial resources.

Since 1816, there existed in Germany a manual for highway engineering, called "Instruction for the Construction, Maintenance and Repair of Artificial Roads."[60] The road had to have, if possible, a straight alignment, two stone-paved lanes for traffic going in opposite directions and a so-called Summer Way made of compacted gravel. Generally, the width of the cross section was 12.6 m. On both sides of the road, shoulders with trees and adjacent ditches were present. These roadside features gave the roadway an avenue character (referred to as "chaussee" in Germany) (see Fig. 5.27).

As the road system in Germany grew, the mobility of the Germans increased too. However, the new freedom on the road was not unlimited because its use was accompanied, in most cases, with payments of road fees. Beginning in 1840 the road fee was canceled in Bavaria, and later by the different states. Between the years 1876 and 1900, the length of the road network nearly doubled to around 100,000 km. Even though the chaussee traffic reduced freight transportation costs by considerable amounts, the emergence of the railroad, however, from around 1855 took precedence over road travel with respect to long-distance transportation.

With the expansion of the railroad network in the second half of the nineteenth century, this tendency toward railroad travel consequently increased. Much of the "through traffic" was shifted from the highway network to the rail network. Roads were still important in moving people locally to and from railway stations, but the long distance travel needs were satisfied by the faster, more comfortable railroad. Advancements in both horizontal and vertical alignment practice were driven for the next several decades by the railroad designers and builders. Figures 5.28 and 5.29 show early railroads in Germany.[386]

5.3.3   The Early Twentieth Century—Questions of the Alignment

Engineers have always been concerned with the geometric alignment of highways since the beginning of road construction. An economic alignment meant avoiding long, steep upgrades in order to reduce the moving resistance and to better utilize the pulling power of animals.

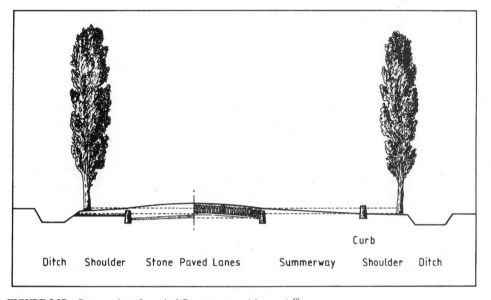

FIGURE 5.27   Cross section of a typical German avenue (chaussee).[60]

FIGURE 5.28   The first train of the Berlin-Potsdam Railroad, Germany (September 22, 1838).[8]

FIGURE 5.29   The first railroad of Munich, Germany (August 25, 1839).[8]

Before the invention of the automobile, the highways were designed for animal-drawn vehicles which rarely exceeded a speed of at most 13 km/h (about 8 mi/h). Thus, speed was not an important design factor and curves were mostly designed as simple sharp bends between two long tangents. Because of the low speeds, the sight distance was also not an important design criterion. Instead, the length and maneuverability of vehicles were the main considerations of horizontal alignment design.[466,617]

The low speed also had consequences with respect to the construction of a road. In the curved sections, the roadways were crowned instead of being banked. The main reason for doing so was to remove water from the road body along a short distance, because no demands existed for banking the curves with respect to vehicle speeds.

The invention and widespread use of the automobile led to considerations with respect to the safety standard of roads which were designed for horse-drawn vehicles. One of the main results of these safety-related questions was the establishment of *speed* as the main design factor in order to increase the safety standard of newly designed roads.[466,617]

Design Elements of the Horizontal Alignment.   In the year 1887, Launhardt wrote his *Theory of the Alignment*[463] to document the accepted technical practices of his day. For example, it was recommended to smooth the break in the alignment caused by a curve. In other words, the transition from a tangent to a circular curve ("circular line" in railroad engineering) could better be accomplished by using compound curves of differing radii or by perhaps a parabola. This concept of smoothing led to combinations of transition curves with double or triple the radius of the circular curve in railroad engineering (Fig. 5.30) and ultimately to the Euler spiral (clothoid), as the transition curve in highway engineering.[386]

The minimum radii of curves for horizontal geometric design were for a long time based on the geometric dimensions of long timber vehicles (Fig. 5.31). These vehicles represented one of the key subjects of textbooks on road construction until World War II, after which the mixed traffic of both horse-drawn carriages and motor vehicles gave way to the exclusively motor-driven traffic of today.

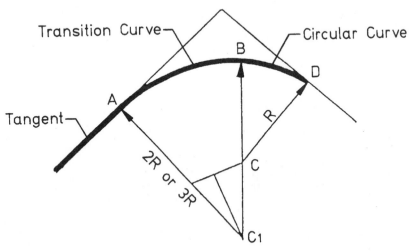

Legend: R   =   Radius of Curve
           C, C₁   =   Center Points
           A, B, D   =   Main Points of the Alignment.

FIGURE 5.30 Transition curve design.[463]

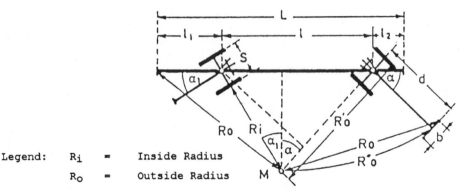

Legend: R_i = Inside Radius

R_o = Outside Radius

FIGURE 5.31   Long timber vehicle for the design of minimum radii of curve.[176]

In his book *The Construction of Rural Roads,* which was published in 1920, Euting reported the following for Europe:[176]

> A minimum radius of curve of 50 m is to recommend for all rural roads with heavy road traffic. Larger radii of curve of up to several hundred meters can be used in areas where topography conditions permit, such as in flat areas or, in some cases, in mountainous terrain. When a minimum radius of curve of 50 m is used, in-depth safety investigations are not necessary because this minimum radius allows all vehicles, including the long timber vehicles, to move easily around the curve.

For the United States, a radius of 15 m is regarded as sufficient for long freight wagons on secondary roads, whereas on primary roads a radius of 25 m is considered large enough. In case a shorter radius is required, then curve widening becomes necessary.[617]

With respect to the cross section, Euting recommended pavement widths of 4 to 8 m. For roads with heavy traffic, pavement widths of at least 5.5 to 6.0 m were suggested.

With respect to gradients, Euting recommended the following:

Plain and flat valleys	3%
Hilly topography	4–5%
Medium-altitude (subalpine) topography	5–6%
High mountainous (alpine) topography	7–8%

He cautioned that these maximum grades should only be used if more modest upgrades could not be reached, and that for main roads the smaller of the indicated values should be applied, wherever possible.[176]

Besides these first, nevertheless exact, indications about curvature relations, cross-sectional design, and grades, that same year, 1920, Graevell[230] published an interesting paper on "Superelevation and Widening of Chaussee-Curves" in the *Journal of Traffic Technique.* The term "chaussee-curve" is an old European expression for wide main roads. Graevell's publication emphasized (perhaps for the first time) the traffic speed issue, pointing out that speeds are important for the *safe* driving through curves. For instance, even at the time when horse-drawn carriages dominated the traffic stream, Graevell proposed the following speeds:

Slow step	1 m/s = 3.6 km/h
Short and stretched trot	6 m/s = 21.6 km/h
Ordinary gallop	7 m/s = 25.2 km/h
Fast gallop	10 m/s = 36.0 km/h

Graevell concluded that the upper speed limit for large transport horse-drawn carriages was between 3 and 4 m/s or, on average, 12 km/h. For this speed, Graevell showed that a minimum radius of curve of 30 m and a maximum superelevation rate of 7 percent were sufficient for the safe operation around curves.

Even though side friction was not regarded in these early investigations of centrifugal acceleration, the driving dynamic issue was pointed out. For motor vehicles, Graevell indicated that a speed of 25 km/h seemed reasonable. In response to the question: "How curves should be passed by motor vehicles?" he suggested the following: "That for fast rides in narrow curves an adequate speed reduction has to be performed, should be a matter of fact for every reasonable person." Therefore, he recommended that for a minimum radius of curve of 30 m and a maximum superelevation rate of 7 percent, the drivers of tractor-trailers and motor buses should reduce their speeds by 9 km/h (from 25 km/h to 16 km/h), in order to drive safely around the curve.[230]

However, for the normal design case the minimum radius of curve was increased to 50 m. Besides driving dynamic considerations, the driving behavior of the faster motor vehicles passing a curve demanded still larger radii of curve. In addition, on sharp curves, the faster vehicles tended to move much of the loose gravel to the road's edges.[176,466]

The concept of "superelevation on curves" was introduced for the first time in the State of New York in the year 1912. Generally, superelevation was applied to curves with radii smaller than 160 m (500 ft). The rate of superelevation depended on the pavement type. For Macadam surfaces, the rate was 1 in/ft (8.33 cm/m), and for concrete surfaces, it was $\frac{5}{8}$ in/ft (5.2 cm/m). These fixed superelevation rates did not take speed and curvature into consideration. The common opinion at that time was that different rates of superelevation were not necessary.[617]

In the year 1920, Ludecke and Harrison developed a mathematical model for representing superelevation:

$$e = \frac{V^2}{15R} \qquad \text{imperial system}$$

$$e = \frac{V^2}{127R} \qquad \text{metric system}$$

where e = rate of roadway superelevation, ft/ft (m/m)
$\quad V$ = vehicle speed, mi/h (km/h)
$\quad R$ = radius of curve, ft (m)

Because this model required up to 5 times the superelevation used in practice, the pavement skid resistance or friction had to take care of the difference.

The General Motors Proving Ground, which was built in the year 1926, was the first fully spiraled roadway in the United States. The first public roadway with spiral curves in Virginia was the Mount Vernon Memorial Highway, which was built in 1929 by the Bureau of Public Roads.[617]

Design Elements of the Vertical Alignment.*   Because of the low hauling power of horses, there was a very critical concern for steep upgrades. If possible, the grades were limited to a maximum of 5 percent; under normal topographical conditions grades up to 3 percent were ideal. Steeper grades led, consequently, to higher operating costs. Operating costs and grades as a function of the number of horses used are graphically shown in Fig. 5.32.

As Fig. 5.32 reveals, grades at that time were the most important design factor, especially when the maximum speed was something around 13 km/h ($\approx$8 mi/h). Thus, with respect to operating costs, engineers aimed for relatively flat grades, even if this yielded an indirect connection between two points. It should be noted that because of the exploding costs associated with steeper grades, the horizontal alignment was of little concern.

Major changes in the design, construction, and layout of roads are attributed to the invention of the automobile. Restrictions with respect to flattened grades no longer mattered because motor-

---

*Elaborated based on Ref. 617.

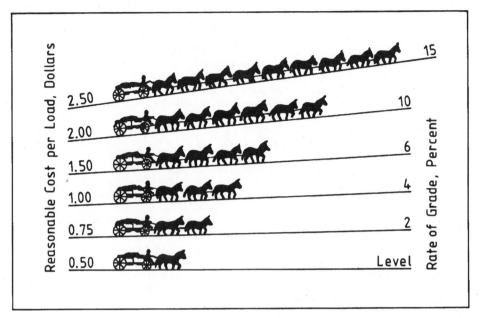

FIGURE 5.32   Effect of grades on hauling power.[755]

ized vehicles had much more power than horse-drawn carriages. Consequently, with reduced restrictions on grades, economic considerations aimed for straight alignments where highway costs were directly proportional to road mileage.[466]

However, with respect to traffic capacity demands, road maintenance, environmental protection, and uniform operating speeds, it was recognized that the flat grades required by horse-drawn carriages provided several advantages, especially with respect to operational issues. Furthermore, increased sight distance and more passing opportunities could be derived from using flatter grades and smoother vertical curves. Thus, the safety of the road could be improved.

Because of the relatively low speeds of horse-drawn carriages, there was little concern about the design of vertical curves. If there were any problems with the intersection of two relatively steep grades, then a small amount of rounding was regarded as sufficient. However, as motor vehicles became more and more common, the speeds increased considerably and questions about driving comfort and available sight distance were taken into consideration in the design of the vertical alignment. These new design issues were already recognized by Wiley in his 1928 book *Principles of Highway Engineering*,[755] in which he discussed the need for an adequate vertical curvature in order to limit the discomfort of vertical acceleration. The other, and even more important, question, which he also discussed, was the demand for a sufficient stopping sight distance. Drivers should be able to see each other and stop the car in such a way so as to avoid a head-on collision. A distance of about 120 m (≈400 ft) would be required in such a case. It should be noted, however, that the relationship between speed and the available stopping sight distance was not considered in his book.

In the early 1920s a suggestion was made for vertical curves to have radii of at least 15 m (≈50 ft) in order to avoid any discomfort aspects. At the same time, the use of parabolic vertical curves was suggested along with specific recommendations for the lengths of vertical curves with respect to the algebraic difference in grades.[466]

In 1929, AASHO established a minimum sight distance of 152 m (500 ft) for vertical curve design. Eye height was taken as 1.40 m (4.5 ft). The same was true for an object. During the same time period, the Germans used an eye height of 1.20 m (3.9 ft) and an object height of 20 cm (8 in). It should be noted that, in the 1930s, German highway engineers used a distance of 370 m (≈1215 ft) as a sight distance criterion for vertical curves in the design of the Autobahnen.

As with the German assumptions, AASHO also considered in the new American Standards of 1939 small objects that could be in the path of vehicles. This became known as the "dead cat" theory.

Note, however, that, besides these pure geometric and driving dynamic demands, the road traffic—expressed by volume, composition, flow, and most importantly by the human being—also play an important role when considering safety aspects (Fig. 5.33).

International Road Congresses. With the development of the motor vehicle in the 1920s and 1930s, the field of highway engineering became very important. By 1920, the United States already had 7 million vehicles in operation, and 1.5 million additional vehicles were produced during the year (Fig. 5.34).

FIGURE 5.33 Traffic and humans. (*Source:* Unknown.)

FIGURE 5.34   Traffic congestion in the 1920s in the United States. (*Source:* Unknown.)

A high percentage of roads at that time still had untreated roadway surfaces (Figs. 5.35 and 5.36). Consequently, dust-free, noise-protected, safe roads became very important to the new automobile owners of this period.

The first international road congress, the product of a joint effort of 30 nations, was held in Paris in 1908. At the fourth international road congress, which was held in 1923, issues like police enforcement, traffic rules and regulations, and the establishment of speed limits were discussed.[100] In the aftermath of this congress, speed limits were advocated, for example, for trucks weighing between 3 and 11 tons, depending on the kind of tires used (iron, solid rubber, pneumatic), as shown in Table 5.1.

For passenger cars, a maximum speed of 40 km/h was proposed, depending on the condition of the road and the motor power of the car.

The growing demands for motor vehicles over the years led to a number of engineering measures for the design of curves, transition curves, longitudinal grades, superelevation rates, cross-sectional elements, and drainage requirements, based mainly on driving dynamic considerations. These engineering measures would later constitute the basis for the geometric design guidelines of different countries, such as, for example, those given in Refs. 1 and 558.

5.3.4   Time Period: 1930–1970 (First Guidelines and Freeway Systems)

The two main aspects of this time period were perhaps the establishment of geometric design guidelines and the introduction of the first exclusive automobile roads. These freeways were called *autobahn* or *interstate* and they significantly influenced mobility and economic growth by promoting a reliable, efficient transportation system.

5.3.4.1   Development of Early Guidelines.   Until the 1930s, road construction was based almost exclusively upon economic and structural considerations, such as bearing capacity, pavement wear and tear, smoothness, drainage, and driving geometry on narrow curves.

FIGURE 5.35 Road full of potholes at the beginning of road construction, Germany. (*Source:* Unknown.)

Consequently, alignment principles were developed by the contractor and/or the supervising agency exclusively for every highway. In 1926, for instance, the German HAFRABA Association planned a freeway which would connect Hamburg, Germany, via Frankfurt with Basel, Switzerland. This freeway was designed with a crown width of 20.50 m and a minimum radius of 500 m in mountainous topography or 2000 m for other conditions (Fig. 5.37).

Additional alignment guidelines were developed in 1927 for a freeway project which would connect three German cities—Munich, Leipzig, and Berlin. According to these guidelines, the horizontal alignment was as straight as possible, and the curve radii were more than 1000 m in flat topography, 500 m in hilly topography, and 300 m in mountainous topography. In order to minimize extensive real estate separation, the alignment of the freeway followed, as far as possible, the railroad tracks. Urban areas were avoided and large cities were connected by means of a direct straight line from the freeway to the city.[215]

In the 1930s, safety considerations, which were based on driving dynamics, began to be regarded in developing criteria for the design of:

- Radii of horizontal curves
- Transition curves
- Superelevation rates
- Crest and sag vertical curves
- Stopping and passing sight distances

In Austria, the first "Guidelines for the Construction of Modern Roads with Mixed Traffic" were published in 1935. In Germany, the first "Preliminary Guidelines for the Design of Rural Roads (RAL 1937)" were published in 1937.[558] In the United States, guidelines were published at the beginning of the 1940s.[1]

FIGURE 5.36   Muddy road at the beginning of road construction, United States. (*Source:* Unknown.)

TABLE 5.1   Maximum Speed Limits for Trucks Established in 1923 at the Fourth International Road Congress[100]

| Total weight of the loaded wagon, kg | Iron wheel | Maximum speed, km/h, for: | | | |
| | | Regular traffic roads | | Specially designed roads | |
		Solid rubber tire	Pneumatic tire	Solid rubber tire	Pneumatic tire
3,001–4,500	12	25	35	30	45
4,500–8,000	8	20	30	25	40
8,000–11,000	5	15	20	20	30
Over 11,000	5	5	10	15	20

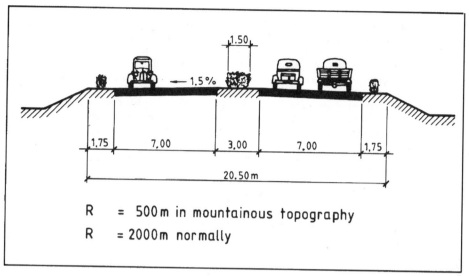

FIGURE 5.37   HAFRABA cross section.[215]

In Germany, the so-called project speed was introduced in order to achieve a sound, comfortable, and safe speed profile with respect to driving dynamic and traffic safety aspects. Limiting values for design parameters like radii of horizontal curves, radii of crest and sag vertical curves, grades, sight distances, and cross-sectional elements were established with regard to the project speed. This project speed was divided between 20 and 100 km/h in increments of 10 km/h.

The term *design speed* (V_d) appeared for the first time in the "Guidelines for the Design of Roads (RAS-L, 1963)."[558] It was regarded as an economic value, and thus it was defined according to an appropriate cost-benefit ratio for a planned new road or with respect to a rehabilitation project. A suitable design speed was selected according to the cross section, average annual daily traffic (AADT), and topographic conditions. The design speed ranged from 20 to 120 km/h. Limiting and standard values for most of the design elements were set according to the design speed concept.

According to Lay,[466] Young introduced for the first time the *design speed concept* in the United States in 1930. Limiting values were set for the design parameters: degree of curve, minimum sight distance and maximum superelevation rate, for design speeds ranging from 30 to 110 km/h (20 to 70 mi/h).

Later, three new speed definitions were introduced by Badlock in 1935. The so-called critical speed was defined as the highest possible speed which could be reached in motor racing or on a generous alignment of the road. The *design speed* was defined as the 80th-percentile level of the critical speed and corresponded to a well-trained driver. Generally, risky driving maneuvers are not expected in this case. This speed can be reached by the average driver under normal traffic flow conditions. The third speed term which was introduced was called *recommended safe speed*. This speed is lower than the design speed. According to Badlock, limiting values for the design parameters—minimum radii, sight distances, lengths of transition curves, and superelevation rates—should be established on the basis of the critical speed. Typical critical speeds were higher than 100 mi/h (160 km/h).

All design parameters were generally discussed during this time period, especially with respect to driving dynamic considerations. However, no explanations were given to describe the interactions between individual design parameters. This fact led to the conclusion that narrow curves are dangerous.

The first classification of U.S. roads appeared in AASHO's guidelines in 1940.[1] An *assumed design speed,* the maximum approximate uniform speed, which can be reached by the majority of drivers was defined. It should be noted that this speed is not equal to the speed of reckless drivers. The assumed design speed ranged from 50 to 110 km/h (30 to 70 mi/h).

In 1946, a comparison between English and American guidelines revealed an overall difference in the design process. The AASHO guideline required a uniform design speed which should be applied to the overall alignment of the planned road. The English guidelines noted additionally that the speed along an alignment is especially affected by geometric and roadside parameters of the road. Consequently, the speed along a roadway section is not constant, and the driver adjusts his speed according to the previously passed alignment and to the geometric features of the curve ahead.[466]

Even with guidelines based upon driving dynamic safety considerations and design speed concepts, highway design often continued to weigh low-cost solutions over safety. Noting this practice, the U.S. Congress responded to a presentation of highway design characteristics in the mid-1940s with the following statement: "We didn't ask you to build it cheap, we asked you to build it right (safe)."

Such a response was regarded as a mandate then, and should continue to guide the practice of highway designers and traffic safety engineers even today. As a result, highway geometric design guidelines were increasingly developed and revised by taking into account driving dynamic safety aspects. By the end of the 1960s, many countries had adopted and employed highway geometric guidelines.

5.3.4.2 Development of Early Freeways.   First considerations of exclusive automobile roads aside from the existing road network appeared in Europe and in the United States soon after the turn of the twentieth century. The predecessors of today's freeway systems, like the autobahnen and interstates, were several racetracks. In 1906, the Long Island Motor Parkway was built on private property as a race course for the Vanderbilt Challenge Cup. Originally, it consisted of a 10-m-wide and 10-km-long tarred track. In 1908, the racetrack became a public toll parkway, and was extended to a length of 80 km with an asphalt surface. It was the first public parkway with interchanges and superelevated, narrow curves. The parkway remained a public toll road until 1937. During the period after the depression, additional parkways were built in the United States as public works programs for unemployed people. One example is the 24-km-long section of the George Washington Memorial Parkway along the cliff-lined banks of the Potomac River near Alexandria, Virginia.[466]

The increased motorized traffic in Europe also led to consideration of a better adoption of the road system with respect to traffic demands. Thus, in 1909 the design of an exclusive automobile road, referred to as AVUS (*automobile traffic test road*), began. The construction of the 9.8-km-long section started in Berlin in 1913. The AVUS was financed by a private joint-stock company. The construction process was interrupted by World War I. It opened for the public in 1921.

In 1929, the pavement, a tarred gravel layer, was replaced by a more resistant bituminous concrete surface.[598] The AVUS highway contained all the modern basic features of a freeway, like median-separated travel lanes, full access control with grade-separated entrance and exit ramps, and a generous, mostly straight alignment (Fig. 5.38).

When Mussolini seized power in Italy in 1922, the idea for a new road network came up. It was developed by Dr. Piero Puricelli. The first Italian road exclusively for automobiles, which ran from Milan via Varese to Lake Como, had a length of 130 km. It was an undivided freeway with edge strips and three driving lanes in each direction. The design speed of the highway was 60 km/h, and the pavement consisted of a concrete surface. The highway, a toll road, was funded and built by a private company. This private company was later renamed the Italian National Road Association [Azienda Nazionale Autonoma delle Strade (ANAS)].[598]

The idea for a comprehensive network of roads exclusively for automobiles in Germany gained momentum with the completion of the freeway between Milan and the upper Italian lake region, and with progressive considerations of the geometric design of roads in both Germany and the Netherlands. In this connection, the previously mentioned HAFRABA Association played a

FIGURE 5.38   AVUS, the first automobile traffic test road in Berlin, Germany, 1921. (*Source:* Unknown.)

key role in the design and construction of roads exclusively for automobiles. The name of this somewhat private association was an abbreviation of the first initials of the three cities **HA**mburg, **FRA**nkfurt, **BA**sel—the cities through which the automobile road was supposed to pass between northern Germany via the south of Germany to Switzerland.[375] The alignment of the planned HAFRABA Autobahn through Germany and its extension through Switzerland to Genoa in Italy is shown in Fig. 5.39.

The planning of the alignment of the HAFRABA Autobahn resulted in many technical papers being written, which discussed questions of modern geometric design, new kinds of pavements, and traffic-related developments. The practical realization of the HAFRABA Autobahn was, however, unsuccessful because of poor economic conditions in Germany at that time.[375]

It should be noted that the dream of the HAFRABA Autobahn was finally realized 50 yr later (1976) when the last two kilometers between south Germany and the border of Switzerland near Basel opened to the public.

In May 1935, Hitler opened the first autobahn section between Frankfurt and Darmstadt. The material applied at that time for the surface of the new autobahn was cement concrete (around 90 percent) and bituminous pavements to a lesser extent.

Between the early 1930s and the end of World War II, 3881 km of autobahnen opened to public traffic in Germany and in the occupied regions.[33]

5.3.4.3   First Design Recommendations for Freeways in Germany.   Design instructions and alignment principles for German freeways (autobahnen or interstates) were recommended in 1943, and were referred to as *BAURAB TG*.[92] On the basis of the BAURAB TG and the

FIGURE 5.39   Course of the HAFRABA Autobahn through Germany and its extension through Switzerland to Genoa in Italy, 1926.[375]

"Preliminary Guidelines for the Design of Rural Roads (RAL 1937),"[558] modern German guidelines for alignment design were developed beginning in the year 1959.

According to the "Building Instructions for National Autobahnen BAURAB TG,"[92] elements of the alignment were already related to safety, comfort, economic principles, as well as to aesthetic considerations. The design elements depended on four autobahn categories, which were differentiated by topography classes.

Table 5.2 provides one of the first listings with respect to design element limitations in plan and profile versus speed for different design (topography) categories.

The standard cross section of the autobahn is shown in Fig. 5.40. The overall width was 28.50 m and consisted of the following cross-sectional elements:

Two unpaved shoulders	2.00 m =	4.00 m
Two paved shoulders	2.25 m =	4.50 m
Two directional travel lanes	7.50 m =	15.00 m
Two inside paved edgestrips	0.50 m =	1.00 m
One median	4.00 m =	4.00 m

With respect to highway geometric design, the following specific recommendations were introduced in the "Building Instructions for National Autobahnen (BAURAB TG)" in 1943:[92]

1. The horizontal alignment should be consistent and well tuned. Spiral curves (clothoids) should be applied as transitions between tangents and curves as well as between curves and curves. Successive curves in the same direction of curvature with a small tangent in between should be avoided.

2. Tangents can be applied and are useful in flat topography. In hilly and mountainous topographies, however, the tangent should be replaced by appropriate curves. Very long tangents lead to drivers' fatigue, and increase the speed level considerably.

3. A radius of $R = 250$ m should be regarded as an absolute minimum for autobahnen (interstates). The length of the circular curve should be at least 300 m. In order to avoid inconsistencies in the horizontal alignment, clothoids should be used as transitions between tangents and curves when the radius of curve is less than 3000 m.

4. The necessary full superelevation rate should be provided over the whole length of the circular curve. To establish a transition between the superelevation rate on the tangent and the superelevation rate on the circular curve, the clothoid should be used for the superelevation runoff. This was realized by rotating the directional travel lanes around the inside edge, or in other words, around the borderline between the travel lane and the edge strip. An exceptional case was rotating the directional travel lanes around the centerlines.

5. Generally, gradients on autobahnen should not exceed 6 percent. Long steep upgrades ($L \geq 1.5$ km) should be interrupted by flat sections with lengths of at least 400 m. When selecting radii of crest vertical curves, appropriate stopping sight distances should be guaranteed. To achieve a consistent vertical alignment, the radii of vertical curves should be taken as large as possible. Very short tangents between two vertical curves should be avoided because they are aesthetically unpleasant.

6. These early recommendations also recognized the importance of a well-tuned three-dimensional alignment, as demonstrated by the following statement[92]:

> Because the alignment of a road has to be fitted into a three-dimensional landscape, the design of the horizontal and of the vertical alignments must be well tuned, and the spatial view must be examined. Measures to do this include models, photographs, or spatial perspectives of the road in order to achieve a design which is aesthetically pleasing and safe.

TABLE 5.2   Design Elements Versus Speed for Different Design (Topography) Categories on Autobahnen, 1942[92]

Autobahn category	Category 1, flat topography, V = 160 km/h*	Category 2, hilly topography, V = 140 km/h*	Category 3, mountainous topography, V = 120 km/h*	Category 4, high mountains, V = 100 km/h*
Minimum radius of curve	2,000 m (1,000 m)†	1,200 m (600 m)†	800 m (400 m)†	500 m (250 m)†
Minimum radius of crest vertical curve	20,000 m	12,000 m	8,000 m	6,000 m (4,000 m)†
Minimum radius of sag vertical curve	10,000 m	8,000 m	6,000 m	6,000 m (4,000 m)†
Maximum permanent grade	4.0%	5.0%	6.0%	6.5%
Minimum sight distance	300 m	250 m	200 m	150 m
Minimum superelevation	1.5%	1.5%	1.5%	1.5%
Maximum superelevation	6.0%	6.0%	6.0%	6.0%

*Basic speed for different design topography categories.

†Applied only in exceptional cases.

  With respect to the information just given, it should be emphasized that these early and quite detailed recommendations will accompany us like an unbroken thread through many of the following chapters. It is quite surprising that fundamental issues of highway geometric design, which are known to us today, were already regarded and acknowledged as important in 1943. During that year, Hans Lorenz published a report on design consistency. He provided the following principles for homogeneous autobahn alignment[485]:

- The transition of two successive circular curves should be consistent; that means, a ratio of up to 1:1.5 between two radii of curve should not be exceeded.
- A long stretched curve should not be interrupted by a tangent.
- Crest and sag vertical curves should not be interrupted by short tangents because they lead to visual breaks or to the broken back effect.
- The reversing points of curves in the horizontal and vertical alignment should be set at approximately the same location in order to provide a consistent three-dimensional alignment.
- Besides consistency in geometric design, consistency in speed is also an important issue.
- Between two sections with different road characteristics there should be an appropriate transition length for speed adaption.

These six statements provide high standards for safety and comfort,[485] as will be shown, for example, in Chap. 16.

  In 1943, Hoffmann[294] conducted the first accident investigation on the new autobahnen (interstates); he found that the relative accident frequency on autobahnen was considerably lower than that on two-lane rural roads. He indicated that the main accident cause was inappropriate speed coupled with technical and physical characteristics of the motor vehicles and the drivers. He further indicated that horizontal curves didn't cause any more accidents than tangent sections. He concluded that a generous horizontal alignment, wide travel lanes, and adequate sight distances may be reasons for such findings, and that special attention should be given to long, steep grades. In this connection, comprehensive accident investigations on the autobahn network revealed that long, steep grades with steadily increasing grades represented the most dangerous sections of the alignment. In the downgrade direction, inappropriate high speeds were often observed at the end

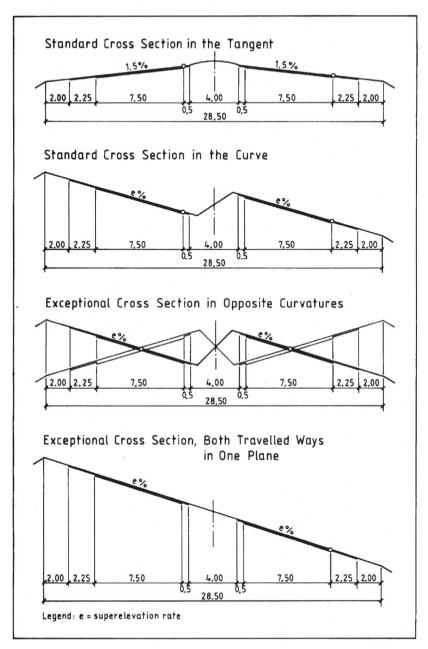

FIGURE 5.40 Cross sections of the national autobahnen, 1942.[92]

of the gradients, whereas in the upgrade direction, heavy trucks or small motorized vehicles experienced problems on the steep grades. This leads to the conclusion that, in order to decrease the accident situation, long, steep grades should be designed with constant slopes and moderately sloped sections in between.

Another interesting result of these early investigations was that a higher accident frequency was experienced on the monotonous portions of the alignment than on the curvilinear ones. The latter is related to the fact that there exist considerable differences in attention levels between monotonous and curvilinear portions. Monotonous sections, like tangents that are too long, often lead to driver's fatigue and, consequently, to a lower attention level, thus increasing the risk of accidents.[294]

All of the aforementioned early considerations, along with practical and research investigations, have led to further development of geometric design standards for freeways as well as for two-lane rural roads, not only in Germany, but also in other countries of the world. As a result, considerably higher traffic quality, safety, and comfort level of the road system were achieved during the last few decades.

5.3.4.4 Development in the United States.*

The main goal of highway design and construction in the United States until the 1920s was to establish a road network which would enable access to the farthest parts of the country. At the beginning of the 1930s, this goal was achieved. At that point in time, economic and mobility demands stood in the forefront, as in Europe, in expanding the highway networks.

The construction of the first city parkway, the Henry Hudson Parkway, which passes through the west side of Manhattan, began in 1934. Soon after, more and more exclusive automobile roads were built, which had a great impact on the driving behavior of motorists and on traffic safety. As the word "parkway" reveals, this type of roads mainly served recreational areas. Thus parkways were built for the purpose of reaching beautifully landscaped, scenic areas, whereas freeways were designed for the speedy, efficient transport of people and goods within the nation. Freeways are comparable to the German autobahnen. The word "freeway" was first introduced by Edward Bisset, chairman of the American National Conference on City Planning, to describe roads which could be used with no tolls. Nowadays, the word freeway refers, in particular, to grade-separated interchanges with full access control.

The 6-km-long Pasadena Freeway between downtown Los Angeles and the suburb of Pasadena was opened for traffic in 1940. It resulted in a general boom of highway design and construction in the United States. In 1939, the councillor of city planning, Lloyd Aldrich of Los Angeles, recommended the planning and construction of about 1000 km of freeways in the Los Angeles metropolitan area. Such technically magnificent, but unrealistic, dreams often came up until the 1950s and 1960s. Fortunately, they were not all realized.

Another very important step in the development of the interstate system was the 270-km-long Pennsylvania Turnpike, which opened for traffic in 1940. The turnpike between Harrisburg and Pittsburgh was the product of an election promise. Its construction took only 2 yr. The turnpike design engineers partially adopted the techniques of the Germans with respect to autobahn design, especially concerning the cross section with respect to the widths of medians and travel lanes. The new freeway cut travel time by half. This turnpike constituted the first portion of today's huge interstate system in the United States.

The National Interregional Highway Committee, which was founded in 1941, recommended in a 1944 report the establishment of a national system of interstates and defense highways. As the name implies, military issues were also regarded in the planning process of a comprehensive interstate network. However, because of low public interest after the end of World War II, the planned interstate system, originally initiated by President Roosevelt, was not realized at that time.

In 1954, President Eisenhower founded a committee to examine the needs of freeways in the United States. The chairman of the committee was General Lucius D. Clay, the former organizer of the Berlin airlift. Another member of the committee was the Secretary of Defense, Charles

---

*Elaborated based on Lay.[466]

Wilson, who reflected the strong influence of the military lobby. The investigation recommended a comprehensive interstate network as a national requirement. In 1956, the Federal Aid Highway Act and Highway Revenue Act was passed as a legal law and a funding plan was provided. The design and planning of the interstate network, based on a 1975 traffic forecast, encompassed the construction of some 70,000 km of interstates which should be finished by 1972. There are, however, still gaps in this network. A governmental decision established a trust fund fed by fuel and motor vehicle taxes in order to finance almost 90 percent of perhaps the world's largest civil engineering project.

While freeway networks had a great impact on the development of efficient transportation systems throughout the world, urban freeways remained, more or less, characteristic of the United States. Perhaps the two main reasons for this development are as follows:

1. In 1956, at the start of establishing the interstate system, there were around 800 km of urban freeways that already existed or were under construction.

2. The chessboard-like pattern of the infrastructure of a typical U.S. city led to freeways that ran directly into or passed through downtown areas.

Unfortunately, most of the urban freeways exceeded reasonable dimensions, for example, corridors that were more than 100 m wide. Consequently, they took up a considerable amount of space in valuable downtown areas, and today still cause heavy noise and air pollution. In addition, they are unpleasant to look at. A number of studies have indicated that huge urban freeways often attract more traffic than they can handle, and cause traffic jams or at least traffic congestion. However, it should be noted that urban freeways are especially successful in suburban areas in cases where the development follows construction of the freeway.[466]

As the previous discussion has revealed, the development of roads exclusively for automobiles, like autobahnen, interstates, or other freeway systems, was an important step forward for economic and mobility developments of nations, and enhanced, to a certain extent, living standards and social lives. In this respect, the introduction of the first design guidelines in Europe and the United States led to standardized rules for the design and construction of roads and freeways.

The road engineering stages in Germany and the United States were selected here as examples from two continents where early approaches for modern highway systems, as well as corresponding design and construction recommendations, have existed since the 1920s. Of course, similar developments existed during that time in several other countries, too.

5.3.5   Time Period: 1970–1980

In 1976, an Office of Economic and Cultural Development (OECD) Symposium, entitled "Methods for Determining Geometric Design Standards,"[556] was held in Helsingoer, Denmark, in order to provide researchers with an opportunity to exchange ideas on how highway geometric design is related to traffic safety, economy, environmental impacts, and energy consumptions.

Topics of discussion at the symposium included:[556]

- Models for driving behavior and traffic flow related to design features
- The impact of design parameters on traffic safety and traffic flow
- The economic consequences of various geometric design guidelines
- The relationship and optimization of certain highway geometric design elements with respect to the road class

Specific attention was given to:

- Driving dynamics and road characteristics with respect to operating speeds
- Influence of design parameters on accident frequency and costs

- Traffic flow on roads outside built-up areas related to certain design elements
- Effects of geometric design guidelines on road construction costs
- Basic considerations about road classification systems

Important outcomes or conclusions drawn from the symposium included the following:[556]

- The establishment of highway geometric design guidelines is an evolutionary process, caused by changing traffic compositions, new insights on driver behavior and attitudes, continuing developments in vehicle design, and improved roadway technology. This requires more intensive, multidisciplined cooperation among diverse professionals.
- Economic considerations must continue to influence highway geometric design. However, it was agreed that, due to the long service life of roadways' geometric parameters, they cannot be unduly compromised to save on initial costs. This does not mean that economic considerations should be neglected, but that the improvement in road safety should also play a decisive role in future highway design.
- Driver characteristics and driving behavior have to be considered more seriously in design.
- The relationships between traffic flow (quality) and highway design parameters have to be given special consideration.
- Road classification systems are necessary to define design standards which are practical. However, the design standards should be based on the actual driving behavior of road users.
- Several countries appeared to be developing new concepts which were directed toward relating design elements to:
  1. Sound network and socioeconomic planning
  2. The quality of traffic flow and traffic goals
  3. Characteristics and requirements of the desired and expected driving behavior of the road users with particular respect to observed operating speeds.

Until the 1970s, economic and qualitative driving dynamic safety aspects were most important in highway geometric design guidelines. However, in the decade from 1970 to 1980, researchers developed:

- Relationships between design elements and the actual driving behavior of motorists (normally expressed by the 85th-percentile speeds of passenger cars under free flow conditions[348,385]
- A more sophisticated, driving dynamic safety-based design procedure[382,383,385]
- Related both issues to observed accident rates[368]

The results of the conceptions of the time period from 1970 to 1980 were partially introduced into several guidelines.[2,35,124,139,241,510,607,686,691] Furthermore, basic requirements about road functions in networks[242] and a sound coordination of horizontal/vertical alignments and cross sections[240] were developed during this time period.

5.3.6 Outlook

Highway design and traffic safety researchers have produced volumes of reports and studies and continue to do so. For instance, the reference list of the German Commentary to the Guidelines for the Design of Rural Roads[124] accounts alone for 1100 references. Therefore, in what follows, only a relatively small number, but important, design fundamentals, which became relevant during the last two decades, are cited:

- The classification of roads into category groups, based on their functions in the road network and the coordination of corresponding design and operational demands for individual road categories

- The definition of relevant speeds, such as design speed and operating speed, and their establishment and tuning
- The evaluation of consistency of the alignment, as well as sight distances
- The sound driving dynamic development of geometric design elements, based on defined friction factors

Before the 1970s, an acceptable design was one in which each design element met the driving dynamic requirements. No specific rules were given for coordinating the actual driving behavior with the design speed concept. No mention was made or guidance given for any design procedure which would achieve consistency between roadway sections and the corresponding operating speed profiles.

As a result of the developments in the 1970s and 1980s, specific design and operational characteristics were assigned to individual road categories. At that time, the design speed concept had the strongest impact on the arrangement of driving spaces. The design speed is a performance speed which should correspond to the desired speed of passenger cars under free flow conditions and can be freely selected within road category limitations. Furthermore, the design speed depends on environmental, topographical and economic conditions, the provided network function of the road, and the required traffic quality for this function in terms of the travel purpose. For safe operation, the design speed has to be coordinated with the actual driving behavior on the roadway section under study.

Operating speeds of free-flowing vehicles are largely governed by the roadway geometry presented to the driver. Horizontal geometric design exerts the most control over vehicular speeds due to the forces acting on a vehicle as it traverses a curve. The design speed of a horizontal curve represents the upper limit of safe operation of the curve. Tighter curves may have lower design speeds, especially on existing (old) alignments. As curves are flattened (radii increased), design speeds become higher until one reaches the "flattest" curve—a tangent section—where the theoretical design speed is infinity.

In developing a geometric alignment, highway designers have often mixed curves of varying design speeds with tangent sections. As presented to the driver, this highway alignment places varying speed controls on the speeds the driver wishes to travel. However, safe design of the highway results when the changing operating speed caused by the alignment is consistent with a driver's desires and expectations.

To achieve safe design, the designer needs to know the impact that various curves have on operating speeds (specifically on 85th-percentile speeds), in order to tune them with the selected design speed.[347,348,350,385] Further design guidelines must be developed which recommend maximum allowable changes in radii of curve sequences, as well as for the transitions between tangents and circular curves. By defining an allowable mix of radii, the designer reduces the probability that a design will attempt to govern the driver's desires in a surprising and unsafe manner. The design has to be adjusted, so that alignment-related speed changes do not lead to risky driving operations.

Detailed research into the connection between design speed and 85th-percentile speed led to reconsideration of skid resistance attitudes between tires and the road surface. Investigations in the 1970s[382,383,385] resulted in equations for sound tangential and side friction factors, which cover 95 percent of the skid resistance values on wet road surfaces in Germany.[747] Incorporating these results, yielded recommendations for

- Minimum radii of curve
- Minimum stopping sight distances and radii of crest vertical curves

with significant safety margins.

It should be noted, however, that, to date, mainly qualitative criteria have been developed for achieving sound alignments, and no accurate procedure has been given to predict quantitatively the traffic safety performance of new designs or redesigns.

Alignment design procedures are influenced primarily by the experience and education of the highway design engineer. According to Fig. 5.41 the development started with simple polygon

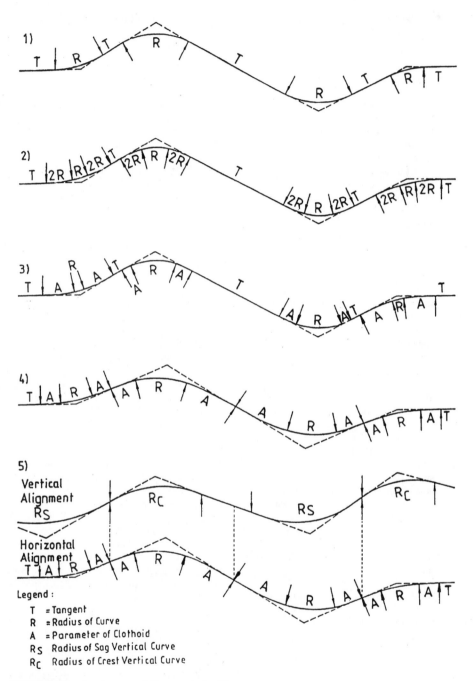

FIGURE 5.41   Development of alignment design.[124]

sections to describe horizontal alignments which were based first on circular curves. Over time, alignments were developed using the standard elements—tangents, transition curves (clothoid or spiral) and circular curves—in horizontal alignment and by the elements—tangents, circular curves, and quadratic or cubic parabolas—in the vertical alignment.[124] Today, an early incorporation of the vertical alignment into highway geometric design and mutual coordination with the horizontal alignment are routinely performed.[657]

Figure 5.41 shows the development over time of alignment design:

1. Tangents and circular curves.

2. Tangents and circular curves with transition curves (circular curve with double radii of curve as transition curve).

3. Tangents and circular curves with transition curves (clothoid or spiral, cubic parabola, etc.).

4. Alignments like in item 3, but without any interim tangent.

5. Three-dimensional alignments with superimposed reversing points as in item 4, but including the vertical alignment. These could be called ideal "curvilinear alignments."[444]

During the past 55 years, many countries have adapted, refined, and updated their policies to reflect national conditions and safety as well as operating experience. Design speed continues to be a cornerstone of alignment design, but interesting differences now exist in how design speed is selected and applied. Several countries have recognized a need to base design speeds more directly upon actual driver speed behavior and to include checks on estimated operating speeds along designed alignments.[361]

Many of the design issues developed during the last two decades are meanwhile, at least partially, introduced into modern international guidelines, for example, see Refs. 5, 35, 37, 139, 246, 453, 663, 686, 693, 700, and 708.

CHAPTER 6
OVERVIEW

6.1 RECOMMENDATIONS FOR PRACTICAL DESIGN TASKS*

The Southern African Transport and Communications Commission (SATCC) gives with respect to their recommendations on road design standards the following introduction about geometric design on rural roads[663]:

> The criteria for formulating geometric design shall be pertinent with local conditions in respect to management, finance and development trends. They shall further be economically feasible, taking into consideration all relevant existing conditions such as the climate, topography and geotechnical conditions encountered throughout the planning region. The recommendations for geometric design have been made for all categories of rural and major suburban roads, but with special emphasis on two-lane rural roads.[663]

6.1.1 Content

Part 2, "Alignment (AL)," contains principles and methods as well as limiting and standard values for new designs, redesigns, and restoration strategies of nonbuilt-up roads outside (rural) and inside built-up areas (suburban).

This chapter deals with the geometric design of horizontal and vertical alignments and cross sections. Furthermore, instructions and principles are provided for the three-dimensional alignment and for the creation of driving space. Special emphasis is given to the relationships and interrelationships between highway geometric design, speed, driving dynamics and safety, and human factors and vehicular safety.

Chapter 6 is divided to a major extent into 2 subchapters: Subchapter 1: Recommendations for Practical Design Tasks; and Subchapter 2: General Considerations, Research Evaluations, Guideline Comparisons, and New Developments (for more information see Preface).

6.1.2 Range of Validity

Roads for public travel can be categorized, according to Part 1, "Network" (Fig. 3.1) by

- Location (outside or inside built-up areas)
- Adjacent area (without buildings or with buildings)
- Relevant design function, such as mobility (connector), access (collector), and local or pedestrian use.

---

*Elaborated based on Refs. 159, 241, 243, 244, 246, 432, 433, and 453.

TABLE 6.1   Range of Validity of the Recommendations for the Design of Highway Facilities*

Location	Adjacent area	Relevant design function	Category group	Applicable design part
Outside built-up areas (rural)	Without buildings	Mobility (connector)	A	Network (NW) Cross sections (CS)
Inside built-up areas (Suburban/urban)	Without buildings	Mobility (connector)	B†	Alignment (AL) Intersections (IS) Interchanges (IC)
	With buildings or built-up capable	Mobility (connector)	C†	Network (NW) Major urban roads (MA)
		Access (collector)	D	Network (NW) Minor urban roads (MI)
		Local or pedestrian use	E	

*Modified based on Ref. 246.

†Note that collector functions may superimpose connector functions, especially for category groups B and C (see Fig. 1.3 of Part 1, "Network").

They are subdivided into the five road category groups A to E according to Table 6.1. Since there are definite differences in the application of category groups A to E in the road network, the range of validity of the individual design parts is also shown in Table 6.1.[246]

According to the respective significance of goals for mobility, access, and local functions in the road network, roads of category groups A to E can be further divided into 14 road categories (designated by the connector function levels I to VI, as developed in Part 1, "Network"). The functional road categories vary from statewide or interstate connections (I), to highways intended to join major and minor activities (II to V), and to serving residential traffic (VI). Important design and operational characteristics are assigned to these 14 road categories in Table 6.2.

Part 2, "Alignment," is especially valid for new designs, redesigns, and rehabilitation strategies for roads of category group A and road categories B II and B III (see Table 6.2).

The elaboration of the parts "Major Urban Roads (MA)" for roads of category group C and "Minor Urban Roads (MI)" for roads of category groups D and E will not be part of this book. The same is true for the parts "Intersections (IS)" and "Interchanges (IC)" (Table 6.1). The contents of the above-mentioned parts are so extensive that they would verify an additional book.

Additionally, important information about the desirable structure of modern highway geometric design is presented in Subchapter 6.2 (Table 6.4).

Furthermore, it is important to consider certain principles for design purposes with respect to different road categories, even in the early stages, when establishing the Part 2, "Alignment." According to Table 6.3, those principles are:

1. Harmonizing the design speed (V_d) with the 85th-percentile speed ($V85_i$) of free-moving passenger cars for longer roadway sections (Col. 3).

2. Harmonizing the 85th-percentile speeds ($V85_i$ and $V85_{i+1}$) between successive design elements to achieve consistency in the alignment, especially on two-lane rural roads,* for example, between a tangent and a curve or between a curve and a curve (Col. 4).

3. Radii relations (Col. 5).

4. Design principle (Col. 6).

Rural Roads are defined as *nonbuilt-up roads* outside built-up areas (Table 6.1), including the road categories A I to A V (Table 6.2).

5. Harmonizing the side friction assumed (f_{RA}) in the existing design guidelines with the actual side friction demanded (f_{RD}) at curved sites, especially on two-lane rural roads (Col. 7).

6. Utilization ratio of the side friction factor (Col. 8).

7. Presence of transition curves (Col. 9).

8. Reaction and perception time (Col. 10).

9. Presence of passing sight distances (Col. 11).

With respect to items 3, 4, and 7, three quantitative safety criteria have been developed for differentiating good, fair (tolerable), and poor design levels. These criteria are particularly applicable for evaluating the design of two-lane rural roads. Improving the safety of two-lane rural roads would address locations which in the United States and western Europe are the sites of at least 50 percent of all fatal and serious injury accidents.

Fundamental and practical considerations for the development and application of the three safety criteria are presented in Chaps. 8 to 10.

6.1.3   Objective

Part 2, "Alignment of Nonbuilt-Up Roads (AL)," is the basis for safe, functional design and construction of roads; it corresponds to today's state-of-the-art. The statements are built upon theoretical considerations of research results and upon practical experience and testing. The application will not only promote the uniformity of highways of the same type but will also show that different categories reveal perceptible differences. Therefore, the design principles of "AL" are adjusted for each road category, as shown in Tables 6.2 and 6.3. Such a functional differentiation of basic principles makes it possible to take into account the different traffic and nontraffic goals of the road. The design rules in "AL" must be balanced against other interests such as:[243,246]

- Regional planning
- Village and municipal planning
- Creation of the road space
- Economy of construction and operation
- Energy conservation
- Prevention of air pollution
- Environmental protection and care for the landscape
- Safety and comfort of motorized and nonmotorized traffic
- Demands for nontraffic-related utilizations

in order to achieve the best solution for an individual roadway segment. Note that Part 2, "Alignment," is first of all related to roads of category group A and road categories B II and B III.

6.1.4   Application

Application of the content of "AL" should be done with care and judgment. Issues of the environment and economy must be weighed along with the design and safety considerations contained in "AL" so that appropriate, balanced solutions can be arrived at. It is not expected or even desirable that "AL" be applied without regard to environmental consequences or cost considerations. The designer's task is to develop alternate solutions and carefully weigh the alternatives in terms of their satisfaction of environmental, economic, and design engineering safety goals.

The design of roads outside built-up areas must take into consideration confined living spaces; therefore, together with the design of roads of category group A, road traffic interests have to be considered carefully along with those of the environment. Specific environmental concerns which are important for these situations include:

TABLE 6.2 Classification of Roads by Group and Categories (Applicable for Most Countries)*

Road Function		Design and Operational Characteristics				
Category group (1)	Road category (2)	Kind of traffic (3)	Permissible speed limit, km/h (4)	Cross section (5)	Intersection access (6)	Design speed (V_d), km/h (7)
A: Low concentration of buildings — Rural areas — Important connector functions	A I: Statewide or interstate connection	Vehicles	≤120	Multilane	Controlled	(130) 120 (110) 100 90 (80)
		Vehicles	≤100	2 + 1 Lanes	(Controlled) Free	100 90 (80)
	A II: Overregional/regional connection	Vehicles	≤110	Multilane,	Controlled (Free)	(120) 110 100 90 (80)
		All†	≤90	2 or 2 + 1 Lanes	Free (Controlled)	(100) 90 80 (70)
	A III: Connection between municipalities	Vehicles	≤90 (≤80)‡	Multilane,	Controlled (Free)	90 80 70
		All†	≤90 (≤80)‡	2 Lanes	Free	90 80 70 (60)
	A IV: Large area accessibility connection	All†	≤80	2 Lanes	Free	80 70 60 (50)
	A V: Subordinate connection	All†	≤50 (≤70)‡	2 Lanes	Free	(70) 60 50
B: Low concentration of buildings — Urban or suburban areas — Important connector functions	B II: Primary arterial	Vehicles	≤90 (≤80)‡	Multilane	Controlled (Free)	(100) 90 80 70 (60)
	B III: Secondary arterial	All†	≤70	Multilane,	Free	(80) 70 60 (50)
		All†	≤70	2 Lanes	Free	70 60 (50)
	B IV: Main collector	All†	≤60	2 Lanes	Free	60 50
C: High concentration of buildings — Urban areas — Important connector functions	C III: Primary/secondary arterial	All†	≤50 (≤70)‡	Multilane,	Free	(70) (60) 50 (40)
		All†		2 Lanes	Free	
	C IV: Main collector	All†	≤50 (≤60)‡	2 Lanes	Free	(60) 50 (40)
		All†	≤50 (≤60)‡	2 Lanes	Free	(60) 50 (40)
D: High concentration of buildings — Urban areas — Important access functions	D IV: Collector	All†	≤50	2 Lanes	Free	None
	D V: Local	All†	≤50	2 Lanes	Free	None
E: High concentration of buildings — Urban areas — Important local functions	E V: Local	All†	≤30 (Walking speed)	2 Lanes	Free	None
	E VI: Pedestrian use	All†	(Walking speed)	2 Lanes	Free	None

* Modified based on Refs. 243 and 453.
† Indicates all types of road user groups combined. In contrast, "vehicles" means exclusive motor vehicle road.
‡ ( ) = exceptional values.

TABLE 6.3 Design Principles for Roads of Category Groups A to C (Applicable for Most Countries)*

Category group (1)	Road category (2)	Design consistency between $V85_i$ and V_d safety criterion I (3)	Operating speed consistency between $V85_i$ and $V85_{i+1}$ safety criterion II (4)	Radii relations (5)	Design principle (6)	Driving dynamic consistency between f_{RA} and f_{RD} safety criterion III (7)	Utilization ratio of side friction factor (8)	Transition curve (9)	Reaction and perception time (10)	Passing sight distance (11)
A: Low concentration of buildings Rural areas Important connector functions	A I: Statewide or Interst. Conn. A II: Overregional/Region. Conn. A III: Connection Between Municipalities A IV: Large Area Accessibility Connection A V: Subordinate Connection	Multilane: $V85_i = V_d + 20$ km/h $V_d \geq 100$ km/h or $V85_i = V_d + 30$ km/h $V_d < 100$ km/h 2 lane (rural): $|V85_i - V_d| \geq \Delta V_{allw}$ $V85_i$ depends on CCR_s	Multilane: Normally not necessary† 2 lane (rural): $|V85_i - V85_{i+1}| \geq \Delta V_{allw}$ $V85_i$ depends on CCR_s	Necessary	Vehicle dynamics	$f_{RA} - f_{RD} \geqq \Delta f_R$	40% for $e_{max} = 7\%$ (hilly/mountainous topography) 45% for $e_{max} = 8\%$ (9%) (flat topography) 10% for $e_{min} = 2.5\%$	Necessary	2.0 s	Necessary for 2 lane rural roads
B: Low concentration of buildings Urban or suburban areas Important connector functions	B II: Primary arterial B III: Secondary arterial B IV: Main collector	$V85 = V_{perm} + 20$ km/h $V85 = V_{perm} + 10$ km/h $V85 = V_{perm}$	Normally not necessary†	Necessary Desirable	Vehicle dynamics	$f_{RA} - f_{RD} \geqq \Delta f_R$ Normally not necessary†	60% for $e_{max} = 6\%$ 30% for $e_{min} = 2.5\%$	Necessary Desirable	2.0 s	Not necessary
C: High concentration of buildings Urban areas Important connector functions	C III: Primary/secondary arterial C IV: Main collector	$V85 = V_{perm} + 10$ km/h $V85 = V_{perm}$	Normally not necessary† Not necessary	Desirable	Vehicle dynamics	Normally not necessary†	70% for $e_{max} = 5\%$ 70% for $e_{min} = 2.5\%$	Desirable	1.5 s	Not necessary

*Modified and completed based on Refs. 243 and 453.

†If in doubt or for extreme alignment conditions, the safety evaluation criteria for two-lane rural roads should be applied.

Note:

$V85_i$	= 85th-percentile speed of the design element i, km/h	f_{RA}	= side friction assumed
V_d	= design speed, km/h	f_{RD}	= side friction demanded
ΔV_{allw}	= allowable speed difference, km/h	Δf_R	= allowable difference in side friction
V_{perm}	= speed limit, km/h	e_{max}	= maximum superelevation rate, %
CCR_S	= curvature change rate of the single curve, gon/km	e_{min}	= minimum superelevation rate, %

6.5

- Protection of natural resources
- Protection of landscape and historic monuments
- Prevention of air pollution.

For roads inside built-up areas, effects of the environment depend strongly on the use of adjacent properties and built-up areas. Therefore, for roads of category group B, it is especially necessary to check the compatibility of the environment and the road with regard to:[243,246]

- The nonmotorized traffic utilization demands
- The effects of noise and emissions
- The utilization of green zones
- The design of municipalities

In the text, tables, and figures, limiting values (minimum or maximum) are given for the design of roads. These values are related to driving dynamics, driving geometry, optical, and drainage requirements which will ensure the safe travel of a single vehicle. In addition, "AL" contains three quantitative safety criteria which enable the highway engineer to evaluate the interaction between highway geometric design, driving behavior, driving dynamics, and the expected accident situation in order to distinguish good, fair (tolerable), and poor design practices; see, for example, Refs. 108, 424, 435, 438, 444, 447, 448, 453, and 455.

Arrangements that result primarily from reasons of uniformity and esthetics are usually recognized as standard values.

6.1.5 Important Goals in Highway Geometric Design

Good highway geometric design is measured according to six important but sometimes conflicting goals:

- Esthetics
- Economy
- Environment
- Function
- Safety
- Traffic quality (capacity)

This alphabetic listing indicates that it is often impossible to maximize one important planning goal without taking into account essential losses from other important goals. Therefore, when balancing these different goals with respect to highway design, an absolute optimum can never be reached. Instead, the best solution is often where an "acceptable compromise" has been found.

Furthermore, human, vehicular, and roadway factors influence, directly or indirectly, the important design issues:

- Network
- Cross sections
- Horizontal alignment
- Vertical alignment
- Sight distance
- Three-dimensional alignment

The flowchart in Fig. 6.1 is an attempt to clarify the relationships between important goals and highway geometric design parts. As can be seen from Fig. 6.1, the important goals are highly cor-

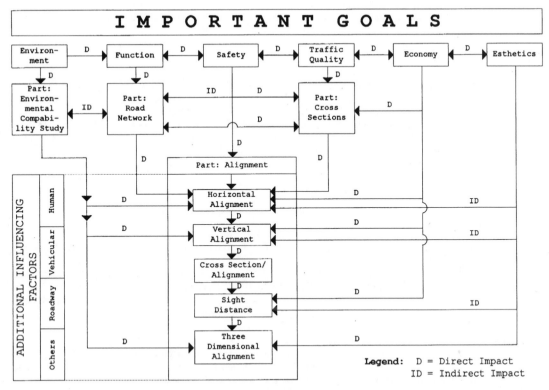

FIGURE 6.1   Overall flowchart for the interrelationships between important goals in highway geometric design and the corresponding design parts.[433,436]

related with each other. Placing them in priority order would require the subjective opinion of the evaluating engineer. The same is true regarding the direct and indirect impacts of these important goals on the individual design parts and their interactions.

Since human, vehicular, and roadway factors influence all important goals, as well as all highway geometric design parts, their impact is not individually presented in the flowchart in Fig. 6.1 (see also Chaps. 19 and 20, especially Sec. 20.3).

Most countries of the European Union have enacted laws to require detailed environmental studies of proposed highways, railroads, waterways, and airports prior to their final design and construction. These studies are an attempt to forecast the environmental consequences of alternate solutions so that the best "acceptable compromise" can be reached (see Chap. 7).

6.2   GENERAL CONSIDERATIONS, RESEARCH EVALUATIONS, GUIDELINE COMPARISONS, AND NEW DEVELOPMENTS

6.2.1   Content

Table 6.1 presents the range of validity of recommendations for the design of highway facilities. In this table, Part 2, "Alignment of Nonbuilt-Up Roads (AL)," is assigned to roads of the category groups A and B and encompasses the development, analysis, and evaluation of design elements and element sequences in plan and profile with special emphasis on driving behavior and traffic safety. Design principles for a sound three-dimensional alignment are also integrated.

6.2.2   Range of Validity

In developing "Recommendations for Practical Design Tasks," consideration was given to including roads of category groups A to E (Table 6.1) and roads of categories A I to E VI, as defined in Table 6.2, or differentiating between roads outside built-up areas (rural) and inside built-up areas (urban). Both possibilities are found in the guidelines of different countries. However, as Tables 6.1 and 6.2 show, overall recommendations for all category groups (A to E) which contain the constraints:

- Outside and inside built-up areas
- Roads without and with buildings
- Roads with different traffic functions (connector, collector, local)

are very complex and difficult to deal with. For example, the consistency and driving dynamic demands of nonbuilt-up roads are very different from the constraints frequently present in municipal road design.

Based on these considerations and on the German experiences with their 1984 "Highway Geometric Design Guidelines"[243] (which included the category groups A to C and partial treatment of D and E[159]), it appears desirable to confine Part 2, "Alignment," strictly to roads of category group A and road categories B II and B III. According to the German guideline framework (which may be recommended as the basis for modern highway geometric design guidelines in the countries of the European Union[159]), it is recommended that for roads of category group C an individual part called "major urban roads" be developed and for the category groups D and E an additional part called "minor urban roads" be developed (see Table 6.1). Of course, interfaces and superimpositions between these parts are always present and desirable.

For a better understanding, a desirable structure of modern highway geometric design guidelines is presented here with brief statements about the contents of individual parts in Table 6.4.[160] Only those parts which have a direct impact on Part 2, "Alignment of Nonbuilt-Up Roads," are listed.

6.2.3   Objective

6.2.4   Application

The recommendations provided in "AL" should not be regarded as fixed rules but rather as overall balanced solutions, and should be used in conjunction with other utilization demands and environmental protection issues.

Section 6.2.5 provides interrelationships between important goals in highway geometric design and the corresponding design parts.[433]

Furthermore, an additional chapter, which describes the basic procedure used in road planning and road design for different design levels, including the environmental protection issues, is introduced, as an overall integrated *road design process* (see Chap. 7). Figures 7.1 and 7.2[160,433] were developed in order to present systematically the interfaces with other parts during the design stages, and to inform the design engineer about how to regard and apply them with special emphasis on environmental compatibility.

TABLE 6.4 Structure of Main Parts for Modern Highway Geometric Design*

Parts	Brief statements
Part: Network (NW)	Fundamentals of road network design Classification of roads into road categories Functional justified design of the road network
Part: Cross sections of nonbuilt-up roads (CS)	Fundamentals for the dimensions of the components of the road's cross section Cross-sectional design Proof of traffic quality

Part: Alignment of nonbuilt-up roads (AL)	Elements of the alignment:
	  Relevant speeds
	  Safety criteria
	  Design elements in plan
	  Design elements in profile
	  Design elements of cross section
	  Design elements of sight distance
	Three-dimensional alignment:
	  Elements of three-dimensional alignment
	  Design of the driving space
	  Procedure and remedies

Part: Intersections (IS)	At-grade intersections:
	  Fundamentals and basic rules for intersection design
	  Principal solutions for intersections
	  Design elements for intersections
	  Equipment
	  Application modes

Part: Interchanges (IC)	Grade-separated intersections (interchanges):
	  Fundamentals of interchange design
	  Interchange systems
	  Design and construction
	  Equipment

Part: Environmental compatibility study (ECS)	Space-related sensitivity investigation Comparison of alternatives

Part: Landscape cultivation (LC)	Section 1: Landscape-justified planning:
	  Integration of the preservation of natural beauty and wildlife and of landscape cultivation into the road planning process
	  Planning fundamentals, relationships between effects
	  Attendant planning for landscape cultivation
	
	Section 2: Realization of landscape cultivation:
	  Planting plan
	  Realizations of road plantings
	  Cultivation of road plantings

Part: Major urban roads (MA)	Recommendations for the design of major urban roads (arterials):
	  Arterials in connection with municipalities
	  Fundamentals and design examples for the design of road spaces

Part: Minor urban roads (MI)	Recommendations for the design of collector roads:
	  Collector planning in connection with municipalities
	  Fundamentals and design examples for the design of road spaces

(Continued)

TABLE 6.4   Structure of Main Parts for Modern Highway Geometric Design*   (*Continued*)

Parts	Brief statements
Part: Drainage (DR)	Planning and design
	Surface-water drainage
	Above-ground installations for water runoff
	Underground installations for water runoff
	Outlet
	Structures
	Dripping installations
	Drainage of engineering structures
	Drainage of roads in water resources areas
	Drainage of roads during construction
	Landscape-justified planning of earth basins
Part: Economy (EC)	Goals and fundamentals of economy investigations
	Value rates for the benefit component
	Network definition, section arrangement, and traffic flow
	Determination of costs
	Determination of benefits
	Determination of the benefit/cost ratio
Part: Public transport streetcar (PT)	Part: Facilities for local public transport—Sec. 1: Streetcar:
	Fundamentals
	Basic dimensions of streetcar track infrastructures
	Track alignment
	Stops
	Examples
Part: Public transport bus (PT)	Part: Facilities for local public transport—Sec. 2: Bus and trolley bus:
	Fundamentals
	Basic dimensions of bus infrastructures
	Stops
	Examples

*Modified based on Ref. 160.

6.2.5   Important Goals in Highway Geometric Design

With respect to the design of road infrastructures, the demands of future roads have to be met with respect to the individual goal criteria shown in Fig. 6.1. The most important goals are:[159,433]

- Safety
- Traffic quality (capacity)
- Economy
- Environmental compatibility

These goals, as well as their interactions, have to be considered together. They depend on the function of the road (connector, collector, local) and the expected use (traffic volume and composition, as well as on the desired traffic qualities). Furthermore, it is necessary to coordinate these goals with other subject areas, such as regional and municipal planning. How these goals affect the individual design parts, shown in Table 6.4, will be discussed in the following section:[159,160,433]

Goal: Safety   The performance of a highway with regard to safety results from problems of human perception and behavioral modes, as well as from the physical rules governing the interaction between the vehicle and the road surface. The following safety issues are included in the individual design parts:

Three newly developed safety criteria for an overall evaluation process	Part: Alignment
Specific safety aspects, such as: Driving dynamics	Part: Alignment
Drainage	Part: Alignment; Part: Drainage
Sight conditions	Part: Alignment; Part: Intersections; Part: Interchanges
Design of driving space	Part: Alignment; Part: Intersections; Part: Interchanges
Road equipment, traffic control	All basic design parts

Goal: Traffic Quality (Capacity)   The traffic quality is determined by traffic composition, cross section of the road (Part: Cross Sections), and horizontal and vertical alignment (Part: Alignment).

Goal: Economy   The economy goal is to choose the largest possible benefit-cost ratio among alternate design approaches including a zero and an RRR* alternative. The same is true for considering a sequence of several road construction projects, according to their priority needs with regard to different benefit-cost ratios (Part: Economy).

Goal: Environmental Compatibility   The effects of road construction on ecology and the environment have to be considered in the planning and design stages (Part: Environmental Compatibility Study; Part: Landscape Cultivation).[159,160]

---

*RRR = restoration, rehabilitation, resurfacing strategies for existing roads.

CHAPTER 7

BASIC PROCEDURES IN ROAD PLANNING AND DESIGN WITH SPECIAL EMPHASIS ON ENVIRONMENTAL PROTECTION ISSUES

7.1 RECOMMENDATIONS FOR PRACTICAL DESIGN TASKS

7.1.1 General

The *French Highway Design Guide*[700] provides several general directives with respect to precautions related to protection of the environment that must be taken into consideration during the planning stages; it also emphasizes that special attention should be given to the best possible way of integrating the road into the environment. The environment should, as far as possible, be protected from any damage; unavoidable damage should be kept to a minimum. Damage correction or mitigation measures should be considered if the environment is damaged. However, taking some design precautions in the early stages of planning or restoration is far better than having to take corrective measures later. Where corrective measures are deemed necessary, they often can provide only compromise solutions, and are usually less effective and considerably more expensive. In addition, the roadside, its appurtenances, and maintenance should also be considered a major issue in protecting the road environment. Not only the natural environment, but also the human environment can suffer from a road. Examples are increased traffic noise, disrupting the urban road network, and interference with agricultural activity caused by new or improved roads.[700]

It is also well known that a road improvement can degrade the natural environment in different ways. Therefore, environmental issues have to be addressed very early during the process of restoration and rehabilitation projects or negative impacts on the environment, such as the following, can be observed:[700]

- Disturbance of the hydrology, especially by modification of the surface drainage pattern flow, which can lead to negative impact on vegetation, slope stability, and groundwater purity, etc.
- Disturbance of the local ecosystem and perhaps an adverse impact on plant and animal life near the road
- Degrading of the landscape, for example, with extensive earthmoving for high embankments or deep cuts[700]

The only way to evaluate an alignment with the least impact on the environment is to study several alternative alignments. In this way, sensitive areas like fragile ecosystems, natural reserves, or sensitive water zones can be avoided. Another issue about such an alternative comparison is the limitation of earthwork and the examination of the planned land use with respect to the different alternative alignment designs.

Furthermore, the following is reported in the *French Highway Design Guide:*[700]

> The visual integration of the road into its environment is also important. Good integration through consistent tuning between topography and alignment (for example, by including esthetic considerations in the design and location of structures such as bridges) leads to good environmental compatibility. Landscape architecture can enhance the value of a road and make traveling along it more pleasant and safer. A further important characteristic of a well-integrated road is early recognition and comprehension of the alignment by the driver.
>
> The quality of the environment can also be affected by the design and management of the roadside. It should be recognized that inappropriate design elements or objects may make maintenance tasks difficult or impossible. Therefore, it is recommended that only elements that allow effective maintenance activities be designed and installed. Such activities on the roadside include grass mowing, trimming of shrubs and trees, repair and cleaning of the drainage system, and collecting litter, waste, etc.[700]

Corresponding to the coordination of the important goals in highway geometric design presented earlier (Fig. 6.1), the elaboration of a roadway design includes several design levels, as shown in Fig. 7.1. With each level, the requirements become more concrete, and consequently the design becomes more detailed. At the same time, the remaining planning levels are concentrated on fewer and fewer alternatives, since through such a selection procedure, useless alternatives will be excluded—for example, based on the criteria environmental compatibility, function, safety, esthetics, and economy as well as traffic quality (capacity) (see Fig. 6.1). In principle, the unity of planning, design, construction, and operation has to be considered during the whole design procedure, that is, the project has to be harmless to the environment, safe, economical, traffic-efficient, and has to ensure sound operation.[159,246]

In parallel with technical requirements, legal procedures must be followed with respect to regional planning and acquisition of rights-of-way and easements (drainage, temporary construction, etc.). The extent of compensation and replacement measures for the new road structure must also be clarified and settled according to local laws and regulatory requirements. Figure 7.1 shows the range of validity, as suggested for modern highway geometric designs, for different design levels with special emphasis on environmental protection issues, in order to clarify the interrelationships that exist with other design parts and legal requirements.[159,246,436] Figure 7.1 reveals the importance that the environment, in its most comprehensive sense, should take, beginning with the very first stages of highway design. Note, however, that environmental questions must be considered until the final stage of the design procedure and throughout the daily tasks of road management and maintenance.

7.1.2   Procedure for Conducting an Environmental Compatibility Study (ECS)*

The numerous relationships that exist between driving behavior, road design, traffic flow, and the environment require an iterative procedure for design development. In other words, preliminary assumptions have to be defined and then, during subsequent design stages, have to be evaluated as being valid or not. In the latter case, the design process must be repeated with better assumptions, according to Fig. 7.1. Furthermore, while a conclusive solution which satisfies all problems is not generally possible, various alternatives (which, as far as possible, cover the overall spectrum of possibilities) should be investigated, in order to arrive at the most appropriate solution.

---

*Elaborated based on Refs. 159 and 679.

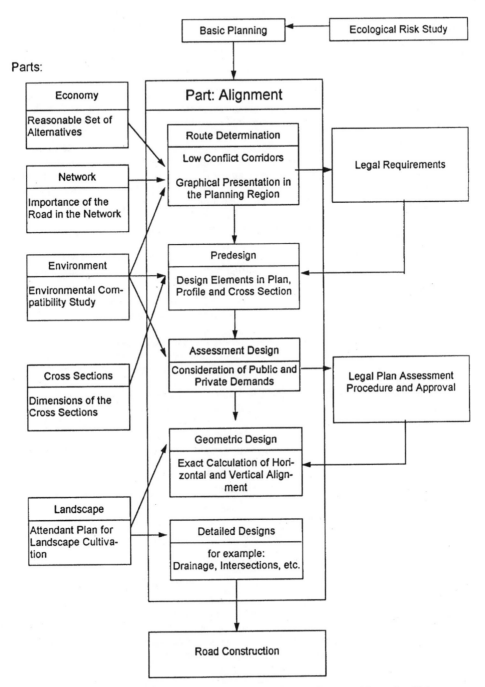

FIGURE 7.1   Design levels and range of validity of Part 2, "Alignment," suggested for modern highway geometric design. (Elaborated and modified based on Refs. 159 and 436.)

In planning major traffic routes, like federal and state routes, an obligation generally exists at all levels of planning to include the competent responsible authorities with respect to nature and landscape protection in the planning process. The basis for this in Europe is the "Law for the Realization of the Guidelines of the Board of June 27, 1985 for the Environmental Compatibility Examination (ECE) for Specific Public and Private Projects (85/337/EWG)" of February 12, 1990.[465]

According to this law, uniform principles for effective environmental protection must be applied, which means the impact on the environment must be identified, described, and evaluated. The results of ECE must be considered in all official decisions for approving a traffic project.[433,436]

Thus, for the construction or alteration of major traffic routes, which have to undergo a legal plan assessment, the law requires the establishment of an Environmental Compatibility Examination (ECE).[465] The ECE procedure is regulated by law; it is a continual and integrated part of the planning process for the determination of the alignment and the legal plan assessment.

ECE is based on an Environmental Compatibility Study, which is controlled and regulated by the "Instructural Guide for the Environmental Compatibility Study in Highway Planning (ECS)."[305] An ECS is normally conducted in two stages (Fig. 7.2):

1. *Space-related sensitivity investigation:* Goal-orientated space analysis and evaluation. Establishment of relatively low-conflict corridor(s) for the alignment and allocation of specific conflict areas.

2. *Comparison of alternatives:* Comparative evaluation of alterations, including the rehabilitation/restoration alternative and the zero (do nothing) alternative.

As an essential part of ECS, the "sensitivity investigation" has to include all relevant environmental issues concerning areas worthy of protection, areas with special sensitivities, as well as the existing and planned land use. Geographical information systems can be used for conducting these studies.

The basic goal of ECS is to identify a highway alternative inside a low-conflict corridor that has the least environmental impact on the surroundings. As mentioned previously, the design engineer also has to consider the rehabilitation and the zero alternative in the evaluation process.

For a better understanding of the individual steps which are involved in an ECS, the flowchart in Fig. 7.2 was developed. In the upper part of the figure, the impact of the space-related sensitivity investigation is presented, while the lower part shows the selection of the most favorable alternative for the planned traffic route. It should be noted that the individual steps listed in Fig. 7.2 are mainly related to the following issues:[679]

- Human, flora, and fauna
- Soil, water, air, climate, and landscape
- Interrelationships between these factors
- Properties, estates, and cultural heritage
- Alignment design requirements

Note that the term "environment" also connotes many social components; that is, life itself, as well as human relations and communications, are included in the term "environment."

Considering traffic route projects, five basic successive steps have to be established in order to conduct an ECS. These principal steps are required by the *Instructural Guide for ECS in Highway Planning.*[305] They are (see Fig. 7.2):

1. Definition and demarcation of the region to be examined.

2. Determination, description, and evaluation of all areas in the region to be examined with relevant functions and impacts on the environment by thematic maps.

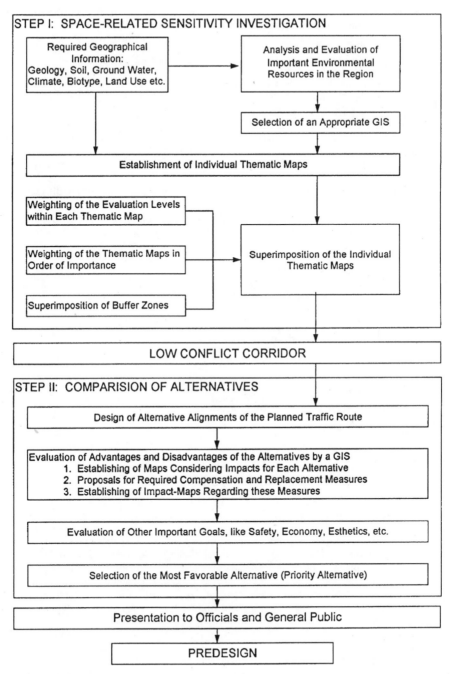

FIGURE 7.2 Flowchart for the design part Environmental Compatibility Study (ECS). (Elaborated based on Refs. 433 and 679.)

3. Development of relatively low-conflict corridor(s) by means of superimposing all areas with relevant functions on the environment by thematic maps.

4. Design of different alternatives for alignments in the low-conflict corridor(s).

5. Evaluation of all advantages and disadvantages for each alignment.

As a result of the described procedure, the determination or the selection of the alignment with the least impact on the environment becomes possible, which means the most compatible alignment with respect to the environment can be made.[679]

Furthermore, it should be noted that other important planning goals, like safety, esthetics, costs, etc., also have to be considered at the same time as the assessment of the most favorable alternative.

The practical procedure for conducting an ECS and the results of a case study are presented in Subchapter 7.2.

In the following, the planning procedure is discussed, as based on the different design levels of Fig. 7.1.[159]

The graphic elaboration is sufficient for the first planning level "determination of alternative routes." For the determination of alternatives in the framework of an ECS, the following evaluation criteria, together with Figs. 7.1 and 7.2, should be considered:

• Traffic-related goals of regional planning
• Environmentally related evaluation criteria, such as:

    Actual use and biotype function

    Present soil and geological functions

    Groundwater and hydrological functions

    Landscape quality

    Dwelling functions

    Recreational functions

    Cultural and other goods worth protecting

    Climate and air (wind)

• Vicinity of settled areas
• Section length, size of radii of curve, and gradients
• Requirements for bridges and walls
• Demolition of buildings
• Construction time and costs

The result of this very comprehensive investigation and analysis process is the selection of a priority alternative, based on the comparison of alternatives (Fig. 7.2, Step II).

The next planning phase is the predesign phase (Fig. 7.1). In this phase, the selected routes for one or perhaps for a few different alternatives are examined and the design elements for the cross section and the alignment are determined. Horizontal and vertical alignments are calculated for at least the main and constraint points of the alignment. More precise information about any adverse effects on the environment and the residents, about construction, operation, and road-user costs are desirable. In many cases, the predesign is sufficient for legal procedures, but often a more detailed study is necessary.

After the predesign has been approved, the geometric design which follows represents a further development of the predesign. Now, horizontal and vertical alignment, drainage, and accurate right-of-way requirements have to be calculated. The geometric design serves as a basis for the acquisition of land, invitations of tenders, and construction purposes (Fig. 7.1). For special designs, like drainage, intersections/interchanges, bridges, etc., or for the whole project, more detailed designs may still be necessary.[159]

Furthermore, during the design phase, presentation of any conflicts with the environment and landscape has to be performed according to the required compensation and replacement measures, as is the case in Germany, which are discussed in the attendant plan for landscape cultivation.[245] If such a plan has not yet been developed, then landscape cultivation measures should nevertheless be very strongly considered in the future (see Fig. 7.1).

7.1.3 Computer-Based Procedure for the Environmental Compatibility Study*

As a result of more and more extensive use of modern, efficient computers and programs, a proposal for conducting an ECS using CAD and GIS tools is discussed in the following.

A very effective software tool for conducting an ECS is a Geographic Information System (GIS), in case a direct data exchange between CAD and GIS is possible. In this way, an integrated traffic route design, which considers all the different impacts, can be realized. The following flowchart (Fig. 7.3) shows a possible highway design procedure, from the space-related sensitivity investigation to the predesign according to Fig. 7.2.

This procedure will be thoroughly described and illustrated in Subchapter 7.2 by means of a case study concerning an environmental compatibility study in southwestern Germany.

7.1.4 Preliminary Conclusion

In conclusion, it should be noted that an examination of environmental compatibility should in the future be considered as an integrated procedure in road planning and design. With an increase in planning accuracy, an increase in the verification of environmental issues can thus take place. In general, the following specific investigation in relation to the environment *has* to be considered for individual design levels according to Fig. 7.1:[668]

Basic planning: Ecological risk evaluation

Alignment design: Environmental compatibility study (ECS)

Detailed design elaboration: Landscape cultivation attendant plan.

Summarizing, in the future it will be necessary to consider environmental protection issues to a large extent on the basis of the flowcharts presented in Figs. 7.1 to 7.3.

7.2 GENERAL CONSIDERATIONS, RESEARCH EVALUATIONS, GUIDELINE COMPARISONS, AND NEW DEVELOPMENTS

7.2.1 General

Generally, AASHTO states the following with respect to environmental issues:[5]

A highway necessarily has wide-ranging effects beyond that of providing traffic services to users. It is essential that the highway be considered as an element of the total environment. Environment as used herein refers to the totality of man's surroundings: social, physical, natural and manmade. It includes human, plant and animal communities as well as the forces that act on all three. The highway can be and should be located and designed to complement its environment and to serve as a catalyst for environmental improvements. The area surrounding a proposed highway is an interrelated system of natural, manmade and sociological variables. Changes in one variable within this system cannot be made without some effect

---

*Elaborated based on Ref. 679.

<u>**S T E P S**</u>   <u>**T O O L S**</u>

1.   First GIS System Steps

I.1	Digitalization of Important Environmental Resources of the Region like Geology, Soil, Climate, Water, Biotype, Recreational, and Land Use-Related Functions, etc.

Digitizer of the GIS System, which Allows the Assigning of Spatial Attributes to the Digitized Areas

I.2	Calculation of Individual Environmentally Related Contour Thematic Maps

MapTool of the GIS System

I.3	Establishment of the Low Conflict Corridor by Superimposing of the Individual Thematic Maps according to the Following Steps:
	a) Weighting of Each Spatial Attribute within Each Thematic Map
	b) Weighting of the Thematic Maps in order of Importance
	c) Superimposition of Defined Buffer Zones
	d) Superimposition of the Thematic Maps

Superimposition and Calculation Tool of the GIS System

Map Tool

Map Tool

Map Tool
Superimposition Tool

I.4	Presentation of the Low Conflict Corridor

Plotter, Printer, etc.

I.5	Export of the Calculated Low Conflict Corridor to the CAD System in an Appropriate Exchange Format

Export Tool of the GIS System

2.   CAD System Steps

II.1	Import of the Data into the CAD System in the Appropriate Data Exchange Format in order to Design Different Alignments

Import Tool of the CAD System

II.2	Design of Different Alignment Alternatives Inside the Low Conflict Corridor

Alignment Tools of the CAD System

II.3	Export of the Alignment Data, or of the Whole CAD Data Depending on the CAD and GIS System Used

Export Tools of the CAD System

FIGURE 7.3 Flowchart of a computer-based environmental compatibility study. (Elaborated based on Ref. 679.)

3. Second GIS System Steps

III.1 Import of the Alignment Data into the GIS System	Import Tools of the GIS System

III.2 Evaluation of the Different Alignment Alternatives according to the Following Steps a) Calculation of Environmental Impact Maps for Each Alignment b) Superimposition of these Maps on Other Environmentally Related Thematic Maps	Analysis and Calculation Tools of the GIS System Analysis and Calculation Tools Superimposition Tool

III.3 Selection of the Most Favorable Alternative Alignment (Priority Alternative)	Calculation and Map Tools of the GIS System

III.4 Presentation of the Selected Alignment Inside the Low Conflict Corridor	Plotter, Printer, etc.

FIGURE 7.3 (*Continued*) Flowchart of a computer-based environmental compatibility study. (Elaborated based on Ref. 679.)

on the other variables. Some of these consequences may be negligible, but others may have a strong and lasting impact on the environment, including the sustenance and quality of human life. Because highway location and development decisions have an effect on adjacent area developments, it is important that environment variables be given full consideration.[5]

As discussed in Subchapter 7.1, the governments of most western European countries have enacted laws for environmental compatibility examinations (ECEs) and require engineers to perform environmental compatibility studies (ECSs) when planning highways, railroads, waterways, and airports. These studies always have to be done *before* starting the design and parallel to the design phases of a project.[305,465]

Thus, for new construction and major reconstruction, rehabilitation, and restoration projects, an environmentally justified compatibility study also has to be developed for the future highway route or location according to the laws enacted in western Europe. These studies have to include all existing natural, ecological, and cultural resources which are important to the integrity of an observed region. These statements fully agree with the American point of view.[5]

The space-related sensitivity investigation of the ECS is based on the superimposition of the individual thematic maps which represent the environmental functions evaluated from the available geographic information. This investigation leads to a presentation that covers the entire area of interest for all relevant functions of the environment with special consideration of the following:

- Protected developed and undeveloped areas or resources and those worth protecting
- Areas with special environmental sensitivities or with specific significance for the environment
- Existing and planned land use

The so-called overall map is the result of the superimposition of the individual thematic maps. The overall map forms the basis for establishing the low-conflict corridor, which is the final result of the first step of an ECS according to Figs. 7.2 and 7.3.[679]

Along with the space-related sensitivity investigation, the highway engineer has to identify alternative alignments that should be evaluated in multidisciplinary cooperation with regard to the low-conflict corridor(s). By comparing the alternatives, the results have to be presented based on numerous steps,[305,465] the most important of which are listed here:

- Evaluating the advantages and disadvantages of the alternatives.
- Revealing the differences.
- Evaluating the alternatives and ranking them from the position of environmental compatibility, in order to reach a sensible decision.

For a better understanding of the individual steps that are involved in an ECS, the flowchart in Fig. 7.2 was developed. In the upper part of this figure, the sequence within the space-related sensitivity investigation is presented. The result of the ECS is the low-conflict corridor. The lower part of Fig. 7.2 reveals the selection of alternatives for the planned roadway. As can be seen, other important planning goals, such as safety, esthetics, and costs, are also evaluated in connection with the assessment of the most favorable alternative. In other words, the result of an ECS is the establishment of the priority alternative for which the predesign will be developed.

The "Instructural Guide for the Environmental Compatibility Study in Highway Planning (ECS),"[305] recommends Geographic Information Systems (GIS) for conducting those examinations.

7.2.2   Procedure for Conducting an Environmental Compatibility Study*

In the varied and intensively used cultural landscapes of nearly all industrialized countries, the sensitivity investigation will identify only a few areas relatively free of conflicts that would allow an easy design of new highways. Rather, a pattern will emerge in which conflicts from competing important environmental goals will vary from area to area, with some areas having more impacts (conflicts) than others. However, that does not mean that in developing countries the sensitivity investigation should be omitted. Just the opposite is true in order to protect unique landscapes and unspoiled countrysides.[436]

To develop a low-conflict corridor, it is necessary to look for all conflicting environmental concerns in order to find those areas with the fewest environmental concerns. A low-conflict corridor is an area with relatively low protection worthiness and low sensitivity for the design of a highway between two goals, A and B—that is, the corridor with the least environmental impact.[436]

The general procedure for conducting an Environmental Compatibility Study in traffic route planning is divided into five hierarchical steps, which have already been developed in Subchapter 7.1. They are repeated here for a more in-depth understanding according to the following detailed discussions:[305]

1. Definition of the planning region, that is, demarcation of the general planning area for the roadway and its functional and operational goals.

2. Determination, presentation, classification, and evaluation of all areas with environmentally relevant functions by thematic maps within the planning region.

3. Development of relatively low-conflict corridor(s) by means of the superimposition of those areas with environmental-relevant functions (thematic maps) in the planning region.

---

*Elaborated based on Refs. 305, 465, and 679.

4. Design of several alternatives for the planned roadway in the low-conflict corridor. The RRR and the zero (no-build) alternatives also have to be considered in a general comparison regarding the environmental impact.

5. Elaboration of the advantages and disadvantages of each individual alternative, and evaluation of the environmental compatibility for the selected alternative.

These five steps are presented in Fig. 7.2 and are graphically shown in Fig. 7.4 where the relationships and interactions with Fig. 7.2 are compared.

Regarding the general procedure of an ECS, each of the five steps is discussed in detail, as follows:

Step 1: Definition of the Planning Region. As the first individual step, the planning region for the new traffic route project has to be defined to achieve a clearly limited, economic, as well as an ecologically appropriate area. This area must allow the design of different road alternatives (variants), including the restoration/rehabilitation and the zero alternative. Note that the planning region is not limited only to the road itself, but also has to cover all areas where the planned roadway may have a specific impact on the environment. Based on the demarcation of the planning region, the potential alignment alternatives (rough design) and/or alignment corridors can be selected. Alignment corridors can be established on the basis of topographical maps (scale of 1:25,000) considering regional development plans, land-use plans, regions worth protection (biotypes, water protection areas, natural recreational areas, national nature reserves, or national parks), as well as the personal knowledge of the area to be examined.

In summary, the planning region must include:

1. The corridor of the planned roadway, that is, the space where the alignment alternatives could probably be designed.

2. Zones of possible environmental impact surrounding the planning corridor(s). These zones have to be wide enough to include all areas on which the planned highway would have an environmental impact.[679]

Step 2: Determination, Presentation, and Evaluation of Environmentally Relevant Functions. Each ECS must include detailed information about the planned traffic route and the resulting cause-and-effect relationship on the environment. It is important to note that not only the long-term effects caused by the operation of the roadway have to be considered, but also the relevant impact caused by the construction and maintenance of the traffic route.

Step 2 requires the determination, description, and evaluation of all areas with relevant environmental functions and impacts in the defined region to be examined (see Fig. 7.4). In order to consider the most important environmental and human functions, multidisciplinary cooperation between specialists of all concerned fields is required. The space-related sensitivity investigation is based on a comprehensive graphical presentation of all relevant environmental functions (called *thematic maps*) with special emphasis on the following features:

- Areas and resources worth protection (prevention)
- Areas with special environmental sensitivities (sensitivity)
- Areas with priorities for specific land use (suitability)
- Areas of existing or planned land use (preutilization)

Thematic maps represent different environmental functions, and have to be established with respect to the previously listed features, like protection worthiness, sensitivity, suitability, and preutilization of land. Relevant environmental functions are, for example, groundwater potential, climatic conditions, biotype distribution, land use, or recreational value of a region (see the case

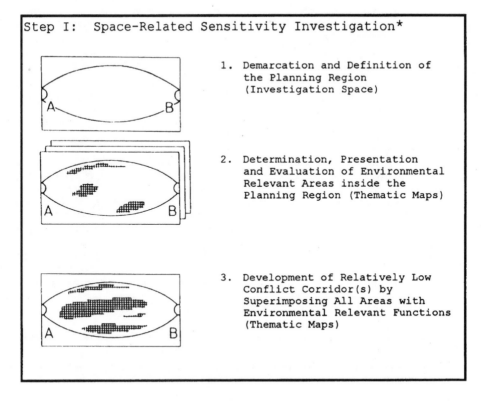

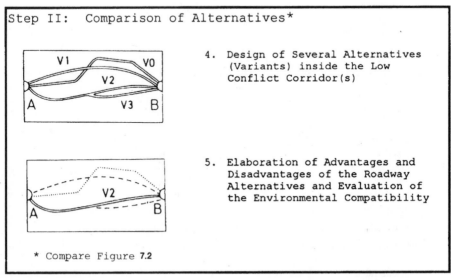

FIGURE 7.4 Graphic presentation of basic steps in conducting an Environmental Compatibility Study. (Elaborated based on Ref. 305.)

study in the next section, Sec. 7.2.3). The evaluation of each feature has to be conducted separately and be provable, for example, by exactly defined evaluation levels and/or scales. For each environmental function, a weighted superimposition with respect to the previously mentioned features is required to establish the individual thematic maps. The weight of the features has to be individually determined and no general rules exist for this process. This procedure has to be repeated for all relevant environmental functions in the investigation area to produce the thematic maps. For example, five thematic maps will result for the case study in Sec. 7.2.3. However, for other planning projects, there may be many more.

Step 3: Development of Low-Conflict Corridor(s).   On the basis of the individual thematic maps, an overall environmental sensitivity map can be established. The analysis is performed by superimposing the before weighted maps. Each of the thematic maps is weighted in order of importance with respect to the environmental functions. The new composite or the resulting overall map shows areas of different levels of environmental impact on the basis of the weighted thematic maps for each environmental function. So far, however, no general rules concerning the weighting process exist. Weighting in each thematic map and the order of importance of the thematic maps depend on the sensitive, prudent decisions of the responsible multidisciplinary team of specialists.

Identifying areas with a relatively high environmental sensitivity may lead to the detection of corridors with relatively low sensitivity. These corridors are called *low-conflict corridors* and represent the best locations for the alignment design of the planned traffic route. In other words, a low-conflict corridor has a relatively high suitability for the design of one or more alternatives. However, it should be noted that there may exist points of special environmental sensitivity even in low-conflict corridors. These have to be clearly identified as negative compulsory alignment areas. That means, if possible, the alignment should be designed in a way to avoid these areas. Furthermore, note that generally a new roadway also has an environmental impact on the surroundings. Since the design should be the best realizable compromise between nature and economic requirements, parallel to the planning of the road, different compensation and replacement measures should be considered to reduce the impact on the environment as far as possible. For those measures, the landscape cultivation attendant plan is provided in Germany. In conclusion, it can be stated that the proposed procedure for detecting low-conflict corridor(s), and for analyzing the environmental impact on the surroundings, provides a very sensitive, sound, and reliable evaluation of all possible environmental interrelationships in the planning region.[305,679]

Step 4: Design of Alternatives (Variants) in the Low-Conflict Corridor(s).   Based on the results of the space-related sensitivity investigation, the highway engineer has to design the horizontal and vertical alignments in the low-conflict corridor(s) for one or more alternatives. Of course, for individual cases, the zero alternative or restoration/rehabilitation projects may also be considered as favorable.

Step 5: Elaboration of Advantages and Disadvantages of the Roadway Alternatives and Evaluation of the Environmental Compatibility.   The last step of the ECS is a comparison of different alternatives as well as an evaluation of the environmental compatibility for each alternative. This step will result in selection of the priority alternative for the planned roadway with respect to environmental compatibility.

The following issues should be addressed in this step:

- Evaluation and description of the adjacent areas affected by the traffic route, including existing negative impacts.
- Evaluation of positive and negative environmental impacts on the areas adjacent to the roadway.
- Evaluation of damaging effects on the environment.

- Development and presentation of measures to reduce or prevent any negative impact on the environment.
- Evaluation of the remaining negative impacts and their compensation.

However, it should be noted that the resulting outcome has to be coordinated with other important planning goals like safety, economic, and esthetic considerations. Therefore, for individual cases, it is quite possible that the most favorable alternative with respect to environmental compatibility may not be suitable for the final priority traffic route alternative.

The following section provides practical recommendations for performing an ECS concerning the application of a Geographical Information System (GIS), based on a case study. GIS systems increase the performance, effectiveness, and accuracy of an Environmental Compatibility Study considerably.

7.2.3   Computer-Based Case Study for the Development of a Low-Conflict Corridor*

The case study area is the town of Staufen im Breisgau, which is in southwestern Germany near the French and Swiss borders. The town is in a scenic, commercially rich region. Its historical town center, with its medieval half-timbered houses and beautiful old vineyards in the surrounding area, attracts thousands of tourists each year. The Black Forest, with its rich biotypes, is in the vicinity.

Increasing numbers of tourists and the growing economy during the past two decades have caused an enormous increase in through traffic in Staufen and an increase in both accident frequency and severity. The important state route (SR 3) runs directly through the residential, shopping, and commercial areas of the town. Therefore, the federal, state, and municipal agencies have decided to provide a bypass around Staufen to alleviate the critical traffic conditions and safety problems.

In-depth environmental studies, which are generally deemed necessary, are even more important for such an attractive and environmentally sensitive region as the Staufen area. To understand the important environmental impacts of the planned bypass, databases of environment functions in the form of thematic maps were developed by experts. It was found that the protection of a number of environmental functions are of specific importance for the Staufen region and would need to be examined by an ECS before the actual design and construction phases of the bypass could begin. The environmental functions to be examined in the space-related sensitivity investigation of the ECS for the Staufen region are

1. Groundwater potential
2. Climate
3. Biotype distribution
4. Land use
5. Recreational value

Not all of the environmental functions may have to be considered during the space-related sensitivity study of an ECS. The decision depends on the importance of the spatial and environmental impact of the planning region. The following environmental functions were not considered in detail in this case study:

---

*Elaborated based on Ref. 436.

1. Landscape because the plain adjacent to the River Rhine was the sole planning alternative in order to avoid a negative impact on the Black Forest region east of Staufen with its areas of natural beauty, wildlife, and recreational significance.

2. Geology and soil type because similar strata and soil types are present in the whole planning area, and they represent a relatively low value of environmental sensitivity. For this reason, the impact of geology and soil type were excluded from further analysis.

3. Noise because from the beginning it was decided to provide as much distance as possible between residential/recreational areas and the planned bypass, and to protect impacted areas by means of noise-protection barriers. Therefore, this impact is not shown on a thematic map.

4. No other relevant issues were found by the multidiscipline team of experts studying the natural resources of the Staufen region.

To protect the five environmental resources discussed here, thematic maps have to be used to decide whether specific areas are suitable for the bypass. The databases represented by the thematic maps in Figs. 7.5 to 7.9 contain current conditions as well as future conditions on residential, commercial, and recreational developments—as far as they were known during the study phases. The thematic maps were developed by AKG Software Consulting GmbH, Ballrechten-Dottingen, Germany.

Thematic maps, which contain complex geographical information about the planning region, are primarily digitized maps. The most suitable tool to perform an ECS is a Geographical Information System.[22,291,327,534,679] A Canadian program known as SPANS (Spatial Analysis System)[352, 437, 673] was used to analyze the complex relationships and to search for a low-conflict corridor. The benefit of a GIS is that data of different georeferenced formats and from different origins can be read in, converted, analyzed, and displayed together.

Normally, thematic maps originated by a GIS are differentiated by means of different colors. Since this was not possible here, different shadings were used to present the results. These may sometimes be of lesser quality than color graphic presentations.

Thematic Map: Groundwater Potential   The thematic map for groundwater potential is shown in Fig. 7.5. Different shadings indicate the three main groundwater levels: low, medium, and high. If the level is low, the risk of fast and uncontrolled flow of pollutants is low. On the contrary, if expressways with high traffic loads (like the bypass in the Staufen region) are planned in areas with high groundwater levels or permeable soils, they might pose a great source of danger for the environment. Traffic accidents involving tanker trucks or other high-risk transports can happen at any time and could threaten the groundwater resources because of pollution.

Therefore, the groundwater level must be regarded as an important issue for the sensitivity investigation. Areas with high groundwater levels and short distances to local drainage systems should be avoided, if possible, when planning a new highway (in this case, a four-lane median-separated bypass).

In connection with groundwater levels, the presence of permeable or impermeable soil is also of great importance. For example, in case of an accident, impermeable polluted soils can be dug out to prevent groundwater pollution. Since in the Staufen region the soils were comparable and mainly impermeable, it was not necessary to include the thematic map for soils in this study. However, for other projects, this thematic map might be of great importance for preventing environmental pollution.

Thematic Map: Climate   Thematic maps of climate may include temperature, weather, and humidity conditions as well as wind speeds and direction. Because the region to be investigated is relatively limited, significant changes related to the first three previously mentioned variables are not to be expected. However, because of the climatic conditions between the Black Forest Mountains and the River Rhine Valley, wind speed and direction play important roles. Therefore, the thematic map of wind conditions was established and is presented in Fig. 7.6.

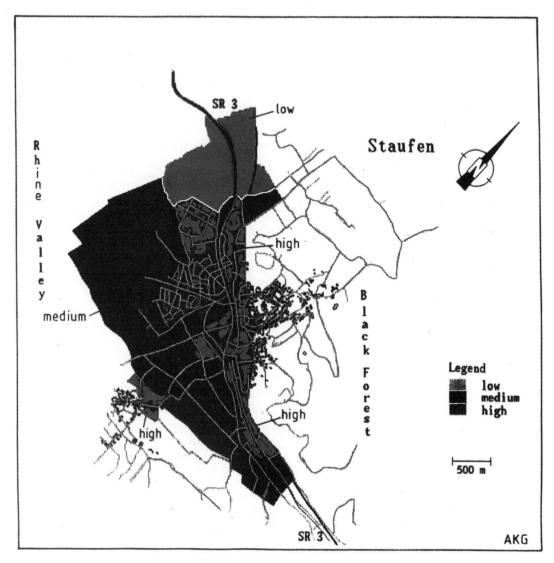

FIGURE 7.5   Groundwater potential.

It is well known that alterations in the morphological contours of the landscape created by building traffic routes can affect local wind systems significantly. In addition, cutting down parts of forests for traffic routes could lead to wind speeds up to 4 times higher than those before the forests were cut. Local experts have shown that the newly planned bypass will not cause these types of major impacts in the Staufen region. Wind speed and direction, however, are always important issues in connection with every major road being planned.

The local wind speeds are arranged on the thematic map of Fig. 7.6 and are again separated into three levels (high, medium, and low). The main wind direction is shown by arrows. In many cases, the emission concentration depends strongly on weather conditions. In particular, unfavor-

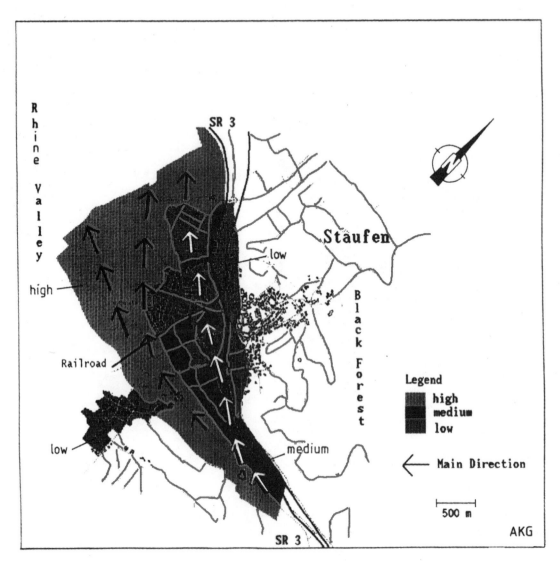

FIGURE 7.6   Climate (wind speed).

able circumstances exist for calm and inversion wind situations. Therefore, for the same amount of emission, emission concentrations can exist that differ by a factor of 5.

Considering the planned bypass project, it is important to note that the wind force is sufficient to blow away and disperse the exhaust emissions of the planned bypass (Fig. 7.10), which consist mainly of health-damaging substances (carbon monoxide, hydrocarbons, and nitrogen monoxide from gasoline engines, as well as soot particles from diesel engines). Consequently, the main wind direction should not blow from the highway corridor to residential or recreational areas.

Thematic Map: Biotype Distribution.   Biotypes are of special interest because many protected species live there in their natural environment. Therefore, biotypes are very important for the eco-

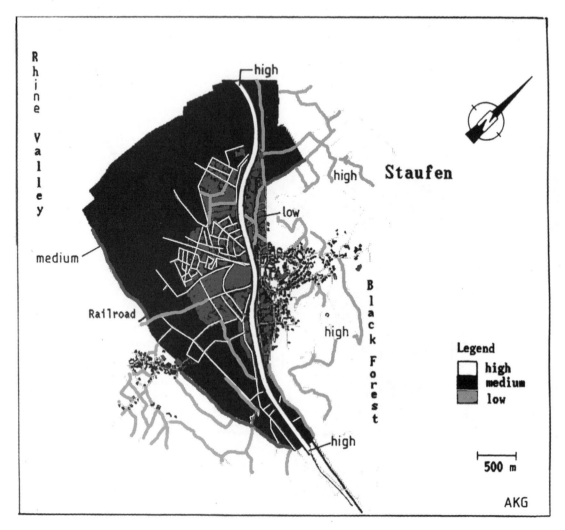

FIGURE 7.7 Biotype distribution.

logical balance of a region. Biotypes include the flora (biotypes of plants) and the fauna (biotypes of animals) of the observed area. Figure 7.7 shows the thematic map of the biotype distribution.

Again, the map is separated into three different levels by using different shadings. "High" means conservation areas with natural beauty and wildlife: in the present case, mostly river, brook, and pond areas, parks and gardens, as well as the Black Forest region at the northeast side of Staufen. In contrast, "low" means relatively low-value agricultural and commercial land for this case study.

The distances between the planned bypass and important biotype areas should be as great as possible (see Fig. 7.7).

Thematic Map: Land Use. The land use of the Staufen region is presented in the thematic map in Fig. 7.8. As can be seen, the land is subdivided into residential and shopping areas, commer-

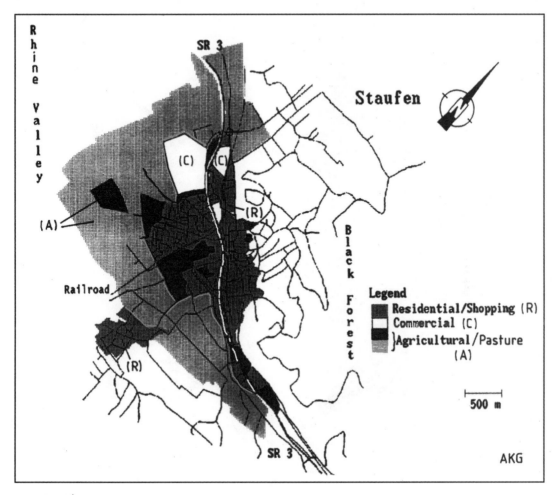

FIGURE 7.8 Land use.

cial, and agricultural areas, and pasture land. Politicians and local groups requested that residential and shopping areas should be given the highest priority for protection. Agricultural and pasture land, as well as commercial areas, could be used for the new planned bypass of Staufen.

Thematic Map: Recreational Value. Based on the thematic map in Fig. 7.9, the recreational value of the area concerned can be evaluated. The map in Fig. 7.9 was based on a poll conducted in the Staufen region in 1990. Thus, the results presented in the thematic map in Fig. 7.9 represent the opinions of Staufen citizens, visitors, and tourists. Only areas with mainly low levels of recreational value should be used for the planned bypass project.

Other Thematic Maps. For other planning projects, additional or other thematic maps may certainly be of importance. The environmental impacts are so complex that almost every project requires a modified selection of thematic maps for conducting environmental compatibility studies in highway planning. This, however, does not change the fundamental procedure of the ECS

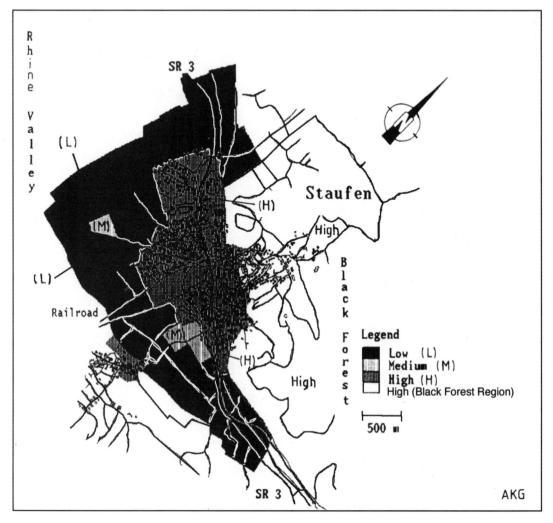

FIGURE 7.9   Recreational value.

in searching for a relatively low-conflict corridor, for example, by using a GIS for superimposing project-specific thematic maps, as discussed in the next section.

Superimposition of Thematic Maps and Buffer Zones.   On the basis of the five thematic maps that were developed, the space-related sensitivity investigation of ECS for the planned bypass of Staufen can be performed. The analysis is done by the superimposition of the thematic maps (Figs. 7.5 to 7.9) by using a computer-supported GIS examination procedure. Since the thematic maps are normally available in digitized form, they are more or less directly usable for the calculation processes provided by an appropriate GIS. The different layers (thematic maps) of such a map superimposition (see Fig. 7.4) should contain all important data of the observed region to differentiate areas that should be protected and less valuable areas, normally distinguished by the levels

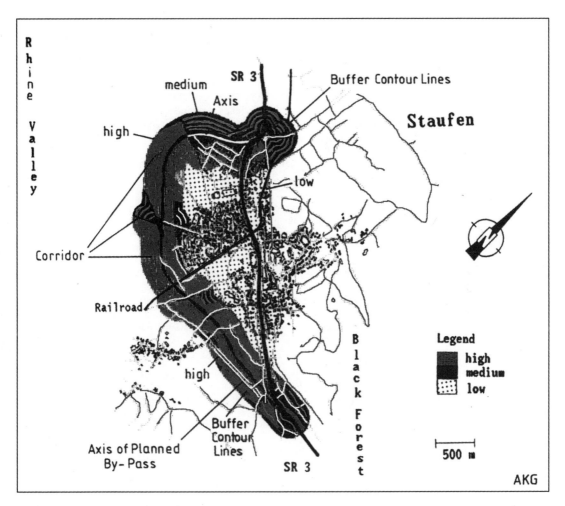

FIGURE 7.10   Suitability for the low-conflict corridor.

"low, medium, and high." These evaluation levels, based on digitized map information, can be presented on a computer screen or printed out using different colors or shadings (Figs. 7.5 to 7.9).

One great advantage of this method of analysis is that computer-supported GISs can provide a weighted examination of the different superimposed digitized thematic maps. A weighted examination is the calculation of the data from the thematic maps in order to create a new map which includes all relevant results of the superimposition process.

In actual practice, the weighting process is a two-step procedure, according to Fig. 7.2 (Step I). First, each thematic map is analyzed by the study team and subdivided into areas of similar environmental impact. An evaluation level (for example, low, medium, or high) is then assigned to each area. The GIS allows the user to assign a weight to each evaluation level of a thematic map. In the present case study, the first step was performed by weighting the "advantageous" evaluation level by the factor 3, the "medium" level by the factor 2, and the "disadvantageous" evaluation level by the factor 1 for each thematic map. This information is entered into the GIS.

Next, the GIS can be applied to superimpose the thematic maps on each other in order of importance. This order is decided by the study team, and reflects the importance of the different environmental functions in terms of local conditions. For example, this can be done sequentially. Different systems also offer a matrix overlay in order to define special dependencies between environmental functions.

For this case study, the second step, weighting of thematic maps in order of importance, was performed in the following way. The most important environmentally justified issue was the groundwater potential, followed by the thematic maps for wind speed and direction, biotype distribution, land use, and recreational value.

According to the two decision criteria, discussed previously

1. Order of evaluation levels within each individual thematic map
2. Sequence of the thematic maps between each other

the computer-supported calculation procedure of the GIS generates an overall map of the results. On the basis of the results, shown in the composite map, areas of different levels of environmental impact can be identified (see Fig. 7.10). By identifying the areas with environmental sensitivity, the suitability for a low-conflict corridor or corridors can be established for the planned bypass. In other words, according to Fig. 7.10, a relatively high level of suitability or low level of environmental sensitivity should be present in the corridor where the alternative(s) for the road project are to be located.

In addition, distance buffers around the corresponding alternative(s) can be superimposed on the new overall thematic map. The distance buffers are provided in steps of 50 m in the present case. Thus, distance requirements can be examined by assessments such as the following:

- The new highway should pass not closer than 250 m to residential or recreational areas.
- The distance between the bypass and a biotype that needs protection should be at least 100 m.

Of course, the input of any other local distance requirement is possible and can be solved clearly and quickly by a GIS.

By assigning weights to the different evaluation levels of thematic maps in a reasonable order, by superimposing the thematic maps in a desired sequence, and by introducing distance buffers, a relatively low-conflict corridor with lateral buffer zones could be established for the planned bypass in the Staufen region, as shown in Fig. 7.10.

The overall width of the buffer zones for the established corridor is 500 m, 250 m on each side of the axis of the planned bypass. By comparing the results of Fig. 7.10 with the evaluation levels in Figs. 7.5 to 7.9, it can be concluded that the corridor developed by the GIS examination procedure excludes at least high-conflict areas.

In addition, according to Fig. 7.2 (Step II), the ECS requires a comparison of alternatives. For the bypass around Staufen, five alternatives were investigated. Consequently, the most favorable ones for the planned four-lane median-separated bypass were located in the relatively low-conflict corridor shown in Fig. 7.10. For each alternative, a sensitivity investigation has to be performed in order to establish the priority alternative. This evaluation must also include other important goals like esthetics, economy, function, safety, and traffic quality. It is important to keep these important goals at all times along with the ECS procedure (compare Figs. 7.1 to 7.4).

7.2.4 Preliminary Conclusion

Finally, two questions should be discussed regarding the effectiveness and the usefulness of an ECS conducted with a computer-based GIS:

1. Do cost and time comparisons exist between current methods of developing an ECS and those using a GIS?

The only answer is as follows: If an appropriate GIS is available, the cost and time factors will be reduced considerably because of the automatic weighting and evaluation processes performed on the basis of quantitative criteria (expressed, for example, by thematic maps). Only a few aspects of an ECS could be dealt with manually, if at all, and only qualitative results should be expected. A cost and time estimation would lead to speculation because of the numerous different relationships and interrelationships that would influence a specific project.

2. Would the GIS database approach change the location of a planned highway over the location that might have been selected by engineers who may have been provided with similar information but without the graphic analysis tools?

The only answer is probably not. However, the question remains: How would the engineer be provided with those complex pieces of information without the graphic analysis tools of a GIS to make unbiased and factually correct decisions regarding the protection worthiness of environmental issues?

Summarizing the procedure of an ECS, the following conclusions can be drawn:

1. One important consequence of the present study is that in highway geometric design an ECS must be conducted before starting any predesign, design, or construction phase of a project. Examination of the other important planning goals can take place after the assessment of the most favorable alternative by an ECS.

2. The case study revealed in this way how areas that need protection can be left intact, and how future highways can be placed in those areas for which it is less important to maintain the integrity of the region being examined. Finally, such a procedure saves money and time in the complex highway geometric design process, because only those alternatives for which mainly low-conflict areas are available and their corresponding alignments must be considered. Thus, the optimum solution is a product of an ECS, which is conducted at the beginning of the planning process.

3. Incorporation of the previous concepts about an ECS into a computerized GIS format is a significant step forward in accomplishing such work and in helping to present it to the public for clearer understanding. And it will probably contribute to general public acceptance and approval.

CHAPTER 8
RELEVANT SPEEDS

8.1 RECOMMENDATIONS FOR PRACTICAL DESIGN TASKS

8.1.1 Definition of Terms*

Part 2, "Alignment (AL)," includes the speeds:

$$V_{\text{perm}} = \text{permissible speed, km/h}$$

$$V_d = \text{design speed, km/h}$$

$$V85 = \text{85th-percentile speed, km/h}$$

The *permissible speed,* V_{perm}, is the general or local maximum speed limit. Recommendations for assigning V_{perm} to the individual road categories are given in Table 6.2, Col. 4, under "permissible speed limit."

Individual design elements are not based on the permissible speed limit. For the alignment, the relevant *design speed* should always be higher than or equal to the permissible speed limit (see Table 6.2).

The *design speed,* V_d, depends on environmental and economic conditions based on the assumed network function of the road and the desired quality of traffic flow, and should be assessed for new designs according to Table 6.2, Col. 7. Limiting and standard values for most of the design elements are defined according to the design speed. The design speed determines

- Maximum tangent lengths
- Minimum radii of curve, R_{min}
- Minimum parameters of clothoids, A_{min}
- Maximum longitudinal grades, G_{max}
- Required parameters for vertical curves

within a specific roadway section. The design speed decisively influences the road characteristics, traffic safety, quality of the traffic flow, as well as the costs, especially for roads of category group A. Therefore, a constant design speed, V_d, should be applied consistently on longer sections or at least on longer connected roadway sections. Furthermore, the design speed is used in the safety evaluation process of highway geometric design according to safety criteria I and III, developed in Chaps. 9 and 10.

For existing alignments, however, the design speed is not often known or was not selected according to the assumptions made in Table 6.2, Col. 7, relative to a specific network function of the road and to a desired traffic quality. Therefore, in order to establish a sound design speed, especially for redesigns and RRR strategies in case of old alignments, a new procedure has been

---

*Elaborated based on Refs. 124, 159, 241, 243, 246, 429, and 453.

developed. The aim of the latter is to coordinate the design speed with the actual driving behavior, expressed by the average of 85th-percentile speeds on an observed roadway section. The procedure is explained in detail in Sec. 9.2.2.1.

It is also recommended that this procedure be used in the examination of the selected design speed for new designs, in order to avoid under- or overdimensioning and to yield economic, environmental, and safety-related sound alignments.

The *85th-percentile speed,* $V85$, represents a driving dynamic value for the geometric design of single parameters in plan, profile, and cross section. For category group A roads, $V85$ corresponds to the speed below which 85 percent of passenger cars operate under free-flow conditions on clean, wet road surfaces; for category group B roads, the 85th-percentile speed corresponds to the permissible speed limit (see Sec. 8.1.2). Based on $V85$,

- Individual tangent lengths
- Superelevation rates in circular curves (e)
- Required stopping sight distances (SSD)
- Required passing sight distances (PSD)
- Required parameters for crest vertical curves (based on sight distance criteria)
- Minimum radii of curve for negative superelevation rates
- Drainage requirements

are determined and evaluated.

The 85th-percentile speed depends on the road geometry, and is also used for the safety evaluation process according to safety criteria I to III.[424,428,434,435,438,444,447,448,453,455,673]

Discussions, analyses, and evaluations with respect to the term "wet" road surfaces can be found in Sec. 8.2.1.2. Based on these investigations, it can be concluded that the 85th-percentile speed on a "dry" road surface does not differ significantly from the 85th-percentile speed on a "wet" road surface, with rainfall from a drizzle to moderately heavy rain and when visibility is not affected appreciably.[416,422] This assumption corresponds to sight distances of about 150 m.

This means, for the critical design case, that high-speed levels are still observed on wet road surfaces, and that the established 85th-percentile speeds are valid for both "wet" and "dry" road surfaces.

8.1.2   General Speed Determinations*

For new roads of category group A, the design speed, V_d, is determined based on the desired road category in Table 6.2. Depending on the traffic flow quality, the difficulty of the topography, and the accumulation of constraint points, the higher or lower values of the design speed (Table 6.2, Col. 7) are selected. For roads of category group B, the design speed, V_d, normally corresponds to the permissible speed limit, V_{perm}.

The 85th-percentile speed, $V85$, is determined in the following way. Because of the very generous alignments of multilane median-separated rural roads of category group A (like interstates or autobahnen), no statistically sound results have been determined for the relationship between 85th-percentile speed and road characteristics. The few existing studies indicate a small decrease in the 85th-percentile speed with a decreasing radius of curve (see Sec. 8.2.4.1). Therefore, the 85th-percentile speed on these roads will be defined as shown in the following (for example, for establishing superelevation rates and stopping sight distances):

$$V85 = V_d + 20 \text{ km/h} \qquad V_d \geq 100 \text{ km/h} \qquad (8.1)$$

$$V85 = V_d + 30 \text{ km/h} \qquad V_d < 100 \text{ km/h} \qquad (8.2)$$

where V_d is the design speed, km/h (see also Table 6.3, Col. 3).

For two-lane rural roads of category group A, the 85th-percentile speed will be determined according to Sec. 8.1.3.

---

*Elaborated based on Refs. 154, 159, 243, and 453.

For roads of category group B, which includes nonbuilt-up roads in the periphery and inside built-up areas (arterials and main collectors, see Tables 6.1 and 6.2), new investigations have not determined any statistically sound relationship between 85th-percentile speed and road characteristics. This is caused, in general, by the lower speed levels which exist on these roads with imposed speed limits, other traffic regulatory measures, and different structural and municipal design features. However, studies also show that on these roads the 85th-percentile speed level is normally higher than the posted speed limit (V_{perm}).

For improving traffic safety, it is recommended that superelevation rates and stopping sight distances be selected according to the following assumptions:

$$V85 = V_{perm} + 20 \text{ km/h (road category B II, for example, primary arterials)} \tag{8.3}$$

$$V85 = V_{perm} + 10 \text{ km/h (road category B III, C III, for example, secondary arterials)} \tag{8.4}$$

$$V85 = V_{perm} \text{ (road category B IV, C IV, for example, main collectors)} \tag{8.5}$$

where V_{perm} is the posted speed limit.

The previous rules will improve safety performance by regarding higher superelevation rates. At the same time, the danger of a further speed increase does not exist because higher superelevation rates are not easily recognized by the motorist.

8.1.3   Determination of the 85th-Percentile Speed for Two-Lane Rural Roads of Category Group A*

For two-lane rural roads of category group A, the 85th-percentile speed, $V85$, is a function of road characteristics, as numerous research investigations have shown. (Also see Sec. 8.2.3.)

The 85th-percentile speeds were originally based on longer roadway sections with similar road characteristics.[241,243] However, new studies have indicated that it would make more sense to focus on the individual design elements (tangent or curve).

In this connection, it was found that the most successful parameter in explaining much of the variability in 85th-percentile speeds, $V85$, and accident rates, AR, is the new design parameter, curvature change rate of the single circular curve with transition curves (CCR_S, Eq. 8.6).[†] This parameter describes the design of a curve through the length-related course of the curvature, which appears to be one of the most important variables in operating speed and accident situations. Furthermore, this new design parameter includes the influence of the transition curves (in front and behind the circular curve, if present), and considers the overall length as well as the angular change of the curve. In the case where a curve site in the same direction of curvature is composed of several circular curves and/or connecting transition curves, it can also be regarded as a single curve site for determining CCR_S, in order to simplify matters see Figs. 8.1 and 8.2.

Tangent sections[‡] ($CCR_S = 0$ gon/km[§]) are considered individually in the process of calculating curvature change rates of the single curve (see Fig. 8.2).

The 85th-percentile speeds, $V85$, can be determined along an investigated roadway section for each individual curved site in relation to CCR_S according to Fig. 8.3. For independent tangents, the 85th-percentile speed is related to $CCR_S = 0$ gon/km.

---

*Elaborated based on Refs. 406, 408, 434, 435, 438, 447, 448, 453, and 455.

† Abbreviated as curvature change rate of the single curve (CCR_S) in the following chapters.

‡ An *independent tangent* is classified as a tangent that is long enough to be considered as an independent design alignment in the curve-tangent-curve design process, while a *short tangent* is called nonindependent and can be ignored. For defining and classifying independent tangents, see Sec. 12.1.1.3.

§ gon is a common designation of the angular unit in the centesimal system, which has found a widespread use in many continental European countries and has certain inherent advantages in various engineering computations. In the anglosaxon countries, this unit is known as grade or grad. In the centesimal system, the circle is divided into 400ᵍ (400 gon or grads) instead of 360° (360 degrees) in the sexagesimal system. The conversion is: 1ᵍ = 0.9°.

The formula for determining the curvature change rate of the single curve is

$$CCR_S = \frac{(L_{cl1}/2R + L_{cr}/R + L_{cl2}/2R)\,63{,}700}{L} \qquad (8.6)$$

where CCR_S = curvature change rate of the single circular curve with transition curves, gon/km

$L = L_{cr} + L_{cl1} + L_{cl2}$ = length of curve, km

L_{cr} = length of circular curve, m

R = radius of circular curve, m

L_{cl1}, L_{cl2} = lengths of clothoids (preceding and succeeding the circular curve), m

$63{,}700 = 200/\pi \times 10^3$

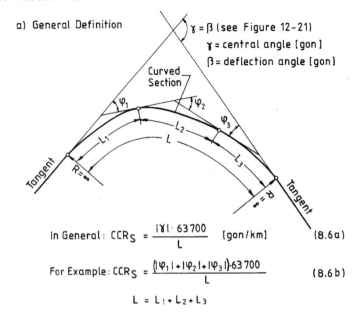

a) General Definition

$\gamma = \beta$ (see Figure 12-21)

γ = central angle [gon]

β = deflection angle [gon]

In General: $CCR_S = \dfrac{|\gamma|\cdot 63700}{L}$ [gon/km] (8.6a)

For Example: $CCR_S = \dfrac{(|\varphi_1| + |\varphi_2| + |\varphi_3|)\cdot 63700}{L}$ (8.6b)

$L = L_1 + L_2 + L_3$

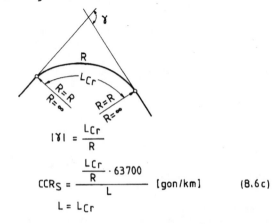

b) Curve Consisting of one Circular Arc with Radius R

$|\gamma| = \dfrac{L_{Cr}}{R}$

$CCR_S = \dfrac{\dfrac{L_{Cr}}{R}\cdot 63700}{L}$ [gon/km] (8.6c)

$L = L_{Cr}$

FIGURE 8.1 Equations for determining the curvature change rate of the single curve, CCR_S, for different design cases.

c) Curve Consisting of a Circular Arc with Radius R and Two
Adjacent Clothoids with Parameters A_1 and A_2

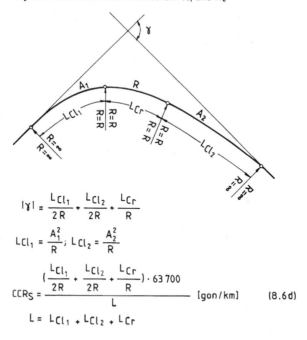

$$|\gamma| = \frac{L_{Cl_1}}{2R} + \frac{L_{Cl_2}}{2R} + \frac{L_{Cr}}{R}$$

$$L_{Cl_1} = \frac{A_1^2}{R}; \quad L_{Cl_2} = \frac{A_2^2}{R}$$

$$CCR_S = \frac{\left(\dfrac{L_{Cl_1}}{2R} + \dfrac{L_{Cl_2}}{2R} + \dfrac{L_{Cr}}{R}\right) \cdot 63\,700}{L} \quad [\text{gon/km}] \qquad (8.6\text{d})$$

$$L = L_{Cl_1} + L_{Cl_2} + L_{Cr}$$

d) Curve Consisting of Three Circular Arcs with Radii R_1, R_2, and R_3

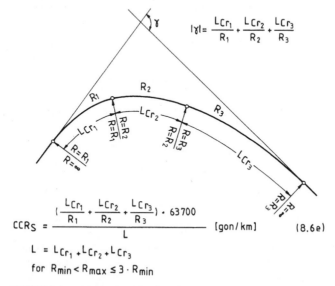

$$|\gamma| = \frac{L_{Cr_1}}{R_1} + \frac{L_{Cr_2}}{R_2} + \frac{L_{Cr_3}}{R_3}$$

$$CCR_S = \frac{\left(\dfrac{L_{Cr_1}}{R_1} + \dfrac{L_{Cr_2}}{R_2} + \dfrac{L_{Cr_3}}{R_3}\right) \cdot 63\,700}{L} \quad [\text{gon/km}] \qquad (8.6\text{e})$$

$$L = L_{Cr_1} + L_{Cr_2} + L_{Cr_3}$$

$$\text{for } R_{min} < R_{max} \leq 3 \cdot R_{min}$$

FIGURE 8.1 (*Continued*)   Equations for determining the curvature change rate
of the single curve, CCR$_S$, for different design cases.

e) Curve Consisting of Two Circular Arcs with Radii R_1 and R_2 ($R_1 > R_2$), Two Clothoids with Parameters A_1 and A_2 and one Egg-Shaped Clothoid with Parameter A_E

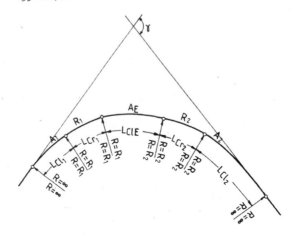

$$|\gamma| = \frac{LC_{l_1}}{2R_1} + \frac{LC_{r_1}}{R_1} + \frac{A_E^2}{2R_2^2} - \frac{A_E^2}{2R_1^2} + \frac{LC_{r_2}}{R_2} + \frac{LC_{l_2}}{2R_2}$$

$$LC_{l_1} = \frac{A_1^2}{R_1}, \quad LC_{l_2} = \frac{A_2^2}{R_2}, \quad LC_{lE} = \frac{A_E^2}{R_2} - \frac{A_E^2}{R_1} \qquad (R_1 > R_2)$$

$$CCR_S = \frac{\left(\dfrac{LC_{l_1}}{2R_1} + \dfrac{LC_{r_1}}{R_1} + \dfrac{A_E^2}{2R_2^2} - \dfrac{A_E^2}{2R_1^2} + \dfrac{LC_{r_2}}{R_2} + \dfrac{LC_{l_2}}{2R_2}\right) \cdot 63\,700}{L} \qquad [\text{gon/km}] \quad (8.6f)$$

$$L = LC_{l_1} + LC_{r_1} + LC_{lE} + LC_{r_2} + LC_{l_2}$$

Legend:

    LC_{ri} = Length of Circular Curve (i)   [m]
    LC_{lj} = Length of Clothoid (j)   [m]
    L = Length of Curve   [km]

FIGURE 8.1 (*Continued*) Equations for determining the curvature change rate of the single curve, CCR_S, for different design cases.

By determining the new design parameter, CCR_S, for a curve site according to Figs. 8.1 and 8.2, the 85th-percentile speed, $V85$, can be directly estimated from the operating speed background of Fig. 8.3 (example: Greece) with respect to the lane width. Additional operating speed backgrounds for numerous countries can be found in Fig. 8.12 and Table 8.5.

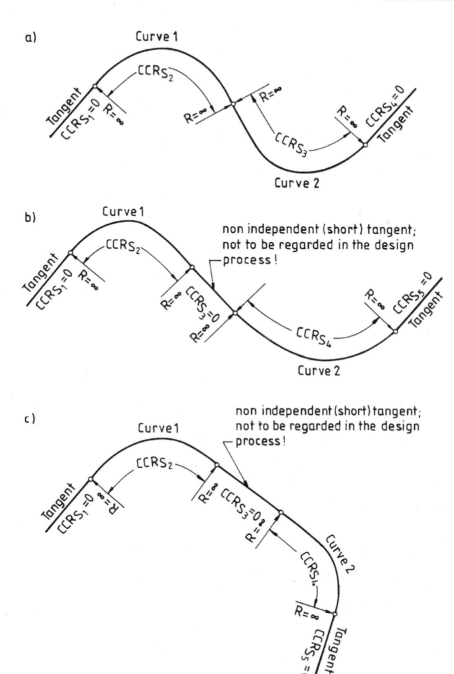

FIGURE 8.2 Systematic sketches for determining the curvature change rate of the single curve, CCR_S. (Independent and nonindependent tangents are defined in Sec. 12.1.1.3.)

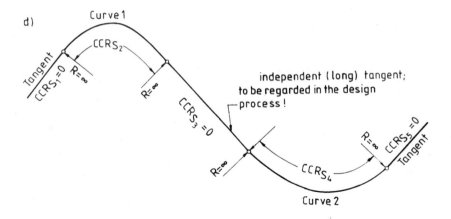

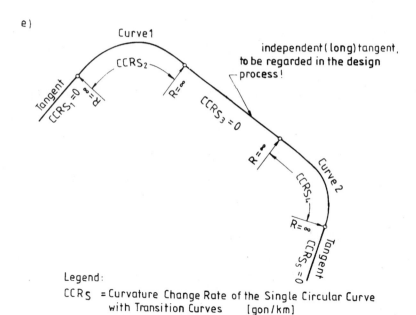

Legend:

CCR$_S$ = Curvature Change Rate of the Single Circular Curve
with Transition Curves   [gon/km]

FIGURE 8.2 (*Continued*) Systematic sketches for determining the curvature change rate of the single curve, CCR$_S$. (Independent and nonindependent tangents are defined in Sec. 12.1.1.3.)

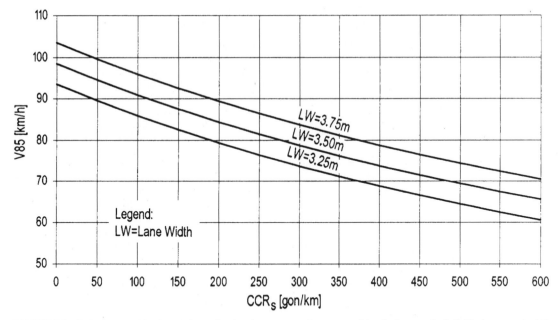

FIGURE 8.3 Operating speed background, as related to the curvature change rate of the single curve for individual pavement widths (two-lane rural roads), example: Greece.[453] (Further examples for operating speed backgrounds are presented in Fig. 8.12 and Table 8.5.)

8.1.4 Evaluation of Other Design Parameters That Influence Operating Speeds

Operating speed characteristics for multilane rural roads, as well as for longitudinal grades $G \geq 6$ percent, are discussed in Sec. 8.2.4 and in Sec. 18.5.1.

8.2 GENERAL CONSIDERATIONS, RESEARCH EVALUATIONS, GUIDELINE COMPARISONS, AND NEW DEVELOPMENTS

8.2.1 Definition of Terms

8.2.1.1 General. Based on the literature review,[109,153,158,159,268,271,325,361,405,433,434] which covers the highway geometric design guidelines of many countries, it can be concluded that two different speeds are used to govern highway geometric design. The design speed is a decisive influence for determining design-related parameters such as horizontal curves, grades, and vertical curves.

The second speed considers the expected actual speed behavior, and is usually expressed as the 85th-percentile speed. In general, this speed is used to determine operating speed-related design parameters such as sight distances, corresponding crest vertical curves, and superelevation rates.

8.2.1.2 Operating Speeds on Dry and Wet Road Surfaces. An inadequate choice of a vehicle speed which is inconsistent with road, traffic, and sight conditions is a prime cause of traffic accidents. Posted speed limits show only maximum permissible speeds, which shouldn't be exceeded and should be driven under ideal conditions. They don't indicate how fast one can actually drive under bad road conditions, for example. Similar considerations are valid for the selected design speed.

In Sec. 8.1.1, the definition of the 85th-percentile speed is based on clean, "wet" road surfaces in order to reflect the impact of lower skid resistance values under "wet" conditions.

It is important to clarify whether operating speeds on dry pavements are significantly different from operating speeds on wet pavements. The results are based on the investigations in Refs. 384, 416, 422, and 705.

Weather conditions should result in modification of vehicular operating speeds because of a reduction in visibility and a possible impairment of surface conditions. The general effect of wet pavement conditions is to reduce friction values or coefficients between the tire and the roadway. The amount of reduction depends upon the presence of moisture, snow, ice, or on the thickness of the film of water covering the pavement. As the thickness of the film of water increases, skid resistance decreases and, in cases of heavy rain, hydroplaning conditions may even occur.[5,401] Note that geometric design deficiencies, as well as drainage and pavement design, may affect the incidence of wet pavement accidents, especially at curved sites. For example, when consistency between successive design elements is not present, or design speed and operating speed are not in harmony at a certain curved site, or an adequate dynamic safety of driving cannot be provided because of a reduction in friction factors due to wet pavement conditions,[420] critical driving maneuvers may occur. Other roadway sections which may be hazardous when roadway surfaces are wet are those sections with insufficient minimum or maximum superelevation rates, superelevation runoffs with insufficient longitudinal slopes, or sag vertical curves which do not satisfy the drainage requirements.

According to a report prepared by the U.S. National Transportation Safety Board (NTSB),[539] approximately 13.8 percent of all fatal highway accidents occurred on highway pavements that were wet. The NTSB cited a study of wet pavement accidents on the West Virginia highway system which revealed that the average rate of accidents on wet pavement was 2.2 times the rate of accidents on dry pavement and that the *maximum* rate was 85 times the accident rate on dry pavement. Forty percent of the accidents on the West Virginia interstate system occurred on wet pavement. The report estimated that the roads in West Virginia were not wet more than 1.5 percent of the time. If analyses of national and international data reflect findings similar to those cited by the NTSB,[539] it may be concluded that the wet pavement problem should be of major concern worldwide and that more resources should be allocated for correcting the sources of the problem.

A review of accidents in Europe and the United States[391,402,413,415,425,426] showed that at least 50 percent of all fatalities in both parts of the world occurred on two-lane rural highways and about 30 percent at curved sites. Analyses of accidents on two-lane rural highways have yielded the following results:

1. "Fatal/injury accidents" accounted for more than 70 percent of the accidents on curves, while property damage accidents greater than $400 represented less than 30 percent.

2. "Wet pavement conditions" contributed to nearly 50 percent of the accidents on curves, even though it is a well-known fact that the vehicle kilometers driven under these conditions are much lower than those driven on dry pavements.

3. Approximately 60 percent of the accidents at curved sites were "run-off-the-road" accidents, normally with only one vehicle involved.

In summary, it may be concluded that a high fatal/injury accident risk *does* exist on curves of two-lane rural roads, especially under wet pavement conditions.

It is widely accepted that because of the lower coefficient of friction on wet pavement as compared with dry, the wet condition more frequently leads to critical driving maneuvers or even accidents.[197,385,396,617] In this connection, it should be recognized that speed is an additional vital contributing factor to many wet weather accidents, because the skid resistance decreases as a vehicle's speed increases.[420] For this reason, most of the studied geometric design guidelines (for example, Refs. 5, 35, 37, 139, 246, 453, 607, 643, 663, 686, 693, and 700) stress the fact that, for driving dynamic considerations in horizontal and vertical alignments, the design speeds, operating speeds (for example, 85th-percentile speeds) and tangential/side friction factors must be considered and selected on the basis of wet pavement conditions.

The purpose of this chapter is to present the results of studies conducted to determine the effect of design parameters, traffic volume, and wet pavement conditions on free operating speeds of passenger cars on curved sections of two-lane rural highways. This chapter will determine whether drivers recognize the fact that wet pavements offer less skid resistance than dry pavements do and that they should adjust their speeds to accommodate wet pavement, particularly on curves, in order to maintain vehicle control similar to that under dry conditions.

In this connection, a short description of research investigations[416,422] and their conclusions are provided here. They are based on experiences in Germany and the United States.

Twenty four sites on horizontal curves of various degrees were selected from an overall database of 261 two-lane rural roadway sections.[406,408] The degree of curve varied from 0 to 27°/100 ft, which corresponds to curvature change rates* from 0 to 985 gon/km. The grades were level or nearly so at the curved sites and for a considerable distance before and after. This minimized the effect of grades on operating speeds.

Observations on wet pavements were taken several weeks after speeds on dry pavements were collected. On all occasions, the surfaces were wet and rain was falling from a drizzle to moderately heavy rain. On no occasion did it rain so hard as to significantly affect visibility. Sight distances were approximately 140 to 170 m, which represent the limiting values of stopping sight distances for design speeds between 90 km/h and 100 km/h for most countries (see Table 15.4). This control was observed because of the results of research experiences[401] which indicated that when sight distance is affected appreciably by heavy rain, drivers tend to reduce their speeds. In other words, drivers reduce their speeds not because of the danger created by lower skid resistance values on wet pavements, but instead because of limited sight distance.

The available sight distance was determined by a member of the study team who drove through the study section at different fixed time intervals. By passing a specified mile marker, the driver had to clearly see a mile marker located at a distance of 0.1 mile (160 m) in front. If this was not possible due to heavy rain, the driver would instruct the observer who was taking the measurements to stop.

The number of vehicles recorded varied considerably from site to site, as the wet pavement studies were dependent on continued rain and adequate visibility. Only speeds of passenger cars under free-flow conditions, with a minimum time gap of at least 6 s between successive vehicles, were recorded.

Cumulative speed distribution curves were plotted for each location studied. Figure 8.4 shows a typical example of the distribution of speeds on dry and wet pavements for the United States[416,422] and Fig. 8.5 for Germany.[384]

The most notable feature of the speed data collected in the study is that there was little difference in the speed distribution of free-moving passenger cars on dry and wet pavements for all curved sites. Similar results were shown by Stohner.[682]

The averages of operating speeds (expressed by the 85th-percentile speeds) on dry and wet pavements for the 24 curved road sections were 76.06 and 75.86 km/h for both directions of traffic (see Fig. 8.4). The observed decrease in 85th-percentile speeds due to wet road surface is only 0.27 percent, which is not statistically significant.

The operating speeds on dry and wet pavements were additionally examined by plotting the dry speeds on the y axis and the wet speeds on the x axis for each test section (Fig. 8.6). It should be noted that: (1) if the operating speed on wet pavement was identical to the operating speed on dry pavement, the data point would fall on the hypothetical 45° diagonal line shown, (2) if the operating speed on wet pavement was greater than the operating speed on dry pavement, the data point would fall below the line, and (3) if the operating speed on dry pavement was greater than the operating speed on wet pavement, the data point would be above the line. As illustrated in Fig. 8.6, the data are random with points distributed both above and below the hypothetical 45° line.

To determine whether or not operating speeds on dry pavements were significantly different from operating speeds on wet pavements, the Kolmogorov-Smirnov (K-S) two-sample statistical test was used; for further information about this statistical test procedure, see Refs. 97, 515, and 647.

---

*For the definition of the U.S. design parameter degree of curve and the German design parameter curvature change rate, see Sec. 8.2.3.6 and for the corresponding conversion factor, see Table 8.4.

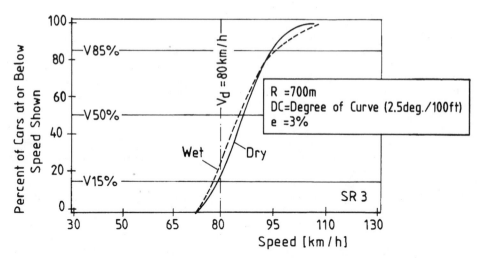

FIGURE 8.4   Typical example illustrating the distribution of speeds on dry and wet pavements on New York State Route 3, United States.[416,422]

It follows from the statistical tests (level of significance $\alpha = 0.05$) that the relationship between operating speed and degree of curve (Fig. 8.7) is valid for both dry and wet pavements in the United States so long as the visibility is not appreciably affected by heavy rain. This conclusion is substantiated by the results of a study conducted in the Federal Republic of Germany,[384] which established that a significant decrease in operating speeds was observed *only* for high precipitation levels combined with greatly reduced visibility conditions.

For passenger cars, the expected operating speed ($V85$) on wet or dry pavements can be determined for the United States by applying the nomogram for the relationship between degree of curve and 85th-percentile speed in Fig. 8.7.

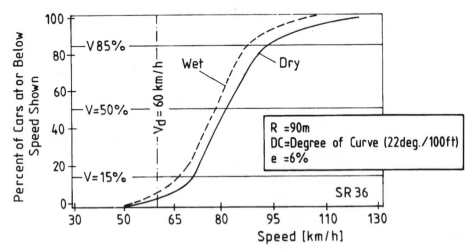

FIGURE 8.5   Typical example illustrating the distribution of speeds on dry and wet pavements on State Route 36, Germany.[384]

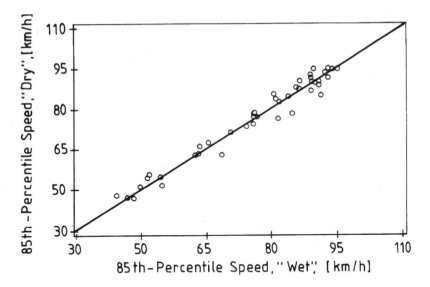

FIGURE 8.6   Plot relating 85th-percentile speeds on dry pavements to those on wet pavements, United States.[416,422]

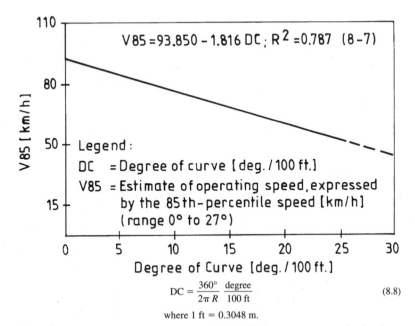

$$V_{85} = 93.850 - 1.816 \, DC; \quad R^2 = 0.787 \quad (8-7)$$

Legend:

DC = Degree of curve [deg. / 100 ft.]

V_{85} = Estimate of operating speed, expressed by the 85th-percentile speed [km/h] (range 0° to 27°)

$$DC = \frac{360°}{2\pi R} \frac{\text{degree}}{100 \text{ ft}} \qquad (8.8)$$

where 1 ft = 0.3048 m.

FIGURE 8.7   Nomogram for evaluating operating speeds of passenger cars as related to degree of curve under wet and dry pavement conditions, United States.[416,422] *Note:* American Design Parameter degree of curve for describing the road characteristics (to a certain extent comparable with the curvature change rate of the single curve, see Sec. 8.2.36).

Thus, it can be concluded that operating speeds on dry pavements were not statistically significantly different from operating speeds on wet pavements and that drivers do not adjust their speeds sufficiently to accommodate for inadequate wet pavement on curves in particular. (If the wet pavement is accompanied by decreased visibility, speed reductions do take place.) It is obvious that the drivers, at least in this research investigation, did not recognize the fact that friction is significantly lower on wet pavements than on dry. Furthermore, it was found that drivers reduce their speeds not so much because of the danger created by lower skid resistance values on wet pavements, but rather because of limited sight distances due to heavy rain. From this finding it may be assumed that, by not adapting to wet roadway conditions, drivers run a higher risk of being involved in a traffic accident.

Therefore, it can be presumed that designing for moderately heavy rain (with sufficient sight distance) is more critical than for heavy rain (visibility affected). In the first case, the operating speeds, $V85$, of drivers are not affected by the "wet" surface condition. Their driving behavior is not different from that under "dry" surface conditions, whereas in the second case, drivers reduce their operating speeds considerably.

A Swiss study[705] confirms these results. Speed measurements were conducted for several weeks at five measurement spots on two-lane rural roads. The rain intensities were grouped as follows:

- *Light rain:* 0.5 to 0.9 mm/h
- *Moderate rain:* 1.0 to 2.9 mm/h
- *Heavy rain:* ≥3.0 mm/h

Table 8.1 indicates that considerable differences in the driving behavior (expressed by the average and 85th-percentile speeds) in the daytime existed for heavy rain only in comparison to dry weather as well as to light and moderate rain conditions. However, for dry weather conditions at night, the average speeds were 5 km/h higher, while the 85th-percentile speeds were 7 km/h higher than those during daylight. For rain and daylight hours, the speeds decrease considerably under heavy rain only, while at night they show a considerable decrease even under light rain conditions. In general, the speed level under rainy conditions does not differ considerably between daytime and nighttime, at least for "light" and "moderate" rain intensities (see Table 8.1).

Thus, the investigations by Thoma[705] support the previous statements, that is, a decrease in the speed level on wet road surface takes place at daytime only under heavy rain and at night even under light rain. That means, only in cases of driving conditions where sufficient sight distances are severely limited can a decrease in speed levels be expected.

Similar results have also been reported in another Swiss study.[91] An 85th-percentile speed reduction of 10 km/h was observed only for passenger cars exposed to heavy rain (greater than

TABLE 8.1   Speed Results of Passenger Cars for Different Weather and Daylight Conditions on Two-Lane Rural Roads (Time Gap > 5 s)[705]

	Parameter	Dry weather	Rain		
			Light	Moderate	Heavy
Day	$\overline{V}$	82	84	81	76
	$V85$	92	94	90	83
	V_{perm}, % 80 km/h	48	54	46	20
Night	$\overline{V}$	87	81	79	—
	$V85$	99	92	91	—
	V_{perm}, % 80 km/h	60	44	38	—

Note: $\overline{V}$ = average speed, km/h; $V85$ = 85th-percentile speed, km/h; V_{perm} = percentage of passenger cars which exceed the permissible speed limit of 80 km/h, %.

0.1 mm/m). For trucks, which generally have lower speed levels than passenger cars, rain did not have any influence on their speeds. If we take into consideration the fact that heavy rains with intensities of more than 0.1 mm/m account for no more than 5 percent of all rainfall, then it can be presumed that the other wet weather conditions, like light/moderate rain, could impose risky driving conditions, since friction is significantly reduced while vehicle speeds remain practically constant.

It can be presumed, therefore, that it is not the "dry" or "wet" road surface which affects the speed choice of the motorist, but only the additional impairment of sight distance that leads to significant speed reductions (in daylight, heavy rain; at night, light rain). Thus, ample evidence exists to indicate that wet pavement does not have a great effect on operating speed and that drivers will not adjust their speeds sufficiently to accommodate for inadequate wet pavement, particularly on curves.

In this connection, Hiersche, et al.[287] have reported the following:

> When observing the speed behavior with regard to rain intensities, the reaction of the driver caused by sight distance impairments stood in the foreground. An increase in rain intensities and an increase in waterfilm thickness led to similar results, i.e. speed decrease. "Light rain" showed only a small influence on driving behavior. "Heavy rain" on the other hand led to a considerable speed reduction because of sight distance impairments.

Since the term "light rain" in this study[287] corresponds to "moderately heavy rain"[416,422] and to "moderate rain,"[705] it can be concluded that this study confirms the results discussed.

Finally, Pfundt[575] reports that the driving behavior on moderately wet pavements is not essentially different than that on dry pavements.

A study by Thoma,[705] which summarized an interesting comparison of risk ratios for dry and wet weather conditions, as well as for day and nighttime conditions, yielded the results shown in Table 8.2, which are valid for rural highways, except freeways.

TABLE 8.2   Risk Ratios under Different Driving Conditions[705]

Driving conditions	Risk ratio
Day versus night	1:2
Dry versus wet	1:2.5
Day and dry versus night and wet	1:6

As can be noted from Table 8.2, driving at night with wet pavement conditions is 6 times more dangerous than driving during the day under dry pavement conditions.

For the following implementation of these results for design purposes, it can be stated that the recommendations for coordinating design speed and 85th-percentile speeds (Secs. 9.1.2 and 9.2.2), as well as the recommendations for achieving operating speed consistency (Secs. 9.1.3 and 9.2.3), are valid for both dry and wet pavement conditions as long as visibility is not reduced by heavy rainfall or even cloudbursts. Only in the latter case, because of affected visibility, will the driver recognize the danger resulting from wet road surfaces and reduce speed accordingly.

It is presumed that the experiences gained in the United States, Germany, and Switzerland are also comparable to those in other countries, but further investigations are recommended.

Finally, it should be noted that according to Krebs and Dieterle,[370] the point-mass model as a simplification for the system vehicle-road surface, is also applicable for wet pavements, based on the driving dynamic assumptions, discussed in Chap. 10.

8.2.2 General Speed Determination

The proposals for the establishment of 85th-percentile speeds for multilane median-separated rural roads of category group A and for roads of category group B in Sec. 8.1.2 are mainly based on European experience.[159,246,320] However, new investigations in Greece[453,586] indicated that the assessments, according to Eqs. (8.1 to 8.5), make sense for other countries as well (see Fig. 8.13).

Backgrounds and new developments for the relationships between 85th-percentile speeds and design parameters for two-lane rural roads of category group A, with special regard to road characteristics, will be discussed, analyzed, and evaluated in the next section.

8.2.3 Determination of the 85th-Percentile Speed for Two-Lane Rural Roads of Category Group A

8.2.3.1 General Considerations. The driving behavior of motorists is based on a number of influencing variables and the driving habits of the motorist. The range of influencing variables is described in Ref. 347, as follows: "The driving behavior and the speed choice are determined both by the optical picture of the driving space as well as by the accumulated experiences of the driver." This observation shows that a determination of the driving behavior can be established only by data collection, reduction, and analysis of the most important influencing variables. Consideration of all influencing variables cannot be guaranteed by any investigation because the individuality of the drivers and their behaviors can never be totally comprehended. Besides the cited variables for analyses of speed behavior, the continuing motor vehicle shift to higher-powered vehicles also has to be considered.

Future analyses may use radar techniques to follow cars to provide excellent databases for the investigations of unaffected speed behavior for both cross-sectional and sectional-related speed analyses.[674] However, this new speed data collection method does not currently represent the state-of-the-art. Most research to date has been based on cross-sectional (local) speed measurements, especially for establishing large databases.[408,431,586]

Many prior publications have been concerned with the problem of the speed choice resulting in basic information on expected operating speeds. Many investigators used the 85th-percentile speed of free-flowing vehicles as a measurement to help to explain driving behavior.

Therefore, the goal of the present research is to describe the speed behavior, $V85$, on two-lane rural roads and to analyze the relevant influencing design variables. Traffic-related and operational-limiting conditions were also considered in these investigations.

It is difficult to analyze and quantitatively evaluate the effects of single influencing variables on operating speed because these variables are not independent. That means they are effective individually, but are partially interrelated to a certain extent. In addition, the data collection of several influencing parameters is difficult or even impossible. For example, the influence of the three-dimensional effect of the driving space of the road certainly influences speed choice, but controlling the various aspects presented to the driver is impossible.

8.2.3.2 Influencing Variables (Road Characteristics).* The significance of the single influencing parameters on the overall system "driver-vehicle-road" will be presented in the following section.

The far-reaching range of influencing variables reveals a distinct definition of the term *road characteristics,* which considerably affects the speed behavior, in addition to the traffic-related influencing variables. Schliesing[627] describes the term *road characteristics* as follows:

> The design features of a roadway which are relevant for the traffic behavior over a continuous roadway section represent the road characteristics. Relevant design features are alignment, cross section, design of intersections and accesses to adjacent properties. Road characteristics must correspond to traffic task, traffic volume, and traffic composition.

According to the German Road and Transportation Research Association[703] road characteristics are defined as: "The entirety of the design features of a roadway which are relevant for the behavior of the drivers on an observed section-interval."

---

*Elaborated based on Ref. 674.

Leins[469] defined this term in this way:

> The task of the road characteristics is to create a balanced arrangement of all perceivable road features and recognizable road types which provide the motorist with an intuitively correct driving behavior, increase traffic safety and improve traffic flow.

This explanation of the term *road characteristics* by Leins[469] was taken as a basis for the present investigation because it reflects the complexity of influencing variables. Based on the previous statement, the difficulty of separating the individual influencing variables in order to investigate their influence on speed behavior becomes apparent.

Koeppel and Bock[347] covered the entire range of influencing variables in the following manner: "Three dimensional picture of the driving space + accumulated experience $\rightarrow$ driving behavior and speed selection." By this "equation," the driver and his or her behavior become more and more significant.

Studies by Dilling,[143] Leutner,[472] and Durth[149] build on these definitions requiring additional consideration of human behavioral modes, as well as the entire road space with its adjacent areas.

More recent investigations have attempted to describe the speed and driving behavior with respect to road space and behavioral analyses. However, all of these multidisciplined studies point out the difficulty of evaluating and quantifying the influencing variable driver and behavioral mode. Additional variables, which should certainly be considered, might be the continuously changing vehicle population and the increased level of the driver's experience. Increases in power, handling, and vehicle safety automatically affect the speed behavior.

In the following sections, results of prior investigations are briefly discussed. The influencing design variables are distinguished by curvature change rate, radius of curve, sight distance, longitudinal grade, and cross-sectional design.[440]

Curvature Change Rate and Radius of Curve.   Several traffic research studies have attempted to analyze the influence of different curve parameters on the driving behavior of the motorist.[476] The studies of Dilling;[143] Koeppel and Bock;[347] and Durth, Biedermann, and Vieth[151] showed that a limitation on the curve itself is not a sufficient speed predictor because for radii of curve of the same size, different speeds were observed. According to Leutzbach and Papavasiliou,[475] better predictive results can be obtained by considering the approach sections to curves in addition to radii.

The design parameter, curvature change rate (CCR), has been found to more accurately relate road characteristics (the entirety of the design features of a roadway) to driving behavior, since many investigations have shown a significant influence of this parameter, CCR, on observed operating speeds (for the definition of CCR, see Fig. 8.10). For example, it was reported by Trapp:[712] "The dominant road-dependent influence on the driving behavior is the CCR-value." It was reported by Dilling:[143] "A statistically significant influence of the CCR-value on the driving behavior was established." It was reported by Trapp and Oellers:[713] "The CCR-value is the strongest variable for the driving behavior. The CCR-value represents the relevant sectional influence for the driving speed on an observed roadway section." Finally, it was reported by Al-Kassar, Hoffmann, and Zmeck:[12] "For the free roadway section it was found that the CCR-value and the pavement width...represent the relevant influencing variables on speed."

In contrast, Lamm[385] limits the influence of curvature change rate on speed behavior. He indicates that, for a range of CCR values, a single value alone cannot adequately describe the influence over a certain length, even if a similar road characteristic is assumed for the different design elements.[674] Lamm states, furthermore, that the CCR value is suitable only for partially evaluating the influence of curvature on driving behavior, particularly for changing ranges of curve radii. In these curve radii ranges, the observed speeds are so different that the driving behavior along a roadway section, even with similar road characteristics, cannot be explained by *one* CCR value alone.

Therefore, along a roadway section, the curvature change rate of the single circular curve, including transition curves CCR_s, is introduced in Sec. 8.2.3.6 to determine different 85th-percentile speeds at different individual curved sites (even with similar road characteristics).

Schlichter[625] also indicates that for the same CCR values for two roadway sections, a different driving and speed behavior has to be expected, since CCR represents only the "average features" for a certain alignment section.

Generally, statistically sound relationships between the speed of the motorist and the curvature change rate, as an essential design parameter for road characteristics, have been established. A study conducted by Krebs and Damianoff[372] indicated that the speed decreases as CCR increases (see Fig. 8.8).

The newest speed measurement results are represented by "curve 7."[431] These measurements reveal a sharp increase in the speed level during the last two decades, caused principally by the

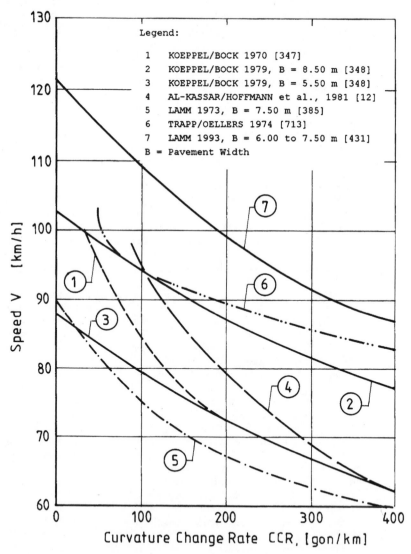

FIGURE 8.8 Relationships between speed, *V*, and curvature change rate, CCR, for different regression models. (Elaborated and completed based on Ref. 372.)

change in vehicle population, the increased level of driver experience,[674] and, at least in Germany, insufficient police enforcement.

Since the curvature change rate is strongly influenced by the radius of curve (Fig. 8.1), similar statements are valid for the radius of curve as for the design parameter, curvature change rate. That means that small radii of curve correspond to high curvature change rates, and vice versa. Therefore, the following statements can be cited, which deal particularly with the radius of curve.

For example, Ruwenstroth[620] reports: "Radii of curve less than 350 m represent an especially strong influence on the speed. However, the influence of curvature change rate is dominant."

Kloeckner and Maier-Strassburg[339] also have the opinion that, on the one hand, the individual design element, radius of curve, influences the driving behavior, and on the other hand, the same radius of curve can be traversed with different speeds on different roadway sections, depending on:

- Curvature change rate of the section in front
- Other visually effective influences in the road space
- Passing needs or pressures of varying strength

Bald, cited by Pfundt,[42] indicates that the highest deceleration is observed at narrow, imperceptible curves both in front of and in the curve itself. He concludes that the curvature change rate of a given roadway section is not as relevant as the radii of individual curves.

Finally, investigations by Dilling,[143] Koeppel and Bock,[348] and Durth, et al.[151] indicated that one should not concentrate exclusively on the radius of curve because for the same radius of curve, different operating speeds have been observed.

These different meanings with respect to curve radii support the introduction of the curvature change rate of the single circular curve by the authors for estimating operating speeds (see Sec. 8.2.3.6).

Sight Distance.   According to Trapp and Oellers,[713] the influence of sight distance (SD) on driving behavior is not uniform. It could be stated that the influence of sight distance should not be considered separately because it is strongly related to curvature change rate. Analogous results were reported in Ref. 712, as well as by Koeppel and Bock:[347] "The influencing variables depend strongly on each other, especially the CCR-value and the sight distance."

The research results of Al-Kassar, Hoffmann, and Zmeck[12] point in the same direction: For the free roadway section, it was found that the CCR value and the cross-sectional width cause the relevant influences on driving speed, while the influence of the longitudinal grade and sight distance are of lesser effect. Therefore, it appears justifiable not to consider the SD (sight distance) in a regression analysis. Trapp and Kraus[714] also do not regard SD as an independent influencing variable, but assume that SD correlates highly with the CCR value and that the influence of SD is already included in the design parameter CCR. Yagar and Van Aerde[768] found that sight distance was not a contributory factor in controlling vehicle speed.

Regression analyses by Lamm and Choueiri[406,408] between a number of design elements and speed yielded similar conclusions with respect to sight distance.

Longitudinal Grade.   In general, it can be assumed that the driving behavior of free-moving passenger cars ($V85$) is not significantly affected up to longitudinal grades of ± 5 to ± 6 percent.[347,406,408,691,713] Up to these grades, the performance characteristics of modern cars have made this geometric attribute unrelated to free-flowing passenger vehicle traffic. Trucks, of course, are a different matter.

The background of the French operating speed is shown in Fig. 8.14 as a function of grade on upgrade sections. This figure indicates that the 85th-percentile speed of passenger cars begins to be affected by longitudinal grades of more than 5 percent.

Based on recent speed measurements by Schulze,[638] it can be concluded that longitudinal grades of up to 6.5 percent do not have a significant influence on the speed behavior of free-moving passenger cars. In the Swiss Norm SN 640 080b (1991 edition), a value of 7 percent was even cited.[691]

However, the longitudinal grade has a strong influence on the speed of vehicles with low power–weight ratios (hp/t), like trucks, as Fig. 8.9 reveals. The curves shown in this figure are

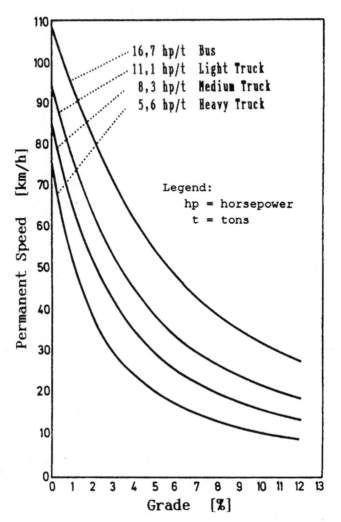

FIGURE 8.9 Permanent speeds of heavy vehicles with different power–weight ratios. (Modified based on Ref. 785.)

based on investigations conducted in 1979.[785] That means the curves should be regarded only as qualitative because modern trucks have essentially higher power–weight ratios (see Secs. 13.1.2 and 13.2.2) and approach today higher permanent speeds.[638]

Cross Section. An examination of the influence of cross-sectional design and lane (pavement) width on speed behavior led to the following conclusions.

According to Trapp:[712] "The pavement width combined with shoulder width affect the speed behavior." According to Schiller:[624] "The investigations revealed, for different pavement width-classes with good road surface conditions, that speeds increase with increasing pavement widths." According to Lamm:[385] "Based on regression analyses, it can be stated that there exists a marginal relationship between 85th-percentile speed and pavement width." According to Trapp and Oellers:[713] "The curvature change rate has a relevant influence on the speed. The lane width alone has no significant influence."[575] According to Al-Kassar, Hoffmann, and Zmeck:[12] "For the free roadway section, it was found that curvature change rate, cross section width, and shoulder width collectively describe the relevant influence on speed." Finally, according to Baumann:[45] "...considering the free moving motor vehicles, no significant difference was found between the relevant percentile speeds for different standard cross sections of two-lane rural roads."

The pavement (or lane) width has been included as an additional parameter for describing the relationship between curvature change rate (respectively degree of curve) and the 85th-percentile speed in the operating speed backgrounds of several countries (see Figs. 8.3, 8.10, and 8.11). This was done to give the highway engineer a better classification system with respect to pavement (or lane) width. However, it should be noted that, based upon all conducted regression analyses in the different countries, only marginal differences were found regarding the influence of this design parameter on operating speeds ($V85$).[241,406,408,453]

8.2.3.3 Developments in Germany.   Since 1973 in the German Guidelines for the Design of Roads, Part: Alignment, there has been an operating speed background (Fig. 8.10) for evaluating the expected 85th-percentile speeds with respect to the design parameters curvature change rate and pavement width. This background is based upon the research work of Koeppel and Bock,[347] Lamm,[385] and the practical experience of the committee in charge. The upper curve in Fig. 8.10 represents the results of speed measurements in the 1990s,[431] and shows the need for reconsidering the existing background because of the significantly increased operating speed level, $V85$, during the last decades. On average, speeds have increased 1 km/h/y.

Despite this fact, it should be emphasized that German designers have evaluated operating speeds ($V85$) with respect to the road characteristics (expressed by the design parameter curvature change rate) since 1973[241] in order to coordinate operating speed and design speed, as well as to consider the different operating speeds between successive design elements.[271,405] In contrast, other countries have only just begun to regard these important design impacts since the beginning of the 1990s. Exceptions include Switzerland[691] and the United States, where a new concept in design speed application was proposed by Leisch and Leisch in 1977.[470]

8.2.3.4 Developments in the United States.*   In contrast to Germany, degree of curve (DC) is used in many countries as the relevant design parameter for describing the driving behavior with respect to road characteristics.

In order to estimate operating speeds with respect to highway design parameters in the United States, the following procedure developed by the authors is presented. Because the speed investigations were conducted in the United States, the design parameter, degree of curve, was used instead of the design parameter, curvature change rate, to describe the driving behavior. The relationship between degree of curve and curvature change rate is shown in Table 8.4. The various steps of the procedure are described here.

Data Collection: Operating Speeds and Design Parameters

Data collection was broken down into three phases: first, the selection of road sections appropriate for the study; second, the collection of as much field data as possible about the road sections; and third, the measurement of operating free speeds at each section.

Selection of appropriate road sections.   Site selection was limited to sections with the following characteristics:

1. A tangent to curve or curve to curve section
2. Removed from the influence of intersections
3. No physical features adjacent to or in the course of the roadway that may create abnormal hazard
4. Grades less than or equal to 5 or 6 percent
5. Average annual daily traffic between 1000 and 12,000 vehicles/day.

---

*Elaborated based on Refs. 104, 108, 402, 406, 408, 409, 410, 412, 417, 418, 421, and 434.

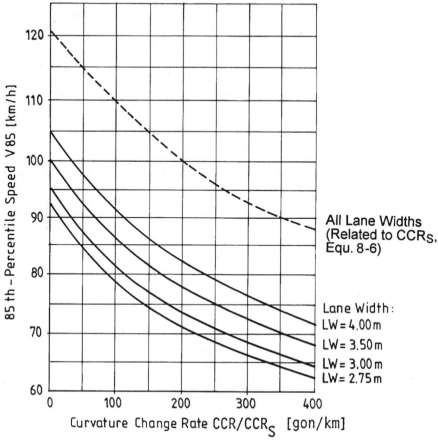

──────── German Operating Speed Background 1973 [241] and 1984 [243]
- - - - - - ISE Speed Measurements, 1993 [431]
            ISE = Institute for Highway and Railroad Engineering, University
            of Karslruhe, Germany

$$\mathrm{CCR} \quad \frac{\displaystyle\sum_{i=1}^{n} |\gamma_i|}{L} \quad \mathrm{gon/km} \tag{8.9}$$

see also Table 8.4.

where CCR = curvature change rate, gon/km

$\Sigma|\gamma_i|$ = angular changes in curves = $\Sigma| \tau_i + \alpha_i |$, gon

τ_i = angular change in transition curve, gon

α_i = angular change in circular curve, gon

L = length of an observed roadway section with similar road characteristics, km

LW = lane width, m

Note: 1 gon = 0.9 degrees

FIGURE 8.10 Operating speed background, as related to the curvature change rate for individual pavement widths, road categories A I to A IV (two-lane rural roads), example: Germany.

It was attempted to maintain a regional distribution and, at the same time, retain the longest road segments in the selection process. The road sections selected provided the widest range of changes in alignment which could be found by observation.

Field data collection.   This phase involved obtaining as much data in the field as possible about the road sections and specifically about the curves or curved sections within the observed road section. Information recorded included degree of curve, length of curve, superelevation rate, sight distance, and gradient.

Speed data collection and reduction.   In order to ensure that the speeds measured represented the free speeds desired by the driver under a set of roadway conditions and were not affected by other traffic on the road, only the speeds of isolated vehicles with a minimum time gap of about 6 s were measured. Speed measurements were made during daytime hours on weekdays under both dry and wet pavement conditions (see Sec. 8.2.1.2).

The basic method used for speed data collection involved the measurement of the time required for a vehicle to traverse a measured course laid out in the center of a curve. Speed measurements were also taken on preceding and succeeding tangents to the curved site. The length of the course was 50 m. The method used for measuring time over the measured distance involved use of transverse pavement markings that were placed at each end of the course and an observer who started and stopped an electronic stop watch as a vehicle passed the markings. The observer was placed at least 5 m from the pavement edge of the road to ensure that his or her presence would not influence the speeds of passing vehicles, but not too far away to prevent accurate measurement. Normally, the observer was located at a position, where she or he could not directly be seen by the drivers of traversing vehicles.

By applying this procedure, satisfactory speed data were obtained for both directions of travel. Because of cost, time, and personal constraints, about 80 to 100 passenger cars under free-flow conditions were sampled at each site for both directions of traffic. Speed data were then analyzed to obtain the operating speed, expressed herein by the 85th-percentile speed—that speed below which 85 percent of the vehicles travel.

Methodology

Many factors affect operating speeds on two-lane rural roads, including, but not limited to, characteristics of the site, characteristics of traffic and road users, characteristics of controls, and characteristics of variable factors. Each of these factors can act in different ways and to varying degrees at a given location. Therefore, the task of determining the influence of each on operating speeds becomes difficult.

In this study, only the effect of the following parameters on operating speeds will be addressed: degree of curve, length of curve, superelevation rate, sight distance, gradient, posted recommended speed, and average annual daily traffic (AADT).

For evaluation of the quantitative effects of design and traffic parameters, the multiple linear stepwise regression technique (max R^2 improvement technique) was used. The stepwise technique consists of adding one independent variable to the regression equation in each step. Thus, the stepwise process produces a series of multiple regression equations in which each equation has one more independent variable than its predecessor in the series.[406,408,505] The following stipulations were used to terminate the stepwise process and to determine the final multiple regression equation:

1. The selected equation had to have a multiple regression coefficient, R^2, that was significant at the 0.05 level.

2. Each of the independent variables included in the multiple regression equation had to have a regression coefficient that was significantly different from 0 at the 0.05 level.

3. None of the independent variables included in the multiple regression equation could be highly correlated with each other.

The superelevation rate and posted recommended speed were withheld from any subsequent regression analysis because they are highly correlated with degree of curve.

The selected multiple regression equation had to fulfill all three stipulations. In addition, the following conditions were assumed to hold:

1. Degree of curve was taken as positive whether a curve turned left or right.
2. An uphill gradient was treated as positive, whereas a downhill gradient was treated as negative.

Outcome of the Data Analyses and Relationships between Variables

The analysis is based on data collected for 261 curved roadway sections under dry pavement conditions in the United States (State of New York).

Various stages of regression analyses found that the most successful equation for explaining much of the variability in 85th-percentile speeds, in terms of statistical significance and overall form, is as follows. The overall equation:

$$V85 = 55.52 - 1.61 \,(DC) + 10.92 \,(LW) + 0.91 \,(SW) + 0.00064 \,(AADT)$$
$$R^2 = 0.842 \tag{8.10}$$

where $V85$ = estimate of the operating speed expressed by the 85th-percentile speed, km/h
   DC = degree of curve (range 0 to 27°)
   LW = lane width, m
   SW = shoulder width, m
   $AADT$ = average annual daily traffic, vehicles/day
   R^2 = coefficient of determination

The large R^2 (0.842) suggests that the relationship represented by Eq. (8.10) is a strong one.

Design parameters, like sight distance, length of curve, and gradient, were not included in the regression model because the regression coefficients associated with these parameters were not significantly different from 0 at the 95 percent level of confidence.

However, in comparing Eq. (8.10) with the following reduced Eq. (8.11), which only includes the design parameter DC, note from the coefficients of determination, R^2, that the influence of LW, SW, and AADT in Eq. (8.10) explains only about an additional 5.5 percent of the variation in the expected operating speeds. In other words, much of the variability in operating speeds is explained by the design parameter degree of curve. The reduced equation is

$$V85 = 93.850 - 1.816 \,(DC)$$
$$R^2 = 0.787 \tag{8.11}$$

The moderately large R^2 value (0.787) suggests that the relationship represented by Eq. (8.11) is also strong, and can be considered a competitor to the relationship represented by Eq. (8.10).

The corresponding regression equations for operating speed with respect to degree of curve for individual lane widths and all lanes combined are presented in Fig. 8.11. They are schematically shown in this figure for individual lane widths. Again, the relatively large regression coefficients, R^2, of Eqs. (8.12 to 8.14) suggest that the relationships are strong ones. In Fig. 8.11, the regression lines are fitted to the data for 3.0-, 3.3-, and 3.6-m lane widths. Their slopes are nearly identical, indicating that a given degree of curve will have, more or less, the same effect on speeds for a 3.6-m two-lane highway as it would have for a 3.0-m one.

The potential usefulness of Fig. 8.11 is obvious. By knowing the degree of curve, the engineer could directly predict the expected operating speed.

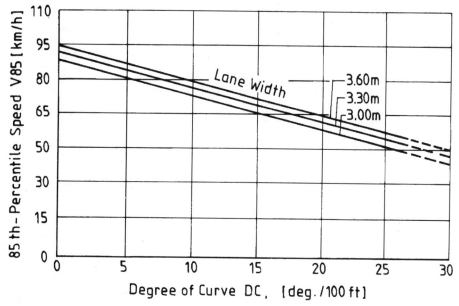

Regression equations:

3.0-m lanes:

$V85 = 89.034 - 1.630\,DC$ $R^2 = 0.753$ $\qquad$ (8.12)

3.3-m lanes:

$V85 = 93.296 - 1.683\,DC$ $R^2 = 0.746$ $\qquad$ (8.13)

3.6-m lanes:

$V85 = 95.594 - 1.597\,DC$ $R^2 = 0.824$ $\qquad$ (8.14)

All lanes:

$V85 = 93.850 - 1.816\,DC$ $R^2 = 0.787$ $\qquad$ (8.11)

where $V85$ = estimate of the operating speed expressed by the 85th-percentile speed, km/h

$\quad\quad DC$ = degree of curve (range 0 to 27°)

$\quad\quad R^2$ = coefficient of determination

FIGURE 8.11 Operating speed background, as related to degree of curve for individual lane widths, two-lane rural roads, example: United States (State of New York).[406,408]

More recent investigations in the United States by Ottensen and Krammes[565] determined the following regression equation between operating speed and degree of curve:

$$V85 = 103.04 - 1.94\,DC \qquad km/h$$
$$R^2 = 0.80 \qquad\qquad (8.15)$$

Equation (8.15) is based on speed measurements and field investigations on the east coast (New York and Pennsylvania), on the west coast (Washington and Oregon), and in the south (Texas). Equation (8.15) may be regarded as valid to describe the impact of the design parameter degree of curve on operating speed, expressed by the 85th-percentile speed, at curved sites in the United States. The influence of the design parameter, lane width, on operating speed was not found to be significant.

8.2.3.5 *Overall Development in Different Countries.** * Two theories have been proposed to predict operating speeds on horizontal curves:

1. Speeds are a function of individual local characteristics (for example, degree of curve or curvature change rate, longitudinal grade, and cross section within the curve, etc.).

---

*Elaborated based on Ref. 565.

2. Speeds are a function of both local characteristics and the general character of the alignment. The general character of the alignment has been reflected in drivers' desired operating speed, which has been estimated as a function of the terrain and overall character of the alignment. Incorporating the desired operating speed into the regression models for speeds on curves increase the proportion of explained variability. The desired speed on a particular highway, however, is not easily defined and is difficult to measure. Normally, it is defined as the 85th-percentile speed.

Table 8.3 lists the various forms of regression equations that have been developed to relate operating speeds on horizontal curves to the design parameter degree of curve. These equations represent, in a general form, the models developed by Glennon, et al.,[217] Lamm and Choueiri,[406,408] and Taragin[699] in the United States; McLean[506] in Australia; Emmerson[169] in England; Gambard and Louah[202] in France; Lamm, et al.[431] in Germany; Kanellaidis, et al.[323] in Greece; Lindemann and Ranft[479] in Switzerland; and Choueiri and Lamm[108] in Lebanon.

Each equation in Table 8.3 uses the 85th-percentile speed at the curve midpoint, $V85$, as the dependent variable and degree of curve, DC, as the independent variable. (Equations using radius of curve were converted to degree of curve to make the comparisons meaningful.)

In general, 85th-percentile speeds on horizontal curves have been estimated with reasonable accuracy (R^2 values between 0.647 and 0.87) using models based on local characteristics alone. R^2 values of approximately 0.92 were reported for models reflecting both local characteristics and desired travel speeds.[565]

Similar model forms for predicting 85th-percentile speeds were found for the German parameter, curvature change rate, as well as using the newly developed design parameter, curvature change rate of the single curve (see Table 8.5).

8.2.3.6 New Design Parameter: Curvature Change Rate of the Single Circular Curve with Transition Curves.*

It can be seen from numerous research studies that strong relationships exist between the U.S. design parameter, degree of curve (DC), the German design parameter, curvature change rate (CCR), and operating speeds ($V85$). But, when applying these two design parameters, several shortcomings are apparent:

1. The design parameter, degree of curve (DC), is solely related to the circular curve itself and does not consider preceding and/or succeeding transition curves (Table 8.4). Thus, describing operating speeds $V85$ by degree of curve only (which means the angular change of the circular curve) would not take into consideration the possible influence of present transition curves on the driving behavior and the accident situations. Since the application of transition curves is required

---

*Elaborated based on Refs. 428, 431, 433, 434, 435, 438, 440, 448, 453, 462, 672, and 673.

TABLE 8.3 Model Forms for Predicting 85th-Percentile Speeds on Horizontal Curves[108,169,202,217,323,406,408,449,506,699]*

Model type	Characteristics considered	Model form $V85 =$
Linear	Local only	$\beta_0 + \beta_1 DC$
Exponential	Local only	$\beta_0 + \beta_1 e^{(\beta_1 DC)}$
Nonlinear/polynomial	Local only	$\beta_0 + \beta_1 \sqrt{DC}$
	Local and general	$\beta_0 + \beta_1 DC + \beta_2 DC + \beta_3 \sqrt{DC} + \beta_4 V_f$
	Local and general	$\beta_0 + \beta_1 DC + \beta_2 DC + \beta_3 DC^2 + \beta_4 V_f$
Inverse	Local only	$\beta_0/(\beta_1 + \beta_2 DC^{\beta_1})$

*Elaborated based on Ref. 565.

Note: $V85$ = 85th-percentile speed, mi/h or km/h; β_n = constants; DC = degree of curve, degrees/100 ft; and V_f = desired travel speed, mi/h or km/h.

TABLE 8.4 Relationship between the U.S. Design Parameter, Degree of Curve, and the German Design Parameter, Curvature Change Rate, for Circular Curves without Transition Curves[434,438,453,672,673]

Degree of curve (DC)	Curvature change rate (CCR)

$$DC_{ft} = \frac{360°}{2\pi R} \quad \left(\frac{degree}{100\ ft}\right)$$

$$= \frac{5729.6}{R} \quad \left(\frac{degree}{100\ ft}\right)$$

where DC_{ft} = radius of curve, ft (1 ft = 0.3048 m)

$$DC_m = \frac{6370}{R} \quad \left(\frac{gon}{100\ m}\right)$$

where R = radius of curve, m

$$DC_m = \frac{63{,}700}{R} \quad \left(\frac{gon}{km}\right)$$

where R = radius of curve, m

The relationship:

$$DC_{ft} \quad \left(\frac{degree}{100\ ft}\right) 36.5 \approx CCR \quad \left(\frac{gon}{km}\right)$$

where 36.5 [−] = conversion factor between DC_{ft}
(imperial system) and CCR (metric system)

$$\text{or } DC_m \quad \left(\frac{degree}{100\ m}\right) 11.13 \approx CCR \quad \left(\frac{gon}{km}\right)$$

where 11.13 [−] = conversion factor between DC_m
and CCR (metric system)

$$CCR = \sum_{i=1}^{n} |\gamma_i / L \quad \left(\frac{gon}{km}\right)$$

where γ_i = angular changes in curves, gon
(gon related to 400°)

L = length of unidirectional curved road section, km

$$CCR = \frac{(L_{cr}/R)\, 63.7}{L} \quad \frac{gon}{km}$$

where L_{cr} = length of circular curve, m
with $L = L_{cr}/1000$ (km)
R = radius of curve, m

$$CCR = \frac{63{,}700}{R} \quad \left(\frac{gon}{km}\right)$$

where R = radius of curve, m

---

in modern highway geometric design, this has to be regarded as a disadvantage for using degree of curve to predict operating speeds.

2. The design parameter, curvature change rate (CCR), includes, according to Fig. 8.10 and Table 8.4, the angular changes in circular and transition curves. *Curvature change rate* is defined as the absolute sum of the angular changes for a longer roadway section with similar road characteristics, divided by the length of this section. Note that an objective definition of the term *longer roadway section with similar road characteristics* does not appear anywhere in the German guidelines. The estimation of the lengths of road sections, on which the calculation of the curvature change rate value is based, relies on the subjective opinion of the highway engineer in charge. Considering that roadway sections can include curves and tangents, the German term *longer road section,* used to assess the design parameter curvature change rate, may be biased because different highway engineers may obtain different curvature change rate values. This biased definition for roadway sections with similar road characteristics is certainly a disadvantage.

To avoid these possible disadvantages, a new design parameter called *curvature change rate of the single circular curve with transition curves* (CCR$_S$)* was developed (according to Eq. [(8.6) and Figs. 8.1 and 8.2], in order to evaluate the operating speed (V85) on a roadway section from

---

*In the following, this term is abbreviated as curvature change rate of the single curve (CCR$_S$).

one design element to the next. This is applicable in situations like independent tangent to curve or curve to curve. This procedure corresponds to that used for degree of curve but also includes the effect of transition curves.

In this way, the design parameter, curvature change rate, is separated from the mathematically biased term, roadway section with similar road characteristics, or homogeneous roadway section, and considers the exact lengths of individual design elements (tangents and circular curves and transition curves). Furthermore, the new parameter is more suitable for data processing systems because computers are not able to estimate geometric parameters based on their optical appearance and can only work based on exact input information.

Based upon the limitations noted here, it appears preferable to use the new design parameter, curvature change rate of the single curve (CCR_S), as the relevant design parameter for describing the relationship between operating speed ($V85$) and road characteristics. This is especially true for two-lane rural roads of category group A. This parameter has already been used for establishing the operating speed background of the Greek Guidelines for the Design of Highway Facilities (Fig. 8.3) and partially for examining the existing German operating speed background (Fig. 8.10).

8.2.3.7 *Operating Speed Backgrounds for Two-Lane Rural Roads in Different Countries.*
In the previous sections, the influence of road characteristics on operating speeds ($V85$) was discussed with respect to the design parameters:

- Degree of curve (DC)
- Curvature change rate (CCR)
- Curvature change rate of the single curve (CCR_S)

All three parameters have proven to produce reasonable results although to varying degrees. For example, the use of the parameter, DC, is limited to curves composed of one circular arc without transition curves; CCR parameters, on the other hand, refer to a certain length of the highway composed of curves and tangents and, therefore, can be considered as an average parametric description of that particular road section. The only parameter which fully considers the impact, which each design element of a highway independently of its type (circular arc with or without transitions or a tangent) imposes on the driver's behavior, is the curvature change rate of the single curve (CCR_S). Consequently, the parameter, CCR_S, enables a road designer to perform a microscopic analysis of alignment consistency and, therefore, provides a safer roadway alignment.

For determining the 85th-percentile speeds ($V85$) with respect to the aforementioned design parameters, existing and new operating speed backgrounds were compiled in Fig. 8.12a and Table 8.5 for various countries. Equations using radius of curve were converted to either of the previously listed design parameters.

By knowing DC, CCR, or CCR_S values of curved roadway sections and/or independent tangents (compare Sec. 12.1.1.3), the expected 85th-percentile speeds can be determined graphically from Fig. 8.12a or mathematically from Table 8.5 for the respective country under study.

Operating speed backgrounds, like those in Fig. 8.12a, should be made available in every country to enable design engineers to provide a good curvilinear alignment in conjunction with sound transitions between independent tangents and curves and consequently consistent road characteristics. In this way, design consistency (safety criterion I), operating speed consistency (safety criterion II), and driving dynamic consistency (safety criterion III) can be achieved, as will be discussed in Secs. 9.1.2, 9.2.2, 9.1.3, 9.2.3, 10.1.4, and 10.2.4.

For those countries where operating speed backgrounds have not been developed to date, the application of a comparable country's operating speed background or the average one in Fig. 8.12b is recommended.

The procedure used to establish operating speed backgrounds, like those in Fig. 8.12a, is described in detail in Sec. 8.2.3.4. However, it should be noted that for future highway geometric design, the application of the new design parameter, CCR_S, is strongly recommended in order to

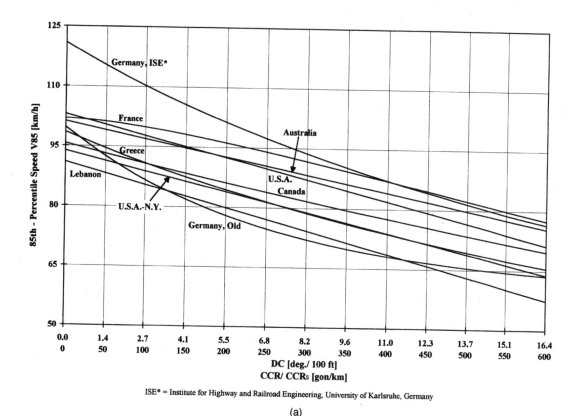

ISE* = Institute for Highway and Railroad Engineering, University of Karlsruhe, Germany

(a)

FIGURE 8.12   (*a*) Operating speed backgrounds for two-lane rural roads in different countries (all lane widths) (based on Table 8.5).

establish superior results compared to the two other parameters (DC and CCR). Normally, about 70 to 100 curved sites with varying CCR_S values between 0 and 600 gon/km are sufficient to establish the operating speed background for the respective country under study.

The effect of wet pavement on 85th-percentile speeds of passenger cars was examined in Sec. 8.2.1.2. Ample evidence exists to indicate that wet pavement does not have a great effect on operating speed and that drivers will not adjust their speeds sufficiently to accommodate slippery wet pavement on curves. The statistical analyses indicate that the relationship between operating speeds on dry pavements is also valid for wet pavement conditions so long as visibility is not affected appreciably by heavy rain. It is obvious that drivers do not seem to recognize the fact that the lower coefficients of friction on wet pavements could lead to critical driving maneuvers or even accidents.

As previously mentioned, the regression equations in Table 8.5 show that the introduction of additional pavement (or lane) width does not significantly influence the relationships between $V85$ and road characteristics (expressed by DC, CCR, or CCR_S) significantly. Despite the statistical insignificance, the influence of the design parameter pavement (lane) width is incorporated

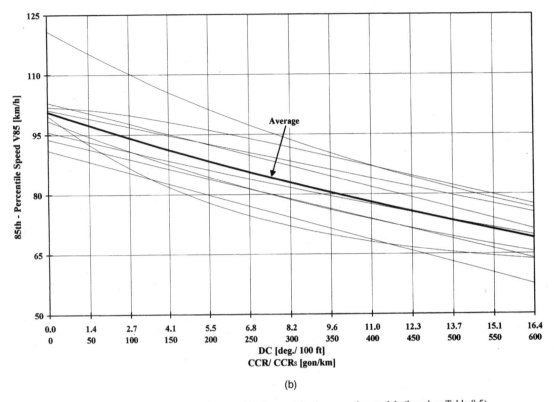

(b)

FIGURE 8.12 (*Continued*) (*b*) Average operating speed background for the regression models (based on Table 8.5).

into the operating speed backgrounds in Figs. 8.3 (Greece), 8.10 (Germany), and 8.11 (United States) to provide the highway engineer with more detailed information.

The relationships for all lane widths combined are shown schematically in Fig. 8.12*a* for a number of countries. Although the 85th-percentile speed levels differ among the countries under study, the overall form is similar. See also the average operating speed background of Fig. 8.12*b*. The studies show that with an increase in the design parameters (DC, CCR, or CCR_S), the operating speeds decrease.

Furthermore, it has been determined that the speed difference between the means of 85th-percentile speeds on the inside and outside lanes of two-lane rural roads are not significant. Therefore, a single regression equation containing information pertinent to the curve (expressed by DC, CCR, or CCR_S) is considered sufficient for both directions of travel.[108,406,408,431,453,565,586]

Finally, an interesting traffic flow phenomenon observed by Steierwald and Buck[674] and Lamm, Hiersche, and Mailaender[431] will be discussed. This finding is based on new local speed measurements for passenger cars under free-flow conditions (normally defined by a time gap of $t \geq 6$ s).

At intersections or within villages, a backup of vehicles caused by traffic lights, pedestrian crossings, etc., is often observed. These backups are often headed by slow-moving vehicles. Because there are no passing possibilities, the vehicle backups are often not able to disperse before the local cross section where speed measurements are taken. These slow-moving vehicles often represent a selection of the slowest drivers, even though they are "free moving." Including these vehicles in speed studies causes the distribution of speeds to become lower.[674] For this reason, the new speed measurements[108,431,565,586] were conducted without including the leaders of

TABLE 8.5 Regression Models for Operating Speed Backgrounds for Two-Lane Rural Roads in Different Countries

---

Germany, ISE[431]

$$V85 = \frac{10^6}{8270 + 8.01 \, CCR_S} \qquad R^2 = 0.73 \tag{8.15}$$

where speed limit = 100 km/h

Greece[453,586]

$$V85 = \frac{10^6}{10150.1 + 8.529 \, CCR_S} \qquad R^2 = 0.81 \tag{8.16}$$

where speed limit = 90 km/h

United States—New York[406,408]

$$V85 = 93.85 - 1.82 \, DC = 93.85 - 0.05 \, CCR_S \qquad R^2 = 0.79 \tag{8.17}$$

where speed limit = 90 km/h

Germany, old[241,243]

$$V85 = 60 + 39.70 \, e^{(-3.98 \times 10^{-3} \, CCR_S)} \tag{8.18}$$

where lane width = 3.50 m
speed limit = 100 km/h

United States[565]

$$V85 = 103.04 - 1.94 \, DC = 103.04 - 0.053 \, CCR_S \qquad R^2 = 0.80 \tag{8.14}$$

where speed limit = 90 km/h

France[643,700]

$$V85 = \frac{102}{1 + 346 \, (CCR_S/63,700)^{1.5}} \tag{8.19}$$

where speed limit = 90 km/h

Australia[505]

$$V85 = 101.2 - 1.56 \, DC = 101.2 - 0.043 \, CCR_S \qquad R^2 = 0.87 \tag{8.20}$$

where speed limit = 90 km/h

Lebanon[108]

$$V85 = 91.03 - 2.06 \, DC = 91.03 - 0.056 \, CCR_S \qquad R^2 = 0.81 \tag{8.21}$$

where speed limit = 80 km/h

Canada[508,527]

$$V85 = e^{(4.561 - 5.86 \times 10^{-3} \, DC)} = e^{(4.561 - 5.27 \times 10^{-4} \, CCR_S)} \qquad R^2 = 0.63 \tag{8.22}$$

where speed limit = 90 km/h

Note: DC_{ft} (degree/100 ft) $36.5 \approx CCR_S$ (gon/km) and DC_m (degree/100 m) $11.13 \approx CCR_S$ (gon/km) (without taking transition curves into consideration).

vehicle backups. This procedure resulted in 85th-percentile speeds which were considerably higher than, for example, the old operating speed backgrounds found in Germany (compare the curves "Germany, ISE" and "Germany, Old") or in the United States, compare the curves "U.S.A." and "U.S.A.-N.Y." in Fig. 8.12a. Of course, additional influencing factors, such as the change in vehicle population, the increased level of experience of drivers during the last two decades, and the lack of police enforcement, at least in Germany, or for example, the poor roadway conditions in Lebanon, also have to be considered in this connection.

In conclusion, it can be stated that the establishment of a sound operating speed background according to Fig. 8.12a and Table 8.5 is the first important step to undertake in modern highway geometric design. In this connection, it is recommended to use the curvature change rate of the single curve as the design parameter for describing the road characteristics and to disregard leaders of vehicle backups.

8.2.4   Evaluation of Other Design Parameters that Influence Operating Speeds

8.2.4.1   Multilane Median-Separated Rural Roads (Interstates and Autobahnen).   In Sec. 8.1.2, the following statement was made:[159]

> For multilane rural roads of Category Group A (Interstates, Autobahnen) no statistically sound results have been determined with respect to the relationship between 85th-percentile speed and road characteristics because of the normally very generous alignments.

Operating speed measurements under free-flow conditions (time gap ≥ 6 s) on multilane roads were conducted in Germany in 1972[320] on sections with CCR_s values of up to 60 gon/km ($R \approx$ 1000 m); these measurements were later repeated by Lamm in 1995. Compared to the 1972 measurements, his results revealed an increase of about 20 km/h in operating speeds.

These latter results (Fig. 8.13) are supported by the findings of Kellermann[331] who reported, for the whole German Autobahn network, an operating speed $V85$ of at least 148.2 km/h for passenger cars under free-flow conditions.

Similar investigations were made in Greece in 1995.[586] A graphical presentation of the results for the average of 85th-percentile speeds on the driving and passing lanes is given in Fig. 8.13.

Figures 8.12a and 8.13 demonstrate that drivers exhibit different speed behavior attitudes on multilane rural roads as compared to two-lane rural roads. First, the 85th-percentile speed differences are, on average, about 30 km/h higher in both Greece and Germany related to tangent sections. Next, the rate of change in the 85th-percentile speed levels between successive design elements can be assumed to be significantly lower on multilane rural roads as compared to those which were observed for CCR_s values, of up to 500 and 600 gon/km on two-lane rural roads (Fig. 8.12a). Last, for CCR_s values of up to 60 gon/km, the average of 85th-percentile speeds in the driving and passing lanes, were relatively constant (Fig. 8.13).

It then follows that the safety assumptions made in Sec. 8.1.2 between $V85$ and V_d hold true since obviously generous alignments, as is normally the case on interstates or autobahnen, have little effect on the driving behavior.

This is especially true for those countries where enforcement by the police is strong, such as in Canada, France (Fig. 8.13), Switzerland, and the United States. However, in the case of Germany where a general speed limit does not exist on multilane median-separated rural roads but a speed of 130 km/h is recommended, and in Greece where a general speed limit of 120 km/h is in effect on multilane rural roads, then the assumptions made in Sec. 8.1.2 do not agree fully possibly due to a lack of police enforcement.

However, special attention must be given to multilane rural roads of road categories A II and A III, where relatively low design speeds are permitted (see Table 6.2). At least on those roadway

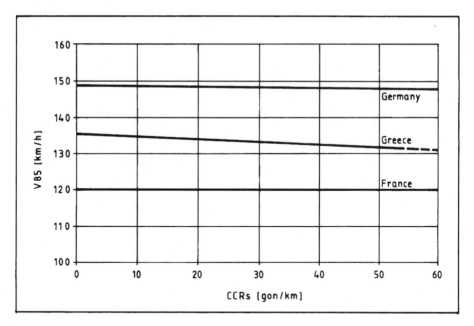

Regression models (average of 85th-percentile speeds in the driving and passing lanes):

Germany: $V85 = 148.75 - 0.0162\ CCR_S$ (8.23)

Greece: $V85 = 135.42 - 0.0755\ CCR_S$ (8.24)

France: $V85 = 120\ \text{Km/h}$

FIGURE 8.13   85th-Percentile speed versus curvature change rate of the single curve for multilane median-separated rural roads (interstates or autobahnen).[320,586,700]

sections with large radii of curve, significant deviations between 85th-percentile speeds and design speeds can be expected. These may result in safety problems related, for example, to super-elevation rate and sight distance.

Therefore, regarding the speed relationships in Fig. 8.13, it is definitely recommended to use a design speed as high as possible for multilane rural roads, even though the speed relationships observed in other countries with significant police enforcement of speed limits may not be as pronounced.

8.2.4.2   Longitudinal Grades on Two-Lane Rural Roads.   The established relationships shown in Fig. 8.12a are valid, in general, for two-lane rural roads with longitudinal grades up to 5 percent (maximum 6 percent).

For grades above 5 or 6 percent, operational speed backgrounds for passenger cars are known from measurements made in France[700] and in Greece.[586] The French speed background in Fig. 8.14 demonstrates a decrease in the 85th-percentile speed with an increase in longitudinal grades on roadway sections greater than 250 m. For the grade range between 5 to 10 percent, a decrease of about 20 km/h in the 85th-percentile speed is noted.

The Greek operating speed background in Fig. 8.15 is based on a large number of curves with CCR_S values ranging from 0 gon/km (tangent section) to 600 gon/km and grades (G) between 5 and 10 percent. Curve 2 in Fig. 8.15 represents the 85th-percentile speeds on downgrade sections, whereas curve 4 represents the corresponding results for upgrade sections (all lane widths includ-

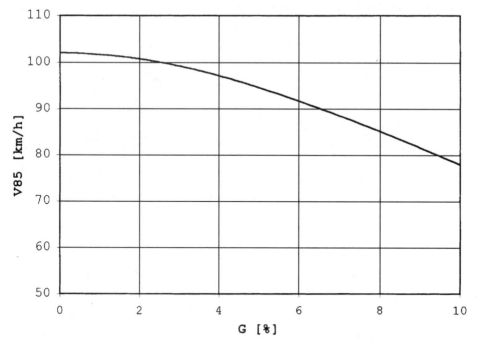

FIGURE 8.14   85th-Percentile speed as a function of grade on upgrade sections (>250 m), example: France.[700]

ed). Because statistical tests show no significant differences between the two operating speed data sets, both data sets were combined into one set for which the resulting regression curve 3 in Fig. 8.15 was derived.[586]

Curve 1 in Fig. 8.15 represents the operating speed background in Greece for grades less than or equal to 5 percent. This curve corresponds to Eq. (8.16) in Table 8.5 and to the curve designated "Greece" in Fig. 8.12a. Comparing curve 1 and curve 3 in Fig. 8.15 reveals for tangent sections and curves with CCR_S values up to 50-gon/km ($R \approx 1300$-m) operating speed differences of $\Delta V85 \approx 25$ km/h for the different grade classes observed. This finding coincides, to a certain extent, with the French result referred to earlier.[586,700] Furthermore, it is very important to note that the superimposition of the horizontal alignment (expressed by the CCR_S value) and the vertical alignment (expressed by the grade) yield for continuously increasing CCR_S values and grades smaller and smaller differences in operating speeds, as indicated by curves 1 and 3 in Fig. 8.15. For example, the operating speed on a very sharp curve with a CCR_S value of 550 gon/km ($R \approx 115$ m) combined with high grades is only about 5 km/h lower than the operating speed in a

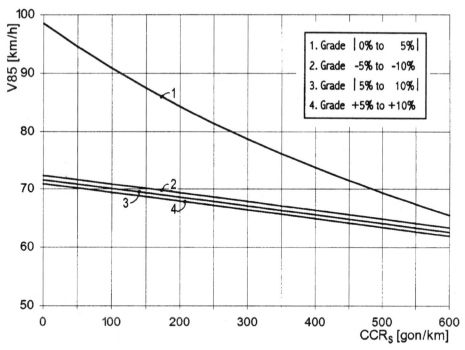

Regression models:

Curve 1: $V85 = \dfrac{10^6}{10150.10 + 8.529\ CCR_S}$ $R^2 = 0.81$ (8.16)

Curve 2: $V85 = 72.36 - 0.015\ CCR_S$ $R^2 = 0.71$ (8.25)
Curve 3: $V85 = 71.58 - 0.015\ CCR_S$ $R^2 = 0.83$ (8.26)
Curve 4: $V85 = 70.91 - 0.015\ CCR_S$ $R^2 = 0.81$ (8.27)

FIGURE 8.15 The influence of longitudinal grades on 85th-percentile speeds on two-lane rural roads, example: Greece.[586]

relatively flat country. Consequently, Fig. 8.15 proves that the speed reduction is overwhelmingly caused by the horizontal alignment (design parameter CCR_S) and to a minor extent by the vertical alignment (design parameter G). These findings apply only to passenger cars. For trucks, the opposite holds true, as will be discussed in detail in Secs. 13.1.2 and 13.2.2.

Recent research work about the relationships and interrelationships between 85th-percentile speeds → high grades → and high CCR_S values were conducted by Schulze (Sec. 13.1.2)[638] and Eberhard (Sec. 18.5.1).[164] Both authors reported that the operating speeds of passenger cars on high grades superimposed with high CCR_S values are influenced considerably more by the horizontal alignment than by the vertical alignment.

CHAPTER 9
SAFETY CRITERIA I AND II

9.1 RECOMMENDATIONS FOR PRACTICAL DESIGN TASKS

9.1.1 Fundamentals

For the safety evaluation of alignments with longitudinal grades of up to 5 or 6 percent, three quantitative safety criteria will be introduced in the following sections. These safety criteria are based on the speed behavior and are specifically relevant to two-lane rural roads of category group A, which are the locations of 50 to 60 percent of all severe accidents in Europe and the United States (see Fig. 9.2). The road characteristics which influence operating speed behavior, expressed by the 85th-percentile speed, are best defined by the design parameter curvature change rate of the single curve (CCR_S) (see Figs. 8.3 and 8.12, and Table 8.5).

Roads other than two-lane rural roads of category group A,

- Multilane median-separated roads of category group A
- Road categories B II, B III, and B IV

in comparison to two-lane rural roads, have fewer serious accidents, and estimates of the 85th-percentile speed can be made, according to Sec. 8.1.2. However, for extreme alignment conditions or where traffic safety is a concern, these roads should also be analyzed using the three safety criteria shown here.

The three safety criteria are

- Achieving design consistency (safety criterion I)
- Achieving operating speed consistency (safety criterion II)
- Achieving driving dynamic consistency (safety criterion III)

The establishment of a classification system for these safety criteria, which is based on accident research, is presented in Sec. 9.2.1.4. In-depth descriptions, analyses, and evaluations of the important fundamentals for the development and understanding of safety criteria I to III are given in Secs. 9.2.2, 9.2.3, and 10.2.4.

The following relationships have been drawn between design and traffic-related variables and accident history, particularly on two-lane rural roads.

Influence of Pavement Width on Safety (see Part 3, "Cross Sections," Sec. 24.2.4 and Part 2, "Alignment," "Cross-Sectional Elements" in Sec. 9.2.1.3)

- A distinct tendency for accident rates* to decrease with increasing lane width up to 3.75 m (12 ft) was established.
- The same is true for paved shoulders up to 2.50 m (8 ft) on high-volume roads and for unpaved shoulders up to 1.50 to 2.00 m (5/8 ft) on both high- and low-volume roads.
- Highways with four or more lanes have to be separated by a median.
- On multilane median-separated roads, the more lanes that are provided in the traveled way, the lower is the accident rate.
- Edge lines increase safety.

Influence of Radius of Curve on Safety (see "Radius of Curve" in Sec. 9.2.1.3)

- A negative relationship between radius of curve on the one hand and accident rate (AR)* and accident cost rate (ACR)* on the other hand was established.
- Large reductions in the accident rate were noted when comparing very large radii of curve with smaller radii [$R \leq 100$ m (330 ft)].
- A curve of a certain radius found in a sequence of properly balanced curves is safer than an identical curve found within an unbalanced sequence of curves.
- For radii less than 200 m (650 ft), the accident rate is at least twice as high as that for radii of 400 m (1300 ft). Increasing curve radii beyond 400 to 500 m results in very small improvements in traffic safety.

Influence of Curvature Change Rate, Degree of Curve, and Curve Radii Ratio on Safety (see "Curvature Change Rate, Degree of Curve, Length of Curve, and Curve Radii Ratio" in Sec. 9.2.1.3)

- With an increasing curvature change rate of the single curve, the degree of curve accident rates (AR) and accident cost rates (ACR) increase.
- Unbalanced curve radii ratios have an unfavorable effect on accident measures.
- The most successful parameters in predicting much of the variability in operating speeds and accident rates were degree of curve and curvature change rate. Therefore, the curvature change rate of the single curve can be used to distinguish good, fair, and poor design practices and to guide designers in achieving sound driving behavior and improved traffic safety.

Influence of Clothoids (or Spirals) on Safety [see "Clothoid (Spiral) Transition" in Sec. 9.2.1.3]

- For passing through a transition curve (clothoid or spiral) from tangents to circular curves, a safety gain could be observed only for radii of curve less than 200 m (650 ft). Clothoids provided no safety improvements over direct tangent to circular curve transitions for curves with radii greater than 200 m. (Because of insufficient databases, further research is needed on this subject.)

---

*Accident rates (AR) and accident cost rates (ACR) are defined in "Relative Accident Numbers" in Sec. 9.2.1.3.

Influence of Grade on Safety (see "Grade" in Sec. 9.2.1.3)

- Grades between 0 and ±2 percent are the safest.
- Grades less than 6 percent have a small impact on accident situation. For grades greater than 6 percent, a sharp increase in the accident rate was noted.
- Downgrade sections are more dangerous than level or upgrade sections.

Influence of Sight Distance on Safety (see "Sight Distance" in Sec. 9.2.1.3)

- As sight distance increases, the accident risk decreases.
- High accident rates were associated with sight distances of less than 100 m (330 ft).
- For sight distances of more than 150 m (500 ft), no major improvement in accident rates was noted.

Influence of Traffic Volume on Safety (see "Traffic Volume" in Sec. 9.2.1.3)

- Between traffic volume and accident rate a U-shaped relationship was established. That is accident rates are highest at the lowest and highest range of traffic volume. In contrast, the accident cost rate decreased continuously with increasing traffic volume.

Influence of Design Speed on Safety (see "Design Speed" in Sec. 9.2.1.3)

- The accident rate decreases with increasing design speed up to 80 km/h (50 mi/h). The accident cost rate, on the other hand, seems to increase with increasing design speed. Because of insufficient databases, further research is needed on this subject.

Superimposition of Design Parameters/Breakpoint in Safety (see "Superimposition of Design Parameters/Breakpoint on Safety" in Sec. 9.2.1.3)

- Strong correlations between a number of design parameters and accident rates were established.
- As a breakpoint in safety, an accident rate of 2.0 accidents per million vehicle kilometers (MVkm) seems to be in order. This breakpoint in safety corresponds to the following limiting maximum or minimum values of design parameters, in the range of design speeds between 80 and 90 km/h:

  Lane width ≥ 3.25 m (11 ft)

  Radius of curve ≥ 350 m (1150 ft)

  Longitudinal grade ≤ 6 percent

  Sight distance ≥ 100 m (330 ft)

These values have to be regarded as approximations due to limitations imposed by the research methodologies.

Failure of a design to fall within these limiting values could mean that the breakpoint in safety is exceeded. This breakpoint in safety should be regarded as the borderline between "safe" and "unsafe" geometry for new designs and rehabilitated roadways.

Design parameters within the recommended values cited here should be used when establishing sound safety criteria and reasonable highway geometric design practices in plan and profile.

9.1.2 Safety Criterion I: Achieving Design Consistency

The design speed should be constant for longer roadway sections. Furthermore, the design speed, V_d, and the 85th-percentile speed, $V85$, should be balanced. The designer must make sure that the road characteristics are adjusted to the driving behavior of motorists.

TABLE 9.1 Recommended Ranges for Good, Fair, and Poor Design Levels for Safety Criterion I (Two-Lane Rural Roads of Category Group A)

Case 1: Good design

$|V85_i - V_d| \leq 10$ km/h

No adaptions or corrections are necessary. A balanced speed behavior can be expected especially at curved sites.

Case 2: Fair design

---

10 km/h $< |V85_i - V_d| \leq 20$ km/h

At these curve locations:
1. The speed behavior should be brought down through speed limits and/or appropriate traffic control devices.
2. The superelevation rates should be related to $V85$ to ensure that side friction assumed would accommodate side friction demanded. For stopping sight distances, $V85$ should be relevant too.

Case 3: Poor design

$|V85_i - V_d| > 20$ km/h

Critical discrepancies between design speed and actual driving behavior of motorists are present. The expected severe accident situation may lead to uneconomic, unsafe operation. Therefore, redesigns are normally recommended. The decision whether or not to institute reconstruction measures should be based on the local situation. If redesigns are not possible, the installation of very stringent traffic control devices like, for example, speed limits combined with chevrons and guardrails or even automatic radar devices may serve sometimes as a surrogate measure.

Note: V_d = design speed, km/h, and $V85_i$ = expected 85th-percentile speed of design element *i,* km/h.

Based on driving behavior and accident research (discussed in detail in Sec. 9.2.1.4 and 9.2.2), safety criterion I was developed. The criterion applies especially to two-lane rural roads of category group A. The quantitative ranges are shown in Table 9.1.

This safety criterion provides appropriate tuning between design speed and 85th-percentile speed for good, fair, and poor design levels. It is applicable for new designs, redesigns, and for restoration, rehabilitation, and resurfacing (RRR) projects on two-lane rural roads (road categories A I to A IV).

Safety criterion I is related to the individual design elements along longer roadway sections, such as curves or independent tangents.*

For the evaluation of two-lane rural roadway sections, the following procedure is recommended:

1. Assess the road section where new designs are contemplated or where redesigns, respectively, RRR projects of existing alignments will be conducted.

2. Determine the design speed for the section under investigation with respect to its road category (network function) according to Table 6.2 for new designs or determine the original design speed for existing sections. If this is not possible, go to step 5. (In many cases it is advisable to examine the selected design speed according to step 5 to avoid design deficiencies from the beginning.)

3. Determine the expected 85th-percentile speed for each curve or independent tangent* in relation to the design parameter curvature change rate of the single curve (and lane width) according to the respective operating speed background of the country under study (for example, Figs. 8.3 and 8.12 and Table 8.5. CCR_S has to be determined from Eqs. (8.6a to 8.6f) in Fig. 8.1 along with the sketches in Fig. 8.2. For independent tangents, the 85th-percentile speed corresponds to $CCR_S = 0$ gon/km. However, for exact design cases, see Sec. 12.1.1.3.

4. Evaluate the difference between $V85_i$ and V_d according to the classification system in Table 9.1 for good, fair (tolerable), and poor design levels.

---

*An *independent tangent* is classified as one that is long enough to be considered in the curve-tangent-curve safety evaluation design process, as an independent design element, while a short tangent is called *nonindependent* and can be neglected.[411] For defining and classifying independent tangents, see Sec. 12.1.1.3.

5. However, for existing alignments, the design speed is often not known and was certainly not selected according to the assumptions of Table 6.2 and Sec. 8.1.1 with respect to a specific network function of the road and a desired traffic quality. Therefore, in examining old alignments, the following procedure is recommended for the estimation of a sound design speed. Calculate the length-related average $\overline{CCR}_S$ value based on all the curves in the observed roadway section (see Sec. 9.2.2.1). Tangent sections should not be included. Determine for this average $\overline{CCR}_S$ value the corresponding average 85th-percentile speed from the operating speed background of the country under study (Fig. 8.12 or Table 8.5). This $\overline{V}85$ value represents a good estimate for the design speed, V_d of existing (old) alignments. For future redesigns or RRR strategies, this speed will, to a certain extent, already optimize the design process from the viewpoint of economic, environmental, and safety-related issues. The evaluation process between $V85_i$ and V_d (Table 9.1) would have to be conducted again according to steps 3 and 4 given previously.

6. Separate the horizontal alignment into good, fair, and poor ranges according to the classification system in Table 9.1. (For case studies, see "Case Studies" in Sec. 12.2.2.4. The examination of existing alignments is thoroughly discussed in Sec. 18.4.2)

New designs of two-lane rural roads of category group A should always be assigned to the *good design level* (Case 1 in Table 9.1).

Redesigns and RRR strategies also can be assigned, in substantiated cases, to the fair design level (Case 2). However, it should be noted that, based on the experiences gained in Sec. 9.2.1.4 and Sec. 18.3, the expected accident rate is at least twice as high as that of Case 1. Also, a higher accident cost rate has to be expected.

In case of speed differences of Case 3, a redesign has to be considered for existing roadway sections. If not, an unfavorable alignment with respect to the actual driving behavior can be expected from safety and economic points of view, because of the normally high accident risk and the resulting high accident costs in Case 3, alignments (see Sec. 9.2.1.4 and Sec. 18.3). In cases where redesigns are not possible, compare comment to case 3 in Table 9.1.

Normally, an examination of safety criterion I is not necessary based on the conservative assessments between design speed V_d, 85th-percentile speed $V85$, and maximum permissible speed limit V_{perm} in Sec. 8.1.2 for:

- Multilane median-separated roads of category group A
- Road categories B II and B III

However, an examination of criterion I can be advantageous in order to reveal discrepancies in extreme alignment conditions.

9.1.3 Safety Criterion II: Achieving Operating Speed Consistency

The design speed, V_d and the operating speed, $V85_i$ should remain constant along longer roadway sections. In this way, the road characteristics are balanced for the motorist along the course of the road section. If, in the case of a longer road section, a definite change in topography occurs, necessitating a change in road characteristics and a corresponding change in the design speed, then the design elements in the transition section must be carefully adjusted to each other so that any change between them is gradual. Likewise, the 85th-percentile speed should be consistent along the road section. For two-lane rural roads of category group A, this can be achieved by applying the recommendations of safety criterion II between successive design elements. Table 9.2 summarizes these recommendations.

A well-balanced operating speed sequence between successive design elements within a road section with the same design speed promotes a consistent, economic driving pattern. This is especially true for two-lane roads of category group A. For roads of category group B, where the driving pattern is strongly influenced by the permissible speed limit, the principle of operation speed consistency between successive design elements should also be considered, if no essential disadvantages originate from the process.[243]

TABLE 9.2   Recommended Ranges for Good, Fair, and Poor Design Levels for Safety Criterion II (Two-Lane Rural Roads of Category Group A)

Case 1: Good design

The range of change in operating speed is

$| V85_i - V85_{i+1} | \leq 10$ km/h

For these road sections, consistency in horizontal alignment exists between successive design elements, and the horizontal alignment does not create inconsistencies in vehicle operating speeds. No adaptions or corrections are necessary.

Case 2: Fair design

The range of change in operating speed is

10 km/h $< | V85_i - V85_{i+1} | \leq 20$ km/h

These road sections may present at least minor inconsistencies in geometric design between successive design elements. Normally, they would warrant speed regulations and/or appropriate traffic control devices but no redesigns, unless a documented safety problem exists.

Case 3: Poor design

The range of change in operating speed is

$| V85_i - V85_{i+1} | > 20$ km/h

These road sections reveal strong inconsistencies in horizontal geometric design between successive design elements, combined with those breaks in the speed profile, that may lead to critical accident frequencies and severities, and therefore to an uneconomic, unsafe operation. Therefore, redesigns normally are recommended. The decision whether or not to institute reconstruction measures should be based on the local accident situation. If redesigns are not possible, the installation of very stringent traffic control devices like, for example, speed limits combined with chevrons and guardrails or even automatic radar devices may serve sometimes as a surrogate measure.

Note: $V85_i$ = expected 85th-percentile speed of design element i, km/h, and $V85_{i+1}$ = expected 85th-percentile speed of design element $i + 1$, km/h.

9.1.3.1 *Classification System.*
The examination of consistency in horizontal alignment is established based on quantitatively developed ranges valid for two-lane rural roads of category group A (Table 9.2). For the evaluation process, the differences between 85th-percentile speeds and successive design elements are used. (Examples could be the sequence independent tangent i to curve $i + 1$ or curve i to curve $i + 1$). The 85th-percentile speeds ($V85_i$ and $V85_{i+1}$) are determined based upon the curvature change rate of the single curve (CCR_{Si} and CCR_{Si+1}), and the operating speed background of the country under study (for example, Fig. 8.12 or Table 8.5). Having determined the magnitude of the ranges for the differences in the 85th-percentile speeds between element i and element $i+1$, Table 9.2 permits the classification into good, fair (tolerable), and poor design levels according to safety criterion II.[408,412,418,444]

A detailed discussion and evaluation of the recommended ranges in Table 9.2 is presented in Sec. 9.2.1.4 and Sec. 9.2.3.1, based upon research results on driving behavior and accident situations.

Safety criterion II is always related to two successive design elements.

For the evaluation of roadway sections of two-lane rural roads, the following procedure is recommended.

1. Assess the road section where new designs or where redesigns, respectively, RRR projects of existing alignments will be conducted.

2. For every curve, the curvature change rate of the single curve and the lengths of (interim) tangents between sequential curves must be determined. CCR_S has to be calculated according to Eqs. (8.6*a* to 8.6*f*) in Fig. 8.1 along with the sketches of Fig. 8.2.

3. Determine the expected 85th-percentile speed for each curved site using the design parameter, CCR_S, according to the operating speed backgrounds shown in Fig. 8.12 and Table 8.5.

4. Decide whether the (interim) tangents between succeeding curves are independent or non-independent tangents. For *independent (long) tangents,* the sequence tangent to curve is used in the design process, whereas for *nonindependent tangents,* the sequence curve to curve without the short interim tangent is used. For independent tangents, the 85th-percentile speed corresponds to $CCR_S = 0$ gon/km. However, for exact definitions, see Sec. 12.1.1.3.

5. Calculate the difference in 85th-percentile speeds between successive design elements (independent tangent i to curve $i + 1$ or curve i to curve $i + 1$).

6. Separate the horizontal alignment into good, fair, and poor ranges according to the classification system in Table 9.2. (For case studies, see Sec. 12.2.4.2. The examination of existing alignments is thoroughly discussed in Sec. 18.4.)

New designs of two-lane rural roads of category group A should always conform to the good design level (Case 1 in Table 9.2).

Redesigns and RRR strategies could conform, in substantiated individual cases, to the fair design level (Case 2). However, it should be noted that, in these cases, the expected accident rate is at least twice as high as that of Case 1, based on the experiences gained in Sec. 9.2.1.4 and Sec. 18.3. Also, a higher accident cost rate has to be expected.

In the situations where 85th-percentile speed differences correspond to Case 3, a new design should be considered for the roadway sections examined. If not, an unfavorable alignment with respect to the actual driving behavior will result, with expected high accident risk and the related high accident costs (see Sec. 9.2.1.4 and Sec. 18.3). For cases when redesigns are not possible, compare to Case 3 in Table 9.2

Normally, an examination of safety criterion II is not necessary, based on the conservative assessments of design speed V_d, 85th-percentile speed $V85$, and maximum permissible speed limit V_{perm} (see Sec. 8.1.2) for:

- Multilane median-separated roads of category group A
- Road categories B II and B III

However, an examination of criterion II can be advantageous in revealing any severe discrepancies in the alignment conditions.

In cases of redesigns and RRR strategies for existing roadway sections, the design parameters of preceding and succeeding sections should always be considered. Where distinct differences in road characteristics are present, the transitions should be planned in a careful manner (definitely at the good design level).

9.1.3.2   *Graphic Presentation: Relation Design.**   *Relation design* means that design element sequences are formed, in which the design elements following one another are subject to specific relations or relation ranges.[444] This concept is the opposite of the practice in which single design elements are put together more or less arbitrarily.

In order to achieve operating speed consistency between two circular curves in the same or in the opposite directions, the radii of these curves should have a well-balanced relationship for roads of category group A (known as *relation design*).[243] The same is true for the sequence independent tangent to a curve.

Relation design is intended to ensure that the operating speed, $V85$, between successive design elements does not change abruptly and coincides with the assumptions underpinning safety criterion II, especially for two-lane rural roads of category group A. This means that for good design practice, the range of change in operating speeds between successive design elements should not exceed 10 km/h (Table 9.2). In other words, this indicates a well-balanced design or that a good relationship exists between successive design elements. The same is true for fair design, where the range of change in operating speeds should not exceed 15 km/h. This indicates a fairly balanced design or a tolerable relation design. However, with respect to poor design practices according to safety criterion II, a change of more than 15 km/h in operating speed would certainly indicate an unbalanced

---

*Elaborated based on Refs. 444, 447, 448, 453, 455, and 462.

design or a poor relation design. (Note that the authors decided to reduce the fair range of safety criterion II to $\Delta V85 \leq 15$ km/h for relation design based on the considerations in Sec. 9.2.3.2.)

The preceding ranges of change in operating speeds were used as the basis for calculating relation design backgrounds, based upon the respective operating speed background of the country under study (see Figs. 8.3, 8.12 or Table 8.5). The calculation procedure is shown in "Calculation of Sound Relation Design Backgrounds" in Sec. 9.2.3.2.

Relation design has previously been limited to considering sequences of curves. A good relation design automatically represented the meaning of a good curvilinear alignment, whereas a poor relation design stood for a noncurvilinear alignment.[159,243,444] However, recent research has shown[628] that curvilinear alignments are not always desirable and that well-balanced sequences between independent tangents and curves can also function well. Therefore, the definition of relation design has been extended here from simply considering "sequences of curves" to the relationships of "independent tangents to curves." Both curve-to-curve and tangent-to-curve sequences should be considered when evaluating roadway sections for good, fair (tolerable), or poor design practices.

The relations for radii between successive circular curves and between independent tangents and curves are given in Fig. 9.1 (example: Greece). Additional relation design backgrounds were

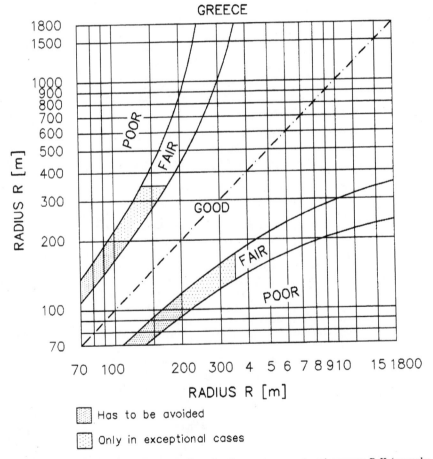

FIGURE 9.1 Relation design background, roads of category group A and category B II (example: Greece).[453] (Additional examples for relation design backgrounds are presented in Figs. 9.36 to 9.40.)

developed for selected countries (Australia, Canada, Germany, Lebanon, and the United States) and are shown in Figs. 9.36 to 9.40.

For new designs of roads of category group A, the radii of curve sequences should always fall into the good range. For roads of at least category B II, good relation designs are also desirable.

In cases of redesigns and of RRR strategies for existing two-lane rural roads, the good relation design range often leads to conflicts with goals related to the protection of the landscape or municipal demands. Therefore, in substantiated individual cases, the fair relation design range is tolerable if especially adverse conditions can be avoided. However, in these cases, the expected accident rate could be at least twice as high as that for the good range of the relation design backgrounds and a higher accident cost rate may result. Therefore, in situations where fair relation design ranges exist, it is recommended that additional optical guidance be given to drivers. Examples include using planting or delineation measures or roads equipped with appropriate traffic warning devices.

In cases where the relation design is at the poor level, a new design should be considered for existing roadway sections. If not, an unfavorable alignment with respect to the actual driving behavior can be expected from safety and economic points of view because of the normally high accident risk and the resulting high accident costs.

Recommendations for the application of relation design backgrounds based on Fig. 9.1 and 9.36 to 9.40 are as follows:

1. Fundamentally, the use of the fair design range cannot be recommended for radii of curve of $R < 200$ m (650 ft). This recommendation is based on work done by Krebs and Kloeckner[368] in Germany and Lamm and Choueiri[408] in the United States, which found that road sections with radii less than 200 m have accident rates that are twice as high as those on sections with radii greater than 400 m (1300 ft). With respect to the design parameter, CCR_S, Table 9.10 also proves that the accident rate of the fair range is at least twice as high as that of the good design range (whereas the poor range was 4 to 5 times higher compared to the good range). Since curves with radii of $R < 200$ m correspond to the fair and especially to the poor design ranges of Table 9.10, it is necessary to avoid using the fair range of radii relations of $R \leq 200$ m for design purposes. Furthermore, according to Table 9.10, the good design range ends with CCR_S values of 180 gon/km, which corresponds to radii of curve of $R = 350$ m (1150 ft). Therefore, it is recommended to use:

The fair range for $R \leq 200$ m under no circumstances

The fair range for $R \leq 350$ m only in exceptional cases

with respect to Fig. 9.1 and Figs. 9.36 to 9.40 because of the expected high accident risk and severity (Table 18.14).[628]

New accident research conducted by Lippold[480] has led to comparable results.

2. For the element sequence independent tangent—clothoid—circular curve, the good range of the relation design backgrounds should always apply. Based upon calculation results and the graphical interpretation of Fig. 9.1 (Greece), Fig. 9.36 (Australia), Fig. 9.37 (Canada), Fig. 9.39 (Lebanon), and Fig. 9.40 (United States), curves with radii of $R \geq 400$ m could be considered sufficient for following independent tangents. The only exception observed to date has been in Germany (Fig. 9.38), where the high operating speed background (see Fig. 8.12, Germany, ISE) leads to curves with radii of $R \approx 500$ m. Therefore, it was decided to assign the following to the transition independent tangent—clothoid—circular curve: $R \geq 400$ m for the countries investigated to date and $R \geq 500$ m (exceptional case so far for Germany) (see Table 12.4) as long as the selected design speed, V_d, does not require larger radii of curve.

3. Based upon the discussion in "Calculation of Sound Relation Design Backgrounds" in Sec. 9.2.3.2, it is recommended that for the direct transition independent tangent to circular curve without transition curves, the following curve radii be selected: $R \geq 800$ m (for the countries investigated to date) and $R \geq 1000$ m (exceptional case so far for Germany) (see Table 12.8).

4. Note that, when adjusting radii of curve sequences, the minimum radius of curve has to at least support the selected design speed of the roadway.

5. Note that, in hilly topography, strong curvilinear alignments, based upon the developed relation design backgrounds, are very favorable and influence traffic safety positively. Curvilinear alignments in such topography are more important than the observance of the limiting values for curve radius.[162]

6. In contrast, note that, in flat or mountainous topography, curvilinear alignments often do not satisfy the demands for a landscape adapted and economic design. However, relation design, as understood here, means more. Relation design is not only directed towards achieving good curvilinear alignments (the main idea so far,[241,243,444] but should also include the sound transitions between independent (long) tangents and curves.

> For example, for flat topography, this could mean the achievement of well-balanced transitions between independent tangents and curves which may or may not be interspersed with curvilinear alignment sections.

> In mountainous topography, limiting design parameters could stand in the forefront in combination with relation design issues; if not, the mountainous topography may even require a curve design with "hairpins."

7. Also, in case of redesigns or RRR projects, curvilinear alignments may not lead to favorable solutions, as revealed by the discussions with respect to the case study presented in Table 9.12.

Thus, for achieving sound relation designs in cases of new designs, major reconstructions, and RRR projects, the highway engineer should examine horizontal alignments using:

- Safety criterion I according to Table 9.1
- Safety criterion II according to Table 9.2
- Relation design backgrounds based on Figs. 9.1 and 9.36 to 9.40 with respect to the country investigated.

If all three evaluation procedures fall into the good design range, it can definitely be said that a good, sound relation design exists. Normally, the results using criterion II and the relation design ranges in Figs. 9.1 and 9.36 to 9.40 correspond to each other, since both evaluation procedures depend on similar assumptions. The results of safety criterion I, however, are independent of the other two.

In the same way, existing two-lane rural roads can also be classified to detect fair and poor design practices in order to evaluate endangered (fair) and dangerous (poor) road sections.

Driving dynamic aspects with respect to permissible tangential and side friction factors, as well as safety criterion III for achieving driving dynamic consistency, are developed in Chap. 10.

In "Evaluation of Tangents in the Design Process" in Sec. 12.1.1.3 the introduction of the tangent as an independent versus nonindependent design element will be introduced.

Case studies, including new designs, redesigns, and existing alignments, are provided for different countries in "Case Studies" in Sec. 12.2.4.2 and Sec. 18.4.2 to show how to conduct safety evaluation processes for practical design tasks from an overall point of view.

9.2   GENERAL CONSIDERATIONS, RESEARCH EVALUATIONS, GUIDELINE COMPARISONS, AND NEW DEVELOPMENTS

The approximately 5.0 million km (3.1 million miles) of two-lane rural highways in the United States represent 97 percent of rural mileage and 80 percent of all U.S. highway miles. Two-lane rural highway travel constitutes an estimated 66 percent of rural highway travel and 30 percent of all U.S. highway travel. Two-lane rural highways have:[82]

- Higher accident rates than all other kinds of rural highways except four-lane undivided roads.

- Higher percentages of head-on collisions than any other kind of rural highway and also higher percentages of single-vehicle accidents, especially "run-off-the-road" accidents.

The probability of an accident on two-lane rural highways is highest at horizontal curves, intersections, and bridges.[82] Similar statements can be made for nearly all highly motorized countries.

9.2.1 Fundamentals

Geometric design standards or guidelines specify appropriate minimum, maximum, and desirable values for visible road elements; these values are usually specified separately although many are interrelated. Since these standards are mainly based on logically derived relationships rather than on safety studies, it is difficult to quantify the safety implications of departures from standards due to environmental or terrain restrictions.[553] Therefore, "Influence of Design and Operational Parameters on the Accident Situation" in Sec. 9.2.1.3 will draw general conclusions from the available international knowledge on the relationships between design parameters and safety.

Fundamentally, highway geometric design guidelines, standards, recommendations, etc., should guarantee the construction of sound roads. However, the fact that accidents occur disproportionately in certain segments of the roadway system suggests that improving our standards will lead to safer systems. Accidents are more than just driver- and vehicle-related; the roadway plays a role, too.

Geometric design guidelines have long been the subject of dispute in the literature. Some argue that the guidelines do not present a clear measure for evaluating the safety level of roadways. For instance, in their discussion of the German design guidelines, Feuchtinger and Christoffers[191] stated:

> When a road goes into operation, the accident experience afterwards is the only indicator of the safety performance of the road. During the planning stage, there is no way to tell what level there is for traffic safety.

Similarly, Bitzl[55] stated:

> Unlike other engineering fields, in road design it is almost impossible to determine the safety level of a road. In other words, the guidelines provide no basic values to describe the safety level of a road in relation to design parameters and traffic conditions; whereas in other engineering fields, such as structural, there exist safety criteria for constructing, for example, bridges or buildings.

Similar statements about safety levels in highway geometric design guidelines were made by Auberlen[32] and Krebs.[364]

In a discussion about the German design guidelines,[241] Krebs and Kloeckner[368] said:

- If the guidelines guarantee the safety of a road, then "no" or "only a few" accidents should occur on that road. When accidents happen, drivers are always the ones who take the blame for the mishap.
- Accidents are not uniformly distributed on the road network. High accident locations are clear indication that, besides driver's error, there exist other influencing parameters which are characterized by the road itself.

Along the same line, Mackenroth stated according to Ref. 390:

- No one is in a position to state whether or not a driver's discipline was in order before a high accident location, but then failed at that location. When a driver fails at a high accident location, it is often said that it was his way of driving which caused the accident.
- When drivers fail a number of times at certain locations, then it becomes obvious that the problem lies, not with the drivers, but mainly with the geometry of the road itself.

The preceding statements indicate that no one is in a position to state whether a road section of considerable length is safe or not, nor can anyone guarantee that a road section will provide a minimum level of safety or a maximum level of endangerment, that is, unsafety.

Despite this knowledge, accident statistics and accident evaluations attribute up to 90 percent of accidents to human failure.[391] Such a description of accident causes is superficial and biased. Of course, there are always traffic participants who do not obey traffic rules, endanger themselves and others, and who end up finding out that the physical rules do not make exceptions for them.[159] But human failure can only partially describe the accident situation.

Durth, et al.[151] concluded that the techniques used in the development of motor vehicles lead to motor vehicles with increased efficiency and performance capability. But in doing so, the possible speed differences between free-moving vehicles become greater and the accident risk consequently increases.

Krebs and Kloeckner[368] argue that the structural and traffic-related conditions of the road are only a part of the cause of accidents. Finally, not all influencing factors are even comprehensible.

Research has revealed that, for those accidents which can be explained by road structural aspects, a safety examination should be provided during the design stages.[349,385] Since the beginning of the 1980s, the demands for the development of quantitative safety criteria have become more important to the design community. In this connection, Lamm[390] in 1980 attempted an evaluation of design elements, in order to estimate a tolerable minimum level of safety or a maximum level of endangerment, that is, unsafety ("Superimposition of Design Parameters/Breakpoint on Safety" in Sec. 9.2.1.3).

9.2.1.1 General Accident Considerations. General impacts on road safety are discussed in Secs. 20.1 and 20.2.

Statistical Viewpoints

Figure 9.2 schematically shows the fatality distributions for different road categories in selected countries. From this illustration it is clear that between 50 and 60 percent of traffic fatalities can be attributed to accidents that occur on two-lane roads outside built-up areas and that somewhat less than half of them (25 to 30 percent) can be attributed to those accidents that occur on curved

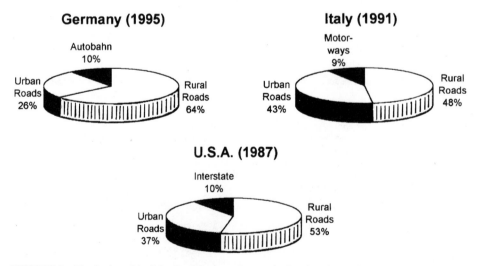

FIGURE 9.2 Distribution of fatalities for different road categories in selected countries.

roadway sections.[106,425,426,438] As an example, 57 percent of the rural fatalities in Germany can be attributed to "run-off-the-road accidents"—the typical accident type at curved sites. It follows then, based on Fig. 9.2, that

$$0.57 \times 0.64 = 0.36 \text{ or } 36 \text{ percent}$$

of the fatal accidents on the rural road network system, which consists mainly of two-lane roads, occur at curved sites and the transitions to curves. Similar results are true for accidents resulting in serious injuries.

These estimates are supported by the results of Brinkman and Smith[82] who reported that horizontal curves are ranked with intersections as the most likely locations for accident concentrations on two-lane rural highways. According to the estimates of the U.S. National Safety Council, about 56 percent of all urban accidents and about 32 percent of all rural accidents occur at intersections.[538] This appears to be consistent with the previous estimate of about 25 to 30 percent of very severe accidents (fatalities) at curved sites.

Thus, curved sites represent one of the most important critical locations for answering the question "Where do people die?" Obviously, measures aimed at reducing accident frequency and severity should be focused there. Therefore, two-lane rural road safety is considered an issue of pressing national concern in both Europe and the United States. It has been identified as the highest priority research need in the area of responsibility by the Transportation Research Board Committee on Geometric Design in the United States. Two-lane roads have the highest accident rate of any class of rural highways, with fatal and injury vehicle kilometer exposure accident rates consistently 4 to 7 times higher than those on multilane median-separated rural highways.[114] With respect to the accident type, an investigation by Pfundt[574] revealed that curved sections of two-lane rural roads are the main location of run-off-the-road accidents.* This applies not only to old alignments but also to newly constructed roadway sections, and leads to the conclusion that driving through curves makes great demands upon a driver (Chap. 19).

Run-off-the-road (ROR) accidents normally result when a vehicle leaves the roadway. However, consequences of ROR accidents can also occur because of collisions with traffic in the opposite direction, for example, in cases with higher traffic densities mostly in combination with wet or even icy road surfaces. Therefore, ROR accidents are not always single-vehicle accidents.

On the other hand, multilane median-separated highways are much safer. For example, the U.S. interstate system and the comparable German autobahn system, with about 10 percent of the total number of fatalities, even though 25 percent of the vehicle kilometers driven are normally done on these roads, represent the safest road class (Fig. 9.2). Multilane highways are normally designed very generously, with relation design aspects prudently included in the design of these roads.[444]

Figure 9.3 schematically shows the distribution of fatalities by age in selected countries. As can be seen from the figure, young drivers between the ages of 15 and 24 years are the most endangered due primarily to excessive speed and lack of driving experience.[106,182,391,398,723] Also see Sec. 20.2.5.

This age group accounts, on average, for about 28 percent of all highway fatalities in both Europe and the United States, even though it represents only about 16 percent of the populations of both continents. This age group experiences almost twice the number of fatalities that would be expected based on its proportion of the total population. Note also that the percentage share of fatalities for the age group over 64 is higher than the percentage that this age group makes up of the total population (Fig. 9.3). Therefore, with respect to the question, "who dies?", it can be concluded that the age groups 15 to 24 and over 64 are the most endangered age groups (see, for comparison, Sec. 20.2.5).

---

*Run-off-the-road accidents occur when a driver loses control of the vehicle because the selected speed did not conform to the course of the roadway, the cross section, the grade, or the surface condition of the pavement or since he or she recognized a critical design inconsistency too late.

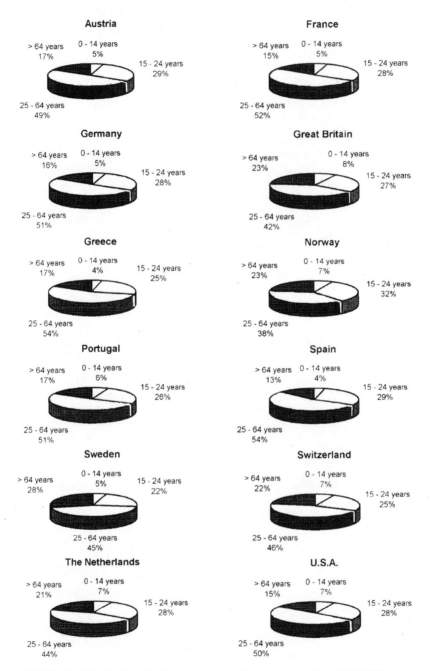

FIGURE 9.3 Distribution of fatalities by age groups for selected countries (1991).[435,438]

Even though significant improvement in traffic safety has been achieved in recent decades, the amount of positive improvement has varied between urban roads, two-lane rural roads, and interstates (autobahnen).[575]

For instance, the number of fatalities on urban roads in West Germany decreased by 75 percent between 1970 and 1990. During the same time period, fatalities on rural, noninterstate roads decreased by only 50 percent. Since the end of 1970s, the fatality rate on two-lane rural roads has been consistently higher than that on urban roads. Similar historical data have been observed in a number of European countries and several states in the United States. This means that two-lane rural roads are the site of a greater share of the traffic fatality total even though fatality rates overall have declined.

This statement is confirmed by research investigations by Pfundt,[575] in Germany:

> In order to describe the development in the last two decades, the accident statistics for the years 1970 to 1972 (Before-Period) and the years 1988 to 1990 (After-Period) were compared.
>
> The fatality rate (number of fatalities per billion vehicle kilometers: 10^9 veh. km) decreased in the following manner:
>
> - Urban roads: from 95.4 to 18.3 (a reduction of 80 percent),
> - Interstate: from 27.4 to 6.3 (a reduction of 77 percent),
> - Two-lane rural roads: from 79.4 to 26.6 (a reduction of 66 percent).
>
> The time period it took for the fatality rate to decrease to 50 percent of its original value is:
>
> - Urban roads: 7.6 years,
> - Interstates: 8.5 years, and
> - Two-lane rural roads: 11.4 years.
>
> The injury rate (number of serious injuries per million vehicle kilometers: 10^6 veh. km) decreased in the following manner:
>
> - Urban roads: from 1.082 to 0.408 (a reduction of 62 percent),
> - Interstates: from 0.206 to 0.060 (a reduction of 72 percent), and
> - Two-lane rural roads: from 0.518 to 0.254 (a reduction of 51 percent).
>
> The time period it took for the injury rate to decrease to 50 percent of its original value is:
>
> - Urban roads: 12.8 years,
> - Interstates: 10 years, and
> - Two-lane rural roads: 17.6 years.
>
> All these numbers clearly indicate that the location cause of the severe accident problem is over time shifting to two-lane rural roads, and that the observed accident reduction in nearly all of the countries is lowest mainly on two-lane rural roads.
>
> Thus, special emphasis should be placed on traffic safety work on two-lane rural roads.

It seems clear from these observations[575] that two-lane rural roads experience the highest accident risks and severities. Therefore, this portion of the road network should be considered most carefully when designing, redesigning, and conducting restoration, rehabilitation or resurfacing (RRR) projects.

General Viewpoints

Many experts believe that abrupt changes in operating speed lead to accidents on two-lane rural roads and that these speed inconsistencies may be largely brought about by abrupt changes in road characteristics.[270,271,286,348,385,418,470,553]

Therefore, providing longer roadway sections with relatively consistent alignment, and thereby achieving a more consistent driving behavior, is an important step toward reducing critical driving maneuvers, obtaining less hazardous road sections, and enhancing traffic safety on two-lane rural highways.

Even though design speed has been used for several decades to determine sound horizontal alignments, it is possible to induce certain inconsistencies into the highway alignment. At low and intermediate design speeds, the portions of a relatively flat alignment, interspersed between the controlling curvilinear portions, may produce operating speed profiles that exceed the design speed in the controlling sections by substantial amounts, especially on two-lane rural roads.[470]

Furthermore, the road cross section, the alignment, and intersections/interchanges are essential parts of road characteristics. They influence traffic safety together. Therefore, they have to be coordinated with each other.

The safety of traffic flow depends on numerous, partially unassessable influencing factors. Besides traffic volume and composition, the design of the cross section of a road is of great significance. Since the moving and safety spaces decrease with decreasing lane width, the risk with respect to traffic in the opposite direction and passing maneuvers increases if the speed is not reduced accordingly. Therefore, the cross-sectional features are very important for the road characteristics, as discussed in detail in Part 3, "Cross Sections," of this book.

One of the strongest impacts on traffic safety has been shown to be whether a road is separated by a median or not. A median separation results in a significant jump for traffic flow and safety. For separated four- or six-lane highways, combined with grade-separated intersections, it has been observed that

- The personal injury accident rate is less than half
- The fatality rate is only a quarter

of the corresponding values for two-lane rural roads with at-grade intersections. These road safety characteristics are particularly important to consider. When based on traffic volume and economic design considerations, a two-lane, three-lane, or four-lane median-separated cross section can be selected (see Part 3, "Cross Sections"). Obviously, the median-separated alternative has clear safety benefits.

From the point of view of highway and traffic engineers, two-lane rural roads represent one of the most important target groups for reducing accident frequency and severity on the roadway system. Consequently, methods to improve highway alignment consistency and to generate more uniform road characteristics have to be developed for modern highway geometric design. For two-lane rural roads focusing on curved sections and tangent curve transition sections holds the most promise for improving safety.

9.2.1.2 Existing Recommendations and Procedures. Design procedures for rural highway alignments, for which the design speed is less than the driver's desired operating speed, cannot guarantee alignment designs that promote uniform operating speed profiles. Typically, a few horizontal curves limit the design speed, while the intervening alignment has a higher design speed which permits higher operating speeds. Drivers who attempt to operate at their desired operating speed are forced to reduce their speed on the limiting curves. Accident analyses indicate that accident rates on curves increase as the magnitude of the speed reduction required from the approach tangent to the curve increases.

For these reasons, Australia and several European countries have incorporated a feedback loop into their design procedures for rural highway alignment to check and to correct for operating speed consistency violations. Two types of violations are of concern: (1) disparities between the design speed, V_d, and the predicted operating speed, $V85$, on individual curves and independent tangents and (2) excessive differences in operating speeds between successive horizontal alignment design elements.[565]

Therefore, the following two safety criteria should be considered for future highway geometric design:[406,408,412,417,418]

1. *Safety criterion I:* Achieving design consistency, that is, harmonizing design speed and operating speed.

2. *Safety criterion II:* Achieving operating speed consistency, that is, harmonizing operating speeds between successive design elements and thereby achieving consistent road characteristics.

The literature review has shown that, within a given roadway section or curve, the difference between operating speeds and/or design speed should not be allowed to exceed certain margins.[405] These are:

- An operating speed concept proposed by Leisch and Leisch[470]
- A theoretical speed model used in the Swiss design standard[691]
- A German design procedure related to the design parameter curvature change rate (CCR)[124,241,243,246]

With respect to safety criterion I, the German guidelines require the following control between 85th-percentile speed $V85$ and design speed V_d on an observed roadway section:

$$V85 - V_d \leq 20 \text{ km/h}$$

With respect to safety criterion II, all three procedures agree that by limiting the change in operating speeds between two road sections to certain ranges, it can be determined whether the break in the speed profile is acceptable or whether it may cause a speed change that could lead to critical driving maneuvers. The severity of the speed break also gives an indication as to the degree of improvement that may be required. For example, the following maximum allowable operating speed changes, ΔV, to prevent abrupt transitions in operating speeds between sections of roadway with dissimilarities in road characteristics, or between two successive design elements, have been recommended:

Leisch method, United States: $\Delta V \leq 10$ mi/n (16 km/h) (average running speed)

German guidelines: $\Delta V \leq 10$ km/h (85th-percentile speed)

Swiss standard: $\Delta V \leq 20$ km/h (project speed) for project speeds ≥ 70 km/h and
   $\Delta V \leq 10$ km/h for project speeds <70 km/h

The meaning of *operating speed* is not the same in these three procedures, but is expressed in terms of the 85th-percentile speed, average running speed, project speed, or speeds related to alignment and layout constraints.[405]

Only the Leisch method[470] uses a speed profile technique to achieve consistency between the horizontal and vertical alignments. It suggests the use of the 10 mi/h (16 km/h) rule as a design principle applied in specific situations as follows:

- Within a given design speed, potential average passenger car speeds generally should not vary by more than 10 mi/h (16 km/h).
- A reduction in design speed, where called for, normally should not be more than 10 mi/h (16 km/h).
- Potential average truck speeds generally should not be more than 10 mi/h (16 km/h) below average passenger car speeds at any time on common lanes.

A comparison of the three procedures for evaluating speed consistency is provided in Ref. 405. Although the differences may be quite substantial at times, the basic conclusions that may be drawn about the investigated road section are the same with all three methods: the critical speed changes occurred at the same spots.

The German CCR method[124] produces the same basic results as those obtained by using the Leisch or the Swiss speed profile methods, and it has several advantages over the graphical techniques.[470,691] The CCR method is the only one which is based on speed measurements and thus reflects the actual driving behavior of motorists. The speed profiles methods are based largely on theoretical considerations.

It would appear that the CCR method would be the most convenient to use in the process of locating inconsistencies in the horizontal alignment. However, the term *section length with simi-*

lar road characteristics would have to be defined more precisely and distinctly. To overcome this weakness in current practice, the new design parameter, curvature change rate of the single curve (CCR_S), was introduced (see Sec. 8.2.3.6).

The need for a method to achieve consistency in highway operation was also emphasized by several findings of an in-depth study team sponsored by the International Road Federation, which surveyed current geometric and pavement design practices in European countries.[271] The report concluded the following:

> Several European countries place great emphasis on achieving consistency among design elements:
>
> • In most cases the effect of individual design elements on operating speed is the mechanism for determining design consistency.
> • The use of design speed as a concept to be applied to individual elements appears to be diminishing in favor of operating speed parameters.

In conclusion, for all three procedures,[124,470,691] the main objective is to design for driver expectations and to comply with inherent driver characteristics to achieve operational consistency and to improve driving comfort and safety.

9.2.1.3   *Influence of Design and Operational Parameters on the Accident Situation.*   There are many factors which exhibit a measurable influence on driving behavior and traffic safety on two-lane rural highways. These include, but are not limited to:[109]

1. Human factors such as improper judgment of the road ahead and traffic, speeding, driving under the influence of alcohol or drugs, lack of driving experience (young people), handicapped (especially for the older segment of the driving population), and sex.[106,110,391,425,426]

2. Physical features of the site such as horizontal and vertical alignments and cross section, combined with the degree of roadside development and access control.

3. Presence and action of traffic such as traffic volume, traffic mix, and seasonal and daily variations.

4. Legal issues such as overall mandatory federal and state laws, type of traffic control devices at the sites, and degree of enforcement.

5. Environmental factors such as weather and pavement conditions.

6. Vehicle deficiencies such as tires, brakes, and vehicle age.

All of these factors, therefore, constitute a complex mix of variables that cause traffic accidents, of which the road itself represents only one (but a very important) factor.

To show to what extent traffic safety is influenced by the road itself, it is necessary to select the design parameters which characterize the road and the traffic volume expected, since they can be categorized by size and number. However, these parameters affect the accident situation collectively rather than independently. Therefore, if conclusions are to be made about the design and traffic conditions of the road with regard to traffic safety, it is necessary to consider these interdependencies. Investigations into the relationship between one or a combination of design and traffic volume parameters and the accident situation may provide valuable information, as long as it is understood that these parameters are only *some* of a variety of influencing factors that are related.

Numerous quantitative and qualitative analyses, appraisals, and discussions of traffic safety have appeared in the literature of highway and traffic engineering in an attempt to provide a better understanding of accident risk and accident severity characteristics. In the planning, design, and operation of a highway transportation system, knowledge of such characteristics is imperative if sound engineering decisions are to be made.[109]

Presented in the following subsections is the result of an extensive international literature review, coupled with the results of research studies by the authors, that was conducted to provide current information on the safety performance of two-lane rural highways, especially for the most critical road categories—A I to A IV.

The studies covered the effects on traffic safety, as measured by the relative accident numbers (accident rates and accident cost rates), of

- Cross-sectional elements
- Radius of curve
- Curvature change rate, degree of curve, and length of curve, curve radii ratio
- Clothoids or spiral curves
- Grade
- Sight distance
- Traffic volume
- Design speed

on two-lane rural highways.

These geometric and operational parameters were chosen for analysis because:

1. It was anticipated that they would exhibit a measurable influence on traffic safety.
2. They can easily be measured.
3. Accident research studies have found statistically measurable impacts of these parameters on traffic safety.

It should be noted that this review is not totally comprehensive and complete. For instance, the authors would have liked to have learned more about the sample sizes and statistical techniques used in the various research studies to give some evaluative comments on the papers investigated or to evaluate the worth of the studies reviewed. Unfortunately, this information was not available or was incomplete in many publications. It appears that until the 1970s, there was a tendency to report only the results of research studies without sufficiently describing the databases or the analytical techniques used. Today, a research paper that does not give information on exact sample sizes and the analysis techniques used is not likely to be accepted by the research community. For these reasons, the reader should understand that the aim of this chapter is, to a certain extent, informative rather than critical.

Relative Accident Numbers

In accident investigations it does not make much sense to compare absolute numbers of accidents because of differing comparative conditions. Under comparative conditions, however, it is understood that section length and traffic volume exhibit an influence on the accident situation. For instance, the longer the roadway section is, the higher the accident possibilities are. Similarly, the higher the traffic volume is, the higher the accident possibilities are. Therefore, the length of an investigated section and the traffic volume on that section (that is, vehicle kilometers traveled) must be considered in comparative accident investigations.

For this reason, relative accident numbers, such as accident rate and accident cost rate, consider the length of a roadway section and the traffic volume to allow a direct comparison of different roadway sections with respect to traffic safety.[286,389]

Accident rate

The *accident rate* relates the number of accidents which occur on an investigated section in a given time period to vehicle kilometers. The accident rate is determined by the following equation:

$$AR = \frac{\text{accidents} \times 10^6}{\text{AADT } 365\ T\ L} \text{ accidents per } 10^6 \text{ vehicle kilometers} \qquad (9.1)$$

$$AR = \text{Accident rate}$$
$$AADT = \text{average annual daily traffic, vehicles/24 h}$$
$$L = \text{length of the investigated section, km}$$
$$T = \text{length of the investigated time period, yr}$$
$$365 = \text{number of days/yr}$$

Accident cost rate

The accident rate evaluates all accidents equally and does not differentiate them by accident severity. Therefore, influences according to accident severity and, especially, accident costs cannot be described by the accident rate. While the accident rate represents the individual risk of being involved in an accident, the *accident cost rate* additionally quantifies and compares the accident severity using cost estimates.[286,338,385,389] Therefore, the accident cost rate provides a measure that quantifies the accident danger (risk) in terms of monetary units.

The accident cost rate, which represents the sum of property and personal damages, is calculated from the following formula:

$$ACR = \frac{S\ 100}{AADT\ 365\ T\ L} \text{ monetary unit per 100 vehicle kilometers} \qquad (9.2)$$

where ACR = accident cost rate
      S = sum of all property and personal damages in the time period T observed (monetary unit of the country under study)

Accident density

The *accident density* represents the relative accident number which is related to the section length of the roadway under consideration. It measures the frequency of road traffic accidents during a certain time period on the observed roadway section. The corresponding formula is

$$AD = \frac{\text{accidents}}{L\ T} \text{ accidents per kilometer per year} \qquad (9.3)$$

where AD = accident density
      L = length of the investigated section, km
      T = length of the investigated time period, yr

The accident density and the accident cost density, which follows, represent economic aspects, as, for example, the number of accidents and the accident costs per observed kilometer and year.

Accident cost density

$$ACD = \frac{S}{L\ T} \text{ monetary unit per kilometer per year} \qquad (9.4)$$

where ACD = accident cost density
      S = sum of all property and personal damages in the time period T observed (monetary unit of the country under study)

Generally, the accident rate is used worldwide. However, in some European countries, the accident cost rate is used in addition. In Sec. 18.1, all of the previously defined relative accident numbers will be combined in a procedure to group roadway sections on a given network according to their safety level.

For determining accident costs, the property and personal damages have to be calculated separately. The property damages are estimated by the police at the accident scene and are listed in the accident report. Based on economic losses, the costs for quantifying personal damages were evaluated by different authors at different time periods. A listing of these costs for Germany can be found in Table 9.3.[550]

Emde, et al.[168] determined average values for property and personal damages for certain accidents. In the following chapters, when analyzing possible differences of costs with respect to design parameters,[476] the police estimates for property damages, as well as the costs for personal damages according to Emde, et al.[168] will be used.

Regarding the accuracy of the police estimates, it should be noted that Kloeckner[338] did not find systematic over- or underestimations. However, Emde, et al.[168] found, through comparing the estimates of the police with the average property damage costs of insurance companies, that police normally underestimated the repair costs. Furthermore, the police registered less than half of the property damage accidents, in comparison to those which had to be regulated by the insurance companies.

Cross-Sectional Elements

The design and the influence of cross-sectional elements on the accident situation is discussed in detail in Part 3, "Cross Sections (CS)." Such elements include lane width, shoulder width and type, roadside features (for example, side slope, clear zone, etc.). For example, see Secs. 24.2.4 and 25.2.3.

Generally speaking, there exists a distinct tendency for accident rates to decrease with increasing lane width (up to 3.50 to 3.75 m) and paved shoulder width (up to 2.00 to 2.50 m) on high-speed and/or high-volume roads.

For low-volume roads, a lane width of at least 3.00 m and an unpaved shoulder width of at least 1.50 m are recommended.

However, it should also be mentioned that lane width alone cannot describe the accident situation satisfactorily and that other influencing design parameters or traffic volumes also need to be considered. The favorable influence of increasing lane width was found to be most prominent when combined with low-traffic volumes.[476,575]

With regard to the accident cost rate, the results so far are contradictory.[286,476]

For multilane highways the presence of a median has the effect of reducing accidents. Furthermore, if medians are supplemented by barriers, the severity of accidents is significantly reduced.

TABLE 9.3 Development of the Costs for Personal Damages in Germany[550]

Reference	Costs for personal damages, dm		
	F	SI	LI
Reinhold (1938)[600]	13.000*	7.000*	500*
Willeke, et al. (1967)[757]	49.412	2.333	63
Niklas (1970)[549]	119.388	4.027	119
Helms (1972)[281]	308.200	11.400	1.800
Kentner (1974)[333]	200.000	10.000	1.000
Emde, et al. (1979)[167]	700.000	75.000	7.000
Krupp and Hundhausen (1983)[376]	1.080.000	50.000	3.800
Emde, et al. (1985)[168]	1.200.000	54.000	4.100
German Federal Highway Research Institute (BASt) (1996)	1.600.000	74.000	7.200

*Costs in RM (Third Reich).

Note: F = fatality, SI = serious injury, LI = slight injury, and dm = German marks.

A continuously decreasing accident rate for increasing median widths up to 3.0 m was found in relevant studies.[296,704]

Radius of Curve

Figure 9.4 depicts the relationship between accident rate and radius of curve as derived from the results of several research studies.[109]

The safe, efficient movement of traffic is greatly influenced by the geometric features of the highway. A review of accident spot maps normally shows that accidents tend to cluster on curves, particularly on very sharp curves. Even though the design engineer possesses detailed information—derived from driving dynamic formulas and standard values—on driving through a curve, accident frequency and severity often do not appear to coincide with the actual driving behavior.

The horizontal alignment of the road may not only be characterized by the radius of curve. The same curve radius in a sequence of similarly aligned curve radii can have effects on the accident situation different from those in a nonaligned sequence of different curve radii, as is usually the case on most old alignments. As Srinivasan[669] noted, an isolated narrow curve in an otherwise straight alignment is more dangerous than a succession of curves of the same radius, and horizontal curves are more dangerous when combined with gradients and surfaces with low coefficients of friction. Similarly, Brenac[71] reported that studies based on detailed accident data show that small radius curves are only dangerous if there is a road alignment anomaly such as a difficult isolated curve in an otherwise easy section.[553]

The general opinion today is that the accident risk decreases as the radius of the curve increases or as the degree of the curve decreases. However, different opinions exist regarding the extent of this influence on accidents. An investigation by Baldwin[43] of U.S. roads with traffic volumes of less than 5000 vehicles/day indicated that the accident rate decreases as the radius of the curve increases. Pfundt[574] studied accidents on low- and high-volume roads in Germany. He indicated that sharply curved, low-volume roads had high accident frequencies. He also indicated that drivers tended to drive faster on low-volume roads than on high-volume roads. Baldwin[43] concluded that the accident rate decreases as curve frequency (radius of less than 600 m) increases. On the

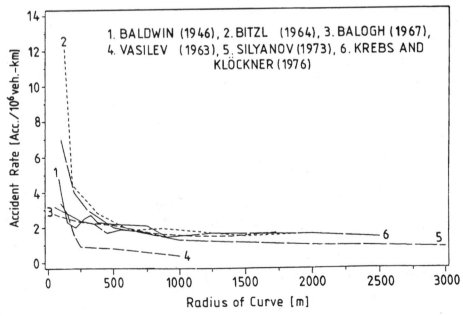

FIGURE 9.4 Examples illustrating the relationship between accident rate and radius of curve.[109,440]

basis of his investigation, a single curve with a small radius should be regarded as more unfavorable than the same curve within a section with a sequence of curves with similar radii.

An investigation of injury accidents by Coburn[115] in the United Kingdom indicated that the accident rate was especially high on curves with radii of less than 175 m. For curves with larger radii, he indicated that the increase in traffic safety was relatively small. In a later publication, Coburn[116] stated that the accident rate on sharp curves was higher than that on gentle curves.

Balogh[44] in Hungary, Raff[591] in the United States, Bitzl[54] in Germany, and Vasilev[40,732] in the former U.S.S.R. also indicated that the accident rate decreases as the radius of the curve increases.

Knoflacher[343] found that, in Germany, for radii of up to 800 m, the percentage of skidding accidents on wet pavements was higher than that on dry pavements; for radii of less than 250 m he found the difference to be statistically significant.

Wilson[761] reported that the accident rate on curves with radii of less than 170 m was about 5 times that on curves with radii of greater than 910 m. He pointed out the danger a single curve poses after a long tangent. The dangers that single isolated curves pose was also mentioned by Babkov[39] in the former U.S.S.R. Because of speed differences before and within the curve, Babkov spoke of:

- "Safe curves," when the change in speeds was less than 20 percent;
- "Relatively safe curves," when the change in speeds was between 20 and 40 percent;
- "Dangerous curves," when the change in speeds was between 40 and 60 percent; and
- "Very dangerous curves," when the change in speeds was greater than 60 percent.

Pfundt[574] indicated that nearly two-thirds of run-off-the-road accidents in Germany occur in curves or near curved sites. For road sections with different road characteristics, he indicated that the risk of having run-off-the-road accidents increases with the increasing complexity of the alignment. He also indicated that road sections with few curves are more dangerous than sections with many curves.

From the accident databases of a number of countries, Silyanov[649] in the former U.S.S.R. established a distinct tendency for the accident rate to decrease with an increasing radius of curve. In a study of accidents in Great Britain, O'Flaherty[554] also concluded that the accident rate decreases as the radius of curve increases.

Rumar[618] analyzed 14,000 accidents on 9000 km of two-lane roads in Sweden. His results showed a reduction in accident rates with increasing radii of horizontal curves.

Statistical analyses by Zegeer, et al.[782] revealed:

1. A significantly higher number of accidents on sharper curves.
2. Accident reductions of up to 80 percent, depending on the central angle and amount of curve flattening. See also "Curve Flattening" in Secs. 12.1.2.5 and 12.2.2.5.

In this respect, a recent French study, quoted by Brenac,[72] showed that accidents at curves depend on two significant variables: the radius and the tangent lengths on both approaches.

Krebs and Kloeckner[368] in Germany and Lamm and Choueiri[408] in the United States determined the following (Fig. 9.5):

1. Accident risk decreases with an increasing radius of curve.
2. Road sections with curve radii of less than 200 m have accident rates that are twice as high as those on sections with curve radii greater than 400 m.
3. A radius of 400 m provides a cross-point in safety.
4. For radii greater than 400 m, the gain in safety is relatively small.

Qualitatively, similar results are reported by Spacek,[665] who found that the average accident rate, AR, for radii of curve classes of $R < 350$ m is about 5 times higher than for radii of curve classes of $R > 400$ m. A significant jump in the accident rate could be observed between the class $R = 200$ to

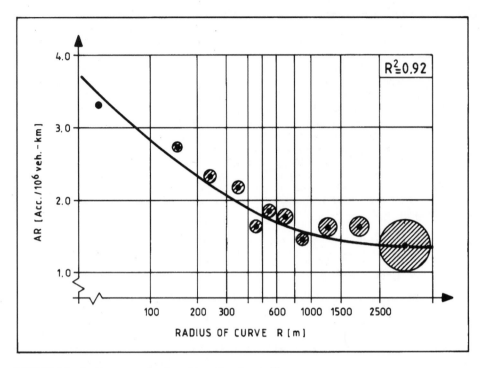

FIGURE 9.5   Accident rate as function of the radius of curve.[368]

350 m ($\overline{AR}$ = 0.61 accidents per 10^6 vehicle kilometers) and R = 400 to 600 m ($\overline{AR}$ = 0.27 accidents per 10^6 vehicle kilometers).

The U.K. Department of Transport publication[139] includes a graph which compared accident rates for horizontal curvature to a base accident rate by means of a multiplier (Fig. 9.6) which agrees closely with the previous results.[368,408] The U.K. data indicates continually increasing accident rate with decreasing radius. This increase in accident rate becomes particularly apparent at curve radii less than 300 m.

The research studies reported here have generally shown that a negative relationship exists between the radius of curve and the accident rate (see Figs. 9.4 to 9.6). A considerable increase in accident risk exists on curves with sharp radii, where run-off-the-road accidents most frequently occur, especially after long tangents. In addition to the magnitude of the curve radius, the road characteristics play an important role. Curves that dictate a significant change in operating speeds and cause inconsistencies in road characteristics are especially dangerous.[410,415,449] Furthermore, the pavement width influences, to a certain extent, the magnitude of the accident rate. Curves that are combined with wide pavements do not affect the accident risk as unfavorably as those curves that are combined with narrow pavements (see Sec. 24.2.4 of Part 3, "Cross Sections").

Based on the research investigation,[476] relationships between accident rates and radius of curve are shown graphically in Figs. 9.7 and 9.8. The data analyzed covered 1300-km-length sections of two-lane rural roads and 27,972 accidents during the time period from 1978 to 1985.

The magnitude of the curve radius influences the accident situation considerably. As the radius increases to about 400 m, the accident rate falls to about 30 to 40 percent of the value corresponding to radii of curve less than 100 m. Larger radii of curve, up to 1000 m, cause a further slight decrease in the accident risk. A slight increase is observed for radii of curve R > 1000 m.[476]

Lamm[385] explained the latter phenomenon in the following manner:

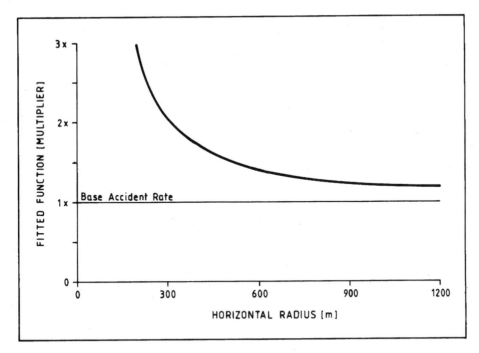

FIGURE 9.6   Accident rate versus horizontal radius, 10-m single carriageway.[139]

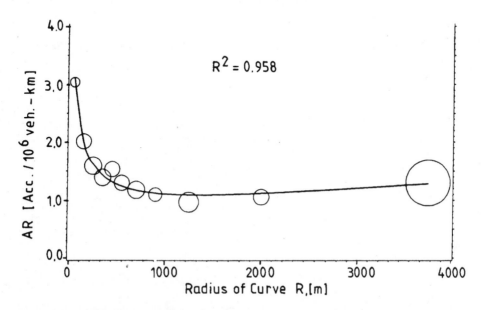

FIGURE 9.7   Accident rate with regard to the radius of curve for all accident types.[476]

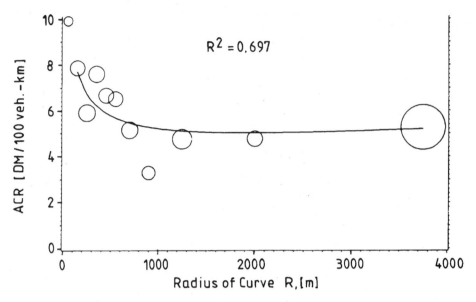

FIGURE 9.8   Accident cost rate with regard to the radius of curve for all accident types.[476]

Large radii are often associated with relatively low design speeds, such as 80 km/h, for which corresponding superelevation rates are between 2.5 and 3 percent. However, actual 85th-percentile speeds on these curves require superelevation rates of at least 5 percent to 6 percent. Such a discrepancy between design speed and actual operating speed could influence the accident situation unfavorably.

Similar relationships, although less pronounced, are shown for the accident cost rate. The relationships determined apply for different daylight and road surface conditions, traffic volumes, and pavement widths.

The results shown in Figs. 9.7 and 9.8 confirm the previous findings that for $R \geq 400$ m, the increase in safety is low. Furthermore, for small radii of curve the accident rate and the accident cost rate are extremely high.

In addition, a number of studies have indicated that horizontal realignment of rural highways is the most efficient way of increasing safety. Reductions in the number of accidents of the order of 80 percent have been reported.[669,677] Table 9.4 shows the prediction model developed from a Swedish study on roads with a 90-km/h speed limit.[85,553]

TABLE 9.4   Accident Reduction Factors for Various Increases in Horizontal Radii of Curve (Proportion of Original Accident Rate)[85]

	To, m		
From, m	500	700	1500
300	0.25	0.35	0.45
500	—	0.10	0.30
700	—	—	0.20

From the foregoing, the following can be concluded:

- Large accident numbers are associated with sharp curves.
- The accident risk (expressed by the accident rate) and the accident severity (expressed by the accident cost rate) decrease with increasing radius of curve.
- Road sections with curve radii less than 200 m have an accident rate which is at least twice as high as that on sections with curve radii greater than 400 m.
- A radius of 400 m provides a cross-point in safety.
- For radii greater than 400 m, the gain in safety is relatively small.
- Accident reductions of up to 80 percent could be obtained, depending on the central angle and the amount of curve flattening.
- The same radius of curve in a sequence of similarly aligned radii can have effects on the accident situation different from those in a nonaligned sequence of different radii, as is usually the case on most old alignments, where "isolated curves" between long tangents are especially very dangerous.

Curvature Change Rate, Degree of Curve, Length of Curve, and Curve Radii Ratio

The design variables, curvature change rate and degree of curve, directly or indirectly depend on the curvature. For single curves without transition curves, the dependencies between these variables are presented in Table 8.4.

Curvature change rate

Several traffic-related scientific investigations attempted to analyze the influence of different curve parameters on the driving behavior of motorists. For example, the studies of Dilling;[143] Koeppel and Bock;[348] Durth, Biedermann, and Vieth[151] revealed that limiting the investigations on the curve itself is not sufficient, since different speeds were observed for equivalent radii. According to Leutzbach and Papavasiliou,[475] more valid results can be established by considering the approach sections to the curves in addition to the curves themselves.

Dilling,[143] Koeppel and Bock,[347,348] Lamm,[385] and Trapp and Oellers[713] established statistically sound relationships between the driving behavior of motorists and curvature change rate, an essential design parameter for road characteristics. A comparison, conducted by Krebs and Damianoff,[372] reveals similar relationships despite different mathematical regression analyses. With increasing CCR, the speed decreases (see Fig. 8.8).

In several investigations, an attempt was made to describe the relationship between accident rate and number of curves per section length (Pfundt[574] and Babkov[40]). It was found that sections with many curves are less dangerous than those with a few single isolated curves.

Krebs and Kloeckner[368] found that, with increasing CCR values, the accident variables also increase. However, these increases are relatively small. Road sections with high CCR values have accident rates about two-thirds higher than those with low CCR values. For exceptionally high CCR values (CCR ≥ 500 gon/km), the accident rate and accident cost rate represent high values. Their research results are presented in Fig. 9.9 for the accident rate and in Fig. 9.10 for the accident cost rate. According to Leutzbach and Zoellmer,[476] the curvature change rate exerts a much smaller influence on traffic safety than radius of curve does. An increase in curvature change rate is accompanied by a slight increase in the accident rate (Fig. 9.11). However, these results stand in contrast to other, more in-depth investigations by the authors and are due to the fact that the design parameter, curvature change rate, is not related to the individual curve site but to a longer roadway section with similar road characteristics, and thereby the individual accident rates are smoothed. Compare Sec. 8.2.3.6.

With regard to the accident cost rate, an increase is noticed up to about 100 gon/km. Above this level, an increase in curvature change rate leads to a reduction in the accident cost rate (Fig. 9.12). This is understandable when one looks at the effects caused by the (design) speed, as shown

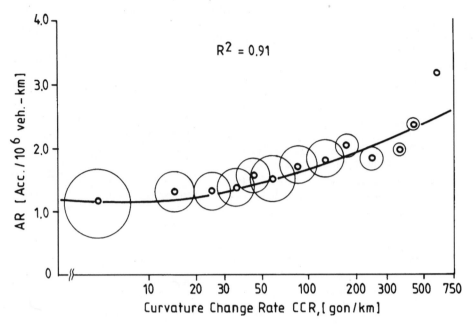

FIGURE 9.9   Accident rate with regard to the curvature change rate for all accident types (1977).[368]

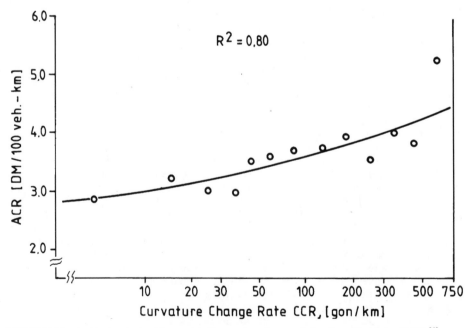

FIGURE 9.10   Accident cost rate with regard to the curvature change rate for all accident types (1977).[368]

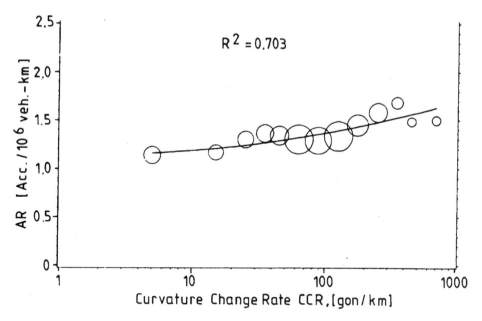

FIGURE 9.11   Accident rate with regard to the curvature change rate for all accident types (1989).[476]

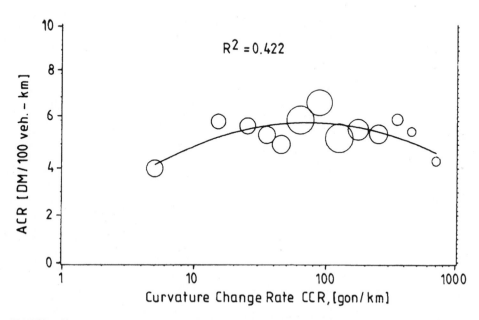

FIGURE 9.12   Accident cost rate with regard to the curvature change rate for all accident types (1989).[476]

in "Design Speed" in this section, since with higher curvature change rates, the speed decreases and the accident severity may also decrease.

However, high levels of curvature change rate have a negative impact on run-off-the-road and passing accidents, as well as on wet/slippery road surfaces. For these differentiated considerations, the accident rate increases by a factor of approximately 2 between the lowest (<10 gon/km) and the highest (>500 gon/km) curvature change rate class.[476]

Hiersche and Lamm, et al.[286] established different relationships between curvature change rate and accident variables for old alignments and redesigned alignments of existing road sections, built according to the "German Guidelines for the Design of Roads," Part: Alignment, 1973 edition.[241] In developing these guidelines, the safety issues of

- Driving dynamics
- Coordinating design speed and operating speed
- Achieving a balanced relation design

took on significant importance for the first time.

The investigation method was based on comparative studies of the accident situation—"before" and "after" the redesign of existing road sections. Absolute, relative accident numbers were considered. In cases, where before investigation time periods were not available or insufficient, the studies were completed by comparative sections with old alignments (with/without investigations).

The study included 28 two-lane rural road sections, designed according to Ref. 241. The overall section length was 86.4 km for the before/after comparison and 92.0 km for the with/without comparison. During the investigated time periods, about 750 million vehicle kilometers traveled were observed on the investigated sections. For the same time periods, an overall number of 1420 accidents were considered for analysis.

Considerable differences were expected between the before/after and with/without sections in terms of more consistent road characteristics and smaller curvature change rate values (called *relation design,* see Secs. 9.1.3.2 and 9.2.3.2). In order to examine the influence of curvature change rate (CCR), accident rates were calculated for certain CCR classes, as shown in Fig. 9.13. In the lower part of this figure, the corresponding percentile for "vehicle kilometers traveled" is also given. On sections of old alignments, CCR values up to 500 gon/km could be observed, but on newly designed roadway sections, values up to only 300 gon/km were reached. Furthermore, high percentages of the vehicle kilometers traveled could be observed in the CCR classes up to 150 gon/km on old alignments, while for new ones 85 percent of the vehicle kilometers traveled were driven on CCR classes less than 100 gon/km. The course of the accident rates in the upper part of Fig. 9.13 clearly shows that, on old alignments, the accident rate increases progressively with respect to CCR, while on new alignments the accident rate also increases but less severely.

Fundamentally, accident rates are lower on the new alignments than on the old ones. This demonstrates, on the one hand, that safety advantages are obtained by the relation design, and, on the other hand, that advantages may be gained through other design parameters like lane width and smooth grades, as well as the appropriate selection of design speeds, sound side/tangential friction factors, and superelevation rates.[382,383,385]

In addition, it is interesting to note that the differences in accident rates between "old" and "new" in the CCR classes between about 50 and 150 gon/km are nearly constant and result in a reduction of at least 30 percent in accident rates for new alignments.

The attempt made by Lamm in Ref. 390 to propose a maximum permissible accident rate of AR = 2.0 accidents per 10^6 vehicle kilometers is substantiated by Fig. 9.13. This value seems to be a reasonable limit for new alignments designed in conformity with modern highway geometric design recommendations (see "Superimposition of Design Parameters/Breakpoint on Safety" in Secs. 9.2.1.3).

Figure 9.14 presents the measured relationships for the accident cost rate observed in the study.[286] Basically, the same tendencies could be noticed. The accident cost rate for new align-

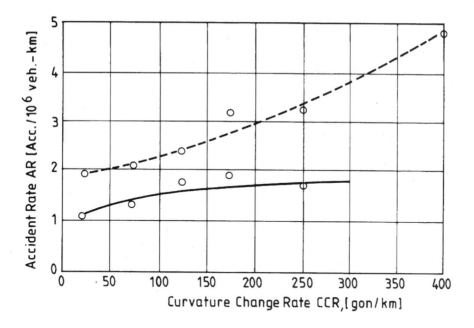

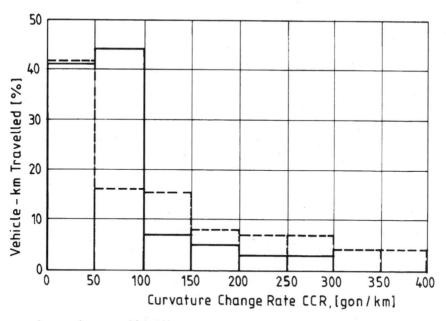

FIGURE 9.13 Accident rate and vehicle kilometers traveled versus curvature change rate.[286]

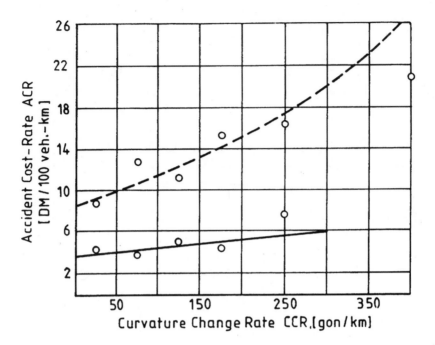

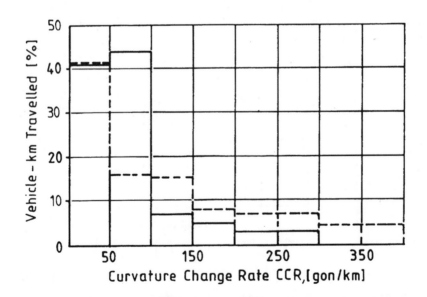

FIGURE 9.14   Accident cost rate and vehicle kilometers traveled versus curvature change rate.[286]

ments is decisively lower than that for old ones. Again, with increasing CCR values the accident cost rate increases, but at a rate much slower for new alignments than for old ones. To clarify the results, two examples are presented, based on Fig. 9.14.

Curvature change rate:	50 gon/km
New alignment:	ACR = 4 German marks/100 vehicle kilometers
Old alignment:	ACR = 10 German marks/100 vehicle kilometers (accident severity = 2.5 times)
Curvature change rate:	250 gon/km
New alignment:	ACR = 5.8 German marks/100 vehicle kilometers
Old alignment:	ACR = 17.5 German marks/100 vehicle kilometers (accident severity = 3.0 times)

That means, for the investigated sections, there exists a tendency for accident severity (expressed by the accident cost rate) to be 2.5 to 3.0 times higher on old alignments than on new ones.[286] In addition, the increase in the accident cost rate is greater on old alignments than on new ones.

Analysis of traffic vehicle kilometers traveled on the new alignments shows that 85 percent of the vehicle kilometers were present on CCR classes less than 100 gon/km. That means, on new alignments the predominant part of vehicle kilometers traveled corresponds to an average accident cost rate of about 4.0 German marks/100 vehicle kilometers.

The comparison between a balanced relation design of new road sections and old alignments reveals that, based on average ratios, a safety gain of about 2 is achieved in the accident rate and a safety gain of about 3 is achieved in the accident cost rate. Furthermore, it could be demonstrated that the safety gain on the new roadway sections investigated was considerably higher than the typical decreasing accident trends noted in most western European countries. Greatest reductions were noted for run-off-the-road and passing accidents.[286]

In conclusion, the design parameter, curvature change rate, revealed significantly lower accident rates and accident cost rates on new alignments than on old ones. Since this parameter is still based on the mathematically biased definition roadway section length with similar road characteristics, it may be expected that the newly developed, mathematically exact, and definable design parameter, curvature change rate of the single curve (CCR_s), will lead to even more favorable safety results.

Furthermore Pfundt[575] reported:

> Local increases of the curvature change rate in an already curved environment may be critical. Dangers can originate in curves with central angles between 15 and 40 gon through "curve cutting." Single curves with large central angles and/or smaller radii can also be critical. Therefore, it is recommended to consider in the future not only a sound relation design but also the magnitude of the central angle with regard to the radius of curve.

These critical design cases can be avoided through the exact application of the curvature change rate of the single curve, CCR_s, in accordance with the recommended ranges of safety criteria I and II for good design levels (see Secs. 9.1.2 and 9.1.3), and through the classification of tangent sections into independent and nonindependent tangents for the design process (see "Evaluation of Tangents in the Design Process" in Sec. 12.1.1).

Degree of curve

Instead of the design parameter, radius of curve or curvature change rate, the design parameter, degree of curve, is used in the United States and several other countries. The relationships between both parameters for circular curves without transition curves are presented in Table 8.4. Research by the authors[410,412,415,434,449,673] yielded the nomograms in Fig. 9.15 for evaluating accident rates in relation to degree of curve for both the United States and Germany (West). A detailed

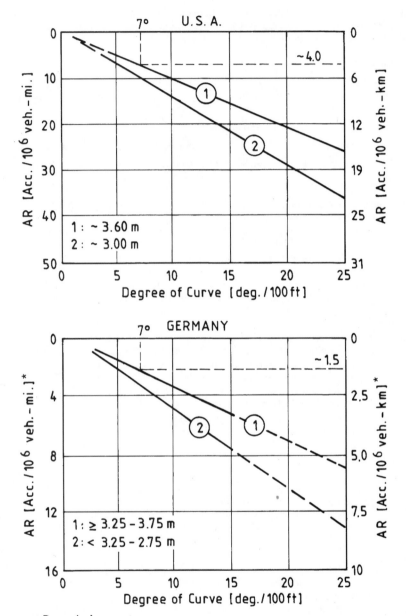

FIGURE 9.15   Nomogram for evaluating accident rates, as related to degree of curve.

TABLE 9.5 Simple Correlation Matrix (3.6-m Lane Width)

	AR	LC	DC	RS	SW	e	G	AADT	SD
AR	1.000	−0.347	0.852	−0.833	−0.514	0.590	0.075	−0.126	−0.316
LC		1.000	−0.321	0.292	0.455	−0.212	−0.022	−0.013	−0.271
DC			1.000	−0.929	−0.561	0.765	0.095	0.022	−0.312
RS				1.000	0.616	−0.721	−0.082	−0.043	0.309
SW					1.000	−0.416	−0.058	−0.101	0.437
e						1.000	0.061	0.158	−0.311
G							1.000	−0.046	−0.145
AADT								1.000	0.034
SD									1.000

discussion of Fig. 9.15 will be provided in "Classification System for Safety Criteria I and II Based on Accident Research" in Sec. 9.2.1.4.

With respect to the U.S. nomogram in Fig. 9.15 (upper part), the following steps were carried out to determine the relationship between accident rate and degree of curve.

Initially, before any models were constructed to predict accident rate, a check of the correlation coefficients was made. (See Table 9.5 for the 3.6-m lane width as an example.) This step was carried out to diagnose any dependencies that may exist between the potential predictor variables used in the study: degree of curve, length of curve, lane width, shoulder width, superelevation rate, sight distance, gradient, posted recommended speed, and average annual daily traffic (AADT). This examination was necessary to detect any strong collinearities between any two independent variables, since their inclusion in the same regression equation can lead to poor estimates; stepwise methods are best when the independent variables are nearly uncorrelated, the condition under which finding a descriptive model is likely to be relevant.[292]

Table 9.5 reveals that some of the independent variables are moderately to highly correlated. For example, the correlation between degree of curve DC and posted recommended speed RS is high ($R = -0.929$), as could be expected. In reality, with increasing degree of curve, posted recommended speeds decrease. In addition, the correlation between degree of curve DC and superelevation rate e is a marked one ($R = +0.765$). This could be expected because of the existing driving dynamic design process. The same holds true for the correlation ($R = -0.721$) between recommended speed RS and superelevation rate e. From corresponding tables, which were developed for 3.0- and 3.3-m lane widths, nearly similar correlations were obtained between these independent variables.

Because of these marked correlations, it was decided to include only one of the three parameters (DC, RS, or e) in the prediction model. For the purpose of this study, the impact of degree of curve on the accident rate was considered.

Related to the current accident database, there are now six potential predictors: degree of curve (DC), length of curve (LC), shoulder width (SW), gradient (G), sight distance (SD), and average annual daily traffic (AADT). Thus, there are now $2^6 = 64$ possible subset models (including the full model and the model containing only the intercept).

An example of such regression models, with the lowest Mallows C_p values, is shown in Table 9.6. The importance of the Mallows C_p statistic is that it measures bias in a regression model; generally, small values of C_p are desirable.[492] To make a decision as to which model should be adopted, an initial step is, therefore, to consider the model with the lowest C_p value. As Table 9.6 indicates, it is the model containing the three independent variables degree of curve (DC), average annual daily traffic (AADT), and length of curve (LC). However, further statistical analyses indicated that the regression coefficient associated with the length of curve (LC) was not significantly different from zero at the 95 percent confidence level ($t = 1.88$). Therefore, it was decided to delete this particular variable from the regression model.

TABLE 9.6   Regression Models for the Dependent Variable Accident Rate (3.6-m Lane Width)

Number of variables in model	Coefficient of determination R^2	C_p values	Variables in model
1	0.726	13.171	DC
2	0.746	3.527	DC, AADT
3	0.753	2.057	DC, AADT, LC
4	0.754	3.454	DC, AADT, LC, SD
5	0.754	5.141	DC, AADT, LC, SD, SW
6	0.755	7.000	DC, AADT, LC, SD, SW, G

The final model now reads, for a 3.6-m lane:

$$AR = 0.968 + 0.674\,DC - 0.0005\,AADT$$

$$R^2 = 0.746 \tag{9.5}$$

Range of degree of curve = 1 to 24°

where AR = estimate of the accident rate, accidents/10^6 vehicle kilometers
　　　DC = degree of curve, degrees/100 ft
　AADT = average annual daily traffic (range = 1000 to 10,000 vehicles/day)
　　　R^2 = coefficient of determination

The regression coefficients associated with the independent variables (DC) and (AADT) in Eq. (9.5) are significantly different from 0 at the 0.05 level. The algebraic signs for the coefficients of the independent variables are output as part of the statistical analysis.

The regression model in Eq. (9.5) has an R^2 value of 0.746. This marked R^2 value suggests that the relationship is a strong one. However, this equation is not really a marked improvement over Eq. (9.6).

$$AR = -0.341 + 0.672\,DC$$

$$R^2 = 0.726 \tag{9.6}$$

Range of degree of curve = 1 to 24°

as can be seen from the third term of Eq. (9.5), which contributes little to the calculated accident rate (an improvement of 2 percent, as related to the R^2 values). Thus, among the design and traffic volume parameters, degree of curve *alone* was able to explain more than 72 percent of the variation in the accident rate [see Eq. (9.6)].

The same observations were made for 3.0- and 3.3-m lane widths. The resulting equations are as follows. For a 3.0-m lane:

$$AR = -0.639 + 0.946\,DC$$

$$R^2 = 0.300 \tag{9.7}$$

Range of degree of curve = 2.3 to 26.9°

and for a 3.3-m lane:

$$AR = -0.161 + 0.859\,DC$$

$$R^2 = 0.462 \tag{9.8}$$

Range of degree of curve = 1.8 to 19°

Generally, it can be noted from Eqs. (9.6, 9.7, and 9.8) that for every 1° increase in degree of curve, an increase of about 0.72 accidents/million vehicle kilometers could be expected for 3.6-m lanes or about 0.86 to 0.95 accidents/million vehicle kilometers for 3.3- to 3.0-m lanes. Thus, lane width appears to have some influence on traffic safety (see Fig. 9.16).

However, a statistical test revealed no statistical differences in the slopes of the three lines over the degrees of curve of interest in Fig. 9.16. A decision was then made to combine the three equations. The least-squares regression technique was used to produce the following formula. For all lanes:

$$AR = -0.55 + 0.881 \, DC$$

$$R^2 = 0.434 \tag{9.9}$$

$$\text{Range of degree of curve} = 1 \text{ to } 26.9°$$

Equations (9.6 to 9.9) are shown schematically in Fig. 9.16.

One should not object to the fact that the intercepts of regression, Eqs. (9.6 to 9.9), are negative in these cases. For the relevant values of degree of curve 1 to 24° for a 3.6-m lane, 1.8 to 19° for a 3.3-m lane, 2.3 to 26.9° for a 3.0-m lane, and 1 to 26.9° for all lanes, the estimated values of AR are positive. Hence, it does not matter here whether the intercept is positive or negative, so long as the predicted value of AR is positive for the relevant range of degree of curve.

The low coefficients of determination ($R^2 = 0.300$) in Eq. (9.7), and the moderate ones ($R^2 = 0.462$) in Eq. (9.8) and ($R^2 = 0.434$) in Eq. (9.9) are not at all surprising. In the field of accident research, the relationships are not simple or direct ones but often very complex, and changes in

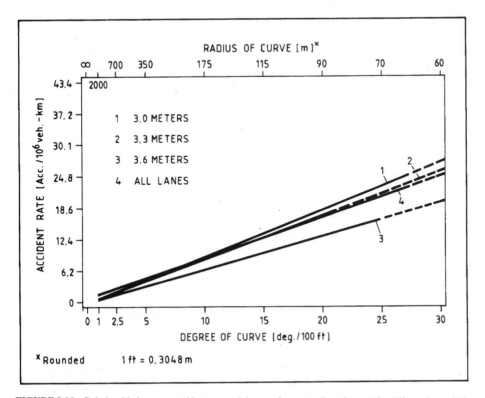

FIGURE 9.16 Relationship between accident rate and degree of curve (radius of curve) for different lane widths and for all lane widths combined.

accidents are often the result of the interplay of many factors in addition to the design parameters and traffic volume.

A cross validation of the models containing degree of curve on a new sample of 61 rural, curved roadway sections of varying degrees of curve, in the State of New York revealed that the regression model, Eq. (9.9) could be used, with a moderate degree of confidence, for prediction purposes.

In general, Eq. (9.9) reveals the following:

1. With an increasing degree of curve (decreasing radius of curve), accident rates increase (see Fig. 9.16).
2. For radii of curve greater than 400 m, an increase in radius leads only to a comparatively low safety gain, that is, a radius of 400 m corresponds to a degree of curve of 4.4°.
3. The accident rate on curves with radii less than 200 m (degree of curve = 8.8°) is more than twice as high as that on curves with radii of 400 m (degree of curve = 4.4°).

These results clearly indicate that, during the design or redesign stages, a designer could already predict accident rates on curves or curved sections of two-lane rural highways from a priori knowledge of degree of curve (radius of curve) and lane width (Fig. 9.16).

Further discussion is presented in "Classification System for Safety Criteria I and II Based on Accident Research" in Sec. 9.2.1.4.

A number of U.S. researchers have attempted to relate changes in accident rates to specific characteristics of curve geometry, usually concentrating on degree (or radius) of curve.[710] Past studies differ considerably in estimates of accidents per vehicle mile (kilometer) as a function of degree of curve, partly because of differences in techniques used for calculating the amount of travel and identifying accidents considered to be curve-related. In addition, some of the accident databases were limited, and influences of other geometric and traffic characteristics on curve-related accidents were not properly treated in some of the analyses.[218] A study sponsored by the Federal Highway Administration (FHWA)[217] succeeded in eliminating many of these problems and, in doing so, assembled the most reliable accident database currently available for horizontal curves in the United States.

Like the data in earlier studies, the FHWA study[217] indicates a strong link between degree of curve and accidents. The link between accidents and other measures of curve geometry, including curve length and central angle, is much weaker. For cost-effective analyses of horizontal curve improvements, the study used a numerical relationship based on the FHWA data. In this relationship, the expected change in accidents resulting from a horizontal curve improvement is based upon the change in degree of curve (Fig. 9.17).[710]

The relationship in Fig. 9.17 between accidents and degree of curve must be regarded as rough in nature. This is especially true for isolated horizontal curves, without consideration of the alignment of adjacent highway segments, because the relationship is not fully correct for interrelated effects of other geometric features (for example, sharp curves occur more frequently on roads with narrow lanes and dangerous roadsides).

As degree of curve decreases, this relationship indicates that the number of accidents at the curve also decreases. On average, about three fewer accidents per degree of curve take place for each 100 million vehicles passing through the curve.[710]

Length of curve *

A 5-year accident investigation of 10,900 curves in Washington State by Zegeer, et al.[781] came to the conclusion that, for the same degree of curve, long curves are associated with a higher number of accidents. While some authors prefer to use the central angle of a horizontal curve as an accident parameter, others rely on the length of curve. This, however, should not alter the assess-

---

*Elaborated based on Ref. 781.

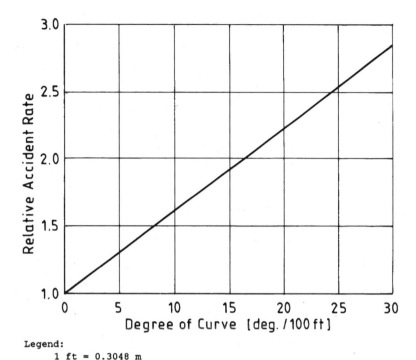

Legend:
    1 ft = 0.3048 m

FIGURE 9.17   Relative accident rate with regard to degree of curve.[710]

ment process of accidents on a horizontal curve, since both parameters are interdependent, as shown in the following equation:

$$\gamma = DC\,L\,0.01 \tag{9.10}$$

where γ = central angle of curve, degrees
    DC = degree of curve, degrees/100 ft
    L = length of curve, ft

In metric units, Eq. (9.10) becomes

$$\gamma = DC\,L\,0.0328 \tag{9.11}$$

where the parameters γ and DC have the same dimensions as shown previously and L is in meters.

The influence of length of curve on traffic accidents is shown in Fig. 9.18. This figure reveals that the number of accidents may almost triple when the length of curve is 4 times longer. This result is valid for each degree of curve and is a very important finding. It provides a clear "picture" of the relationship between curve geometrics and safety. According to Zeeger, et al.[781] it is not sufficient to relate accident potential only to degree of curve; it must also be related to length of curve. A long curve thus presents a more difficult task for the driver than a short curve. It follows then that sharp curves that, for example, have to be flattened, require longer curved sections of highway, on the one hand, and a decrease in degree of curve on the other hand. That means a net accident decrease might not be as great as expected.[781]

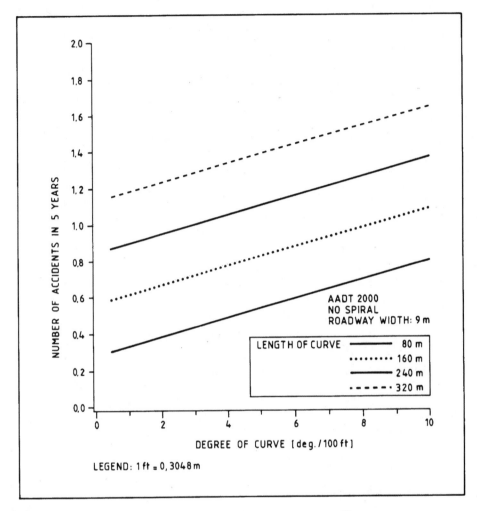

FIGURE 9.18 Predicted accidents in 5 yr for degree of curve and curve length.[781]

It should be clear that, for a given amount of curve flattening, the percent reduction in accidents is slightly larger for smaller central angles than for greater central angles. For example, flattening a 10° central angle curve with DC = 20°/100 ft to a curve with DC = 10°/100 ft will result in a reduction of 48 percent in accidents; for a 50° central angle, the reduction will be only 41 percent. However, it should be remembered that a 50° central angle curve would be expected to have a greater number of total accidents than a 10° central angle for a curve with the same degree of curve. In other words, the *net* number of accidents reduced may be greater on a 50° central angle than on a 10° central angle for the same flattening improvement.

Therefore, in laying out new roadway alignments, redesigns, or RRR projects, designers should strive to avoid situations where large central angles are necessary. Central angles greater than 30° may result in safety problems; those greater than 45° should be avoided whenever possible; that is, curve flattening should not result in curves that are too long.[781]

Curve radii ratio

Leutzbach and Zoellmer[476] developed an interesting approach by coming up with the so-called curve radii ratio. This ratio represents the quotient of the radius of the circular curve of interest divided by the radius of the preceding circular curve (the calculation is performed for both driving directions). The accident experience is then analyzed in relation to this value. Separate evaluations are performed for:

1. The ratio related to all radii of curve, which means all radii of curve were included, even those which have a short tangent or a transition curve in between. The ratio is the quotient of the radius of the circular curve of interest and the radius of the preceding circular curve in the observed driving direction.

2. The ratio related only to directly succeeding radii of curve, which means the circular curve of interest is the one being preceded by a circular curve in the observed direction with nothing in between. Again, the ratio is the quotient of these two radii of curve.

The principal results of the analyses are presented in Figs. 9.19 to 9.22. As can be seen from the figures, curve radii ratios < 0.8 produce a noticeable increase in the accident rate; ratios < 0.15 result in a pronounced jump in the accident rate to about double the rate for ratios > 0.8. For ratios > 0.8, only marginal reductions in the accident rate are noticed. This relationship applies equally to all radii and to directly succeeding radii of curve, and illustrates the favorable effect on traffic safety which can be made by the use of relation design according to Secs. 9.1.3.2 and 9.2.3.2.

Thus, it could be proved that a sound tuning of radii of curve sequences leads to low accident rates (ratio > 0.8). Intermediate accident rates can be expected for ratios between 0.8 and 0.15, while high accident rates and accident cost rates can be expected for ratios < 0.15. These results support the relation design backgrounds of Figs. 9.1 and 9.36 to 9.40.

Contrary to the opinion of many experts, the presence of short tangents and/or transition curves between succeeding circular curves did not demonstrate measurable positive impacts on the accident situation compared to the direct sequence of two circular curves. This raises a question as to whether the extensive and often very distinguished use of transition curves is overem-

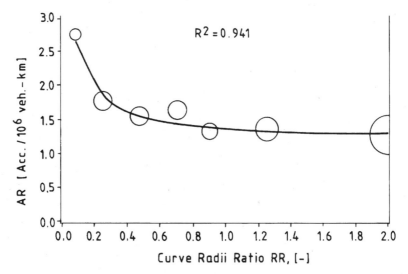

FIGURE 9.19   Accident rate with regard to the curve radii ratio for all accident types and all radii of curve.[476]

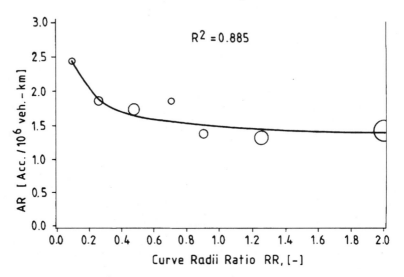

FIGURE 9.20   Accident rate with regard to the curve radii ratio for all accident types and *only* directly succeeding radii of curve.[476]

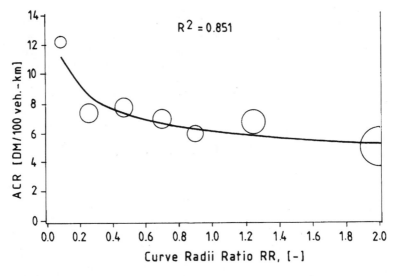

FIGURE 9.21   Accident cost rate with regard to the curve radii ratio for all accident types and all radii of curve.[476]

phasized in several European countries. Compare also the following section on "Clothoids" and Sec. 12.1.1.3 about the evaluation of nonindependent (short) tangents in the design process.

In conclusion, it can be stated that with an increasing curvature change rate or degree of curve values, the investigated accident variables increase too. This result is confirmed by the overwhelming majority of the investigated research studies.

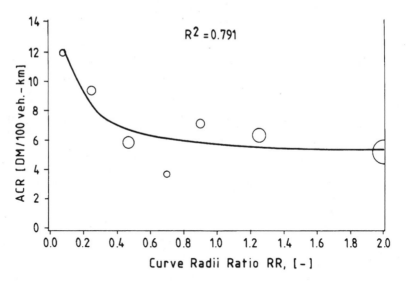

FIGURE 9.22   Accident cost rate with regard to the curve radii ratio for all accident types and *only* directly succeeding radii of curve.[476]

Unbalanced curve radii ratios have an unfavorable effect on traffic safety, especially in the case of low traffic volume and/or narrow pavement width. They also contribute to critical accident developments in combination with relatively high curvature change rates and degrees of curve, particularly on isolated curves.

Clothoid (Spiral Transition)

A 1983 FHWA study by Glennon, et al.[217] found a measurable operational benefit of clothoids. Drivers were found to position themselves in advance of the curve to effect a spiral transition. Based on computer simulation, the authors concluded that adding spiral transitions to highway curves dramatically reduces the friction demands of the critical vehicle traversals.[728]

These findings were supported by the 1991 FHWA study, which represents the first successful documentation of safety effectiveness for spiral transitions on high-speed horizontal alignments. The study showed that spiral transitions reduced curve accidents by 2 to 9 percent, depending on the degree of curve and central angle.[781] The researchers determined that a reduction of 5 percent in the total number of accidents was most representative of the effect of adding spiral transitions at both ends of a curve on two-lane rural highways.

Providing a spiral transition curve to an existing curve may be accomplished in conjunction with a routine 3R (resurfacing, restoration, and rehabilitation) project, particularly where a curve-flattening and/or curve-widening improvement is proposed.[728]

In contrast, some studies reported by O'Cinneide[553], have concluded that transition curves are dangerous because of driver underestimation of the severity of the horizontal curvature.[72,677] Stewart[678] quotes a California Department of Transportation study involving a rigorous comparison of over 200 bends, both with and without transitions curves; those with transitions had, on average, 73 percent more accidents with injuries (probability < 0.01). The Department's report "Accidents on Spiral Transition Curves in California" also recommends against any use of these curves. However, it is understood that recent studies in Germany and the United Kingdom have concluded that the impact of transition curves on safety is neutral.[553]

An interesting approach for evaluating the influence of spiral transitions (clothoids) was given by Leutzbach and Zoellmer.[476]

The tangent, circular curve, and transition curve (clothoid) are the design elements of the horizontal alignment. The transition curve allows a gradual change of the centrifugal acceleration in the transition from one curvature to the other. The transition, tangent to circular curve, has to be designed according to most design guidelines by the element sequence, tangent—transition curve—circular curve, especially for roads of category group A and for road categories B II and B III. However, several guidelines allow the immediate transition, tangent to circular curve, for larger radii of the circular curve, for example, 1500 m (for design speeds up to 80 km/h), and 3000 m (for design speeds of more than 80 km/h).[241] This is currently considered too conservative as revealed in Secs. 9.1.3.2 and 9.2.3.2 (see also Tables 12.4 and 12.8).

But the overwhelming majority of roadway sections in many countries are not built according to these standards. In practice, smaller radii of curve are often immediately combined with tangents. The impact of how the insertion of a clothoid would affect the accident situation is examined here.

Roadway sections including the design element, circular curve, were chosen for investigation. A distinction was made as to whether a tangent or a clothoid was in front of the circular curve (for both driving directions). The radii of curve were arranged into 11 classes. The last class encompassed all radii of curve of $R \geq 2500$ m.[476]

Analyses of the accident rates (Fig. 9.23) show distinct reductions with increasing radii of curve for both element sequences. In the range up to 200 m, the sequence clothoid to radius of curve is more favorable than the sequence tangent to radius of curve. The values are significantly lower at the 95 percent level of confidence. For radii of curve greater than 200 m, no systematic differences are present with respect to the element sequences. Because of the small sample size, the deviations are higher, for the sequence clothoid to radius of curve (coefficient of determination $R^2 = 0.763$) than for the sequence tangent to curve ($R^2 = 0.959$).

For accident cost rates, the sample sizes were too small to provide statistically sound results. For both sequences, a general reduction trend was observed with increasing radius of curve.

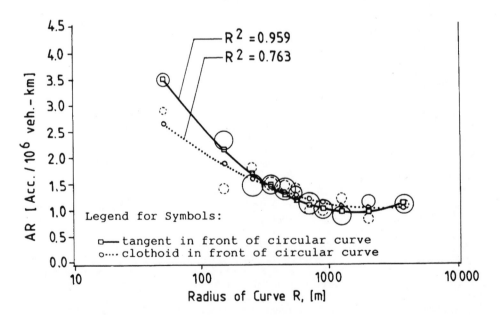

FIGURE 9.23   Accident rate with regard to the radius of curve (element sequences) for all accident types.[476]

In conclusion, we can state the following. With regard to the element sequence for the transition from tangents to circular curves, a safety gain is observed only for radii of curve less than 200 m with transition curves (clothoids) when compared to the direct sequence, tangent to circular curve. Accident rates were significantly lower for curves <200 m when using transition curves. However, for curve radii >200 m, no systematic differences were observed. Figure 9.23 does not exactly represent the statistical outcome.

The present results support the findings of the previous section, which discussed the curve radii ratio. Obviously, the introduction of transition curves between tangents and circular curves leads to statistically measurable accident rate reductions only for relatively small radii of curve. This conclusion does not contradict the statements made in Refs. 728 and 781, since no distinctions were made there between different radii of curve or degree of curve classes.

Grade

The operating speed of a vehicle is influenced by the characteristics of the vertical alignment. Trucks and buses suffer the most on grades, especially on upgrades where speed reductions may become significant (Rotach[615]). On downgrades, trucks and buses are often driven at crawl speeds to compensate for the effect of longer braking distances. On longer downgrade sections, with steep longitudinal grades, brakes may not adequately slow down a heavy vehicle traveling at a high speed and bring it to a stop. For passenger cars, longitudinal grades also lead to a variation in operating speeds but not in a manner that is as pronounced as that for trucks. It may be concluded that, with increasing longitudinal grades, an increase in the nonhomogeneity of traffic flow and consequently in passing maneuvers could increase the risk of accidents.[109]

A number of studies concerning the relationship between vertical alignment and accident risk have been done.

An investigation by Bitzl, which is cited by Pucher,[589] for German two-lane rural roads established a positive relationship between gradient and accident rate. In other words, the accident rate increases as the gradient increases. In another study related to German expressways, Bitzl[58] found a marked relationship between grade and accident rate. He indicated that steep grades of 6 to 8 percent produce over 4 times the number of accidents as gradients of less than 2 percent.

Vasilev[732] determined in the former U.S.S.R. that accident rates were especially high on steep grades. In a study that evaluated databases from Germany, Great Britain, and the former U.S.S.R., Silyanov[649] indicated that the accident rate increased as the gradient increased. Similar results were reported by Babkov[39,40] in the former U.S.S.R.

Studies cited by Pignataro[578] showed that steeper grades increase the accident rates and skidding accidents on two-lane rural curved sections.

Krebs and Kloeckner[368] analyzed accident data for two-lane rural roads in Germany. They indicated that the accident rate showed a slight increase up to grades of about 6 percent. For grades of more than 6 percent, a sharp increase in the accident rates was noted (see Fig. 9.30). Studies by the authors[406,408] indicated that grades of up to 5 percent did not have any particular effect on the accident rate.

For two-lane rural highways, the research studies reported previously have generally shown that

1. Grades of less than 6 percent have relatively little effect on accident rate.

2. A sharp increase in accident rate occurs on grades of greater than 6 percent.

Figure 9.24 illustrates the relationship between accident rate, accident cost rate, and gradient[286] for new designs and redesigns executed according to the German design guidelines.[241] From Fig. 9.24 it can be seen that

1. Longitudinal grades of between 0 and +2 percent show the most favorable results. With increasing upgrades, the accident rate gradually increases, whereas with increasing downgrades, the risk of being involved in an accident increases exponentially.

2. Between upgrades of +7 percent and downgrades of −7 percent, the accident cost rate gradually increases, which is understandable, since operating speeds are highest on steep downgrades.[286]

Recent research in the United States reveals the following:[728, 781, 782]

The vertical alignment of a highway is described by both vertical lines or grades and vertical curves which include sags and crests. In general, the three major factors which affect the design of vertical alignment of a highway are safety, terrain, and construction cost.

Studies have shown that the accident rate on downgrades is 63 percent higher than that on upgrades, assuming that upgrades have as much vehicular traffic as downgrades. Table 9.7 shows

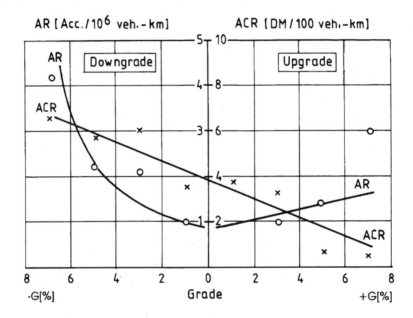

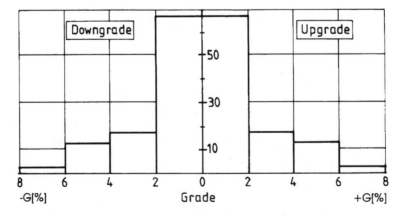

FIGURE 9.24   Accident rate and accident cost rate versus grade.[286]

TABLE 9.7   Accident Frequency and Severity by Vertical Alignment[80]

Vertical alignment	Number of accidents	Percent of total accidents	Percent injured	Percent killed
Level	2001	34.6	53.6	4.7
Upgrade	943	16.3	55.6	3.9
Downgrade	1533	26.5	58.4	5.1
Up on crest	373	6.5	59.5	6.0
Down on crest	461	8.0	62.6	5.9
Up on sag	258	4.5	57.8	6.3
Down on sag	211	3.7	61.7	6.8
Total known	5780	100.0		
Total unknown	2192			
Total	7972			

that downgrade accidents are more frequent and result in higher percentages of injuries and fatalities than upgrade accidents. Also, injury and fatality rates on vertical curves are higher than on level or upgrade locations.[80]

The same conclusion was made by Glennon,[218] who examined the results of a number of studies in the United States. They concluded that grade sections have higher accident rates than level sections, steep grades have higher accident rates than mild grades, and downgrades have higher accident rates than upgrades.

The standards of the United Kingdom[139] include a graph which compares accident rates on grades to a base accident rate by means of a multiplier. It is reported that studies on the effect of gradients on accidents[255] have shown that the accident rate can be expected to increase with steeper gradients. The fitted functions of accident rate versus gradient are shown in Fig. 9.25. It can be seen that there is a negligible increase in accidents in the uphill direction but a significant increase is indicated in the downhill direction. However, the accident rates shown are calculated as accidents per vehicle kilometer and, on that basis, short steep gradients will be subject to higher accident rates for shorter lengths of gradient (see Fig. 9.26). This suggests that the total effect of the adoption of a steeper gradient is relatively minor, although this may be an oversimplification of causal factors. It should be noted that the studies referred to were made on road sections without accesses or junctions which would be likely to exacerbate the accident risks of steep gradients.[139]

The results of Fig. 9.25 confirm, to a large extent, the investigations of Hiersche and Lamm[286] presented in Fig. 9.24, as well as Glennon's conclusions.[218]

From the foregoing the following conclusions can be drawn:

- Grades of between 0 and ±2 percent are the most favorable.

- Grades of less than 6 percent have relatively little effect on the accident rate. For grades greater than 6 percent, a sharp increase in accident rate was noted. This increase is obviously more pronounced on downgrade sections, where accident severity plays a decisive role in comparison to level and upgrade sections.

- The same negative trend seems to exist when comparing downgrade and upgrade crest and sag vertical roadway sections.

- The geometry of vertical curves does not affect accident severity, as long as adequate stopping sight distance is provided.

In addition, National Cooperative Highway Research Program (NCHRP) reports[709,710] concluded that the accident rate increases with gradient on curves.

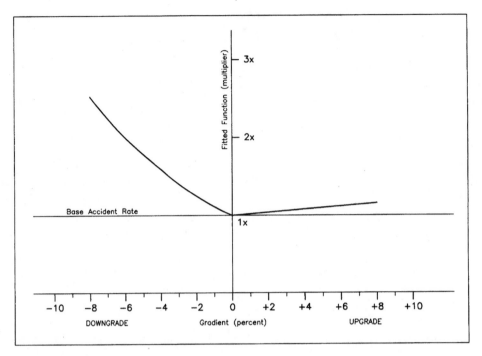

FIGURE 9.25 Accident rate versus gradient for a 10-m single carriageway.[139]

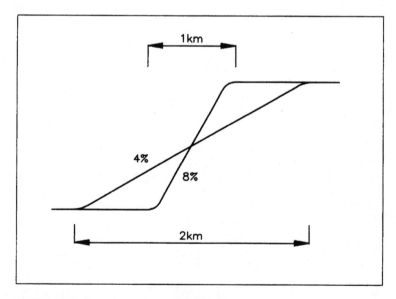

FIGURE 9.26 Alternative gradient options.[139]

Further information can be found in "Safety Considerations" in Sec. 13.2.1.1 and "Driving Behavior and Accidents" in Sec. 13.2.4.2.

Sight Distance

Sight distance, which is dependent on both horizontal and vertical alignments, is of great importance to traffic safety. In Germany, Hiersche[285] pointed out that sight distance is the most important criterion in the design of highway alignments. Krebs and Kloeckner[368] did not fully agree with that statement, but said that insufficient sight distances are the cause of many accidents. Meyer, et al.[513] stated that about one-quarter of all rural accidents result from overtaking maneuvers for which passing sight distances were not sufficient. Similar results were reported by Netzer[544] in Germany, who determined that passing maneuvers accounted for about 21 percent of all traffic accidents.

An analysis of accidents on U.S. roads by Young[770] showed that the accident rate correlated negatively with sight distance. For a sight distance of less than 240 m, the accident rate was twice as high as that for a sight distance of more than 750 m.

Sparks[666] also established a negative relationship between stopping sight distance and accident rate in the United States. Similar results were reported by Silyanov[649] in the former Soviet Union and Kunze[378] in Germany.

A U.K. study[255] reported that there is little erosion of safety resulting from sight distances below absolute minimum design standards on "clean" sites (no accesses, intersections, etc.). It was also noted that accident rates rise steeply at sight distances below 100 m.

Transportation Research Board (TRB) Special Report 214[710] stated that a study carried out in the United States under carefully controlled conditions found that accident frequencies were 52 percent higher at sites with sight reductions than at the control sites.

A Swedish study has shown that accident rates decrease with increasing average sight distance, especially for single-vehicle accidents in darkness.[23]

Another study of accidents on two-lane rural roads in Germany by Krebs and Kloeckner[368] determined the following:

1. As sight distance increases, the accident risk decreases.

2. High accident rates were associated with sight distances of less than 100 m.

3. With sight distances of between 100 and 200 m, accident rates were about 25 percent lower than those associated with sight distances less than 100 m.

4. For sight distances more than 200 m, no major decreases in accident rates were noted.[109]

Studies by the authors[104,406,408] determined that sight distances of more than 150 m did not have any particular effect on accident rates.

The research studies reported here have established that a negative relationship exists between sight distance and accident risk. This is especially true for run-off-the-road accidents and to a lesser extent for passing accidents.[476]

In a German investigation, Bitzl and Stenzel[56] reported that the frequency of accidents related to improper passing maneuvers sharply increased for sight distances of less than 400 to 600 m.

However, it can be hypothesized that other influencing parameters, such as wide pavements and gentle radii of curve, also play a part in the observed positive effect of greater sight distances on the accident situation. For narrow road sections, an increase in sight distances could favorably affect traffic safety.

Traffic Volume

The traffic volume strongly influences the traffic flow and consequently the frequency and severity of accidents. A detailed, controlled investigation of the direct effect of traffic volume would be difficult, since the specific traffic volume at the time of an accident is generally unknown. Therefore, only the average annual daily traffic (AADT) can be related to accident history.

When analyzing the relationship between AADT and accident rate, it must be noted that road sections with high traffic volumes normally have good designs (that is, wide pavements, gentle curvilinear alignments, low gradients, etc.). For low-volume roads, these influencing parameters often play a contradictory role. This fact plays an important role in the scientific investigation of the relationship between traffic volume and accidents.

Goldberg[223] investigated accidents on rural two-lane highways in France. For traffic volumes of up to 20,000 vehicles/day, he established a U-shaped distribution between accident rate and traffic volume.

Paisley[567] studied the effect of traffic volume on accident severity in Great Britain. For fatal accidents, he indicated that the accident rate decreased as traffic volume increased. However, for accidents with injuries, he indicated that the accident rate increases as the traffic volume increases. Paisley's study was based on traffic volumes of up to 10,000 vehicles/day.

Roosmark and Fraeki[614] analyzed accident types on roads in Sweden with traffic volumes of up to 11,000 vehicles/day, and they established the following results:

1. For single-vehicle accidents, the accident rate decreased as traffic volume increased.

2. For multiple-vehicle accidents, the accident rate increased as traffic volume increased.

An investigation of accidents on two-lane rural roads in Austria by Knoflacher[344] established a U-shaped distribution between accident rate and traffic volume. Accident rate was at a minimum for traffic volumes of between 6000 and 6500 vehicles/day. For traffic volumes of less than 6000 to 6500 vehicles/day, single-vehicle accidents dominated. For traffic volumes of more than 6000 to 6500 vehicles/day, multiple-vehicle accidents prevailed.

For traffic volumes of up to 10,000 vehicles/day, Lamm and Kloeckner[388] in Germany reported that the accident rate decreased as the traffic volume increased. They also indicated that the level of design correlated highly with traffic volume, a result that could explain the favorable trends in accident rates on high-volume roads.

Generally speaking, several investigations have established a U-shaped relationship between traffic volume and accident rate for multiple lane highways. For example, see Gwynn,[248] Pfundt,[574] Leutzbach, et al.,[473] and Kloeckner.[337] This relationship was also confirmed by Leutzbach, et al.[474] in studies which established, for different two-lane road categories, a U-shaped relationship between accident rate and traffic volume, based on hourly traffic volume data. The accident rate experiences a relatively high value of about 1.5 accidents/10^6 vehicle kilometers for very low traffic volumes, then decreases in the range of traffic volumes (400 to 1000 vehicles/h) to a nearly constant value of about 0.5 accidents/10^6 vehicle kilometers. This is followed by a gradual increase to about 1.0 accidents/10^6 vehicle kilometers for very high traffic volumes (1400 to 1600 vehicles/h).[476]

For multiple lane roads, Brilon[74] indicated that the U-shaped relationship is caused by the superimposition of the decreasing accident risk for run-off-the-road accidents and the increasing risk for rear-end accidents with increasing traffic volume. Furthermore, his investigation established a distinct increase of head-on accidents with increasing traffic volume on two-lane roads.

The same functional relationship was shown by Ceder,[102] based on accidents with one motorist involved or with more than two motorists involved. Carre, Lassare, and Liger[99] pointed out:

> The safety problem…is represented by the limiting ranges of the U-shaped curve at both sides: for low traffic volumes, "run off the road accidents" dominate, caused primarily by inappropriate speeds, alcohol consumption and fatigue, and for high traffic volumes, "rear-end accidents" predominate in addition to "head on accidents" in the case of two-lane rural roads.

In conclusion, the research study of Leutzbach and Zoellmer[476] summarizes the following results for two-lane rural roads. The relationship between accident rate and traffic volume was found to be a U-shaped curve. A fall in the accident rate up to a traffic volume range of 10,000 to

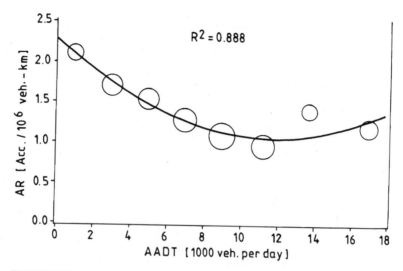

FIGURE 9.27   Accident rate with regard to the AADT for all accident types.[476]

12,000 vehicles/day was followed by an increase for higher traffic volume ranges (Fig. 9.27). In contrast, there is a constant reduction in the accident cost rate with increasing AADT (Fig. 9.28). These relationships can also be illustrated in a similar manner for run-off-the-road and passing/rear-end accidents, as well as for different daylight and roadway conditions.

In summary, accident rates decrease with traffic volume increases until the volume reaches about 11,000 vehicles/day. Volumes in excess of 11,000 vehicles/day drive rates upward. Accident cost rates, however, decrease consistently as volumes increase.

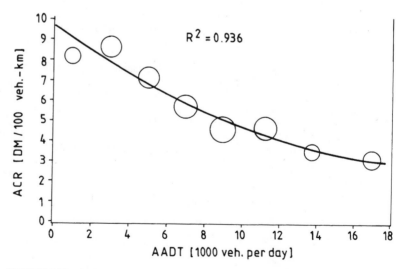

FIGURE 9.28   Accident cost rate with regard to the AADT for all accident types.[476]

TABLE 9.8   Accident Rate and Accident Cost Rate versus Design Speed[286]

Design speed, km/h	60	70	80	100
Vehicle kilometers (absolute), 10^6 vehicle kilometers	52.3	18.0	234.5	52.0
Vehicle kilometers traveled, %	14.6	5.2	65.7	14.5
Accident rate, accident/10^6 vehicle kilometers	2.12	1.78	1.15	1.11
Accident cost rate, dm/100 vehicle kilometers	3.62	3.85	4.21	5.21

Design Speed

In addition to studying the effects of design and traffic volume-related variables on traffic safety, Hiersche and Lamm[286] also studied the effect of design speed on accident rates and accident cost rates. Table 9.8 presents an overview of their results for new alignments designed according to the German design guidelines.[241] An examination of Table 9.8 reveals that the accident rate decreases as design speed increases from 60 to 80 km/h. However, for design speeds greater than 80 km/h, the accident rate did not decrease.

On the other hand, the accident cost rate increased as the design speed increased throughout the range. This is understandable, assuming that for higher design speeds a more generous road design capable of supporting operating speeds was in place. These higher operating speeds led to more severe accidents.

Therefore, an accident rate decrease does not automatically mean an improvement in the overall accident situation. Similar effects with regard to increasing operating speeds were observed for the accident cost rate:

- In Fig. 24.10 of Part 3, "Cross Sections," with respect to increasing pavement widths
- In Fig. 9.24 with respect to increasing downgrades

These findings are interesting and decisively contradict the conventional wisdom of today. But further research, including larger databases, is clearly needed on these subjects before definite conclusions can be drawn.

With respect to speed in general and its relationship with traffic safety, O'Cinneide[553] reported the following: Speed is one of the major parameters in geometric design, and safety is synonymous with accident studies. For example, Finch, et al.[194] recently concluded that a reduction of 1.6 km/h (1 mi/h) in the average speed reduces the incidence of injuries by about 5 percent. It is also generally accepted that there are substantial safety benefits from lower speed limits. For example, reducing rural speed limits from 100 km/h to 90 km/h has been predicted to reduce casualties by about 11 percent.[192] It is interesting to note that the relationship between the design speed and the speed limit is not referred to in the geometric design standards of many countries.[552]

With respect to the latter statement, the German guidelines,[243,246] the Greek guidelines,[453] and the present Part 2: Alignment (Table 6.2) provide a classification table of roads in order to adjust, among other things, permissible speed limits and design speeds. It is safe, however, to say that no one has been able to establish a concrete relationship between the design parameter, design speed, and the operational parameter, speed limit, at least from a traffic safety point of view.

*Superimposition of Design Parameters/Breakpoint on Safety**

A number of researchers have used regression analyses to obtain quantitative estimates of the effects produced by design and traffic volume parameters on accident rates.[132,198,272, 531,634,660,735,766]

---

*Elaborated based on Ref. 390.

However, their findings do not provide any practical applications, and do not yield any clue to the level of design parameters, such as curvature change rate, above which improvements in traffic safety become particularly important.

Research by Lamm and Choueiri[104,406,408] to analyze the joint effect of several parameters—degree of curve/curvature change rate of the single curve, length of curve, superelevation rate, gradient of up to 6 percent, sight distance, lane width, shoulder width, and AADT—on operating speeds and accident rates, determined that DC and CCR_S explained most of the variation in the expected operating speeds and accident rates on curved sections of two-lane rural highways. On the basis of this research, recommendations for good, fair, and poor design practices are presented in this book, based on three safety criteria.

Studies conducted at the Institute for Highway and Railroad Engineering at the University of Karlsruhe, Germany[368,385,390] yielded the results shown in Figs. 9.29 to 9.32. The database covered 4 years of accidents (14,200 accidents) on 1162 km of two-lane roadways in western Germany. Figures 9.29 and 9.30 show that there is a definite correlation between accident rate and curvature change rate, radius of curve, pavement width, gradient, and sight distance. However, the results in Figs. 9.29 and 9.30 do not yield a clue to the level of design parameters above which improvements in traffic safety become particularly important. In other words, the results do not present a minimum level of safety or a maximum level of unsafety.[390]

A number of studies conducted in Europe and the United States have shown that curvature change rate correlates highly with operating speeds and accident rates (see "Curvature Change Rate and Radius of Curve" in Sec. 8.2.3.2 and "Curvature Change Rate, Degree of Curve, Length of Curve, and Curve Radii Ratio" in Sec. 9.2.1.3). As shown in Fig. 9.31, an increase in the curvature change rate leads to a decrease in operating speeds. Furthermore, for new designs, Lamm[385] indicated that curvature change rates greater than or equal to 250 gon/km produced relatively high accident rates. It should be noted that for new designs and redesigns in Germany, curvature change rates greater than 250 gon/km are used in very few cases (see Figs. 9.13 and 9.31).

Lamm[390] concluded: If a curvature change rate of 250 gon/km is the highest acceptable level of curvature in modern geometric design guidelines, then the accident rate corresponding to this curvature change rate should be regarded as the breakpoint between levels of safety and unsafety. On the basis of the data in Figs. 9.13 and 9.29, an accident rate of 2.0 accidents/million vehicle kilometers corresponds to a curvature change rate of approximately 250 gon/km. Furthermore, Lamm suggested that design parameters, such as radius of curve, lane width, and gradient, should be laid out during the design stages (for new designs or redesigns) in such a way that when the road is in full operation, the accident rate should not be allowed to exceed 2.0 accidents/million vehicle kilometers; this corresponds to a level of safety/unsafety of $1 - (2 \times 10^{-6}) = 0.999998$, or a 99.9998 percent chance that an accident will not occur.

Despite this high percentage, Lamm indicated that hazardous situations could still be expected. The following example illustrates the point. For a safety level of 100 percent, AADT of 10,000 vehicles/day, it can be estimated that 3,650,000 ($365 \times 10,000$) vehicles/y would pass a section kilometer without being involved in accidents. However, according to the proposed level of safety/unsafety (99.9998 percent), only 3,649,993 vehicles would pass safely; that means, 7 vehicles/y would be involved in collisions on the observed section kilometer.

From Figs. 9.29 and 9.30, an accident rate of 2.0 accidents/million vehicle kilometers corresponds to

- Pavement width of about 6.5 m
- Radius of curve of about 350 m
- Longitudinal grade of about 6.5 percent
- Sight distance of about 100 m

Falling short of or exceeding these values could result in an accident rate greater than 2.0 accidents/million vehicle kilometers. Therefore, it should be noted that modern highway geometric design considerations seem to meet minimum safety requirements, when they recommend radii of

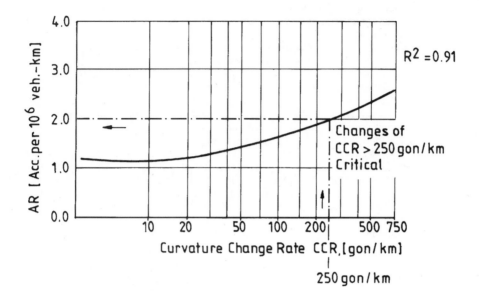

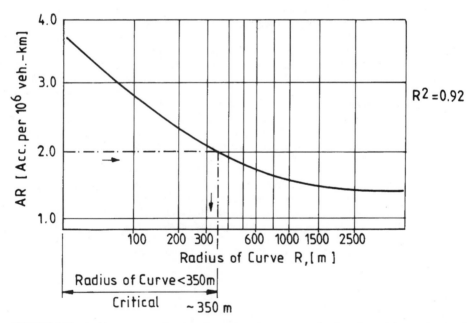

FIGURE 9.29 Accident rate as a function of curvature change rate and radius of curve for two-lane rural roads.[368,390]

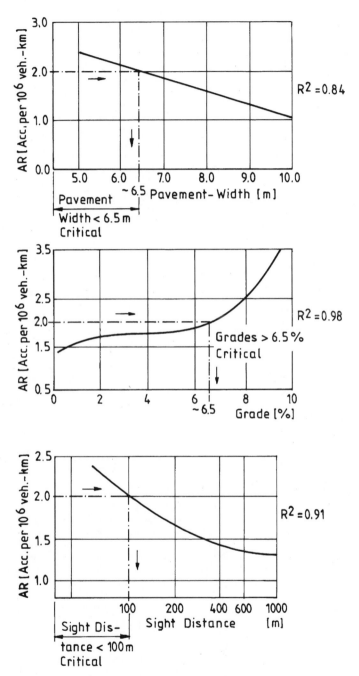

FIGURE 9.30   Accident rate as a function of pavement width, grade, and sight distance for two-lane rural roads.[368,390]

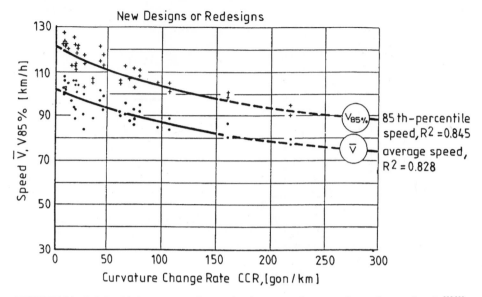

FIGURE 9.31 Relationship between operating speed and curvature change rate for two-lane rural roads.[385,390]

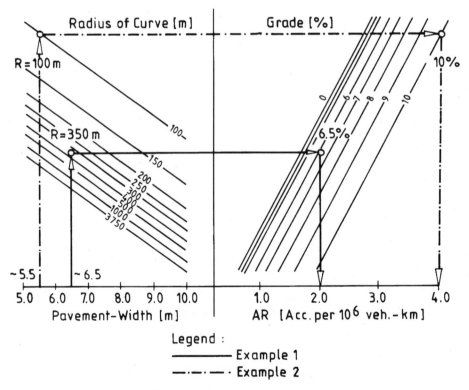

FIGURE 9.32 Nomogram to determine the accident rate by pavement width, radius of curve, and longitudinal grade for two-lane rural roads.[368,390]

curve of >350 m, pavement widths ≥6.5 m, maximum grades of ≤6 percent, and stopping sight distances ≥100 m for design speeds of between 80 km/h and 90 km/h. These values should be regarded as approximations.

For future designs, the use of radii of curve of less than 350 m should be carefully considered. For safety reasons, curvature change rates greater than 250 gon/km, pavement widths less than 6.5 m, gradients more than 6.0 percent, and sight distances less than 100 m should be avoided. For responsible agencies, policy makers, and designers, the question here becomes: Which is more important—safety, economics, or the environment? This decision must be made on a case-by-case basis.

Up to now, the risk of an accident, as measured by the accident rate, was regarded as a function of single design parameters. Since the accident situation cannot be described by just one design parameter, as noted previously, Krebs and Kloeckner[368] studied the joint impact of several design parameters, including radius of curve, lane width, gradient, sight distance, and traffic volume on accident rate. Sight distance was later removed from the analyses because it correlated highly with radius of curve. The same was true for curvature change rate. Traffic volume was also excluded because it did not affect the accident rate significantly. As a result of the study, Fig. 9.32 was developed. Examination of Fig. 9.32 leads to the following interesting results:

- Use of the design parameters recommended here (Example 1 in Fig. 9.32), which should at least be present in cases of new designs or redesigns from a safety point of view, results in an accident rate of 2.0 accidents/million vehicle kilometers.

- Use of minimum design standards (Example 2 in Fig. 9.32) results in an accident rate twice as high as that recommended.

Once again, the proposed breakpoint in safety, as related to the accident rate of 2.0 accidents/million vehicle kilometers, was obtained even when several design parameters were superimposed.

Originally, this value was based on a curvature change rate of 250 gon/km for new designs and redesigns (see Fig. 9.29). Converting a CCR value of 250 gon/km into a degree of curve value, according to Table 8.4, leads to DC ≈ 7°/100 ft. As can be seen from Fig. 9.15, this value corresponds to an AR ≈ 4.0 accidents/million vehicle kilometers for the U.S. database and to an AR ≈ 1.5 accidents/million vehicle kilometers for the German database. Considering that both databases consist of large portions of old alignments and the German database represents only run-off-the-road accidents, the recommended breakpoint in safety of 2.0 accidents/million vehicle kilometers can be considered as reasonable for modern highway design practices.

It should be noted that one unfavorable design parameter should not be superimposed on others. By correctly applying Fig. 9.32, a high accident rate resulting from one unfavorable design parameter could at least be taken care of partially by proper selection of the other design parameters.

The authors do not intend[109,390] to give the impression that the limiting values proposed here are generally valid, since the study was mainly related to the influence of the road itself on traffic safety. As noted earlier, there are many factors that may exhibit a measurable influence on driving behavior and traffic safety on two-lane rural highways. These include, but are not limited to human factors, physical features of sites, presence and action of traffic, legal issues, environmental factors, and vehicle deficiencies. All of these constitute a complex mix of various causes of traffic accidents, of which the road itself represents only one factor, but a very important one.

Because accident cost rate, which is an indicator of accident severity, was shown in the present study to increase with increasing design speed and pavement width as well as with increasing downgrade, the authors propose that further studies with large databases should be conducted in this field.

9.2.1.4 Classification System for Safety Criteria I and II Based on Accident Research.
Research that evaluated the impact of design parameters (degree of curve/curvature change rate, length of curve, superelevation rate, lane width, shoulder width, sight distance, gradient, and posted speed) and traffic volume on operating speeds and accident variables on two-lane curved high-

TABLE 9.9 Regression Equations for 85th-Percentile Speeds and Accident Rates[408,412,449,673]

Germany (West)
For 3.50 m: $V85 = 60 + 39.70\ e^{(-3.98 \times 10^{-3}\ CCR_S)}$
For 3.00 m: $V85 = 60 + 36.85\ e^{(-5.21 \times 10^{-3}\ CCR_S)}$
For ≥3.25 m: $AR^* = -0.18 + 6.4 \times 10^{-3}\ CCR_S$; $R^2 = 0.33$
For <3.25 m: $AR^* = -0.31 + 9.4 \times 10^{-3}\ CCR_S$; $R^2 = 0.35$

United States
For 3.60 m: $V85 = 95.60 - 0.0438\ CCR_S$; $R^2 = 0.82$
For 3.00 m: $V85 = 89.04 - 0.0448\ CCR_S$; $R^2 = 0.75$
For 3.60 m: $AR = -0.341 + 0.0185\ CCR_S$; $R^2 = 0.73$
For 3.00 m: $AR = -0.639 + 0.0259\ CCR_S$; $R^2 = 0.30$

*only ROR accidents

Note: $V85$ = expected 85th-percentile speed, km/h; CCR_S = curvature change rate of the single curve, gon/km; R^2 = coefficient of determination; AR = expected accident rate, accidents/10^6 vehicle kilometers.

way sections in New York State and Germany demonstrated that the most successful parameters in explaining much of the variability in operating speeds $V85$ and accident rates AR were the design parameters, degree of curve and curvature change rate. Based on this knowledge, the new design parameter, curvature change rate of the single curve (CCR_S), was developed in Sec. 8.2.3.6.

For the following investigation the U.S. database* consisted of 261 two-lane rural curved highway sections while the German database* consisted of 204 sections, some of which contained tangents between two curves.

The relationships between operating speed $V85$, accident rate AR, and curvature change rate of the single curve CCR_S are quantified for different lane widths by the regression models presented in Table 9.9, and schematically shown in Fig. 9.33 for Germany (West). Similar results were found in the United States (Table 9.9 and Fig. 9.34).

Generally speaking, the results of Figs. 9.33 and 9.34 show a certain degree of agreement. It should be noted that the U.S. accident rates are related to all accidents, whereas the German ones are related only to run-off-the-road (ROR) accidents.

This difference helps to explain the lower accident rates in Germany. Related to research by the main author, normally an additional length of 100 to 150 m in front and behind the curve is included in the accident analysis and evaluation of curved sites, since increased ROR accidents could be observed on these section lengths, which are caused by the curve. In contrast, while ROR accidents in tangents can be more or less neglected, there are other types of major importance.[628,632]

Thus, once more it could be shown by Figs. 9.33 and 9.34, as conclusion to the previous sections, that operating speeds decrease with increasing CCR_S values, while accident rates increase with increasing CCR_S values.

When searching for reliable safety ranges, it should be noted that differences in the CCR_S values of about 180 gon/km between successive curves and between long (independent) tangents and curves correspond more or less to differences in the 85th-percentile speeds of about 10 km/h, as the average operating speed background of Fig. 8.12b reveals. In this respect, for good pavement

---

*The databases contain roadway sections with grades up to 5 to 6 percent and traffic volumes between 1000 and 10,000 vehicles/day.

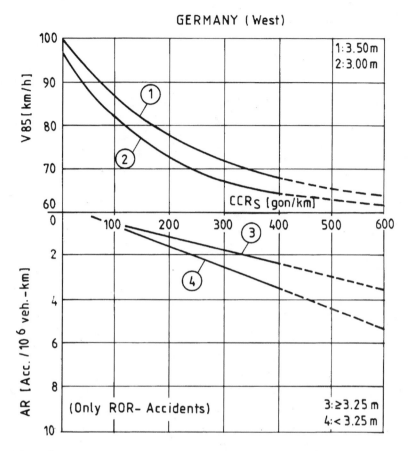

FIGURE 9.33 Nomogram for evaluating operating speeds and ROR accident rates as related to the curvature change rate of the single curve, Germany (West).[243,449,673]

conditions and grades of up to 4 percent, Al-Masaeid, et al.,[18] in a study conducted in Jordan, indicated that the CCR_S value corresponding to a speed reduction of 10 km/h is 185 gon/km.

For differences in CCR_S values of about 360 gon/km, differences in the 85th-percentile speeds of about 20 km/h can be expected. These findings are confirmed by Fig. 8.12b in particular, and also approximately by Fig. 9.34.

These relationships were taken as the basis for the speed ranges of safety criteria I and II in the following sections (see Tables 9.11 and 9.13).

As mentioned previously, it is evident from Figs. 9.33 and 9.34 that accident rates increase with increasing CCR_S values. Thus, it can be concluded that with increasing differences in the

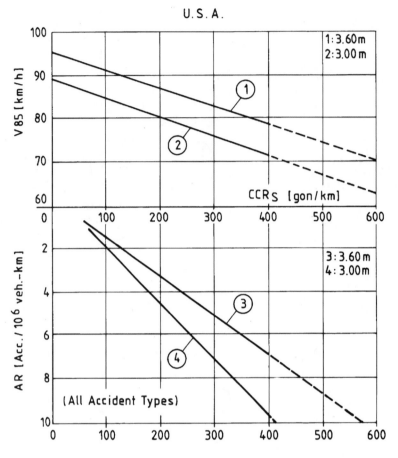

FIGURE 9.34   Nomogram for evaluating operating speeds and accident rates as related to the curvature change rate of the single curve, United States.[408,412,449,673]

CCR_S values between successive design elements, greater differences in the accident rates can be expected.

In order to get a better overview of the accident situation, the curvature change rate of the single curve was broken down in Table 9.10 into different design classes (CCR_S classes) for the existing databases from Germany[449,673] and the United States.[408,412] For every design class, a mean accident rate was calculated. The selected ranges of the CCR_S classes from 180 to 360 gon/km and from 360 to 550 gon/km go back to the original investigations in the United States,[406,408] which were related to the U.S. design parameter degree of curve. The conversion of the original ranges of the DC classes (ΔDC = 5 to 10 deg./100 ft. and ΔDC = 10 to 15 deg./100 ft.) leads, according to Table 8.4, to the selected CCR_S classes in Table 9.10.

TABLE 9.10 t-Test Results of Mean Accident Rates for Different CCR_S Classes for Germany (West) and for the United States[408,412,449,673] (see also Tables 10.12 and 18.14)

CCR$_S$/design classes, gon/km	Mean AR	$t_{calc.}$	$t_{crit.}$	Significance	Remarks
Germany (204 two-lane rural test sections)					
tangent (0)	0.35*				Considered as:
		5.20 > 1.99		Yes	
35–180	0.68*				Good design
		6.9 > 1.99		Yes	
>180–360	2.28*				Fair design
		—			
>360–550	3.84*†				Poor design
		—			
>550–900	6.39*†				Poor design
United States (261 two-lane rural test sections)					
tangent (0)	1.17				Considered as:
		4.00 > 1.96		Yes	
35–180	2.29				Good design
		7.03 > 1.96		Yes	
>180–360	5.03				Fair design
		6.06 > 1.99		Yes	
>360–550	10.97				Poor design
		3.44 > 1.99		Yes	
>550–990	16.51				Poor design

*Including run-off-the-road accidents only.

†No mean accident rate because of an insufficient database, estimated from the overall regression models for Germany according to Table 9.9.

Note: AR = accident rate, accidents/10^6 vehicle kilometers.

As shown in Table 9.10, the t-test results indicate significant increases (at the 95 percent level of confidence) in the mean accident rates between the different CCR$_S$ classes compared. This means that higher accident rates can be expected with higher CCR$_S$ classes, despite the stringent traffic warning devices often installed at curve sites.

The results of Table 9.10 are fully supported by Glennon[218] who reported that the average accident rate for highway sections which include horizontal curves is at least 3 times that for horizontal tangents.

The findings in Table 9.10 indicate, based on the observed accident situation of two databases in different continents:

1. Gentle curvilinear horizontal alignments consisting of tangents or transition curves, combined with curves up to CCR$_S$ ≤ 180 gon/km (R > 350 m), experienced the lowest average accident risk. They are classified here as "good design."

2. The accident rate on sections with a change in CCR$_S$ values between 180 and 360 gon/km (R ≈ 175 to 350 m) was at least twice as high as that on sections with a change in CCR$_S$ values up to 180 gon/km. They are classified here as "fair (tolerable) design."

3. The accident rate on sections with a change in CCR$_S$ values between 360 and 550 gon/km was about 4 to 5 times higher than that on sections with a change in CCR$_S$ values up to 180 gon/km. They are classified here as "poor design."

4. For changes in CCR$_S$ values greater than 550 gon/km (R < 115 m), the average accident rate was even higher, based on the available databases (Table 9.10).

Note that the previously recommended breakpoint in safety of 2.0 accidents/10^6 vehicle kilometers lies somewhere in the range of good design (Table 9.10), considering that the German database contains ROR accidents only. This means that this value is fully confirmed by the results of Table 9.10. Thus, it seems reasonable that this value should not be exceeded in cases of new designs and redesigns of existing alignments.

Furthermore, the CCR_S classes of Table 9.10 and the resulting radii of curve (converted according to Table 8.4 without considering transition curves) agree well with the knowledge about accidental developments with respect to "Radius of Curve" in Sec. 9.2.1.3. For instance, Krebs and Kloeckner[368] found that road sections with radii of curve less than 200 m have an accident rate which is at least twice as high as that on sections with radii greater than 400 m. For radii greater than 400 m, the gain in safety is relatively small. The same is true regarding the results of Table 9.10, issues 1 and 2.

These findings are once again substantiated by a German research report in this field.[476] The research agrees with conclusions previously drawn (see Figs. 9.7 and 9.8) for both accident rates and accident cost rates.

Furthermore, according to the U.K. Department of Transport,[139] horizontal curves have higher accident rates than tangent sections. This difference becomes apparent for CCR_S values greater than 65 gon/km (or radii less than 1000 m). This corresponds to the lower boundary (35 gon/km) of the good design range shown in Table 9.10. In other words, the U.K. findings also support the results of this table.

Based on the presented results of accident research, it can be assumed that the proposed CCR_S ranges of Table 9.10 represent a sound classification system for the arrangement of good, fair, and poor design practices in modern highway geometric design.

9.2.2   Safety Criterion I: Achieving Design Consistency

All reviewed highway geometric design guidelines indicate that the design speed should be constant along longer roadway sections. Research investigations have shown that the driving behavior on an observed road section often exceeds the design speed, on which the original design of the road section was based, by substantial amounts. This is particularly true for lower design speeds. Therefore, harmonizing design speed and operating speed is an important safety criterion that should be considered in the design, redesign, and/or rehabilitation processes of two-lane rural highways and networks.

To achieve this goal, the 85th-percentile speed, $V85$, of every independent tangent* or curve must be tuned with the existing or selected design speed, V_d, in the course of the observed roadway section according to the recommended design speed criteria presented in Table 9.11.

Safety criterion I is always related to the individual design elements (curves or independent tangents). On the basis of the logically defined ranges for good, fair, and poor designs, as shown in Table 9.10 for the new design parameter, CCR_S, the classifications shown in Table 9.11 are proposed as safety criterion I for a sound coordination of operating speed $V85$ and design speed V_d. Table 9.11 is valid for new designs, as well as for examining existing alignments in the case of redesigns or RRR projects.

The recommended ranges for good, fair, and poor designs in Table 9.11 were used as the basis for Table 9.1, although to simplify matters, only the ranges of the 85th-percentile speeds and the design speed are designated. This is possible, since the investigations in the previous section and many case studies in Sec. 12.2.4.2 and Sec. 18.4.2 have proven that the CCR_S ranges and the speed ranges in Table 9.11 correspond to one another within relatively narrow limits.

---

*An independent tangent is classified as one that is long enough to be regarded in the curve-tangent-curve design process, as an independent design element, while a short tangent is called nonindependent tangent and can be ignored.[411] For defining and classifying independent tangents, see Sec. 12.1.1.3.

TABLE 9.11 Classification System for Safety Criterion I

Safety criterion I: Design consistency

Case 1: Good design level

The permissible differences are

$| CCR_{Si} - \overline{CCR_S} | \le 180$ gon/km

$| V85_i - V_d | \le 10$ km/h

Case 2: Fair design level

The tolerated differences are

180 gon/km $< | CCR_{Si} - \overline{CCR_S} | \le 360$ gon/km

10 km/h $< | V85_i - V_d | \le 20$ km/h

Case 3: Poor design level

The nonpermissible differences are

$| CCR_{Si} - \overline{CCR_S} | > 360$ gon/km

$| V85_i - V_d | > 20$ km/h

Note: $CCR_{Si} = CCR_S$ value of the design element i, gon/km; $\overline{CCR_S}$ = average curvature change rate of the single curve for the observed roadway section without regarding tangents, gon/km, based on "Existing (Old) Alignments" in Sec. 9.2.2.1, Eq. (9.12); $V85_i$ = 85th-percentile speed of the design element i, km/h; and V_d = design speed for the observed roadway section, km/h.

9.2.2.1 *Determination of Sound Design Speeds.*

As noted earlier, a number of researchers have indicated that the design speed and the 85th-percentile speed have to be coordinated with each other, especially on wet road surfaces. This requires a fine-tuning between the variables: road characteristics, driving behavior, and driving dynamics. Because there is a tendency for a considerable number of drivers to drive faster than the assigned design speed, dangerous situations may arise, especially on roadway sections that were planned with low design speeds. This is particularly true in case of redesigns or RRR projects resulting in widened cross sections but keeping unfavorable existing alignment designs. In these cases, structural or geometric inconsistencies could impair the driver's behavior and could negatively affect traffic safety.

In order to avoid these types of dangerous situations, three safety criteria are developed in Chaps. 9 and 10 in order to enable the design engineer to achieve a uniform road characteristic, in the case of new designs or redesigns, by avoiding future accident black spots during the predesign stages.

New Alignments

The design speed, V_d, depends on environmental and economic conditions based on the assumed network function of the road and the desired traffic quality, and must be selected for a specific road category from Table 6.2. Such an adjustment is recommended for an interactive design process, and is considered for new designs regarding the interrelationships between the design parts, "Network, Alignment, and Cross Sections." However, to avoid design deficiencies, the procedure proposed in the next section is also recommended for examining new designs.

Existing (Old) Alignments

Case studies have revealed, for existing alignments, that the assumptions of the previous chapter are difficult to achieve because old alignments were normally not constructed for an exactly defined design speed. Thus, in the case of redesigns and RRR strategies, achieving fine-tuning between network function, traffic quality, design speed, and 85th-percentile speed can often lead to unfavorable results with respect to economic, environmental, and safety issues. This conclusion is based on new experiences at the Institute for Highway and Railroad Engineering, University of Karlsruhe, Germany.[460,462,628] Therefore, a new procedure for determining sound design speeds for existing (old) alignments had to be developed.

In this connection, it is possible to estimate for existing alignments a justifiable average curvature change rate of the single curve, $\overline{CCR}_S$, based on a length-related calculation of the average of the CCR_{Si} values for the individual curves along the observed roadway section without considering tangent sections. Thus

$$\overline{CCR}_S = \frac{\sum\limits_{i=1}^{i=n} (CCR_{Si} L_i)}{\sum\limits_{i=1}^{i=n} L_i} \tag{9.12}$$

where $\overline{CCR}_S$ = average curvature change rate of the single curve for the observed roadway section without considering tangents, gon/km

CCR_{Si} = curvature change rate of the single curve i, gon/km

L_i = length of curve i, m (see the case study in Table 9.12)

This average $\overline{CCR}_S$ value can be taken as the basis for determining an average 85th-percentile speed, $\overline{V}85$, based on the respective operating speed background of the country under study in Fig. 8.12 or Table 8.5. This so-called average $\overline{V}85$ will be considerably exceeded in the case of large radii of curve or independent tangents. However, in the case of small radii of curve, $V85$ will seldom be achieved. Since the design speed, V_d, should be constant on longer sections, or at least on longer connected roadway sections, it makes sense to regard the estimated average 85th-percentile speed of the observed existing alignment as the basis for the selection of a reliable design speed for future redesigns or RRR strategies.

In doing so, it can be assumed that, for this design speed, over- and underdimensioning of existing elements can be avoided and that even they can be adapted to each other to a certain extent. This is one of the reasons that the ranges for safety criterion I between $V85$ and V_d in Tables 9.1 and 9.11 were set in absolute symbols, in contrast to the German considerations, which assume that the 85th-percentile speed normally exceeds the design speed. This case is normally only true for new alignments.

That means when taking the average 85th-percentile speed as an estimation for the design speed, it can be assumed that redesign of existing alignments can be optimized at least up to a certain extent from the viewpoints of economic, environmental, and safety-related issues.

For a better understanding, a case study is presented in Table 9.12a. The old existing alignment in Greece consists of three curves combined with two independent tangents. The average $\overline{CCR}_S$ value was found to be about 250 gon/km; this corresponds to an average $V85 \approx 82$ km/h from the Greek operating speed background in Fig. 8.12 or Table 8.5.

It was decided to select a design speed of $V_d = 90$ km/h because of the long independent tangents, where high operating speeds can be expected. In cases of more curvilinear alignment designs, the selection of the lower design speed level of $V_d = 80$ km/h would have been more appropriate.

As a result of the evaluation process of safety criterion I (Table 9.12a, Col. 7), it can be shown that the speed ranges of the independent tangents fall into the range of good design practices and are well coordinated with the selected design speed of 90 km/h (see Tables 9.1 and 9.11).

With respect to safety criterion II (which is discussed in detail in Sec. 9.2.3.1; see also Table 9.13), the operating speed transitions between curve element 1 and independent tangent 2 fall into the fair range, whereas between independent tangent 4 and curve element 5, they fall into the poor range (see Table 9.12a, Col. 8). As possible solutions, for example, in the case of future redesigns, the following measures are proposed:

TABLE 9.12 Estimation of Reliable Design Speeds and Safety Evaluation Processes (Example: Greece)[460]

(a) Existing (old) alignment							
Element no.	Design parameters,* m	Length L_i, m	CCR_{Si}, gon/km	$V_d \approx \overline{V85}$, km/h	$V85_i$,** km/h	Safety criterion I $\lvert V85_i - V_d \rvert$, km/h	Safety criterion II† $\lvert V85_i - V85_{i+1} \rvert$, km/h
(1)	(2)	(3)	(4)	(5)	(6)	(7)	(8)
1	$R = 245$	155	260	90	81	9 (good)	
							17 (fair)
2	$R = \infty$	510‡	0	90	98	8 (good)	
							10 (good)
3	$R = -425$	195	149	90	88	2 (good)	
							10 (good)
4	$R = \infty$	555‡	0	90	98	8 (good)	
							26 (poor)
5	$R = 145$	100	439	90	72	18 (fair)	

(b) Redesign of the existing alignment							
Element no.	Design parameters, m	Length§ L_i, m	CCR_{Si}, gon/km	$V_d \approx \overline{V85}$, km/h	$V85_i$, km/h	Safety criterion I $\lvert V85_i - V_d \rvert$, km/h	Safety criterion II $\lvert V85_i - V85_{i+1} \rvert$, km/h
(1)	(2)	(3)	(4)	(5)	(6)	(7)	(8)
1	$R = 400$ $A = 150$	225 56	143	90	88	2 (good)	
							10 (good)
2	$R = \infty$ $A = 150$	405 53	0	90	98	8 (good)	
							8 (good)
3	$R = -425$ $A = 150$	142 53	118	90	90	0 (good)	
							8 (good)
4	$R = \infty$ $A = 150$	409 56	0	90	98	8 (good)	
							10 (good)
5	$R = 400$	248	144	90	88	2 (good)	

*No transition curves present.

** $V85_i$ based on the operating speed background of Greece in Fig. 8.12 or Eq. (8.16) in Table 8.5: $V85_i = 10^6/(10150.1 + 8.529\ CCR_{Si})$.

†According to Tables 9.2 or 9.13.

‡Independent tangents according to Case 2 in "Evaluation of Tangents in the Design Process" in Sec. 12.1.1.3, $V85_i \rightarrow \overline{CCR}_{Si} = 0$.

§Rounded.

Note: For part *a:*

$$\overline{CCR}_S = \frac{155\ 260 + 195\ 149 + 100\ 439}{155 + 195 + 100} \approx 250\ \text{gon/km} \rightarrow \overline{V85} \approx 82\ \text{km/h} \Rightarrow V_d = 80\ \text{km/h or} \Rightarrow V_d = 90\ \text{km/h}.$$

and for part *b:*

$$\overline{CCR}_S = \frac{143\ (225 + 56) + 118\ (53 + 142 + 53) + 144\ (56 + 248)}{281 + 248 + 304} \approx 136\ \text{gon/km}$$

$$\rightarrow \overline{V85} \approx 89\ \text{km/h} \Rightarrow V_d = 90\ \text{km/h}$$

1. A curvilinear alignment to replace the independent tangents. In this case, the radius of element 3, for example, may be taken as a basis to select favorable radii of curve relations from Figs. 9.1 and 9.36 to 9.40 with respect to the country under study. However, this solution may be too expensive since large portions of the alignment have to be changed.

2. Select a design speed of $V_d = 90$ km/h, and accordingly larger radii of curve for the design elements 1 and 5 in order to achieve sound operating speed transitions between these elements and the independent tangents. As discussed in Sec. 9.1.3.2, at least radii of curve $R \geq 400$ m combined with transition curves have to be selected after independent tangents for the majority of investigated countries (see also Table 12.4).

Of course, other possibilities exist to improve the existing alignment in terms of safety-related issues. Administrative, economic, and environmental demands may play an important additional role here.

The results presented in Table 9.12a allow quantitative statements for the individual safety criteria about good, fair, and poor design practices for the investigated (old) existing alignment, which support the search for a favorable redesign.

Once fair or poor designs along the length of an existing alignment are recognized (Table 9.12a), the question now becomes "Which appropriate redesigns are relevant?" This question is directly related to the favorable selection of a reliable design speed for the overall observed roadway section which normally consists, besides poorly designed segments, of good and/or fair ones. The responsible Highway Department in Greece required the designers to maintain at least portions of the independent tangents and to flatten the critical circular curves, including transition curves, in such a way that a sound alignment, which represents good design levels according to safety criteria I and II would be achieved. For the redesign, the design speed of $V_d = 90$ km/h was assumed to be appropriate and radii of curve of $R = 400$ m were selected for the design elements 1 and 5.

Accordingly, as shown in Table 9.12b, the redesign of the old alignment was developed. As can be seen, the safety evaluation process does provide good design levels for safety criteria I and II. That means, a well-tuned speed behavior can be expected on individual curved sites, in addition to balanced operating speed transitions between successive design elements. Furthermore, the redesigned alignment confirms the selection of a design speed of $V_d = 90$ km/h.

Based on a questionnaire which was sent to international colleagues, it has been estimated that at least between 70 and 80 percent of the two-lane rural networks worldwide consist of (old) alignments for which no clear indication exists about the design speed. That means the majority of the two-lane rural roads do correspond to the previous considerations—a magnitude which should not be underestimated.

It is recommended that this procedure be used even for examining the selected design speed in case of new designs, in order to avoid any design deficiencies from the beginning and to yield sound economic, environmental, and safety-related alignments.

Further case studies are presented in Sec. 12.2.4.2 and in Sec. 18.4.2.

9.2.3 Safety Criterion II: Achieving Operating Speed Consistency

9.2.3.1 Classification System. Based on the preceding discussions, analyses, and evaluations about the driving behavior, expressed by the 85th-percentile speeds, and the accident situation, expressed by the accident rates, the ranges of Table 9.13 are considered reasonable for safety criterion II for the classification of good, fair, and poor design levels. The comparison between the assumed CCR_S classes, based on accident research (Table 9.10), and the corresponding average 85th-percentile speed ranges, based on Fig. 8.12b reveals that the relationships in Table 9.13 are well coordinated with each other. This criterion was developed for achieving operating speed consistency in horizontal alignment, and is related to the transition between two succeeding design elements.

The calculation process for the CCR_{Si} values is described in Sec. 8.1.3. (Compare also Figs. 8.1 and 8.2.) Based on the CCR_{Si} values, the corresponding $V85_i$ values can be determined for the

TABLE 9.13 Classification System for Safety Criterion II

Safety criterion II: Operating speed consistency

Case 1: Good design level

The permissible differences are

$| \text{CCR}_{S_i} - \text{CCR}_{S_{i+1}} | \leq 180$ gon/km

$| V85_i - V85_{i+1} | \leq 10$ km/h

Case 2: Fair design level

The tolerated differences are

180 gon/km $< | \text{CCR}_{S_i} - \text{CCR}_{S_{i+1}} | \leq 360$ gon/km

10 km/h $< | V85_i - V85_{i+1} | \leq 20$ km/h

Case 3: Poor design level

The nonpermissible differences are

$| \text{CCR}_{S_i} - \text{CCR}_{S_{i+1}} | > 360$ gon/km

$| V85_i - V85_{i+1} | > 20$ km/h

Note: $\text{CCR}_{S_i} = \text{CCR}_S$ value of design element i, gon/km; $\text{CCR}_{S_{i+1}} = \text{CCR}_S$ value of design element $i + 1$, gon/km; $V85_i = $ 85th-percentile speed of design element i, km/h; and $V85_{i+1} = $ 85th-percentile speed of design element $i + 1$, km/h.

respective operating speed background of the country under study (see Fig. 8.12 or Table 8.5). If no operating speed background is available, then one may select the one corresponding to a comparable country, the average one according to Fig. 8.12b, or develop a new one.

The differences in the CCR_{S_i} values and $V85_i$ values between the two observed successive design elements reveal, in comparison with the ranges of Table 9.13, the classification of the investigated alignment section into good, fair, or poor design levels. Table 9.13 is valid for new designs, redesigns, or RRR projects, or for examining existing (old) alignments, as will be shown in Sec. 18.4. (See also case studies in Secs. 12.2.4.2 and 18.4.2.)

The recommended ranges for good, fair, and poor design practices in Table 9.13 were used as the basis for the development of Table 9.2. To simply illustrate the scientific background of safety criterion II, especially in terms of the expected accident situation (see Table 9.10), the ranges for good, fair, and poor design levels for the design parameter, curvature change rate of the single curve, are incorporated into Table 9.13. This is not necessary for practical design work, since the CCR_S ranges and the $V85$ ranges correspond to one another within relatively narrow limits, as previously discussed. Therefore, in order to simplify matters, only the ranges of the 85th-percentile speeds between successive design elements are referred to in Table 9.2 for distinguishing good, fair, and poor design practices. A case study for the evaluation process of safety criterion II was already presented in Table 9.12.

9.2.3.2 Graphic Presentation: Relation Design*

Background

In the TRB Special Report 214 "Designing Safer Roads,"[710] it is reported:

> Safety relationships attempt to capture at least partially situational influences present in the roadway environment that contribute greatly to roadway hazards. Illustrative of these particular hazards are sharp horizontal curves following long segments of generally straight alignment, compound curves—contiguous horizontal curves turning the same way—in which a flat curve precedes a much sharper one and high-volume intersections in isolated

---

*Elaborated based on Refs. 444, 447, 448, 455, 461, and 462.

rural settings. Common to such situations is the violation of driver expectancy: the unfamiliar or inattentive driver, lulled to complacency by the gentleness of the approaching roadway, is surprised by the sudden appearance of a potential hazard. The response is uncertain and slow, leading possibly to inappropriate maneuvering and an increase in accident potential.

In addition to inconsistencies that occur at point or spot locations, such as illustrated previously, other situations are found in which clues from the physical environment belie the nature of the roadway hazard. Perhaps the most important arises from possible incompatibilities between the roadway cross section and its horizontal and vertical alignment. In the case of a roadway improvement, for example, upgrading cross-sectional elements without corresponding upgrading of alignment can result in an erroneous and potentially hazardous illusion of safety and the selection of operating speeds excessive for the critical alignment conditions.[710] Compare Part 3, "Cross Sections (CS)" and Chap. 16.

In general, the degree of hazard inherent in a specific feature, such as a sharp curve, narrow bridge, or a roadway without shoulders, depends not only on the feature itself but also on the nature of the nearby roadway environment. Although the safety effects of such interactions have not been quantified satisfactorily, it is quite likely that essential safety gains are achievable if roadway inconsistencies are eliminated as a part of the following improvements. Useful techniques for eliminating inconsistencies or compensating for their potentially adverse effects include:[710]

- Provision of gradual geometric transitions appropriate to the anticipated vehicle operating speed (safety criterion II).
- Improvement of sight distance for early detection of the presence of the critical feature (safety criteria I and II, and Chap. 15).
- Provision of gentle sideslopes with few roadside obstacles at critical locations (compare "Roadside Conditions" in Secs. 25.1.3.1 and 25.2.3.1 of Part 3, "Cross Sections").
- Installation of appropriate traffic control devices.[710]

In order to satisfy a major portion of the previously stated safety issues, consistent characteristics of a roadway, including horizontal/vertical alignment, cross section, and road adjacent environment, are therefore of prime importance.

With regard to consistency issues, it is recommended in Ref. 5, as an example, and in the guidelines of many other countries that:

- All of the pertinent features of the highway should be related to the design speed to obtain a balanced design (safety criterion I).
- Changes in design speed should be in increments no greater than 10 km/h (safety criterion I).
- The use of greater sight distance or flatter horizontal curves is encouraged (safety criteria I and II).
- Winding alignments composed of short curves should be avoided because they are usually a cause of erratic operation (safety criterion II).
- Sharp curves should not be introduced at the end of long tangents (safety criterion II).
- Sudden changes from areas of flat curvature should be avoided (safety criterion II).
- In alignments predicated on a given design speed, use of maximum curvature for that speed should be avoided wherever possible. The designer should attempt to use generally flatter curves, retaining the maximum for the most critical conditions (safety criteria I and II, and "Curve Flattening" in Secs. 12.1.2.5 and 12.2.2.5).
- Caution should be exercised in the use of compound curves. Compound curves with large differences in curvature introduce the same problems that arise in a tangent approach to a circular curve (Safety criterion II and "Compound Circular Curves" in Secs. 12.1.2.4 and 12.2.2.4).
- Abrupt reversals in alignment should be avoided (safety criterion II).[5]

With respect to human factors, Lunenfeld[490] made the following statements at the 1992 Transportation Research Board meeting, which, according to him, should be carefully considered in highway geometric design:

> Consistency is an often violated aspect of geometric design. Driver expectancies are developed through experience and knowledge gained by driving a facility and is directly related to the geometric consistency of that facility. Consistency affects how drivers perceive and react to the information provided, by means of signing and pavement markings. Geometric consistency reinforces driver expectancies, which aids the driver in making quick and correct responses to decisions.

In addition, he indicated the following:[490]

> …incompatibility in geometric and operational requirements may be caused by trying to fit together geometric components conveniently and economically rather than trying to satisfy operational requirements. Therefore, design consistency should be maintained in addition to standardizing roadside features such as concrete barrier walls, aluminum guardrail, signing, pavement marking, traffic control devices and the like.[490]

In conclusion, many countries regard acceptable design as one in which each design element, such as radius or degree of curve, superelevation rates, vertical curves, and sight distances, meets the minimum or maximum requirements for the individual design elements for the selected design speed(s). No specific guidelines are given for the relationships between design elements that occur in sequence, that is, no mention is made or guidance is given through a design procedure to achieve a better consistency and improve safety on two-lane rural roads.

With respect to the German guidelines, Lippold[480] reported the following: In connection with a consistent road characteristic, the German guidelines for the design of roads[241,243] have required, beginning in 1973, well-aligned element sequences in the horizontal alignment in addition to sound design elements in plan, profile, and cross section. This is to be achieved primarily by using the required sequences of curves shown in Fig. 9.35. This figure, known as relation design, defines the ranges where the radii of two successive circular curves in the same or in opposite directions should fall in order to achieve a well-balanced relationship for safety reasons. Relation design should ensure that motorists adapt their speed to the present geometry of the road, and that only low speed differences with low deceleration rates should exist in front of or within curves themselves. The radii relations of Fig. 9.35 were based on accident investigations conducted in 1953 and between 1963 and 1966, and on investigations related to driving behavior in 1968.[346,347] The speed differences were based only on assumptions about possible speed changes. Today, it can be concluded that the radii relations in Fig. 9.35 have been used successfully for rural roads[154] and have contributed considerably to the safety of new alignments.[286,480]

Without a doubt, the relation design fundamentals in Fig. 9.35 have greatly improved consistency experiences in highway geometric design during the last two decades, even though the exact fundamentals for calculating relation design backgrounds were not available at the time of the development of Fig. 9.35.

The application of Fig. 9.35 for coordinating radii of curve sequences leads to more or less strong curvilinear alignments. The term *curvilinear alignment* is usually considered to mean a long curve–short tangent–long curve type of alignment, as opposed to the more common long tangent–curve–long tangent type[444] found in many countries.

In this connection it was found, from the evaluation of case studies related to new designs and redesigns of existing alignments,[628] that curvilinear alignments are very favorable, especially in hilly topography, and improve traffic safety as well. Furthermore, curvilinear alignments and coordination between horizontal and vertical alignments are recognized as a means of achieving an esthetically pleasing three-dimensional highway alignment (see Chap. 16). Thus, curvilinear alignment can be seen as a tool to achieve highway esthetics as well as a tool to achieve increased safety.

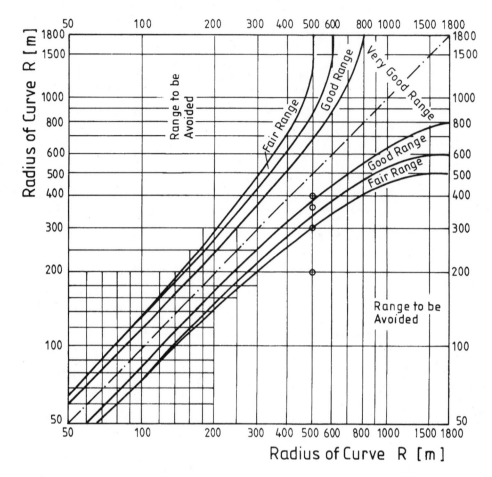

FIGURE 9.35 Tuning of radii of curve sequences for roads of category group A and road category B II in Germany.[241,243]

However, the case studies also revealed that in flat or mountainous topography, or in cases of redesigns, curvilinear aspects often do not lead to optimum solutions. These findings are supported by Durth and Lippold.[162]

Therefore, the term *relation design* is extended here in such a way that, besides the advantages of curvilinear alignments, a well-balanced design between independent (long) tangents and curves is also included. This is very important for future application, depending on topographic, landscape-related, and economic demands whether to use, either the long curve–short tangent–long curve or the long tangent–curve–long tangent alignment type or to combine both types.

Calculation of Sound Relation Design Backgrounds

Safety criterion II was developed for use in achieving operating speed consistency. The assessed operating speed ranges for good, fair, and poor design practices of safety criterion II are fully based on accident research according to Tables 9.10, 10.12, and 18.14. The same is true for the development of the relation design backgrounds in Fig. 9.1 and Figs. 9.36 to 9.40, since they are to be understood as the graphical presentation of safety criterion II.

However, the authors decided to reduce the fair design range of safety criterion II (Table 9.2 or 9.13) from

$$10 \text{ km/h} < | V85_i - V85_{i+1} | \leq 20 \text{ km/h}$$

to

$$10 \text{ km/h} < | V85_i - V85_{i+1} | \leq 15 \text{ km/h}$$

for the establishment of the relation design backgrounds. The reasons for this more conservative consideration are

1. Relation design backgrounds will mainly be used for new alignment designs and redesigns. It is expected that they will be applied only to a minor extent for the safety evaluation process of existing (old) alignments, which are examined by safety criteria I to III according to Sec. 18.4.

2. Regarding Tables 9.10 and 10.12, it is obvious that the expected accident rates for the "fair range" are at least 2 to 3 times higher, as compared to the "good range." Therefore, it may be beneficial to reduce the speed range of the fair design level from 20 km/h to 15 km/h with respect to the development of relation design backgrounds in order to limit the accident risk and accident severity (Table 18.14), in order to make relation design between successive design elements much safer.

The development of relation design diagrams will be described in the following, using Germany as an example (Fig. 9.38).

According to Table 8.5, the equation of the operating speed background for Germany (ISE) is

$$V85 = \frac{10^6}{8270 + 8.01 \text{ CCR}_S} \tag{8.15}$$

Step 1. Set, for example, $R = 1000$ m.

Step 2. Calculate CCR_S with respect to R from Eq. (8.6) or Table 8.4 without regarding transition curves:

$$\rightarrow \text{CCR}_S = \frac{63,700}{R}$$

$$\rightarrow \text{CCR}_S = \frac{63,700}{1000} = 63.7 \text{ gon/km} \tag{8.6}$$

Step 3. Determine $V85$ from Eq. (8.15) or graphically from Fig. 8.12 (Curve: Germany, ISE):

$$\rightarrow V85 = 114 \text{ km/h}$$

Step 4. Subtract 10 km/h from $V85$ in step 3 to reflect good design or 15 km/h to reflect fair design:

$$V85 = 104 \text{ km/h} \quad \text{for good design}$$

$$V85 = 99 \text{ km/h} \quad \text{for fair design}$$

Step 5. For $V85$ in step 4, determine CCR_S from Eq. (8.15) and R from Eq. (8.6). For good design:

$$\text{CCR}_S = 169 \text{ gon/km}$$

or

$$R = 377 \text{ m}$$

and for fair design:

$$\text{CCR}_S = 230 \text{ gon/km}$$

or

$$R = 277 \text{ m}$$

Step 6. The intersections of the lines drawn horizontally or vertically from radii of curve of 1000 and 377 m, and from 1000 and 277 m, respectively, indicate the points which should fall on the relation design curves for good and fair design (see Fig. 9.38).

Step 7. Repeat steps 1 through 6 for radii of curve of less than and greater than 1000 m with increments of 100 m.

The calculation procedure was conducted in the same way for all relevant operating speed equations in Table 8.5, in order to establish the relation design backgrounds shown in Fig. 9.1 (Greece), Fig. 9.36 (Australia), Fig. 9.37 (Canada), Fig. 9.38 (Germany, ISE), Fig. 9.39 (Lebanon), and Fig. 9.40 (United States) for roads of category group A and road category B II.

Based on the relation design curves, the designer in the country under study could immediately decide whether certain radii of succeeding curves fall into the ranges of good, fair, or poor design. For instance, from the relation design background of the United States (Fig. 9.40) a radius of curve of 500 m, combined with the following radii of curve, leads to

$R = 100$ m	poor design
$R = 180$ m	fair design
$R = 300$ m	good design
$R = 1500$ m	also good design

However, note that when tuning radii of curve sequences the minimum radius of curve has to at least support the selected design speed of the roadway.

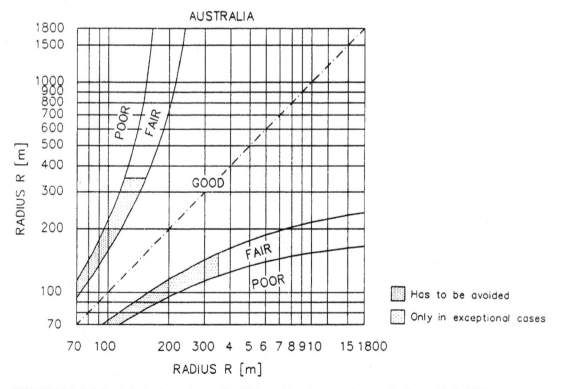

FIGURE 9.36   Relation design background, Australia. (Elaborated based on operating speed background in Ref. 505.)

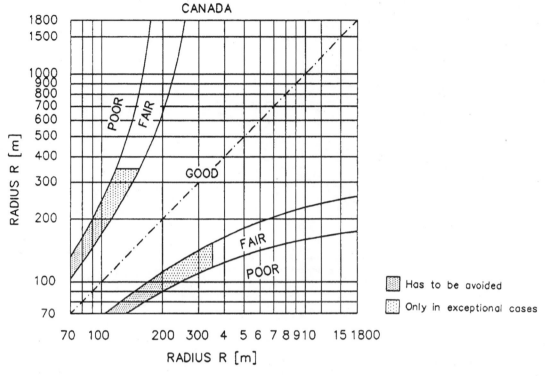

FIGURE 9.37   Relation design background, Canada. (Elaborated based on operating speed background in Ref. 508.)

For the transition independent tangent–clothoid–circular curve, the good design range should always apply. That means, according to the previously cited figures and corresponding calculations, that radii of curve of at least

- *Australia:* $R \geq 300$ m (rounded), Fig. 9.36
- *Canada:* $R \geq 300$ m (rounded), Fig. 9.37
- *Germany:* $R \geq 500$ m (rounded), Fig. 9.38
- *Greece:* $R \geq 400$ m (rounded), Fig. 9.1
- *Lebanon:* $R \geq 350$ m (rounded), Fig. 9.39
- *United States:* $R \geq 350$ m (rounded), Fig. 9.40

should follow independent tangents.

Similar conclusions were established in several other studies. For instance, Krebs and Kloeckner[368] indicated, in a study conducted in 1977, which was based on a large database of traffic accidents in Germany, that radii of curve of 400 to 500 m provide a certain cross point in safety on circular curves. This result was later confirmed by Lamm, Mailaender, and Choueiri in a study conducted in the United States and in Germany in 1989.[418] Another 1989 study conducted by Leutzbach and Zoellmer[476] compared the transitions between tangents and circular curves with or without transition curves (clothoids) in Germany. They indicated the following: With respect to the element sequence for the transition from tangents to curves, a safety gain from the inclusion of a transition curve (clothoid) could only be observed for radii of curve less than 200 m,

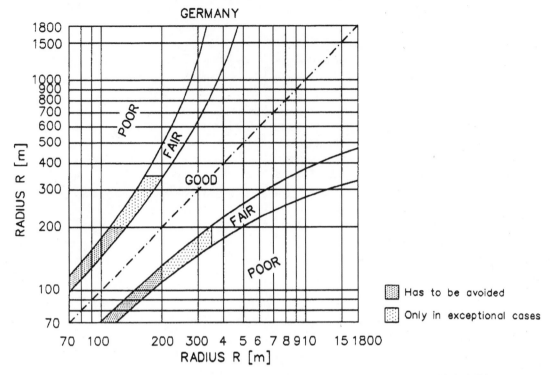

FIGURE 9.38 Relation design background, Germany. (Elaborated based on operating speed background in Ref. 431.)

where accident rates were significantly lower in comparison to the direct sequence tangent-circular-curve. No systematic differences were detected for radii of curve of $R > 200$ m; this finding also applies to the accident cost rate for the entire range of the investigated radii of curve.

Based on these findings and research results, it can be expected that the introduction of a minimum radius of curve of 400 to 500 m for the transition tangent to circular curve with and without transition curves will not affect the accident situation negatively. Of course, one should not forget the importance of other design impacts that transition curves provide, besides accident-related issues, such as:

- Gradual increase or decrease of the centrifugal force
- Convenience in application of superelevation run-off
- Improvement in esthetic appearance

Therefore, it was decided that curves with radii of $R \geq 400$ m (an exception is so far Germany with $R \geq 500$ m), in conjunction with transition curves, should follow an independent tangent in order not to create inconsistencies in vehicular operating speeds. In cases where the selected design speed requires larger radii of curve, these values have to be used.

Comparing the results shown in the relation design backgrounds of Figs. 9.1 and 9.36 to 9.40 with the corresponding operating speed backgrounds of the individual countries in Fig. 8.12 or Table 8.5, it can be concluded that the 85th-percentile speed on an independent tangent $(V85_T)$* which corresponds to a radius of curve of $R \approx 400$ m is $\overline{V85}_T < 105$ km/h. For Germany, where $R \approx 500$ m, the corresponding 85th-percentile speed is $V85_T \geq 105$ km/h. On the basis of these find-

---

* $V85_T$ = 85th-percentile speed on an independent tangent according to "Evaluation of Tangents in the Design Process" in Sec. 12.1.1.3. It can be determined from Fig. 8.12 or Table 8.5 for $CCR_S = 0$ gon/km.

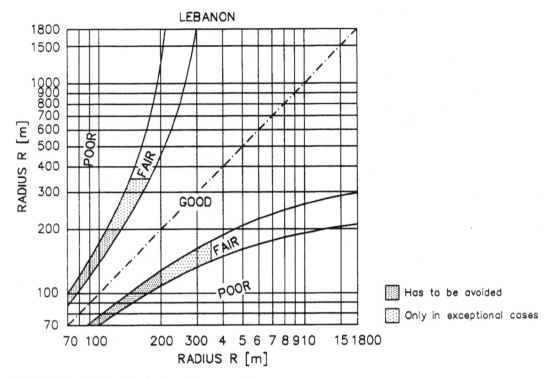

FIGURE 9.39   Relation design background, Lebanon. (Elaborated based on operating speed background in Ref. 108.)

ings, it was decided to assign to the transition independent tangent–clothoid–circular curve the following radii of curve with respect to the 85th-percentile speed, $V85_T$, on independent tangents:

$$R \geq 400 \text{ m} \qquad \text{for } V85_T < 105 \text{ km/h}$$

and

$$R \geq 500 \text{ m} \qquad \text{for } V85_T \geq 105 \text{ km/h (see Table 12.4)}$$

For the direct transition independent tangent to curve, one can find in the guidelines of the different countries different assumptions related to minimum radii of curve of up to 3000 m for design speeds of $V_d > 80$ km/h.[241,243] However, previous research results revealed that these assumptions are far too conservative.[476] Therefore, it is recommended, based on the experiences of the authors (unconfirmed so far by any known research), that radii of curve of

$$R \geq 800 \text{ m}$$

should follow independent tangents for operating speed backgrounds ($V85_T < 105$ km/h), and

$$R \geq 1000 \text{ m}$$

for operating speed backgrounds ($V85_T \geq 105$ km/h) (see Table 12.8).

For the first time, relation design backgrounds of radii of curve sequences in combination with sound transitions between independent tangents to curves with and without transition curves have been developed, based on actual operating speed backgrounds and accident research on two continents for evaluating good, fair (tolerable), and poor design practices.

Operating speed backgrounds (like those in Fig. 8.12) and relation design backgrounds (like those in Figs. 9.1 and 9.36 to 9.40) provide essential information for modern highway geometric

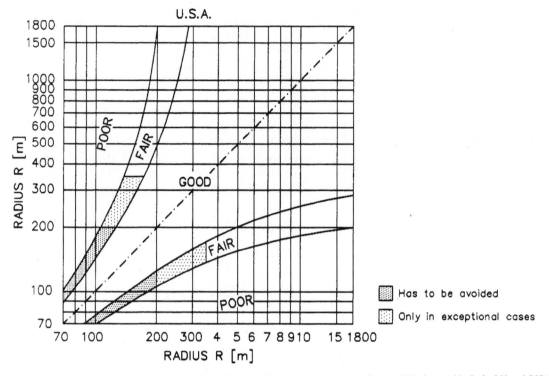

FIGURE 9.40   Relation design background, United States. (Elaborated based on operating speed background in Refs. 360 and 565.)

design tasks. If these backgrounds are not available and cannot be established, it is recommended that the designer select one which may best fit the operational requirements of the country under study.

Relation design, as described herein, will allow the designer to quantitatively determine if an alignment is consistent or whether an alignment change which is necessary for the required consistency meets driver expectancy in order to achieve safer operation. This quantification of consistency, in terms of criteria I and II in Tables 9.11 and 9.13, as well as the proposed relation design ranges according to Figs. 9.1 and 9.36 to 9.40, will allow the designer to evaluate the effects of fitting together geometric components conveniently and economically and to satisfy operational requirements.

Relation design should be considered a useful tool to achieve a more consistent road design (and thereby avoid potential safety errors) than is generally attained by the individual element/maximum-minimum design approach. This is true for new alignments, upgraded highway alignments, or full-blown RRR projects of two-lane rural roads.

The proposed relation design process is based on quantifiable and sophisticated criteria for evaluating operating speed changes between successive design elements (curves to curves and independent tangents to curves, see "Evaluation of Tangents in the Design Process" in Sec. 12.1.1.3) and for coordinating operating speeds and design speeds of individual design elements with each other.

CHAPTER 10
DRIVING DYNAMICS AND SAFETY CRITERION III

10.1 RECOMMENDATIONS FOR PRACTICAL DESIGN TASKS

10.1.1 Fundamentals*

This chapter concentrates on the examination of the frictional behavior between tire and road surface which depends, in general, on the speed.

A comparison of a number of highway geometric design guidelines in Europe, Australia, South Africa, and the United States, for example, and the introduction of new research results reveal that considerable differences exist between permissible tangential and side friction factors. The differences have decisive consequences on the design of minimum radii of curve, required sight distances, minimum radii or lengths for crest and sag vertical curves, and so on, with respect to design speed or 85th-percentile speed.

A sufficient driving dynamic safety margin is nearly always required. However, this margin is only rarely verifiable, since, in many countries, the exact information about the theoretical background, on which the determination of permissible friction factors is based, is not available.

To overcome this weakness in current practice, a driving dynamic safety criterion will be introduced in Secs. 10.1.4 and 10.2.4.

The investigations of the forces acting on the vehicle which are relevant for the selection of design parameters regard the vehicle as a rigid body and assume that the resulting forces act on the center of gravity only. In this way, the vehicle is idealized as a point-mass model, and no information is given about the actual distribution of forces, for example with respect to the individual wheels of the vehicle. This procedure still has to be considered as typical for all modern highway geometric design guidelines today. Therefore, the considerations about the frictional behavior between vehicle and road surface must be confined to these idealizations.

For the driving dynamic distribution of forces, related to the vehicle as point-mass, the following formula is basically valid:

$$\text{friction force} = \text{friction factor} \times \text{wheel weight}$$

The basic driving dynamic formula for curve design is

$$f_R + e = \frac{V^2}{127\,R} \tag{10.1}$$

---

*Elaborated based on Refs. 124, 159, 385, 433, 434, and 453.

where f_R = side friction factor

e = superelevation rate, %/100

V = speed, km/h

R = radius of curve, m

The basic driving dynamic formula in the tangential direction is given in Sec. 15.1.2.

10.1.2   Establishment of Permissible Friction Factors*

10.1.2.1   Skid Resistance.   As an essential component of traffic safety on a highway, the skid resistance of the pavement under wet conditions and the resulting coefficients of friction have to be considered.

The coefficient of friction between the tire and the pavement specifies a condition which consists of adhesion, friction, and interconnection, whereby the two surfaces—the tire and the pavement—which act together must be considered as equal partners.

The coefficient of friction varies considerably because of the influences of many physical elements, such as the air pressure of the tires, composition of the tires, the tire tread pattern and the depth of the tread, the type and the condition of the pavement surface, and the presence or absence of moisture, mud, snow, or ice.[5]

Skid resistance, better known as the frictional properties of a pavement, also includes the influence of the pavement surface, through its material properties and its geometric microtexture, on the magnitude of acceleration, braking, and radial forces that are maximally transmitted from the tire to the road. The skid resistance of a pavement is characterized, in most European countries and the United States, for example, by the skidding value under confined test conditions. The skidding value, μ_G, is defined as the coefficient of friction of a retarded wheel (100 percent slippage) (see Fig. 10.10).

Measurements of the skidding values on representative road surfaces in some countries already provide the evaluation backgrounds for the distribution of skid resistance values needed to establish permissible tangential and side friction factors. Normally, it is required that between 80 and 95 percent of the skid resistance values of road surfaces must be covered by the permissible tangential friction factors in order to provide a relatively high level of driving dynamic safety. (Germany, for instance, uses the 95th-percentile distribution curve in Figs. 10.25 and 10.26 as the basis for assessing maximum permissible tangential friction factors.)

10.1.2.2   Tangential Friction Factor.   In most countries, skid resistance evaluation backgrounds do not exist yet. Furthermore, the procedure presented in Subchapter 2 for developing skid resistance evaluation backgrounds, in order to establish country-specific maximum permissible tangential friction factors, is not always possible because existing roadway networks sometimes present relatively low skid resistance values, such as, for example, in Greece.[453] Assigning the maximum permissible tangential friction factors to the skid resistance values that are reached by at least 80 percent of existing road surfaces results in uneconomical and irresponsible environmental solutions in some countries (see, for example, Fig. 10.30). Therefore, another solution must be found for establishing reasonable tangential friction factors.

Assuming that the permissible tangential friction factors correspond to the current state of the art with respect to driving dynamic aspects, the specifications in acknowledged highway geometric design guidelines (Germany, France, Sweden, Switzerland, and the United States) were taken as a basis for developing an overall relationship between permissible tangential friction factor and design speed (see Fig. 10.27).

The overall tangential friction equation, Eq. (10.2), developed can be considered as reasonable for safety, economic, and environmental demands. It may be used for determining maximum permissible tangential friction factors for modern highway geometric design tasks, also see Eq. (10.40):

---

*Elaborated based on Refs. 241, 382, 383, 385, 396, 414, 420, 428, 435, 438, 453, and 686.

$$f_{T\text{perm}} = 0.59 - 4.85 \times 10^{-3} \, V_d + 1.51 \times 10^{-5} \, (V_d)^2\}$$

$$R^2 = 0.731 \tag{10.2}$$

where $f_{T\text{perm}}$ = maximum permissible tangential friction factor
V_d = design speed, km/h
R^2 = coefficient of determination

The present relationship will be introduced for the calculation of maximum permissible tangential friction factors with respect to speed and in Sec. 15.1.2 for determining reliable stopping sight distances.

With regard to future driving dynamic safety demands, the required tangential friction factors resulting from Eq. (10.2) should be reached and maintained as the necessary skid resistance values of pavement surfaces in cases of new designs, redesigns, and RRR projects.

Normally, skid resistance measurements are conducted for speeds of 40, 60, and 80 km/h. The required skid resistance values of pavement surfaces that should guarantee the tangential friction factors of Eq. (10.2)

0.42	for	$V = 40$ km/h
0.35	for	$V = 60$ km/h
0.30	for	$V = 80$ km/h

have to be reached for new construction and be maintained until resurfacing projects are deemed necessary.

10.1.2.3 Side Friction Factor.   Following the establishment of the relationship between the maximum permissible tangential friction factor and the design speed, the question that arises is from which range should the utilization ratio, *n,* of the maximum permissible side friction factor be selected? International experience indicates that *n* varies between 40 and 50 percent for rural roads. This means that 92 and 87 percent, respectively, of friction in the tangential direction is still available when driving through curves for acceleration, deceleration, braking, or evasive maneuvers (see Fig. 10.13 and Table 10.4).

According to Table 10.1, the general equation for the maximum permissible side friction factor is

$$f_{R\text{perm}} = n \, 0.925 \, f_{T\text{perm}} \tag{10.3}$$

The reduction factor of 0.925 corresponds to tire-specific influences.

10.1.3 Arrangements for Maximum Permissible Side Friction Factors*

10.1.3.1 Roads of Category Group A (Rural Roads).   Based on topography conditions in many countries (flat, hilly, and mountainous topography), different utilization ratios were considered as reasonable for the side friction factors for this category group.

Flat Topography, Maximum Superelevation Rate (Rural).   In flat topography, a utilization ratio of $n = 45$ percent is justified for the application of a maximum superelevation rate of $e_{\text{max}} = 8$ percent on the basis of practical experience. As an exceptional case, a maximum superelevation rate of $e_{\text{max}} = 9$ percent applies in areas with no ice and snow. The relationship between maximum permissible side friction factor and design speed is shown as curve 3 in Fig. 10.1. The

---

*Elaborated based on Refs. 124, 159, 241, 435, 438, 448, 453, and 458.

TABLE 10.1   Recommended Equations for the Relationships Between Maximum Permissible Side Friction Factors and Design Speed for Different Road Category Groups, Topography Classes, and Maximum and Minimum Superelevation Rates[435,453]

The maximum permissible tangential friction factor is

$$f_{Tperm} = 0.59 - 4.85 \times 10^{-3} V_d + 1.51 \times 10^{-5} (V_d)^2 \tag{10.2}$$

The general form of the equation for the maximum permissible side friction factor is

$$f_{Rperm} = n\, 0.925\, f_{Tperm} \tag{10.3}$$

Roads of category group A (rural roads)

Flat topography, maximum superelevation rate $e_{max} = 8$ to 9%:

$$n = 45\%$$

$$f_{Rperm} = 0.45\, 0.925\, f_{Tperm} = 0.416\, f_{Tperm} \tag{10.4a}$$

$$f_{Rperm} = 0.25 - 2.02 \times 10^{-3} V_d + 0.63 \times 10^{-5} (V_d)^2 \tag{10.4b}$$

Hilly and mountainous topography, maximum superelevation rate $e_{max} = 7\%$:

$$n = 40\%$$

$$f_{Rperm} = 0.40\, 0.925\, f_{Tperm} = 0.37\, f_{Tperm} \tag{10.5a}$$

$$f_{Rperm} = 0.22 - 1.79 \times 10^{-3} V_d + 0.56 \times 10^{-5} (V_d)^2 \tag{10.5b}$$

Minimum superelevation rate $e_{min} = 2.5\%$:

$$n = 10\%$$

$$f_{Rperm} = 0.10\, 0.925\, f_{Tperm} = 0.0925\, f_{Tperm} \tag{10.6a}$$

$$f_{Rperm} = 0.05 - 0.45 \times 10^{-3} V_d + 0.14 \times 10^{-5} (V_d)^2 \tag{10.6b}$$

Roads of category group B (suburban roads)

Maximum superelevation rate $e_{max} = 6\%$:

$$n = 60\%$$

$$f_{Rperm} = 0.60\, 0.925\, f_{Tperm} = 0.555\, f_{Tperm} \tag{10.7a}$$

$$f_{Rperm} = 0.33 - 2.69 \times 10^{-3} V_d + 0.84 \times 10^{-5} (V_d)^2 \tag{10.7b}$$

Minimum superelevation rate $e_{min} = 2.5\%$:

$$n = 30\%$$

$$f_{Rperm} = 0.30\, 0.925\, f_{Tperm} = 0.2775\, f_{Tperm} \tag{10.8a}$$

$$f_{Rperm} = 0.16 - 1.34 \times 10^{-3} V_d + 0.42 \times 10^{-5} (V_d)^2 \tag{10.8b}$$

---

corresponding equations, Eqs. (10.4a) and (10.4b), are found in Table 10.1. For comparative reasons, the relationship between the maximum permissible tangential friction factor and design speed is presented as curve 1 in Fig. 10.1.

Hilly and Mountainous Topography, Maximum Superelevation Rate (Rural).   For hilly and mountainous topography, a utilization ratio of $n = 40$ percent for a maximum superelevation rate of $e_{max} = 7$ percent is regarded as reasonable in order to compensate, from a driving dynamic safety standpoint, for the decrease in the superelevation rate by the application of a lower utilization ratio. The relationship between the maximum permissible side friction factor and design speed is presented as curve 4 in Fig. 10.1. The corresponding equations, Eqs. (10.5a) and (10.5b), are found in Table 10.1.

Using the utilization ratios of side friction of $n = 45$ percent for flat topography and of $n = 40$ percent for hilly/mountainous topography, 90 and 92 percent, respectively, are still available for friction in the tangential direction when driving through curves (Fig. 10.13).

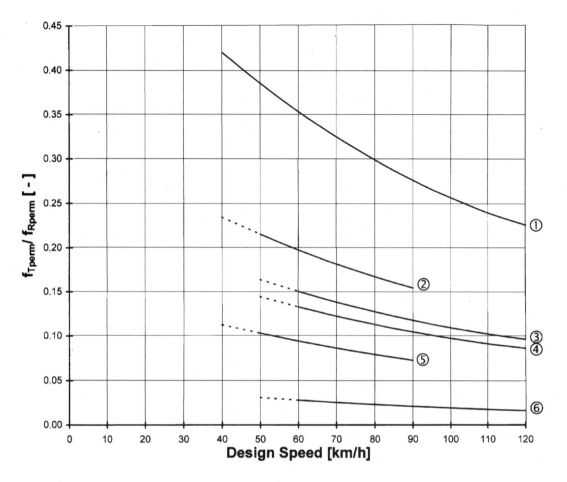

Legend:

1 f_{Tperm} for "RR" and "SR"

2 f_{Rperm} for "SR" n=60%, e_{max}=6%

3 f_{Rperm} for "RR" Flat Topog.
   n=45%, e_{max}=8% (9%)

4 f_{Rperm} for "RR" Hilly/Mount. Topog.
   n=40%, e_{max}=7%

5 f_{Rperm} for "SR" n=30%, e_{min}=2.5%

6 f_{Rperm} for "RR" All Topog. Classes
   n=10%, e_{min}=2.5%

f_T/f_{Rperm} = maximum permissible
tangential/side fric-
tion factors

RR = Rural Roads

SR = Suburban Roads

n = utilization ratio of side
friction

e_{max}/e_{min} = maximum/minimum
superelevation rates

FIGURE 10.1   A graphical presentation of the relationships between the side friction factor and the design speed (see also Tables 10.1 and 10.2).[435]

Minimum Superelevation Rate (Rural).   Generally, the establishment of utilization ratios, which are used for determining permissible side friction factors, is regarded as useful for implementing maximum and minimum superelevation rates. For low superelevation rates, the utilization ratio, *n,* can be assigned to a low level,[124] on the basis of considerations regarding danger classes.

The minimum superelevation rate in the countries investigated is usually selected to be equal to that in tangents because of drainage and driving dynamic safety demands. In the majority of countries, it is equal to e_{min} = 2.5 percent and the utilization ratio of n = 10 percent is considered reasonable.[243] The relationship between the side friction factor and design speed is presented as curve 6 in Fig. 10.1 and Eqs. (10.6*a*) and (10.6*b*) in Table 10.1 for all topography classes.

10.1.3.2   *Roads of Category Group B (Suburban Roads).*   For this category group, the speed level is normally lower than that on rural roads. Therefore, a higher utilization ratio of side friction factor can be tolerated with respect to driving dynamic safety considerations, and is demanded because of economic, environmental, and municipal development reasons.

Maximum Superelevation Rate (Suburban).   For suburban roads with e_{max} = 6 percent, a maximum utilization ratio of n = 60 percent is justified.[124,159] In this case, 80 percent of friction is still available in the tangential direction when driving through curves. Curve 2 in Fig. 10.1 and Eqs. (10.7*a*) and (10.7*b*) in Table 10.1 represent this case.

Minimum Superelevation Rate (Suburban).   Because of lower speed levels on suburban roads compared with those on rural roads, for the application of minimum superelevation rates e_{min} = 2.5 percent, a high utilization ratio of n = 30 percent for the side friction factor appears to be justified.[241,243] The relationship between the side friction factor and the design speed is shown as curve 5 in Fig. 10.1, and is given by Eqs. (10.8*a*) and (10.8*b*) in Table 10.1.

Maximum permissible tangential and side friction factors in relation to design speed for the road categories group A "rural" and group B "suburban," for different topography classes, and for recommended maximum and minimum superelevation rates are given in Table 10.2.

The minimum radii of curve in Table 10.2 were calculated on the basis of the driving dynamic formula for curve design [Eq. (10.9)] with respect to the newly developed side friction factors and the recommended maximum and minimum superelevation rates.

Preliminary Conclusion.   The results of Table 10.2 for flat, as well as for hilly/mountainous, topography agree well with the minimum radii of curve used in most other countries, and result in balanced designs with respect to economic, environmental, and safety-related issues.

A differentiation of utilization ratios of side friction for individual road category groups, topography classes, and maximum and minimum superelevation rates appears to be justifiable because it provides logical driving dynamic safety reserves where they are needed most. In addition, through considering maximum permissible side friction factors for minimum superelevation rates, the overall driving dynamic design process becomes complete.

In this study, maximum and minimum superelevation rates were selected on the basis of international experience. The values of superelevation rates of e_{max} = 10 percent and e_{max} = 12 percent[5] were consciously not considered because of construction limitations, maintenance difficulties, and the operation of heavy vehicles at low speeds, especially under snow and ice conditions.

Fundamentally, the developed maximum tangential and side friction factors consider important safety aspects and enable the designer to effect the proper dimensioning of design elements knowing that certain qualitative safety reserves are present. But no one has been able so far to examine quantitatively the level of "safety" or "unsafety" at curved sites. For this reason, the development of a quantitative driving dynamic, safety criterion, is deemed necessary.

10.1.4   Safety Criterion III: Achieving Driving Dynamic Consistency*

Based on the information presented so far and the driving dynamic safety considerations elaborated in Sec. 10.2.4, it follows that a number of safety aspects for the establishment of reliable tan-

---

*Elaborated based on Refs. 424, 428, 435, 438, 447, 448, 449, 453, 455, 458, 632, and 673.

TABLE 10.2 Maximum Permissible Tangential and Side Friction Factors and Recommended Minimum Radii of Curve with Respect to Design Speed, Road Category Groups, Topography Classes, and Maximum and Minimum Superelevation Rates*

Design speed, km/h		40	50	60	70	80	90	100	110	120
Category groups rural roads (RR) and suburban roads (SR)										
f_{Tperm}		0.420	0.385	0.353	0.324	0.299	0.276	0.256	0.239	0.225
Category group RR (flat topography, e_{max})										
f_{Rperm}	$n = 0.45$ $e_{max} = 8$ to (9%)	—	0.160	0.147	0.135	0.124	0.115	0.107	0.100	0.094
R_{min}, m		—	85	125	180	250	330	425	530	650
		—	(80)	(120)	(170)	(235)	(310)	(400)	(500)	(620)
Category group RR (hilly and mountainous topography, e_{max})										
f_{Rperm}	$n = 0.40$ $e_{max} = 7\%$	—	0.143	0.131	0.120	0.110	0.102	0.095	0.089	0.083
R_{min}, m		—	95	140	200	280	370	480	600	740
Category group RR (all topography classes, e_{min})										
f_{Rperm}	$n = 0.10$ $e_{min} = 2.5\%$	—	0.036	0.033	0.030	0.028	0.026	0.024	0.022	0.021
R_{min}, m		—	325	490	700	960	1250	1600	2000	2500
Category group SR (e_{max}, e_{min})										
f_{Rperm}	$n = 0.60$ $e_{max} = 6\%$	0.233	0.214	0.196	0.180	0.166	0.153	—	—	—
R_{min}, m		45	70	110	160	225	300	—	—	—
f_{Rperm}	$n = 0.30$ $e_{min} = 2.5\%$	0.117	0.107	0.098	0.090	0.083	0.077	—	—	—
R_{min}, m		90	150	230	335	470	630	—	—	—

*Modified and completed based on Refs. 159 and 453.

Note: RR = rural roads; SR = suburban roads.

The driving dynamic formula for curve design is

$$R = \frac{V_d^2}{127\,(f_R + e)} \tag{10.9}$$

where R = radius of curve, m
f_R = side friction factor
e = superelevation rate, %/100
V_d = design speed, km/h

gential and side friction factors have already been considered. However, on the basis of known highway geometric design guidelines, an overall quantitative driving dynamic safety analysis is not available.

For the safety-related evaluation of the interplay of geometric design, driving behavior, driving dynamics, and the accident situation, three safety criteria will be proposed here. For coordinating the design speed with the expected 85th-percentile speed for the individual design element, safety criterion I was established. For coordinating the 85th-percentile speeds between successive design elements, such as between tangent to curve or curve to curve, safety criterion II was developed. Thus, the important questions regarding speed, which are relevant for the design of a road and the resulting driving behavior of free-moving passenger cars, are evaluated by safety criteria I and II, which consider quantitative ranges for good, fair (tolerable), and poor design practices on the basis of accident research.

The only question left now is the development of safety criterion III, in order to evaluate the driving dynamic aspects of driving through curves. Safety criterion III is also related, like safety criterion I, to the individual design element which, in the present case, is the circular curve.

Safety criterion III compares side friction assumed for curve design (f_{RA}) for the different design speeds, V_d, with the actual side friction demand, f_{RD}, required at curved sites for the expected 85th-percentile speeds of the country under study (see Fig. 8.12 and Table 8.5).

Side friction assumed f_{RA} has to be determined based on Eq. (10.3). However, new research has revealed that the utilization ratios of $n = 40$ percent and $n = 45$ percent for maximum permissible side friction factors in Table 10.1 and Fig. 10.1, providing as much driving dynamic safety as possible with respect to new designs and redesigns, are too conservative for a generally valid safety evaluation process according to criterion III. Note that the procedure should not only regard new designs but should also be applicable for redesign and RRR projects of existing alignments. Therefore, it is recommended that a utilization ratio of $n = 60$ percent for the calculation of side friction assumed f_{RA} be used. That means that there will still be 80 percent of friction available in the tangential direction when driving through curves (Table 10.4). Thus, the recommended formula for side friction assumed f_{RA} for redesigns and RRR strategies of existing alignments reads:

$$f_{RA} = 0.60 \ 0.925 \ f_{T\text{perm}} \tag{10.10a}$$

for $f_{T\text{perm}}$, see Eq. (10.2), and

$$f_{RA} = 0.33 - 2.69 \times 10^{-3} \ V_d + 0.84 \times 10^{-5} \ (V_d)^2 \tag{10.10b}$$

where f_{RA} = side friction assumed
V_d = design speed, km/h

As stated in Table 10.1, Eq. (10.4) is valid for new designs in flat country and Eq. (10.5) in hilly or mountainous topography. In cases of upgrading existing alignments or full-blown RRR projects, the decision is basically left to the highway engineer, whether to select Eqs. (10.4), (10.5), or (10.10), based on economic, environmental, or safety-related issues. However, normally a utilization ratio of $n = 60$ percent is recommended.

Side friction assumed f_{RA} remains constant along the investigated roadway section for achieving driving dynamic consistency.

Another problem arises when regarding the design speed, V_d, with respect to safety criterion III and to Eq. (10.10b). While the selected design speed is known for new designs, this is often not true for existing alignments being evaluated for intended redesigns and RRR projects. Therefore, according to the detailed discussions for safety criterion I, it is recommended in those cases, for the estimation of a sound design speed, to calculate the length-related average, $\overline{CCR_S}$ value, based on all the curves in the observed roadway section according to Eq. (9.12) (see "Existing (Old) Alignments" in Sec. 9.2.2.1). Tangent sections should not be considered. From this length-related average $\overline{CCR_S}$ value, the average 85th-percentile speed, $\overline{V}85$, has to be determined from Fig. 8.12 or Table 8.5 for the country under study. This average $\overline{V}85$ represents a good estimate of the new design speed, V_d, for future redesigns and RRR strategies, since this speed will, to a certain extent, optimize the design process from the viewpoints of economic, environmental, and safety issues at least for two-lane rural roads of category group A. (For determining $\overline{CCR}_S \rightarrow$ $V85 \rightarrow V_d$, see the case study in Table 9.12.)

Therefore, this design speed, V_d, has to be introduced with respect to the evaluation process of safety criterion III into the recommended equation, Eq. (10.10b), for determining side friction assumed f_{RA} of existing alignments in cases of redesigns and RRR strategies, if the design speed is unknown.

The actual side friction demanded is calculated from the following equation:

$$f_{RD} = \frac{V85^2}{127\,R} - e$$

(10.11)

where f_{RD} = side friction demanded (actually needed)

$V85$ = 85th-percentile speed, km/h, with respect to CCR_S according to Fig. 8.12 or Table 8.5

R = radius in the observed circular curve, m

e = superelevation rate, %/100

Knowing the side friction assumed, f_{RA}, and side friction demanded, f_{RD}, one can conduct the safety evaluation process according to safety criterion III (see Table 10.3).

For the evaluation of curved roadway sections, the following procedure is recommended:

1. Assess the road section where new designs are contemplated or where redesigns or RRR projects of existing alignments shall be conducted.

2. Determine the design speed for the investigated section with respect to road category (network function) according to Table 6.2 for new designs or find out the original design speed for existing sections. If this is not possible, determine the design speed of an existing two-lane roadway section of category group A according to the previous recommendations.

3. Determine the 85th-percentile speed for each curved site with respect to the design parameter curvature change rate of the single curve, CCR_S, on the basis of the operating speed background for the country under study (Fig. 8.12 or Table 8.5). CCR_S is calculated from Eq. (8.6) or the equations in Fig. 8.1.

4. For multilane roads of category group A and the road categories B II, B III, and B IV, the 85th-percentile speeds can be determined according to the information presented in Sec. 8.1.2.

5. The difference between side friction assumed f_{RA} and side friction demanded f_{RD} represents a quantitative measure of driving dynamic safety or unsafety:

$$f_{RA} - f_{RD} \lessgtr \Delta f_R$$

(10.12)

In this respect, a good design level exists when side friction assumed f_{RA} exceeds the side friction demanded f_{RD} because friction reserves are still available in this case, at least theoretically.

On the other hand, when side friction demanded f_{RD} exceeds side friction assumed f_{RA}, we can say that a poor design level exists. In this case, it can be assumed, at least theoretically, that there is a high probability that accident risk will increase because of critical driving situations.

On the basis of current research results, which are discussed in detail in "Classification System for Safety Criterion III" in Sec. 10.2.4.3, safety criterion III provides quantitative ranges for good, fair, and poor design levels (see Table 10.3). Safety criterion III should be used for examination of two-lane and *multilane* rural roads (category group A), as well as for *suburban roads* (road categories B II, B III, and B IV).

New designs for category groups A and B must correspond to the good design level (Table 10.3, case 1).

In cases of spatial constraints, the fair design level (case 2) can be assigned to redesigns and RRR projects. However, on the basis of the experience gained by accident research (Tables 9.10 and 10.12), one can expect the accident rate in this case to be at least twice as high as that of case 1.

TABLE 10.3   Recommended Ranges for Good, Fair, and Poor Design Levels for Safety Criterion III (Category Groups A and B)

Recommended Driving Dynamic Consistency Ranges (Safety Criterion III

Case 1: Good design:

$$f_{RA} - f_{RD} \geq + 0.01 \qquad\qquad (10.12a)$$

There exists a high probability that sufficient friction reserves are available at curved sites. No adaptations or corrections are necessary.

Case 2: Fair design

$$-0.04 \leq f_{RA} - f_{RD} < +0.01 \qquad\qquad (10.12b)$$

At curved sites:

1. The speed behavior must be calmed down by local speed limits and/or traffic warning devices.
2. Superelevation rates must be related to $V85$ to ensure that side friction assumed will better accommodate side friction demanded, especially for redesigns and/or RRR projects.
3. In case of resurfacing, high demands should be placed on skid resistance.

Case 3: Poor design:

$$f_{RA} - f_{RD} < - 0.04 \qquad\qquad (10.12c)$$

There exists a high probability that insufficient friction conditions are present at those curved sites, especially on wet road surfaces, which may lead to an increase in accident frequencies and severities and to uneconomic, unsafe operation. The decision whether or not to institute reconstruction measures must be based on the local accident situation. For recommended countermeasures, besides redesigns, compare comment d in Sec. 18.4.1.

Sometimes a cost-effective accident reduction can be achieved by reconstructing the pavement (for example, increasing the skid resistance values and/or the superelevation rates).

For curved sites which correspond to case 3 in Table 10.3, a redesign or a reconstruction (like resurfacing) is recommended. Compare, for example, the case studies in Sec. 12.2.4.2 and comment d in Sec. 18.4.1.

In conclusion, the utilization ratio of $n = 60$ percent for side friction assumed may be used for the safety evaluation process of criterion III in Table 10.3 for the examination of existing alignments in conjunction with redesigns or RRR projects. For new designs, a more generous use of the maximum permissible side friction factors, f_{Rperm}, according to Table 10.1 and Fig. 10.1, is required.

10.2   GENERAL CONSIDERATIONS, RESEARCH EVALUATIONS, GUIDELINE COMPARISONS, AND NEW DEVELOPMENTS

10.2.1   Fundamentals

Driving dynamics deals with the physical laws that govern a vehicle's motion with respect to the vehicle's properties and the properties of the road (alignment and pavement).

To describe thoroughly the performance of a vehicle on a road, very complex (and, therefore, difficult to handle) mathematical models are required. These models are also used for the design and construction of the vehicles themselves. Figure 10.2 shows the various basic driving situations that may arise when a motor vehicle is moving on a highway.[522,611]

Including the wheels and the chassis, a vehicle is composed of five individual parts that are connected in a movable way. Every movable body is considered a rigid body with six levels of

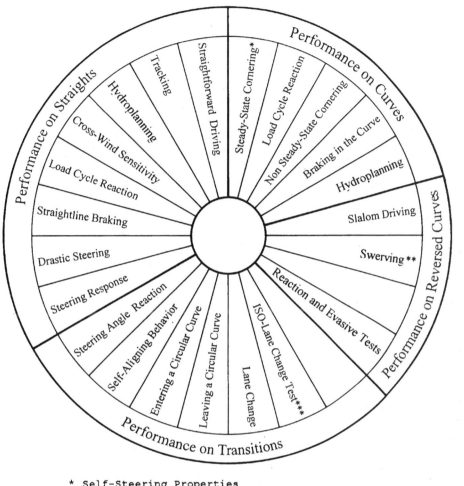

* Self-Steering Properties
** Drastic Steering and Acceleration
*** ISO = International Organization for Standardization

FIGURE 10.2   Basic driving performance of a motor vehicle.[522,611]

freedom—three rotational and three translatory—resulting in 5 × 6 = 30 degrees-of-freedom for the whole vehicle (Fig. 10.3).

These movements can be described thoroughly by a number of differential equations. The number of these differential equations increases considerably if other components of the vehicle, such as the motor (engine), the load, the occupants, and so on, are also taken into consideration.

During the motion of a vehicle, different forces are transmitted between the tires and the pavement. These forces act normal to the road surface (vertical forces) as well as parallel to it (horizontal forces). The horizontal forces can be projected to the tangential direction, that is, parallel to the travel path of the vehicle, and to the radial direction, that is, normal to the travel path of the vehicle, in order to analyze vehicle movements in these two characteristic directions, which are of interest in highway geometric design.

Furthermore, the actions of the driver result in reactions of the vehicle, which act again on the driver, who is forced to take new actions, and so forth. Additionally, the driver and vehicle are subject to different disturbances and road environmental factors such as alignment (curves,

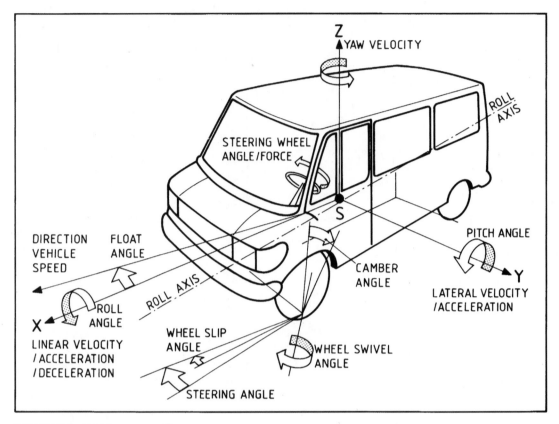

FIGURE 10.3   Vehicle movements.[62]

grades, and so on), pavement characteristics, and weather conditions. The driver, the vehicle, and the road represent a closed-loop control system (Fig. 10.4).

Each tire on the vehicle contributes individually to the transmission of forces while the vehicle is in motion. From the point of view of vehicle mechanics, it is important to calculate the slip angle and the available tangential force based on the tire type if the radial and vertical forces are known. A review, however, of international standards, guidelines, and norms has shown that this approach has not yet been used. On the contrary, all driving performance analyses regarding path stability, acceleration and deceleration, and the forces associated with these are based on the assumption that the vehicle is a rigid body and all forces act on the center of gravity. From this simplification, the well-known point-mass model of the vehicle has resulted.[382,383,385]

For this simple point-mass model, the following compromises are met:[159]

- The whole vehicle mass is concentrated in the center of gravity.
- All forces act in the center of gravity.
- The hypothetical center of gravity of the vehicle lies on the road surface.
- The contact area between tires and surfaces is generalized and simplified.
- Influences caused by sudden movement changes are neglected.
- The three-dimensional movement is reduced to three two-dimensional movements, which are examined separately—on a horizontal plane, on a vertical plane, and on a cross-sectional plane.

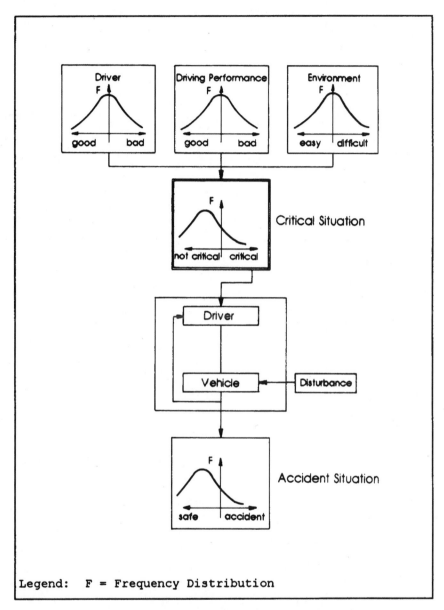

FIGURE 10.4   Common driver-vehicle-road interactions generating critical driving situations.[613]

The previously mentioned assumptions are not to be objected to as long as the actual friction values are significantly greater than those used in practice (that is, in the design policies).

10.2.1.1   Friction.   Forces are transmitted between tires and pavement due to friction. The maximum transmissible friction force depends on the characteristics of the tire and the pavement and on the presence of any substance on the contact area between tires and pavement.

Figure 10.5 shows the factors influencing the friction potential between tire and pavement.[222]

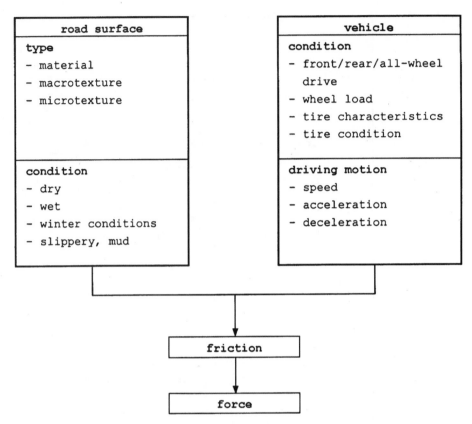

FIGURE 10.5   Factors affecting the friction potential. (Elaborated and completed based on Ref. 222.)

Friction forces depend on the friction coefficient and on the load normal to the transmitting contact patch between tire and road surface. Pressure on the contact area is not only unsymmetrical but also depends on the type of tire. Radial-ply tires have a better pressure distribution at their contact area with the road surface than bias-ply tires (Fig. 10.6).

Driving possibilities are limited to the wheel loads and friction availability. Therefore, friction values supplied by the interaction between tire and pavement are not constant but depend considerably on tire elasticity.[93] Furthermore, it becomes evident from the interaction between wheel and pavement that friction plays a significant role in driving safety and comfort of a vehicle. The construction (design) of faster cars and the need for more safety on the road network make a thorough understanding of the physical laws governing the wheel/pavement interaction absolutely necessary. Effective utilization of this knowledge has so far led to the employment of the antilock braking systems (ABS), which prevent the wheel from locking up during braking and to the electronic traction control systems, which do not allow a wheel to spin when accelerating, such as the vehicle dynamic control system of Bosch which counteracts the vehicle spin phenomenon.[307]

In this connection, the newest development in Germany is being presented by Mercedes-Benz and Bosch through a driving safety system—the electronic stability program (ESP). This system gives the driver a maximum level of driving stability and traction at critical driving dynamic limits without influencing good driving characteristics during normal rides.[530]

The most important forces active in the contact area between wheel and pavement are the longitudinal (tangential) and the radial (side) forces. The longitudinal force in a straight vehicle path causes a driving tractive or braking movement of the wheel, which in turn causes a difference in

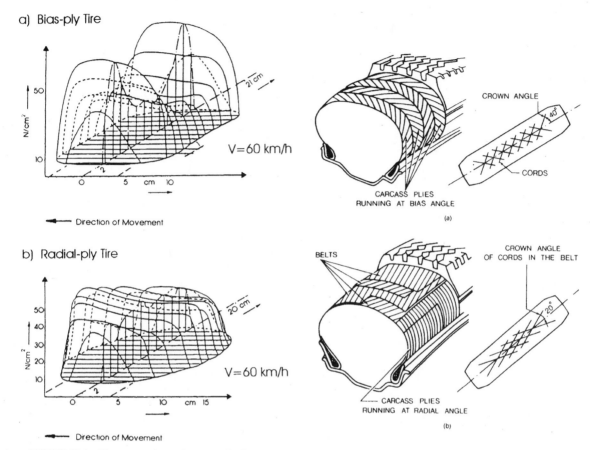

FIGURE 10.6 Tire construction and pressure distribution on the contact area between free-rolling tires of the bias-ply and the radial-ply types and the road surface.[239,764]

velocity between the vehicle velocity itself and the relative velocity between the wheel and the pavement. This velocity difference or wheel slip is decisive for the friction phenomenon. During cornering, on the other hand, the front wheels are at an angle to the longitudinal axis of the vehicle, thereby causing the development of a radial (side) force that is responsible for radial friction. When both the longitudinal and the radial forces are present, their resultant force should not exceed an upper limit (maximum friction force, F_{max}) (see Fig. 10.7).

Exceeding this force can lead to an accident. For example, during panic braking on a curve, the maximum longitudinal force is almost attained and the remaining radial component of the resultant friction force is not sufficient to keep the vehicle in its path, thus causing the vehicle to slide off the road. In Fig. 10.8, the various influencing factors of the radial force are illustrated:[632]

a. A curve with a maximum available friction (in general, an elliptic curve).

b. For a slip angle of $\alpha = 0°$ the radial force is not necessarily zero.

c. So long as maximum longitudinal forces are generated despite wheel slip, the curves with constant slip angles fall back into the characteristic curve field.

d. When a wheel locks up, an inner circle with a minimum friction value is generated on the braking side.

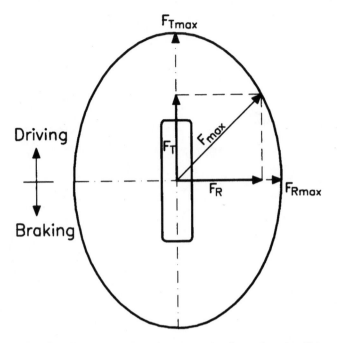

FIGURE 10.7   Horizontal forces between road surface and motor vehicle (tire).[632]

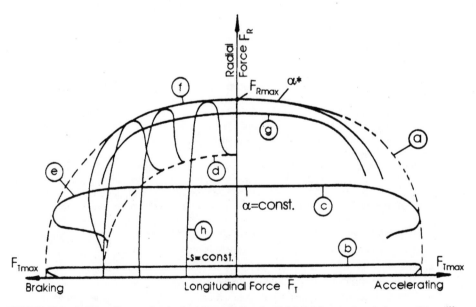

FIGURE 10.8   Relations between longitudinal (tangential) and radial (side) force under varying conditions.[632]

e. For small slip angles, that is, the radial (side) force, F_R, is less than $F_{R\max}$, the longitudinal (tangential) force, $F_T < F_{T\max}$, does not have much impact on the radial force.

f. In the area of maximum radial force, the longitudinal force has a strong impact on the radial force.

g. Curves with constant slip angles (greater than α^*) fall back into the characteristic curve field if the radial force has a maximum according to α^*.

h. Curves with a constant slip have the same form as curves with a constant slip angle ($\alpha = $ const) but rotated by 90°.

Beyond the friction force, which is of merit to the physical phenomenon, other disturbing forces, such as side-wind forces, can act on a vehicle.[64,149,738] These forces act on the vehicle itself and are then transferred to the wheels. Analysis of all the forces acting on a vehicle is a very complex task and in most cases requires the introduction of simulation techniques (see "Simulation Models" in Sec. 10.2.1.3).

In order to initiate the design of a highway, the designer needs to use *design values* as input in the various models considered in the design process. In the case of friction, these design values are expressed in terms of maximum permissible values that are directly associated with the driving and braking limits of a vehicle and, consequently, with the built-in safety of a highway.

Driving and braking limits of a vehicle are defined as the maximum longitudinal (tangential) and lateral (radial/side) accelerations (decelerations) that can be tolerated by the vehicle under steady-state conditions. In Fig. 10.9, the friction potential is illustrated in terms of vehicle longitudinal and lateral acceleration.

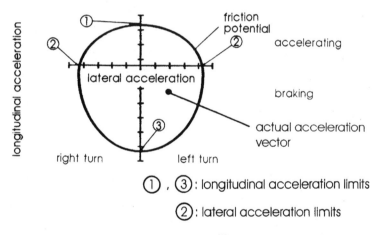

FIGURE 10.9   Definition of the friction potential.[222]

For road design purposes, a rolling vehicle in the longitudinal direction on a wet but clean road surface is considered. The friction values arising from the interaction between tires and pavement are those resulting at the moment of impending skidding. The forces acting on tire/pavement (Fig. 10.7) fulfill the equation:

$$F \le F_{\max} = \mu_P\, Q \tag{10.13}$$

where $F_{\max}$ = maximum possible (transferable) friction force, N
  μ_P = peak friction factor
  Q = weight force, N

This resultant force can be separated into the two perpendicular forces, F_T and F_R, acting in the two major directions that are of importance in highway design—the longitudinal and the radial direction of travel (Fig. 10.7).

For these two forces, the following relationship is approximately valid:

$$F^2 = F_R{}^2 + F_T{}^2 \tag{10.14}$$

Mathematically, the friction factor, f, is expressed as the ratio:

$$f = \text{horizontal force/vertical force}$$

It is almost impossible to give general statements about the friction potential between a tire and a road surface for use as design values, since the following factors have to be thoroughly considered and evaluated:[124]

- The pavement structure, the road surface features due to material properties, the different traffic demands, the time elapsed since construction and the time period of the year
- Weather variations ranging from summer dry, humid and wet with different waterfilm depths to winter dry, wet, slush, snowy, and icy road surface conditions
- Vehicle parameters such as wheel-load, slip angle, position of center of gravity, steering wheel angle, and so on
- The tire type, width, profile, rubber synthesis, and inflation pressure
- The speed and the driving behavior

In highway geometric design, the problem is approached by examining the independent friction factors for the radial (side) and tangential direction of travel and assuming approximate Newton friction values. It follows:

$$f_T = \frac{F_T}{Q} \tag{10.15}$$

$$f_R = \frac{F_R}{Q} \tag{10.16}$$

$$f^2 = f_T^2 + f_R^{2*} \tag{10.17}$$

$$f \le f_{\max} \tag{10.18}$$

where * = friction circle assumed
f = friction factor
F_T = tangential force (horizontal force), N
F_R = radial force (horizontal force), N
f_T = friction factor in tangential direction
f_R = friction factor in radial direction
Q = weight force (vertical force), N

One may differentiate between friction demanded, permissible (assumed) friction, and maximum friction values. Friction demanded is the friction required under specific vehicle movements, for example, during cornering, in order to allow stable vehicle motion. This demand stems from physical relationships, taking into consideration the radius of the curve and the vehicle's speed. The maximum friction values are the values available under specific, extreme conditions. Finally, permissible friction values are those assumed for highway design purposes (design values), such that an adequate safety potential can be assumed in highway geometric design. Permissible friction values are less than the maximum values.

The relationship between friction factor in the tangential direction, f_T, and wheel slip is shown in Fig. 10.10. The permissible friction factor in the tangential direction is defined at 100% wheel slip (wheel lockup), which equals the skidding coefficient μ_G.

In order to define the term *wheel slip,* the following assumptions have to be considered:

When a driving or braking torque is applied to a tire, the tread elements are compressed or stretched accordingly prior to entering the tire-pavement contact area.

Consequently, the tire rotates without the equivalent translatory progression (vehicle/movement). This phenomenon is referred to as (longitudinal) wheel slip.

This wheel slip causes the distance that the tire travels when subject to a torque to be different from the distance it travels when rolling free. As a matter of fact, if the torque is a driving one, then this distance will be less than the corresponding free-rolling distance (positive slip); the opposite is true if the torque is a braking one (negative slip). According to Fig. 10.11, if a tire is rotating at a certain angular speed, ω, but the linear speed, v, of the tire center (chassis) is 0, then Eq. (10.19) implies that the wheel slip, ζ_A, will be 100 percent. This case is often observed on an icy pavement, where the driven tires are spinning at high angular speeds but the vehicle does not move forward.

The opposite situation is observed when the brakes are applied to a moving vehicle. In this case, if the driving wheels are locked, then the angular speed, ω, of the tires becomes 0, whereas the linear speed of the tire center is not 0. Under this condition, the wheel slip, ζ_B (decelerating condition), is denoted 100 percent from Eq. (10.20) and the wheel is skidding.

The same physical conditions as in the tangential direction also apply in the radial direction. The radial friction force is activated on the road surface through a radial skidding motion of the wheel.

Based on measurements on different (wet but clean) road surfaces with standardized tires, a distribution of the skidding friction factors in relation to speed was developed (Fig. 10.12). For

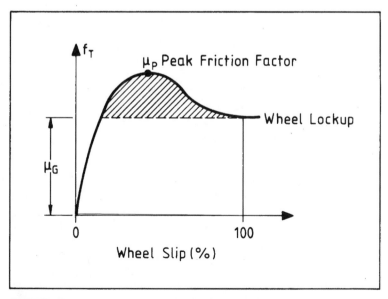

FIGURE 10.10 Definition of the maximum permissible tangential friction factor with regard to the skidding value ($f_{T\max} = \mu_G$).[382,383]

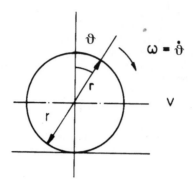

$$\text{slip} = \zeta_A = r \cdot \omega - v/r, \cdot \omega, r \cdot \omega > v \qquad \text{accelerating condition} \qquad (10.19)$$

$$\text{slip} = \zeta_B = v - r \cdot \omega/v, v > r \cdot \omega \qquad \text{braking (decelerating) condition} \qquad (10.20)$$

where ω = tire angular speed, rad/s
v = vehicle translatory speed, m/s
ϑ = rotational angle, rad
r = rolling radius, m

FIGURE 10.11   Definition of longitudinal wheel slip.

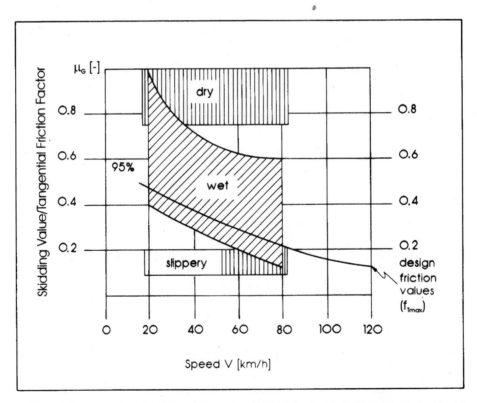

FIGURE 10.12   Distribution of skidding values μ_G (tangential friction factor, f_T).[160] (Originally developed by Wehner.[747])

design purposes, that percentile distribution curve of friction is normally used which falls short of only 5 to 10 percent of the investigated road surfaces. This boundary line is described, for example, by the following equation for German road surface conditions at the 95th-percentile distribution level:[382,383,385,396]

$$\mu_G = f_{Tmax} = 0.214 \left(\frac{V}{100}\right)^2 - 0.640\frac{V}{100} + 0.615 \tag{10.21}$$

where f_{Tmax} = maximum (permissible) friction factor in tangential direction
μ_G = skidding value
V = speed, km/h

The maximum permissible friction factor in radial (side) direction cannot be measured. It is calculated based on the friction factor in tangential direction as follows:

$$f_{Rmax} = 0.925\,f_{Tmax} \tag{10.22}$$

where f_{Rmax} = maximum (permissible) friction factor in radial (side) direction.

Equation (10.22) is derived from f_{Tmax} (friction ellipse) shown in Fig. 10.13, which describes the friction potential more accurately than the friction circle based on Eq. (10.17) (see also Fig. 10.7). In Eq. (10.22), the factor 0.925 sums up all tire-specific influences.[374]

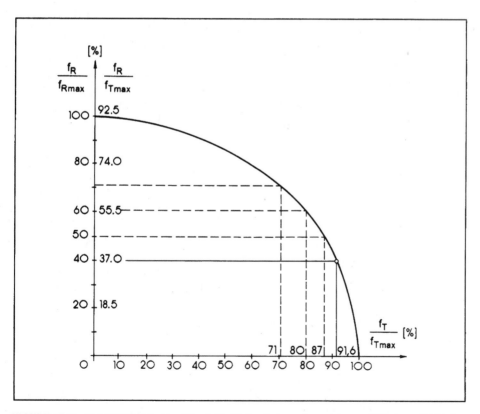

FIGURE 10.13 Relationship between utilized side friction factor and remaining available tangential friction factor.[382,383,396]

For Fig. 10.13, the following ellipse equation is valid:

$$\left(\frac{f_R}{f_{Rmax}}\right)^2 + \left(\frac{f_T}{f_{Tmax}}\right)^2 \leq 1 \tag{10.23}$$

where f_R = available friction factor in the radial direction
f_T = available friction factor in the tangential direction

Equation (10.23) was developed by Krempel.[374] Numerical values of the distribution of friction values in the radial and tangential direction, respectively, are given in Table 10.4. From Eq. (10.23), the available friction factor in the tangential direction can be determined for any given side friction factor or vice versa.

The maximum permissible side friction factor, f_{Rperm}, has to be determined in such a way that friction reserves as large as possible are available, both in the radial and in the tangential direction, so that no skidding off the road can occur when braking during cornering. For example, if a ratio of about 90 percent of friction in the tangential direction is available, thus warranting a considerable driving dynamic safety reserve, a maximum permissible utilization ratio of side friction of about $n = 40$ percent results according to Table 10.4. On the basis of this utilization ratio, the design values of the maximum permissible side friction factors can be calculated, for example, for German road conditions[241] according to Eqs. (10.21) and (10.22):

$$f_{Rperm} = 0.40 \, 0.925 \, f_{Tperm}$$

$$= 0.37 \left[0.214 \left(\frac{V}{100}\right)^2 - 0.640 \, \frac{V}{100} + 0.615 \right] \tag{10.24}$$

The utilization ratio of $n = 40$ percent is only used here as an example for demonstration purposes. In modern highway geometric design guidelines, the utilization ratio selected for determining minimum radii of curve is normally between 40 and 50 percent for rural roads.

In the following section the dynamics of a vehicle moving on a roadway will be discussed.

10.2.1.2   Vehicle Dynamics.   As already mentioned, the vehicle dynamics in highway geometric design are examined by reducing the vehicle to a point mass. To follow the course of the point mass on a road, two discrete motions have to be considered. The first is the *tangential direction*

TABLE 10.4   Distribution of Utilized Side Friction Ratios and Remaining Available Tangential Friction Ratios[382,383,396]

Radially	**Tangentially**
$n = (f_R/f_{Rmax}) \, 100$, %	$(f_T/f_{Tmax}) \, 100$, %
0.0	100.0
10.0	99.5
20.0	98.0
30.0	95.4
40.0	91.7
50.0	86.6
60.0	80.0
70.0	71.4
80.0	60.0
90.0	43.6
100.0	0.0

Note: n = utilization ratio of side friction, %.

of the road, for example, parallel to the direction of the vehicle speed, and the second one is the *radial direction,* for example, perpendicular to the direction of the vehicle speed. The vehicle speed is assumed to remain unchanged (steady-state motion) along the road.

The forces acting on a vehicle during its motion in the tangential and in the radial direction are shown in Figs. 10.14 and 10.15.

Tangential Direction.   With regard to the friction factors, the following principle is valid: skidding in the tangential direction is not possible, if according to Fig. 10.14:[160]

resistant forces ≥ tractive forces

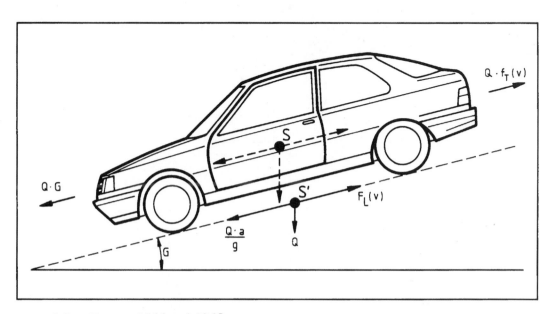

Legend for Figures 10.14 and 10.15:

v	=	Vehicle Speed [m/sec]
a	=	Deceleration/Acceleration [m/sec^2]
$f_T(v)$	=	Tangential Component of the Friction Potential [-]
$f_R(v)$	=	Radial Component of the Friction Potential [-]
R	=	Radius of Curve [m]
$F_L(v)$	=	Aerodynamic Drag Force [N]
F	=	Centrifugal Force [N]
Q	=	Vehicle Weight [N]
G	=	Longitudinal Grade [%/100]
g	=	Acceleration of Gravity [m/sec^2]
e	=	Superelevation Rate [%/100]
S	=	Center of Gravity [-]
S'	=	Center of Gravity Projection on the Road Surface [-]
m	=	Vehicle Mass [kg]
α	=	Angle of Superelevation Rate, tanα = e [%/100]
b	=	Half Vehicle Width [m]
b'	=	Horizontal Projection of Half Vehicle Width (b), [m],
h	=	Height of Center of Gravity from Road Surface [m]

FIGURE 10.14   Forces acting on a vehicle in the tangential direction. (Elaborated based on Ref. 124.)

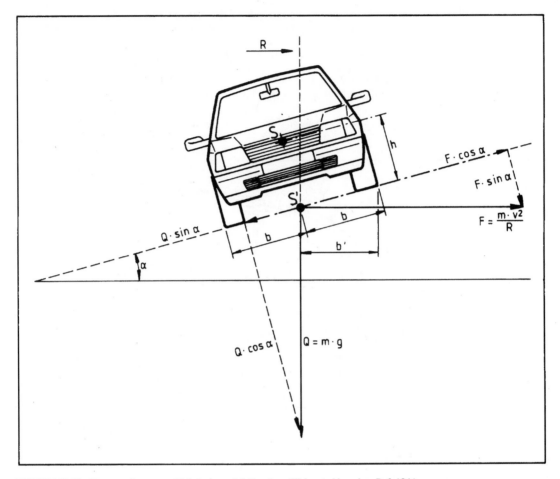

FIGURE 10.15 Forces acting on a vehicle in the radial direction. (Elaborated based on Ref. 124.)

The resistant forces are primarily composed of the aerodynamic drag force, $F_L(v)$, and grade resistance, $Q \cdot G$, if present. Other possible resistant forces, such as rolling resistances, transmission resistances, or drawbar loads, are neglected in highway design. Tractive forces arise from the friction which occurs when the wheel slip is less than 100 percent and are active only on the driven wheels of the vehicle. These tractive forces are equal to the driven wheel load multiplied by the tangential friction factor which corresponds to a specific wheel slip. For a two-wheel drive passenger car, the maximum tractive force that can be developed is

$$\mu_p \cdot \frac{Q}{2}$$

where μ_p is the peak friction factor according to Fig. 10.10 and Q is the vehicle weight.

During braking, there are no tractive forces. Instead there are braking (retarding) forces due to the application of the brakes to *all* wheels of a vehicle.

The *minimum* braking force that can be applied to a vehicle is equal to $\mu_G \cdot Q$, where μ_G is the skidding value according to Fig. 10.10, which is actually developed when the wheels are locked up.

In highway geometric design, the braking and not the driving (tractive) motion of a vehicle is considered to be critical. Therefore, the equations which follow refer to the braking performance of a vehicle (passenger car).

Summing up the forces in the tangential direction provides the following basic equation of motion according to Fig. 10.14:

$$Q f_T(v) + F_L(v) = Q G + \frac{Q}{g} a \tag{10.25}$$

By rearranging the equation for deceleration, it follows:

$$a = g \left[f_T(v) + \frac{F_L(v)}{Q} \pm G \right] = \frac{dv}{dl} v \qquad v \text{ (m/s)} \tag{10.26}$$

where $G = +$ for upgrades and $-$ for downgrades.

By solving Eq. (10.26) for dl and by integrating from an initial speed, v_0, to a final speed, v_1, the distance, required to reduce the vehicle speed from v_0 to v_1 can be calculated. For $v_1 = 0$ m/s, the distance traversed by a vehicle until it stops can be determined. This braking distance will be used in Sec. 15.1.2 to establish stopping sight distances. Compare Eq. (15.3) and Table 15.1.

Radial Direction.   For the radial direction, it is stated in Ref. 5 that:

> A vehicle moving in a circular path is forced radially outward by centrifugal force. This force is resisted by the vehicle weight component due to the superelevation of the roadway and by the side friction between the tires and the roadway surface.

Taking into account the forces acting on a vehicle in the radial direction, the following equilibrium state can be observed according to Fig. 10.15.[160]

Skidding in the radial direction is not possible if

$$\text{resistant forces} \geq \text{tractive forces}$$

Resistant forces consist of the portion of the friction force available in the radial direction parallel to the road surface, while the tractive forces stem from the centrifugal force.

For the forces acting on the center of gravity, the following relationship exists:

$$\text{friction force} = \text{friction factor} \times \text{sum of all wheel loads}$$

Summing up the forces in the radial direction and assuming equal side friction factors on all wheels results in the following equation according to Fig. 10.15:

$$Q \sin\alpha + f_R(v) \, (Q \cos\alpha + F \sin\alpha) \geq F \cos\alpha \tag{10.27}$$

where

$$F = \frac{m \cdot v^2}{R} \qquad Q = m \cdot g$$

and, by dividing through $\cos\alpha$ and substituting

$$\tan\alpha = e \frac{\%}{100} \tag{10.28}$$

the following basic equation for the cornering performance of a vehicle results:

$$\frac{v^2}{g \, R} \leq \frac{f_R(v) + e}{1 - f_R(v) \, e} \approx f_R(v) + e \qquad v \text{ (m/s)} \tag{10.29}$$

Equation (10.29) provides a cornering prediction model because it describes the interrelationships existing among the vehicle speed on the curve, v; the radius of the curve, R; the superelevation rate, e; and the radial portion of the friction, $f_R(v)$. Because $f_R(v)$ is limited to f_{Rmax} according to Eq. (10.22), it follows that the maximum centripetal acceleration, $(v^2/R)_{max}$, that can be applied by a vehicle before an impending skidding is equal to $f_{Rmax} \cdot g$ on a nonsuperelevated highway.

Figure 10.16 gives a general overview of the maximum values of the left part of Eq. (10.29) which can be attained under various weather conditions.

According to Fig. 10.16, the maximum centripetal acceleration for normal tires on dry pavements varies between 0.7 and 1.0g, while for icy pavements a maximum value of 0.15g can be attained during braking maneuvers (temperature about 0°C). Compare also Fig. 10.12.

Table 10.5 contains maximum values of centripetal acceleration observed on dry pavements.

By comparing Table 10.5 and Fig. 10.16, it becomes obvious that the observed centripetal accelerations on dry pavements for common motor vehicles, that is, excluding racing cars, lie below the maximum possible values shown in Fig. 10.16. Actually, this is due to the fact that the general assumptions that led to Eq. (10.29) (that is, point-mass model, negligible wheel-slip angles, equal friction potential to all wheels, etc.) all lie on the "safe side" of the cornering prediction model.[522]

Furthermore, it can be concluded from Eq. (10.29) that the superelevation rate should not exceed the value of 15 percent ($e = 0.15$). This results from the fact that, for icy pavement conditions ($\sim$0°C), the minimum side friction factor can be estimated to be $\sim$0.15 (Figs. 10.12 and 10.16).

Under these conditions, the vehicles tend to move very slowly ($V \approx 0$ km/h). Thus, it may be concluded, at least theoretically, from Eq. (10.29):

$$e \leq f_{Rmin} \approx 0.15 \qquad (10.30)$$

However, since a highway in general also has a longitudinal grade, the 15 percent value is actually referred to as the maximum composite three-dimensional slope of the road surface. Modern editions of highway geometric design guidelines use lower values of the composite slope, for example, 10 percent to accommodate other design criteria as well.

Minimum Radius of Curve.   Solving Eq. (10.29) for the radius, R, it follows that

$$R \geq \frac{v^2}{g\,(f_R + e)} \qquad \text{m} \qquad (10.31a)$$

or substituting the value of g and converting the speed unit from meters per second to kilometers per hour

$$R \geq \frac{V^2}{127\,(f_R + e)} \qquad \text{m} \qquad (10.31b)$$

Equation (10.31b) allows the determination of the minimum radius of curve for a given design speed, V_d. By substituting in Eq. (10.31b) the maximum permissible side (radial) friction factor, f_{Rperm}, and the maximum permissible superelevation rate, e_{max}, it follows:

$$R_{min} = \frac{V_d^2}{127\,(f_{Rperm} + e_{max})} \qquad (10.32, 10.9)$$

where R_{min} = minimum radius of curve, m

V_d = design speed, km/h

f_{Rperm} = maximum permissible friction factor in the radial (side) direction

$\quad$ ($f_{Rperm} = n\,0.925\,f_{Tperm}$)

f_{Tperm} = maximum permissible friction factor in the tangential direction

n = utilization ratio of side friction

e_{max} = maximum allowable superelevation rate, %/100

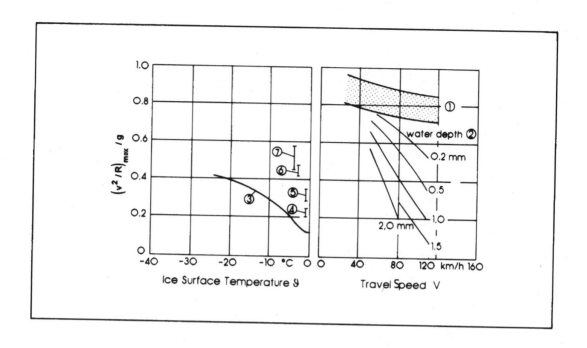

Pavement	Tire Dimension	Profile	Driving Performance
(1) dry concrete and asphalt fine concrete	6.40 - 13 to 6.00 - 15	bias-ply summer	braking
(2) wet	5.60 - 15	bias-ply summer	braking
(3) ice	5.60 - 15	bias-ply M+S*	braking
(4) ice (5) (6) new snow 10 cm (7) packed snow	5.60 - 15	summer M+S* M+S* summer M+S* M+S*	starting from zero

Legend:

M+S* = mud + snow (winter profile)

FIGURE 10.16 Possible maximum centripetal acceleration $(v^2/R)_{max}$, expressed in g's for various weather conditions. (Elaborated based on Ref. 522.)

TABLE 10.5   Observed Maximum Centripetal Acceleration $(v^2/R)_{max}$, Expressed in g's on Dry Pavements[522]

Vehicle	$(v^2/R)_{max}/g$	Remarks
Passenger car	0.73...0.8	Measured in circle with
Racing car	2.5	$R \approx 100$ m limited by
Tractor trailer	0.34...0.36	rollover
Heavy truck	0.35...0.4	
Light truck	0.5 ...0.6	

Equation (10.32) represents the basic driving dynamic equation for the cornering performance of a vehicle. Based on Eq. (10.32) and the assumptions of Table 10.1, the minimum radii given in Table 10.2 were calculated.

If larger radii of curve are combined with smaller superelevation rates, the f_{Rperm} values are often not fully utilized. The relationship between 85th-percentile speed $V85$, radius of curve R, and superelevation rate e is illustrated in Figs. 14.2 to 14.4.

Fundamentally, the radius of a curve should be as large as possible and the radii of successive curves, as well as the transitions between tangents and curves, should be well balanced. For consistency in highway design, all geometric design elements should provide, as far as economically feasible, sound and continuous operation at a certain design or operational speed. To allow the designer to achieve this, the following design, operating, and driving dynamic controls have been developed:

Safety criterion I: Secs. 9.1.2 and 9.2.2

Safety criterion II: Secs. 9.1.3 and 9.2.3

Relation design backgrounds: Secs. 9.1.3.2 and 9.2.3.2

Safety criterion III: Secs. 10.1.4 and 10.2.4

Influence of Grade and Vehicle Parameters.   Equation (10.32) provides the basis for the design of horizontal curves. Because the minimum radius of curve according to this equation is independent of any parameter of the vehicle configuration and of the longitudinal grade of the road, one could immediately conclude that any vehicle, whether it is a small, medium, or large size passenger car, a two-axle truck, or a multiaxle semitractor trailer, is governed by the same vehicle dynamics at curved sites in order to move safely. While this is true to a large extent, there are cases, however, where the point-mass model and its inherent simplification in describing the cornering performance of a vehicle according to Eq. (10.32), can lead to erroneous decisions concerning the selection of the appropriate horizontal curve radius for a specific design speed.[353,588]

This fact has encouraged researchers to attempt to extend the classical vehicle road model by a number of three-dimensional road parameters and other operational features of the vehicle.[148,149]

In this connection, Psarianos, et al.[588] carried out a study to determine the influence of vehicle characteristics (including 52 passenger cars and five heavy trucks) on horizontal curve parameters (radius, superelevation rate, and friction). The vehicle data consisted of vehicle weight, position of the center of gravity (horizontally and vertically), frontal area, aerodynamic drag coefficient, and wheel drive type, that is, front versus rear wheel drive. Thirty-nine of the 52 passenger cars were representative of the European car manufacturers; 13 represented the U.S. passenger car manufacturers.

The results of the study revealed that, on a level roadway section and for speeds of up to 100 km/h, the calculated R_{min} values for all vehicles (passenger cars and trucks), based on the assumptions of Eq. (10.32) were lower than those given in modern guidelines like the American,[5] German,[246] and the Greek.[453] By superimposing a longitudinal grade on the curve, Eq. (10.32) proved to gradually result in unsafe results, especially for grades of more than 5 percent (or 4 percent for some vehicle types) on roadways designed with design speeds of 80 km/h or more. Trucks experienced a greater chance of loss of control on wet pavements when traveling on a sloped curve, since trucks require an additional 10 percent of side friction values in order to corner a curve, as compared to the assumed friction values of passenger cars.[265,267]

For higher speed levels (that is, 120 km/h), larger differences could be identified between the safety requirements of the vehicles when negotiating a curve and those yielded by Eq. (10.32). At high speeds and limiting values of grades (that is, >5 to 6 percent), passenger cars generally require higher R_{min} values than those proposed in the guidelines.

In general, the study revealed that, at high speeds, passenger cars with heavy loads and front wheel drive require higher friction values to perform a cornering process than the same vehicles loaded only with the driver, since the position of gravity is transferred to the rear wheels. For trucks, the study indicated that the safety conditions are even more critical because the tangential friction demand alone could reach or even exceed the total available maximum permissible friction, f_{Tmax}, and, consequently side friction is no longer available.

However, further research work is needed on this issue, mainly due to its stochastic dimension (for example, selection of a representative car of the car manufacturer of a specific country or continent, loading assumptions, etc.), before the vehicle parameters and the three-dimensional configuration of the road become part of the road design process to a proper extent, thus giving rise to a three-dimensional vehicle dynamics curve design.

With respect to the three-dimensional alignment itself, numerous qualitative recommendations for good design practices are presented in Chap. 16.

Rollover.[601]   Rollover due to centrifugal force in a direction opposite to curvature cannot occur as long as the tractive moment is smaller than the resistant moment (according to Fig. 10.15), that is:

$$(F \cos \alpha - Q \sin\alpha) h \le (F \sin\alpha + Q \cos\alpha) b \tag{10.33}$$

It follows

$$F \le \frac{Q (b + h \tan\alpha)}{h - b \tan\alpha} \tag{10.34a}$$

By rearranging terms and substituting

$$F = \frac{m v^2}{R} \qquad Q = m g \qquad \tan\alpha = e \tag{10.34b}$$

the critical cornering speed and radius of curve for impending rollover is obtained for

$$v \le \sqrt{g R \frac{b + h e}{h - b e}} \qquad \text{m/s} \tag{10.35}$$

or

$$R \ge \frac{v^2}{g} \frac{h - b e}{b + h e} \qquad \text{m} \tag{10.36}$$

For the designations, compare Figs. 10.14 and 10.15.

In most highway design guidelines, criteria for horizontal curve design do not explicitly consider vehicle rollover thresholds. The rollover thresholds for passenger cars may be as high as $1.2g$, so a passenger car will normally skid off a road long before it would roll over.[265,502] Thus, the consideration of rollover thresholds is not critical for passenger cars.

However, tractor trailer combinations have relatively high centers of gravity and rollover thresholds lie in the range from 0.27 to $0.40g$.[171,172,265,502] Therefore, the margin of safety for trucks with low rollover thresholds on horizontal curves is not great, if unstable trucks travel even a little faster than the design speed.[267] This is a particular concern on freeway ramps and on minimum radius, low-speed horizontal curves.

10.2.1.3   Simulation Models*

General Development.   Schoch[632] examined new developments in vehicle engineering and simulation techniques in Germany, in order to decide whether or not the point-mass model can and should be replaced in future highway geometric design.

---

*Elaborated by J. A. Reagan and W. A. Stimpson.[595]

As previously discussed, the numerous factors involved in vehicle dynamics lead to many individual and combined movement equations which make an overview about the driving characteristic interrelationships between motor vehicles and roads very difficult. However, these characteristics should be clarified by performing, for example, simulation models for highway geometric design tasks.

Most known simulation models neglect important influencing factors and deal with individual problems and specific questions. There is no doubt that high demands are made on newly developed motor vehicles and vehicle components with regard to driving safety and comfort. Simulation techniques are widely used in the automobile industry.

However, simulation has yet not found wide application in highway engineering, although models do exist for simulating

- The traffic flow on up-grade sections[293]
- The traffic flow on two-lane rural roads[67,76]
- Passing maneuvers[329,330]
- The theoretical evaluation of design elements.[520]

Furthermore, the U.S. two-lane traffic simulation model ROADSIM,[523] the vehicle dynamic model SGUI,[622] and VDANL,[13–16] presented in the next chapter, are worth mentioning here.

Simulation models also exist for accident reconstruction, for example:

- CARAT (computer-aided reconstruction of accidents in traffic)[170]
- HVOSM (highway vehicle object simulation model)
- SMAC (simulation model of automobile collision)[310]

Volkswagen has developed a system that allows simulated rides in the laboratory to provide "real" driving data. The system offers the possibility of modeling a landscape with roads, houses, other vehicles, and persons on a computer. Based on the model, a ride through the simulated traffic world in real time is possible.[623]

The most well-known research project in Europe in this field is PROMETHEUS (program for a European traffic with highest efficiency and unprecedented safety). Thirteen European car manufacturers, and many industrial partners, as well as private and official research institutes, are participating in this project.[295] In the area of safe driving, Porsche is participating in the development of a system for friction surveillance, and in the area of cooperative driving with an automatic distance regulation system.[358,763]

Most of the previous models are related to one or several individual vehicles, which means that the results cannot be transferred directly to other vehicle types. However, in highway geometric design, many different vehicle types have to be combined in one model. Such a general model, including the human being and the often discussed limiting vehicle-, tire-, environmental-, and design-specific considerations does not so far exist.

This may be the reason why all international standards, guidelines, and norms reviewed still use the point-mass model as a simplification; this model is based on the assumption that the vehicle is a rigid body and all forces act in the center of gravity.

There are no objections to the point-mass model as long as the assumed friction factors in highway design are significantly higher than those demanded in practice. Therefore, in order to include driving dynamic considerations, at least for curve design, it becomes necessary to develop a safety criterion III for achieving driving dynamic consistency for new designs, redesigns, and RRR strategies (see Sec. 10.2.4).

Development of the IHSDM Vehicle Dynamics Module. The U.S. Department of Transportation's Federal Highway Administration (FWHA) is in the process of creating the interactive highway safety design model (IHSDM). The IHSDM will consist of a series of CAD-interactive computer software modules (Fig. 10.17) which enable designers and design reviewers to evaluate the safety of alternative geometric designs for both new construction and reconstruction.

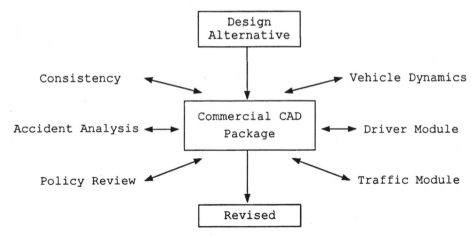

FIGURE 10.17 Interactive highway safety design model (IHSDM).[594]

The first module of the IHSDM to reach a prototype stage suitable for demonstration to the design community is the vehicle dynamics module (VDM). This section reviews the functional objectives and capabilities of this module, plus some of the key issues involved in its application.

Initial concept.   The VDM was conceived as a user-friendly means of providing design engineers with safety-related information on any alternative alignment. The VDM would allow a designer to "drive" the design vehicle through the alternative design and develop a speed profile and data on lateral accelerations. This ability to travel through the design will give the designer a visual method for looking for poor design situations. Concurrently, the VDM would predict the speed-dependent lateral accelerations which largely determine a vehicle's probability of staying safely on the road. Excessive values or rapid changes of lateral acceleration may indicate an incompatible or inconsistent geometric design in need of further evaluation.

The VDM will also play a major future role in research activities. A study has been programmed for 1997 in which the VDM will be used to evaluate the effects of slope and ditch design on rollover accidents. Finally, the VDM will be a key component in human factor studies of the human-vehicle-roadway environment.

Development of VDM prototype.   In September 1993, the FHWA awarded two contracts to develop alternative prototype versions of the VDM. One contract was with the University of Michigan Transportation Research Institute (UMTRI) and the other contract was with Systems Technology, Inc. (STI). UMTRI and STI have had extensive experience in the development of models for simulating vehicle dynamics. Both organizations have also developed closed-loop driver models that allow the vehicle to follow the roadway.

The feasibility of the VDM depends on two main factors. The first deals with the module being able automatically to extract and use a mathematical description of the driving surface from a computer-aided design (CAD) representation of the roadway. The second deals with the visualization of the roadway from the driver's perspective as the vehicle moves along the roadway.

The goal is to make the IHSDM highly interactive, using pull-down menus providing a comfortable look and feel to the experienced CAD user. In this interactive work environment, various IHSDM modules (such as the VDM) would be called up by a user from within the CAD package.

STI approach.   The STI effort built upon past successes in developing, testing, and using the vehicle dynamic analysis nonlinear (VDANL) model. VDANL is a comprehensive vehicle dynamics simulation program configured to run on personal computers. The model was initially intended for the analysis of passenger cars, light trucks, and multipurpose vehicles, but as part of STI's VDM contract, it was upgraded to handle articulated vehicles such as tractor semitrailers.

A recent paper[16] provides an overview of VDANL and its proposed use within the IHSDM (see

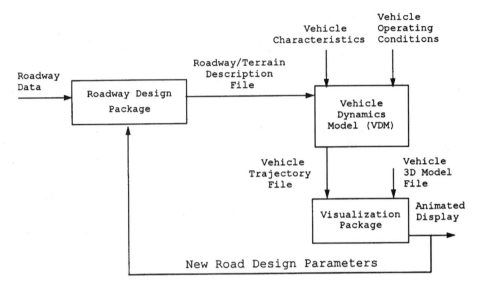

FIGURE 10.18   VDM role within IHSDM.[16]

Fig. 10.18, where the VDM is VDANL). Due to the underlying complexity of the software and its detailed inputs, STI was compelled to develop a simplified user interface for highway designers. This interface receives as inputs the user's choices of roadway design file, vehicle type, and desired maneuver (for example, lane following). It then initiates the simulation using vehicle and driver characteristics stored in hidden parameter files. Users can view time traces of various simulated vehicle performance measures, as well as vehicle roadway scenes produced by a third-party graphics program.

UMTRI approach.   In this effort, the contractor first developed a comprehensive vehicle simulation capability using AUTOSIM software accessed through a simulation graphical user interface.[622] To maximize its efficiency of application in highway design evaluation, the simulation was then used by the researchers to develop a series of vehicle performance signatures. Each "signature" is a plot of a given performance measure versus speed (or a speed derivative). For example, Fig. 10.19 shows body roll (an indicator of rollover potential) versus lateral acceleration, with rollover for this vehicle occurring at $0.36g$.

When a highway designer executes UMTRI's VDM, he or she causes the VDM to combine its prestored vehicle performance signatures with IHSDM-supplied data on speed, horizontal radius of curvature, and superelevation to compute lateral acceleration and related performance measures by station. As with VDANL's application, the user can then examine the performance profiles to identify questionable locations along the proposed alignment.

Currently, this version of the VDM (called *vehicle dynamics models for road analysis and design* or *VDM ROAD*) uses a fairly rudimentary wire-frame visualization program developed by UMTRI. On the other hand, VDM ROAD (unlike VDANL) now runs on either DOS-based PCs or UNIX-based workstations.

VDM application issues.   Numerous application issues are likely to arise as the two versions of prototype VDM are used, refined, and further tested. A few of the issues already identified are discussed briefly in the following.

Sources of vehicle speed profile.   Previous papers on the IHSDM and VDM have indicated different sources for the vehicle speed profile that so strongly influences other measures of vehicle dynamics and safety. It now appears unlikely that any mechanism within the VDM itself will realistically determine speeds, since this module predicts vehicle responses to driver-induced maneuvering, not roadway-dependent driver speed preferences. The more likely sources for vehicle

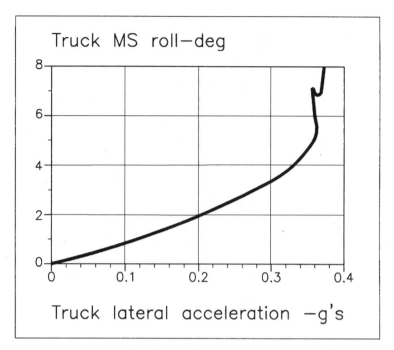

FIGURE 10.19   Vehicle performance "signature."[622]

speeds are the consistency module (for macroscopic project-level evaluations) and the driver module (for microscopic situation-specific evaluations). For example, for the consistency module, the regression models for operating speed backgrounds of Table 8.5 could be used as a basis with respect to the country under study.

User access to vehicle and driver parameters.   It seems unlikely that many highway designers or design reviewers will want to specify or experiment with detailed vehicle and driver descriptive parameters. It might also be somewhat risky to allow persons who are not experts in vehicle dynamics or human factors to change values for parameters they might not adequately understand. One approach that has been suggested would be to limit access to such parameters to a "keeper of the model" within each agency. FHWA will seek additional suggestions on this issue in the near future.

Evaluation scenarios.   Much has been accomplished in the way of developing powerful CAD-interactive VDM simulation/visualization tools. Less has been accomplished in the way of identifying suitable specific uses for those tools. Allen, et al.[16] give curve-following and (lane-changing) crash-avoidance maneuvers as "simple examples of VDM application"; however, IHSDM researchers need to make more systematic, ambitious efforts to identify roadway and driving situations of greater criticality and interest to designers and (potentially) less interest to vehicle dynamics experts. A start on addressing this need has been made in two FHWA contracts aimed at developing specifications for the IHSDM driver module.

Relevant performance measures.   As mentioned previously, the primary measure initially sought from the VDM was lateral acceleration. Since designers do not customarily refer to this measure directly, however, it may be better to proceed as Lamm, et al. have done and compute instead the lateral friction demanded by the driver's speed choice and compare it to the lateral friction assumed by the designer.[424,428,435,447,448,458]

Another key vehicle dynamics measure, particularly related to the operation of large vehicles, is rollover potential. Prediction of this hazard would be especially helpful in evaluating interchange loop ramps and intersection turning roadways, where truck drivers sometimes fail to recognize the limitations of speed-change lanes designed for faster-decelerating passenger cars.

Although not strictly a vehicle dynamics variable, sight distance is an important highway safety characteristic well suited to display in a drive-through visualization (as well as in a separate plot by station). This being the case, the two VDM research contractors were asked to include a calculation of vertically limited sight distance in their prototype software.

Visual displays.   There are well-recognized trade-offs between the number of polygons rendered on a computer monitor and the speed of rendering them. Hence, to reduce hardware/software needs and provide smoother, more responsive animation, the rendering of graphical details not essential to a particular design evaluation task should be minimized.

On the other hand, users of IHSDM/VDM visualization tools need to be cautioned on the practical limitations imposed by computer displays. Of special concern are situations where unwarranted conclusions might be inadvertently drawn regarding highly visual issues, such as obstacle visibility, sight distance to simulated roadway hazards, and sign legibility. It is important to realize that most computer-driven displays do a mediocre job at best in replicating relevant real-world visual conditions, such as surface texturing, color rendition, and brightness contrast.

Additional refinements to the STI and UMTRI prototype versions of the VDM are likely to be made by staff in the FHWA Geometric Design Laboratory.

Closure.   The FHWA approach has been to treat the vehicle dynamics module as an independent, major component of the IHSDM. The initial goal in the development of the VDM is to develop a tool to assist the designer. However, accident analysis has identified driver error as the major cause of highway accidents. Future design procedures must take into account driver perception of the roadway. The FHWA has awarded two contracts to begin the preliminary development of a driver module for use with the vehicle dynamics module. The VDM, coupled with an open-loop driver module, will allow U.S. researchers to study how the roadway environment influences the driver's risk evaluation, speed selection, effects of fatigue, and other factors related to accidents.

Additional information on software for CAD interactive tools with respect to an overall safety evaluation process, also including driving dynamic issues, is presented in Sec. 18.2.

Preliminary Conclusion.   An in-depth literature review about vehicle dynamic considerations and simulation models revealed that those tools can be helpful in clarifying the complex interrelationships between vehicles, roads, and human beings. Simulation models already support modern vehicle development and are also used with success for specific highway engineering and accident reconstruction models. But no model could be found, to date, that combined in an overall sense highway geometric design tasks with safety-related issues in connection with the numerous vehicle-, tire-, road-, and environmental-specific influencing components. Therefore, it was decided to continue using the simple point-mass model with its well-known disadvantages.

10.2.2   Establishment of Permissible Friction Factors

10.2.2.1   Skid Resistance

General.   Skidding accidents are of major concern in the field of highway safety. It would be wrong to attribute skidding accidents *only* to driver error or driving too fast for existing conditions because the roadway must provide a level of skid resistance to safely accommodate braking and steering maneuvers that can be expected at a particular site.

Skid research investigations by Mason and Peterson[496] have indicated that sufficient friction supply should be regarded as an important safety issue. Brinkman[81] found that resurfacing did not alone have a significant effect on the mean skid number. He indicated that skid resistance should be regarded as a major safety issue when resurfacing roadways. Glennon, et al.[217] indicated that the probability that a highway curve may become an accident black spot increases with decreasing pavement skid resistance.

Therefore, a skid-resistant pavement surface is a key element in the prevention of skidding accidents. Most dry pavements provide adequate skid resistance. However, special consideration must be given to the development of skid resistance of wet pavements. The four main causes of poor skid resistance on wet pavements are rutting, polishing, bleeding, and dirty pavement. Rutting causes water accumulation in the wheel tracks. Polishing reduces, and bleeding covers, the microtexture. In both cases, the harsh surface features needed for penetrating the thin water film are diminished. Dirty pavement or pavement contaminated by oil drippings or layers of dust or organic matter will lose its skid resistance.[5]

Skid-resistance research investigations in Europe and the United States[437] have indicated that sufficient friction must be considered an important safety demand, a demand which has been considered in Germany since the 1970s.[382,383,385] Note that reliable friction coefficients have been developed for the German highway geometric design guidelines[124,241,243] and other countries,[91,447,453,686,689,692] or are in preparation.[435,438,443,448]

In this connection, it is interesting to note that Europeans and Americans always require good skid-resistance values for resurfacing projects (especially in curves), since obviously in the last decades on both continents often only insufficient skid-resistance values could be provided. This fact led to numerous accident black spots, especially in curved roadway sections, since with decreasing skid resistance, the risk of accident greatly increases.

Sufficient skid-resistance measurements for the establishment of an evaluation background for the relationship between skid resistance (friction) and speed are not available in many countries.

Therefore, a reliable procedure for the development of such a relationship will be presented in "Tangential Friction Factor" in Sec. 10.2.2.2. In order to achieve this goal, a comparative analysis of the tangential and side friction factors in the highway geometric design guidelines of the United States and of four western European countries—France, Germany, Sweden, and Switzerland—was conducted. Relationships between friction factors and design speed were developed for each of the subject countries. On the basis of the friction and speed data of all the countries, an overall regression relationship between friction and speed was established. This overall relationship was later compared with actual pavement friction inventories, in order to examine its reliability.[420] The reason for doing so was to provide those countries that did not have satisfactory skid-resistance values or simply had insufficient skid-resistance measurements with current state-of-the-art recommendations for the establishment of reliable tangential and side friction factors.

*Measurement Procedure.** The skid resistance (frictional properties) of a pavement's surface is one of the major factors in determining the safety of a highway. The critical aspect of skid resistance is the available friction when a pavement is wet, since almost all pavements have more than adequate friction for safe vehicle maneuvering during dry conditions. Previous studies have clearly demonstrated that the friction levels developed by a pavement in contact with a given tire are largely dependent on two characteristics: surface macrotexture and surface microtexture.

Depending on the road surface type and the speed, the resulting skidding values for individual surface types with different macrotexture and microtexture are shown in Fig. 10.20.

Macrotexture is defined as those surface textural features that are greater than 0.5 mm in height and therefore provide a drainage system for water on the pavement surface, thus preventing a buildup of water between the tire and the pavement and the resulting hydroplanning. Additionally, the macrotexture provides the hysteresis component of the tire-pavement friction, that is, the energy loss as the tire deforms around the macrotexture asperities.

Microtexture is defined as those surface features less than 0.5 mm in height. Its role in friction development is to penetrate the thin water film present on a wet pavement so that the intimate tire-pavement contact is maintained.

Basic assumptions. *Skid resistance* is the force developed when a tire that is prevented from rotating slides along the pavement surface. It is the antonym of slipperiness. The term *coefficient of friction* is used in the mechanical sense as the ratio of the frictional resistance to motion in the plane of the interface between two bodies to the load perpendicular or normal to the plane.

When a tire rolls, slips, or slides on pavement, however, various conditions influence the amount of friction developed, and most of them are difficult to describe and measure. This is particularly true when water is present at the interface. In this case, the preferred term, instead of the coefficient of friction, is *friction factor*, which is also called the *tangential friction factor, f_T*:

---

*Elaborated based on Refs. 5, 130, 131, 228, 288, 401, and 443.

Type	Presentation	Description
1		Neither macrotexture nor microtexture.
2		Only microtexture.
3		Macrotexture (through rifts and hollows) and microtexture.
4		Macrotexture and microtexture as for type 3, with low density of the macrotexture.
5		Macrotexture (through interim spaces between the coarse grains) and microtexture (through the edge sharpness of the coarse grains).
6		Macrotexture (through interim spaces between the coarse grains), but no microtexture.
7		Macrotexture (through rifts and hollows), but no microtexture.

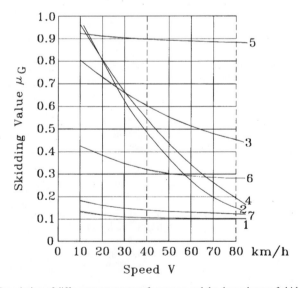

FIGURE 10.20   Description of different pavement surface types and the dependence of skidding values on speed for wet road surface and different surface types.[597]

$$f_T = \frac{F_T}{Q} \qquad \text{see Eq. (10.15)}$$

When describing pavement characteristics, it is technically wrong to say that a pavement has a certain friction coefficient since the coefficient always involves two bodies, each of which may have a large number of variables contributing to the friction. It is imprecise to say a particular tire on a given pavement produces a certain friction factor, unless forward or sliding speed, inflation pressure load, temperature, water-film thickness, and other details are specified. To overcome the resulting problems, standards have been developed which prescribe all variables influencing the friction factor. The current standards and vehicles for determining friction are designed to provide a relative indicator for comparing different pavement surfaces.

Skid-resistance measurements establish the effect of the coarseness (roughness) of the surface texture by measuring the friction resistance for certain test conditions in a strictly defined manner. These test conditions depend on the respective applied investigation method and the country where they were developed. Because of the great number of existing test procedures, this study will discuss only some common test methods. Important parameters for these test methods are

1. Tire type (traction tire type, for example, Phoenix P3 or PIARC standard tire)
2. Wheel load (for example, 3500 N)
3. Inflation pressure load (for example, 1.5 atü)
4. Moistening level of the road surface (for example, 1-mm water film)
5. Speed (for example, 40, 60, and 80 km/h)
6. Friction condition

Skid-resistance measurements have been conducted in Germany and other European countries since the 1960s by applying the retarded wheel procedure (100 percent slippage) for confined test conditions (see Fig. 10.10).

Consultations with responsible transportation agencies in Europe and the United States have revealed that skid-resistance measurements are usually conducted by test vehicles in the following way:

1. An automotive vehicle with one or more test wheels incorporated into it or forming part of a suitable trailer towed by a vehicle.

2. A transducer, instrumentation, water supply, and a proper dispensing system and actuation controls for the brake of the test wheel. The test wheel is equipped with a standard test tire. (The test tire is different in different countries.)

3. The test apparatus is brought to a desired test speed. (The test speeds are different in different countries as they differ in different states of the United States.) For example, in the Federal Republic of Germany a road section is tested at speeds of 40, 60, and 80 km/h. For evaluating skid resistance, the standard procedure is to compare the measured values with the standard values of the 90th-percentile distribution curve of the evaluation background according to Fig. 10.25 (Germany), for example. In this case, the standard threshold values for skid resistance are:[208]

0.42 for V = 40 km/h

0.33 for V = 60 km/h

0.26 for V = 80 km/h

Similar recommendations exist in several other European countries and the United States.

4. Water is delivered ahead of the test tire and the braking system is actuated to lock the test tire. (For the test, a water-film thickness of 1 mm is widely used.[401])

5. The resulting friction force acting between the test tire and the pavement surface and the speed of the test vehicle are recorded with the proper instrumentation. The skid resistance of the paved surface is determined from the resulting force torque record and reported as the coefficient of friction, friction factor, skid number, or friction number. These values are determined from the force required to slide the locked tire at a stated speed divided by the effective wheel load. The wheel load depends on the weight of the test trailers used in the different countries.

Vehicles were developed for the measurement of the skid resistance. These vehicles drive with constant speeds and carry along one or two additional wheels as "measurement wheels" between the front or rear wheels or at the rear end of the test vehicle.

For moistening, the measurement vehicle carries along a water-supply tank, from which the water is delivered immediately ahead of the measurement wheel. The level of moistening is determined by the computational water-film depth. This depth is determined from the water-strip width multiplied by the distance length per time unit. Normally, a water-film depth of 0.5 or 1.00 mm is selected; it should be noted that, with increasing water-film depth, the tangential friction factor decreases. The moistening of the surface is speed-dependent, and for different speeds a constant water-film depth has to be reached. The most common alternatives of the test apparatuses are, in Europe:

1. The locked wheel SRM (Stuttgarter tribometer)
2. The slanted wheel SCRIM (side force coefficient routine investigation machine)
3. The rolling wheel with constant slippage (retarding wheel)

The locked wheel SRM (Stuttgarter tribometer).   Since the 1950s, skid-resistance measurements with the locked test wheel—the Stuttgarter tribometer—have been routine in Germany. The measurement wheel is mostly carried along as a towed wheel (fifth wheel) or in a trailer. The test wheel will be braked by an independent braking system until the wheel is locked and after a skidding distance of, for example, 20 or 40 m, it is released again. With the locked wheel, the danger braking of a motor vehicle in heavy rain is simulated. The test procedure is conducted intermittently, which means the free rolling phase follows the braking phase. By this method, the skid-resistance force is measured.

Figure 10.21 shows the Stuttgarter tribometer (SRM).

The slanted wheel SCRIM.   Since the 1970s, measurements with the English SCRIM (side-force coefficient routine investigation machine) apparatus have been conducted in different countries (Fig. 10.22). In this test procedure, the side friction coefficient is determined by a test tire without traction for a slanted driving angle of 20 degrees. In this procedure, the breaking out of a motor vehicle in light rain is simulated.

The SCRIM procedure offers a good alternative to the Stuttgarter friction apparatus because of costs, capacity, required measurement personnel, and the transfer of the measurement results into practical decisions.

Swedish skiddometer.   Besides the procedures already presented, there is also the Swedish skiddometer, which is able to measure the skidding value alternatively at the locked wheel or the braking force value for constant slippage (Fig. 10.23).

Other methods.   We should also mention the rolling wheel with constant slippage procedure, where the measurement wheel is forced to roll with constant slippage in comparison to the rolling wheels of the vehicle.

In addition, some countries use a hand-carried device, the SRT (skid-resistance tester, see Fig. 10.24).

Finally, it should be noted that other countries, such as the United States and Greece, use the standard test method known as ASTM (American Society for Testing and Materials) E 274.[228,377,640]

The working procedure of ASTM E 274 follows the general description (issues 1 to 5) presented earlier in this chapter. The skid resistance is reported as the skid number, which is determined from the force required to slide the locked test tire at a stated speed divided by the effective wheel load and then multiplied by 100.

Measurements made in accordance with this standard are reported as skid number (SN) or friction number (FN) as follows:

1) Water run off instrumentation and digital data collection
2) Utility box
3) Water supply tank
4) Water valve
5) Measurement wheel

FIGURE 10.21   The Stuttgarter tribometer (SRM).[131]

$$SN = FN = 100\,f_T = 100\,\frac{F_T}{Q} \tag{10.37}$$

where F_T = tangential force, N
Q = weight force, N
f_T = tangential friction factor

The F_T value in Eq. (10.37) is obtained by sliding a locked, standardized tire at a given speed on an artificially wet pavement.

Of course, other testing procedures for measuring skid resistance also exist but we cannot describe all of them here.

FIGURE 10.22   The slanted wheel (SCRIM).[131]

FIGURE 10.23   Swedish skiddometer. (*Source:* Unknown.)

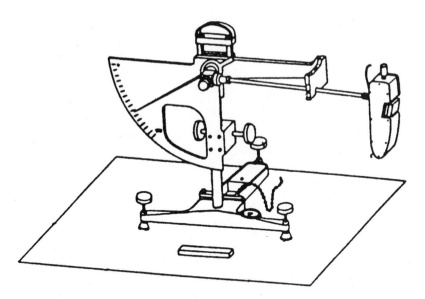

FIGURE 10.24   Skid-resistance tester (SRT). (*Source:* Unknown.)

Evaluation Backgrounds.   Since the 1970s, the Permanent International Association of Road Congresses (PIARC) has been striving to standardize the skid-resistance measurement procedures in the European countries. For several countries, including Germany, that meant that the previously used measurement tires (Germany used diagonal type PHOENIX, Fig. 10.25) had to be replaced by the radial PIARC standard tire (Fig. 10.26). In this respect, the development of a new skid-resistance evaluation background became necessary. In addition to increased doubts about whether or not the measurements conducted prior to the 1970s are still valid, the possibility also exists that the skid-resistance level may have changed overall because many new road surfaces have been built.[130]

The skid-resistance evaluation background should show the range of values which cover the most slippery road surface and the most skid-resistant road surface, and the corresponding frequency distribution with respect to the road network. The evaluation background should encompass a representative body of road surfaces of different ages, different construction methods, and roads of different categories. The evaluation backgrounds in Figs. 10.25 and 10.26 were established by skid-resistance measurements using the Stuttgarter tribometer (SRM), based on the normal distribution of skidding values for speeds of 20, 40, 60, and 80 km/h. The curves in these figures correspond to frequency limits that indicate the percentage of roads with higher skid resistance.

Exemplarily, Fig. 10.25 shows that 80 percent of the roads, which are limited by the 10th- and the 90th-percentile level distribution curves, lie in a relatively narrow band of skid resistance. The wide range of the overall skid-resistance background for wet road surfaces shown in Fig. 10.25 results from roads with extremely good skid-resistance values and roads with extremely bad skid-resistance values. These cases represent only 20 percent of the overall road surfaces. For instance, the standard threshold values of skid resistance,[208] presented in the last subsection, were assigned to the 90th-percentile level distribution curve shown in Fig. 10.25, which means 90 percent of roads would lie above the assessed standard values.[159]

The new German evaluation background for wet road surfaces was introduced in 1980. Evaluations of dry and snowy/icy road surfaces were completed by 1992 (Fig. 10.26).[130,131]

According to Ref. 130, the increase in the skid-resistance level was not significant, and Eq. (10.38), developed by Lamm and Herring,[382,383,385] could still apply. Equations (10.38) and (10.39) are based on the 95th percentile distribution curves.

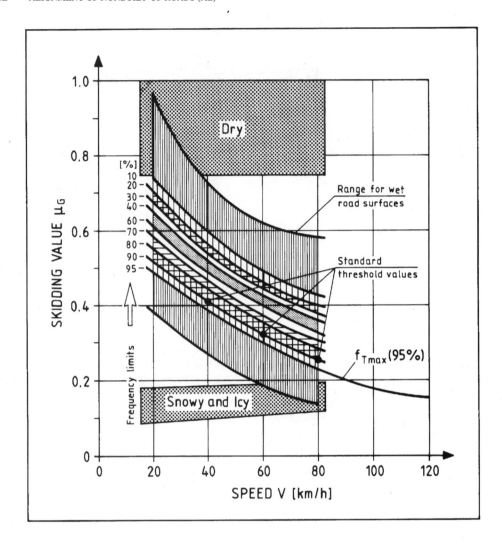

$$\mu_G = f_{Tmax} = 0.214 \left(\frac{V}{100} \right)^2 - 0.640 \left(\frac{V}{100} \right) + 0.615 \qquad (10.38)$$

where μ_G = skidding value

f_{Tmax} = tangential friction factor

V = speed, km/h

Equation (10.38) was developed in Refs. 382 and 383.

FIGURE 10.25 Old evaluation background up to the 1970s for skid-resistance measurements with the SRM (Phoenix P3 standard tire), Germany.[131,208,349,640,747]

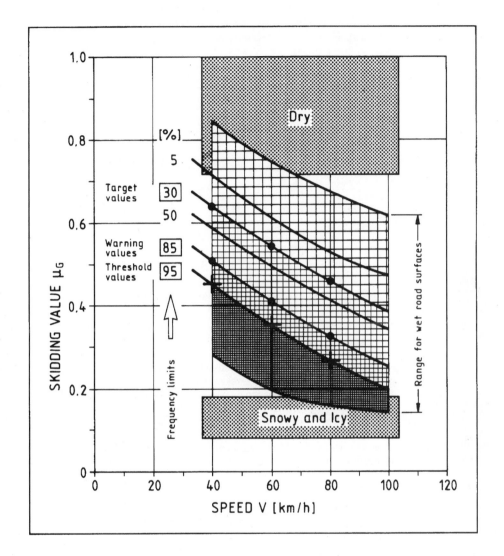

$$\mu_{G\text{new}} = f_{T\text{max}} = 0.241 \left(\frac{V}{100}\right)^2 - 0.721 \left(\frac{V}{100}\right) + 0.708 \qquad (10.39)$$

where the variables are the same as for Eq. (10.38) in Fig. 10.25.

FIGURE 10.26   New evaluation background with completions until 1992 for skid-resistance measurements with the SRM (PIARC Europe standard tire), Germany.[130,131,159]

However, because the PIARC Europe standard tire results in higher skid-resistance values, the standard threshold values in Germany have been increased. They are now related to the 95th-percentile distribution curve of Fig. 10.26, which means that only 5 percent of road surfaces are falling short of the new standard values.

Consequently, the new threshold values for skid resistance (Fig. 10.26) in Germany are

$$0.45 \quad \text{for} \quad V = 40 \text{ km/h}$$

$$0.35 \quad \text{for} \quad V = 60 \text{ km/h}$$

and

$$0.27 \quad \text{for} \quad V = 80 \text{ km/h}$$

For comparison, the skid-resistance values, based on Eq. (10.2) and proposed in "Tangential Friction Factor" in Sec. 10.1.2.2 are repeated here:

$$0.42 \quad \text{for} \quad V = 40 \text{ km/h}$$

$$0.35 \quad \text{for} \quad V = 60 \text{ km/h}$$

and

$$0.30 \quad \text{for} \quad V = 80 \text{ km/h}$$

A comparison of the German values with the ones proposed by the authors reveals that the recommended values are very reasonable. The value of 0.30 corresponding to the speed of 80 km/h appears to be in order to force contractors and highway agencies to use good aggregates for the construction of road surfaces of high-speed roads and to provide reserves for the "aging process." In this respect, we should note that the aggregates used in many countries are less skid resistant than those of Germany and the United States (for information on the State of New York, see the next chapter).

Extensive comparative skid-resistance measurements which were conducted by using the SRM (Stuttgarter tribometer) and the SCRIM (the slanted wheel), in order to determine the suitability of these two systems for future routine measurements in the field of road maintenance,[131] have revealed that both procedures are suitable for

- Evaluating road surfaces with respect to skid resistance
- Evaluating roadway sections with respect to the accident risk on wet road surfaces
- Conducting measurements for pavement management

Table 10.6 shows the numerical differences for warning and threshold values resulting from the two procedures.

Overall, a very good agreement could be observed, and the threshold values shown agree well with those proposed in "Tangential Friction Factor" in Sec. 10.1.2.2 by the authors, especially

TABLE 10.6 Comparison of the Required Values for Road Skid Resistance from SRM and SCRIM Measurement Results[131]

	Frequency, %	V = 40 km/h		V = 60 km/h		V = 80 km/h	
		SRM	SCRIM	SRM	SCRIM	SRM	SCRIM
Warning value	85	0.51	0.52	0.41	0.44	0.33	0.37
Threshold value	95	0.45	0.45	0.35	0.37	0.27	0.30

those of SCRIM for higher speeds.

For comparative purposes, skid numbers from the United States are presented in the following. The State of Ohio, for example, uses the following skid numbers (SN):[228]

Above 40: Normally provides adequate skid resistance

30–40: Dependent on other roadway factors

Below 30: Warrants further investigation

These skid numbers divided by 100 according to Eq. (10.37) for SN also agree well with the recommended skid-resistance values in Sec. 10.1.2.2.

A Kentucky Department of Transportation (DOT) report, cited by Goyal,[228] analyzed skid numbers (SN) as they relate to accident frequencies and types of pavement on two-lane roads. The report presents extensive analyses, including types of aggregate, pavement ages, traffic volumes, and the costs relating to skid resistances. The report concluded that a mature road surface should have an SN of at least 32 on roads having an AADT of more than 3500 vehicles/day. It was noted that burlap-dragged concrete and most normal asphalt pavements with an AADT of more than 3500 vehicles/day must be surface treated to maintain an SN of at least 32. On lower-volume roads, the lowest acceptable SN drops to 28 for a mature pavement. Lower-volume roads (still AADT > 1000) do not deteriorate as rapidly, allowing most normal pavement to meet the lower SN value of 28.

Because of variations in testing procedures (for example, test tires), one should keep in mind that the friction data resulting from using different procedures may be biased. But the fact remains that the basic methods employed to measure skid resistance are, to a certain extent, comparable between countries. Therefore, a comparison of skid-resistance and friction values of different countries should be allowed from a research standpoint, because differences exist in every research field, such as medicine, engineering, and so on, in testing as well as in reporting procedures. In performing comparative analyses of data from different countries, the possibility always exists that the data may be biased.

How the issue of friction and, equally important, the issue of speed are being applied in relation to geometric design in several European countries and in the United States will be the subject of discussion in the following section.

Because of the lower coefficients of friction on wet pavements than on dry, the wet condition governs in determining stopping sight distances and radii of curve, as revealed in the studied design guidelines. Furthermore, the countries in this study assume that the coefficients of friction used for design purposes should represent not only wet pavements in good condition but also surfaces approaching the ends of their useful lives. The maximum allowable friction values should encompass nearly all significant pavement surface types and the likely field conditions.

10.2.2.2   *Tangential Friction Factor**

Comparison of Different Countries.   The data in Table 10.7 represent the maximum permissible tangential friction factors for wet pavements with respect to the design speed contained in the highway design guidelines of the United States (U.S.),[5] Federal Republic of Germany (FRG),[241,243] France (F),[510] Sweden (S),[686] and Switzerland (CH).[691]

Figure 10.27 gives an overview of the maximum permissible tangential friction factors of the European guidelines studied and of the United States. Note that, with the exception of France, all relationships in Fig. 10.27 are quadratic. It should be noted that the European countries in this study were considered typical European countries by a survey of current geometric and pavement design practices in Europe conducted by the International Road Federation in 1985.[271]

Figure 10.27 reveals that: (1) as design speeds increase, friction factors decrease; (2) the friction speed curves for Switzerland and Germany are nearly parallel, with the friction values of

---

*Elaborated based on Refs. 414, 420, 428, 435, 447, 448, and 458,

TABLE 10.7 Maximum Permissible Tangential Friction Factors for Different Design Speeds in Selected Countries

Design speed, km/h	Tangential friction factors f_T (rounded values)				
	U.S.	FRG	F	S	CH
30	0.40	0.43		0.46	0.54
40	0.38	0.39	0.37	0.44	0.50
50	0.35	0.36		0.41	0.45
55	0.34	0.32			0.40
60	0.31	0.30	0.37	0.39	0.39
65	0.32	0.29			0.37
70	0.31	0.27		0.36	0.35
80	0.30	0.24	0.33	0.34	0.32
90	0.30	0.22			0.30
95	0.29	0.20			0.29
100	0.29	0.19	0.30		0.28
105	0.29	0.18			0.27
110	0.28	0.17			0.26
120		0.16	0.27		0.25

Note: Values are converted from miles per hour to kilometers per hour, since the basic investigations were conducted in the United States.[228]

Switzerland higher by about 0.1; (3) the tangential friction values of Sweden are limited because of a maximum design speed of 80 km/h on two-lane rural roads in Sweden; and (4) U.S. values intersect the German curve at a design speed of approximately 55 km/h and the Swiss curve at a design speed of approximately 90 km/h.

In comparison with the other countries, the United States and, to some extent, France have the lowest differences in friction values in the design speed range between 50 and 110 km/h according to Table 10.7 and Fig. 10.27. These small differences in the friction values, or these low-speed gradients of tangential friction, clearly contradict the worldwide research experience which shows that friction values should substantially decrease with increasing speeds. If this experience is not met, critical driving maneuvers may occur, especially when operating speed exceeds design speed by considerable amounts under wet pavement conditions at curved sites.[348,385,470]

Overall Regression Curve. Based on the data of the five countries studied, the following overall regression equation between tangential friction factor f_T and design speed V_d was developed:

$$f_T = 0.59 - 4.85 \times 10^{-3} V_d + 1.51 \times 10^{-5} (V_d)^2$$

$$R^2 = 0.731$$

$$\text{SEE} = 0.044$$

(10.40)

where f_T = tangential friction factor
    V_d = design speed, km/h
    R^2 = coefficient of determination
  SEE = standard error of estimate

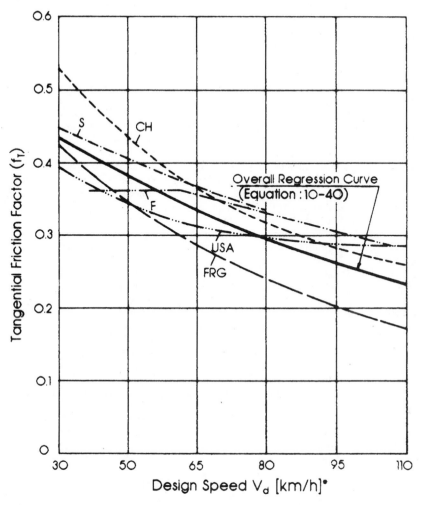

FIGURE 10.27   Relationships between maximum permissible tangential friction factor and design speed for different countries along with the overall regression curve.[228,414,420,428]

The high R^2 value and the low SEE value of Eq. (10.40) indicate that the relationship between tangential friction and design speed is a strong one. *Equation (10.40) was taken as the basis for the overall tangential friction equation, Eq. (10.2).*

Figure 10.27 shows the calculated values of the tangential friction factor [Eq. (10.40)] as a solid line superimposed on the curves of the countries under study. This figure indicates that: (1) the Swiss and Swedish tangential friction values are higher than the tangential friction values of the overall regression curve; (2) for design speeds greater than 55 km/h, the French values are higher, whereas the German values are lower; and (3) the U.S. tangential friction values intersect the overall regression curve at a design speed of about 80 km/h. For design speeds greater than 90 km/h, the French and U.S. tangential friction values are higher than the tangential friction values of the other countries.

The overall regression curve in Fig. 10.27 can be used as a reference for the countries under study to decide where their assumed tangential friction factors lie, and, if necessary, use the values of this curve [Eq. (10.40)] to determine reliable tangential friction factors for future highway design tasks as will be shown in the following section.

Comparison with Actual Pavement Friction Inventories.   In order to show the extent to which the developed overall regression curve (Fig. 10.27) agrees with actual pavement friction inventories, friction inventories from New York State (United States), developed by Goyal;[228] from Germany, developed by Wehner, et al. (see, for example, Refs. 640 and 747); and from Greece, developed by Kanellaidis, et al.[324] were used (Figs. 10.28 to 10.30). For all three friction inventories, skid-resistance values of up to 80 km/h were included.

The relationship in Fig. 10.28 indicates that the overall regression curve [Eq. (10.40)] clearly coincides with the 90th-percentile level distribution curve of New York State. That means 90 percent of wet pavements could be covered by using the overall regression curve as a driving dynamic basis for design purposes. Figure 10.29 shows that the overall regression curve could cover about 80 percent or more of wet pavements in Germany.

For example, stopping sight distances and minimum radii of curve in Germany are assigned to the 95th-percentile level distribution curve of the skid-resistance background shown in Fig. 10.29.[241,243,382,383,385]

However, it was found that, compared with the measured skid-resistance values in Greece (Fig. 10.30), only approximately 20 percent of wet pavement could be covered by using the overall regression curve. This may also occur in other countries when one attempts to establish the corresponding skid-resistance backgrounds, which means that use of the 95th-, 90th-, or even 80th-percentile level distribution curves used so far for the arrangements of maximum permissible tangential friction factors may, for some countries, lead to irresponsible designs from economic and environmental viewpoints. To respond to safety demands and to fill the gap between design values and actual friction potential, higher quality standards for skid resistance should be set when designing and constructing pavements in these countries.

Based on these considerations, possible solutions for the selection of sound friction factors are discussed in Sec. 10.2.3.

10.2.2.3   Side Friction Factor.*   Based on international experiences, it can be assumed that the side friction factor on rural roads should correspond to a utilization ratio of $n = 45$ percent with respect to the tangential friction factor in predominantly flat topography, as well as in snow-free areas. This assumption was made consciously, since this value (45 percent) comes nearest to the assumptions in the highway geometric design guidelines of Germany,[241,243,246] Greece,[453] Switzerland,[691] and Sweden.[686] In areas with hilly or mountainous topography, a utilization ratio of $n = 40$ percent is considered reasonable.[447,453] Using utilization ratios of side friction of $n = 45$ percent or $n = 40$ percent, according to Fig. 10.13, 90 percent and 92 percent, respectively, are still available for friction in the tangential direction when driving through curves; that means that from a driving dynamic viewpoint, decisive friction reserves are present. Those reserves are necessary in order to make possible a safe curve cornering in such a way that in spite of centrifugal forces the vehicle can keep its track during acceleration and deceleration maneuvers. Thus, safety is guaranteed if the side friction assumed is higher than the actual side friction demanded, the basic consideration for the development of safety criterion III in Sec. 10.2.4.

Overall Regression Curve.   The data presented in Table 10.8 give the maximum allowable side friction factors for wet pavement with respect to the design speed applied in the highway design guidelines of the five countries studied.

Figure 10.31 provides an overview of the maximum allowable side friction factors in the European guidelines and for highway design in the United States with respect to design speed. Note that, with the exception of the United States, all relationships in Fig. 10.31 are quadratic.

---

*Elaborated based on Refs. 414, 420, 428, 435, 447, 448, and 458.

SKID RESISTANCE BACKGROUND

(U.S.A.: NEW YORK STATE)

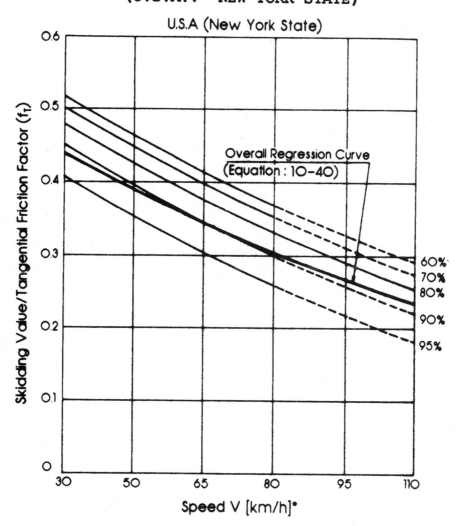

FIGURE 10.28 Percentile distribution curves for the relationship between tangential friction factor and speed for 93 wet pavements in the United States (New York State) along with the overall regression curve.[228,414,420,428]

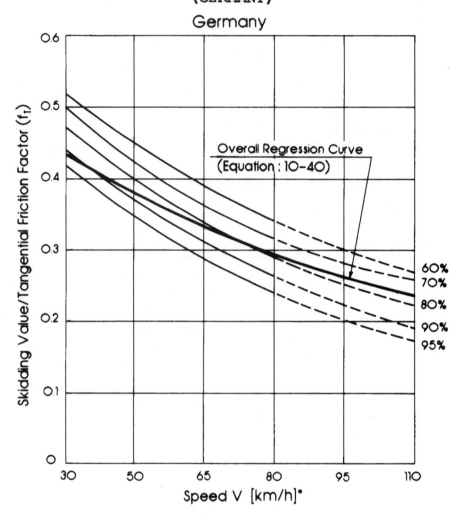

FIGURE 10.29 Percentile distribution curves for the relationship between tangential friction factor and speed for 600 wet pavements in Germany along with the overall regression curve.[414,420,428,640,747]

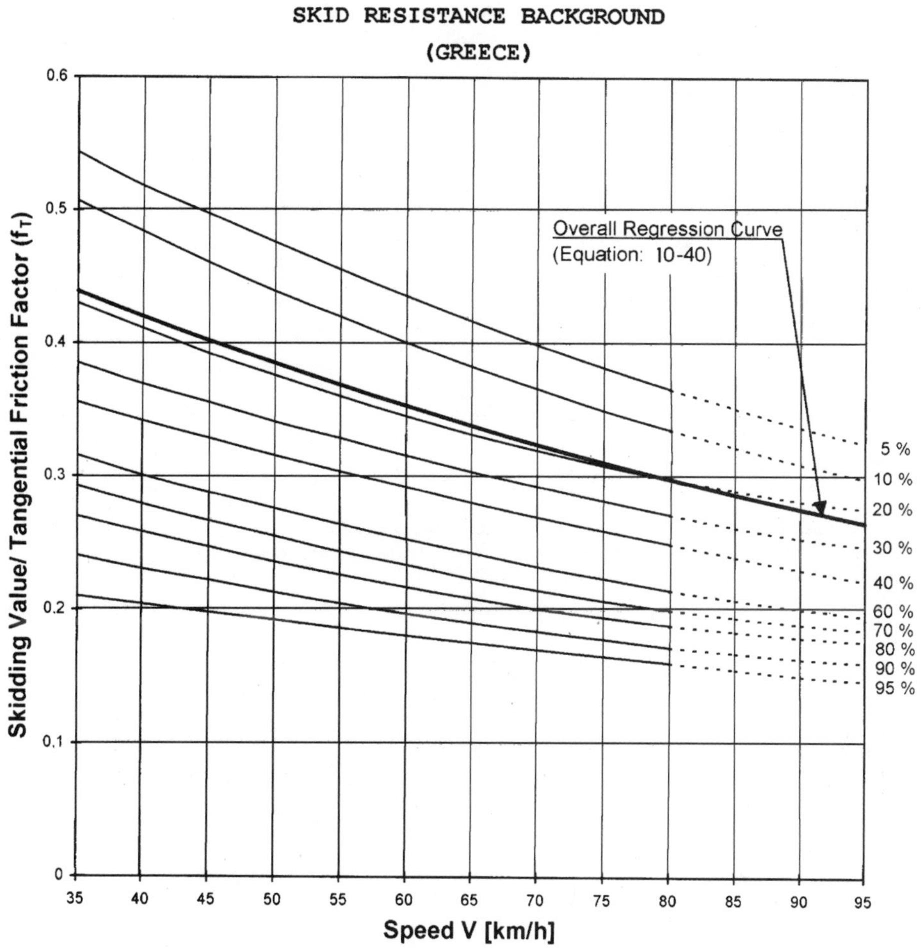

SKID RESISTANCE BACKGROUND
(GREECE)

FIGURE 10.30 Percentile distribution curves for the relationship between tangential friction factor and speed for 107 wet pavements in Greece along with the overall regression curve.[324]

Based on the data of the five countries in Table 10.8, the following overall regression equation between side friction factor f_R and design speed V_d was developed:

$$f_R = 0.27 - 2.19 \times 10^{-3} V_d + 5.79 \times 10^{-6} (V_d)^2$$

$$R^2 = 0.799$$

$$SEE = 0.018 \tag{10.41}$$

where f_R = side friction factor.

The high coefficient of determination, R^2, and the low standard error, SEE, of Eq. (10.41) indicate that the side friction–design speed relationship is a strong one.

Figure 10.31 shows the calculated values of the side friction factor [Eq. (10.41)] as a solid line superimposed on the curves of the countries in this study. With respect to the overall regression

TABLE 10.8 Maximum Permissible Side Friction Factors for Different Design Speeds in Selected Countries

Design speed, km/h	Side friction factors f_R (rounded values)				
	U.S.	FRG	F	S	CH
30	0.170	0.200		0.210	
40	0.165	0.180	0.250	0.190	0.220
50	0.160	0.170		0.180	0.200
55	0.155	0.150		0.170	0.180
60	0.152	0.140	0.160	0.160	0.170
65	0.150	0.130			0.160
70	0.145	0.120		0.150	0.150
80	0.140	0.110	0.130	0.140	0.140
90	0.130	0.100			0.130
95	0.120	0.090			0.130
100	0.115	0.085	0.110		0.125
105	0.110	0.080			0.120
110	0.100	0.075			0.110
120		0.070	0.100		0.110

Note: Values were converted from miles per hour to kilometers per hour because the basic investigations were conducted in the United States.[228]

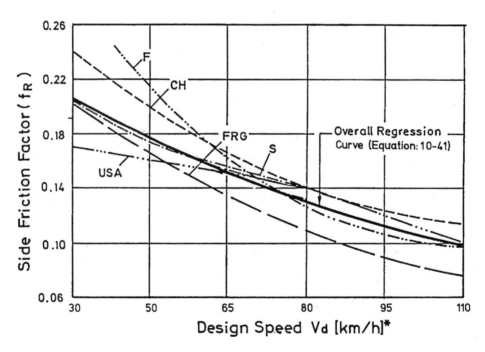

FIGURE 10.31 Relationship between maximum permissible side friction factor and design speed for different countries along with the overall regression curve.[228,414,420,428]

curve, Fig. 10.31 reveals that: (1) the side friction factors of Switzerland are higher; (2) the German side friction factors are lower; and (3) the regression curves for France, Sweden, and the United States intercept the overall regression curve.

Comparison with Actual Pavement Friction Inventories.   Based on the previous investigations, a utilization ratio of $n = 45$ percent was selected to compare the side friction factors resulting from actual pavement friction inventories with the overall regression curve [Eq. (10.41)] of Fig. 10.31. This utilization ratio was selected because, on the average, it was considered reasonable by the Swiss norms,[689,692] the Swedish standard specifications,[686] and the German design guidelines,[241,243,246] and because a portion of friction, about 90 percent, is still available in the tangential direction (see Table 10.4 and Fig. 10.13).

It is interesting to note that, according to Fig. 10.32, the overall regression curve for the side friction factor [Eq. (10.41)] nearly coincides once more with the 80th-percentile distribution curve of Germany and the 90th-percentile distribution curve of New York State. In this connection, we should emphasize the fact that the overall regression curve, which resulted from the maximum permissible side friction factors of the five countries investigated is, at least partially, independent of the utilization ratio, n, whereas the other side friction percentile distribution curves in Fig. 10.32 were established by multiplying the utilization ratio of $n = 45$ percent by the respective tangential distribution curves in Figs. 10.28 to 10.30. The resulting close agreement with the previous findings proves the reliability of the considerations and the correctness of the assumptions.

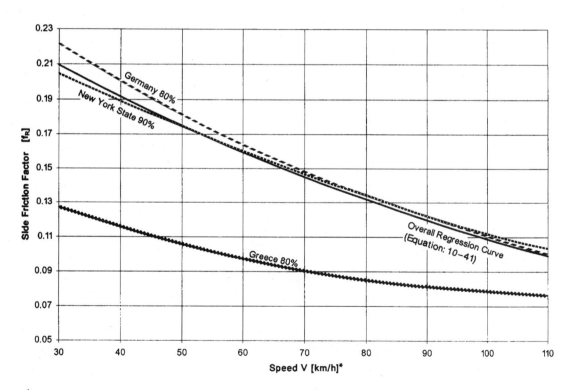

*   Rounded
Legend: see Table 10.8

FIGURE 10.32   80th- and 90th-Percentile distribution curves of different countries for the relationship between side friction factor and speed along with the overall regression curve.[324,414,420,428]

Figure 10.32 also shows that the 80th-percentile distribution curve of Greece lies far below the distribution curves of the other countries studied.

10.2.3 Arrangements for Maximum Permissible Side Friction Factors

Three possibilities exist for the establishment of reliable maximum permissible side friction factors:

1. Let us assume that a skid-resistance background exists for the country under study. Then one should select the nearest percentile distribution curve (95, 90, 85, or a minimum of 80 percent) with respect to the overall regression curve for side friction factors [Eq. (10.41) and Fig. 10.32].

The selected percentile distribution curve then expresses the relationship between side friction factor f_R and speed V, and guarantees sound f_R values which cover 95, 90, 85, or a minimum of 80 percent of wet pavements in the country under study. The same applies for the establishment of the relationship between tangential friction factor f_T and speed V according to the overall regression curve in Figs. 10.28 and 10.29 [Eq. (10.40)].

In order to ensure the condition that friction assumed should exceed friction demanded most of the time, it is appropriate to apply the side (tangential) friction factor that corresponds to the highest percentile level distribution curve.

However, this procedure is not always feasible because reliable skid-resistance backgrounds do not yet exist in many countries, or if they do exist, the 80th-percentile distribution curve lies below the overall regression curve, such as is the case in Greece (see Figs. 10.30 and 10.32). Using the 80th-percentile distribution curve in this case leads to impractical outcomes such as relatively large minimum radii of curve, stopping sight distances, and minimum radii of crest vertical curves. Such outcomes should be rejected, based on economic and environmental reasons.

2. In order not to be too conservative, it is recommended that the tangential friction factors produced by the overall regression curve [Eq. (10.40)] should be used for highway design (Fig. 10.27). Since this curve is based on the maximum permissible tangential friction factors of five acknowledged highway geometric design guidelines, it can be assumed that this relationship represents reliable skid-resistance values which are state of the art. The authors decided to use this solution, and the corresponding Eq. (10.40) was introduced as Eq. (10.2) in "Tangential Friction Factor" in Sec. 10.1.2.2.

The maximum permissible side friction factors depend on the selected utilization ratio, n. Sound recommendations for n, along with corresponding equations and graphical presentations, are given in Table 10.1 and Fig. 10.1 for different design speeds, road category groups, topography classes, and maximum and minimum superelevation rates.

The basic assumption for this procedure, however, is that the responsible highway administration has to include in its regulations directions pertaining to the skid-resistance values that must be used for new designs, redesigns, and rehabilitation strategies. This means in order to be able to provide the tangential friction factors which result from Eqs. (10.2) or (10.40), at least the minimum skid-resistance values given in Sec. 10.1.2.2 *must* be guaranteed. Once more, these values are given below:

$$0.42 \quad \text{for} \quad V = 40 \text{ km/h}$$
$$0.35 \quad \text{for} \quad V = 60 \text{ km/h}$$

and

$$0.30 \quad \text{for} \quad V = 80 \text{ km/h}$$

Based on the evaluation backgrounds for skid resistance in Figs. 10.25 and 10.26, these values proved to be reasonable.

If these skid-resistance values are not observed, the assumed tangential and side friction factors, and the resulting minimum radii of curve, minimum stopping sight distances, and minimum radii of crest vertical curves, would not correspond to the assumed driving dynamic safety demands. This could mean an increase in accident risk and accident severity.

To obtain sufficient skid resistance, the following measures have been recommended:[5]

> Skid resistance should be "built in" on new construction and major reconstruction. Vertical and horizontal alignment and pavement texture can be designed to provide a high initial skid resistance. Measures taken to correct or improve skid resistance should result in the following characteristics: high initial skid resistance, durability to retain skid resistance with time and traffic, and minimal decrease in skid resistance with increasing speed. On roadways of portland cement, tinning has proved to be effective in reducing the potential for hydroplaning. The use of surface courses or overlays constructed with polish-resistant coarse aggregate is the most widespread method for improving the surface texture of bituminous pavements. Overlays of open-graded asphalt friction courses are quite effective because of their frictional and hydraulic properties.[5]

For further discussions, readers may wish to consult Refs. 4 and 208.

3. Finally, the regression model [Eq. (10.41)] for the overall regression curve in Fig. 10.31 could also be used to achieve sound side friction factors. Since this model nearly coincides with the tangential relationship represented by Eq. (10.40) multiplied by an average utilization ratio of $n = 45$ percent, it could also be used for the establishment of reliable side friction factors.

However, in order to guarantee a mathematically correct solution for the relationship between tangential and side friction factor, it was decided to go ahead with the second possibility in order to derive all further relationships.

Maximum permissible side friction factors (or maximum coefficients of side friction or lateral acceleration rates) are specified for driver safety and/or comfort. Table 10.9 summarizes and Fig. 10.33 illustrates the values used in various countries worldwide. Several countries do not report the values used or derive them exactly. Where possible, values were calculated from their minimum radius for a given design speed and their maximum supereleration rate according to Krammes and Garnham.[361]

According to Table 10.9, for a 60-km/h design speed, for example, most countries' side friction factors are between 0.14 and 0.17. For an 80-km/h design speed, maximum permissible side friction normally ranges between 0.11 and 0.14, and for a 100-km/h design speed, between 0.09 and 0.13. Austria, Germany, Greece, Sweden, Switzerland, and Part 2, "Alignment (AL)" in this book specify side friction as a percentage of tangential friction used for stopping sight distance for a given design speed on main rural roadways. The normally used utilization ratio n of side friction is between 40 and 50 percent, which means, according to Table 10.4 or Fig. 10.13, about 87 to 92 percent of friction are still available in the tangential direction for accelerating/decelerating or evasive maneuvers, thus providing a considerable driving dynamic safety reserve.

On the other hand, Australia uses values between 0.35 and 0.26 for 50- to 80-km/h roadways; these are the values drivers were observed to be accepting based upon their 85th-percentile speeds on curves.[361] Overall, the lowest side friction values are used in Germany, Japan, and "AL" for hilly/mountainous topography.

The developed maximum permissible friction values for "AL" are, in comparison with those for other countries, somewhat lower than the average values in cases of flat topography, whereas in cases of hilly/mountainous topography, minimum values are provided to guarantee high side friction supplies for safety reasons (compare Table 10.9 and Fig. 10.33).

In conclusion it can be stated that the developed maximum permissible tangential and side friction factors in "AL" are based on important driving dynamic safety aspects. To establish the following design parameters, they will be effective in such a way that certain safety reserves will be present. We should note, however, that besides friction factors, speed also plays an important role in establishing different design elements. The *speed* is expressed in the design process according to Sec. 8.1.1 as design speed V_d or 85th-percentile speed $V85$. If within the scope of numerous design combinations, the driving dynamic safety reserves are actually present with respect to individual design elements, such as

TABLE 10.9 Maximum Permissible Side Friction Factors as a Function of Design Speed in Various Countries (Worldwide)*

Design speed, km/h	Maximum permissible side friction factors									
	Australia	Austria	Belgium	Canada	France	Germany	Ireland	Italy	Japan	Luxembourg
50	0.35			0.16					0.10	
60	0.33	0.16	0.16	0.15	0.16	0.14	0.15	0.17	0.09	0.16
70	0.31	0.15		0.15		0.12				0.15
80	0.26	0.14	0.13	0.14	0.13	0.11	0.14	0.13	0.08	0.14
85										
90	0.18	0.13	0.11	0.13		0.10				0.13
100	0.12	0.11		0.12	0.11	0.09	0.13	0.11	0.07	0.12
110	0.12			0.10						0.11
120	0.11	0.10	0.10	0.09	0.10	0.07	0.12	0.10	0.06	0.10

	The Netherlands	Portugal	South Africa	Spain	Sweden	Switzerland	U.K.	U.S.	AL + Greece†	AL + Greece‡
50			0.16		0.18	0.19	0.10	0.16	0.164	0.145
60	0.17	0.16	0.15	0.16		0.16	0.10	0.15	0.150	0.133
70	0.15	0.15	0.15	0.15	0.15	0.15	0.10	0.14	0.138	0.122
80		0.14	0.14	0.14		0.14		0.14	0.127	0.113
85							0.10		0.122	0.108
90			0.13	0.14	0.12	0.13		0.13	0.117	0.104
100	0.12	0.12	0.13	0.13		0.12	0.10	0.12	0.109	0.097
110			0.12		0.10	0.11		0.11	0.102	0.091
120	0.08	0.10	0.11	0.10		0.10	0.10	0.09	0.096	0.086

*Elaborated, modified, and completed based on Refs. 153 and 361.

†Flat topography.

‡Hilly/mountainous topography.

Note: AL = abbreviation for Part 2, "Alignment."

- Radius of curve
- Superelevation rate
- Grade
- Sight distance
- Radius of crest and sag vertical curves, and so on

then, certainly they can be described *qualitatively,* but not *quantitatively* by number and measure, for example, in the form of a safety analysis or a safety evaluation process.

One can assume, therefore, that qualitative safety considerations and aspects are more or less present in modern highway geometric design, for example, by the establishment of reliable tangential and side friction factors.

Following the introduction of safety criteria I and II with quantitative ranges for design consistency (Secs. 9.1.2 and 9.2.2) and operating speed consistency (Secs. 9.1.3 and 9.2.3), it appears then that the introduction of a safety criterion III for driving dynamic consistency, with corresponding ranges for good, fair, and poor design practices, is necessary.

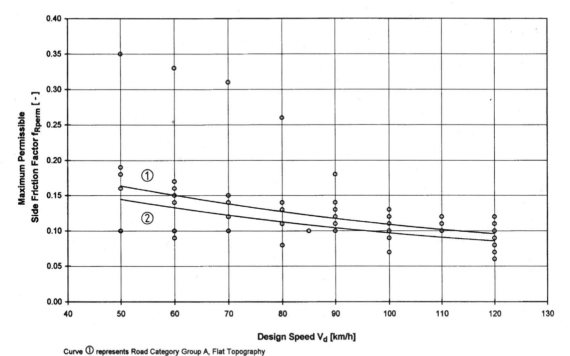

Curve ① represents Road Category Group A, Flat Topography

Curve ② represents Road Category Group A, Hilly/ Mountainous Topography, compare Table 10-1 and Figure 10-1

FIGURE 10.33   Graphical presentation of maximum permissible side friction factors as a function of design speed in various countries and "AL." (Elaborated, modified, and completed based on Refs. 153 and 361.)

10.2.4   Safety Criterion III: Achieving Driving Dynamic Consistency*

10.2.4.1   Providing Adequate Dynamic Safety of Driving in Curves.   One of the most important goals for the development of recommendations for the safety evaluation of planned or existing two-lane rural roads is to reduce accident risk and accident severity by increasing available friction.

In this connection, the safety investigations in most countries attempt to clarify the questions related to the improvement of geometric road design. On the contrary, questions related to the improvement of skid resistance—tangential and side friction factors—are not as often considered even though many publications indicate that sufficient friction represents an important safety aspect.[217,322,385,496] Studies show that the likelihood of a highway curve becoming an accident black spot increases with decreasing skid resistance.[81] Consequently, modern highway geometric design guidelines should clearly emphasize the need for sufficient friction between tire and road surface, *especially on curved roadway sections.* In addition, the consistent upward trend of driving speeds and the increase of traffic densities will unquestionably continue.[431,674] In this connection, accidents caused by skid resistance (friction) that is too low should be taken seriously, since there is no doubt that low skid resistance represents an impairment of safety on wet road surfaces at high speeds.[705]

---

*Elaborated based on Refs. 424, 428, 435, 446, 447, 448, 587, and 672.

The goal of this investigation is to determine the extent of precautions which modern highway geometric design guidelines, such as, for example, in the United States,[5] Germany,[243] and Greece,[453] have taken in order to provide adequate driving dynamic safety in highway geometric design and for road maintenance purposes.

Note that, in comparison to modern guidelines, large portions of the existing road network have to be considered as *not being* in conformity with the guidelines. In these cases, with even higher driving dynamic uncertainties, accident risks have to be expected.

10.2.4.2   *Side Friction: Assumed Versus Demanded.*

For decades it has been known that the original design speed concept does not realistically reflect the actual driving behavior on curved roadway sections and that the assumed design speeds are often considerably exceeded.[270,271,348,349,385,470] This may result in a significantly higher side friction demanded as compared to that originally assumed for curve design.

For the establishment of the driving dynamic backgrounds of Figs. 10.34 to 10.36, the following procedure was used. The curvature change rate of the single curve, CCR_S, had to be calculated for every curved section according to Eqs. (8.6a) to (8.6f) in Fig. 8.1. Also see the explanatory sketches in Fig. 8.2.

Assumed side friction f_{RA} was calculated on the basis of Eq. (10.9a) with respect to the design speed, V_d, while the actual side friction demanded, f_{RD}, was calculated from Eq. (10.11) with respect to the 85th-percentile speed, V85 (see Fig. 8.12 or Table 8.5). For clarity, Eqs. (10.9a) and (10.11) are repeated here:

$$f_{RA} = \frac{V_d^2}{127\,R} - e \qquad\qquad (10.9a)$$

$$f_{RD} = \frac{V85^2}{127\,R} - e \qquad\qquad (10.11)$$

where f_{RA} = side friction assumed

$\quad\ \ f_{RD}$ = side friction demanded

The results that follow are based on investigations on two-lane rural roads of 204 curved roadway sections in Germany,[428,672] 197 curved roadway sections in the State of New York in the United States,[424] and 107 curved roadway sections in Greece.[587] The results of the investigations confirm the fact that the side friction factors for curve design that are assumed for different design speeds, f_{RA}, in the geometric design guidelines of Germany,[243] the United States,[5] and Greece[453] are often exceeded by those that are demanded, f_{RD}, by the 85th-percentile speeds under real-world conditions. The relationships are presented with respect to the design parameter curvature change rate of the single curve (CCR_S) in Figs. 10.34 to 10.36. These figures show the following.

Side friction demanded f_{RD} exceeds side friction assumed f_{RA} for CCR_S values greater than 180 gon/km ($R < 350$ m) in Germany (Fig. 10.34). The point of intersection for the U.S. case in Fig. 10.35 is at about 225 gon/km ($R < 290$ m). For Greek conditions, Fig. 10.36 shows that the curves intersect at about 180 gon/km.

The results of Figs. 10.34 to 10.36 clearly show points of intersection between side friction assumed f_{RA} and side friction demanded f_{RD}. It appears logical, therefore, in connection with driving dynamic considerations, to regard the design range where side friction assumed f_{RA} exceeds side friction demanded f_{RD} as good design, while the design range where side friction demanded f_{RD} exceeds side friction assumed f_{RA} should be regarded as poor design.

Between good and poor design levels lies the tolerated fair design level with respect to the design parameter curvature change rate of the single curve (CCR_S), as was agreed upon based on previous accident research results and arrangements for safety criteria I and II (see Table 9.10). The fair CCR_S class, which ranges from 180 to 360 gon/km, is shown in Figs. 10.34 to 10.36. Based on these figures, recommendations for evaluating good, fair, and poor design levels will be used to develop safety criterion III.

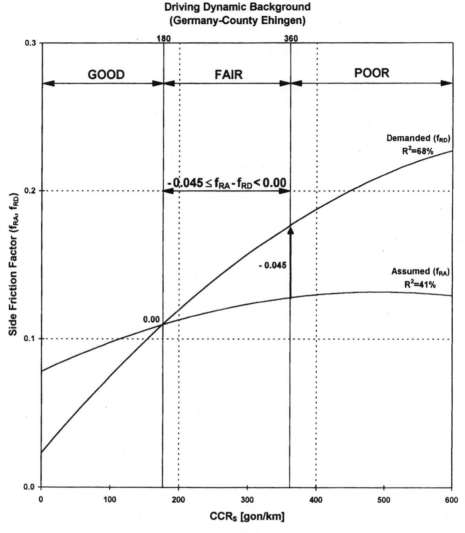

$$f_{RA} = 0.078 + 2.18 \times 10^{-4} \, CCR_S - 2.21 \times 10^{-7} \, (CCR_S)^2 \qquad (10.42)$$

$$f_{RD} = 0.023 + 5.55 \times 10^{-4} \, CCR_S - 3.58 \times 10^{-7} \, (CCR_S)^2 \qquad (10.43)$$

FIGURE 10.34 Relationship between side friction assumed/demanded and the curvature change rate of the single curve for Germany (based on 204 two-lane rural test sections).[428,672]

In addition, note from Figs. 10.34 to 10.36 that the differences between the regression curves to the right side of the points of intersection increase with increasing CCR_S values, which means the relation between side friction assumed and side friction demanded is becoming more and more unfavorable.

These considerations are supported by the relationship between side friction assumed/demanded and accident rates for the United States (see Fig. 10.37). A point of intersection is also shown in this figure. This point of intersection indicates that there are design ranges that are less dangerous (that

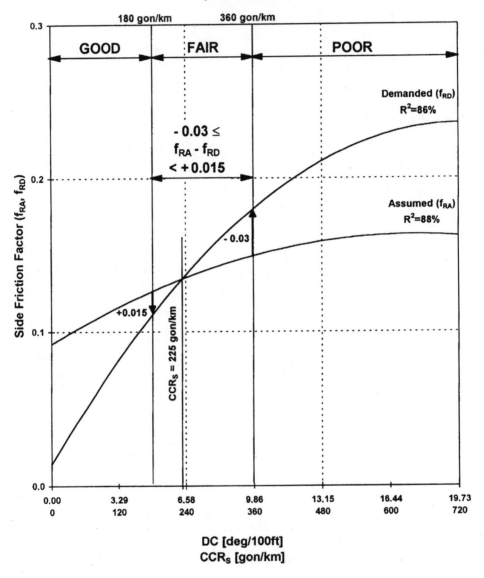

$$f_{RA} = 0.092 + 2.22 \times 10^{-4} \, CCR_S - 1.73 \times 10^{-7} \, (CCR_S)^2 \qquad (10.44)$$

$$f_{RD} = 0.014 + 6.16 \times 10^{-4} \, CCR_S - 4.28 \times 10^{-7} \, (CCR_S)^2 \qquad (10.45)$$

FIGURE 10.35   Relationship between side friction assumed/demanded and the curvature change rate of the single curve for the United States (based on 197 two-lane rural test sections).[424]

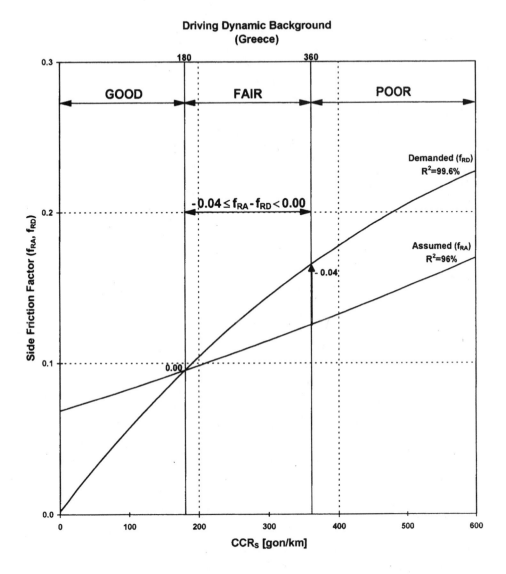

$$f_{RA} = 6.83 \times 10^{-2} + 1.41 \times 10^{-4}\, CCR_S + 4.66 \times 10^{-8}\, (CCR_S)^2 \tag{10.46}$$

$$f_{RD} = 1.95 \times 10^{-3} + 5.90 \times 10^{-4}\, CCR_S - 4.12 \times 10^{-7}\, (CCR_S)^2$$
$$+ 8.82 \times 10^{-11}\, (CCR_S)^3 \tag{10.47}$$

FIGURE 10.36   Relationship between side friction assumed/demanded and the curvature change rate of the single curve for Greece (based on 107 two-lane rural test sections).[587]

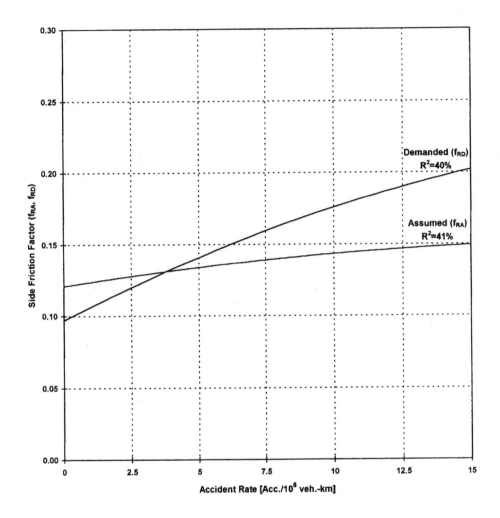

$$f_{RA} = 0.121 + 2.97 \times 10^{-3}\,AR - 5.14 \times 10^{-5}\,(AR)^2 \qquad (10.48)$$

$$f_{RD} = 0.097 + 9.68 \times 10^{-3}\,AR - 1.79 \times 10^{-4}\,(AR)^2 \qquad (10.49)$$

FIGURE 10.37   Relationship between side friction assumed/demanded and accident rates for the United States.[424,428]

is, to the left of the point of intersection) and exhibit smaller accident rates. For these good or relatively good design ranges, the assumed side friction factor exceeds the side friction demand actually needed. Figure 10.37 also indicates that there are design ranges that are more dangerous (that is, to the right of the point of intersection) and exhibit higher accident rates. For these poor or relatively poor design ranges, the side friction demanded exceeds the assumed side friction (see also Figs. 10.34 to 10.36). In addition, Fig. 10.37 shows that the accident risk, expressed by the accident rate, increases with an increase in the utilization of friction when driving through curves.[424,428]

The relatively small coefficients of determination, R^2, of Eqs. (10.48) and (10.49) are not at all surprising because accident research relationships are not simple or direct, but are often complex, and changes in the frequency of accidents are often the result of many factors in addition to the driving dynamic aspects, expressed herein by side friction assumed and side friction demanded. These results clearly support the opinions expressed by several researchers who argue that, in recognition of safety considerations, insufficient dynamic safety of driving has a direct impact on accident rate.

The results are confirmed by Table 9.10 and Figs. 10.34 to 10.36 with respect to the curvature change rate of the single curve, CCR_S.

However, these results are in clear contradiction to the opinion of many practitioners and researchers who argue that the margin of safety against skidding (especially for passenger cars), that is, the difference between assumed friction and available actual pavement friction, is large enough to provide an adequate dynamic safety of driving. Related to good skid-resistant pavements, this margin of safety may reach a factor of 2 for wet pavements and a factor of 4 for dry pavements or even higher. Related to vehicular and human issues, there may be another margin of safety against skidding, that is, the margin based on the fact that in nearly all highway design guidelines assumed friction values are derived from locked-wheel friction measurements. These assumed friction values are lower than the peak friction coefficients that may be reached by experienced drivers or by the presence of ABS (antilock braking) systems. However, even those margins of safety do not alter the fact that higher accident risks do exist on poorly designed roadways, which exhibit inconsistencies in horizontal alignment and lack of coordination between design speeds and operating speeds as compared to those roadways which exhibit good designs or even fair designs, as revealed in Table 9.10 and Figs. 10.34 to 10.38.

It makes sense, then, to develop a driving dynamic safety criterion with certain ranges or limiting values for the evaluation of good, fair, and poor design practices.

10.2.4.3 Classification System for Safety Criterion III

Investigations up to 1993. Investigations on mean accident rates in Germany and the United States (Tables 9.10, 10.12, and 18.14) were taken as the basis for establishing classes of curvature change rate of the single curve, CCR_S, in order to evaluate good, fair, and poor design levels.
The following CCR_S classes,

<180 gon/km (good design)

180 to 360 gon/km (fair design)

>360 gon/km (poor design)

on which the design ranges for safety criteria I and II were based (see Tables 9.11 and 9.13), will also be used here for the development of the driving dynamic safety criterion III.

By incorporating these CCR_S classes into Figs. 10.34 to 10.36, the following differences between side friction assumed/demanded could be observed for the fair (tolerated) design range of safety criterion III:
For German conditions (Fig. 10.34):

$$-0.045 \le f_{RA} - f_{RD} < 0$$

For U.S. conditions (Fig. 10.35):

$$-0.03 \le f_{RA} - f_{RD} < +0.015$$

For Greek conditions (Fig. 10.36):

$$-0.04 \le f_{RA} - f_{RD} < 0$$

That means the differences ($\Delta f_R = f_{RA} - f_{RD}$)

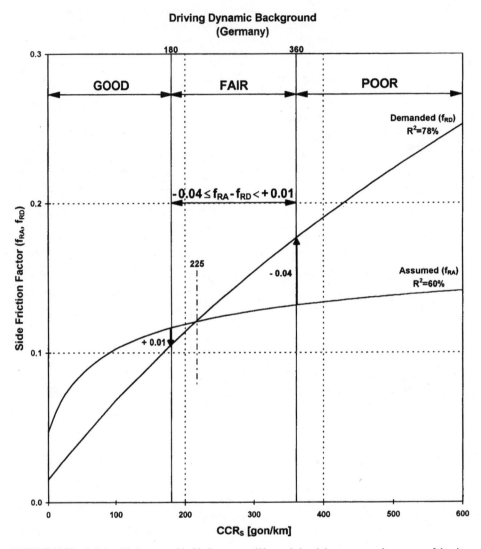

FIGURE 10.38   Relationship between side friction assumed/demanded and the curvature change rate of the single curve for Germany (based on 657 two-lane rural test sections).[632]

$$\Delta f_R > 0 \qquad \text{German conditions}$$

$$\Delta f_R > + 0.015 \qquad \text{U.S. conditions}$$

$$\Delta f_R > 0 \qquad \text{Greek conditions}$$

correspond to good design practices, whereas the differences

$$\Delta f_R < - 0.045 \qquad \text{German conditions}$$

$$\Delta f_R < - 0.03 \qquad \text{U.S. conditions}$$

$$\Delta f_R < - 0.04 \qquad \text{Greek conditions}$$

correspond to poor design practices.

Recent Research Results. *   For the final arrangement of statistically sound driving dynamic consistency ranges, a large new database was used in order to: (1) analyze the relationship between side friction assumed/demanded f_{RA}, f_{RD} and the curvature change rate of the single curve, CCR_S, and (2) compare its results with those of the previous sections.

This database consisted of 657 curved roadway sections of two-lane rural roads in the counties of Ehingen, Enzkreis, and Karlsruhe in southwestern Germany which were distributed with about 50 percent in flat topography and with about 50 percent in hilly topography. Only curved sections with a constant pavement width over the entire investigated roadway section, a minimum length of 50 m, grades between ±5 percent, AADT values between 1000 and 10,000 vehicles/day, and distances between intersections longer than 500 m were taken into consideration.

Again, the curvature change rate of the single curve, CCR_S, was calculated for every curved section according to Eqs. (8.6a) to (8.6f) in Fig. 8.1. The calculation of side friction assumed/demanded f_{RA}, f_{RD} was based on Eqs. (10.9a) and (10.11).

Regression relationships between side friction assumed/demanded f_{RA}, f_{RD} and CCR_S were determined for each of the counties as well as for all the counties combined. The resulting overall regression equations for f_{RA} and f_{RD} are

$$f_{RA} = 0.267 - \frac{0.813}{\ln (CCR_S + 40)} \tag{10.50}$$

and

$$f_{RD} = -2.179 + 0.343 \, [\ln (CCR_S + 600)] \tag{10.51}$$

Equations (10.50) and (10.51) are schematically shown in Fig. 10.38.

As shown in Fig. 10.38, the point of intersection of the two regression curves corresponds to a CCR_S value of 225 gon/km. Without considering clothoids, this CCR_S value corresponds to a radius of curve of about 290 m. To the right side of the intersection point, the differences $f_{RA} - f_{RD}$ increase with increasing CCR_S values, which means the relation between side friction assumed and side friction demanded in this range can be considered unfavorable. This result, which is based on a large database of 657 curved test sections, confirms the results previously shown in Figs. 10.34 to 10.36.

On the basis of the discussions in the previous sections, one can conclude, then, that the differences

$$\Delta f_R = f_{RA} - f_{RD}$$

between side friction assumed and side friction demanded are relevant for the establishment of reliable ranges for the driving dynamic safety criterion III.

---

*Elaborated based on Refs. 435, 438, 447, 448, 453, 455, 458, and 632.

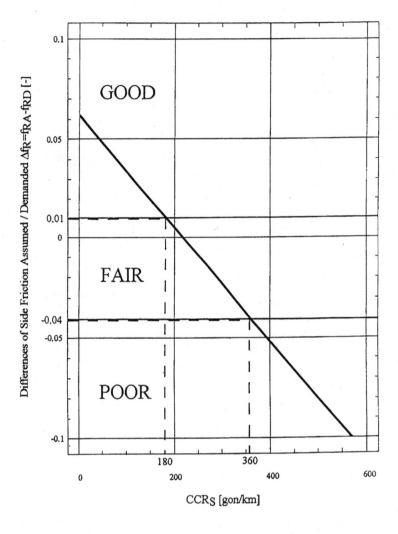

$$\Delta f_R = f_{RA} - f_{RD} = -2.831 \times 10^{-4}\ CCR_S + 0.062 \qquad (10.52)$$

$$R^2 = 71\%$$

FIGURE 10.39   Relationship of the differences between side friction assumed/demanded and the curvature change rate of the single curve for Germany (based on 657 two-lane rural test sections).[632]

Figure 10.39 was developed in order to examine, in detail, the relationships between the differences of side friction assumed/demanded and the curvature change rate of the single curve, CCR_S, and to confirm the graphical results in Fig. 10.38.

As another check, by incorporating the fair CCR_S class, which ranges from 180 to 360 gon/km, into Eq. (10.52) of Fig. 10.39, the following ranges according to the driving dynamic safety criterion III were determined:

Fair design: $-0.039916 \leq f_{RA} - f_{RD} < +0.01102\}$

Rounded: $-0.04 \leq f_{RA} - f_{RD} < +0.01$

$$(10.53)$$

These values are confirmed once more by Figs. 10.38 and 10.39. Design ranges for good, fair, and poor design levels are also incorporated into these figures.

For a final examination of safety criterion III, the results of the investigations conducted so far in Germany, the United States, and Greece were compiled in Table 10.10. Column 8 of this table indicates that the differences agree well with each other. The fair design range on line ($\Sigma 4$) or Eq. (10.53) could then be regarded as adequate and reasonable for future design tasks, at least based on the results of the investigations presented here. Note that the established differences between side friction assumed/demanded were the result of a large database and that for calculating the CCR_S values, they include not only circular curves but also the connecting transition curves, as opposed to the U.S. and Greek results.

Table 10.11 gives the classification system for safety criterion III. For the illustration of the scientific background, especially related to the expected accident situation (compare Tables 9.10 and 10.12), the corresponding ranges for good, fair, and poor design levels with respect to the design parameter curvature change rate are incorporated in Table 10.11. For practical design work, this is not necessary, since the CCR_S ranges and $f_{RA} - f_{RD}$ ranges correspond to each other within relatively narrow limits (see Table 10.10, cols. 2 to 7). Therefore, to simplify matters, only the $f_{RA} - f_{RD}$ ranges are listed in the comparable Table 10.3 of Sec. 10.1.4.

With respect to Table 10.11, it is recommended that side friction assumed f_{RA} should be calculated from Eq. (10.10b) ($n = 60\%$) for redesigns and RRR projects in conjunction with existing alignments depending on the assumed design speed ["Existing (Old) Alignments" in Sec. 9.2.2.1]. For new designs, Eq. (10.4) ($n = 45\%$) is valid in flat topography and Eq. (10.5) ($n = 40\%$) is valid in hilly and mountainous topography (Table 10.1). However, in cases of redesigns, the basic decision about the selected equation is left to the highway engineer, since he or she can

TABLE 10.10   Development of a Classification System for Safety Criterion III[632]

State/county (1)	Side friction assumed f_{RA}		Side friction demanded f_{RD}		$\Delta f_R = f_{RA} - f_{RD}$		Safety criterion III tolerated difference fair design $f_{RA} - f_{RD}$ (8)	Number in database (9)
	360 gon/km (2)	180 gon/km (3)	360 gon/km (4)	180 gon/km (5)	360 gon/km (6)	180 gon/km (7)		
1 United States	0.150	0.125	0.180	0.110	−0.030	0.015	$-0.030 \leq f_{RA} - f_{RD} < 0.015$	197
2 Germany SB	0.130	0.110	0.175	0.110	−0.045	0.000	$-0.045 \leq f_{RA} - f_{RD} < 0.000$	204
3 Greece	0.125	0.095	0.165	0.095	−0.040	0.000	$-0.040 \leq f_{RA} - f_{RD} < 0.000$	107
4a Ehingen	0.130	0.115	0.175	0.105	−0.045	0.010	$-0.045 \leq f_{RA} - f_{RD} < 0.010$	324
4b Karlsruhe	0.130	0.115	0.175	0.105	−0.045	0.010	$-0.045 \leq f_{RA} - f_{RD} < 0.010$	181
4c Enzkreis	0.135	0.120	0.175	0.105	−0.040	0.015	$-0.040 \leq f_{RA} - f_{RD} < 0.015$	152
$\Sigma 4$ Germany LB	0.130	0.115	0.175	0.105	−0.040	0.010	$-0.040 \leq f_{RA} - f_{RD} < 0.010$	657

Note: SB = small database (county: Ehingen); LB = large database (counties: Ehingen, Karlsruhe, and Enzkreis); line 1 based on Ref. 424 and Fig. 10.35; line 2 based on Refs. 428 and 672 and Fig. 10.34; line 3 based on Ref. 587 and Fig. 10.36; line 4 based on Refs. 446 and 632 and Figs. 10.38 and 10.39 and Eq. (10.53).

TABLE 10.11 Classification System for Safety Criterion III[446]

Safety criterion III: Driving dynamic consistency

Case 1: Good design level:

The permissible differences are

$CCR_{S\,i} \leq 180$ gon/km

$f_{RA} - f_{RD} \geq + 0.01$

Case 2: Fair design level:

The tolerated differences are

$180 < CCR_{Si} \leq 360$ gon/km

$-0.04 \leq f_{RA} - f_{RD} < + 0.01$

Case 3: Poor design level:

The nonpermissible differences are

$CCR_{Si} > 360$ gon/km

$f_{RA} - f_{RD} < - 0.04$

Note: CCR_{Si} = CCR_S value for curved roadway section i, gon/km; f_{RA} = side friction assumed; f_{RD} = side friction demanded.

best decide the present economic, environmental, and driving dynamic safety concerns of the investigated roadway section.

Side friction demanded f_{RD} is calculated from Eq. (10.11) with respect to the 85th-percentile speed, $V85$, for the general safety evaluation process of safety criterion III.

Equation (10.9a) in "Side Friction: Assumed Versus Demanded" in Sec. 10.2.4.2, which was also applied for determining side friction assumed f_{RA}, was used only to establish the driving dynamic backgrounds of Figs. 10.34 to 10.36 and 10.38, as precisely as possible, based on the specific design procedures of the U.S.,[5] German,[241,243] and Greek[234] highway geometric design guidelines.[424,428,446,587,632,672,673]

The observed relationship between mean accident rates and the CCR_S classes for different design levels in Table 9.10 was examined once more in Table 10.12 on the basis of the extended

TABLE 10.12 t-Test Results of Mean Accident Rates for Different CCR_S Classes Based on an Extended Database (Germany Based on 657 Two-Lane Rural Test Sections)[632] (See also Tables 9.10 and 18.14)

Design/CCR_S classes, gon/km	Mean AR	t_{calc} t_{crit}		Significance; remarks
Tangent (0)	0.35*			Considered as
		—†		
35–180	0.51		—	Good design
		10.70 > 1.96	Yes	
>180–360	1.72		—	Fair design
		2.64 > 1.98	Yes	
>360–550	2.78		—	Poor design
>550 gon/km (no sufficient data base).				

* From Refs. 672, 673

†Not calculated because of different databases

Note: CCR_S = curvature change rate of the single curve, gon/km, and AR = accident rate [accidents per 10^6 vehicle kilometers], including run-off-the-road accidents only.

data base of 657 curved test sections.[632] Again, the evaluation reveals significant increases in the mean accident rates between good and fair design practices (accident risk factor>3), and between good and poor design practices (accident risk factor >5) for the 95 percent level of confidence.

The results of Table 10.12, therefore, confirm the significant increasing trend of the accident rate with increasing CCR_S classes. In addition, the results reveal once more that the selected CCR_S classes for evaluating good, fair, and poor design levels should be considered reliable for the developed safety criteria I to III.

On the basis of the driving dynamic safety criterion III, the examination of new designs and existing highways is urgently needed. New designs should always correspond to good design practices. The same is true for redesigns and RRR strategies, although this is not always possible because of economic and environmental protection issues. However, note that the introduction of the fair design range on two-lane rural roads leads to an accident risk rate that is 2 or 3 times higher than that of good design practice (see Table 10.12). Similar statements were previously made for the United States (Table 9.10), which indicate that the knowledge gained here has already been confirmed by results from two continents, at least based on limited areas (New York State and southwestern Germany).

Conclusive remarks about the safety classification system with respect to good, fair, and poor design levels are made in Sec. 18.3.

CHAPTER 11
GENERAL ALIGNMENT ISSUES WITH RESPECT TO SAFETY

11.1 RECOMMENDATIONS FOR PRACTICAL DESIGN TASKS

Alignment design consists of the following levels:

- Horizontal alignment
- Vertical alignment
- Cross section/alignment
- Sight distance
- Three-dimensional alignment

The flowchart for alignment design with special emphasis on the newly developed safety criteria I to III, in addition to other important safety aspects, is presented in Fig. 11.1.

The arrangement of the CCR_S classes in Table 11.1 is based on the accident research of four databases in the United States and Germany (Tables 9.10, 10.12, and 18.14). According to these classes, the design consistency ranges of safety criterion I (Table 9.11), the operating speed consistency ranges of safety criterion II (Table 9.13), and the driving dynamic consistency ranges of safety criterion III (Table 9.11) were coordinated in order to be able to distinguish between good, fair, and poor design levels.

Achieving design consistency is of special interest in modern highway geometric design. That means the design speed, V_d, should remain constant on longer roadway sections and should also be coordinated with the actual driving behavior, expressed by the 85th-percentile speed, $V85$. This is guaranteed by safety criterion I, achieving design consistency (see Table 11.1, good design practice).

In this way, the road characteristic is well balanced for the motorist along the course of the road section. If, in the course of a longer road section—for example, by definite changes in the topography—a change in the road characteristic and a corresponding change in the design speed are necessary, then the design elements in the transition section must be carefully adjusted for each other, so that they change only gradually.

The 85th-percentile speed should also be consistent along the roadway section. This is guaranteed by the good design practice of safety criterion II, achieving operating speed consistency (Table 11.1), between two successive design elements. In this connection, the *tangent* is considered here for the first time as a *dynamic design element,* as will be discussed in detail in "Evaluation of Tangents in the Design Process" in Sec. 12.1.1.3. Tangents exist that are long enough to accelerate up to the top 85th-percentile speed, $V85_i$ (for example, see Fig. 8.12 for

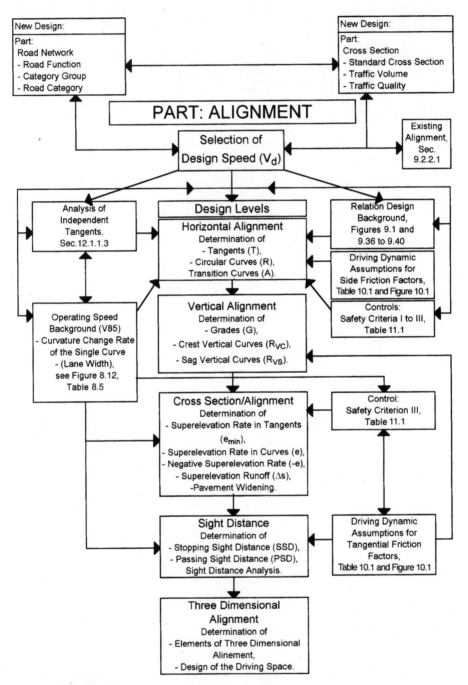

FIGURE 11.1   Flowchart for alignment design with special emphasis on the newly developed safety criteria I to III and other additional important safety aspects.

TABLE 11.1 Quantitative Ranges for Safety Criteria I to III for Good, Fair, and Poor Design Levels

Safety criterion/ CCR_S class	Good ($\leq$180 gon/km)	Fair (>180 gon/km $\leq$360 gon/km)	Poor (>360 gon/km)
I*	$|V85_i - V_d| \leq 10$ km/h	10 km/h $< |V85_i - V_d| \leq 20$ km/h	$|V85_i - V_d| > 20$ km/h
II†	$|V85_i - V85_{i+1}| \leq 10$ km/h	10 km/h $< |V85_i - V85_{i+1}| \leq 20$ km/h	$|V85_i - V85_{i+1}| > 20$ km/h
III‡	$f_{RA} - f_{RD} \geq +0.01$	$-0.04 \leq f_{RA} - f_{RD} < +0.01$	$f_{RA} - f_{RD} < -0.04$

*Related to the individual design elements, i (independent tangent or curve), in the course of the observed roadway section.

†Related to two successive design elements, i and $i+1$ (independent tangent to curve or curve to curve).

‡Related to one individual curved roadway section.

Note:

CCR_S = curvature change rate of the single curve, gon/km [Eq. (8.6)]

V_d = design speed, km/h, from network functions for road categories A I to A IV for new designs (Table 6.2); for redesigns or existing alignments, see Sec. 9.2.2.1

$V85_i$ = expected 85th-percentile speed of design element i, km/h

$V85_{i+1}$ = expected 85th-percentile speed of design element $i + 1$, km/h (according to Fig. 8.12 or Table 8.5) with respect to the design parameter curvature change rate of the single curve

f_T = tangential friction factor for modern highway geometric design:

$$f_T = 0.59 - 4.85 \times 10^{-3} V_d + 1.51 \times 10^{-5} V_d^2 \tag{10.2}$$

f_{RA} = side friction "assumed":

$$f_{RA} = n\, 0.925 \cdot f_T \tag{10.3}$$

n = utilization ratio of side friction, %/100:

n = 0.40 for hilly/mountainous topography, new designs   (10.5)

n = 0.45 for flat topography, new designs   (10.4)

n = 0.60 for redesigns or existing (old) alignments   (10.10*b*)

f_{RD} = side friction "demanded":

$$f_{RD} = V85^2/127\, R - e \tag{10.11}$$

R = radius in the observed circular curve, m

e = superelevation rate, %/100

$CCR_S = 0$), or to decelerate down to the 85th-percentile speed, $V85_{i+1}$, on the succeeding curved section. Those long tangents are called *independent tangents* and must be regarded in the curve-tangent-curve design process as independent design elements. Short tangents, where critical acceleration and deceleration maneuvers are not possible, are called *nonindependent tangents* and can be ignored in the speed-related design process.

In addition, the ranges of safety criterion II are graphically presented as relation design backgrounds for a proper alignment of radii of curve sequences with or without independent tangents for good and fair design, as well as for detecting poor design practice (see Figs. 9.1 and 9.36 to 9.40).

A well-balanced driving dynamic sequence of individual design elements within a road section with the same design speed promotes a consistent, economic driving pattern. This is guaranteed by safety criterion III, achieving driving dynamic consistency, for good design practice in Table 11.1. This safety criterion relies heavily on sound driving dynamic assumptions for tangential and side friction factors.

As can be seen from Fig. 11.1, four of the five "alignment" design levels are either controlled by the three quantitative safety criteria (Table 11.1) or by additional direct or indirect safety-related aspects. These include:

- The selection of an appropriate design speed for new and existing (old) alignments
- The analysis of independent or nonindependent tangents
- The establishment of an operating speed and relation design background for the country under study
- The introduction of sound driving dynamic assumptions for tangential and side friction factors

Thus, it may be concluded that, from the design level horizontal alignment to the design level sight distance, qualitative and quantitative safety-related issues are considered (Fig. 11.1). It may be expected, therefore, that the sensitive combination and superimposition of these design levels will lead, along with the information that will be presented in Chap. 16, to a sound three-dimensional alignment, for which so far, no satisfying safety evaluation processes exist.

Figure 11.2 clarifies the design flow of Part 2, "Alignment," presented in Fig. 11.1, from another graphical point of view. This figure attempts to present the methodology needed to accomplish a consistent design flow with respect to a number of design and safety-related issues, interactions, and interrelationships.

As discussed in detail in the previous chapters, safety criteria I and II and the corresponding operating and relation-design backgrounds are of special importance for two-lane rural roads of category group A.

Because of the conservative assessments for the design speed, the 85th-percentile speed and the maximum permissible speed limit in Sec. 8.1.2, normally there is a control, based on safety criteria I and II, not necessary for:

- Multilane median-separated roads of category group A
- Road categories B II, B III, and B IV

However, an examination of criteria I and II is always useful in revealing discrepancies in the alignment.

In contrast, safety criterion III must be examined for two-lane and multilane rural roads (category group A) as well as for suburban roads (road categories B II, B III, and B IV).

Finally, for redesigns or RRR projects of existing roads, the design elements of the sections that follow the one to be reconstructed must be examined to determine if definite differences in the road characteristic exist. If they do, the transitions must be created very carefully according to the three safety criteria discussed here.

Based on Chaps. 8 to 10, reliable ranges for the new design parameter curvature change rate of the single curve, CCR_S, could be developed for good, fair, and poor design levels based on accident research (see Tables 9.10 and 10.12). These CCR_S classes agree well according to Table 11.1 to the corresponding:

1. Differences between 85th-percentile speeds and design speed for individual design elements (safety criterion I, Table 9.11)
2. Differences of 85th-percentile speeds between successive design elements (safety criterion II, Table 9.13)
3. Relation design demands regarding curve-to-curve and tangent-to-curve transitions (graphical presentation of safety criterion II, Figs. 9.1 and 9.36 to 9.40)
4. Differences between side friction assumed and side friction demanded at curved sites (safety criterion III, Table 10.11)

Thus, a quantitative safety evaluation process has been developed to examine horizontal alignments with respect to new designs, redesigns, and RRR strategies.

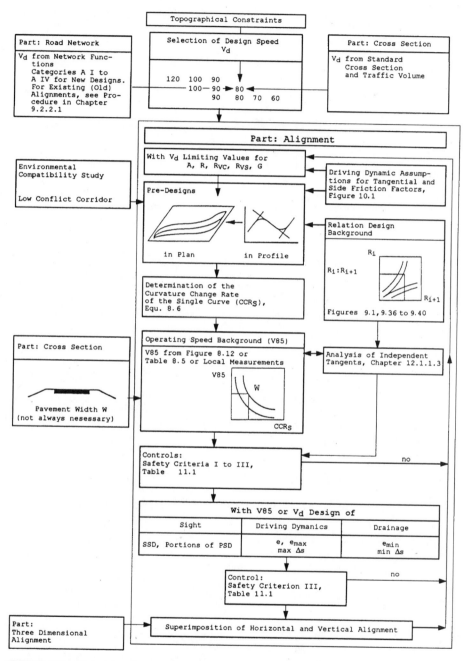

FIGURE 11.2   Methodology for a consistent design flow with special emphasis on safety-related issues. (Elaborated, modified, and completed based on Ref. 159.)

The process is completed by providing sound maximum permissible tangential and side friction factors (see Table 10.1 and Fig. 10.1). For existing alignments, see Eq. (10.10b).

Since the procedure has been tested and found to be reliable for grades of up to 5 to 6 percent and AADT values of about 10,000 (12,000) vehicles/day, it is valid by far for the largest portion of two-lane rural road networks.

11.2 GENERAL CONSIDERATIONS, RESEARCH EVALUATIONS, GUIDELINE COMPARISONS, AND NEW DEVELOPMENTS

11.2.1 Conclusive Remarks

From 1940 to 1970, the only direct safety criterion in geometric design guidelines available to highway engineers in most western European countries and the United States was mainly directed toward evaluating the dynamic safety of driving, such as calculating, for a given design speed, minimum radii of curve, superelevation rates, required stopping sight distances, minimum parameters of crest vertical curves, and so on.[1,558]

Since the 1960s, many experts have recognized the fact that abrupt changes in operating speeds lead to accidents, particularly on two-lane rural roads, and that these speed inconsistencies may be largely attributed to abrupt changes in horizontal alignment.[349,385,470] Since the 1970s, two additional indirect design criteria, related to traffic safety, have been provided in the geometric design guidelines of some European countries. German, Swedish, and Swiss designers, for instance, are partially provided with design criteria to help ensure design consistency between design elements and to coordinate design speed and operating speed.[241,405,686,691]

The design of highways primarily involves three main geometric design levels: the horizontal alignment, the vertical alignment, and the cross section. The design speed, V_d, controls the horizontal and vertical alignment of a highway, which should be based on sound radii of curve and adequate sight distances. Therefore, the horizontal and vertical alignments control the operating speed, $V85$, on the highway. A correct combination of horizontal alignment, vertical alignment, and cross-sectional elements promotes a uniform speed for the motorist traveling on the highway and contributes to a sound design.[728]

Safety research has, so far, focused mainly on horizontal and vertical alignment features. In this connection, the influence of most highway design parameters on accidents has been separately discussed, as shown in "Influence of Design and Operational Parameters on the Accident Situation" in Sec. 9.2.1.3, and will be shown in the following sections.

However, generally speaking, any evaluation of road safety, such as in the driving dynamic field, has, so far, been conducted more or less qualitatively. It is safe to say, from a traffic safety point of view, that no one can predict with great certainty, or prove by measure or number, where traffic accidents might occur or where accident black spots might develop.[109]

However, everyone agrees that a relationship exists between traffic safety and geometric design consistency. There is no doubt that alignment consistency is a key issue in modern highway geometric design. A consistent alignment allows most drivers to operate safely at their desired speed along the entire alignment. But existing design speed–based alignment policies permit the selection of a design speed that is less than the desired speed of a majority of the drivers.[347,359,396,412,470]

Previous research on rural two-lane highway operations and safety has concluded that horizontal curves whose design speed is less than drivers' desired speed exhibit operating-speed inconsistencies that increase accident potential.[362,410] Accident research has consistently found that accident rates on horizontal curves are 1.5 to 4 times the accident rates on tangent sections of rural two-lane highways.[195,415]

Keeping this in mind, a practical procedure, which considers safety rules and criteria for the safety evaluation of new designs, redesigns, and RRR projects, is presented in this book. The procedure is based on statistical investigations of speeds and design parameters in Europe, the Middle East, and North America.[108,431,435,438,453,461] The main components of the procedure are

- Operating speed backgrounds (Fig. 8.12 and Table 8.5)
- Relation design backgrounds (Figs. 9.1 and 9.36 to 9.40)
- Skid resistance backgrounds (Figs. 10.28 to 10.30)
- Driving dynamic backgrounds (Figs. 10.34 to 10.36 and 10.38)

for tangents, curves, and transition curves for different roadway types and topography classes. Based on these backgrounds, three safety criteria were developed.

For a sound balance in highway design, *all* geometric elements must be selected, if economically and environmentally feasible, to provide safe and continuous operation at a speed that the general conditions of that highway or street. For the most part, this can be achieved by a sensible tuning of design speed and operating speed, as recommended by safety criteria I to III, with respect to the individual design elements, the successive design elements, and the whole roadway section (see Figs. 11.1 and 11.2, and Table 11.1).

Criteria I to III were the subject of a number of basic research reports, publications, and guideline proposals by the authors.[406,408,411,420,422,424,434,448,460] These investigations included:

1. Processes for evaluating design speed and operating speed differences
2. Processes for evaluating operating speed differences between successive design elements
3. Processes for evaluating the differences between side friction assumed and side friction demanded on curved roadway sections

The procedure presented provides interrelationships between design parameters, driving behavior, and driving dynamics in order to determine sound roadway alignments and/or to detect poor ones and to influence in a positive manner the accident situation.

Therefore, significant improvements in safety cannot automatically be assumed for new design or redesign projects; safety must be systematically incorporated into every project. Highway designers must deliberately seek opportunities specific to each project and apply sound safety and traffic engineering principles.

As already mentioned, the geometric form of a road is a three-dimensional alignment which is presented in two projections—the horizontal and the vertical alignment. The horizontal alignment consists of three elements: the tangent, the circular curve, and the transition curve. The vertical alignment consists of two elements: the tangent grade and the vertical curves (crest and sag). Other elements of the alignment are sight distances and superelevation rates.[663] Furthermore, the horizontal and vertical alignments have to be combined in a way that results in a safe, esthetically pleasing design. Safety evaluation processes to control alignment design are presented in Figs. 11.1 and 11.2.

Horizontal and vertical alignment are permanent design elements for which a thorough study is warranted. It is extremely difficult and costly to correct alignment deficiencies after the highway is constructed.

Thus, compromises in the alignment design must be weighed carefully, because any initial savings may be more than offset by the economic loss to the public in the form of accidents and delays.[663]

Horizontal and vertical alignment should not be designed independently. They complement each other, and poorly designed combinations can spoil the good points and aggravate the deficiencies of each. Therefore, excellence in the design of the horizontal alignment and the vertical profile, and in the design of their combination, increases usefulness and safety, encourages uniform speed, and improves appearance—almost always without additional cost[5] (see Chap. 16).

Finally, the safety criteria presented in Table 11.1 will constitute the core of an overall safety module[449] (see Sec. 18.2) in order to

1. Examine the expected operating speed in relation to the design speed

2. Examine consistency or inconsistency between successive design elements

3. Examine the dynamic safety of driving on curved roadway sections

It is recommended that road networks and/or roadway sections, existing or planned, be evaluated by the overall safety module or by the three individual safety criteria, mainly in relation to good, fair, and poor design practices.

Numerous case studies are presented in Secs. 12.2.4.2 and 18.4.2 to reveal the safety evaluation processes.

11.2.2   Alignment Design Policy and Practice Worldwide*

In order to familiarize the reader with the following chapters concerning "Horizontal Alignment" (Chap. 12), "Vertical Alignment" (Chap. 13), "Design Elements of Cross Section" (Chap. 14), "Sight Distance" (Chap. 15), and "Three-Dimensional Alignment" (Chap. 16) general observations about alignment design policy and practice worldwide are discussed by Krammes and Garnham[361] as a review for the International Symposium on Highway Geometric Design Practices in Boston, Massachusetts, in August 1995.

The study reviews highway alignment design policies and practices in a sample of countries throughout the world. The goal is to broaden our understanding and perspective of alignment design by highlighting similarities and differences between design philosophies and quantitative guidelines. It is hoped that accomplishing this goal will stimulate the continued improvement of highway geometric design.

The information presented combines and builds upon recent studies in the United Kingdom, United States, and Germany and included reviews of alignment design policies and practices in a sample of countries.[139,159,360] It was generally observed that there are many similarities in fundamental alignment design principles and philosophies and in quantitative guidelines on basic design parameters. These similarities may reinforce the reasonableness of countries' guidelines that fall within the norm of worldwide practice. What may be most interesting and important, however, are differences in policy emphasis and concern that have led to more advanced guidelines on certain geometric elements, and differences in local conditions and experience that have led to deviations from apparent worldwide norms for certain quantitative guidelines. It is the intent of this chapter to gain insight by understanding the reasons for differences rather than to make judgments about deviations from the norm.

Some countries combine their policies for rural and urban streets, whereas other countries have separate policies. This review focuses on alignment design for rural roadways, which is also the main object of this book.

To illustrate the similarities and differences in alignment design philosophy throughout the world, a sample of 12 countries' alignment design policies were reviewed:

- Australia[37]
- Belgium[516]
- Canada[607,708]
- France[510,700]
- Germany[243,246,399,432]
- Greece[453]
- Italy[126]

---

*Elaborated by R. Krammes and M. Garnham.[361]

- South Africa[125, 663]
- Sweden[47, 686]
- Switzerland[687–693]
- United Kingdom[139]
- United States[5]

Guidelines, standards, and norms of additional countries are incorporated in the discussions, elaborations, comparisons, and evaluations of Chaps. 12 to 16.

All countries use design speed as a basis for establishing limits for basic parameters (for example, minimum radius of horizontal curvature and maximum vertical grade). A fundamental difference among countries is the speed used to establish other alignment parameters, including superelevation rates, sight distance, and rate (or radius) of vertical curvature. Some countries (for example, Canada, South Africa, and the United States) follow the approach described by AASHTO, wherein the design speed is selected (based upon road type, land use, and terrain) and used as the basis for all other alignment parameters. This approach presumes that drivers will not exceed the design speed and, therefore, that no formal checks of actual speed behavior are required. Other countries (for example, Australia, France, Germany, Greece, Switzerland, and the United Kingdom) give more formal and explicit consideration to operating speeds and speed consistency among successive alignment features. Although the details vary, these countries estimate operating speeds (typically 85th-percentile speeds) or a surrogate for operating speed (project speed in Switzerland) along the alignment, check for excessive differences between successive features, and iterate to reduce these differences to acceptable levels. They also typically use this operating speed measure (when it is greater than the design speed) for establishing superelevation rates and sight distance requirements (and corresponding vertical curvature parameters). The United Kingdom has a structured system of design speeds that is explicitly related to 99th-, 85-, and 50th-percentile speeds and uses an iterative approach to ensure that operating speeds and design speeds are coordinated.

Minimum radius of horizontal curvature for a given design speed varies among countries (Table 12.9 and Fig. 12.18). This range results from differences in maximum superelevation rates (Table 14.4) and maximum permissible side friction factors (Table 10.9). Most countries' maximum superelevation rates for rural roadways fall between 6 and 8 percent but some are as high as 10 percent (or 12 percent for exceptional cases). Countries apply margins of safety to different aspects of their design guidelines. For example, Japan has a relatively high maximum superelevation rate (10 percent) but relatively low side friction coefficients. Australia's minimum radii for design speeds ≤90 km/h are lower than for most countries, but these radii are based upon relatively precise estimates of 85th-percentile speeds and observed side friction coefficients. Values for individual parameters must be evaluated within the context of a country's overall design policy, which demands considerable care in making comparisons.

Most, but not all, countries specify superelevation rates for curves with above-minimum radii. Several countries use a linear relationship between superelevation and radius. Canada, South Africa, and the United States use a more complex parabolic relationship. Sweden uses only three superelevation rates.

A common concern is the relative dimensions of successive horizontal alignment elements. Several countries (Australia, Germany, Greece, and Switzerland) estimate speed profiles along alignments and have guidelines based upon acceptable speed reductions between successive features. Several countries provide quantitative guidelines on the relationship between the radii of successive horizontal alignment elements. Most countries have guidelines on the radii of compound curves; a ratio of 1.5 to 1 is common. Other guidelines for the radii of compound curves are related to speed. France, Germany, Greece, and Italy have guidelines on the minimum radii following long tangents. Germany and Greece have comprehensive guidelines indicating acceptable and unacceptable ranges of radii for successive features.

Most countries require the use of transition curves (clothoids) from tangents to most curves and between successive curves. Exceptions are made for certain curves following tangents. These exceptions are stated in various ways: for example, curves not requiring superelevation (France), curves requiring superelevation less than 60 percent of the maximum rate (South Africa), or curves with radii longer than specified values (various countries). Some countries, such as the United States, encourage but do not require the use of transition curves. In most countries, transition curve lengths decrease with increasing radius of the subsequent circular curve. France uses a different philosophy, in which the length decreases with decreasing radius, such that a higher rate of change of centripetal acceleration alerts the driver to a sharper curve.

With respect to vertical alignment, maximum gradient guidelines vary in structure but result in similar maximum values (Table 13.10). For higher-type roadways (motorways or freeways) with higher design speeds (100 to 120 km/h), maximum gradients of 3 to 4 percent are typical. For lower-type roadways (two-lane or single carriageways) with lower design speeds (60 to 80 km/h), maximum rates of 6 to 8 percent are typical. In several countries, gradients in more rolling and mountainous terrain may be 1 to 2 percent steeper.

Vertical curves are typically parabolic in shape. Crest (convex) vertical curve radii (K values in some countries) are based upon stopping sight distance requirements (Table 13.22). Two different criteria for minimum sag (concave) vertical curve radii are prevalent; some countries use stopping sight distance, whereas other countries use less stringent comfort criteria (Table 13.24).

For freeways (motorways) and other multilane divided highways (dual carriageways), curvilinear alignments are preferred to conform with the terrain for cost and environmental reasons. For rural two-lane roadways (single carriageways), some countries (for example, Germany) call for curvilinear alignments to ensure operating speed consistency; whereas other countries place greater emphasis on passing (overtaking), which generally leads to segments with longer tangents (straights). Several countries (including France, Germany, Greece, and the United Kingdom) have observed safety problems associated with marginally adequate passing (overtaking) sight distance and have adapted their alignment guidelines to avoid this condition. The United Kingdom avoids certain ranges of horizontal and vertical curve radii, so that passing sight distance is either adequate or clearly inadequate.

Various provisions are made for dealing with exceptional cases. For example, several countries permit higher maximum superelevation rates. Several countries integrate consideration of climbing lanes, for example, Austria (Table 13.10), as an alternative in vertical alignment design to permit go-with-the-ground designs that avoid costly earthwork but maintain desirable traffic operations.

The United Kingdom has perhaps the most systematic approach for dealing with departures from standards (design exceptions), wherein a given design speed corresponds to the 85th-percentile speed on a roadway with that design speed, the 99th-percentile speed on a roadway with the next lower design speed, and the 50th-percentile speed on a roadway with the next higher design speed. As considerations involving impacts on natural and man-made environments become more important, so too will policies for dealing with exceptions.

In closing, several issues seem particularly fertile for fruitful discussions among a worldwide audience:

- Considering the move toward more detailed checking of design for both individual elements and for consistency between adjacent elements: What are the most effective methods to predict and accommodate actual operating speeds along proposed alignments? What factors influences, which methods work best in a particular country?

- Considering the interrelationships among horizontal and vertical alignment and roadway cross section: What are the tradeoffs between flowing alignments for operating speed consistency and requirements for overtaking (passing) sight distance? What are effective methods for considering tradeoffs between alignment (for example, maximum longitudinal grade) and cross section with respect to the cost and operational efficiency of designing to satisfy overtaking demands and to minimize the operational effects of heavy vehicles on grade?

- Considering the variability among maximum superelevation rates and maximum side friction coefficients: What are the safety and operational impacts of alternative maximum superelevation rates and maximum side friction coefficients?

- Considering differences in transition curve length design among countries: What insight does worldwide safety and operational experience provide concerning when transition curves should be used and the extent to which transition curve length increases with decreasing radius?

- Considering the increasing constraints within which roadway geometry is designed: What are appropriate and effective elements of policies for considering exceptions to design policy?[361]

An attempt has been made in Chaps. 8 to 10 to address several of these questions through an analysis of new research and the design practices of various countries for establishing quantitative safety evaluation processes. Chapters 12 to 16 make comparisons and evaluate and discuss the issues with particular emphasis on safety.

Finally, unique combinations of topography, climate, driving behavior and culture, motor vehicle rules and regulations, vehicle characteristics, and traffic volumes preclude a single set of parameter values or policies working equally well in all countries. In dealing with these issues, however, individual countries can benefit from an understanding and appreciation of the practices and experiences in other countries. An ongoing interchange of ideas, policy evaluations, and research results among countries is recommended.[361]

CHAPTER 12
HORIZONTAL ALIGNMENT

12.1 RECOMMENDATIONS FOR PRACTICAL DESIGN TASKS

The design elements of the horizontal alignment are the tangent, the circular curve, and the transition curve (clothoid).

12.1.1 Tangent

12.1.1.1 Application.* As a design element, the tangent can be beneficial

- For roads of category group A:

  For specific topographic relations, for example, in plains or in wide valleys

  At intersections and interchanges

  To achieve passing sight distances on two-lane highways

  For adapting the alignment to railroad sections, canals, and other man-made constraints

- For roads of category group B:

  In case of specific municipal requirements

  At intersections

  However, long tangents with constant grades have the following disadvantages, especially for roads of category group A:

- They usually lead to excessive speeding.
- They make it difficult to estimate the distances and speeds of oncoming and following vehicles.
- They increase the danger of glare from oncoming vehicles at night.
- They cause the drivers to become tired.
- They can only be adapted to the structure of the landscape in hilly topography with great difficulty.

Therefore, for new designs of category group A roads, long tangents with constant grades should be avoided. Furthermore, short tangents between curves in the same direction of curvature should also be avoided ("broken back" effect). If short tangents must be used, the unsatisfying appearance may be improved by the introduction of a sag vertical curve (see Chap. 16).

Elaborated based on Refs. 124, 159, 243, 453, and 663.

*12.1.1.2 Standard Values.** Because of the glare effect at night and the danger of drowsiness, the maximum lengths of tangents with constant grades, max L, m, for roads of category group A should not exceed 20 times the design speed, V_d, km/h (rule of thumb based on German experience). Minimum tangent lengths should not exceed the values for nonindependent tangents in Tables 12.1 and 12.2. Desirable tangent lengths should be >600 m but <1000 m to provide sufficient passing sight distances.

For roads of category group A, tangents between curves in the same direction of curvature should be avoided. If this is not possible, the minimum length, in meters, should be approximately 6 times the design speed to maintain consistency of the optical guidance.

Tangents for roads of category group A should be combined with circular and transition curves in such a way that, taking into consideration the design elements of the vertical alignment, a good three-dimensional alignment can be achieved (see Chap. 16).

12.1.1.3 Evaluation of Tangents in the Design Process.† The following presented tangent theory represents a modification and further development of the Swiss Norm SN 640 080b.[691]

For the first time, the tangent will be considered here as a "dynamic design element," by taking into account the longitudinal acceleration and deceleration movements observed on tangents. In contrast to the tangent, the *circular curve* has been considered since the 1920s as a dynamic design element with respect to the lateral (centrifugal) acceleration as a driving dynamic input.[385] For the safety evaluation of circular curves with or without transition curves, criteria I and III (see Table 11.1) are of prime importance. Safety criterion II, achieving operating speed consistency (Table 11.1), is significant for the safety evaluation of tangents, in order to distinguish good, fair, and poor design levels for a tangent-to-curve transition, especially on two-lane rural roads.

---

*Elaborated based on Refs. 124, 159, 243, 453, and 663.

†Elaborated based on Refs. 104, 411, 412, 417, 434, 452, 453, and 455.

TABLE 12.1   Relationship between Tangent Lengths and 85th-Percentile Speed Changes for Sequences: Tangents-to-Curves ($V85_T < 105$ km/h)

$V85$ in curve, km/h	$V85_T$ in tangent, km/h						
	70 (1)	75 (2)	80 (3)	85 (4)	90 (5)	95 (6)	100 (7)
50	110	140	175	215	255	295	340
55	—	120	155	190	230	270	315
60	—	—	125	165	205	245	290
65	—	—	—	135	175	220	260
70	—	—	—	—	145	185	235
75	—	—	—	—	—	155	200
80	—	—	—	—	—	—	165

Nonindependent tangent:

☐ = short tangent lengths TL_S, m, the maximum allowable lengths of tangents regarded as nonindependent design elements

$V85$, $V85_T$ = 85th-percentile speed, km/h, in curves or tangents, depending on the CCR_S value according to the operating speed background for the respective country under study (see, for example, Fig. 8.12); for tangents: $CCR_S = 0$ gon/km

CCR_S = curvature change rate of the single curve, gon/km, according to Figs. 8.1 and 8.2

Independent tangents:

Long tangent lengths TL_L, m: For those tangent lengths [(col. 7), the maximum operating speed in tangents of $V85_{Tmax} < 105$ km/h is reached for most of the countries under study, as shown in Fig. 8.12 (that is, Australia, Canada, France, Germany (Old), Greece, Lebanon, and the United States (see also Table 8.5)]. The use of col. 7 is recommended for determining long tangent lengths TL_L.

TABLE 12.2 Relationship between Tangent Lengths and 85th-Percentile Speed Changes for Sequences: Tangents-to-Curves ($V85_T \geq 105$ km/h)

$V85$ in curve, km/h	$V85_T$ in tangent, km/h						
	90 (1)	95 (2)	100 (3)	105 (4)	110 (5)	115 (6)	120 (7)
70	145	185	230	280	325	380	430
75	—	155	200	245	295	345	400
80	—	—	165	210	260	310	365
85	—	—	—	170	220	270	325
90	—	—	—	—	180	235	285
95	—	—	—	—	—	190	245
100	—	—	—	—	—	—	200

Nonindependent tangents:

☐ = short tangent lengths TL_S, m, the maximum allowable lengths of tangents regarded as nonindependent design elements

$V85$, $V85_T$ = 85th-percentile speed, km/h, in curve or tangent, depending on the CCR_S value according to the operating speed background for the respective country under study (see, for example, Fig. 8.12); for tangents: $CCR_S = 0$ gon/km

CCR_S = curvature change rate of the single curve, gon/km, according to Figs. 8.1 and 8.2

Independent tangents:

Long tangent lengths TL_L, m: Exceptional cases which may occur according to Fig. 8.12 in some countries (for example, Germany, ISE). That means the maximum operating speeds in tangents $V85_T$ are equal to or greater than 105 km/h. In those cases, which may also occur in countries not investigated in addition to Germany, the use of cols. 4 to 7 is recommended for determining long tangent lengths TL_L. In the present case, for Germany [$V85_{T\max} = 120$ km/h (see Fig. 8.12)], long tangent lengths would correspond to col. 7.

For the following discussion, two definitions of tangents are relevant:

1. *Nonindependent tangents* are tangents that are too short to exceed the possible 85th-percentile speed differences of safety criterion II (Table 11.1) for good design levels ($\Delta V85 \leq 10$ km/h) or even for fair (tolerable) design levels ($\Delta V85 \leq 20$ km/h) during the acceleration and/or deceleration maneuvers. In this case, the element sequence curve-to-curve, and not the interim tangent, controls the safety evaluation design process.

2. *Independent tangents* are tangents that are long enough to permit a driver to exceed the 85th-percentile speed difference of safety criterion II (Table 11.1) for fair design levels ($\Delta V85 > 20$ km/h) during the acceleration and/or deceleration maneuvers. In this case, the element sequence tangent-to-curve should control the design process.

Based on car-following techniques,[411] an average acceleration or deceleration rate of $a = 0.85$ m/s^2 was established. Consequently, the formula for the evaluation of the (theoretical) transition length between two successive curves according to Fig. 12.1 becomes:

$$TL = \frac{V85_1^2 - V85_2^2}{2 \times 3.6^2 a} \tag{12.1}$$

$$TL = \frac{V85_1^2 - V85_2^2}{22.03} \rightarrow TL > 0 \tag{12.1a}$$

where $V85_{1,2}$ = 85th-percentile speeds in curves 1 and 2, km/h

TL = (theoretical) transition length between two successive curves, m

a = acceleration/deceleration rate, m/s^2

In order not to be too conservative, tangent lengths between two successive curves that fall into the ranges of fair design levels will be considered as nonindependent design elements. The corresponding tangent lengths are represented by the values within the outlined boxes shown in Tables 12.1 or 12.2. For tangent lengths equal or shorter than these, changes in the 85th-percentile speeds between two successive curves may be calculated directly without considering the tangent between them as an independent design element for the safety evaluation process according to criterion II (Table 11.1). With this assumption, the most critical case for fair design levels ($\Delta V85 = 20$ km/h) is addressed, especially during a deceleration process. In all the other cases ($\Delta V85 < 20$ km/h), the tangent lengths are not sufficient for the average driver to decelerate or accelerate in such a way that the assumed boundaries for operating speed changes for fair or good design are exceeded.

However, tangent lengths between successive curves that exceed the critical values within the outlined boxes of Tables 12.1 or 12.2 have to be considered as independent design elements. In these cases, a driver is able to accelerate or decelerate in such a way that even the maximum allowable operating speed changes for fair design levels ($\Delta V85 \leq 20$ km/h) according to Table 11.1 may be exceeded. This means that critical driving maneuvers may have already occurred.

Based on the practical and research experience of the authors, the following three cases have to be distinguished when dealing with tangent lengths in the highway design process in order to simplify the design procedure:

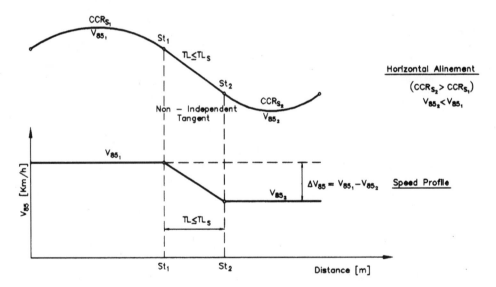

Legend:

St	=	Station,
TL	=	Existing Tangent Length [m],
TL$_S$	=	Short Tangent Length, Acceleration or Deceleration Distance Between Curve 1 and Curve 2 [m],
CCR$_{S1,2}$	=	Curvature Change Rates of the Single Curves [gon/km],
V85$_{1,2}$	=	Operating Speeds in Curves [km/h].

FIGURE 12.1 Systematic sketch of horizontal alignment and speed profile for case 1 (nonindependent tangent). (Modified and completed based on Ref. 691.)

Case 1 (Fig. 12.1).   The existing tangent length, TL, between two successive curves is smaller than the short tangent length, TL_S, given in Tables 12.1 or 12.2, which corresponds to the nearest 85th-percentile speed of the curve with the higher CCR_S value according to the operating speed background of the country under study (see Fig. 8.12 or Table 8.5). From this it follows that the tangent is to be considered as nonindependent and can be assumed to be negligible in the design process, that is, the curve-to-curve sequence controls the design process (see Fig. 12.1 and the case study under item 4 in the "Design Procedure" section that follows).

Case 2 (Fig. 12.2).   The existing tangent length, TL, is at least twice as long as the long tangent length, TL_L, given in col. 7 of Table 12.1, related once again to the nearest 85th-percentile speed of the curve with the higher CCR_S value ($TL \geq 2TL_L$). In this case, it can be assumed without any calculations that the tangent is independent and that operating speeds in tangents $V85_T < 105$ km/h are good estimates for most countries. In exceptional cases ($V85_T$ in tangents ≥ 105 km/h), the application of cols. 4 to 7 of Table 12.2 is recommended. For case 2, the tangent-to-curve sequence controls the design process (see Fig. 12.2 and the case study under item 5 in the "Design Procedure" section that follows).

Case 3 (Fig. 12.3).   The existing tangent length lies somewhere between case 1 and case 2 ($TL_S < TL < 2TL_L$). In this case, the operating speed in the independent tangent has to be calcu-

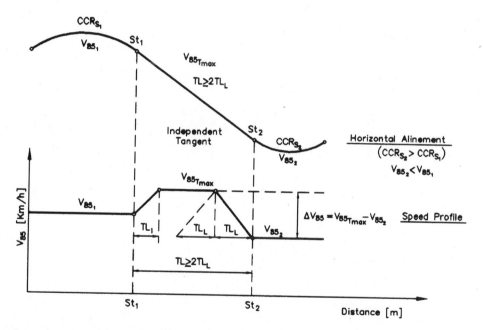

Legend: see also Figure 12-1

TL_L =   Long Tangent Length, Critical Acceleration or Deceleration
                Distance Between Independent Tangent and Curve 2 [m],

TL_i =   Distance Travelled for Section "i" during Acceleration or
                Deceleration [m],

$V85_{Tmax}$ =   Maximum Operating Speed in Tangents [km/h]
                (depending on the country under study).

FIGURE 12.2 Systematic sketch of horizontal alignment and speed profile for case 2 (independent tangent). (Modified and completed based on Ref. 691.)

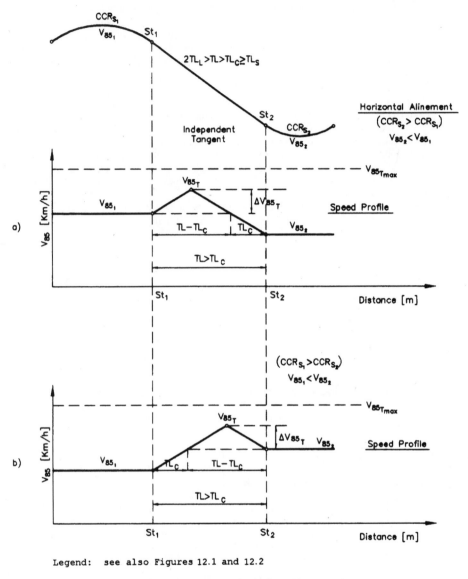

FIGURE 12.3 Systematic sketches of horizontal alignment and speed profiles for case 3 (independent tangent). (Modified and completed based on Ref. 691.)

lated individually according to Fig. 12.3 and Table 12.3. The tangent-to-curve sequence controls the design process. "Example Application" in Sec. 12.2.1.3 reveals an example for this case.

Note that the statements for cases 1 to 3 have to be applied for both driving directions.

Design Procedure. For the safety evaluation process of criterion II, the changes in the 85th-percentile speeds, $\Delta V85$, between successive design elements (tangent-to-curve or curve-to-curve) have to be determined according to Table 11.1.

For evaluating tangents in the design process, the following procedure is recommended:

1. Determine the tangent length, TL, between the observed two successive elements of interest.

2. Calculate the CCR_S values for curves 1 and 2 according to Eq. (8.6) and Figs. 8.1 and 8.2 and determine the corresponding 85th-percentile speeds, $V85_1$ and $V85_2$, by applying the operating speed background for the country under study (for example, use Fig. 8.12 or Table 8.5).

3. Compare the existing tangent length, TL, between the two successive curves with the short tangent length, TL_S, and the long tangent length, TL_L, given in Tables 12.1 or 12.2 in order to distinguish nonindependent tangents from independent tangents. (Note that, for determining TL_S and TL_L section lengths in Table 12.1 or Table 12.2, the nearest 85th-percentile speed of the curve with the higher CCR_S value is the controlling speed in order to simplify the procedure.)

4. *Case 1:* Case 1 is presented in Fig. 12.1. In this case, the driver accelerates (or decelerates) uniformly. If the observed tangent length, TL, is smaller than the maximum allowable for short tangent lengths TL_S according to Tables 12.1 or 12.2, then the tangent is considered nonindependent and not relevant for the design process. That means the change in operating speeds for evaluating good, fair (tolerable), and poor design practices according to safety criterion II (Table 11.1) is related only to the two successive curves

$$\Delta V85 = |V85_i - V85_{i+1}| \qquad (12.2)$$

Example for Greece according to curve "Greece" in Fig. 8.12 or Eq. (8.16) in Table 8.5: Figure 12.1 and Table 12.1 are relevant:

$$\text{TL} \quad = 120 \text{ m} \qquad \text{existing tangent length}$$

$$CCR_{S1} = 90 \text{ gon/km} \qquad V85_1 = 92 \text{ km/h}$$

$$CCR_{S2} = 240 \text{ gon/km} \qquad V85_2 = 82 \text{ km/h}$$

The 85th-percentile speed in Table 12.1 which is closest to 82 km/h in the curve with the higher CCR_S value is 80 km/h. (This simplification was done for an easier application of Table 12.1.) For 80 km/h, the maximum tangent length which is regarded as nonindependent is $TL_S = 165$ m. Since

$$\text{TL} < \text{TL}_S$$

$$120 \text{ m} < 165 \text{ m}$$

the tangent has to be classified as a nonindependent design element, and the sequence curve-to-curve with the corresponding operating speeds, $V85_1$ and $V85_2$, becomes relevant for evaluating operating speed consistency according to safety criterion II (Table 11.1):

$$|V85_1 - V85_2| = |92 - 82| = 10 \text{ km/h} = \text{good design practice}$$

5. *Case 2:* Case 2 is schematically shown in Fig. 12.2. In this case, the tangent is long enough for a driver to accelerate and for some distance to sustain the maximum 85th-percentile speed of the tangent, $V85_{Tmax}$, before decelerating again. As discussed previously, this speed was found to be approximately $V85_{Tmax} \approx 100$ km/h for most of the investigated countries (see Fig. 8.12).

However, new research results reveal that the 85th-percentile speed in tangents may be as high as 120 km/h (such as in Germany, according to Fig. 8.12). There may be other countries, such as Italy, where personal observations seem to support these findings.

Therefore, as a result of comparison to former investigations,[104,411,417,434,453,565] Table 12.2 was developed in order to be able to consider higher operating speed backgrounds in the course of a sound design process.

Figure 12.2 reveals that if the existing tangent is at least twice as long as the long tangent length, TL_L, which is presented in col. 7 of Tables 12.1 or 12.2 (depending on the country under study), then the tangent can be considered independent and no further calculation is necessary. For the safety evaluation process of criterion II (Table 11.1), the sequence tangent-to-curve or curve-to-tangent controls the design process. For the expected critical speed change, the following equation applies:

$$\Delta V85 = |V85_{Tmax} - V85_{i+1}| \tag{12.3}$$

Tables 12.1 and 12.2 were developed in order to provide quick estimates of nonindependent and independent tangent lengths. Of course, the relevant tangent lengths, TL_S and TL_L, can also be calculated from the formulas in Table 12.3 for design cases 1 and 2.

Example for Germany according to curve "Germany, ISE" in Fig. 8.12 or Eq. (8.15) in Table 8.5: Figure 12.2 and Table 12.2 are relevant:

$$TL = 850 \text{ m} \qquad \text{existing tangent length}$$

$$CCR_S = 0 \text{ gon/km} \qquad V85_{Tmax} = 120 \text{ km/h}$$

$$CCR_{S1} = 180 \text{ gon/km} \qquad V85_1 = 103 \text{ km/h}$$

$$CCR_{S2} = 520 \text{ gon/km} \qquad V85_2 = 80 \text{ km/h}$$

The 85th-percentile speed in Table 12.2 that is closest to 80 km/h in the curve with the higher CCR_S value is 80 km/h. To accelerate or decelerate from 80 km/h to the highest operating speed of $V85_{Tmax} = 120$ km/h in the tangent, a distance of $TL_L = 365$ m is needed according to Table 12.2. It follows:

$$TL > 2 \, TL_L$$

$$850 \text{ m} > 2 \times 365 \text{ m} = 730 \text{ m}$$

Thus, it can be concluded that the tangent is independent. For the safety evaluation process, the transitions between the independent tangent and the preceding and the succeeding curve have to be examined according to safety criterion II (Table 11.1):

$$|V85_{Tmax} - V85_2| = |120 - 80| = 40 \text{ km/h} > 20 \text{ km/h} = \text{poor design practice}$$

$$|V85_1 - V85_{Tmax}| = |103 - 120| = 17 \text{ km/h} \rightarrow 10 < 17 \leq 20 \text{ km/h} = \text{fair design practice}$$

6. *Case 3:* Case 3 is schematically shown in Fig. 12.3. In case 3, the tangent is long enough for some acceleration but not sufficiently long enough for drivers to attain the maximum speed, $V85_{Tmax}$, on independent tangents at normal acceleration and deceleration rates. Figure 12.3 reveals that the existing tangent length between curves 1 and 2 is longer than the acceleration or deceleration distance between curves 1 and 2 ($TL > TL_C$). As limiting conditions, it may be possible that

$$TL_S = TL_C$$

or

$$TL_C = TL_L$$

Therefore, the tangent has to be considered independent. For this case, estimates based on Tables 12.1 or 12.2 are not sufficient, and exact calculations must be performed.

The equations for conducting the computation processes according to Fig. 12.3 are given in Table 12.3. For the evaluation of safety criterion II (Table 11.1), the sequence tangent-to-curve and curve-to-tangent controls the design process. For the expected change in the 85th-percentile speeds, the equations now read:

$$\Delta V85 = V85_T - V85_2 \qquad (12.4a)$$

for Fig. 12.3a, and

TABLE 12.3 Equations for the Definition of Various Types of Tangents in the Design Process

Case	Condition	Equation	Result on $V85_{Tmax}$
1 (NIT)	$TL \le TL_S$ (Fig. 12.1)	$$TL_S = \frac{V85_1^2 - V85_2^2}{25.92\,a} \qquad (12.5)$$ $$V85_1 > V85_2$$	$V85_{Tmax}$ not met
2 (IT)	$TL > 2TL_L$ (Fig. 12.2)	$$TL_L = \frac{V85_{Tmax}^2 - V85_2^2}{25.92\,a} \qquad (12.6)$$ $$TL_i = \frac{V85_{Tmax}^2 - V85_1^2}{25.92\,a} \qquad (12.7)$$ $$V85_1 > V85_2$$	$V85_{Tmax}$ met and held, respectively, met and not held
3 (IT)	$TL < 2TL_L$ $TL > TL_C \ge TL_S$ (Fig. 12.3a and Fig. 12.3b) (replace $V85_1$ through $V85_2$)	$$TL_C = \frac{V85_1^2 - V85_2^2}{25.92\,a} \qquad (12.8)$$ $$V85_T = V85_1 + \Delta V85_T \qquad (12.9)$$ $$\Delta V85_T = \frac{-2V85_1 + [4V85_1^2 + 44.06\,(TL - TL_C)]^{1/2}}{2} \qquad (12.10)*$$ $$V85_1 > V85_2$$ Note that when calculating $V85_T$, the curve with the lower CCR_S value must be selected.	$V85_{Tmax}$ not met

*If $TL_C > TL \to \Delta V85_T = 0$, km/h.

Note: See Figs. 12.1 to 12.3. NIT = nonindependent tangent, IT = independent tangent, and a = acceleration/deceleration rate = 0.85 m/s².

$$\Delta V85 = V85_T - V85_1 \qquad\qquad (12.4b)$$

for Figure 12.3*b*.

For case 3, a detailed example is given in "Example Application" in Sec. 12.2.1.3, in addition to background information about the research results that led to the previous definitions.

Equations for the definitions of various types of tangents according to design cases 1 to 3 are given in Table 12.3.

From a dynamic safety point of view, the tangent, regarded here as an operating speed–dependent design element, allows a sound transition between independent tangents and curves. Nonindependent tangents do not need to be considered in the safety evaluation process according to criterion II.

Several different case studies are discussed for various countries in Secs. 12.2.4.2 and 18.4.2.

With respect to the previous discussions and the experiences in Sec. 12.2.1.2 for maximum tangent lengths (independent tangents), the German and South African recommendations[243,663] were taken as a basis:

$$L_{max} = 20\, V_d$$

where L_{max} = maximum tangent length, m
       V_d = design speed, km/h

However, because of passing distance requirements, a length of >600 m but <1000 m (compare Table 15.11) would represent more reliable solutions for independent tangents and also support relation design issues.

Minimum tangent lengths should not exceed the values within boxes in Tables 12.1 or 12.2 for nonindependent tangents.

In this way, the relation design requirements of Secs. 9.1.3.2 and 9.2.3.2 are considered, and well-balanced designs between successive curves, as well as sound transitions between independent tangents and curves, can be expected. With respect to the latter case, with and without transition curves, refer to Tables 12.4 and 12.8.

Finally, note that tangent sections have lower accident rates than horizontal curves. Compare Tables 9.10 and 10.12.

With respect to recent developments a modified procedure for determining tangent speeds and lengths in the safety evaluation process is presented in Sec. 18.5.2.

12.1.2 Circular Curve*

12.1.2.1 Application.   The radii of circular curves of category group A roads should be as large as possible, especially for small deflection angles. The same applies to circular curves with or without transition curves that follow independent tangents. Furthermore, the radii of successive curves, as well as the transitions between independent tangents and curves, should have a well-balanced relationship according to Secs. 9.1.3.2 and 9.2.3.2.

By selecting large radii of curve, sufficient sight distances and a consistent driving behavior should be provided. Otherwise the radii of curve should only be as large as needed so they are in harmony with:

- The topography
- The landscape
- The vertical alignment

---

*Elaborated based on Refs. 35, 37, 124, 243, 429, 453, 663, and 688.

In addition, horizontal curves should represent a well-balanced relationship between the design speed, V_d, and the 85th-percentile speed, V85, in order to adhere to the requirements of good design practices according to safety criteria I and III in Table 11.1.

When selecting very large radii of curve, it should be remembered that the same advantages and disadvantages that apply to long independent tangent sections are also relevant to these roadway sections.

Short curves between long independent tangents appear optically as a kink. Therefore, they should be avoided. If the deflection angle between two tangents is 8° or less, then the length of the horizontal curve should be at least 200 m in order to avoid the impression of a broken line. Australia even requires lengths of at least 500 m to ensure that curves with small deflection angles do not appear as kinks in flat terrain.

For new designs and redesigns of existing roads of category group A, it is especially important to consider the protection of the landscape. For new designs, redesigns, or restoration strategies of category group B roads, municipal conditions require a careful review of the demands of adjacent utilized areas when assessing radii of curve. This is especially true for suburban roads and roads through villages and smaller towns.[441,450,457]

12.1.2.2 Limiting and Standard Values.   Between two circular curves in the same or in opposite directions, the radii of these curves should be in a well-balanced relationship in order to promote safety on roads of category group A and road category B II (known as relation design). Radii relations between successive circular curves are shown in Figs. 9.1 and 9.36 to 9.40 for selected countries. Detailed information on relation design between independent tangents and curves is also provided in Secs. 9.1.3.2 and 9.2.3.2 for new designs and redesigns of existing alignments.

Based on this section, the minimum radii of curve shown in Table 12.4 should be applied for the element sequence: independent tangent–transition curve–circular curve when the selected design speed, V_d, does not require larger curve radii.

Once again, it should be noted here that the radii of curve in Table 12.4 correspond to good design practices according to Table 11.1 for safety criterion II.

The size of the minimum radius of curve is determined by the relationship between design speed, maximum permissible side friction factor, and maximum superelevation rate. This relationship is expressed by the driving dynamic formula:

$$R_{\min} = \frac{V_d^2}{127\,(f_{R\mathrm{perm}} + e_{\max})}$$ (10.9, 12.11)

where $R_{\min}$ = minimum radius of curve, m

V_d = design speed, km/h

$f_{R\mathrm{perm}}$ = maximum permissible side friction factor

$e_{\max}$ = maximum superelevation rate, %/100

[For the derivation of Eq. (12.11), see "Vehicle Dynamics" in Sec. 10.2.1.2.]

TABLE 12.4 Minimum Radii of Curve for the Element Sequence Independent Tangent–Clothoid–Circular Curve*

Operating speed background	$R_{\min}$ of the circular curve
$V85_T < 105$ km/h	$R_{\min} \geq 400$ m
$V85_T \geq 105$ km/h	$R_{\min} \geq 500$ m

*When the selected design speed does not require larger curve radii (Table 12.5).

Note: $V85_T$ = 85th-percentile speed in independent tangents, km/h; refer to the operating speed backgrounds, V85, in Fig. 8.12 or Table 8.5 (CCR_S = 0 gon/km).

In order to establish minimum radii of curve, the assumptions provided in Table 10.1 with respect to maximum permissible side friction factors in relation to design speed, road category group, topography, and maximum superelevation rates should apply. A detailed listing of the parameters which influence the calculation of minimum radii of curve is given in Table 10.2.

For practical design purposes, the minimum radii of curve are given in Table 12.5 for roads of category group A (rural roads) and in Table 12.6 for roads of category group B (suburban roads). In practice, it is normal to select radii of curve larger than those shown in Tables 12.5 and 12.6 in order to reduce superelevation rates and side friction factors below their maximum values. The corresponding relationships between curve radii and superelevation rates are presented in Figs. 14.2 to 14.4.

The circular curve should be long enough to ensure that driving through the curve at the design speed will last for more than 2 s. The corresponding minimum lengths of circular curves are given in Table 12.7.

The values for R_{min} are based upon driving dynamic considerations and do not automatically ensure adequate sight distances in curves.

For instance, sight obstructions in the median (that is, guardrails and fauna) on multilane highways frequently lead to inadequate stopping sight distances. In these cases, a sufficient stopping sight distance must be provided for the left lane. If adequate stopping sight distances cannot be

TABLE 12.5   Minimum Radii of Curve for Roads of Category Group A in Flat and Hilly/Mountainous Topography

	R_{min} for roads of category group A (rural)	
	Flat topography	Hilly and mountainous topography
V_d, km/h	e_{max} = 8 to 9%; n = 45%	e_{max} = 7%; n = 40%
50	85 (80)	95
60	125 (120)	140
70	180 (170)	200
80	250 (235)	280
90	330 (310)	370
100	425 (400)	(480)*
110	530 (500)	(600)*
120	650 (620)	(740)*
(130)	790 (740)	(890)*

Note: n = utilization ratio of side friction; ( ) = exceptional values; and ( )* = for high grades, values are only of theoretical character in practical alignment design. For redesigns or RRR strategies of existing alignments, a utilization ratio of n = 60 percent may be appropriate in case of administrative, financial, and environmental constraints.

TABLE 12.6   Minimum Radii of Curve for Roads of Category Group B

	R_{min} for roads of category group B (suburban)
V_d, km/h	e_{max} = 6%; n = 60%
40	45
50	70
60	110
70	160
80	225
90	300

TABLE 12.7 Minimum Lengths of Circular Curves*

V_d, km/h	L_{min}, m
50	30
60	35
70	40
80	45
90	50
100	55
120	65
(130)	75

*Based on Ref. 246.

guaranteed, then it might be necessary to consider speed limits for wet road surface conditions, to either widen the median in order to increase the distance of the guardrail from the pavement, or to select larger radii of curve.[243]

If, in rare cases, the minimum radii of curve given in Tables 12.5 and 12.6 or the suggested radii of curve sequences shown in Figs. 9.1 and 9.36 to 9.40 cannot be provided, then the resulting decrease in safety must be alleviated by certain measures, such as by improving the visibility of the curve or by easing existing side obstructions. In addition, it is appropriate, in case minimum radii of curve and desired radii of curve sequence cannot be maintained, to take care of these disadvantages by undertaking measures such as planting, delineation, or traffic warning devices.[243]

*12.1.2.3 Safety Considerations.** With respect to the safety evaluations of radii of curve, results of studies on two-lane rural roads of category group A show the following (refer to the literature review in "Radius of Curve" in Sec. 9.2.1.3):

• Accident risk decreases with increasing radius of curve.

• Road sections with radii of curve less than 200 m have an accident rate which is at least twice as high as that on sections with radii greater than 400 m.

• A radius of 400 m provides a cross point in safety.

• For radii greater than 400 m, the gain in safety is relatively small.

• The safety of a winding alignment is usually not seriously affected by a smaller curve, whereas isolated sharp curves on an otherwise flowing alignment are dangerous.

These findings provide the highway engineer with an overview of what he can expect from a safety point of view when using different ranges of radii of curve.

With respect to the three developed safety criteria

• Safety criterion I in Secs. 9.1.2 and 9.2.2

• Safety criterion II in Secs. 9.1.3 and 9.2.3

• Safety criterion III in Secs. 10.1.4 and 10.2.4

it should be noted that the design parameter radius of curve is included in each of the three safety criteria, as will be shown in the following:

Based on safety criterion I, a comparison of the design speed and the 85th-percentile speed is made for different design classes in Table 11.1. With respect to the design speed, the radii of curve

---

*Elaborated based on Refs. 435, 445, 448, 453, and 455.

of category group A roads are given in Table 12.5, at least for determining the minimum radii of curve. The 85th-percentile speed depends on the design parameter curvature change rate of the single curve CCR_S and can be determined from Fig. 8.12 or Table 8.5. CCR_S is calculated from Eq. (8.6), which also considers the design parameter radius of curve (compare Fig. 8.1).

Based on safety criterion II, a comparison of the 85th-percentile speeds between successive design elements is made in Table 11.1. As noted previously, the design parameter radius of curve is considered in the determination of these speeds.

Based on safety criterion III, a comparison of side friction assumed and side friction demanded is made in Table 11.1. Side friction assumed depends on the design speed and thereby, at least indirectly, on the corresponding radius of curve (Table 10.2). On the other hand, side friction demanded directly regards the radius of curve according to Eq. (10.11).

From the foregoing, it can be concluded that all three developed safety criteria are directly or indirectly influenced by the design parameter radius of curve.

The development of the three safety criteria is based on mean accident rates for individual design classes, expressed by the design parameter curvature change rate of the single curve CCR_S (see Tables 9.10 and 10.12) in order to distinguish good designs from fair designs as well as poor designs. The established design classes correspond somewhat to radii of curve classes. For example, the results in Tables 9.10 and 10.12 clearly indicate that the mean accident rate for a fair design (CCR_S = 180 to 360 gon/km; $R \approx$ 175 to 350 m) is at least twice as high as that for a good design (CCR_S < 180 gon/km; R > 350 m), and that the mean accident rate for a poor design (CCR_S > 360 gon/km; R < 175 m) is at least 4 to 5 times higher than that of a good design. These findings also apply to the accident cost rate, as will be shown later in Sec. 18.3.

By transforming the CCR_S classes (design classes) to operating speed classes (Table 9.13), as well as to side friction classes (Table 10.11), it follows that an overall relationship between curve design (radius of curve), traffic safety, operating speed, and driving dynamics could be obtained in order to distinguish good designs from fair designs as well as poor designs.

If all three safety criteria fall into the good design range, it can certainly be said that a good, sound alignment design can be expected with respect to individual curve designs and transitions to succeeding elements.

In a similar manner, existing curved sites and their corresponding transitions can be examined for detecting fair and poor design practices in order to improve or redesign these sites.

In addition, the safety-related discussions mentioned earlier indicate that the limiting values of Table 12.4 for the element sequence tangent to circular curve with transition curves and of Table 12.8 for the element sequence independent tangent to circular curve without transition curves can be regarded as reliable from the safety point of view.

Thus, it can be concluded that the appropriate selection of the radius of curve plays an important role in the development of safety evaluation processes.

12.1.2.4   Compound Circular Curves.*

Compound curves consist of two or three contiguous unidirectional circular curves of differing radii, and they are a primary exception. To minimize any effect of possible nonrecognition of the tightening curvature by vehicles approaching from the longer radius of curve, the sight distance should be maximized so it is at least greater than the stopping sight distance.

For roads of category group A and road category BII, the change in radii of curve should always be related to the good relation design range of Figs. 9.1 and 9.36 to 9.40. In order to establish permissible sequences of radii of curve, only radii of curve where the arc lengths are long enough to allow a vehicle traveling at the design speed to pass in 2 s (rule of thumb) should be considered. More than three circular sections should not be joined together.

For compound curves on open highways, it is generally accepted that the ratio of the flatter radius of curve to the sharper one should not exceed 1.5 to 1. Where feasible, a smaller difference in radii should be used; a desirable maximum ratio is 1.75 to 1. When the ratio is more than 2 to 1, a suitable length of spiral or a circular arc of intermediate radius should be inserted between the two curves.

Elaborated based on Refs. 5, 37, 243, and 688.

The Swiss Norm SNV 640 100a[688] provides a graphical presentation of the previous statements for the following design cases (see Fig. 12.4). The ranges of radii in Fig. 12.4 are also more or less valid for the design of ramps at grade-separated intersections.

12.1.2.5 Curve Flattening.*

According to "Operating Speeds on Dry and Wet Road Surfaces" in Sec. 8.2.1.2, it was concluded that operating speeds on dry pavements were not statistically significantly different from operating speeds on wet pavements and that drivers did not seem to recognize the fact that friction on wet pavements is significantly lower than on dry pavements. These results indicate that these drivers run the risk of being involved in a traffic accident.

Similar conclusions were made by Talarico and Morrall[697] who indicated that, in relation to speed, the margin of safety decreases at a higher rate on wet pavements than on dry.

They further indicated that wet pavements provide a margin of safety of approximately 0.50 less than dry pavements for any given speed and radius of curvature. Therefore, redesigns or RRR improvements of sharp curves should concentrate on curve flattening.

In summary, Talarico and Morrall[698] reported the following:

1. Sharp curves may not provide an adequate margin of safety at normal operating speeds, while flatter curves provide very high margins of safety. Therefore, motorists have a greater chance of exceeding the frictional capacity of a pavement at normal operating speeds on sharper curves than on flatter curves.
2. The margin of safety provided on sharp curves is too small to accommodate evasive driving maneuvers under wet pavement conditions, even if maximum superelevation is provided. For flatter curves, the margin of safety is large enough to accommodate these types of maneuvers, even if lower superelevation rates were provided. This suggests that alternate forms of superelevation rates and side friction factors could be explored.
3. Increasing the superelevation rates on sharper curves is unlikely to yield acceptable margins of safety. Therefore, RRR improvements to these types of curves should concentrate on curve flattening.[698]

In order to quantitatively improve those roadway sections which are often present in the existing (old) alignments by means of redesigns or RRR projects, the driving dynamic safety criterion III was developed. Safety criterion III compares side friction assumed [Eq. (10.10b)] with side friction demanded [Eq. (10.11)] with respect to good, fair (tolerable), and poor design levels (Table 11.1). If, however, despite providing maximum superelevation rates of, for example, e_{max} = 8 percent, the critical ranges for fair design or even poor design have been achieved or exceeded, then *curve flattening* represents the one and only way of obtaining sound combinations of superelevation rates and side friction factors.

A typical case study is shown in Sec. 12.2.4.2 (example: United States) for a dangerous curve of R = 150 m (Fig. 12.26) which has been flattened to a radius of R = 500 m (Fig. 12.28). According to Table 12.16, with respect to the old alignment case, col. 13 reveals poor design according to safety criterion III. Following curve flattening, the interim solution case in Fig. 12.28 represents good design practices according to the established three safety criteria.

The safety effect of flattening sharper horizontal curves is of particular interest on existing two-lane rural roads in case of redesigns and RRR projects. When a sharp curve is improved, transitions from the tangent to the curved portions of the highway are smoother, the length of the curved portion of the roadway is increased, and the overall length of the highway is lightly reduced. In this respect, however, the relationships between the central angle of curve, degree of curve, length of curve, and accidents should also be considered, as presented in "Curvature Change Rate, Degree of Curve, Length of Curve, and Curve Radii Ratio" in Sec. 9.2.1.3, Fig. 9.18.

Curve flattening refers to reconstructing an existing horizontal curve in order to make it less sharp, that is, longer with a lower degree of curve or curvature change rate of the single curve. Curve flattening is highly effective for reducing accident rates on sharp or poorly designed inconsistent curves.

---

*Elaborated based on Refs. 698, 710, and 728.

Two contiguous unidirectional circular curves

Conditions:

    $R_2 > R_1$

    R_2/R_1 must be located in the dotted area.

For open highways, radii of curve smaller than $R_2 < 350$ m should not be allowed.

Three contiguous unidirectional circular curves

Conditions:

    $R_{21} > R_1 < R_{22}$

    R_{2i}/R_1 must be located in the dotted area.

The design cases

        $R_{21} < R_1$ or

        $R_{22} < R_1$

are not allowed.

For open highways, radii of curve smaller than $R_{2i} < 400$ m should not be allowed.

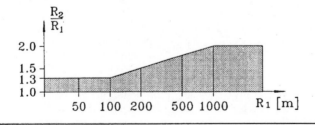

FIGURE 12.4   Application ranges for two or three contiguous unidirectional circular curves. (Elaborated based on Ref. 688.)

A number of studies have indicated that horizontal realignment of rural highways is the most efficient way of increasing safety. Accident reductions of up to 80 percent could be obtained depending on the central angle and the amount of curve flattening.[669,677] Accident reduction factors, which correspond to various increases in horizontal radii of curve or decreases in degrees of curve, are given in Tables 9.4[85] and 12.13,[781] respectively.

When flattening nonisolated curves, the good range of the relation design backgrounds shown in Figs. 9.1 or 9.36 to 9.40 must be used.

When flattening isolated curves, the ranges in Table 12.4 are recommended for the element sequence independent tangent–clothoid–circular curve. On the other hand, for the direct sequence tangent–circular curve, the ranges in Table 12.8 are recommended.

12.1.3 Transition Curve*

12.1.3.1 Application.
The transition curve is represented by the Euler spiral curve (clothoid) applied to effect the transition between two circular curves or between a circular curve and a tangent.

The transition curve will:

- Provide a linear gradual increase or decrease of the centrifugal acceleration for the transition from one design element to the other when passing through curves
- Serve as a transition section for a convenient desirable arrangement for the superelevation runoff
- Make possible through the gradual change of the curvature a consistent alignment and through this a consistent operating speed
- Create a satisfactory optical appearance of the alignment.

The application of transition curves is required for roads of category group A and road categories BI and BII.

The transition curve has to be formed as a clothoid. For this kind of curve, the curvature changes linearly with the arc length. The formula for the clothoid is:

$$A^2 = R\,L \tag{12.12}$$

where A = parameter of the clothoid, m
  R = radius, m (the radius at the end of the clothoid)
  L = length of the clothoid, m (until the radius R is reached)

The clothoid parameter, A, expresses the rate of change of curvature along the clothoid. Large values of A represent slow rates of change of curvature, whereas small values of A represent rapid rates of change of curvature.

Based on the experiences gained in Secs. 9.1.3.2 and 9.2.3.2 (Relation Design), for the minimum radii of curve in Table 12.8, transition curves are usually not necessary for the independent tangent to circular curve sequence. In these cases, the application of clothoids is left to the discretion of the highway engineer.

Once again, and for safety reasons, the radii of curve in Table 12.8 represent values that are double those calculated for good design practices, as based on safety criterion II in Table 12.4.

Furthermore, transition curves can be neglected if the deflection angle of the overall curve is less than 10 gon (9°). In this case, the minimum arc length of the circular curve should correspond to the design speed, in meters (rule of thumb).

---

*Elaborated based on Refs. 35, 124, 243, 453, 663, and 688.

TABLE 12.8   Minimum Radii of Curve for the Element Sequence Independent Tangent–Circular Curve without Transition Curves

Operating speed background	R_{min} of the circular curve
$V85_T < 105$ km/h	$R_{min} \geq 800$ m
$V85_T \geq 105$ km/h	$R_{min} \geq 1000$ m

Note: $V85_T$ = 85th-percentile speed on independent tangents, km/h; compare the operating speed backgrounds $V85$ in Fig. 8.12 or Table 8.5 (CCR_S = 0 gon/km).

12.1.3.2   Limiting Values.

In order for the transition curve to be optically perceivable, and for esthetic reasons, the clothoid should consume an angle of at least $\tau = 3.5$ gon (3°) from the origin (see Fig. 12.5). For roads of all categories, it follows then that the minimum parameter of the clothoid is given by:

$$A_{min} = \frac{R}{3} \qquad (12.13)$$

where A_{min} = minimum parameter of the clothoid, m

R = radius at the end of the clothoid, m

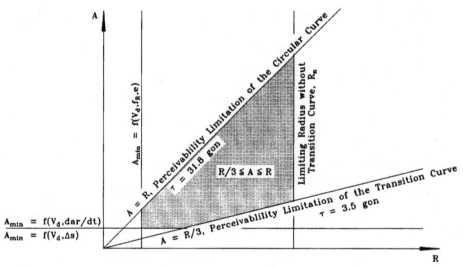

Legend:

V_d = design speed [km/h],

dar/dt = rate of change of radial acceleration [m/sec³],

Δs = relative grade of superelevation runoff [%], see Equation 14-14a, limiting values in Secs. 14.1.3.2 and 14.2.3.2,

f_R = side friction factor [-],

e = superelevation rate [%/100],

R_n = limiting radius without transition curve [m].

FIGURE 12.5   Acceptable ranges for the parameter, A, of the clothoid. (Modified and elaborated based on Ref. 688.)

For safety reasons, and in order for the circular curve to be optically perceivable, the clothoid should not consume an angle larger than $\tau = 31.8$ gon (29°) from the origin (see Fig. 12.5):

$$A_{max} = R \qquad (12.14)$$

where A_{max} = maximum parameter of the clothoid, m
$\qquad R$ = radius at the end of the clothoid, m

Acceptable ranges for the parameter, A, of the clothoid are schematically shown in Fig. 12.5. This figure indicates that the parameter, A, of the clothoid should be selected from the dotted area. This area is subject to the following conditions:

- The change of radial acceleration falls within acceptable ranges
- The circular curve is quite perceivable, that is, $A \leq R$
- The clothoid is quite perceivable, that is, $A \geq R/3$.

With respect to radius of curve, the parameter, A, of the clothoid can be graphically determined from Fig. 12.6, in the range from the maximum to the minimum parameter of the clothoid, by additionally taking into account the requirements of the superelevation runoff according to Eq. (14.21b) in "Limiting Values" in Sec. 14.2.3.2. The design speed–dependent curves in Fig. 12.6 complete the selection of the minimum parameter of the clothoid for the critical design case between a superelevation rate of $e_b = -2.5$ percent at the beginning (crown section) and a super-

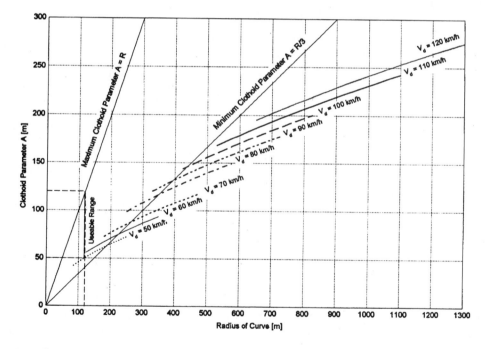

Legend:

V_d = Design Speed [km/h]

FIGURE 12.6 Determination of the parameter, A, of the clothoid including, additionally, the requirements of the superelevation runoff.

elevation rate of e_e = 8 percent at the end of the superelevation runoff. As can be noted, a larger minimum parameter of the clothoid than that required by Eq. (12.13) could result for structural reasons.

For example:

Design speed: V_d = 50 km/h

Initial superelevation rate: e_b = −2.5% (for example, normal crown)

Full superelevation rate: e_e = 8%

Radius of circular curve: R = 120 m

The usable range of the clothoid parameter, A, for the given values can be assumed between A_{min} = 50 m and A_{max} = 120 m according to Fig. 12.6.

For large radii of curve the parameter, A, may be smaller than R/3 by taking into account the speed-dependent arrangements in Fig. 12.6. However, the tangent offset, ΔR, should at least be 0.25 m. See the definition and explanation of the term, ΔR, in Eq. (12.24) and Fig. 12.7, respectively.

12.1.3.3 Geometry of Transition Curves.

It is known today, based on the experience of the past 50 yrs, that the alignment of modern highways has to be consistent and efficient from a driving dynamic and convincing from a driving psychological point of view. In this connection the clothoid, as a transition curve, offers good solutions. The clothoid satisfies esthetical solutions and enables, by being flexible, a good adaptation to the topography and existing local constraints. The clothoid guarantees for motor vehicles an economically efficient ride, and saves, through its appropriate insertion into the local environment, considerable construction costs.

The clothoid, as a transition curve, provides the best adaptation of the steering course, when entering a circular curve. The clothoid is used in all of the guidelines studied. Figure 12.7 shows the geometry of the clothoid.

Geometric properties of the clothoid and its numerical calculations are based on the following equations and standardized clothoid tables which were developed by Kasper, Schuerba, and Lorenz:[326]

General formula:

$$A^2 = R L \tag{12.15}$$

Total length:

$$L = \frac{A^2}{R} \tag{12.16}$$

Deflection angle:

$$\tau = \frac{L^2}{2 A^2} \tag{12.17}$$

Tangent distance:

$$X = \int_0^L \cos \left(\frac{L^2}{2 A^2} \right) dL \tag{12.18}$$

$$X = L - \frac{L^5}{40 A^4} + \frac{L^9}{3456 A^8} - \frac{L^{13}}{599,040 A^{12}} + \cdots \tag{12.19}$$

Tangent offset:

$$Y = \int_0^L \sin \left(\frac{L^2}{2 A^2} \right) dL \tag{12.20}$$

$$Y = \frac{L^3}{6 A^2} - \frac{L^7}{336 A^6} + \frac{L^{11}}{42,240 A^{10}} - \frac{L^{15}}{9,676,800 A^{14}} + \cdots \tag{12.21}$$

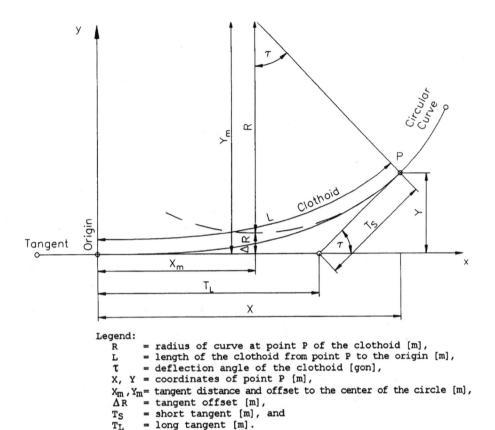

Legend:
```
R      = radius of curve at point P of the clothoid [m],
L      = length of the clothoid from point P to the origin [m],
τ      = deflection angle of the clothoid [gon],
X, Y   = coordinates of point P [m],
Xm,Ym  = tangent distance and offset to the center of the circle [m],
ΔR     = tangent offset [m],
Ts     = short tangent [m], and
TL     = long tangent [m].
```

FIGURE 12.7 Geometry of the clothoid.[326]

Tangent distance to the center of a circle:

$$X_m = X - R \sin (\tau) \qquad (12.22)$$

Tangent offset to the center of a circle:

$$Y_m = Y + R \cos (\tau) \qquad (12.23)$$

Tangent offset:

$$\Delta R = Y_m - R \approx \frac{L^2}{24\,R} \qquad (12.24)$$

Long tangent:

$$T_L = X - Y \cot (\tau) \qquad (12.25)$$

Short tangent:

$$T_S = \frac{Y}{\sin (\tau)} \qquad (12.26)$$

Further information is provided in Ref. 326.

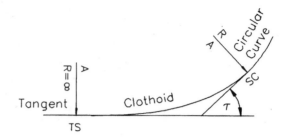

Legend:
General nomenclature for Figures 12-8 to 12-12 and Figure 12-23.
    Points of Transitions:
    TS = Tangent to Spiral (Tangent to Clothoid),
    SC = Spiral to Circular Curve (Clothoid to Circular Curve),
    CS = Circular Curve to Spiral (Circular Curve to Clothoid),
    SS = Spiral to Spiral (Clothoid to Clothoid),
    ST = Spiral to Tangent (Clothoid to Tangent).

FIGURE 12.8   Transition between tangent and circular curve. (Elaborated based on Refs. 35 and 243.)

12.1.3.4   *Types of Transition Curves (Acknowledged as Favorable).*   The application of different types of transition curves is presented in the following figures.

Simple Clothoid.   The *simple clothoid* adjusts the transition between a tangent and a circular curve (see Fig. 12.8).

Based on the relation design assumptions, it becomes obvious that the smaller the upcoming radius of curve is, the larger the clothoid parameter should be. Doing so guarantees a longer and optically smoother transition. Furthermore, the longer the tangent section in front of a circular curve is and the wider the cross section is, then the transition curve should lead gradually and gradually to the upcoming circular curve.

Reversing Clothoid.   The *reversing clothoid* consists of two clothoids with opposing curvature, which are joined at their origins (see Fig. 12.9). For each of the clothoid branches, the conditions of the simple clothoid are valid. For consistent alignment reasons and for the sake of a uniform superelevation runoff, the same parameters should be closely selected for both clothoid

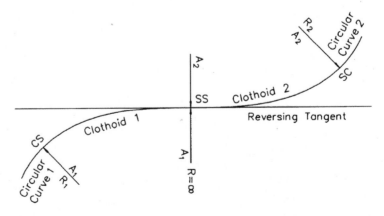

FIGURE 12.9   Transition between two reversing circular curves. (Elaborated based on Refs. 35 and 243.)

branches. For unequal parameters ($A_2 \leq 200$ m) for roads of category group A and road categories BII and BIII, the following conditions should be regarded:[35,243]

$$A_1 \leq 1.5\, A_2 \qquad (12.27)$$

where A_1 = larger parameter of the clothoid, m
A_2 = smaller parameter of the clothoid, m

For relation design purposes, the good design range must be applied for the two circular curves which are joined by the reversing clothoid (see Figs. 9.1 or 9.36 to 9.40).

In case of a symmetrical reversing clothoid, the common clothoid parameter, A_R, can be approximated from the following formula:[326]

$$A_R = \sqrt[4]{24\, d\, R_R{}^3} \qquad (12.28)$$

where d represents the distance between the two circular curves, as shown in Fig. 12.10. It is given by

$$d = \overline{C_1 C_2} - R_1 - R_2 \qquad (12.29)$$

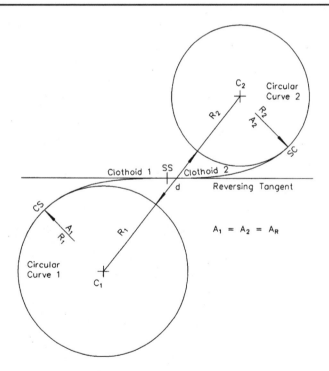

Legend:

A_R = common clothoid parameter [m],
d = shortest distance between the two circular curves [m],
$R_{1,2}$ = radii of the circular curves [m],
$C_{1,2}$ = center points of the circular curves.

FIGURE 12.10   Geometry of the reversing clothoid.

and the surrogate radius, R_R, is equal to

$$R_R = \frac{R_1 R_2}{R_1 + R_2} \tag{12.30}$$

where R_1 and R_2 represent the radii of the circular curves 1 and 2 shown in Fig. 12.10.

Egg-Shaped Clothoid.   The *egg-shaped clothoid* is a clothoid section which connects two circular curves having the same direction of curvature (Fig. 12.11).

With respect to the radii of curve, the relation design backgrounds of Figs. 9.1 or 9.36 to 9.40 must be considered for the individual country under study. An angular change of at least $\tau \geq 3.5$ gon should be provided. By doing so, the egg-shaped clothoid then becomes optically perceivable.

The smaller circular curve must be located on the inside of the larger curve. The circular curves are not allowed to intersect each other or to have the same center point (Fig. 12.12).

The parameter of the egg-shaped clothoid, A_E, can be approximated from the following formula:[326]

$$A_E = \sqrt[4]{24\, d\, R_E^{\,3}} \tag{12.31}$$

where d represents the distance between the two circular curves in Fig. 12.12. It is given by

$$d = R_1 - R_2 - \overline{C_1 C_2} \qquad \text{for } R_1 > R_2 \tag{12.32}$$

and the surrogate radius, R_E, is equal to

$$R_E = \frac{R_1 R_2}{R_1 - R_2} \qquad \text{for } R_1 > R_2 \tag{12.33}$$

where R_1 and R_2 correspond to the radii of the circular curves 1 and 2 in Fig. 12.12.

12.1.3.5 Safety Considerations.   The basic formula of the clothoid

$$A^2 = R\,L \tag{12.12}$$

shows the relationship between the parameter of the clothoid, the radius of curve, and the length of the clothoid.

In "Safety Considerations" in Sec. 12.1.2.3, a detailed discussion has shown that the design parameter radius of curve influences, directly or indirectly, the three developed safety criteria (Table 11.1) with respect to the design speed, the 85th-percentile speed, and side friction assumed

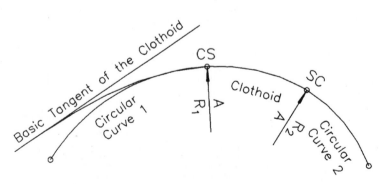

FIGURE 12.11   Egg-shaped clothoid. (Elaborated based on Refs. 35 and 243.)

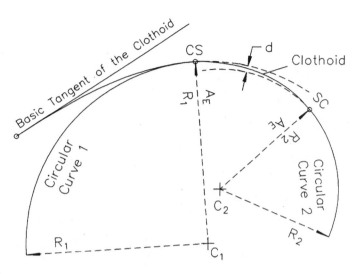

Legend:

A_E = parameter of the egg-shaped clothoid [m],

d = distance between the two circular curves [m],
$R_{1,2}$ = radii of the circular curves [m],
$C_{1,2}$ = center points of the circular curves [-].

FIGURE 12.12   Geometry of the egg-shaped clothoid.

and side friction demanded. Consequently, with respect to design cases where the element sequences are joined together by clothoid(s), the parameter of the clothoid, *A*, also has an influence on the three safety criteria because this parameter is directly related to the radius of curve *R*, as revealed by Eq. (12.12).

Furthermore, all three safety criteria are influenced by the length(s) of the clothoid(s) according to the design parameter curvature change rate of the single curve, CCR_S [Eq. (8.6); see also Fig. 8.1], which is used to determine the 85th-percentile speed(s), as shown in Fig. 8.12 or Table 8.5 and to assess the design classes based on accident rates in Tables 9.10 and 10.12.

Thus, it can be concluded that the influence of the design element transition curve (clothoid) is directly or indirectly included in the three safety criteria in order to distinguish good designs from fair designs (tolerable) and poor design practices.

12.2   GENERAL CONSIDERATIONS, RESEARCH EVALUATIONS, GUIDELINE COMPARISONS, AND NEW DEVELOPMENTS

12.2.1   Tangent

12.2.1.1   Application.   In the highway geometric design process, tangents and horizontal curves with or without transition curves are regarded as *design elements.*

From an esthetic point of view, the tangent may often be beneficial in flat country but rarely in rolling or mountainous terrain. As a design element, the tangent can seriously affect a proper alignment. On the other hand, the placement of tangents is suitable for passing sections, weaving sections, in the area of intersections, and so forth. Therefore, no specific rules are given for the applications of tangents, since a sensitive placement is influenced by several components. It is left open to the design engineer to place tangents where they are appropriate and to avoid them where they interrupt the harmony of the alignment.[35]

12.2.1.2   Standard Values.   Most of the reviewed highway geometric design guidelines give qualitative and, to a lesser extent, quantitative recommendations for maximum or minimum tangent lengths. For example, in Germany[243] and South Africa,[663] tangent lengths between curves are limited by the design speed. The maximum length of tangent sections between two curves is not permitted to exceed 20 times the design speed of that roadway. In this way, long tangents are controlled and a relation design is encouraged.

Desirable tangent lengths should be at least 6 times the design speed. For a typical design speed of 100 km/h, this would correspond to tangent lengths between 600 and 2000 m.[124]

To avoid fatigue as a cause of accidents, it is recommended in France[510] that tangent sections should be limited to a maximum of 40 to 60 percent of long roadway sections, with maximum single tangent lengths between 2000 and 3000 m.

Swiss highway officials[691] also limit tangent lengths to reduce driver fatigue. Designs which permit driving more than 1 min on a straight section are not permitted. That means maximum tangent lengths would be confined to 1670 m for a design speed of 100 km/h. Minimum tangent lengths are related to *project speeds* or roughly translated as *theoretical operating speeds.* For example, for a project speed of 100 km/h, a minimum tangent length of 165 m would be permitted. This value corresponds well to the length of a nonindependent tangent in Table 12.1 for an operating speed change from 80 to 100 km/h between curve and tangent.

In the 1990 AASHTO "Policy on Geometric Design of Highways and Streets,"[5] specific values for maximum or minimum tangent lengths are not given, but the following statement is listed under "General Controls for Horizontal Alignment": "Although the esthetic qualities of curving alignment are important, passing necessitates long tangents on two-lane highways with passing sight distance on as great a percentage of the length of highway as feasible." This statement clearly supports the application of long tangents, especially for the design of two-lane rural highways.

In the "Commentary to the German Guidelines,"[124] it is fundamentally stated that

- The assessment of the maximum or minimum length of a tangent with a constant grade depends on the physiological abilities of the driver, as well as on the optical impression.

- The reasons are: the preservation of a safe ride and an optically consistent picture sequence.

- The criteria are: fatigue symptoms of the driver, the danger of glare, the ability to estimate the speed of opposing (and succeeding) traffic.

The assessment of the previous limiting values is based on the consideration that a roadway section of 6 times the design speed in meters plus the connecting transition curve sections corresponds approximately to the passing sight distance. The assessment of 20 times the design speed for limiting the maximum tangent length appears reasonable as passing time for the motorist to avoid fatigue symptoms.

A short tangent between curves in opposite directions is frequently desirable for reasons of computation practicality. From previous experience, this interferes neither with the reversing transition curves nor with driving dynamic, nor drainage, nor optical aspects, especially if the superimposition of the reversing points in plan and profile is placed according to Chap. 16.

On this subject, Durth and Lippold[159] stated:

> While in the past years numerous research works dealt with the relationships "radius of curve vs. speed" and "radius of curve vs. accident situation," the effect of the placement of tangents on speed and traffic safety was hardly the subject of investigations. Obviously the statements in the German Guidelines [243, 246] are satisfactory for the application of tangents, as usual. Therefore, the given standard values don't need a change.

This explanation is not fully satisfying. Both the Swiss Norm SN 640 080a (1981) and SN 640 080b (1991)[691] considered the tangent as an important dynamic design element in the speed profile as acceleration or deceleration section between two successive curves. The Swiss idea was further developed by Lamm, Choueiri, and Hayward in the publication "Tangent as an Independent Design Element,"[411] where tangents were defined as independent or nonindependent. Independent tangents may cause critical changes in the operating speed profile, $V85$, while nonindependent tangents do not. In this connection, the consideration of tangents as dynamic (speed-dependent) elements similar to curves is very important for the evaluation of (speed) transitions between successive design elements (for example, curve to tangent to curve). Therefore, the introduction of the *dynamic* design element tangent into modern recommendations for practical design tasks is an essential assumption for distinguishing new designs, redesigns, and RRR projects by the quantitative safety criterion II into good, fair, and poor design practices, as discussed in Secs. 12.1.1.3 and 12.2.1.3.

In conclusion, the statements so far about the tangent in highway geometric design cannot be considered as sufficient. Therefore, the tangent as an independent dynamic design element is introduced for the first time here, especially for two-lane rural roads of categories A I to A IV.

12.2.1.3 Evaluation of Tangents in the Design Process.*   The only method developed to evaluate acceleration or deceleration movements between sequences of curve-to-curve or tangent-to-curve are found in the geometric guidelines of Switzerland.[405,691] The Swiss have developed a formula for calculating the *transition length (tangent length)* or the distance required for acceleration or deceleration of a vehicle as it approaches or leaves a curve based on the project (operating) speed difference between two curves or between a tangent and a curve. Unallowable ranges or those that should be avoided for these transition lengths (tangent lengths) are tabulated.

Background and Objective.   The majority of transitions between curves on two-lane rural highway networks in many countries consist of tangents, with the exception of good designs where transition curves are applied and operating speed changes which exceed the limits for good design or even fair design normally do not exist.

Abrupt changes in operating speeds created by the horizontal alignment are among the leading causes of accidents on two-lane rural roads. Primarily at lower design speed levels, the changing alignment may cause variations in operating speeds that may in turn increase the accident risk by substantial amounts.[470] In this connection, the transition from a tangent to a curve (especially an isolated curve) has to be considered as one of the most critical design cases. Therefore, one of the important tasks in modern highway geometric design and rehabilitation strategies for two-lane rural roads in many countries is to ensure design consistency and to detect critical inconsistencies in the horizontal alignment.[359]

A design that conforms well with the driver's expectations is considered, from a geometric highway design point of view, as consistent. This requirement means that drivers are able to operate their vehicles safely at reasonably uniform speeds and with a relatively uniform mental effort along rural highway alignments. Large reductions in speed or significant increases in mental effort may lead to speed selection, as well as path selection, problems. This is the reason why consistency must be measured with respect to both speed and driver mental workload. The two main measures of consistency are, according to Ref. 359, reductions in operating speeds and increases in driver mental workload between successive alignment elements (see Chap. 19).

Most geometric guidelines qualitatively advocate consistent alignment, but provide little objective guidance to ensure that consistency is achieved. Therefore, the *tangent* will be considered in the following as a dynamic design element when examining the recommended speed changes according to safety criterion II, "Achieving Operating Speed Consistency," with respect to the necessary deceleration and acceleration movements.

In this connection, it should once again be noted that safety criterion II is provided to distinguish good, fair, and poor design practices for the transitions between successive design elements (for example, tangent-to-curve). In other words, a sufficient transition length between successive design elements must be present to provide a safe, gentle change in operating speeds.

---

*Elaborated based on Refs. 104, 411, 412, 417, 434, 452, 455, 628, and 691.

To illustrate the critical situation which may result from combining a tangent with a curve, the following sample calculations for three representative countries—United States,[565] Germany,[431] and Greece[453]—are presented.

Suppose a curve with a curvature change rate of the single curve of $CCR_S = 500$ gon/km follows a long tangent ($CCR_S = 0$ gon/km). Then, according to Fig. 8.12 or Table 8.5, the expected operating speeds for the tangent and the curve can be derived from graphical presentations or from the following equations:

United States:
$$V85 = 103.04 - 0.053 \ CCR_S \tag{8.14}$$

Germany:
$$V85 = 10^6/8270 + 8.01 \ CCR_S \tag{8.15}$$

Greece:
$$V85 = 10^6/10,150.1 + 8.529 \ CCR_S \tag{8.16}$$

For the preceding sequence of design elements, the expected (rounded) changes in operating speeds are

United States:

Tangent:	$CCR_S = 0$ gon/km	$V85 = 103$ km/h
Curve:	$CCR_S = 500$ gon/km	$V85 = 77$ km/h
Speed change:		$\Delta V85 = 26$ km/h

Germany:

Tangent:	$CCR_S = 0$ gon/km	$V85 = 121$ km/h
Curve:	$CCR_S = 500$ gon/km	$V85 = 81$ km/h
Speed change:		$\Delta V85 = 40$ km/h

Greece:

Tangent:	$CCR_S = 0$ gon/km	$V85 = 98$ km/h
Curve:	$CCR_S = 500$ gon/km	$V85 = 69$ km/h
Speed change:		$\Delta V85 = 29$ km/h

Therefore, it can be concluded that the speed changes from the tangent to the curve are higher than the maximum allowable change in operating speeds, according to safety criterion II, even for fair design levels for all three countries (Tables 9.13 and 11.1) and that they correspond to poor design levels:

$$|V85_i - V85_{i+1}| > 20 \ km/h$$

However, this statement would be true only for a relatively long tangent. The tangent has to at least be long enough to allow a driver to reach the top 85th-percentile speed for $CCR_S = 0$ gon/km in Fig. 8.12. For short tangents between succeeding curves, it may be expected that the average driver in a typical vehicle would not be able to accelerate or decelerate in such a way that the boundaries of safety criterion II for good design practices ($\Delta V85 \leq 10$ km/h) or even for fair design practices ($\Delta V85 \leq 20$ km/h) are exceeded (Table 11.1). In these cases, operating speed changes would be related to the two successive curves, and the relatively short tangent in between could be ignored in the design process for achieving operating speed consistency or inconsistency. Therefore, the task of this research is to provide recommendations for transition lengths (tangent lengths) between successive curves for:

- Tangents that should be regarded as nonindependent design elements, where the sequence curve-to-curve controls the design process
- Tangents that should be regarded as independent design elements, where the sequence tangent-to-curve controls the design process.

Acceleration and Deceleration Rates.   The transition length, TL, is that road section where the operating speed between two design elements is changing from $V85_1$ to $V85_2$, as shown in Fig. 12.13.

When the CCR_S values of two successive curves are known, the expected 85th-percentile speeds can be determined, based on the operating speed background of the individual country, by Fig. 8.12 or Table 8.5. In order to be able to calculate the transition length (tangent length), a value for the parameter, *a,* that is, the acceleration/deceleration rate between two successive design elements, must be determined.

To determine an estimate of the coefficient, *a,* in Eq. (12.1), typical accelerations and decelerations were studied between tangents and specific curved sections of two-lane rural highways.

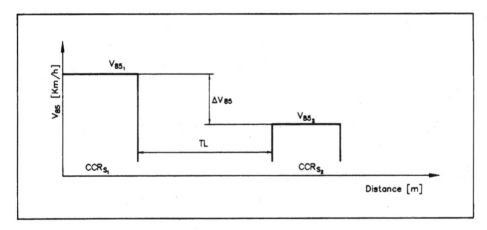

$$TL = \frac{\overline{V85}}{3.6^2 \, a} \, \Delta V85 \quad m$$

$$\overline{V85} = (V85_1 + V85_2)/2 \tag{12.34}$$

$$\Delta V85 = V85_1 - V85_2$$

The transition length is given by the relationship:

$$TL = \frac{V85_1{}^2 - V85_2{}^2}{2 \times 3.6^2 \, a} \tag{12.1}$$

$$TL = \frac{V85_1{}^2 - V85_2{}^2}{25.92 \, a}$$

where $V85_{1,2}$ = operating speeds in curves, km/h
$\overline{V85}$ = average value of $V85_1$ and $V85_2$, km/h
$\Delta V85$ = difference between $V85_1$ and $V85_2$, km/h
a = acceleration/deceleration rate, m/s^2
TL = transition length (tangent length), m

FIGURE 12.13   Determination of transition length (tangent length). (Elaborated and modified based on Ref. 691.)

Because of financial and time constraints, acceleration and deceleration measurements from tangents-to-curves or curves-to-tangents were conducted at curves where speeds of 50 km/h (three sections), 55 km/h (two sections), and 65 km/h (one section) were recommended. The study sites were located in the State of New York.[411] (Some of the speeds discussed in this section were converted and rounded because the original investigations were conducted in the United States.)

The procedure used to record the speeds of individual vehicles consisted of the following: An investigation car (the "follow car"), a car observed in the field (the "test car"), and a tape recorder on which to place any relevant information. Note that two persons, a driver and an observer, were required in the "follow car" to allow observation of the situation while speed data were being recorded.

Measurements of travel speeds were made at particular spots along the routes. The measurement spots were uniform in characteristics:

- Sections were horizontal (longitudinal grades less than 1.5 percent).

- Intersections and places where an influence on traffic flow might be expected through changes in the highway surroundings were not present in the sections.

- Cross sections were representative with regard to the width of the roadway. Three sections with 3.0-m lane widths and three sections with 3.3-m lane widths were selected.

- Sight conditions at measuring spots were adequate.

- Locations for measurements were equipped so as not to be recognizable as such by drivers but obvious enough to be seen by the observers in the follow car.

In all cases, 11 locations (from the beginning of the curve into the tangent section) marked with driveway reflectors were set up along the routes investigated on both sides of the roadway. The distance between marked locations was 75 m; thus, the measurement sections were about 800 m long. On the average, the recommended speed signs were located about 150 m from the curves in the deceleration direction, while in the acceleration direction at this location the normal speed limit of 90 km/h (55 mi/h) was posted.

The car speeds were measured during off-peak traffic periods on weekdays, under dry conditions, and in daylight. The traffic flows were light, and cars were capable of attaining the speeds they desired under the conditions of the site; in other words, a car was selected for the speed survey if it had sufficient headway to be considered traveling at its own free speed.

With regard to the analysis process, the observer in the follow car observing the cars crossing his or her field of view had to select the cars to be sampled. Once a car was spotted under free-flow conditions, an initial acceleration by the driver of the follow car was made in order to catch up and adjust his speed to that of the test vehicle. Then, at each of the marked locations along the highway, the observer in the follow vehicle would record the speed of the test vehicle by reading the speed from the speedometer of the follow car. Other relevant information, such as the gender and approximate age of the driver of the test vehicle and the type and manufacturer of the test vehicle, were also recorded, but their effect was not considered in this study. A distance of at least 1.6 km was necessary for the follow car to accelerate and adjust its speed to that of the test car.

All tests in which evasive action was taken, such as turning maneuvers into driveways before the end of the speed measurements, were recorded, but those measurements were not considered in the analysis.

Normally, the speeds of at least 20 passenger cars (test cars) were recorded on the tape recorder at each of the 11 test spots along the routes investigated from the tangent to the curve (deceleration) and from the curve to the tangent (acceleration). The data on the tape recorder was later analyzed, and the 85th-percentile speed at each of the test locations was determined.

Regression equations relating the 85th-percentile speeds to distances traveled are as follows: *Acceleration (Rounded):*

Recommended speed in curve: 50 km/h

$$V85 = 59.57 + 0.26 \, DT - 0.0003 \, DT^2 \qquad (12.35a)$$

Recommended speed in curve: 55 km/h

$$V85 = 67.62 + 0.21\ DT - 0.0003\ DT^2 \qquad (12.35b)$$

Recommended speed in curve: 65 km/h

$$V85 = 75.67 + 0.21\ DT - 0.0003\ DT^2 \qquad (12.35c)$$

Deceleration (Rounded):

Recommended speed in curve: 50 km/h

$$V85 = 53.13 + 0.26\ DT - 0.0003\ DT^2 \qquad (12.36a)$$

Recommended speed in curve: 55 km/h

$$V85 = 61.18 + 0.21\ DT - 0.0003\ DT^2 \qquad (12.36b)$$

Recommended speed in curve: 65 km/h

$$V85 = 69.23 + 0.21\ DT - 0.0003\ DT^2 \qquad (12.36c)$$

where $V85$ = estimate of 85th-percentile speed, km/h
  DT = distance traveled, m

These equations are plotted in Figs. 12.14 and 12.15. They clearly indicate that the acceleration end and the deceleration processes begin at about 210 to 230 m from the end of the observed curved sections. This means that any reaction from the driver to decelerate begins nearly 70 to 80 m from the recommended speed signs, which are normally posted 150 m in front of a curve or a

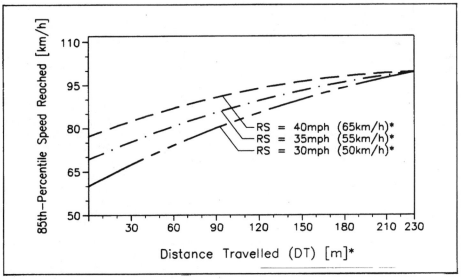

Legend:
 * Rounded
RS = Recommended Speed Sign (Normally Posted at about 150 m from a Curve)
DT = Distance Travelled [m]
DT = 0 m (Begin or End of Curve)

FIGURE 12.14 85th-percentile speed versus distance traveled: passenger cars (acceleration).

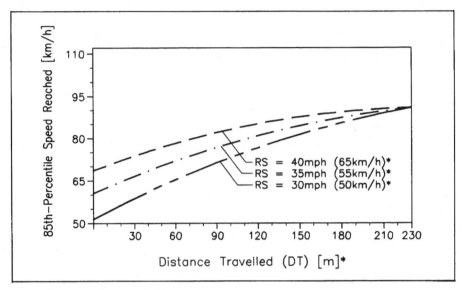

Legend: see Figure 12-14

FIGURE 12.15   85th-percentile speed versus distance traveled: passenger cars (deceleration).

curved section. Another finding is that the operating speeds at the beginning of a curve in the deceleration direction are nearly 6.5 to 8 km/h lower (Fig. 12.15) than those at the end of the curve in the acceleration direction (Fig. 12.14).

Related to the distance of 230 m, the average deceleration and acceleration rates ranged between 0.85 and 0.88 m/s² for the six tested road sections consisting of tangents (length of at least 800 m) followed by curves with recommended speeds between 50 and 65 km/h. Since the differences between deceleration and acceleration rates are negligible, an average acceleration or deceleration rate of 0.85 m/s² was selected for the following analysis. This value agrees well with the deceleration and acceleration rate of 0.8 m/s² on which the design of transition lengths in the Swiss guidelines is based.[691] Furthermore, this value agrees relatively well with the values in the AASHTO design policy, Table III-4,[5] where average acceleration rates of about 0.63 m/s² for passing maneuvers in the speed ranges 50 to 65 km/h and 65 to 80 km/h are tabulated.

Similar results are also found by Steierwald and Buck[674] and Kockelke and Steinbrecher,[345] who investigated the deceleration in front of signs with the names of places. (In Germany, beginning at those sign locations, a general speed limit of 50 km/h is valid.) In the first case, deceleration rates of 0.4 to 0.6 m/s², and in the second case deceleration rates < 1.0 m/s², were reported.

Finally, it should be mentioned that the acceleration and deceleration rate of 0.85 m/s² developed here has been used by Krammes, et al.[362] for speed profile models.

Determination of Necessary Transition Lengths (Tangent Lengths).   For traffic safety reasons, driving behavior during the deceleration process is a particularly important factor.

As noted in "Evaluation of Tangents in the Design Process" in Sec. 12.1.1.3, operating speed differences $\Delta V85$ between two successive design elements of more than 10 km/h for good designs and of more than 20 km/h for fair (tolerable) designs should be avoided, according to safety criterion II (Table 11.1). An illustration of the previous statement is shown in Fig. 12.16.

With an average acceleration or deceleration rate of $a = 0.85$ m/s², the transition length from Eq. (12.34a) becomes according to Fig. 12.13:

(Continues on page 12.34)

$$TL = \frac{\overline{V}85 \; \Delta V85}{11.016}$$ (12.34a)

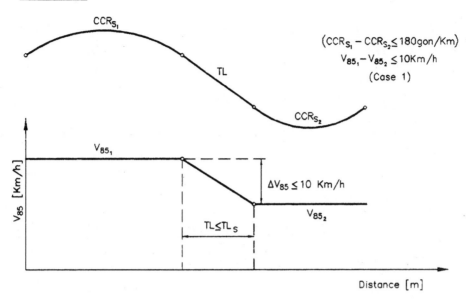

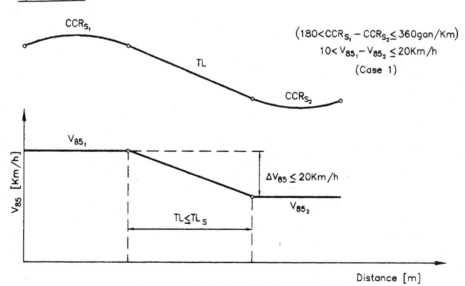

FIGURE 12.16   Transition lengths between successive design elements that correspond to nonindependent tangents (safety criterion II). (*Note:* For the determination of reasonable CCR$_S$ and $V85$ ranges, see Table 11.10.)

$$TL = \frac{V85_1{}^2 - V85_2{}^2}{22.03}$$

(12.1a)

Based on Eq. (12.1a), required transition lengths (tangent lengths) for good and fair design practices are given in Tables 12.1 and 12.2. The values within boxes represent fair design practices, which means that a driver is able to decelerate within the range of operating speed changes of up to 20 km/h (see Fig. 12.16). Thus, from the point of view of reasonable changes in CCR_S values and the corresponding changes in 85th-percentile speeds, the tangent lengths (values within boxes) in Tables 12.1 and 12.2 should represent the maximum boundaries for good and fair design practices, and are considered according to "Evaluation of Tangents in the Design Process" in Sec. 12.1.1.3 to be nonindependent tangents (case 1). The sequence curve-to-curve controls the design process. As explained in Sec. 12.1.1.3, the good and fair ranges were considered here as a unit in order not to be too conservative.

In all the other cases (for example, see the unmarked values in Tables 12.1 or 12.2), a driver is able to exceed the recommended operating speed changes ($\Delta V85 > 20$ km/h), that is, critical driving maneuvers may occur, especially during the deceleration process. These tangent types are defined as independent tangents (cases 2 and 3 of Sec. 12.1.1.3). The sequence tangent-to-curve controls the design process.

An illustration of the latter statement is given for the United States in Fig. 12.17 for a sequence of two curves ($CCR_S = 600$ gon/km) linked by an independent tangent ($CCR_S = 0$ gon/km, $L = 600$ m). The 85th-percentile speeds for this case were determined from Fig. 8.12 (curve "U.S.A.") or Eq. (8.14) of Table 8.5.

As can be seen from Fig. 12.17, a driver is able to accelerate within the tangent from an operating speed of 72 km/h in the curve ($CCR_S = 600$ gon/km) to the highest operating speed of 103 km/h in the tangent ($CCR_S = 0$ gon/km), for which, according to Table 12.1, a tangent length of 235 m is needed for this acceleration.

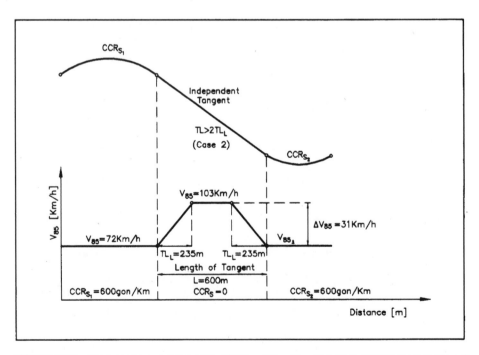

FIGURE 12.17   Example of poor design practice. [$\Delta V85 = 103 - 72 = 31$ km/h > 20 km/h (poor design, safety criterion II, Table 11.1.]

$$\text{TL} > 2\,\text{TL}_L \qquad \text{case 2}$$

$$600\text{ m} > 2 \times 235 = 470\text{ m}$$

(It is interesting to note that, according to Table 12.2, even a tangent length of 280 m could be regarded as justified. However, this would not change the overall statements.)

Thus, it can be concluded that the tangent is independent. In this example, the maximum allowable operating speed change, even for fair designs of $\Delta V85 \leq 20$ km/h, has been exceeded, which is a clear indication of poor design practices (see Table 11.1). A redesign of the existing alignment is recommended.

Example Application.   Case 3 is shown in Fig. 12.3 and the corresponding equations are given in Table 12.3.

An example for the United States, according to curve "U.S.A.-N.Y." in Fig. 8.12 and Eq. (8.17) in Table 8.5 is as follows:

Case 3a (Fig. 12.3a):

$$\text{TL} = 320\text{ m} \qquad \text{existing tangent length}$$

$$\text{CCR}_{S1} = 219\text{ gon/km} \qquad V85_1 = 83\text{ km/h}$$

$$\text{CCR}_{S2} = 818\text{ gon/km} \qquad V85_2 = 53\text{ km/h}$$

The 85th-percentile speed in Table 12.1 that is closest to 53 km/h in the curve with the higher CCR_S value is 55 km/h. For 55 km/h, the maximum length of tangent that is considered to be non-independent is 120 m. Since TL = 320 m > 120 m, the tangent has to be evaluated as an independent design element.

Thus, the sequence tangent-to-curve plays an important role in the design process for evaluating horizontal design consistency or inconsistency for both directions of travel on this road section.

The 85th-percentile speed in the tangent, $V85_T$, can be estimated as follows (see Fig. 12.3a).

Equation (12.8) in Table 12.3 is used to calculate the acceleration or deceleration distance, TL_C, between curve 1 and curve 2. This implies that

$$\begin{aligned}
\text{TL}_C &= \frac{V85_1{}^2 - V85_2{}^2}{22.03} \\
&= \frac{83^2 - 53^2}{22.03} = 186\text{ m}
\end{aligned} \tag{12.8}$$

Then, the remaining tangent length is

$$\text{TL} - \text{TL}_C = 320 - 186 = 134\text{ m}$$

along which a driver is able to perform additional acceleration or deceleration maneuvers. (Exceptional case: $\text{CCR}_{S1} = \text{CCR}_{S2} \rightarrow V85_1 = V85_2 \rightarrow \text{TL}_C = 0$; perform the calculations in the same way with $\text{TL}_C = 0$.) The formula used to calculate $\Delta V85_T$ is given in Table 12.3, Eq. (12.10). See also Fig. 12.3a:

$$\Delta V85_T = \frac{-2\,(83) + [4\,(83)^2 + 44.06\,(320 - 186)]^{1/2}}{2}$$

$$= 8\text{ km/h}$$

Thus, the operating speed in the independent tangent for evaluating the sequences tangent-to-curve in both directions of travel is

$$V85_T = V85_1 + \Delta V85_T = 83 + 8 = 91 \text{ km/h}$$

For the example presented here (Fig. 12.3a), the following changes in operating speeds can be expected between

Tangent-to-curve 1: $\Delta V85 = |91 - 83| = 8$ km/h

Tangent-to-curve 2: $\Delta V85 = |91 - 53| = 38$ km/h

The changes in operating speeds reveal that the sequence independent tangent-to-curve 1 corresponds to good design practices ($\Delta V85 < 10$ km/h), whereas the sequence independent tangent-to-curve 2 corresponds to poor design practices ($\Delta V85 > 20$ km/h) (see Table 11.1).

Case 3b (Fig. 12.3b): (Same operating speed background as for Case 3a.)

$$\text{TL} = 240 \text{ m} \qquad \text{existing tangent length}$$

$$\text{CCR}_{S1} = 880 \text{ gon/km} \qquad V85_1 = 50 \text{ km/h}$$

$$\text{CCR}_{S2} = 680 \text{ gon/km} \qquad V85_2 = 60 \text{ km/h}$$

The 85th-percentile speed in Table 12.1 that is closest to 50 km/h in the curve with the higher CCR_S value corresponds exactly to 50 km/h. For this speed, the maximum length of tangent that is considered to be nonindependent is 110 m. Since TL = 240 m > 110 m, the tangent has to be evaluated as an independent design element.

The 85th-percentile speed in the tangent, related to Fig. 12.3b, can be estimated in the same way as discussed in the previous example. According to Eq. (12.8) in Table 12.3, the acceleration or deceleration distance between curve 1 and curve 2 is

$$\text{TL}_C = \frac{V85_2{}^2 - V85_1{}^2}{22.03}$$

$$= \frac{60^2 - 50^2}{22.03} = 50 \text{ m} \tag{12.8}$$

Therefore, the remaining tangent length becomes

$$\text{TL} - \text{TL}_C = 240 - 50 = 190 \text{ m}$$

According to Eq. (12.10) in Table 12.3, the difference between the operating speed in the curve with the lower CCR_S value and the operating speed in the tangent now becomes

$$\Delta V85_T = \frac{-2(60) + [4(60)^2 + 44.06(240 - 50)]^{1/2}}{2}$$

$$= 16 \text{ km/h}$$

Note that, for the example in Fig. 12.3b, curve 2 is the curve with the lower CCR_S value. It follows then that the operating speed in the independent tangent is

$$V85_T = V85_2 + \Delta V85_T = 60 + 16 = 76 \text{ km/h}$$

For the example presented here (Fig. 12.3b), the following changes in operating speeds can be expected between

Tangent-to-curve 1: $\Delta V85 = |76 - 50| = 26$ km/h

Tangent-to-curve 2: $\Delta V85 = |76 - 60| = 16$ km/h

The changes in operating speeds reveal that the sequence independent tangent-to-curve 1 corresponds to poor design practices ($\Delta V85 > 20$ km/h), whereas the sequence independent tangent-to-curve 2 can still be evaluated as fair design (10 km/h $< \Delta V85 \leq 20$ km/h) (see Table 11.1).

Preliminary Conclusion. The procedure presented here is a rational method that provides recommendations for transition lengths (tangent lengths) between successive curved roadway sections in modern highway geometric design.

Reviews of several design guidelines for rural roads reveal that highway designers rely on controls on maximum and minimum or desirable lengths of tangents between successive curves. Minimum tangent lengths are prescribed to promote operating speed consistency and maximum lengths are suggested to combat driver fatigue. None of the guidelines to date (except for Switzerland) have considered tangents as dynamic (speed-dependent) design elements similar to curves. According to safety criterion II, the approach presented uses operating speed differences between successive horizontal geometric elements (curves and tangents) and acceleration or deceleration profiles derived from car-following techniques to establish limits on tangent lengths.

In conclusion, recommendations are provided for tangent lengths between successive curved roadway sections for

- Tangents that should be regarded as nonindependent design elements, that is, the sequence curve-to-curve is relevant in the design process
- Tangents that should be regarded as independent design elements, that is, the sequence tangent-to-curve is relevant in the design process.

With respect to recent developments, a modified procedure for evaluating tangent speeds and lengths in the safety evaluation process is presented in Sec. 18.5.2.

12.2.2 Circular Curve

12.2.2.1 Application. Up to the 1970s, minimum radii of curve were determined solely as a function of design speed, as revealed in the highway geometric design guidelines of most countries. Information relative to the relationship between design elements was rarely provided. However, following the introduction of relation design, the possibility now exists to achieve a sound, more balanced, and more consistent alignment in curves with respect to permissible radii of curve relations and reliable transitions between independent tangents and curves ("Graphic Presentation: Relation Design" in Secs. 9.1.3.2 and 9.2.3.2).

12.2.2.2 Limiting and Standard Values. Table 12.9 and Fig. 12.18 summarize minimum radii of horizontal curvature as a function of design speed for various countries. These values are a product of maximum superelevation rates and maximum permissible side friction factors [Eq. (12.11)]. For a 60-km/h design speed, for example, most countries' minimum radii are between 120 and 130 m. For an 80-km/h design speed, the minimum radius normally ranges from 230 to 280 m. For a 100-km/h design speed, the minimum radius is between 400 and 500 m.[361]

The values in Sec. 12.1.2.2 (Table 12.5) agree well with the minimum radii of curve used by most countries shown in Table 12.9 for flat and for hilly/mountainous topography. The values for flat topography, which are based on a maximum superelevation rate of $e_{max} = 8$ percent and a utilization ratio for side friction of $n = 45$ percent, are somewhat lower than the average values of other guidelines, which results in balanced designs, as a function of economic, environmental, and safety-related aspects. (See "AL" in Table 12.9 and Fig. 12.18.)

The larger values for hilly/mountainous topography, which are based on a superelevation rate of $e_{max} = 7$ percent and a utilization ratio for side friction of $n = 40$ percent, are intended to com-

TABLE 12.9   Minimum Radii of Curve as a Function of Design Speed for Rural Roads in Various Countries (Worldwide) and AL*

	Design speed, km/h									
Country	50	60	70	80	90	100	110	120	130	140
Australia (flat)	—	—	105	160	270	440	530	670	785	—
Australia (mountainous)	45	65	90	135	215	—	—	—	—	—
Austria	80	125	180	250	—	450	—	700	—	1000
Belgium	—	120	—	240	—	425	—	650	—	—
Canada	80	120	170	230	300	390	530	670	950	—
Denmark	—	120	200	280	380	500	—	800	—	—
France	—	120	—	240	—	425	—	665	—	—
Germany	80	120	180	250	340	450	—	720	—	—
Great Britain	127	180	255	325	410	510	—	720	—	—
Ireland	—	130	—	240	—	400	—	600	—	—
Italy	—	120	—	260	—	400	—	650	—	1000
Japan	100	150	—	280	—	430	—	610	—	—
Luxembourg	—	120	175	215	320	370	525	600	—	—
The Netherlands	—	130	—	260	—	450	—	750	—	—
Portugal	—	120	180	250	—	450	—	700	—	—
South Africa	80	125	180	250	335	440	560	710	—	—
Spain	—	120	180	250	—	450	—	650	—	—
Sweden	160	—	350	—	500	—	625	—	—	—
Switzerland	75	120	175	240	320	420	525	650	—	—
United States	80	125	175	230	305	395	500	665	—	—
AL† (flat)	85	125	180	250	330	425	530	650	790	—
AL† (hilly/mountainous)	95	140	200	280	370	(480)‡	(600)‡	(740)‡	—	—

*Modified and completed based on Refs. 153, 325, 361, and 363.

†AL and Greece agree.

‡( ) For high grades, values are only of theoretical character in practical alignment design.

pensate for the increased possibility of endangerment in areas of high grades and slippery roads during the winter. (See "AL" in Table 12.9 and Fig. 12.18.)

Finally, the recommendations for minimum radii of curve shown in Tables 12.4 or 12.8 must be considered for the transitions between independent tangents and curves with or without transition curves.

12.2.2.3   Safety Considerations.*

Accidents are more likely to occur on horizontal curves than on straight segments of roadways because of increased demands placed on the driver and the vehicle and because of friction between tires and pavements. The safety effect of an individual curve is not only influenced by the curve's geometric characteristics but also by the geometry of adjacent highway segments. The hazard is particularly intense when the curve is unexpected, such as following a long tangent approach or when it is hidden from view by a hill crest.[710] In other words, a road alignment anomaly, such as an isolated narrow curve in an otherwise straight alignment, is more dangerous than a succession of curves of the same radius. Also, horizontal curves are more dangerous when combined with gradients and surfaces with low coefficients of friction.[553]

---

*Elaborated based on "Safety Effectiveness of Highway Design Features, Volume II: Alignment."[728]

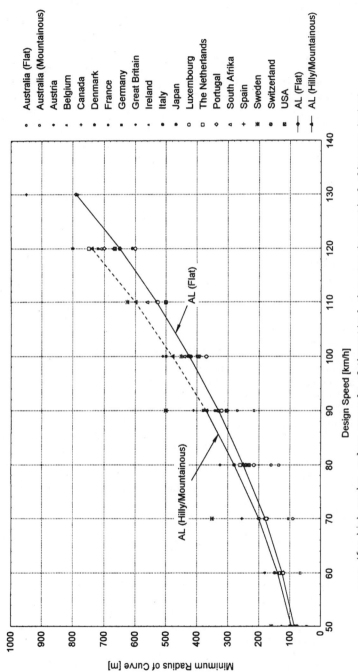

FIGURE 12.18 Minimum radius of horizontal curvature as a function of design speed in various countries and AL. (Elaborated, modified, and completed based on Refs. 153, 325, 361, and 363.)

---- (for high grades, values are only of theoretical character in practical alinement design).

Legend:
- Australia (Flat)
- Australia (Mountainous)
- Austria
- Belgium
- Canada
- Denmark
- France
- Germany
- Great Britain
- Ireland
- Italy
- Japan
- Luxembourg
- The Netherlands
- Portugal
- South Afrika
- Spain
- Sweden
- Switzerland
- USA
- AL (Flat)
- AL (Hilly/Mountainous)

Minimum Radius of Curve [m]

Design Speed [km/h]

AL (Flat)

AL (Hilly/Mountainous)

Instead of the design parameter radius of curve, the degree of curve is used in several countries, such as in the United States. The relationships between both parameters for circular curves without transition curves are presented in Table 8.4. The relationships between accident rates and radii of curve, as well as degree of curve, are discussed in detail in "Radius of Curve" and "Curvature Change Rate, Degree of Curve, Length of Curve, and Curve Radii Ratio" in Sec. 9.2.1.3.

With respect to the accident situation in horizontal curves, the following statements have been made in Ref. 728. Accident studies indicate that horizontal curves experience a higher accident rate than tangents, with rates ranging from $1\frac{1}{2}$ to 4 times greater than tangent sections (also see Tables 9.10 and 10.12). Past research has identified a number of traffic, roadway, and geometric features which are related to the safety of horizontal curves.

These factors include:[134,135,217–219,407,658,701,772,781]

- Traffic volume on the curve and traffic mix (for example, percent of trucks)
- Curve features (degree of curve, length of curve, central angle, superelevation, presence of spiral, or other transition curves)
- Cross-sectional curve elements (lane width, shoulder width, shoulder type, and shoulder slope)
- Roadside hazard on the curve (clear zone, sideslope, rigidity, and types of obstacles)
- Stopping sight distance on the curve (or on the approach to the curve)
- Vertical alignment on the horizontal curve
- Distance to adjacent curves
- Presence/distance from curve to the nearest intersection, driveway, bridge, etc.
- Pavement friction
- Presence and type of traffic control devices (signs and delineation)

In terms of accident characteristics on curves, a 1991 study by Zegeer, et al., which was prepared for the Federal Highway Administration (FHWA), has identified accident factors overrepresented on curves compared to tangents, based on 3427 curve/tangent pairs in Washington State.[781] Groups of accidents generally found to have higher percentages on curves compared to tangents included more severe (fatal and serious injury) crashes, head-on and opposite-direction sideswipe crashes, fixed-object and rollover crashes, crashes at night, and those involving drivers who were drinking.

Identification of Problem Curve Sites. A 1983 study by Glennon, et al.[217] developed a discriminant model for use in identifying horizontal curve sites which were potentially high accident, based on geometric, traffic, and roadside conditions. The identification of such sites will allow investigation and possible correction of such sites before a more serious accident problem develops. The database used for the discriminant model included 330 curve sections which were either high-accident or low-accident sites.

The best derived discriminant function was as follows:

$$D = 0.0713 \, (DC) + 1.8402 \, (LC) + 0.1074 \, (RR) - 0.0352 \, (PR) - 0.4756 \, (SW) - 1.5454 \quad (12.37)$$

where D = discriminant function
  DC = degree of curve, degrees, basis 360°
  LC = length of curve, km
  RR = roadside rating (a measure of roadside hazard)
  PR = pavement rating (a measure of the pavement's skid resistance)
  SW = shoulder width, m

Curves with sharper curvature, greater length, more hazardous roadsides, lower skid resistance, and/or narrower shoulders had higher discriminant scores. A higher discriminant score, D, indicates a higher likelihood that a curve site will be a high-accident location. In fact, the equa-

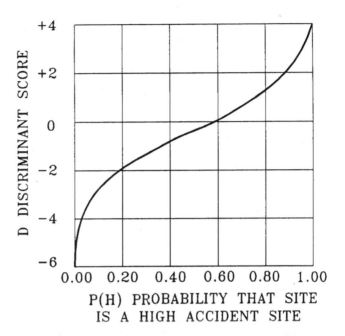

FIGURE 12.19   Relationship between discriminant score and the probability that a site is a high-accident site.[217]

tion correctly classifies 75.9 percent of the high-accident sites, 60.2 percent of low-accident sites, and 69.1 percent of all sites.

The probability of a site being a high-accident site is shown in Fig. 12.19 for various discriminant scores. For example, a curve section with a discriminant score of +2 will have about a 90 percent chance of being a high-accident site.

The *roadside hazard rating* is the probability that an accident with injuries will occur given a roadside encroachment. The roadside hazard rating depends on side slope, coverage factor, and lateral clear zone width from the road to roadside objects. Values of roadside hazard between 24 and 53 percent can be selected from Table 12.10. The study found that hazardous roadside designs represent the largest contributor to a high number of accidents at highway curves.[217]

Accident reduction factors given in Table 12.11 correspond to increasing the clear roadside recovery distance on a horizontal curve. Such roadside improvements include removing trees, relocating utility poles, providing traversable drainage structures, and flattening roadside slopes and other obstacles further from the roadway. Thus, an increase in recovery distance of 1.50 m would be expected to reduce total curve accidents by 9 percent. Providing 6.00 m of additional roadside recovery distance should reduce total curve accidents by 29 percent.[781]

The percent reduction in total curve accidents due to flattening side slopes on curves is given in Table 12.12. The side slope in the "before" condition is found in the left column and the "proposed" side slope is in the next four columns. The number in the table corresponding to these two values yields the expected percent reduction in the total number of curve accidents. For example, flattening a roadside slope from 2:1 to 6:1 on a horizontal curve would be expected to reduce the total number of curve crashes by approximately 12 percent.[781]

The impact of individual design elements and operational parameters, like radius of curve, degree of curve (or CCR_S, curvature change rate of the single curve), length of curve, lane width, shoulder width, gradient, sight distance, design speed, and average annual daily traffic on accident rates in curve design has been discussed, analyzed, and evaluated in detail in Sec. 9.2.1.3.

TABLE 12.10   Roadside Hazard Ratings[217]

Side slope	Coverage factor	Lateral clear zone width, m (rounded)						
		9.0	7.5	6.0	4.5	3.0	1.5	0
6:1 or flatter	90	24	28	32	34	42	46	47
	60	24	27	29	30	35	38	39
	40	24	27	27	27	32	34	34
	10	24	24	24	24	25	26	26
4:1	90	35	37	39	41	44	48	49
	60	35	36	38	39	40	43	44
	40	35	36	37	37	39	41	41
	10	35	35	35	35	36	37	37
3:1	90	41	42	42	43	44	48	49
	60	41	42	42	42	43	45	46
	40	41	42	42	41	41	44	45
	10	41	42	42	41	41	42	42
2:1 or steeper	90	53	53	53	53	45	49	50
	60	53	53	53	53	46	49	50
	40	53	53	53	53	48	50	50
	10	53	53	53	53	50	50	50

Note: Coverage factor is the probability that a vehicle reaching the clear-zone width will impact a fixed object.

TABLE 12.11   Accident Reduction Factors for Increasing Roadside
Clear Recovery Distance on Curves[781]

Amount of increased roadside recovery distance, m*	Percent reduction in total number of curve accidents
1.50	9
2.50	14
3.00	17
3.75	19
4.50	23
6.00	29

*Rounded

TABLE 12.12   Accident Reduction Factors (Percent Reduction in Total
Number of Curve Accidents) for Flattening Side Slopes on Curves[781]

Sideslope in "before" condition on curve	Sideslope in "after" condition			
	4:1	5:1	6:1	7:1 or flatter
2:1	6	9	12	15
3:1	5	8	11	15
4:1	—	3	7	11
5:1	—	—	3	8
6:1	—	—	—	5

12.2.2.4 Compound Circular Curves. Compound circular curves, which consist of two or more contiguous, unidirectional curves of differing radii, should be avoided unless the curves consist of large radii. If economically or physically feasible, a single curve is preferred. If compound curves are unavoidable, the 85th-percentile speed of the flatter curve desirably should not be more than 10 km/h greater than the 85th-percentile speed of the sharper curve (safety criterion II, Table 11.1, "good design"). To minimize any effect of possible nonrecognition of the tightening curve by approaching from the longer radius of curve, the superelevation on the sharper curve should be adjusted in such a way that the friction demanded, when the curve is negotiated at the 85th-percentile speed of the flatter curve, does not significantly (say 20 to 25 percent) increase above that for the flatter curve.[37]

It should be noted that the new German "Guidelines for the Design of Roads"[246] call for the total avoidance of compound curves. The Germans argue that compound curves do not provide sufficient distances to perform the necessary steering efforts on the smaller radius of curve. In place of compound curves, the interspersing of a transition curve between the circular curved sections (egg-shaped clothoid) is recommended.

In this connection, Durth/Weise, et al.,[163] who examined the German relation design background shown in Fig. 9.35 in relation to accident experiences (see Fig. 12.20), indicated, at least graphically, that in the range to be avoided (lower right part of Fig. 12.20) a strong accumulation of accidents exists. Based on these results, they called for a serious review of compound curves in the future, and suggested replacing compound curves with egg-shaped clothoids, wherever possible.

The results of the previous investigation have been interpreted by Hanke[258] in the following manner:

> It is interesting to note that the size of the radius of curve plays a minor role in an accident situation, as compared to the radii of curve relation. Narrow radii of curve can certainly be used if the relation between the observed radius of curve and that of the preceding radius of curve lies into the good design range. Therefore, from a traffic safety point of view, relation design has to be acknowledged as a reliable tool in highway design.

The graphic accident results shown in Fig. 12.20 support the assumptions made in Figs. 9.1 and 9.36 to 9.40 to avoid in the fair range radii relations of $R \leq 200$ m and to use in this range radii relations of $R \leq 350$ m only in exceptional cases.

12.2.2.5 Curve Flattening. Previous studies clearly show that sharper curves are associated with higher accident rates than are milder ones.[135,217,218,408,410,455,781] In this connection, Talarico and Morrall reported the following:[698] Since curved sections increase the demands placed on drivers, these sections of highway have significantly higher accident rates than tangent sections, as many studies indicate.[396,710,782] In fact, accident rates on curves have been found to be 1.5 to 4 times greater than those on similar tangent sections for similar highway sections.[217] These results are confirmed by the authors in Tables 9.10, 10.12, and 18.14.

As a curve becomes flatter, a smoother transition between the tangent and curved sections of a highway is provided, and the length of the curve increases while the total length of the highway decreases. Research indicates that curve flattening is one of the most effective means of decreasing accident rates on horizontal curves.[782] Therefore, flatter curves increase the level of safety of a highway.[698]

For obtaining a sound or relation design on existing highways, all inconsistencies must be eliminated (for example, greater speed reduction in approaching a sharp curve, as is also required by safety criterion II for good design). Therefore, *curve flattening* refers to reconstructing a horizontal curve to make it less sharp, that is, longer with a lower degree of curve.[728]

Figure 12.21 shows graphically the curve-flattening effect. It can be directly seen that through curve flattening (case 2), the radius of curve, R_2, and the length of curve, L_2, become greater, as compared with the existing alignment (case 1); at the same time, degree of curve DC_2 becomes smaller than DC_1. Furthermore, the graph and the corresponding definitions explain the

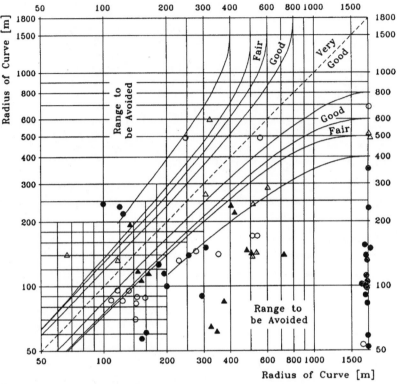

Legend:
```
9 test-sections/325 overall accidents, 69 run-off-the-road accidents,
o accident in element sequence with reverse curvature or tangent to curve,
∆ accident in element sequence with unidirectional curvature,
  empty: ∆V ≤ 5 km/h, V85 < 95 km/h; filled black: ∆V > 5 km/h, V85 < 95 km/h;
  filled gray: ∆V > 5 km/h, V85 ≥ 95 km/h.
```

FIGURE 12.20 The accident situation in relation to the radii of curve sequence.[163,258]

interrelationships between the central angle and degree of curve for those readers who are quite used to using the design element radius of curve in place of degree of curve. Corresponding relationships are also valid for the design parameter curvature change rate of the single circular curve (see Fig. 12.21).

Expected accident reduction factors associated with flattening curves for various degrees of curve before and after improvement and for central angles of 10° to 50° are given in Table 12.13. The table shows percent reductions for isolated curves and nonisolated curves; an *isolated curve* is defined as one having tangents of 200 m or more on both ends of the curve.[728,781]

According to "Evaluation of Tangents in the Design Process" in Sec. 12.1.1.3 (Tables 12.1 and 12.2), the tangent lengths defined here can be classified as independent tangents.

For example, consider a project that would flatten the curve from 10° to 5° with a central angle remaining at 30°. From Table 12.13, 48 percent of the total number of curve accidents would be expected to be reduced from the project for the case "isolated curve." The authors of the study point out that, due to the high construction costs of curve flattening, such improve-

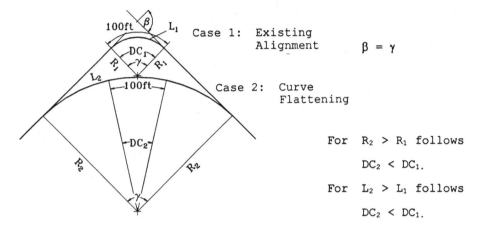

Case 1: Existing
         Alignment $\beta = \gamma$

Case 2: Curve
         Flattening

For $R_2 > R_1$ follows

$DC_2 < DC_1$.

For $L_2 > L_1$ follows

$DC_2 < DC_1$.

Legend:

$R_{1,2}$	=	radii of curve [ft or m],
γ	=	central angle of curve [degrees, basis 360° or 400 gon],
β	=	deflection angle [degrees, basis 360° or 400 gon],
DC_1/DC_2	=	degree of curve [degrees, basis 360° or 400 gon],
$L_{1,2}$	=	lengths of curve [ft or m],

The degree of curve is[5]

$$DC = \frac{100 \times 360}{2\pi R} = \frac{5729.6}{R} \qquad \text{degrees/100 ft for U.S. units} \qquad (12.38a)$$

$$DC = \frac{100 \times 360}{2\pi R} \approx \frac{5729.6}{R} \qquad \text{degrees/100 m} \qquad (12.38b)$$

and

$$DC = \frac{100 \times 400}{2\pi R} \approx \frac{6370}{R} \qquad \text{gon/100 m for metric units} \qquad (12.39)$$

Correspondingly, for circular curves without transitions, the CCR_S values can be calculated from the following formula (see also Table 8.4):

$$CCR_S = \frac{L_i/R}{L}\frac{400}{2\pi} \times 10^3 = \frac{1000 \times 400}{2\pi R} \approx \frac{63,700}{R} \qquad \text{gon/km} \qquad (12.40)$$

where L = length of unidirectional curved road section, km (corresponding, in the present case without transition curves, to the length of curve L_i).

FIGURE 12.21 Interrelationships between central angle, radii, lengths, and degrees of curve in case of curve flattening.

ments are more practical when a sharp or poorly designed curve has an abnormally high number of accidents.[728,781]

With respect to the cost-effectiveness of curve flattening, the following has been reported by Talarico and Morrall.[698] Generally, curve flattening costs depend on the original curve geometry, and costs increase with increases in either central angle or the change in degree of curve.[710] The following equations allow the calculations of all length-related relationships with regard to curve-flattening considerations. As a curve becomes flatter, the curve length increases, as the following equation indicates (see Fig. 12.21):

TABLE 12.13   Accident Reduction Factors for Flattening Horizontal Curves[781]

| Degree of curve | | Percent reduction in related accident types for the central angle, degrees | | | | | | | | |
| | | 10° | | 20° | | 30° | | 40° | | 50° | |
Original	New	Non-isolated	Isolated	Non-isolated	Isolated	Non-isolated	Isolated	Non-isolated	Isolated	Non-isolated	Isolated
30	25	16	17	16	17	16	17	15	16	15	16
30	20	33	33	32	33	31	33	31	33	30	33
30	15	49	50	48	50	47	50	46	50	46	50
30	12	59	60	57	60	56	60	55	60	55	60
30	10	65	67	64	66	63	66	62	66	61	66
30	8	72	73	70	73	69	73	68	73	68	73
30	5	82	83	80	83	79	83	78	83	78	83
25	20	19	20	19	20	18	20	18	20	17	20
25	15	39	40	38	40	36	40	36	40	35	40
25	12	50	52	49	52	48	52	46	52	46	51
25	10	58	60	56	60	55	60	54	59	53	59
25	8	66	68	64	68	62	68	61	67	60	67
25	5	77	80	75	80	74	79	72	79	72	79
20	15	24	25	23	25	22	25	21	25	20	24
20	12	38	40	36	40	35	40	34	39	33	39
20	10	48	50	45	50	44	49	42	49	41	49
20	8	57	60	54	60	52	59	51	59	50	59
20	5	71	75	68	74	66	74	64	74	64	74
15	10	30	33	28	33	26	33	25	32	24	32
15	8	43	46	40	46	37	46	35	45	34	45
15	5	61	66	56	66	53	65	51	65	50	65
15	3	73	79	68	79	64	78	63	78	63	78
10	5	41	49	36	48	32	48	29	47	28	47
10	3	58	69	50	68	45	67	43	66	42	66
5	3	22	37	15	35	13	33	11	32	11	31

Note: The *central angle* refers to the angle which would be formed by extending the tangents on either end of the curve.

$$L_2 = \frac{DC_1}{DC_2} L_1 \tag{12.41}$$

where L_1 = the length of the old curve, m
L_2 = the length of the new curve, m
DC_1 = the degree of curve of the old curve, degrees/100 m
DC_2 = the degree of curve of the new curve, degrees/100 m

In addition, a portion of the tangent section on either end of the old curve will now become circular.

Assuming that the new curve will have the same tangents as the old curve, the reduced tangent length is

$$T = \frac{5729.6}{1000} \left(\frac{1}{DC_2} - \frac{1}{DC_1} \right) \tan\left(\frac{\gamma}{2}\right) \tag{12.42}$$

where T = the tangent length eliminated due to curve flattening, km, and γ = the central angle, degrees. Therefore, the length of the old alignment, L_0, km, is

$$L_0 = L_1 + 2\,T \tag{12.43}$$

and the length of the new alignment, L_n, km, is

$$L_n = L_2 \tag{12.44}$$

When a horizontal curve is flattened, the total length of an alignment is decreased by an amount, ΔL, given by

$$\Delta L = \left[11.4592 \tan\left(\frac{\gamma}{2}\right) - 0.10\,\gamma \right]\left(\frac{1}{DC_2} - \frac{1}{DC_1}\right) \tag{12.45}$$

where ΔL = the change in length, km, and γ = the central angle, degrees. This decreased alignment length can result in benefits, like reductions in the number of accidents, maintenance savings, travel time savings, and vehicle operating savings.

Based on in-depth investigations of annual operating and travel time savings, Talarico and Morrall concluded the following:[698]

1. The cost-effectiveness of flattening a curve increases exponentially with AADT.
2. Care must be taken when selecting an accident reduction model upon which accident savings are based. Some of these models may predict unrealistic savings.
3. An optimum degree of flattening exists for curves with central angles less than or equal to 45 degrees.
4. The cost-effectiveness of flattening curves with central angles greater than 60 degrees increases with the central angle.

12.2.3 Transition Curve

*12.2.3.1 Application.** Parallel to, and in addition to, the statements made in Sec. 12.1.3.1, the following statements have been given in the AASHTO policies with respect to general considerations of transition curves.[5] A motor vehicle follows a transition path as it enters or leaves a circular horizontal curve. The steering change and the consequent gain or loss of centrifugal force cannot be effected instantly. Therefore, all studied countries use transition curves between tangents and circular curves for obtaining a sound horizontal alignment with an optically and esthetically satisfying alignment design and for driving dynamic considerations [at least up to large radii of curve (see, for example, Table 12.8)]. Normally, the Euler spiral, which is also known as the clothoid, is used.[5,35,37,139,243,429,453,663,686,688,700]

The radius of curve varies from "infinite" at the tangent end of the clothoid to the radius of curve at the beginning of the circular curve. By definition, the radius of curve at any point of the clothoid varies directly with the distance measured along the spiral. In the case when a clothoid combines two unidirectional circular curves having different radii, there is an initial radius of curve rather than an infinite value.

The principal advantages of transition curves in horizontal alignments are:

1. A properly designed transition curve provides a natural, easy-to-follow path for drivers, such that the centrifugal force increases and decreases gradually as a vehicle enters and leaves a circular curve. This transition curve minimizes encroachment on adjoining traffic lanes and tends to promote uniformity in speed.[5]

Driving comfort criterion: Considering $dar/dt = 0.5$, m/s^3, as the maximum rate of change of radial acceleration, the minimum value of A is

$$A_{min} = \sqrt{2\,v^3} \tag{12.46}$$

---

*Elaborated based on Refs. 5 and 663.

where v = vehicle speed, m/s

2. The transition curve length provides a convenient desirable arrangement for superelevation runoff. The transition between the normal cross slope or superelevation rate on the tangent and the fully superelevated section on the curve can be effected along the length of the transition curve in a manner closely fitting the speed-radius relation for the vehicle traversing it.[5]

Superelevation runoff criterion: The clothoid must have sufficient length to accommodate the superelevation runoff, that is:

$$A_{min} = \sqrt{R L_{e\,min}} \qquad (12.47)$$

where $L_{e\,min}$ = minimum length of superelevation runoff, m, calculated from Eq. (14.14b) in Sec. 14.1.3.2.

In contrast, where superelevation runoff is effected without a transition curve, usually partly on curve and partly on tangent, the driver approaching a curve may have to steer opposite to the direction of the curve ahead when on the superelevated tangent portion in order to keep his vehicle on tangent.

3. The appearance of the highway or street is enhanced by the application of clothoids. Their use avoids the noticeable breaks at the beginning and ending of circular curves (see Fig. 12.22).[5]

Appearance criterion: For esthetic reasons, the clothoid should consume an angle of at least $\tau = 3.5$ gon. It follows from this requirement that

$$A_{min} = \frac{R}{3} \qquad (12.13)$$

where R = radius of the circular curve, m.

4. Furthermore, the spiral facilitates the transition in width where the pavement section is to be widened around a circular curve. Use of spirals provides flexibility in the widening of sharp curves.[5]

Nearly all of the investigated highway geometric guidelines directly or indirectly assume that the clothoid should fulfill the previously mentioned criteria in the transition section. In this connection, it should be noted that:

- The limiting values of the driving comfort criteria differ among the guidelines, but have only an inferior effect on the determination of the minimum parameter of the clothoid (see Fig. 12.5).
- Almost every guideline contains the length criterion of the superelevation runoff (see Fig. 12.6).
- In most design cases, the appearance criterion becomes relevant, as Fig. 12.5 reveals.

12.2.3.2 Limiting Values.   In the guidelines studied so far, it has been assumed that the transition curve is perceivable for an angular change of at least $\tau = 3.5$ gon, and is limited by a maximum angular change of $\tau = 31.8$ gon (see Fig. 12.5). Based on these assumptions, the parameter of the clothoid has to fulfill the following conditions:

$$\frac{R}{3} \leq A \leq R \qquad (12.13, 12.14)$$

where A = parameter of the clothoid, m
R = radius of the curve, m

However, problems may still exist with the choice of the parameter A. For instance, with respect to very large radii of curve, on the basis of the appearance criterion, very long transition curves which are not necessary from a driving dynamic, comfort, or drainage point of view, have to be partially fitted in.

FIGURE 12.22 Curve with and without clothoid. (Elaborated and modified based on Ref. 5.)

The following considerations are based on the results of Table 12.8 in order to avoid long clothoid sections. According to this table, the application range for clothoids can be confined to radii of curve of $R \leq 800$ m (exceptional case: $R \leq 1000$ m) for the element sequence independent tangent–clothoid–circular curve. For larger radii of curve, the application of clothoids is not usually necessary and is left to the discretion of the highway engineer.

In addition, for circular curves with reversing curvature, three design possibilities exist, beginning with the radii of curve shown in Table 12.8:

1. Apply the superelevation runoff criterion for determining the parameter, A, of the reversing clothoid [Eq. (12.47)] instead of the appearance criterion.

2. Add a nonindependent tangent between the reversing circular curves in order to fit in the length of the superelevation runoff (Tables 12.1 or 12.2).

3. For large radii of curve, negative superelevation (adverse crossfall) may be applied on multilane median-separated highways (Table 14.1).

***12.2.3.3   Geometry of Transition Curves.**   In addition to the Euler spiral (clothoid), other forms of transition curves exist.

- *Cubic parabola:* The cubic parabola fulfills the demands of transition curves in case of large radii of curve and large transition curve lengths. However, for the small radii of curve often used in highway design, a kink occurs between the end of the transition curve and the beginning of the circular curve. In railroad engineering, the cubic parabola as transition curve was applied for the first time in 1871. Because of favorable riding conditions and ease of field output, cubic parabolas of the third or the fourth degree are still used today as transition curves in railroad engineering.

- *Sinus line:* The sinus line represents a consistent course of the curvature. It is applicable in the range between 0 and 100 gon as a transition curve. However, the sinus line has attained no practical meaning because it is difficult to tabulate and to stake out.

- *Different steering curves:* In an effort to find an ideal driving line, driving tests were conducted and distance coordinates were measured. These curves revealed consistent driving courses. However, the practical applicability failed since the attempts did not succeed in deriving valid mathematical formulations.

***12.2.3.4   Types of Transition Curves (to Avoid).†**   Unlike the favorable types of clothoids, which have been discussed in Sec. 12.1.3.4, the types presented in Fig. 12.23 have been partially used but should be avoided in the future:

- *Compound Clothoid (Fig. 12.23a):* This type consists of a sequence of clothoid sections in the same direction of curvature, which have equal radii of curve and a common tangent at their intersection point, R_S. For traffic safety reasons, compound clothoids should be avoided because they interrupt a consistent alignment flow and cause nonuniform driving behavior. However, for deceleration ramp design at interchanges, they may be of advantage.

- *C Clothoid (Fig. 12.23b):* The intersection of two clothoids in the same direction of curvature at their origins (C clothoid) should be avoided. A C clothoid optically effects a flattening, like a short tangent between two curves in the same direction of curvature (broken-back curve) (see Chap. 16).

- *Peak Clothoid (Fig. 12.23c):* The peak clothoid consists of two simple clothoids with the parameters, A_1 and A_2. At the intersection point, R_S, the radii are equal ($R_1 = R_2$), but no circular curve exists between the two clothoids; the circular curve is reduced to a single point. The peak clothoid should be avoided, since theoretically at this peak point, a jerk-blind handling of the steering wheel would be necessary, and from experience it is known that this may result in inconsistent driving behavior or even critical maneuvers.

***12.2.3.5   Safety Considerations.**   As already discussed in "Clothoid (Spiral Transition)" in Sec. 9.2.1.3, Leutzbach and Zoellmer[476] found, based on accident investigations, that up to $R = 200$ m, the element sequence tangent–clothoid–circular curve is more favorable than the direct sequence tangent–to–circular curve. However, for radii of curve greater than 200 m, no systematic differences were detected between the two element sequences with respect to the accident rate, a finding that is also valid for the accident cost rate.

Furthermore, they indicated that, contrary to the opinion of many experts, the presence of short tangents and/or transition curves between succeeding circular curves did not have a measurable positive impact on the accident situation, as compared to the direct sequence of two circular curves. However, relation design issues were of significant importance with regard to the

---

*Elaborated based on Ref. 748.

†Elaborated based on Refs. 241 and 243.

a) COMPOUND-CLOTHOID

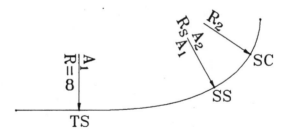

b) C-CLOTHOID

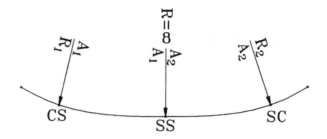

c) PEAK-CLOTHOID

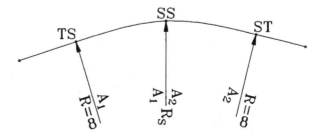

FIGURE 12.23 Undesirable types of transition curves. (Elaborated based on Ref. 243.) (*Note:* For general nomenclature, see Fig. 12.8; R_S = intersection point ($R_1 = R_2$); it follows that the length of the circular curve is $L = 0$ m.)

observed accident development ("Curvature Change Rate, Degree of Curve, Length of Curve, and Curve Radii Ratio" in Sec. 9.2.1.3).

In addition, more recent accident research work suggests that the impact of transition curves on traffic safety is neutral.[553]

Other researchers[781] indicated that a reduction of 5 percent in total accidents was achieved when spiral transitions were added to both ends of a curve on two-lane rural highways.

Generally speaking, with respect to safety effects, the application of clothoids should not be overemphasized in the design process, as it has been done so far in several countries.

Of course, one should not forget the importance of other design impacts that transition curves provide, besides accident-related issues, such as:

- Gradual increase or decrease of the centrifugal force
- Convenient desirable arrangement for the superelevation runoff
- Improvement of the optical appearance
- Curve widening

12.2.4 Practical Procedure for Detecting Errors in Alignment Design and Consequences for a Safer Redesign

12.2.4.1 Background. A procedure is introduced for evaluating horizontal alignment of two-lane roads on the basis of three individual safety criteria. Based on these criteria, design practices are classified into three groups: good, fair, and poor. The procedure can be used to detect the safety deficiencies of existing roadways as well as to identify potential safety errors in new designs that are still in the planning stages.

The safety evaluation process discussed provides a reliable assessment of horizontal alignment based on quantitative criteria. Consequently, safety impacts can be reviewed along with the normally considered local, environmental, esthetical, and economical aspects in making decisions relative to the project.

The procedure consists of three safety criteria. Safety criterion I (harmonizing design speed and operating speed) and safety criterion II (achieving consistency in operating speed) have been discussed thoroughly in Secs. 9.1.2, 9.1.3, 9.2.2, and 9.2.3.

Criteria I and II are based on design and/or operating speed changes for individual design elements, as well as between successive design elements, in order to achieve good designs, classify fair designs, and detect poor designs. Quantitative ranges for the safety evaluation processes are given in Table 11.1.

The third safety criterion (providing adequate driving dynamic safety) compares geometrically assumed side friction, which is dependent on the design speed, with side friction demanded, which is dependent on the 85th-percentile speeds under real-world conditions at individual curve sites, as discussed in detail in Secs. 10.1.4 and 10.2.4. The quantitative ranges of safety criterion III are also listed in Table 11.1.

However, on the basis of the different safety aspects, the results of the three safety criteria do not always agree. For example:

- A curved section may be classified by safety criteria I and II as good. That would mean the absolute difference between the 85th-percentile speed and the design speed in the curve itself (safety criterion I), as well as the absolute difference between the 85th-percentile speeds of preceding and succeeding design elements (safety criterion II), would fall into the range of ≤10 km/h (see Table 11.1). Despite these positive results, safety criterion III may reveal driving dynamic deficiencies, since the superelevation rate in the observed curve is too low or the radius of curve has to be flattened, for example:
- It is possible that criterion II represents a good safety level for a longer roadway section, whereas safety criterion I reveals fair or even poor design practices because of differences between expected 85th-percentile speeds and the selected design speed, which might be too large.

Because of these discrepancies, an individual examination of specific roadway sections, on the basis of the three safety criteria, makes sense. This is especially true when the highway engineer has information about the planned or existing highway, the safety quality (good or fair) to be strived for, and local conditions and available funds. For example, the designer may be able to improve the alignment where failure of only one safety criterion occurred in such a way that the safety deficiency can be eliminated without affecting the other criteria and their impacts on the design.

Note that besides the normal case of a design speed that is too low, it is quite possible to select a design speed that is too high. In such a case, the superior goal "safety" is of minor importance. However, the function of the highway in the road network or the desired traffic quality may cause the design to be uneconomical or unjustifiable because of environmental impacts.[436]

If the safety evaluation process does not reveal errors or deficiencies, then the examined roadway section represents a sound solution. If one or more of the three individual criteria are not fulfilled, then various design alternatives must be evaluated until a satisfactory alignment design is established.

The procedure provides interrelationships between design parameters, driving behavior, and driving dynamics in order to detect poor design practices, develop good designs, and have a positive influence on the accident situation.

On the basis of the databases tested so far, the procedure is applicable to two-lane rural roads with longitudinal grades of up to 5 to 6 percent and average annual daily traffic (AADT) volumes of up to 12,000 vehicles/day.[408,449,628,632,673]

The following case studies are based on Refs. 437, 445, and 595. More recent research results, however, which were developed here in Part 2, "Alignment," are also incorporated. Additional case studies can be found in Sec. 1.8.4.2.

12.2.4.2   Case Studies.   The existing horizontal alignment in Fig. 12.24 shows a two-lane rural state route in southwestern Germany in the plain of the Rhine River. Accident analysis indicates

AXIS No. 1 (Germany)

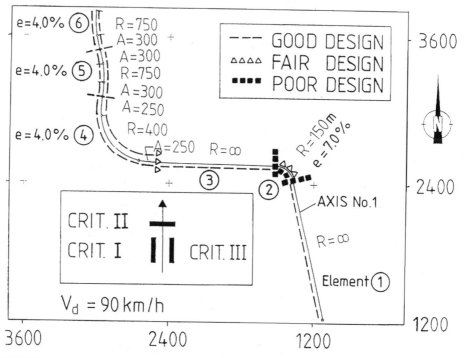

FIGURE 12.24   Graphic presentation of the safety evaluation process, based on the German operating speed background (old alignment).

a high-accident frequency and severity at element 2. The longitudinal grades are less than 2 percent and the AADT values corresponded to 8200 vehicles/day in 1995. The old alignment should be improved, and the new alignment should represent the level of good design practice for all three individual safety criteria. Between the old and new stages, an interim solution should also be planned in the event that federal funds cannot be provided in full.

Germany: Old Alignment.   Figure 12.24 shows the existing old alignment (axis no. 1), which was designed in the 1930s. The lane width is 3.50 m, and the original design speed is unknown.

A serious accident situation exists in the curve of design element 2 ($R = 150$ m), which is situated between two long tangents (elements 1 and 3). Sixteen run-off-the-road type accidents occurred from 1989 to 1991; these included three fatalities, six seriously injured individuals, and 13 slightly injured individuals. The main accident cause, recorded by the police, was "improper speed estimation" in the transition sections and in the curve itself. The main goal, therefore, had to be a reduction in accident severity by appropriately redesigning the old alignment. It is interesting to note that for the long tangent sections 1 and 3, no relevant accidents were recorded (for example, because of passing vehicles).

All necessary information about design elements 1 to 6 are listed in Table 12.14, like stations (col. 2), radius of curve and parameters of clothoids (col. 3), length of the design element (col. 4), curvature change rate of the single curve (col. 5), and superelevation rate (col. 8).

In order to perform the safety evaluation processes of criterion I (col. 11), criterion II (col. 12), and criterion III (col. 13), the following basic calculations have to be made, which are discussed in detail under issues 1 to 5:

1. *Curvature change rate of the single circular curve (Table 12.14, col. 5):* According to Sec. 8.1.3, the formula for CCR_S is

$$CCR_S = \frac{\left[\dfrac{L_{c11}}{2R} + \dfrac{L_{cr}}{R} + \dfrac{L_{c12}}{2R}\right] 63{,}700}{L} \tag{8.6}$$

It follows then for element 4, as an example:

$$L_{cr} = 502 \text{ m} \qquad L_{c11} = 160 \text{ m} \qquad L_{c12} = 160 \text{ m} \qquad L = 822 \text{ m} \qquad R = 400 \text{ m}$$

$$CCR_{S4} = \frac{\left[\dfrac{160}{800} + \dfrac{502}{400} + \dfrac{160}{800}\right] 63{,}700}{822}$$

$$= 128 \text{ gon/km}$$

(see Table 12.14, col. 5).

Accordingly, the CCR_{Si} values for elements 2, 5, and 6 were determined. For tangents, $CCR_S = 0$ gon/km.

2. *85th-percentile speed (Table 12.14, col. 6):* For the old German operating speed background,[241,243] the formula for $V85_i$ is given in Table 8.5 as follows:

$$V85_i = 60 + 39.70 \, e^{\,(-3.98 \times 10^{-3} \, CCR_{Si})} \tag{8.18}$$

(The relationship is schematically shown as curve "Germany, Old" in Fig. 8.12.)

For element 4 as an example, it follows:

$$CCR_{S4} = 128 \text{ gon/km} \rightarrow V85_4 = 84 \text{ km/h}$$

(see col. 6).

TABLE 12.14 Numerical Data for the Safety Evaluation Process, Based on the German Operating Speed Background (Old Alignment)

Element no. (1)	Station, km (2)	Parameters, m (3)	L_r, m (4)	CCR_{Sr}, gon/km (5)	$V85_r$, km/h (6)	V_d, km/h (7)	e, % (8)	f_{RA} (9)	f_{RD} (10)	Criterion I $|V85_i - V_d|$, km/h (11)	Criterion II $|V85_i - V85_{i+1}|$, km/h (12)	Criterion III $f_{RA} - f_{RD}$ (13)
1	0.000 1.190	IT ∞	1190	0	100	90	2.5			10 +	33 –	
2	1.190 1.390	R −150	200	425	67	90	7.0	0.15	0.17	23 –	33 –	−0.02 o
3	1.390 2.374	IT ∞	984	0	100	90	2.5			10 +	16 o	
4	2.374 3.196	A 250 R 400 A 250	822	128	84	90	4.0	0.15	0.10	6 +	7 +	0.05 +
5	3.196 3.586	A 300 R −750 A 300	390	59	91	90	4.0	0.15	0.05	1 +	1 +	0.10 +
6	3.586 3.907	A 300 R 750	321	69	90	90	4.0	0.15	0.05	0 +		0.10 +

Note: + = good design; o = fair design; − = poor design. IT = independent tangent.

12.55

3. *Assessment of an appropriate design speed:* The design speed of the existing (old) alignment in Fig. 12.24 is unknown. Thus, in order to come up with an estimation of an appropriate design speed, a procedure was developed in "Existing (Old) Alignments" in Sec. 9.2.2.1, which is based on the average of the CCR_{Si} values for the individual curves along the observed roadway section without regarding tangent sections:

$$\overline{CCR_S} = \frac{\sum\limits_{i=1}^{i=n} CCR_{Si} L_i}{\sum\limits_{i=1}^{i=n} L_i} \tag{9.12}$$

For the existing alignment, it follows according to Table 12.14:

$$\overline{CCR_S} = \frac{425 \times 200 + 128 \times 822 + 59 \times 390 + 69 \times 321}{200 + 822 + 390 + 321}$$

$$= 136 \text{ gon/km}$$

From this average $\overline{CCR_S}$, it follows from Eq. (8.18) that the average 85th-percentile speed is

$$\overline{V}85 = 83 \text{ km/h}$$

Based on this average 85th-percentile speed, a design speed of

$$V_d = 90 \text{ km/h}$$

(see col. 7) was selected, and was regarded as a reliable estimation for the existing alignment. Of course, a design speed of $V_d = 80$ km/h could also have been chosen. However, the somewhat generous elements 4 to 6 support the selection of the higher V_d value.

4. *Evaluation of tangents:* Elements 1 and 3 of the existing alignment are tangents. The evaluation of tangents in the design process is discussed in detail in "Evaluation of Tangents in the Design Process in Sec. 12.1.1.3. Since the tangent lengths for element 1 with 1190 m and for element 2 with 984 m (col. 4) are relatively long, it is assumed that case 2 in Fig. 12.2 in Sec. 12.1.1.3 is relevant. Figure 12.2 reveals that wherever the existing tangent length is at least twice as long as the long tangent length, TL_L, which is presented in col. 7 of Table 12.1, then the tangent can be regarded as independent, and consequently no additional calculation is deemed necessary.

For *tangent element 1:*

$$TL = 1190 \text{ m}$$

$CCR_{S1} = 0$ gon/km (col. 5) $V85_1 = 100$ km/h (col. 6, Table 12.14)

$CCR_{S2} = 425$ gon/km (col. 5) $V85_2 = 67$ km/h (col. 6, Table 12.14, see issue 2)

The 85th-percentile speed which is closest to 67 km/h in Table 12.1 is 65 km/h. In order to accelerate or decelerate from 65 km/h to the highest operating speed of 100 km/h in the tangent, a distance of $TL_L = 260$ m is needed, as revealed in col. 7 of Table 12.1. It follows then that

$$TL > 2 \times TL_L$$

$$1190 > 2 \times 260 = 520 \text{ m}$$

Thus, it can be concluded that the tangent is independent, and the sequence tangent (element 1)-to-curve (element 2) is relevant for the safety evaluation process of criterion II (col. 12, Table 12.14).

For *tangent element 3:*

$$TL = 984 \text{ m}$$

$$CCR_{S2} = 425 \text{ gon/km} \qquad V85_2 = 67 \text{ km/h}$$

$$CCR_{S3} = 0 \text{ gon/km} \qquad V85_3 = 100 \text{ km/h}$$

$$CCR_{S4} = 128 \text{ gon/km} \qquad V85_4 = 84 \text{ km/h}$$

The 85th-percentile speed which is closest to 67 km/h in the curve with the higher CCR_S value is again 65 km/h (see Table 12.1). Thus, the same long tangent length of $TL_L = 260$ m is needed:

$$TL > 2 \times TL_L$$

$$984 > 2 \times 260 = 520 \text{ m}$$

Consequently, element 3 also represents an independent tangent, and the sequence curve (element 2)-to-tangent (element 3), as well as the tangent (element 3)-to-curve (element 4) is relevant for the safety evaluation process of criterion II (col. 12, Table 12.14).

5. *Side Friction Assumed/Demanded:* The equations for side friction assumed/demanded were developed in Sec. 10.1.4.

Accordingly, the formula for side friction assumed on existing alignments is

$$f_{RA} = 0.60 \times 0.925 \, f_{Tperm} \tag{10.10a}$$

$$f_{RA} = 0.33 - 2.69 \times 10^{-3} \, V_d + 0.84 \times 10^{-5} \, (V_d)^2 \tag{10.10b}$$

For the selected design speed of $V_d = 90$ km/h (issue 3), the side friction assumed is

$$f_{RA} = 0.15$$

see col. 9 (Table 12.14).

The actual side friction demanded is calculated from the following equation:

$$f_{RD} = \frac{V85^2}{127 \, R} - e \tag{10.11}$$

For element 2, as an example, it follows that

$$f_{RD} = \frac{67^2}{127 \times 150} - 0.07 = 0.17$$

(see col. 10, Table 12.14).

The calculations for issues 1 to 5 have been presented here in detail in order to understand the development of Tables 12.14 to 12.19. The calculations will be repeated only in case of important or interesting deviations.

From the listings and/or calculations of the input data in Table 12.14, the safety evaluation processes for criterion I (col. 11), criterion II (col. 12), and criterion III (col. 13) can be easily made in conjunction with the quantitative ranges of Table 11.1 in order to distinguish good design from fair design and poor design practices. For independent tangents, safety criterion III is not relevant, since no centrifugal force exists.

An analysis of the critical curve (element 2 in Table 12.14) indicates that the absolute difference between $V85$ and V_d exceeds 20 km/h and corresponds to poor design according to the ranges of safety criterion I presented in Table 11.1. The same is true for safety criterion II with respect to the absolute $V85$ differences between elements 1 and 2 as well as between elements 2 and 3; these also fall into the ranges of poor design presented in Table 11.1. With respect to safety criterion III, regarding the difference between side friction assumed f_{RA} and side friction demanded f_{RD}, a fair design level could be noticed for element 2 (Table 11.1).

It should be noted that the numerical data presented in Table 12.14 is difficult to describe and those listings—as valuable as they are for an exact evaluation overview—may be too complex for fast, easy comprehension.

Therefore, a graphical presentation of the numerical results in Table 12.14 was developed and is presented in Fig. 12.24. In this way the different design levels, based on individual safety criteria I to III, can be recognized visually by using discriminating colors or symbols. For a better understanding, it should be mentioned that the colors or graphic symbols (as in the present case) for safety criterion II are arranged vertically to the road axis, whereas the symbols for safety criterion I are located on the left side and those for safety criterion III are located on the right side, parallel to the axis.

By evaluating the graphic layout of Fig. 12.24, it can be recognized at once that the critical curve (element 2) corresponds to poor design practices in terms of safety criteria I and II, and to fair design in terms of safety criterion III. This result supports the previous statements about the serious accident situation at this curve site and the corresponding transitions.

In addition, it can be seen that the curve with the radius of 400 m (element 4) can be evaluated only as a fair design for safety criterion II when considering the transition between elements 3 and 4. The accident situation at this site consisted of one serious, three light, and two property damage accidents during the time period investigated, which supports the findings of the safety evaluation process.

The other road sections of the existing alignment reveal good design practices and do not require any changes in future redesigns.

Greece: Old Alignment.   For comparative purposes, the existing (old) alignment will be examined based on the Greek operating speed background,[453,586] which corresponds to the following formula from Table 8.5:

$$V85_i = \frac{10^6}{10,150.1 + 8.529 \, CCR_{Si}} \qquad (8.16)$$

(The relationship is graphically presented as curve "Greece" in Fig. 8.12.)

Accordingly, from the CCR_{Si} values of col. 5 in Table 12.15, the $V85_i$ values in col. 6 can be determined.

The selection of an appropriate design speed is again based on the average curvature change rate of the single circular curve (see issue 3 of the previous chapter):

$$\overline{CCR_S} = 136 \text{ gon/km} \rightarrow \overline{V}85 = 88 \text{ km/h}$$

That means a design speed of $V_d = 90$ km/h can be recommended (Table 12.15, col. 7).

An evaluation of tangents revealed with respect to element 1 cross the following:

$$TL = 1190 \text{ m}$$

$$CCR_{S1} = 0 \text{ gon/km} \rightarrow V85_1 = 99 \text{ km/h}$$

TABLE 12.15 Numerical Data for the Safety Evaluation Process, Based on the Greek Operating Speed Background (Old Alignment)

Element no. (1)	Station, km (2)	Parameters, m (3)		L_t, m (4)	CCR_{SF}, gon/km (5)	$V85_t$, km/h (6)	V_d, km/h (7)	e, % (8)	f_{RA} (9)	f_{RD} (10)	Criterion I $|V85_i - V_d|$, km/h (11)		Criterion II $|V85_i - V85_{i+1}|$, km/h (12)		Criterion III $f_{RA} - f_{RD}$ (13)
1	0.000 1.190	IT	∞	1190	0	99	90	2.5			9	+			
													26	−	
2	1.190 1.390	R	−150	200	425	73	90	7.0	0.15	0.21	17	o			−0.06 −
													26	−	
3	1.390 2.374	IT	∞	984	0	99	90	2.5			9	+			
													10	+	
4	2.374 3.196	A R A	250 400 250	822	128	89	90	4.0	0.15	0.12	1	+			0.03 +
													5	+	
5	3.196 3.586	A R A	300 −750 300	390	59	94	90	4.0	0.15	0.05	4	+			0.10 +
													1	+	
6	3.586 3.907	A R	300 750	321	69	93	90	4.0	0.15	0.05	3	+			0.10 +

$$CCR_{S2} = 425 \text{ gon/km} \rightarrow V85_2 = 73 \text{ km/h}$$

According to Table 12.1 (col. 7), it follows that $TL_L = 200$ m.

$$TL > 2 \, TL_L$$

$$1190 > 2 \times 200 = 400 \text{ m}$$

Thus, element 1 can be regarded as an independent tangent. The same is true for element 3:

$$984 > 2 \times 200 = 400 \text{ m}$$

By knowing the input data of Table 12.15, the safety evaluation process can be conducted.

The results are shown in col. 11 for safety criterion I, col. 12 for safety criterion II, and col. 13 for safety criterion III (see Table 12.15). A graphical presentation is given in Fig. 12.25. The critical curve (element 2) attracts attention with its poor design levels with respect to safety criteria II and III and a fair design level with respect to safety criterion I.

All the other road sections reveal good design practices based on the operating speed background of Greece.

AXIS No. 1 (Greece)

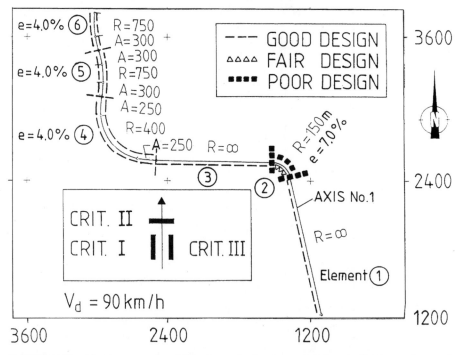

FIGURE 12.25 Graphic presentation of the safety evaluation process, based on the Greek operating speed background (old alignment).

United States: Old Alignment. Finally, the existing alignment will be evaluated based on the operating speed background of the United States.[565] The relationship between CCR_{Si} and $V85_i$ is given by the following equation from Table 8.5:

$$V85_i = 103.04 - 0.053 \, CCR_{Si} \qquad (8.14)$$

(The relationship is graphically presented as curve "U.S.A." in Fig. 8.12.)

Based on Eq. (8.14), the 85th-percentile speeds in col. 6 of Table 12.16 are determined based on the CCR_{Si} values in col. 5.

For an average $\overline{CCR}_S = 136$ gon/km, an average 85th-percentile speed of $\overline{V}85 = 96$ km/h results from Eq. (8.14). Therefore, contrary to the German and Greek results, a design speed of $V_d = 100$ km/h was selected (Table 12.16, col. 7).

Elements 1 and 3 are again classified as independent tangents.

The outcome of Table 12.16 and Fig. 12.26 reveal results that are similar to those of Greece with respect to the critical curve (element 2). According to safety criteria II and III, the results are poor design levels, whereas according to safety criterion I, a fair design range is obtained. Contrary to Germany and Greece, a fair design practice could also be noticed, according to safety criterion III, in the curve with a radius of $R = 400$ m (element 4).

Once again, the other roadway sections reveal good design practices based on the U.S. operating speed background.

In conclusion, the safety evaluation processes of the three investigated countries show strong similarities. For the critical element 2, two of the safety criteria always reveal poor design levels, while one criterion shows a fair design range.

Generally speaking, two "poor" and one "fair" imply a poor design practice, as will be shown later in Sec. 18.2 with respect to the development of a safety module for highway design.

The three safety criteria have proven to be able to detect errors or deficiencies in alignment design and provide a basis for safer redesigns or RRR strategies in the future in case of severe design deficiencies, as for example for element 2, where all three safety criteria call for special attention. On the other hand, less stringent deficiencies exist, where only one criterion stands out. A typical example of the latter case is presented in Fig. 12.24. This figure shows a fair design range according to safety criterion II with respect to the design inconsistency case between the independent tangent (element 3) and the following curve with a radius of $R = 400$ m (element 4), based on the German operating speed background. Before a reconstruction project can be considered for this design case, it would first be appropriate to check if traffic control devices, like signs, chevrons, delineations, etc., or local speed limits, might be sufficient to alleviate the problem. Another example is presented in Fig. 12.26, which shows a fair design according to safety criterion III with respect to a minor driving dynamic safety problem in the curve with a radius of $R = 400$ m (element 4), based on the U.S. operating speed background. For this case, an increase in the superelevation rate of about $e = 2$ to 3 percent could prove to be sufficient for changing the fair design range to the good design range according to safety criterion III.

Germany: Interim Solution. A fair design practice, as a minimum requirement for an interim solution, was requested for the present case study in order to keep the reconstruction costs down. Related to the critical radius of the curve (element 2), it was found that in order to combine a tangent and a curve in the fair design range, the least possible radius is $R = 500$ m according to the old German relation design background in Fig. 9.35.[241,243] At the time of the investigation in 1993,[445] the new considerations about sound relation design backgrounds in Secs. 9.1.3.2 and 9.2.3.2 had not yet been developed. Furthermore, the authors decided to apply the exact superelevation rates provided in the German guidelines.[243] In this way, axis no. 2 in Fig. 12.27 was developed, and the design input data was listed in Table 12.17.

The selection of a sound design speed is again based on the average curvature change rate of the single circular curve without regarding tangent sections, as previously discussed.

For the interim solution, it follows, according to Table 12.17, that

TABLE 12.16 Numerical Data for the Safety Evaluation Process, Based on the U.S. Operating Speed Background (Old Alignment)

Element no. (1)	Station, km (2)		Parameters, m (3)		L_r, m (4)	CCR_{Sr}, gon/km (5)	$V85_r$, km/h (6)	V_d, km/h (7)	e, % (8)	f_{RA} (9)	f_{RD} (10)	Criterion I $|V85_i - V_d|$, km/h (11)		Criterion II $|V85_i - V85_{i+1}|$, km/h (12)		Criterion III $f_{RA} - f_{RD}$ (13)	
1	0.000	1.190	IT	∞	1190	0	103	100	2.5			3	+				
														22	−		
2	1.190	1.390	R	−150	200	425	81	100	7.0	0.14	0.27	19	o			−0.13	−
														22	−		
3	1.390	2.374	IT	∞	984	0	103	100	2.5			3	+				
														7	+		
4	2.374	3.196	A	250	822	128	96	100	4.0	0.14	0.14	4	+			0.00	o
			R	400													
			A	250										4	+		
5	3.196	3.586	A	300	390	59	100	100	4.0	0.14	0.07	0	+			0.07	+
			R	−750													
			A	300										1	+		
6	3.586	3.907	A	300	321	69	99	100	4.0	0.14	0.06	1	+			0.08	+
			R	750													

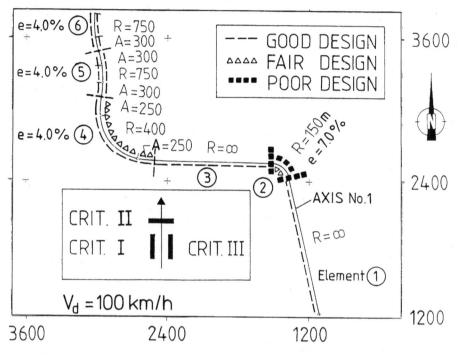

FIGURE 12.26   Graphic presentation of the safety evaluation process, based on the U.S. operating speed background (old alignment).

$$\overline{CCR}_S = \frac{107 \times 791 + 128 \times 822 + 59 \times 391 + 69 \times 320}{791 + 822 + 391 + 320}$$

$$= 101 \text{ gon/km}$$

From the German operating speed background [Eq. (8.18)], it follows that

$$\overline{V}85 = 87 \text{ km/h}$$

Based on this average 85th-percentile speed, a design speed of

$$V_d = 90 \text{ km/h}$$

(see col. 7, Table 12.17) was regarded as a reliable estimate of the interim design.

Elements 1 and 3 were again classified as independent tangents.

For the same safety evaluation procedure discussed before, but this time based on the input data in Table 12.17, it can be seen that the interim solution produces fair design practices between elements 1 to 2, 2 to 3, and 3 to 4. That means according to safety criterion II, the absolute differences in $V85$ for these element sequences lie somewhere in the range between 10 and 20 km/h according to Table 11.1 (compare the results in col. 12 of Table 12.17). Safety criteria I and III reveal good design practices in all cases.

AXIS No. 2 (Germany)

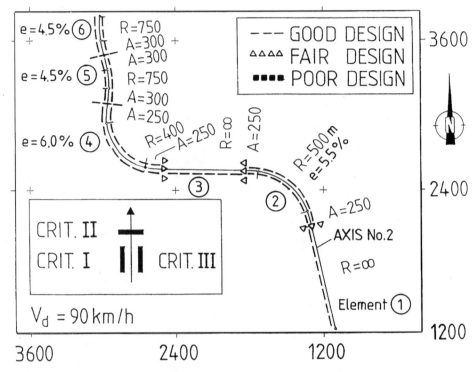

FIGURE 12.27 Graphic presentation of the safety evaluation process, based on the German operating speed background (interim solution).

From an economic point of view, the alignment of axis no. 2 can be evaluated as favorable because of low construction costs (at least 50 percent less than the cost of the final curvilinear alignment). However, it is difficult to determine to what extent the remaining transition sections, which correspond to fair design levels, may have an unfavorable impact on the accident situation. As a matter of fact, higher accident risks can be expected on sections which correspond to fair design than on sections which correspond to good design (see Tables 9.10 and 10.12).

Greece and the United States: Interim Solution. The selection of a sound design speed is once again based on the average curvature change rate of the single circular curve. In the previous section, this parameter was calculated for axis no. 2 to be:

$$\overline{CCR}_S = 101 \text{ gon/km}$$

From the Greek operating speed background [Eq. (8.16), Table 8.5], it follows that

$$\overline{V}85 = 91 \text{ km/h} \rightarrow V_d = 90 \text{ km/h} \qquad \text{Greece}$$

With respect to the U.S. operating speed background [Eq. (8.14), Table 8.5], it follows that

$$\overline{V}85 = 98 \text{ km/h} \rightarrow V_d = 100 \text{ km/h} \qquad \text{United States}$$

TABLE 12.17 Numerical Data for the Safety Evaluation Process, Based on the German Operating Speed Background (Interim Solution)

Element no. (1)	Station, km (2)		Parameters, m (3)		L_t, m (4)	CCR_{St}, gon/km (5)	$V85_t$, km/h (6)	V_d, km/h (7)	e, % (8)	f_{RA} (9)	f_{RD} (10)	Criterion I $|V85_i - V_d|$, km/h (11)		Criterion II $|V85_i - V85_{i+1}|$, km/h (12)		Criterion III $f_{RA} - f_{RD}$ (13)	
1	0.000	0.852	IT	∞	852	0	100	90	2.5			10	+	14	o		
2	0.852	1.643	A R A	250 −500 250	791	107	86	90	5.5	0.15	0.06	4	+			0.09	+
3	1.643	2.288	IT	∞	645	0	100	90	2.5			10	+	14 16	o o		
4	2.288	3.110	A R A	250 400 250	822	128	84	90	6.0	0.15	0.08	6	+			0.07	+
5	3.110	3.501	A R A	300 −750 300	391	59	91	90	4.5	0.15	0.04	1	+	7	+	0.11	+
6	3.501	3.821	A R	300 750	320	69	90	90	4.5	0.15	0.04	0	+	1	+	0.11	+

AXIS No. 2 (Greece and U.S.A.)

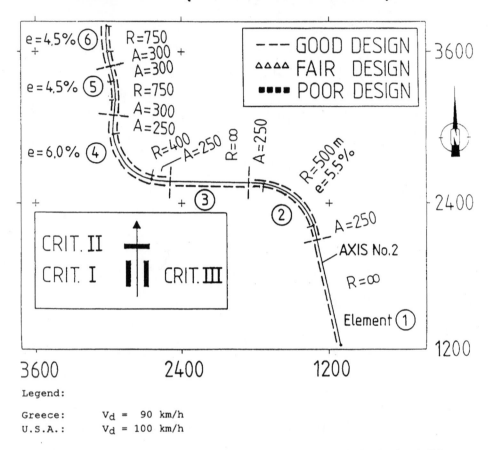

FIGURE 12.28 Graphic presentation of the safety evaluation process, based on the Greek and on the U.S. operating speed backgrounds (interim solution).

Despite different design speeds, a graphical presentation of the safety evaluation processes in Fig. 12.28 reveals good design practices for both countries with respect to all design elements and corresponding transitions. That means further investigation of curvilinear alignment, for example, as will be shown in the next section, is not necessary for Greece and the United States.

Germany: Curvilinear Alignment.   For safety reasons, good design practices should always be strived for if no other superior goals are of relevant importance. This is true for the new design of multilane, as well as two-lane, rural roads. Besides the individual safety criteria I to III discussed here, a tool for achieving good designs is introduced by the term *relation design* or *curvilinear alignment*. This means that single design elements should no longer be put together; rather sound design element sequences should be formed.[444] To support this idea, relation design backgrounds were developed in Fig. 9.1 for Greece and in Figs. 9.36 to 9.40 for Australia, Canada, Germany, Lebanon, and the United States. The selected element sequences for the curvilinear alignment in Fig. 12.29 satisfy the demands of the good range for all of the relation design backgrounds investigated so far.

AXIS No. 3 (Germany)

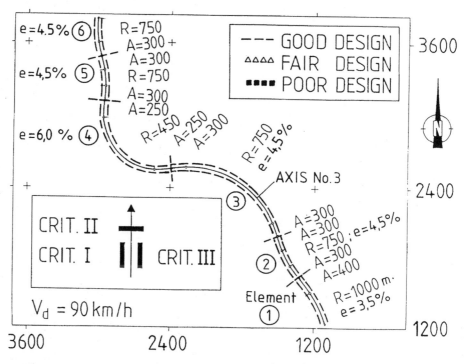

FIGURE 12.29 Graphic presentation of the safety evaluation process, based on the German operating speed background (curvilinear alignment).

The following calculations and evaluations are done based on the German operating speed background [Eq. (8.18), Table 8.5]. The input data for the curvilinear alignment are listed in Table 12.18 and are graphically presented in Fig. 12.29.

Based on the road function, the design speed of a new alignment normally has to be determined according to Table 6.2. The state route observed here mainly represents regional connections in the network and corresponds to road category A II, for which design speeds of 90 and 80 km/h (col. 7, Table 6.2) should normally be provided for two-lane rural roads.

In order to examine these recommendations, the design speed is once again determined for axis no. 3 (Table 12.18) based on the average curvature change rate procedure:

$$\overline{CCR}_s = \frac{52 \times 449 + 58 \times 385 + 76 \times 1148 + 121 \times 952 + 58 \times 367 + 69 \times 321}{3622}$$

$$= 80 \text{ gon/km}$$

From the German operating speed background [Eq. (8.18), Table 8.5], it follows that

$$\overline{V}85 = 89 \text{ km/h} \rightarrow V_d = 90 \text{ km/h}$$

Therefore, a design speed of $V_d = 90$ km/h, which is in agreement with the recommendations of Table 6.2, was selected for the network function of the observed state route. As could be noted,

TABLE 12.18 Numerical Data for the Safety Evaluation Process, Based on the German Operating Speed Background (Curvilinear Alignment)

Element no. (1)	Station, km (2)	Parameters, m (3)		L_r, m (4)	CCR_{Si}, gon/km (5)	$V85_r$, km/h (6)	V_d, km/h (7)	e, % (8)	f_{RA} (9)	f_{RD} (10)	Criterion I $|V85_i - V_d|$, km/h (11)	Criterion II $|V85_i - V85_{i+1}|$, km/h (12)	Criterion III $f_{RA} - f_{RD}$ (13)
1	0.000 0.449	R A	−1000 400	449	52	92	90	3.5	0.15 0.11*	0.03	2 +		0.12 + 0.08* +
												1 +	
2	0.449 0.834	A R A	300 750 300	385	58	91	90	4.5	0.15 0.11*	0.04	1 +		0.11 + 0.07* +
												1 +	
3	0.834 1.982	A R A	300 −750 300	1148	76	90	90	4.5	0.15 0.11*	0.04	0 +		0.11 + 0.07* +
												5 +	
4	1.982 2.934	A R A	250 450 250	952	121	85	90	6.0	0.15 0.11*	0.07	5 +		0.08 + 0.05* +
												7 +	
5	2.934 3.301	A R A	300 −750 300	367	58	92	90	4.5	0.15 0.11*	0.04	2 +		0.11 + 0.07* +
												2 +	
6	3.301 3.622	A R	300 750	321	69	90	90	4.5	0.15 0.11*	0.04	0 +		0.11 + 0.07* +

*Utilization ratio of side friction assumed n = 45 percent.

the $\overline{CCR}_S$ procedure for determining sound design speeds is not only valid for existing (old) alignments but is also valid for new designs or redesigns, as is the case here.

The results of the safety evaluation process according to Table 12.18 and Fig. 12.29 show no safety errors or deficiencies on the basis of safety criteria I to III. All three criteria confirm good design practices for the curvilinear alignment along the entire observed two-lane rural roadway section. Thus, it can be expected that the final alignment presented in Fig. 12.29 is a sound one.

Based on the assumptions made in Sec. 10.1.4, the calculation of side friction assumed is based on a utilization ratio of $n = 60$ percent according to Eqs. (10.10a) and (10.10b) for existing (old) alignments (see also issue 5 in "Germany (Old) Alignment" in this section). However, the present redesign case, which corresponds to axis no. 3, no longer represents an old alignment or an interim solution. Therefore, the assumption made for side friction assumed for new designs ($n = 45$ percent) was additionally tested here according to Eqs. (10.4a) and (10.4b) in Table 10.1. With respect to safety criterion III, it can be seen from Table 12.18 (cols. 9 and 13) that the reduced assumed side friction factor also leads to good design levels for all investigated elements.

However, relation design, based on Secs. 9.1.3.2 and 9.2.3.2, means more. Relation design should not only be directed toward achieving good curvilinear alignments (the main theme so far,[241,243,444]) but should also include sound transitions between independent tangents and curves. For example, for flat topography, it could mean achieving well-balanced transitions between independent tangents and curves interspersed (or not) between curvilinear alignment sections, as was shown for the interim solution. Note that for Greece and the United States, the interim solution revealed an already sound alignment design (Fig. 12.28), whereas for Germany, the existing fair ranges of safety criterion II could be avoided by slightly increasing the radii of curve of elements 2 and 4 (Fig. 12.27). Thus, both curvilinear alignments or improved alignments according to the interim solution can lead to good results regarding the three safety criteria.

Other Aspects.   For the alignment in Fig. 12.29, it can be observed that by eliminating the tangent sections, a well-balanced curvilinear alignment would result, and the risk of run-off-the-road accidents would certainly be reduced.

However, by eliminating the tangents, the risk of critical passing maneuvers may increase. Safe passing maneuvers require minimum passing sight distances (PSDs). Therefore, a PSD analysis was conducted on the basis of a minimum PSD of 575 m, which is required for $V85 = V_d = 90$ km/h in Table 15.2. This analysis involved the roadway from point 1 to point 2 in Fig. 12.30, where the main redesign measures will take place. The rest of the alignment more or less remains unchanged. The result of the PSD analysis for the observed road section revealed that the minimum PSD always exists because of the presence of large radii of curve between 750 and 1000 m of axis no. 3. It could even be proved that the PSD requirements are improved decisively by the curvilinear alignment in comparison with the old alignment of axis no. 1, where the radius of $R = 150$ m influenced the PSD unfavorably. This is an additional positive aspect resulting from the analysis of road sections by using the three safety criteria. Therefore, negative impacts on traffic safety are not to be expected for the final curvilinear alignment resulting from PSD considerations.

The lateral displacement of axis no. 3, in comparison with that of axis no. 1 (see Fig. 12.30), is of minor importance because an environmental compatibility study, performed as described in Sec. 7.2.3,[436] revealed that all three axes are located in a low-conflict corridor. Regarding land use, sufficient agricultural land and pasture are available for the new corridor (classified as being worthy of a low level of protection), whereas the topography of the present case study, located in the plain of the Rhine River, plays a minor role.

Preliminary Conclusion.   A practical procedure that enables the highway engineer to evaluate horizontal alignments of two-lane rural roads by means of the three individual safety criteria was presented in this chapter.

The procedure was tested by changing the alignment of an existing two-lane rural roadway, which revealed poor design practices, via a fair, but economical solution into a sound curvilinear alignment representing only good design levels.

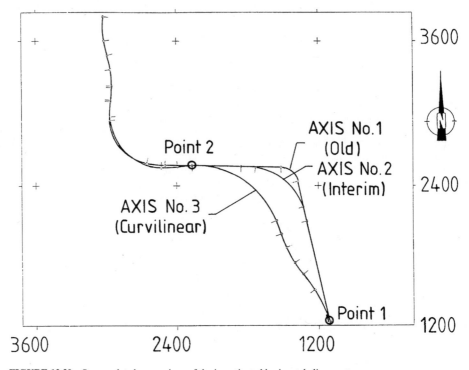

FIGURE 12.30   Space-related comparison of the investigated horizontal alignments.

The next research step should be to examine the validity of the results from the proposed safety model in relation to the actual accident situation, for example, by extending the model to hazard rating or to estimating the number of accidents (measured by rate or severity) for the road segment being considered. First efforts in this direction will be made in Sec. 18.2 and will reveal good agreement. A statistically sound analysis and evaluation, however, has not so far been possible because of the insufficient accident databases, especially with respect to single roadway sections with relatively low numbers of accidents. It should not be forgotten, however, that the ranges of validity shown in Table 11.1 were, for safety criteria I to III, established on the basis of the mean accident rates presented in Tables 9.10 and 10.12 in order to distinguish good designs from fair and poor designs. These tables were developed in order to supplement the ranges of validity for safety criterion I in Table 9.11, safety criterion II in Table 9.13, and safety criterion III in Table 10.11.

In addition, more recent research investigations[628] confirm, by and large, based on 30 individual case studies, the developed close relationships between design, operating speed, driving dynamic consistencies or inconsistencies, and the actual accident situation (Sec. 18.3).

12.2.4.3   *Fundamentals for Computer-Aided Design*

General.   In order to be effective, the safety evaluation process must be integrated into modern highway design tools which are available to today's highway design engineers. These tools consist of computer-aided design (CAD) systems for highway geometric design, and usually contain a component for the design of the horizontal alignment. In order to incorporate the safety evaluation process into the horizontal alignment component of a commonly used CAD system, a subprogram for safety computations was developed on the basis of the three individual criteria. Consequently, safety impacts are integrated into what is normally considered the local, environmental, esthetic, and economic aspects in making decisions on a given project.

Modern data processing systems for traffic routes should consist of at least the following components:[236–238,423,437,445]

- Environmental compatibility study
- Geometric surveying
- Horizontal alignment
- Vertical alignment
- Cross section
- Graphic layouts (as direct derivations of the computations)
- Three-dimensional evaluation (perspective view)
- Different structural components
- Construction accounting
- Automatic management of land acquisition

For the present study, the horizontal alignment component is of special interest. Programs for the numerical computation of road axes for horizontal alignments have existed since the 1960s. They were first developed by IBM.[61,300] These programs were related to mainframe computer applications that computed whole systems of roads and interchanges on the basis of descriptive data by using provided input data explicitly. The input data, coordinates of certain fixed points, and basic information about circular and transition curves according to a predesign of the horizontal alignment are provided by the highway engineer and are based on preliminary work at the location. The computer then prints out all necessary numerical design data for establishing the future road axes.

However, numerical printouts are difficult to work with, and the examination of the results is nearly impossible. Therefore, modern data processing systems have the capability of providing information on both the numerical level and the graphic level, which permits an immediate change of computational results into graphic layouts or vice versa on the PC screen or on printouts, thus allowing exact computations and information control graphics to appear side-by-side.[236,237]

The horizontal alignment component of a commonly used computer-aided design (CAD) system in Germany was selected[10,237] for the possible integration of the new subprogram for evaluating horizontal alignments with the three individual safety criteria. With this component, the axes of horizontal alignments can be computed and displayed on a PC screen for various alternatives. This is important not only for studying topographical and local conditions (see, for example, the development of a low conflict corridor in Sec. 7.2.3) but also for making the necessary alignment changes required by the safety evaluation process. Furthermore, all necessary design data for the axis are available in a computer (digital) mode for future processing steps.

It makes sense, therefore, to develop an additional subprogram for the new safety evaluation process on the basis of the three individual criteria and to integrate this into the horizontal alignment component of an overall CAD system. Figure 12.31 shows the interactive flow of information. In this way, the future axis of a specific roadway section can be evaluated automatically with special regard for traffic safety.

Development of a Subprogram for Safety Calculations.   Because the subprogram for safety calculations needs only the information about the geometry of the road in a computer (digital) mode, this subprogram can be integrated into any CAD system for highway geometric design. The only assumption is that the system can provide a clear data interface for the output of the horizontal design data in digital mode. The flowchart in Fig. 12.31 shows that the input of the descriptive design data is possible for planned or existing roadways.

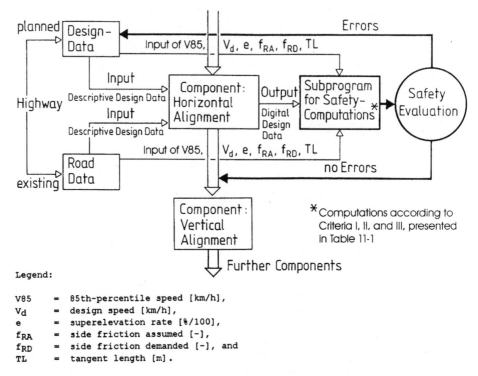

FIGURE 12.31 Flowchart for the subprogram: safety evaluation process.

As input data, the safety computation subprogram needs the geometric output data and the elements of the horizontal alignment component, which are as follows. The descriptive design data are

- Kind of design elements (circular curve, clothoid, tangent)
- Lengths of elements
- Parameters of design elements (radius of curve R, parameter of clothoid A)
- Stations
- Curvature change rates of the single curve CCR_S, Eq. (8.6)

The output data for the previously discussed case studies are presented numerically in Tables 12.14 to 12.18 (cols. 1 to 5) and graphically in Figs. 12.24 to 12.29.

For safety calculations, the following input data are required (see Table 11.1):

- 85th-percentile speed: $V85 = f(CCR_S)$ for the country under study (see Table 8.5)
- Design speed V_d from Table 6.2 or from Eq. (9.12) in "Existing (Old) Alignments" in Sec. 9.2.2.1 for the average $\overline{CCR}_S$
- Superelevation rate e
- Side friction assumed f_{RA}, Eq. (10.10b) for existing alignments or Table 10.1 for new designs
- Side friction demanded f_{RD}, Eq. (10.11)
- Evaluation of the tangent length, TL, according to the formulas in Table 12.3 or according to the limiting ranges in Tables 12.1 or 12.2. (See also Sec. 18.5.2.)

All pertinent equations, as well as the ranges for the three safety criteria in Table 11.1 for evaluating good, fair, and poor design practices, are contained in the safety computation subprogram (Fig. 12.31). The process described in this figure is interactive. Therefore, an automatic safety evaluation process with regard to the input data listed here is possible for planned or existing road axes. Corresponding output data for the previously discussed case studies are presented numerically in Tables 12.14 to 12.18 (cols. 6 to 13) and graphically in Figs. 12.24 to 12.29.

If the evaluation process does not reveal errors or deficiencies, the highway geometric design procedure can be pursued by means of the vertical alignment component. If one or more of the three individual safety criteria are not fulfilled, various design alternatives can be evaluated until a satisfactory road axis is established, as demonstrated for the previous case studies.

Even though Tables 12.14 to 12.18 were manually created in order to demonstrate in detail the individual calculation steps, the safety evaluation subprogram provides the numerical output data automatically. For instance, Table 12.19 shows such an output in comparison to the manual listing of Table 12.14. The graphic presentations in Figs. 12.24 to 12.29, however, fully represent the outcome of the safety evaluation subprogram.

In conclusion, to recognize safety errors in new designs or redesigns in the planning stages or necessary improvements in RRR projects before implementation, modern planning tools, like CAD systems for highway geometric design, have to be available. In this connection, for the horizontal alignment component of the overall CAD system, an additional subprogram for a new safety evaluation process was developed. The new subprogram allows the evaluation of horizontal alignments of planned or existing roadways on the basis of good, fair, and poor design practices.

In this way, it is possible to evaluate safety impacts for future establishment of horizontal alignment alternatives. This allows a change, not only from a design point of view, but also from a safety point of view.

Note that the data processing and CAD backgrounds developed to date are mainly usable by experts for scientific applications. Future developments for every day use is intended.

Further information on safety-related CAD procedures will be provided in Sec. 18.2.

TABLE 12.19   Numerical Output Data of the Subprogram: Safety Evaluation Process (Old Alignment) Elaborated and modified based on Refs. 445 and 595.

AXIS : 1

ELEM. : 1	STATION		CLOTHOIDS				SUPER-
RADIUS	FROM	TO	BEFORE	BEHIND	CCR_S	V85	ELEVATION
0	0.00	1190.42	0.00	0.00	0.00	99.70	2.5

CRIT. I : $|V85_1 - V_d| = 9.70$ => GOOD DESIGN

Transition 1-2 for Crit. II: $|V85_1 - V85_2| = 32.98$ => POOR DESIGN

ELEM. : 2	STATION		CLOTHOIDS				SUPER-
RADIUS	FROM	TO	BEFORE	BEHIND	CCR_S	V85	ELEVATION
-150	1190.42	1390.00	0.00	0.00	424.67	67.32	7.0

CRIT.   I : $|V85_2 - V_d| = 22.68$ => POOR DESIGN
CRIT. III : $f_{RA} - f_{RD} = -0.02$ => FAIR DESIGN

Transition 2-3 for Crit. II: $|V85_1 - V85_3| = 32.98$ => POOR DESIGN

ELEM. : 3	STATION		CLOTHOIDS				SUPER-
RADIUS	FROM	TO	BEFORE	BEHIND	CCR_S	V85	ELEVATION
0	1390.00	2373.79	0.00	0.00	0.00	99.70	2.5

CRIT. I : $|V85_3 - V_d| = 9.70$ => GOOD DESIGN

Transition 3-4 for Crit. II: $|V85_3 - V85_4| = 15.95$ => FAIR DESIGN

ELEM. : 4	STATION		CLOTHOIDS				SUPER-
RADIUS	FROM	TO	BEFORE	BEHIND	CCR_S	V85	ELEVATION
400	2373.79	3195.87	250.00	- 250.00	128.08	83.75	4.0

CRIT.   I : $|V85_4 - V_d| = 6.25$ => GOOD DESIGN
CRIT. III : $f_{RA} - f_{RD} = +0.05$ => GOOD DESIGN

Transition 4-5 for Crit. II: $|V85_4 - V85_5| = 7.66$ => GOOD DESIGN

ELEM. : 5	STATION		CLOTHOIDS				SUPER-
RADIUS	FROM	TO	BEFORE	BEHIND	CCR_S	V85	ELEVATION
-750	3195.87	3586.17	300.00	- 300.00	58.82	91.41	4.0

CRIT.   I : $|V85_5 - V_d| = 1.41$ => GOOD DESIGN
CRIT. III : $f_{RA} - f_{RD} = +0.10$ => GOOD DESIGN

Transition 5-6 for Crit. II: $|V85_5 - V85_6| = 1.25$ => GOOD DESIGN

ELEM. : 6	STATION		CLOTHOIDS				SUPER-
RADIUS	FROM	TO	BEFORE	BEHIND	CCR_S	V85	ELEVATION
750	3586.17	3906.89	300.00	0.00	69.04	90.16	4.0

CRIT.   I : $|V85_6 - V_d| = 0.16$ => GOOD DESIGN
CRIT. III : $f_{RA} - f_{RD} = +0.10$ => GOOD DESIGN

CHAPTER 13
VERTICAL ALIGNMENT

13.1 RECOMMENDATIONS FOR PRACTICAL DESIGN TASKS

According to AASHTO,[5] the topography of the land traversed has an influence on the alignment of roads and streets. Although topography affects horizontal alignment, its effect is more evident on vertical alignment. To characterize variations, engineers generally separate topography into three classifications based on the terrain (see also Sec. 14.1.2.2).

Flat (level) terrain is that condition where highway sight distances, as governed by both horizontal and vertical restrictions, are generally long or could be made to be so without construction difficulty or major expense.

Hilly (rolling) terrain is that condition where the natural slopes consistently rise above and fall below the road or street grade and where occasional steep slopes offer some restriction to normal horizontal and vertical roadway alignment.

Mountainous terrain is that condition where longitudinal and traverse changes in the elevation of the ground with respect to the road or street are abrupt and where benching and side hill excavations are frequently required to obtain acceptable horizontal and vertical alignment.

Terrain classification is the general character of a specific route corridor. Routes in valleys or passes or mountainous areas that have all the characteristics of roads or streets traversing flat or hilly topography should be classified as flat or hilly. In general, hilly terrain has steeper grades, causing trucks to reduce speed below that of passenger cars, and mountainous terrain aggravates the situation, resulting in some trucks operating at crawl speeds.[5]

The design elements of the vertical alignment are tangent grades and vertical curves.

13.1.1 Grades

It is seldom practical to construct roads with grades sufficiently flat so as to permit all vehicles to operate at the same speed. Therefore, it is necessary to adopt a more modest design standard, in which limits are placed on steepness and length of grade which may be used.

13.1.1.1 Maximum Grades. For safety reasons, as well as with respect to operating costs, energy savings, pollution control, and quality of traffic flow, grades shall be designed as low as possible. On the other hand, in order to protect the landscape and municipalities, as well as to reduce construction costs, grades shall be adapted as close as possible to the existing topography.[246]

Maximum grades in relation to design speed and topography, which are recommended for roads of category group A as well as for roads of categories B I and B II, are given in Table 13.1. For safety reasons, the regular values should be used for major roads (high percentage of trucks), whereas the exceptional values could be used for minor, low-traffic roads.

Accident research shows ("Grade" in Sec. 9.2.1.3), that: (1) grades of less than 6 percent have relatively little effect on the accident rate and (2) a sharp increase in the accident rate was noted on grades of greater than 6 percent, at least for two-lane rural roads (see Figs. 9.30 and 13.14 and 13.15).

TABLE 13.1   Maximum Grades, as a Percentage

Design speed, V_d km/h	Category group A (rural roads) topography			Road categories B I, B II (suburban roads)
	Flat	Hilly	Mountainous	All
50	—	—	—	8 (12)
60	6 (8)	7 (9)	9 (10)	7 (10)
70	5 (7)	6 (8)	8 (9)	6 (8)
80	4 (6)	5 (7)	7 (9)	5 (7)
90	4 (5)	5 (6)	7 (8)	—
100	3 (5)	4 (6)	6 (8)	—
120	3 (5)	4 (6)	—	—
(130)	3 (4)	—	—	—

( ) = exceptional values.

On the basis of these latter findings, Table 13.1 was developed. This table shows that grades are permitted to be higher than 6 percent only in rare cases, such as low design speeds and mountainous topography. Table 13.1 also shows, especially for high-speed roads (flat and hilly topography), that the maximum grades for roads of category group A are considerably lower than 6 percent.

In the area of at-grade intersections, grades over 4 percent must be avoided, because of design (connection with minor roads) and traffic-related (limited stopping sight distance) reasons.[243]

In tunnel sections, grades must be limited to G_{max} = 4 percent for roads of category group A. For long roadway sections, a G_{max} = 2.5 percent is desirable. The disadvantages of higher grades in tunnel sections are as follows:[243]

- Higher pollution
- Higher potential of accidents
- Faster spreading of flammable liquids
- Reduction of truck speeds

Critical lengths of upgrades is a term used to indicate the maximum length of a designated upgrade on which a loaded truck can operate without an unreasonable reduction in speed below 50 km/h. The critical lengths of upgrades shown in Table 13.2 are based on a design truck with a power/weight ratio of 10 hp/ton and an acceptable speed reduction of 15 km/h, as derived from new speed measurements by Schulze[638] for trucks on upgrades, assuming an approach speed of 65 km/h.

These critical upgrade lengths have to be considered a first indicator for examining if either the alignment or the cross section should be changed, for example, by adding an auxiliary lane in the upgrade direction (climbing lane) (see also Secs. 13.1.2 and 13.2.2).

TABLE 13.2   Critical Lengths of Upgrades[638]

Upgrade, %	4	5	6	7	8
Length, m	750	350	260	240	220

13.1.1.2   Minimum Grades in Distortion Sections.* Generally, the minimum longitudinal grade of the road has to be at the least as large as the relative grade of the superelevation runoff, Δs. Thus, in order to provide sufficient drainage of the road surface, the following relation [Eq. (13.1)] applies:

$$G - \Delta s \geq 0.5\% \tag{13.1a}$$

$$\geq 0.2\% \quad \text{(in exceptional cases)} \tag{13.1b}$$

where G = longitudinal grade of the road, %

Δs = relative grade of the superelevation runoff, % (see Sec. 14.1.3.2)

For curbed roadway sections, Eq. (13.1a), which always ensures sufficient drainage in gutter areas, applies.

The assumptions of Eq. (13.1) guarantee that the relative grade of the superelevation runoff will never exceed the minimum longitudinal grade of the road.

For the most critical case, which means a distortion section between opposite superelevation rates, minimum longitudinal grades of $G_{min} \geq 0.7$ percent (or higher $G_{min} \geq 1$ percent) are recommended to solve drainage problems.

13.1.2   Auxiliary Lanes on Upgrade Sections of Two-Lane Rural Roads†

13.1.2.1   General Considerations. According to AASHTO,[5] freedom and safety of operation on two-lane highways are adversely affected by heavily loaded vehicles operating on grades of sufficient length, resulting in speeds that could impede vehicles that are following. In the past, provision of extra auxiliary lanes in upgrade directions to improve safe operation has been rather limited because of extra construction costs. However, because of the increasing number of serious accidents occurring on grades, these lanes are more commonly included in original construction plans and additional lanes on existing highways built as safety improvement projects. The accident potential created by this condition has been dramatically illustrated in Fig. 13.16.[5]

Auxiliary lanes on upgrade sections of two-lane highways should reduce or eliminate the impairment of the traffic flow caused by heavy truck traffic on longitudinal grades. The impairment consists of passing needs behind the heavy truck traffic, especially in the upgrade direction but also in the downgrade direction due to limited passing sight distances, such as in the case of highly curved roadway sections. Auxiliary lanes have the effect of separating the slower and faster traffic. They also create passing possibilities without the corresponding passing sight distances. In general, alignments with auxiliary lanes adapt well to the topography. Interference with the landscape is low because deep cuts and high embankments on an otherwise stretched, smooth alignment can be avoided.

In conclusion, the design of auxiliary lanes on upgrade sections makes it possible to pass slow-moving vehicles and, thus, contributes to smooth traffic flow and to increased traffic safety.[443,664,715]

Examination of existing guidelines for many countries, as well as international research investigations, have revealed that the design of auxiliary lanes in the upgrade direction is normally based on a speed evaluation background for a defined (heavy) design vehicle. Even though the underlying assumptions differ considerably from country to country, the basic procedure used by the different countries is, however, comparable. The procedure makes use of a truck speed evaluation background (see the examples in Figs. 13.2, 13.6, and 13.18). Based on the experience, investigations, and research results presented in Subchapter 13.2, this design practice, in relation

---

*Elaborated based on Refs. 124, 243, and 453.

†Elaborated with the support of C. Schulze based on Refs. 638 and 639.

to a truck speed evaluation background, is logical and, therefore, will be applied in this chapter for developing an international procedure for a sound design of auxiliary lanes in the upgrade direction. In order to accomplish this, it was important first to select a reliable design truck, and second to establish an appropriate speed evaluation background that describes best the actual truck speed measurements, which were conducted for this purpose by the Institute for Highway and Railroad Engineering, University of Karlsruhe, Germany.[459]

13.1.2.2   Consequences of Existing Guidelines.   In this section, the dimensions for the design of auxiliary lanes in the upgrade direction on two-lane rural roads will be assessed on the basis of the in-depth reviews of several guidelines, which were presented in Sec. 13.2.2.2, in addition to the investigations and resulting experiences of Schulze.[638]

As a result, the dimensions for the design of upgrade auxiliary lanes (two-lane rural roads) follow:

Minimum length:	500 m (according to the findings and discussions in Table 13.13).
Minimum interim distance:	800 m. The *interim distance* is the length between two succeeding auxiliary lanes. If this distance is shorter than the minimum distance listed previously, then an uninterrupted auxiliary lane must be provided in the upgrade direction; consult also the findings and discussions relevant to Table 13.13.

Tapered Sections in Front of and Behind Auxiliary Lanes.   Tapered sections are designed with or without hatchings in different countries. On the basis of British, French, and German experiences, it was decided not to use hatchings at the beginning of an auxiliary lane, but hatchings at the end of it. The corresponding design is shown in Fig. 13.1.

The design in Fig. 13.1 corresponds to the French recommendations. It was selected on purpose because France is well experienced with auxiliary lanes and because of safety considerations

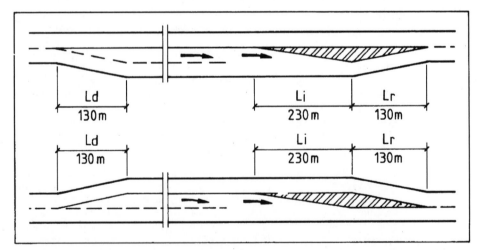

Legend: See also Figure 25-17 of Part 3: Cross Sections (CS)

Ld: diverge taper length (lane addition) [m],
Li: merge taper length (lane drop) [m],
Lr: lateral displacement taper length [m].

FIGURE 13.1   Design of auxiliary lanes on two-lane rural roads. (Elaborated based on Ref. 700.)

with respect to deceleration maneuvers. For instance, the merge taper length of the auxiliary lane, *Li,* is the longest of the guidelines investigated. The selected dimensions are:

Diverge taper length (*lane addition*): 130 m.

Merge taper length (*lane drop*): 230 m (160 m), depending on *V*85 according to Fig. 25.17 of "CS" in Part 3.

Lateral displacement taper length: 130 m (refer to the findings and discussions of Table 13.14 and Fig. 13.1).

With respect to the lane width of auxiliary lanes, Table 13.15 shows that most guidelines use a lane width which corresponds to the standard cross section of the country under study. Because this reflects a sound solution, it will be handled here in the same manner, see Fig. 25.1, "Standard Cross Sections," and Fig. 25.3, "Intermediate Cross Sections," in Part 3, "Cross Sections (CS)."

13.1.2.3   Design Truck.   Based on the investigations presented in Sec. 13.2.2.3, it was determined that a semitrailer or full trailer combination, with a power/weight ratio of 10 hp/ton, represents the best motor distribution available today. Thus, it was considered here as a heavy design vehicle.

This design truck (10 hp/ton) was also selected for determining the truck speeds on upgrades based on the Swiss speed evaluation background (see Fig. 13.2). As can be seen from this figure, truck speeds normally are expressed as *15th-percentile speeds*. The other design trucks, which were discussed in "Consequences of Existing Guidelines" in Sec. 13.2.2.2, had a lower power/weight ratio (see also Table 13.16).

13.1.2.4   Actual Speed Measurements and Comparisons with Existing Speed Evaluation Backgrounds for Trucks.   As discussed in detail in Sec. 13.2.2.4, eight test sections with auxiliary lanes in the upgrade direction were selected (see Table 13.19) in order to describe the driving behavior of trucks and passenger cars, as based on 24 local speed measurements. The main results of the speed measurements are summarized in the following so that the highway engineer knows what to expect when designing an auxiliary lane in the upgrade direction of two-lane rural roads. (The results are based on the investigations of Schulze[638] for Germany.)

Longitudinal Grades Superimposed by Horizontal Alignments with CCR$_s$ Values of <700 gon/km (*see, for example, Fig. 13.19*).   The following statements are valid for two-lane rural roads with an auxiliary lane in the upgrade direction:

• The difference between the 15th-percentile speed of trucks and the 85th-percentile speed of passenger cars ranges from 40 to 50 km/h in plain topography.

• The difference between the 15th-percentile speed of trucks and the 85th-percentile speed of passenger cars increases on upgrade sections, in relation to the longitudinal grade and the length of the upgrade section, and can reach about 80 km/h.

• The speed behavior of passenger cars on upgrades of up to 6.5 percent is independent of the longitudinal grade and the length of the upgrade section.

• The 85th-percentile speed level of passenger cars in the downgrade direction is considerably lower (about 20 km/h) than that in the upgrade direction for high grades.

• The 15th-percentile speed level of trucks in the downgrade direction falls between 50 and 60 km/h, and is independent of the longitudinal grade of between 4 and 8 percent and of the length of the downgrade section.

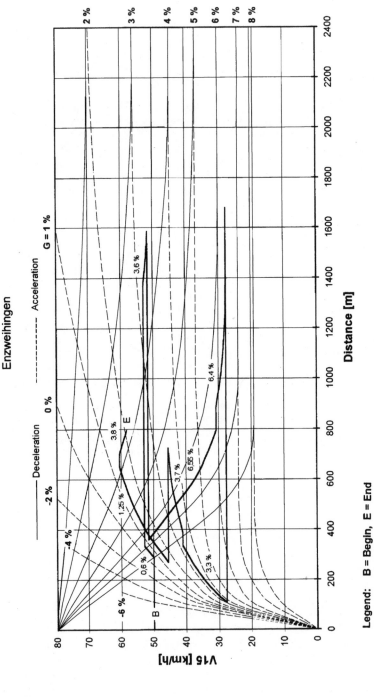

FIGURE 13.2 Truck speed evaluation background of Switzerland (design truck = 10 hp/ton) in addition to the truck speed profile with respect to the vertical alignment of Fig. 13.3 (test section SR 10 Enzweihingen). (Elaborated based on Ref. 693, SN 640, 138a.)

Longitudinal Grades Superimposed by Horizontal Alignments with CCR$_S$ Values of >700 gon/km (see, for example, Fig. 13.20)

- The difference between the 15th-percentile speed of trucks and the 85th-percentile speed of passenger cars ranges from 20 to 25 km/h. This difference is significantly lower than that in plain topography or on upgrades with CCR$_S$ values, <700 gon/km.
- Generally, the curvature change rate of the single curve influences the driving behavior of passenger cars (see Fig. 8.12) and the driving behavior of trucks especially for CCR$_S$ of more than 700 gon/km.
- With increasing CCR$_S$ (for example, for narrow curves), the speed distributions of passenger cars in the upgrade and downgrade directions are nearly comparable. The same can be said about the speed distributions of trucks.

In conclusion, the detailed comparisons and analyses made in Sec. 13.2.2.4 have indicated that high curvature change rates reduce truck speeds considerably on longitudinal grades.

Based on these experiences, it should be noted that the studied speed evaluation backgrounds for the trucks of today, as presented in "Consequences of Existing Guidelines" in Sec. 13.2.2.2 (see, for example, Fig. 13.18), depend on the longitudinal grade and do not consider the influence of the important design parameter curvature change rate of the single curve when designing auxiliary lanes in the upgrade direction. Therefore, according to Schulze,[638] the existing backgrounds allow limited applications on roadway sections with high curvature change rates of the single curve. Thus, in order to develop a new international procedure for designing auxiliary lanes in the upgrade direction, it is important to consider the design parameter longitudinal grade with respect to the vertical alignment and the design parameter curvature change rate of the single curve with respect to the horizontal alignment, especially when CCR$_S$ values exceed 700 gon/km on two-lane rural roads.

The next important step was to determine which of the existing truck speed evaluation backgrounds would best fit the truck speed measurements conducted on the eight upgrade auxiliary lanes in the German Black Forest area (see Table 13.19). Consequently, in "Consequences of Existing Guidelines" in Sec. 13.2.2.2 a comparison is made between the measured 15th-percentile truck speeds with those resulting from the seven existing truck speed evaluation backgrounds of several countries. It will be shown that the Swiss background best fits the driving behavior of trucks.

13.1.2.5 International Design Procedure with Case Studies.
The following procedure is related to two-lane rural roads.

As discussed in Sec. 13.1.2.3, a semitrailer/full trailer combination with a power/weight ratio of 10 hp/ton was selected as the design truck on which the Swiss speed evaluation background is also based (Fig. 13.2). Judging by the truck speed measurements discussed in Sec. 13.2.2.4, it could be shown additionally that the Swiss speed evaluation background, as compared to other backgrounds under study, agrees well with the actual driving behavior of trucks on upgrade sections.

Introduction.   In the following subsections, a new design procedure, which is partially based on the Swiss design practice for the design of auxiliary lanes in the upgrade direction, will be developed.[693] The procedure considers a comparison between the 15th-percentile speed behavior of trucks ($V15$) and one-half the 85th-percentile speed behavior of passenger cars ($V85$) under free-flow conditions, as described here:

$$V15 \text{ (TR)} \rightarrow 0.5 \; V85 \text{(PC)}$$

where $V15$ (TR) = 15th-percentile speed of trucks, km/h

$V85$ (PC) = 85th-percentile speed of passenger cars, km/h

According to the Swiss norm SN 640 138a,[693] auxiliary lanes should be provided when the 15th-percentile speed of trucks falls below one-half the 85th-percentile speed of passenger cars (0.5 $V85$) on a minimum section length of 500 m on two-lane rural roads. Thus, the auxiliary lane should begin at the point where the truck speed falls short of the 0.5 $V85$ value of passenger cars and should end at the point where the truck speed exceeds the latter limit again. If the distance between two succeeding auxiliary lanes is shorter than 800 m, then an uninterrupted auxiliary lane must be provided (compare the case studies in Figs. 13.3 and 13.7). The design of auxiliary lanes should always correspond to the graphical presentation shown in Fig. 13.1. In the downgrade direction, a no-passing zone should be indicated.

Upgrade Driving Behavior of Trucks with CCR$_S$ Values (<700 gon/km, Case Study I).   For this design case, the 15th-percentile speed behavior of trucks on upgrades with CCR$_S$ values of <700 gon/km is generally influenced by the grade size on the upgrade sections; the curvature change rate of the single curve does not have to play a role there. In order to determine the speed of the design truck (V15), the speed evaluation background of Switzerland, as presented in Fig. 13.2, was used. Figure 13.2 shows the construction of the truck speed profile on the SR 10 Enzweihingen test section (Table 13.19, no. 2) in relation to the vertical alignment shown in Fig. 13.3. (The determination of truck speeds on grades works similar to that presented in Fig. 13.18 for Australia. The only difference is that the deceleration and acceleration curves in the Swiss evaluation background are superimposed.)

The truck speed profile established in Fig. 13.2 is presented in Fig. 13.3 as V15 (TR). The truck speed profile is mainly influenced by the longitudinal grade because the truck speeds shown in Fig. 13.3 always fall short of the speed relations presented in Fig. 13.5, in conjunction with the CCR$_S$ values of the horizontal alignment. For a better understanding, the following example is provided.

Figure 13.5 describes the relationship between the 15th-percentile speeds of trucks and the curvature change rate of the single curve. As an example, with respect to case study I, the highest truck speed from Fig. 13.3 is about 60 km/h at station 0 + 400 with a CCR$_S$ value of 82 gon/km (see dotted line with arrow). The intersection point of these two values, as shown in Fig. 13.5, lies below the developed relationship between V15 (TR) and CCR$_S$. (Similar conclusions result from combining the truck speeds with the other CCR$_S$ values of the horizontal alignment shown in Fig. 13.3.)

Thus, it can be concluded that the curvature change rate of the single curve is not relevant for the truck speed profile of case study I.

Upgrade Driving Behavior of Passenger Cars.   According to "Longitudinal Grade" in Sec. 8.2.3.2 as well as in Secs. 8.2.4.2 and 18.5.1, it was shown that the operating speeds, V85, of free-moving passenger cars were not affected by longitudinal grades of up to ±6 percent, but they were significantly affected by the horizontal alignment (design parameter CCR$_S$), as revealed by the operating speed backgrounds of several countries shown in Fig. 8.12 and by the corresponding regression models given in Table 8.5.

Furthermore, the speed measurements of passenger cars conducted by Schulze[638] on upgrade sections superimposed with CCR$_S$ values of more than 700 gon/km were best represented by the regression model of the German operating speed background (ISE) given by Eq. (8.15) in Table 8.5. In this connection, it was found that in tight curves (R < 90 m; CCR$_S$ > 700 gon/km) the 85th-percentile speeds of passenger cars in the studied countries strongly adapt to each other, since aspects of traffic laws, engine performances, and driving dynamics become less important, whereas driving geometric influences more often determine the speed behavior of passenger cars. Therefore, the regression model of Eq. (8.15) in Table 8.5 was used as a basis for describing the 85th-percentile speed of passenger cars on upgrade sections (G ≤ 6 percent) with respect to CCR$_S$. This relationship is schematically shown in Fig. 13.4 for CCR$_S$ values of up to 1500 gon/km (R ≈ 40 m). Because the highest operating speeds of passenger cars of about 100 km/h (except in Germany) can be reached, according to Fig. 8.12, on independent tangents (CCR$_S$ = 0 gon/km), the operating speeds in Fig. 13.4, therefore, were limited to this value for CCR$_S$ values between 0 and 220 gon/km.

With reference to the horizontal alignment of case study I (CCR$_S$ = 82 to 97 gon/km) in Fig. 13.3, it follows that 85th-percentile speeds V85(PC) of 100 km/h, according to Fig. 13.4, can be expected. However, because of lower existing local speed limits, like 50, 60, and 80 km/h, these become relevant for the actual passenger car speed profile in Fig. 13.3.

According to the previously described procedure, an auxiliary lane should begin at the point where V15(TR) falls short of the 0.5 V85(PC) level and should end at the point where the truck speed exceeds this limit again. For case study I in Fig. 13.3, the first intersection point coincides with station 2 + 340 m, whereas the second coincides with station 1 + 160 m. Consequently, an

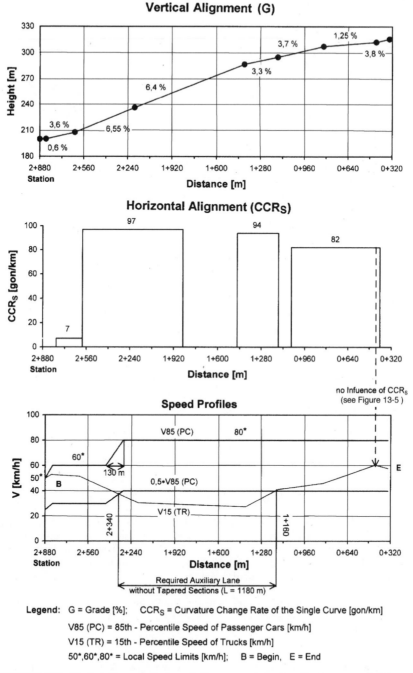

FIGURE 13.3   Vertical alignment, horizontal alignment, and speed profiles for case study I (test section SR 10 Enzweihingen, Table 13.19).

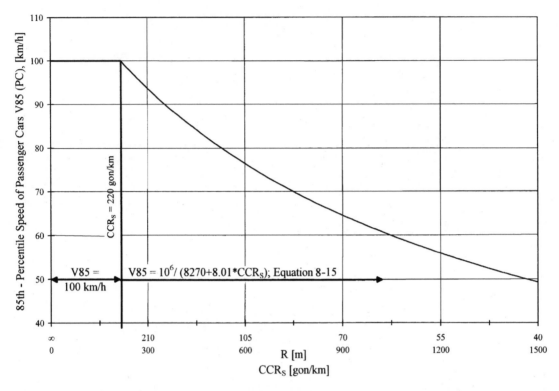

FIGURE 13.4   85th-percentile speeds of passenger cars on upgrade sections ($G \leq 6$ percent) with respect to the curvature change rate of the single curve for two-lane rural roads.

auxiliary lane of 1180 m must be introduced between these two stations in the upgrade direction because the minimum design length of 500 m was exceeded. In addition, the lengths of the tapered sections at the beginning and at the end of the auxiliary lane should correspond to Fig. 13.1. It follows then that the overall length of the auxiliary lane, including tapered sections, is

$$130 \text{ m} + 1180 \text{ m} + 230 \text{ m} + 130 \text{ m} = 1670 \text{ m}$$

Besides the influence of CCR_S, consideration must also be given to the fact that the 85th-percentile speeds of passenger cars are additionally influenced by steep longitudinal grades. In this respect, research investigations by Schulze[638] and other international design experiences[693,700] yielded the results presented in Table 13.3, in order to estimate the 85th-percentile speed of passenger cars on steep upgrade sections. With respect to case study I (Fig. 13.3), upgrades of $G = 6.5$ percent exist, which correspond, according to Table 13.3, to $V85(PC)$ of 95 km/h. Because the existing local speed limit of 80 km/h is lower than 95 km/h, it then follows that the speed reduction as a result of steep upgrades does not have an influence on the speed profile of passenger cars for case study I.

TABLE 13.3   85th-Percentile Speeds of Passenger Cars on Steep Straight Upgrade Sections ($G > 6$ percent)

G, %	≤6	6.5	7	8	9	10
$V85(PC)$, km/h	100	95	90	85	80	70

A comparison of Fig. 13.4 and Table 13.3 reveals that on steep upgrades superimposed with high CCR_S values of more than 700 gon/km, the operating speeds of passenger cars are influenced more by the horizontal alignment than by the vertical alignment. Thus Fig. 13.4 also becomes valid in those design cases for $G > 6$ percent.

The *proof* is as follows:

Horizontal alignment:

$$CCR_S \geq 700 \text{ gon/km} \rightarrow V85(PC) \leq 70 \text{ km/h; Fig. 13.4 (approximately).}$$

Vertical alignment:

$$G \leq 10\% \rightarrow V85(PC) \geq 70 \text{ km/h; Table 13.3.}$$

Furthermore, Table 13.4 was developed in order to determine the transition lengths needed by passenger cars to perform speed changes between successive design elements.[691] For acceleration or deceleration maneuvers, the following formula was developed on the basis of the car-following techniques[411,452] presented in "Evaluation of Tangents in the Design Process" in Sec. 12.1.1.3:

$$TL = \frac{V85_1{}^2 - V85_2{}^2}{2 \times 3.6^2 \times a} \tag{12.1}$$

$$TL = \frac{V85_1{}^2 - V85_2{}^2}{22.03} \tag{12.1a}$$

where $V85_{1,2}$ = 85th-percentile speeds of passenger cars on successive design elements (tangent-to-curve or curve-to-curve), km/h
   TL = transition length between two successive design elements, m
   $a = 0.85$ = acceleration/deceleration rate, m/s^2

Table 13.4 corresponds to a large extent to Table 12.1 which is used to assess the lengths of nonindependent tangents as well as independent tangents; it was extended here in order to include the transition lengths that are shorter than the lengths of nonindependent tangents (see Sec. 12.1.1.3).

With respect to case study I in Fig. 13.3, the transition length needed to accelerate from the local speed limits of 60 to 80 km/h is $TL = 130$ m (see Table 13.4). This distance is shown on the $V85(PC)$ speed profile.

TABLE 13.4 Relationship between Transition Lengths and 85th-Percentile Speed Changes between Successive Design Elements

$V85_1, V85_2,$ km/h	$V85_1, V85_2$, km/h									
	50	55	60	65	70	75	80	85	90	95
	Transition lengths TL, m									
55	25									
60	50	25								
65	80	55	30							
70	110	85	60	30						
75	145	120	95	65	35					
80	180	155	130	100	70	35				
85	215	195	165	140	105	75	40			
90	255	230	205	175	145	115	80	40		
95	300	275	250	220	190	155	120	85	45	
100	345	320	295	265	235	200	165	125	90	45

Case study I was selected on purpose in order to show, based on local constraints, that speed limits often govern the driving behavior of passenger cars on upgrade sections. In these cases, the 85th-percentile speeds of free-flowing passenger cars, based on Fig. 13.4 for the horizontal alignment and on Table 13.3 for the vertical alignment, do not become relevant. Therefore, for case study II, a roadway section without any local speed limits and with low to high CCR_S values was selected in order to describe the influence of the horizontal alignment superimposed with high grades on the driving behavior of upgrade sections.

Upgrade Driving Behavior of Trucks with CCR_S Values (>700 gon/km, Case Study II). For CCR_S values of more than 700 gon/km, the curvature change rate of the single curve can influence, to a large extent, the 15th-percentile speed of trucks. Figure 13.5, which is based on truck speed measurements,[638] was developed for pairs of data between $CCR_S = 780$ and 1480 gon/km (solid line), in order to show the relationship between $V15(TR)$ and CCR_S. The dotted line in Fig. 13.5 was established in the following way, since no statistically sound database was available in this CCR_S range. Beginning with $CCR_S = 780$ gon/km, the corresponding (statistically sound) value of $V15(TR) = 36$ km/h was connected linearly with the speed for trucks in tangent sections ($CCR_S = 0$ gon/km), which corresponds to a 15th-percentile speed of $V15(TR) \approx 65$ km/h for trucks for grades of up to 2 percent, based on the investigations of Schulze.[638]

In this way, a relationship between the driving behavior of trucks, as expressed by the 15th-percentile speeds, and the curvature change rate of the single curve was established (Fig. 13.5). It should once again be noted that, according to the investigations made in the previous chapters,

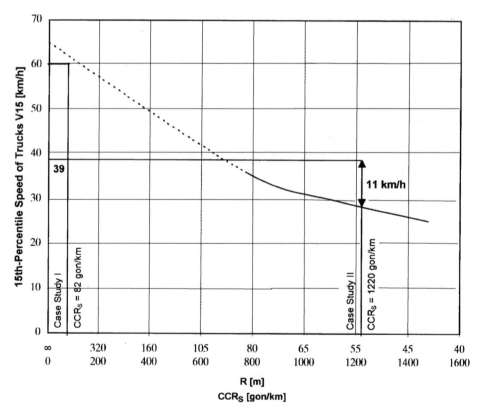

FIGURE 13.5 15th-percentile speed of trucks with respect to the curvature change rate of the single curve for two-lane rural roads.

that the new design parameter CCR_S is the best indicator for describing the influence of road characteristics on driving behavior ("New Design Parameter: Curvature Change Rate of the Single Circular Curve with Transition Curves" in Sec. 8.2.3.6). For trucks, Schulze[638] succeeded, for the first time, in establishing a relationship between $V15(TR)$ and CCR_S (Fig. 13.5).

Nevertheless, it should be pointed out that Fig. 13.5 only reveals *approximate* values because of limited truck speed measurements. Thus, further research is needed in this field.

The relationships shown in Fig. 13.4 for $V85(PC)$ versus CCR_S, in Table 13.3 for $V85(PC)$ versus grade, and in Fig. 13.5 for $V15(TR)$ versus CCR_S will be further explained when discussing case study II.

In order to determine the speed of the design truck, the speed evaluation background of Switzerland as presented in Fig. 13.2, is once again used for case study II. Figure 13.6 shows the construction of the truck speed profile for the test section SR 314 Epfenhofen (Table 13.19, no. 6) in relation to the vertical alignment of Fig. 13.7.

The truck speed profile established in Fig. 13.6 is also shown in Fig. 13.7 as $V15(TR)$. The truck speed profile depends on the longitudinal grade (for speed profile construction, see Fig. 13.6) and on the design parameter, CCR_S. The influence of CCR_S can be taken from the newly developed relationship between $V15(TR)$ and CCR_S shown in Fig. 13.5 in the following manner. The highway engineer simply compares the values of the truck speed profile and the corresponding superimposed values of CCR_S as shown, for example, in Fig. 13.7. If the intersection point of these two values should fall below the curve in Fig. 13.5, then the highway engineer can conclude that the curvature change rate of the single curve has no influence on the truck driving behavior in the upgrade direction. On the other hand, if the intersection point should fall above the curve in Fig. 13.5, then, besides the upgrade, CCR_S also has an additional influence. For a better understanding, the following example is provided.

According to Fig. 13.7, the truck speed, as represented by $V15(TR)$, which is superimposed by a CCR_S value of 1220 gon/km at station 7 + 560 (see dotted line with arrow) is 39 km/h. Plugging these two values into Fig. 13.5 would result in an intersection point which clearly lies above the curve. This means that the curvature change rate of the single curve is responsible for an additional speed reduction of 11 km/h at that curved section. The influence of this speed reduction is clearly indicated in Figs. 13.6 and 13.7. A comparison of the other data pairs of the relationship between $V15(TR)$ and CCR_S in Fig. 13.7 with Fig. 13.5 shows that CCR_S results in no or in marginal speed reductions of 1 to 2 km/h.

The construction of the speed profile for passenger cars in Fig. 13.7 depends on Fig. 13.4 for the horizontal and vertical alignment (≤ 6 percent) and on Table 13.3 for the vertical alignment (≥ 6 percent). On the first part of the test section, relatively low CCR_S values are present, resulting in $V85(PC)$ of 100 km/h (for example, $CCR_S = 225$ gon/km $\rightarrow V85 = 100$ km/h from Fig. 13.4). However, because the vertical alignment in that roadway section lies between 6.8 and 6.6 percent, a decision was made to approximate $V85(PC)$ as 95 km/h based on Table 13.3. The transition length needed to accelerate from the approach speed of 70 to 95 km/h is 190 m (see Table 13.4).

On the second part of the same test section which begins at station 6 + 900, the relationship between $V85(PC)$ and CCR_S (Fig. 13.4) plays the more important role in establishing the passenger car speed profile. For example:

$$CCR_S = 833 \text{ gon/km} \rightarrow V85(PC) = 67 \text{ km/h [Fig. 13.4, Eq. (8.15)]}$$

$$CCR_S = 1255 \text{ gon/km} \rightarrow V85(PC) = 55 \text{ km/h}$$

$$CCR_S = 721 \text{ gon/km} \rightarrow V85(PC) = 71 \text{ km/h}$$

The speed profile for passenger cars was thus constructed by additionally considering the transition lengths needed for acceleration and deceleration maneuvers between successive design elements (see Table 13.4).

According to the procedure discussed earlier, an auxiliary lane in the upgrade direction should begin at the point where $V15(TR)$ falls below the 0.5 $V85(PC)$ level and should end at the point

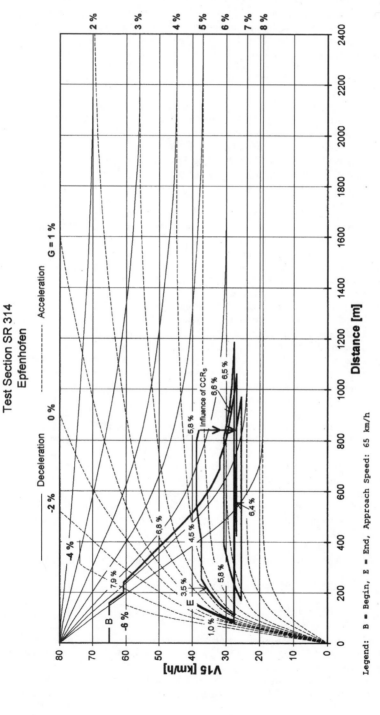

FIGURE 13.6 Truck speed evaluation background of Switzerland (design truck = 10 hp/ton) in addition to the truck speed profile with respect to the vertical and horizontal alignment of Fig. 13.7 (test section SR 314 Epfenhofen). (Elaborated based on Ref. 693, SN 640, 138a.)

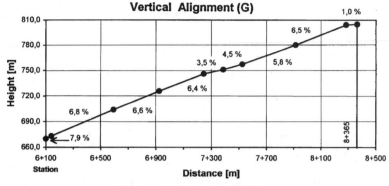

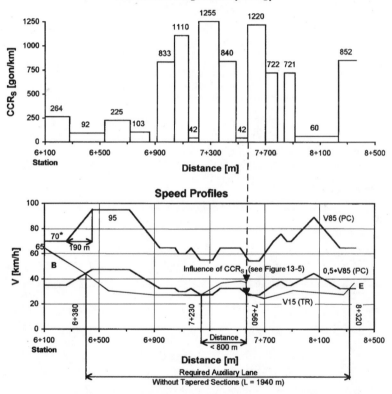

FIGURE 13.7 Vertical alignment, horizontal alignment, and speed profiles for case study II (test section SR 314 Epfenhofen, Table 13.19).

where the truck speed exceeds this limit again. For case study II, shown in Fig. 13.7, the first inter-section point coincides with station 6 + 380 m, whereas the second coincides with station 8 + 320 m. Between these two stations, however, a section exists (station 7 + 230 to 7 + 560) where an auxiliary lane is not needed. Because this section length of 330 m is less than 800 m according to the design rules of Sec. 13.1.2.2, then an uninterrupted auxiliary lane in the upgrade direction with a length of 1940 m must be provided. In addition, the lengths of the tapered sections at the beginning and at the end of the auxiliary lane must correspond to Fig. 13.1, which means the over-all length of the auxiliary lane is

$$130 \text{ m} + 1940 \text{ m} + 230 \text{ m} + 130 \text{ m} = 2430 \text{ m}$$

Supplementary Requirements.   In addition to the design and speed-dependent decision crite-ria discussed so far, traffic-related aspects are also important for the design of auxiliary lanes on upgrade sections. In this respect, very complex and difficult procedures have been proposed by some countries, for example, in Refs. 5 and 183, in order to handle this issue. The Swiss norm SN 640 138a,[693] however, provides an interesting approach to tackle this problem. This approach will be the focus of the present procedure.

Figure 13.8 depicts the relationships between upgrades, traffic volumes, truck percentages, and sight distances and is subdivided into three ranges—A to C. According to this figure, auxil-iary lanes in the upgrade direction are desirable when the intersection point of the hourly traffic volume and truck percentage falls into range A and for restricted sight distance into range B. A restricted sight distance refers to the fact that, for more than about 50 percent of the section length, the sight distances are below 600 m on upgrades of between 3 and 6 percent and 400 m for upgrades of more than 6 percent (including the sections with no passing zones).[693]

Multilane Rural Roads.   The procedure described so far with respect to designing auxiliary lanes in the upgrade direction is especially pertinent to two-lane rural roads. With respect to the driving behavior of trucks on multilane rural roads, it can be stated that the evaluation background of Switzerland with respect to the vertical alignment is still relevant and that the influence of the curvature change rate of the single curve can be ignored here because of the generous alignments of these roads, in general.

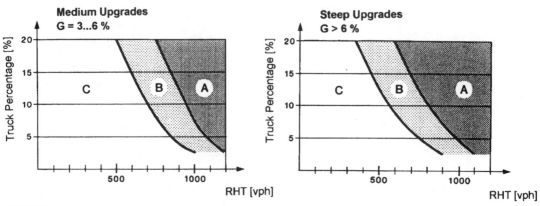

Legend:

RHT: relevant hourly traffic volume for both directions [veh. per hour],
A: auxiliary lane desirable,
B: auxiliary lane desirable, if the sight distances are restricted,
C: auxiliary lane is not necessary.

FIGURE 13.8   Supplementary requirements for assigning auxiliary lanes in the upgrade direction on two-lane rural roads (Ref. 693, SN 640, 138a).

With respect to the driving behavior of passenger cars, the general speed limits on multilane roads are regarded as $V85(PC)$ to comply with the values given in Table 13.3 for steep upgrades. Otherwise, the proposed design procedure for auxiliary lanes in the upgrade direction remains unchanged.

Based on the in-depth reviews of the design guidelines in Sec. 13.2.2.2, the authors propose the following dimensions for the design of auxiliary lanes in the upgrade direction on multilane rural roads:

Minimum length: 1000–1500 m

Minimum interim distance: 2000–2500 m

(on the basis of the findings and discussions for Table 13.13).

For the layout in Fig. 13.1, it is recommended that, besides an extra lane in each direction, the two driving directions be separated by a median and not just with a solid line.

Preliminary Conclusion. A new procedure for the design of auxiliary lanes in the upgrade direction, which is partially based on the Swiss design practice, that proved to be the best of all the studied guidelines was developed here. The Swiss procedure is based on a design truck with a power/weight ratio of 10 hp/ton. This design truck adapted well to the driving behavior on upgrade sections, as can be deduced from new truck speed measurements[459,638]; this design truck was also considered here.

The new design procedure entails the following important assumptions for adapting the driving behavior of trucks and passenger cars, in order to establish sound design lengths of auxiliary lanes as well as to pinpoint their exact beginnings and endings.

The main design assumptions are as follows:

- Modern truck speed evaluation background (Switzerland)
- Relationship between truck speeds and the curvature change rate of the single curve (Fig. 13.5)
- Relationship between passenger car speeds and the curvature change rate of the single curve on upgrade sections (Fig. 13.4)
- Relationship between passenger car speeds and steep upgrades (Table 13.3)

Sound design dimensions for auxiliary lanes in the upgrade direction were assessed based on in-depth reviews of numerous existing guidelines. (More detailed information can be found in Sec. 13.2.2.)

13.1.3 Safety Facilities on Steep Downgrade Sections

13.1.3.1 Emergency Escape Ramps.* A review of several highway geometric guidelines revealed that emergency escape ramps are mainly used in Australia and in the United States. However, it is recommended that these special facilities be provided in other countries in order to bring runaway vehicles to a safe, controlled stop on steep downgrades. The best discussion about emergency escape ramps is found in AASHTO, on which the following statements, regarding typical design cases for these facilities, are based.

Where long descending grades exist or where topographic and location could require such grades on new alignment, the design and construction of an emergency escape ramp at an appropriate location is desirable for the purpose of slowing and stopping an out-of-control vehicle away from the main traffic stream. Considerable experience with ramps constructed on existing highways in Australia and the United States has led to the design and installation of effective ramps that are saving lives and reducing property damage. Furthermore, the experience with existing ramps indicates that their operational characteristics are providing acceptable deceleration rates and affording good driver control of the vehicle on the ramp.

---

*Elaborated based on Ref. 5.

1. SANDPILE

2. DESCENDING GRADE

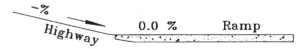

3. HORIZONTAL GRADE

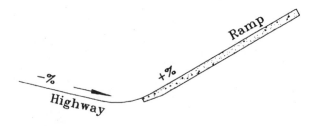

4. ASCENDING GRADE

(Note: Profile is along the baseline of the ramp)

FIGURE 13.9   Basic types of emergency escape ramps.[5]

There are four basic types of emergency escape ramps currently in use—the sandpile, descending grade, horizontal grade, and ascending grade. The four types are illustrated in Fig. 13.9:

1. Sandpiles composed of loose, dry sand are usually no more than 120 m in length. The influence of gravity is dependent on the grade of the sandpile. The increase in rolling resistance is supplied by the loose sand.

2. and 3. The horizontal grade and descending ramps can be rather lengthy because the gravitational effect does not help reduce the speed of the vehicle. For the horizontal grade ramp, the force of gravity is 0; for the descending grade ramp, the force of gravity acts in the direction of vehicle movement. The increase in rolling resistance is supplied by an arresting bed composed of loose aggregate.

4. The ascending grade ramp uses both the arresting bed and the effect of gravity, in general reducing the ramp length necessary to stop the vehicle. The loose material in the arresting bed increases the rolling resistance, as in the other types of ramps, while the force of gravity acts downgrade, opposite to the vehicle's movement. The loose bedding material also serves to hold the vehicle in place on the ramp grade after it has come to a safe stop.

TABLE 13.5   Rolling Resistance of Roadway Surface Materials[5]

Surfacing material	Rolling resistance, lb/1000 lb GVW	Equivalent grade, %*
Portland cement concrete	10	1.0
Asphalt concrete	12	1.2
Gravel, compacted	15	1.5
Earth, sandy, loose	37	3.7
Crushed aggregate, loose	50	5.0
Gravel, loose	100	10.0
Sand	150	15.0
Pea gravel	250	25.0

*Rolling resistance expressed as equivalent gradient.

Note: GVW = gross vehicle weight, lb, and lb (pounds) = 0.454 kg (kilograms).

Each one of these ramp types is applicable to a particular situation where an emergency escape ramp is desirable and must be compatible with established location and topographic control at possible sites.

One important influencing parameter for designing escape ramps is the rolling resistance. It is influenced by the type and displacement characteristics of the surface material of the roadway. Each surface material has a coefficient, expressed in lb/1000 lb of gross vehicle weight in Table 13.5, which determines the amount of rolling resistance of a vehicle.

The values shown in Table 13.5 for rolling resistance were obtained from various sources throughout the United States and the best available estimate.

To determine the distance required to bring the vehicle to a stop while considering the rolling resistance and gradient resistance, the following simplified equation may be used:

$$L = \frac{V^2}{252\,(r \pm G)}$$
(13.2)

where L = distance to stop, that is, length of the arrester bed, m
   V = entering velocity, km/h
   G = longitudinal grade divided by 100, %/100
   r = rolling resistance expressed as equivalent percent gradient divided by 100, %/100, according to Table 13.5

As an example, assume that topographic conditions at a site selected for an emergency escape ramp limit the ramp to an upgrade of 10 percent ($G = +0.10$). The arrester bed is to be constructed with loose gravel. According to Table 13.5, r is determined to be 0.10. The ramp is to be designed for an entering speed of 145 km/h. From Eq. (13.2), it follows that the length of the arresting bed should be 420 m.[5]

Another very interesting approach to reduce the accident risk and severity on steep downgrade sections through the use of radar-activated cameras is presented in the following section, and with respect to a long-term investigation on Elzer Mountain, as discussed in detail in Sec. 13.2.3.2.

13.1.3.2   Effective Speed Control by Automatic Radar Devices.*   Experiences in many countries have indicated that the introduction of speed limits often has only a short-term effect on reducing speeds, and consequently the number of accidents, unless police regularly enforce the speed limits. Posted speed limits alone will not guarantee compliance. Only when backed up by strict police enforcement can speed limits both reduce speed and alleviate accidents.

---

*Supported by J. Wacker, Ministry for Economy, Traffic, and Regional Development, State of Hessen, Germany.

A long-term investigation of the relation between driving behavior and accidents when speed limits are strictly enforced by the police with automatic radar devices was conducted at a steep downgrade section of the Autobahn (interstate) A 3 with three lanes in each direction between Cologne and Frankfurt in Germany (Fig. 13.21). The following comprehensive statements about this long-term investigation from 1970 to 1996 can be made based on the in-depth analyses and evaluations presented in Sec. 13.2.3.2.

General Information.   On steep downgrade sections with gradients of up to 5 percent (Fig. 13.22), the operating speeds on all three lanes were quite high, with great variations between them, especially between passenger cars and trucks. Consequently, there were numerous passing and lane-changing maneuvers with high, dangerous risks, caused mainly by trucks.

The "before" investigation conducted in 1971 indicated that the maximum design speed of 100 km/h for grades of up to 5 percent was exceeded by about 95 percent of passenger cars in the left lane and by about 80 percent of passenger cars in the middle lane. The general speed limit of 80 km/h for trucks in Germany was exceeded by 15 percent of the heavy vehicles (Fig. 13.23). The personal injury accident rates in 1971 for the downgrade direction at Elzer Mountain were about 4.5 times higher than the comparable values for the upgrade direction; the same was true in comparison to the whole German Autobahn network (Fig. 13.32). The high number of accidents and their severity were mainly caused by rear-end collisions of trucks and from the resulting congestion accidents.

In 1972, lane-related speed limits (100 km/h in the left and middle lanes and 40 km/h in the right lane) and additional **DO NOT PASS** signs for trucks were introduced. In 1973, automatic radar devices to control the speed limits were installed together with a psychologically striking warning sign at the beginning of Elzer Mountain (see Figs. 13.25 to 13.28).

Results and Consequences.   Since the permanent operation of the automatic radar devices in 1973 at the steep downgrade section of Elzer Mountain, the following speed and accident developments were observed for more than two decades in the downgrade direction:

- Operating speeds were significantly reduced. For passenger cars the 85th-percentile speeds were reduced by 45 km/h in the left lane (see Fig. 13.29) and 35 km/h in the middle lane (see Fig. 13.30). For trucks, the reductions were 35 km/h to 40 km/h (see Fig. 13.31). (It should be noted that Germany recommends speeds of 130 km for passenger cars on interstates (autobahnen). For trucks, general speed limits of 80 km/h exist.

- The variability of speeds decreased significantly (as measured by the standard deviation) and led to a uniform traffic flow in the left and middle lanes for passenger cars and in the right lane for trucks (see Figs. 13.29 to 13.31).

- Significant reductions in the number of accidents, fatalities, and injuries were observed since the mid-1970s (see Table 13.21).

- With respect to personal injuries, the accident rates were reduced by a ratio of about 9 to 1 between the early 1970s through the 1980s and 1990s (see Fig. 13.32).

- The average number of between seven and eight fatalities per year until 1972 was reduced to none or one fatality per year during the last decades (see Table 13.21).

- Since 1976, there were no significant differences between the accident rates on the downgrade and upgrade directions at Elzer Mountain and the average accident rate of the entire German autobahn network (Fig. 13.32).

In conclusion, it can be stated, with reference to steep downgrade sections on multilane median-separated motorways, that the two-pronged impact of reasonable lane-related speed limits and strict surveillance by police with automatic radar devices has reduced the number and severity of accidents to a level that can be considered today as normal. The investigation period of about 25 years appears to be long enough to verify that the improvement in the speed behavior and accident situation is permanent. A similarly high absolute or relative accident decrease could not be observed on any other multilane roadway section in Germany.

The most recent solution in this field combines the ideas of Sec. 13.1.3.1, "Emergency Escape Ramps," with those of Sec. 13.1.3.2, "Effective Speed Control by Automatic Radar Devices."

A steep, dangerous downgrade section of a multilane median-separated motorway (Autobahn A 7, Germany) was equipped with lane-related speed limits surveyed by automatic radar devices together with a psychologically striking warning sign and additional flashing speed limit signs for trucks (see Fig. 13.10). The downgrade section is about 5 km long with grades of up to 8 percent. As can be seen from Fig. 13.10, lane-related speed limits (100 km/h for passenger cars in the left and middle lanes and 60 km/h for trucks in the right lane) were installed on traffic sign bridges together with automatic radar devices. Furthermore, **DO NOT PASS** signs for trucks were added between the right and middle lanes.

The last traffic sign bridge was combined with an emergency escape ramp as shown in Fig. 13.11 for the purpose of slowing down and stopping an out of control vehicle away from the main traffic stream, such as in the case of a brake failure along the long steep downgrade section. The emergency escape ramp somewhat corresponds to type 3 in Fig. 13.9.

The operational and structural design, according to Figs. 13.10 and 13.11, represents so far the best solution for avoiding risky driving maneuvers and for improving the accident situation on long, steep downgrade sections of multilane median-separated motorways in terms of the interaction of trucks and passenger cars.

The influence of automatic radar devices on speed behavior and accidents is discussed in detail for a long-term investigation in Sec. 13.2.3.2.

Finally, it is recommended to improve the poor design level of safety criteria I and II through the introduction of stationary automatic radar devices, for example, at dangerous curve sites of two-lane rural roads, if redesigns cannot be achieved because of economic or environmental and/or political constraints (see Tables 9.1 and 9.2).

Warning Sign at the Beginning and in the
Course of the Downgrade Section

(a)

FIGURE 13.10 Equipment of the downgrade section of Autobahn A 7 in combination with lane-related speed limits, surveyed by automatic radar devices: (*a*) warning sign at the beginning and in the course of the downgrade section.

Traffic Sign Bridge with Lane-Related Speed Limits

(b)

Automatic Lane-Related Radar Devices

(c)

FIGURE 13.10 (*Continued*) Equipment of the downgrade section of Autobahn A 7 in combination with lane-related speed limits, surveyed by automatic radar devices: (*b*) traffic sign bridge with lane-related speed limits; and (*c*) automatic lane-related radar devices.

**Traffic Sign Bridge with Emergency Escape Ramp
Placed Parallel to the Main Traffic Stream**

(a)

**The Emergency Escape Ramp Begins to
Move away from the Main Traffic Stream**

(b)

FIGURE 13.11   Combination of lane-related speed limits, automatic radar devices, and emergency escape ramp: (*a*) traffic sign bridge with emergency escape ramp placed parallel to the main traffic stream; (*b*) emergency escape ramp begins to move away from main traffic stream.

**The Lateral Displacement from the
Main Traffic Stream Becomes Obvious**
(c)

**Arresting Bed Composed of Loose Aggregate; the
Brake Marks Indicate that the Emergency Escape
Ramp has already been Used Several Times**
(d)

FIGURE 13.11   (*Continued*)   Combination of lane-related speed limits, automatic radar devices, and emergency escape ramp: (*c*) lateral displacement from main traffic stream becomes obvious; and (*d*) arresting bed composed of loose aggregate; brake marks indicate the emergency escape ramp has already been used several times.

13.1.4 Vertical Curves

With respect to vertical curves, general considerations are best described in AASHTO.[5]

Vertical curves which affect gradual change between tangent grades may be any one of the crest or sag types depicted in Fig. 13.12. Vertical curves should be simple in application and should result in a design that is safe, comfortable in operation, pleasing in appearance, and adequate for drainage. The major control for safe operation on crest vertical curves is the provision of ample sight distances for the design speed. Minimum stopping sight distance should be provided in all cases. Wherever economically and physically feasible, opposing or passing sight distances should be provided on two-lane rural roads.

Consideration of motorists' comfort requires that the rate of change in grade be kept within tolerable limits. This consideration is most important in sag vertical curves where gravitational and vertical centrifugal forces act in the same direction. Appearance should also be considered. A long curve has a more pleasing appearance than a short one, which may give the appearance of a sudden break in the profile due to the effect of foreshortening.

Drainage of curbed pavements on sag vertical curves, type III (Fig. 13.12), requires a careful profile design to retain a gradeline of not less than 0.5 percent.[5]

The general geometry, as well as the different types of vertical curves relative to changes and variations of tangent grades, are presented in Fig. 13.12. For the general application of vertical curves the following rules are given in Ref. 243.

Crest and sag vertical curves are connected by tangents but they can also be joined together directly. For the connection of two crest vertical curves or two sag vertical curves by a short interim tangent, the statements of the three-dimensional alignment are valid for roads of category group A (see Chap. 16).

The crest and sag vertical curves should be selected in such a way that, together with the horizontal design elements, they

- Result in a balanced three-dimensional alignment
- Guarantee a maximum level of safety by applying favorable sight distances, as far as possible
- Preserve the landscape
- Adapt well to the topography
- Have low construction costs

These requirements are less important for roads of categories BI and BII. It is of major importance here to preserve municipal realities. If in the case of redesign and RRR projects of existing roads of categories BI and BII, the crest vertical curves are lower than the standard values and cannot be improved because of municipal constraints, then a speed limit combined or not combined with additional traffic warning devices should be considered for safety reasons (perhaps a speed limit for wet pavement conditions should be posted).[243]

13.1.4.1 Application.   In general, the radii of vertical curves selected should be as large as possible.

Vertical curves can be circular or parabolic. The latter is preferred in highway design because it provides a constant rate of change of curvature. It closely approximates to a circular curve and is simple to apply to the relatively flat grades of highway profiles where the assumption of a vertical axis introduces no practical error.

The form generally adopted for vertical curves is the circle or the simple parabola. The equation for the circular vertical curve is approximate (for the definition of symbols, see Fig. 13.12):

$$(y^1 + R_V)^2 = R_V^2 - x^2 \tag{13.3}$$

---

*Elaborated based on Refs. 139 and 663.

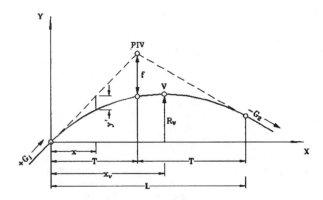

Legend:

PIV	=	point of intersection of tangent grades
G_1, G_2	=	tangent grades [%]
R_v	=	radius of vertical curve [m]
T	=	tangent length [m]
y'	=	tangent offset (vertical) [m]
x	=	horizontal length in plan [m]
f	=	tangent offset at PIV [m]
L	=	length of vertical curve [m]
V	=	vertex

General Geometry of Vertical Curves

	Convex: Crest Vertical Curve	Concave: Sag Vertical Curve
Changes of Tangent Grades $\dfrac{\lvert G_1 - G_2 \rvert}{100}$	Type I	Type III
Variations of Tangent Grades $\dfrac{\lvert G_1 - G_2 \rvert}{100}$	Type II	Type IV

Sign Rules:

    Upgrade: positive $(+G_1, +G_2)$
    Downgrade: negative $(-G_1, -G_2)$

FIGURE 13.12 Geometry and types of vertical curves. (Elaborated and modified based on Refs. 124 and 663.)

For small values of x relative to R_V, the circular vertical curve may be approximated by the parabola having the equation:

$$y^1 = \frac{x^2}{2\,R_V}$$ (13.4)

The tangent length of the vertical curve is calculated from

$$T = \frac{R_V}{2}\,\frac{|G_1 - G_2|}{100}$$ (13.5)

where $|G_1 - G_2|$ is the absolute value of the algebraic difference of tangent grades as a percentage (see Fig. 13.12).

The minimum values of radius of curvature, R_{Vmin}, for vertical curves are determined from the required sight distances. For crest vertical curves, the absolute minimum depends on the stopping sight distance (SSD), whereas the desirable minimum depends on the passing sight distance (PSD). For sag curves, the minimum is determined from the headlight sight distance and the comfort criterium.[663]

13.1.4.2   Crest Vertical Curves*

Sight Distance Control.   Figure 13.13 illustrates the effect of vertical curvature on visibility. There are two design cases to consider when determining the radius of crest vertical curves:

a. The sight distances (SSD or PSD) are shorter than the length of the vertical curve

b. SSD or PSD are longer than the length of the vertical curve

The basic formulas for the minimum radius of a crest vertical curve, with respect to the algebraic difference between sight distances and tangent grades, are as follows:
For situation a:

$$S \leq L \qquad R_{VCmin} = \frac{S^2}{2\left(\sqrt{h_1} + \sqrt{h_2}\right)^2}$$ (13.6)

and for situation b:

$$S > L \qquad R_{VCmin} = \frac{200\,S}{|G_1 - G_2|} - \frac{20{,}000\left(\sqrt{h_1} + \sqrt{h_2}\right)^2}{|G_1 - G_2|^2}$$ (13.7)

$$
\begin{aligned}
\text{where } S &= \text{sight distance (SSD or PSD), m} \\
\text{SSD} &= \text{stopping sight distance, m} \\
\text{PSD} &= \text{passing sight distance, m} \\
L &= \text{length of vertical curve, m} \\
|G_1 - G_2| &= \text{absolute algebraic difference in tangent grades, \%} \\
R_{VCmin} &= \text{minimum radius of crest vertical curve, m} \\
h_1 &= \text{height of eye above the road surface, m} \\
h_2 &= \text{height of object above the road surface, m}
\end{aligned}
$$

The following discussions and calculations are first of all related to situation a and are based on Eq. (13.6), which is relevant for most design cases with respect to stopping and passing sight distances. Situation b is dealt with in Table 13.25. As revealed in this table, situation b occurs only where changes in grades are minor.

Situation a for $S \leq L$.   The minimum radii for crest vertical curves in Table 13.6 were calculated with respect to stopping sight distance (SSD) and passing sight distance (PSD) on the basis

---

*Elaborated based on Refs. 159, 246, and 663.

Situation (a)

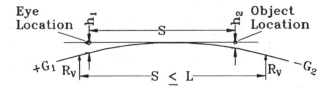

Situation (b)

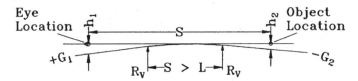

FIGURE 13.13   Stopping and passing sight distances for situations when $S \leq L$ and $S > L$.

TABLE 13.6   Recommended Ranges for Radii of Crest Vertical Curves as Related to Stopping Sight Distance, as Well as to Half and Full Passing Sight Distance for $S \leq L$*

V_{d}, km/h	Permissible range 1, R_{VC} (SSD$_{min}$ ... 0.5 PSD), m	Permissible range 2, R_{VC} (to be avoided) (0.5 PSD ... 1.0 PSD), m	Permissible range 3, R_{VC} (>PSD), m
50	1,500	—	—
60	2,500 ... 7,800	7,800 ... 30,000	>30,000
70	3,200 ... 8,600	8,600 ... 35,000	>35,000
80	4,300 ... 10,300	10,300 ... 40,000	>40,000
90	5,700 ... 12,200	12,200 ... 48,000	>48,000
100	7,400 ... 13,000	13,000 ... 52,000	>52,000
110†	≥11,000	—	—
120†	≥15,000	—	—
(130)†, ‡	≥20,000	—	—

*Elaborated, modified and completed based on Refs. 157, 159, 453, and 477.

†"Passing" only makes sense for design speeds of up to 100 km/h for two-lane rural roads.

‡( ) = exceptional cases.

Note: R_{VC} = radius of crest vertical curve, m.

of Eq. (13.6) and the corresponding sight distance model shown in Fig. 13.13 for situation a. For safety reasons, the relevant SSDs and PSDs are determined in Chap. 15 with respect to the 85th-percentile speed. Thus, despite the fact that Table 13.6 is related to the design speed, the 85th-percentile speed is indirectly included in the sight distance assumptions presented in Fig. 15.1 for SSD and in Table 15.2 for PSD. The same is true for the height of the object above the road surface, h_2, in Eq. (13.6), which also depends on $V85$ according to Table 15.3. Therefore, when designing crest vertical curves based on sight distance control, one should always strive to achieve the good design range of safety criterion I given in Table 11.1.

$$V85 \leq V_d + 10 \text{ km/h}$$

As can be seen from Table 13.6, the minimum radii of crest vertical curves are based on three design ranges:[159]

- Range 1 encompasses the necessary stopping sight distance up to one-half the passing sight distance
- Range 2 extends from one-half to full passing sight distance
- Range 3 describes radii of crest vertical curves which provide passing sight distances longer than the required ones

For future design tasks, radii of crest vertical curves, which correspond either to range 1 or range 3, should be selected. Range 2 must be avoided because it increases the probability of passing accidents.[159] This accident type is mainly represented by head-on collisions, which normally result in serious injuries or fatalities. If, because of some constraints, passing sight distances of range 2 must be applied, then a nonpassing zone should be posted for safety reasons.

Situation b for S > L.   As can be seen from Table 13.25, situation b is only relevant for minor changes in grade.

Minimum radii of crest vertical curves, as determined from sight distance requirements, are generally satisfactory from the standpoints of safety, comfort, and appearance.

However, when examining existing alignments in cases of redesigns and RRR projects, the criteria given in Sec. 13.1.4.3 for comfort and appearance should also generally be adhered to.

13.1.4.3   Sag Vertical Curves.   At least four different controls for establishing radii of sag vertical curves are recognized. These are headlight sight distance, rider comfort, drainage control, and general appearance.[5]

Sight Distance Control.   For appearance reasons, sag vertical curves should normally *not* be smaller than one-half the corresponding radii of crest vertical curves. By applying the radii of sag vertical curves given in Table 13.7, sufficient sight distance for underpasses (clearance = 4.5 m and trucker's eye height = 2.5 m) is guaranteed for headlight sight distance at night and for comfort control, based on new research experience in Europe.[157,477]

Comfort Control.   The comfort effect of change in the vertical direction is greater on sag than on crest vertical curves because gravitational and centrifugal forces are combining rather than opposing forces. For comfort reasons, it is generally accepted that the centrifugal acceleration on vertical curves should not exceed 0.5 m/s². The general expression for such control is

$$R_{VS\text{min}} = \frac{V_d^2}{3.6^2 \times 0.5} = \frac{V_d^2}{6.5} \tag{13.8}$$

where $R_{VS\text{min}}$ = minimum radius of sag vertical curve, m

V_d = design speed, km/h

How comfort control influences the values of Table 13.7 is the subject of discussion in "Sag Vertical Curves" in Sec. 13.2.4.3, Table 13.23.

TABLE 13.7 Recommended Minimum Radii of Sag Vertical Curves for Headlight Sight Distance and Comfort Control

$V_{d'}$ km/h	$R_{VSmin'}$ m
50	750
60	1,000
70	1,250
80	1,550
90	2,400*
100	3,800*
110	6,300†
120	8,800*
(130)	10,000†

*Values based on Ref. 157.

†Values based on Ref. 453.

Note: R_{VSmin} = minimum radii of sag vertical curves, m.

13.1.4.4 Appearance Control and Drainage Control

Appearance Control. For very small changes in the grade, a vertical curve (crest or sag) has little influence other than the appearance of the profile and thus may be omitted. For significant changes in grade, however, short vertical curves affect the appearance. This is particularly evident on sag curves of high standard roads.

Table 13.8 gives ranges of radii of vertical curves for satisfactory appearance, as developed in Sec. 13.2.4.4 (Table 13.25). Larger radii or longer curves can be used where they better fit the topography.

Drainage Control. Drainage affects the design of type I and type III vertical curves shown in Fig. 13.12. In the area of the level point of crest and sag vertical curves, longitudinal grades of $G \leq 0.5$ percent exist. The length of this area corresponds to Eq. (13.9) according to:[243]

$$L_D = \frac{R_V}{100} \qquad (13.9)$$

where L_D = length of the area with small longitudinal grades, m

R_V = radius of crest/sag vertical curve, m

Therefore, the gutter areas must represent minimum longitudinal grades of $G \geq 0.5$ percent for vertical curves of type I and type III (Fig. 13.12) in case of curbed roadway sections.

13.1.4.5 General Controls for Vertical Alignment.
In addition to the preceding specific controls for vertical alignment, there are several controls, as shown in the following, that should be considered in design, according to AASHTO[5] and the South African Road Design Standards:[663]

1. A smooth gradeline with gradual changes, consistent with the type of highways, roads, or streets and the character of terrain, should be strived for in preference to a line with numerous breaks and short lengths of grade.

2. Normally, crest vertical curves should be placed in cuts and sag vertical curves on embankments. Sag vertical curves should be avoided in cuts unless adequate drainage can be provided.

3. Vertices of horizontal and vertical curves should, as far as possible, coincide (see Chap. 16, "Three-Dimensional Alignment").

4. The "roller coaster" or the "hidden dip" profile type should be avoided. Such profiles generally occur on relatively straight horizontal alignment where the roadway profile closely follows a rolling natural ground line (see Chap. 16).

5. A broken-back gradeline (two vertical curves in the same direction separated by a short section of tangent grade) generally should be avoided, particularly in sags where the full view of both vertical curves is not pleasing (see Chap. 16).

6. In flat or gently rolling terrain, a sag vertical curve combined with a long horizontal tangent may give the impression of a kink (see Chap. 16).

7. Short sag vertical curves between long constant grades are not acceptable, particularly where the horizontal alignment is straight (see Chap. 16).

8. On long grades it may be preferable to place the steepest grades at the bottom and lighten the grades near the top of the ascent or to break the sustained grade by short intervals of lighter grades instead of a uniform sustained grade that might be only slightly below the allowable maximum.

9. Where intersections at grade occur on roadway sections without moderate steep grades, it is desirable to reduce the gradient through the intersection.[5,663]

TABLE 13.8 Length Ranges and Radii of Vertical Curves for Appearance Control for Very Small Changes of Grade*

Design speed, V_d, km/h (1)	$|G_1 - G_2|$,† % (2)	L,‡ m (3)	R_v,§ m (4)
40	1.0	40–80	4,000–8,000
50	0.9	50–100	5,500–11,000
60	0.8	60–120	7,500–15,000
70	0.7	70–140	10,000–20,000
80	0.6	80–160	13,000–26,000
90	0.5	90–180	18,000–36,000
100	0.4	100–200	25,000–50,000
110	0.3	110–220	36,000–73,000
120	0.2	120–240	60,000–120,000

*Elaborated and modified based on Ref. 139.

†Maximum algebraic difference in grades without need for vertical curves. In practice, vertical curves are provided for all changes of grade.

‡Length ranges of vertical curves for satisfactory appearance.

§Radius ranges of vertical curves for satisfactory appearance.

13.2 GENERAL CONSIDERATIONS, RESEARCH EVALUATIONS, GUIDELINE COMPARISONS, AND NEW DEVELOPMENTS

Roads and streets should be designed to encourage uniform operation throughout. Use of a selected design speed, as previously discussed, is a means towards this end by correlation of various geometric features of the road or street. Design values have been determined and agreed upon for many highway features but few conclusions have been reached on roadway grades in relation to design and operating speed.[5]

13.2.1 Grades

On upgrade and downgrade sections, considerable differences are frequently observed between the speeds of passenger cars and those of trucks. This fact leads to inconsistencies in traffic flow and affects traffic safety negatively. In addition, the slow-moving heavy vehicles, which cause higher amounts of pollution and noise emission, can do substantial damage to the environment. These statements indicate that it would be desirable to build roads only with low grades in order to achieve uniform traffic flow. However, this leads to extremely high construction costs, especially in hilly/mountainous topography. Therefore, higher grades cannot always be avoided as a result of adapting the road to the topography.

13.2.1.1 Maximum Grades. According to "Longitudinal Grade" in Sec. 8.2.3.2, it can be assumed that the driving behavior of free moving passenger cars, $V85$, is not significantly affected up to longitudinal grades of ± 5 to ± 6 percent;[347,406,408,691,700,713] for higher grades, considerable decreases can be expected, as discussed in "Longitudinal Grades on Two-Lane Rural Roads" in Sec. 8.2.4.2 (Figs. 8.14 and 8.15).

Critical lengths of upgrades from various countries are listed in Table 13.9, which reveals the maximum upgrade length on which a loaded truck can operate without an unreasonable speed reduction. The values in Table 13.9 vary considerably, depending on the selected design trucks (Table 13.16), on the existing criteria for the permissible truck speed behavior on upgrades (Table 13.12), and on the speed evaluation backgrounds of trucks for individual countries (for example, Figs. 13.2 and 13.18).

TABLE 13.9 Critical Lengths of Upgrades for Various Countries and AL

		Length, m								
		Grade, %								
Country	Approach speed, km/h	2	3	4	5	6	7	8	9	10
Australia	80	—	—	630	460	360	300	270	230	200
	60	—	—	320	210	160	120	110	90	80
Ireland	80	1,500	700	500	350	250	—	—	—	—
South Africa	80	—	500	350	275	225	200	175	150	150
United States	88	750*	430*	300*	230*	200*	160*	140*	120*	—
AL (see Table 13.2)	65	—	—	750	350	260	240	220	—	—

*Rounded.

As can be seen from Table 13.9, the critical lengths of Part 2, "Alignment (AL)," represent maximum values, with the exception of one design case for Australia. Based on research work carried out by Schulze,[638] the critical lengths of upgrades in "AL" can be regarded as reasonable for the truck speed behavior of today, especially in Europe. In this respect, the widely used approach speed of 80 km/h in many countries, which is quite high, and the selected power/weight ratios of design trucks (7 to 8 hp/ton), which are quite low, often result in unrealistic evaluation backgrounds of truck speed behavior and accordingly to the variations in critical lengths of upgrade (Table 13.9).

Safety Considerations. The following statements have been made by Krebs and Kloeckner:[368]

> The driving speed of a motor vehicle depends on longitudinal grades, besides other influencing parameters. Truck speeds are especially affected by the length of the upgrade section which often leads to wide deviations in the speed distributions between the speeds of passenger cars and those of trucks. Also, on downgrade sections trucks often ride at reduced speeds, in order to compensate for the effects of greater stopping distances. It can be assumed that the strong speed decrease of trucks on upgrade sections can cause traffic congestions which may lead to an increase in rear-end and passing accidents. To avoid these critical design cases, the provision of auxiliary (climbing) lanes is thus an important issue in modern highway geometric design.

Figures 13.14 and 13.15 show the relationships between accident rate/accident cost rate and longitudinal grade. As can be seen from these figures, the statements made in "Grade" in Sec. 9.2.1.3 are once again confirmed here:

- Grades of less than 6 percent have relatively little effect on accident rate and accident cost rate
- A sharp increase in accident rate and accident cost rate was noted on grades of more than 6 percent

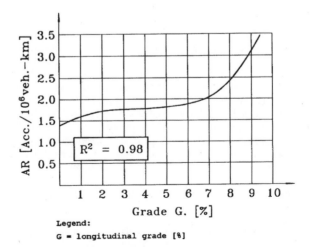

Legend:

G = longitudinal grade [%]

FIGURE 13.14   Accident rate with respect to the longitudinal grade for all accident types.[368]

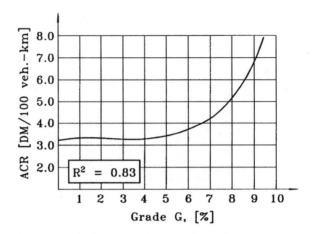

FIGURE 13.15   Accident cost rate with respect to the longitudinal grade for all accident types.[368]

Furthermore, it is reported in Ref. 368 that, for run-off-the-road accidents, the accident rate on grades of 7 percent was already twice as high as that for grades ≤6 percent. For grades >7 percent, the ROR accident rate was doubled again. Similar results were found for the ROR accident cost rate.

It is interesting to note that, with respect to the horizontal alignment, similar conclusions were noted for the design parameter curvature change rate of the single curve (CCR_S) (see Tables 9.10 and 10.12). That is, with increasing CCR_S classes (decreasing radii of curve classes), the accident rates for fair and poor design practices were at least double and quadruple, respectively, the rates for good design.

Based on these results, it can be assumed that when superimposing high CCR_S values (small radii of curve) in plan with high grades in profile, high accident rates have to be expected, as was already demonstrated in Fig. 9.32.

According to U.S. practice,[5] the basis for determining critical lengths of upgrades is a reduction in truck speeds to below the average running speed. An ideal situation would be for all traffic to operate at this speed. However, this is not practical. In the past, the common practice was to use a speed reduction of about 10 mi/h (16 km/h) in truck running speed to below the average running speed of all traffic. As shown in Fig. 13.16, the accident involvement rate increased significantly when the truck speed reduction exceeded this value; the involvement rate was 2.4 times higher for a 16-mi/h (25-km/h) reduction than for a 10-mi/h (16-km/h) reduction. On the basis of these findings, it is recommended for the United States to use the 10-mi/h (16-km/h) reduction criterion as a general guide for determining the critical lengths of upgrades (see Table 13.12). Parallel to this criterion, a 15-km/h speed reduction was used in "AL" for the development of Table 13.2.

Finally, two interesting statements by McGee, et al.[503] about vertical alignment design deserve to be mentioned here:

- Steep (that is, greater than 4 percent) upgrades and downgrades pose a greater hazard, especially with respect to accidents involving trucks.

- The accident potential is much greater for horizontal curves on or immediately after grades steeper than 3 percent.

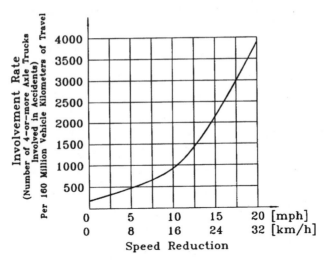

FIGURE 13.16 Accident involvement rate of trucks for which running speeds are reduced to below the average running speed of all traffic.[216]

In conclusion, and as discussed in "Grade" in Sec. 9.2.1.3, it can be stated that accidents increase with gradients, with downgrades having considerably higher accident rates than upgrades.

Worldwide Comparison. Guidelines for maximum gradients vary in complexity between countries and consider some or all of the following factors: road type (or functional classification), design speed, and terrain.

For a worldwide comparison, maximum grades in relation to design speed, road type, and topography classes were compiled in Table 13.10 for rural roads for selected countries on different continents. It should be noted that the maximum gradient in Switzerland is mainly a function of design speed. In Germany and Greece, the maximum gradient is a function of roads of category groups A (rural) and B (suburban) and design speed. In Australia and South Africa, the maximum gradients are based on design speed and topography. Besides the latter factors, the United States relies on road type. In "AL," maximum gradients are based on the road category group, topography, and design speed.

Austria alone considers maximum gradients for climbing lanes and differentiates between roads with or without medians.

As can be seen from Table 13.10, the maximum grades developed for "AL" for different topography classes agree well with the values of most of the other countries. Once again, it should be noted that the maximum grades of "AL" are, to a large extent, based on accident research experience.

Finally, the United Kingdom specifies desirable absolute maximum gradient values for different road types:[139]

	Desirable maximum gradient, %
Motorways	3
Dual carriageways	4
Single carriageways (two-lane rural roads)	6

TABLE 13.10   Maximum Longitudinal Grades in Relation to Design Speed, Road Type, and Terrain for Rural Roads in Various Countries (Worldwide) and AL

Country	Design speed, km/h										
	40	50	60	70	80	90	100	110	120	130	140
	G_{max}, %, for rural roads										
Austria with median	—	—	—	—	—	—	3	—	3	—	3
+Climbing lane	—	—	—	—	—	—	6	—	5	—	4
Without median	9	8	7	6	5	—	4	—	3	—	—
+Climbing lane	12	11	10	9	8	—	6	—	5	—	—
France	—	—	7 (R)	—	6 (T/R)	—	5 (T)	—	—	—	—
Germany (A)	—	—	8	7	6	5	4.5	—	4	—	—
Greece (A)	—	11	10	9	8	7	5	4.5	4	3	—
Switzerland	12	—	10	—	8	—	6	—	4	—	—
Italy	10	10	7	7	6	5	5	5	5	5	5
( ) Minor roads	(12)	—	(10)	—	(7)	(6)	(6)	—	—	—	—
Canada	7	7	6, 7	6	4-6	4, 5	3, 5	3	3	3	—
Minor roads	11	11	10, 11	9	7,8	6, 7	5-7	5, 6	5	5	—
United States											
Level	—	—	—	5*	4	—	3*	3*	—	—	—
Rolling	—	—	—	6*	5	—	4*	4*	—	—	—
Mountainous	—	—	—	8*	7	—	6*	5*	—	—	—
South Africa											
Flat	—	—	—	5	4	3.5	3	3	3	—	—
Rolling	—	7	6	5.5	5	4.5	4	—	—	—	—
Mountainous	10	9	8	7	6	—	—	—	—	—	—
Australia											
Flat	—	—	6-8	—	4-6	—	3-5	—	3-5	—	—
Rolling	—	—	7-9	—	5-7	—	4-6	—	4-6	—	—
Mountainous	—	—	9-10	—	7-9	—	6-8	—	—	—	—
Japan	7	6	5	—	4	—	3	—	2	—	—
AL (A)											
Flat	—	—	6 (8)	5 (7)	4 (6)	4 (5)	3 (5)	3 (5)	3 (5)	3 (4)	—
Hilly	—	—	7 (9)	6 (8)	5 (7)	5 (6)	4 (6)	4 (6)	4 (6)	—	—
Mountainous	—	—	9 (10)	8 (9)	7 (9)	7 (8)	6 (8)	—	—	—	—

Note: A = category group A (rural roads), T = single carriageway, R = interurban arteries, *Rounded. ( ) = exceptional values.

These values are intended to be used as a general guide for maximum gradients below which economic effects with respect to user costs and construction costs are not significant. They have been found to suit the majority of design situations and to lead to profiles which produce economical earthwork designs. In hilly terrain where the cost/environmental implications of working to desirable maximum gradients are severe, an assessment could be made to review the trade-off between construction/environmental costs and user costs.[139]

The maximum longitudinal grades of "AL" agree well with the previously mentioned desirable maximum gradients for both motorways with typical design speeds of 100 to 130 km/h and single carriageways (two-lane rural roads) with typical design speeds of 70 to 100 km/h in flat and hilly topographies (see Table 13.10).

13.2.1.2 Minimum Grades in Distortion Sections. Superelevation runoff should possibly be constructed along stretches where the longitudinal grade is between 0.5 and 3 percent in order to ensure good water runoff and to reduce the risk of hydroplaning. Superelevation runoff on longitudinal grades lower than 0.5 percent can result in water pools accumulating on the roadway.[686] In order to avoid these design cases, the assumptions of "Minimum Grades in Distortion Sections" in Sec. 13.1.1.2 and "Drainage Considerations" in Sec. 14.1.3.3 should be considered.

13.2.2 Auxiliary Lanes on Upgrade Sections of Two-Lane Rural Roads*

13.2.2.1 General Considerations. On roadway sections with high longitudinal grades, the speed-dependent differences between passenger cars and trucks become evident. The impediments of the slower trucks impair the traffic quality of the motor vehicle collective. Besides, the nonhomogeneity of the traffic flow has negative effects on the accident situation. As shown in "Grade" in Sec. 9.2.1.3 and "Safety Considerations" in 13.2.1.1, two-lane rural roads with longitudinal grades of at least 5 percent are more dangerous than those with lower longitudinal grades. In addition, impairments with respect to environmental quality should be considered, because driving in groups of slow-moving vehicles on upgrade sections using low gears could lead to increased exhaust and noise emissions.

Generally, steep longitudinal grades cannot be avoided in alignment design. For this reason, many countries increasingly use auxiliary lanes on upgrades in order to separate for safety reasons, slow-moving traffic from fast-moving traffic. The influence of auxiliary lanes (on upgrade sections) on the reduction in accident rates is presented in Fig. 13.17.[686] Generally speaking, on longitudinal grades of more than 3 percent, the construction of auxiliary lanes in the upgrade direction produces a considerable reduction in the accident rate.

Furthermore, St. John and Harwood[670] have indicated that the safety gains through the use of auxiliary lanes on upgrades increase with increasing longitudinal grades, sections of upgrades, and percentages of heavy vehicles. With increasing truck percentages, an exponential safety gain was noted.

*Elaborated with the support of C. Schulze based on Refs. 638 and 639.

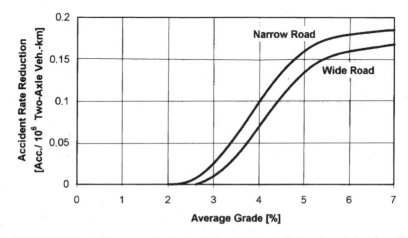

FIGURE 13.17   Accident rate reduction in relation to the longitudinal grade as derived from the construction of auxiliary lanes on two-lane rural roads.[686]

Furthermore, O'Cinnéide has indicated the following:[553] "A climbing lane is an extra travel lane provided on uphill gradients for slow moving vehicles." Hedman[273] quotes a Swedish study which concluded that climbing lanes on rural two-lane roads reduced the total accident rate by an average of 25 percent, 10 to 20 percent on moderate upgradients (3 to 4 percent), and 20 to 40 percent on steeper gradients. It was also observed that additional accident reduction can be obtained within a distance of about 1 km beyond the climbing lane. In earlier studies, Jorgensen[316] found no change in accident experience in the United States due to the provision of climbing lanes, while Martin and Voorthees[495] found a 13 percent reduction in accidents in the United Kingdom.[553]

From the limited information which is available, it appears that climbing lanes can significantly reduce accident rates.

Based on an in-depth literature review, it was found that the speed behavior of trucks in eight countries is determined from speed evaluation backgrounds corresponding to defined design trucks with *specific* engine performances (engine performance/vehicle weight). These countries are:

- Australia[37]
- Austria (Fiolic[196])
- Germany[183,247,443]
- Hungary[765]
- Ireland[504]
- Sweden[686]
- Switzerland[693]
- The United States of America[5]

In the following section, the various design procedures of the countries just mentioned will be analyzed, compared, and evaluated in order to determine the one which is most appropriate for describing the speed behavior of trucks and passenger cars on upgrade sections.

13.2.2.2   Consequences of Existing Guidelines.   The speed evaluation backgrounds of the majority of the investigated countries were derived from the 15th-percentile speeds ($V15$) of selected heavy-design vehicles. A general overview of how individual countries assess the use of auxiliary lanes on upgrade sections (climbing lanes) is provided in the following.

Australia.[37]   In the "Guide to the Geometric Design of Rural Roads," a semitrailer with a power/weight ratio of 8 hp/ton is selected as a heavy-design vehicle. An auxiliary lane in the upgrade direction is provided when the speed of the design truck falls below 40 km/h.

In order to understand the procedure used to determine the speed of the design truck on upgrades for different section lengths, the Australian speed evaluation background is schematically shown in Fig. 13.18 along with an example application. The form and use of such a background is typical for all the countries under study. From the example application presented in Fig. 13.18, note that the longitudinal profile does not require an auxiliary lane because the truck speeds on the given grades do not fall below 40 km/h.

In addition to the speed factor, the Australian guidelines look at opportunities for one vehicle to overtake another, current-year design volume (AADT), and percentage of slow vehicles when considering the construction of an auxiliary lane in the upgrade direction.

Depending on design speeds, the normal maximum length of auxiliary lanes should be 1200 m and the minimum should be 200 m.

The auxiliary lane has to be provided on the *inside* of the original driving lane.

Austria.[196]   The requirement for auxiliary lanes is based on design diagrams with regard to traffic volume, percentage of heavy traffic, longitudinal grade, and sight distance.

Minimum lengths, minimum distances between two succeeding auxiliary lanes, and taper lengths depend on the design speed (see Tables 13.13 and 13.14). In addition, the speed of a designated design truck with a power/weight ratio of 8 hp/ton has to be determined on the basis of a speed evaluation background, in order to locate the beginning and the end of auxiliary lanes.

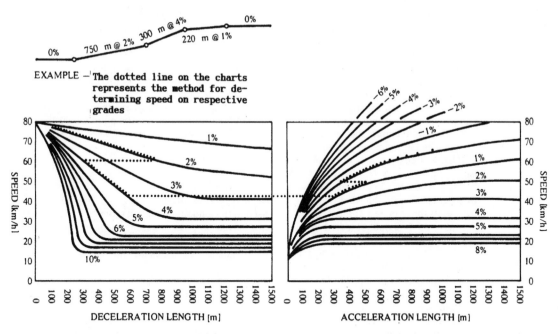

FIGURE 13.18 Determination of truck speeds on grades (Example: Australia). (© Austroads 1989. Produced under license from Austroads Incorporated, Australia.)

Germany.[183]   In the "Guidelines for the Design of Auxiliary Lanes on Upgrade Sections," a hypothetical truck is used as a design vehicle, based on speed measurements conducted by Brannolte.[66] If the truck speed ($V15$), as determined from a speed evaluation background, does not fall short of 70 km/h, then an auxiliary lane is not necessary. However, if the truck speed falls short of 30 km/h on multilane roads and 20 km/h on two-lane roads, then an auxiliary lane must be provided for traffic safety reasons. In case the speed should lie between these two limiting values, then it is necessary to examine, by means of a very complex procedure which regards curvature change rate, traffic volume, and percentage of heavy truck traffic, whether or not an auxiliary lane is needed.

The auxiliary lane should begin at the point where the design truck falls short of 70 km/h and should end at the point where the design truck once again exceeds 70 km/h. The minimum length of an auxiliary lane on multilane roads is 1500 m and 500 m on two-lane rural roads. If the distance between two succeeding auxiliary lanes is shorter than 2500 m on multilane roads and 800 m on two-lane roads, then an uninterrupted auxiliary lane must be provided. The auxiliary lane has to be assigned to the left of the original driving lane(s); in the downgrade direction, a no-passing zone should be established.

Hungary.[765]   A design truck with a power/weight ratio of 7.8 hp/ton is selected. If the truck speed, which again is determined from a speed evaluation background, falls short of 50 km/h (assuming an approach speed of 70 km/h), then an auxiliary lane is necessary. It should be noted that the auxiliary lane does not end at the point where the design truck once again exceeds 50 km/h, because the Hungarians add an additional 200-m safety section before the end of the auxiliary lane. In case of a distance of less than 300 m between two succeeding climbing lanes, then an uninterrupted construction has to be provided.

Contrary to the design procedures discussed earlier, the auxiliary (climbing) lane has to be provided (as in the United States) on the right side of the original driving lane and is regarded as a crawling lane.

Ireland.[504]   A semitrailer with a power/weight ratio of 7.9 hp/ton is regarded as a heavy-design vehicle. If the truck speed ($V15$), depending on a speed evaluation background, decreases by more than 25 km/h, then a climbing lane should be provided. Furthermore, the traffic volume with respect to certain truck percentages plays an additional role in requiring climbing lanes.

The climbing lane should begin at the point where the speed of the design vehicle decreases by more than 15 km/h and should end at the point where the truck speed ($V15$) reaches 50 km/h. The climbing lane is regarded, as in the United States, as a crawling lane.

Sweden.[686]   An auxiliary lane should be provided at the point where the design vehicle's speed (7.2-hp/ton semitrailer) falls short of 65 km/h and should end at the point where the truck speed reaches 60 km/h; this distance should be longer than 400 m. The truck speed is determined from a speed evaluation background.

The auxiliary lane has to be provided, as in Germany, on the left side of the original driving lane.

Switzerland.[691,693]   In the Swiss norm "Alignment—Auxiliary Lanes in Upgrades and Downgrades," a design truck with a power/weight ratio of 10 hp/ton is used. This design truck represents the strongest of the design trucks used in the guidelines studied (compare Table 13.16). Furthermore, it should be noted that the Swiss design procedure not only depends on the speed behavior of a design truck but also on the driving behavior of passenger cars and on their super-imposition. This is one reason why the Swiss approach is regarded as the soundest in the proposed procedure for the design of auxiliary lanes on upgrade sections.

Whereas the truck speed ($V15$) is once again determined from a speed evaluation background, the driving behavior of passenger cars is represented by a project speed, V_p. The project speed is given in Table 13.11 in relation to radius of curve.

Because two-lane rural roads in Switzerland have a general speed limit of 100 km/h, the project speed is limited to this general speed which corresponds according to the Swiss standard[691] to radii of curve of $R \geq 420$ m. For multilane rural roads, the general speed limit of 120 km/h is associated with radii of curve of $R \geq 650$ m.

Whenever the 15th-percentile speed of the design truck falls below 50 percent of the project speed of passenger cars (0.5 V_p) on a minimum section length of 200 m on two-lane rural roads, then an auxiliary lane is required. For multilane roads, the following limiting values are relevant:

- Forty percent of the project speed (0.4 V_p)
- A section length of 500 m

Typical diagrams for evaluating the previously defined speed-dependent criteria for trucks and passenger cars are presented in Figs. 13.3 and 13.7, in which the project speed of the Swiss guidelines was replaced by the more general 85th-percentile speed.

The auxiliary lane should begin at the point where the truck speed falls short of 50 percent of the project speed and should end at the point where the truck speed exceeds this limit again (or 0.4 V_P for multilane roads). If the distance between two succeeding auxiliary lanes on two-lane rural roads is shorter than 600 or 2,000 m on multilane roads, then an uninterrupted auxiliary lane has to be provided. The auxiliary lane is assigned to the left side of the original driving lane(s); in

TABLE 13.11   Project Speed in Curves for Passenger Cars[691]

Two-lane rural road												
Radius of Curve R [m] 45	60	75	95	120	145	175	205	240	280	320	370	≥420 - ∞
V_p [km/h] 40	45	50	55	60	65	70	75	80	85	90	95	100

Multilane rural road			
	470	525	580	≥650 - ∞
	105	110	115	120

the downgrade direction, a no-passing zone should be established, like in Germany.

The procedure takes into account additional requirements like the relationship between traffic volume, percentage of heavy truck traffic, longitudinal grades, and sight distances on two-lane rural roads. These requirements are presented in two additional design diagrams, and were also introduced in the developed design procedure in "AL" (see Fig. 13.8).

For multilane roads, the previous issues are subject to the requirement that auxiliary lanes are deemed necessary for traffic volumes of more than 1500 vehicles/h and direction and a truck percentage of at least 15 percent.

United States.[5]   According to AASHTO,[5] the following three conditions and criteria, which reflect economic considerations, should be satisfied in order to justify a climbing lane:

1. Upgrade traffic flow rate is in excess of 200 vehicles/h per direction.
2. Upgrade truck flow rate is in excess of 20 vehicles/h per direction.
3. One of the three following conditions exists:

A 10-mi/h (16-km/h) or greater speed reduction is expected for the design truck (7.5 hp/ton), based on a speed evaluation background.

Level of service E or F exists on the upgrade section.

A reduction of two or more levels of service is experienced when moving from the approach segment to the grade.

The level of service is determined through a very complex procedure, which considers a number of influencing factors like passenger car/truck-proportion, directional distribution, ratio of flow rate to ideal capacity for level-of-service D, adjustment factor for narrow lanes, restricted shoulder width, etc.

Tne climbing lane should begin at the point where the speed of the design truck decreases by more than 10 mi/h (16 km/h) below the operating speed of 55 mi/h (88 km/h) of the traffic flow and should end at the point where the latter speed is once again reached.

The climbing lane has to be provided on the right side of the original driving lane. It is regarded as a crawling lane for slow vehicles.

Preliminary Conclusions.   Speed criteria for selected design trucks in various countries, with respect to auxiliary lanes (climbing lanes) in the upgrade direction, are listed in Table 13.12. As can be seen from this table, criteria exist for the general provision of auxiliary lanes, which depend on absolute speeds (col. 1) and speed reductions (col. 2). The speeds of design trucks on auxiliary lanes are defined in most countries as the 15th-percentile speeds. Furthermore, speed criteria exist for the beginnings and ends of auxiliary lanes (cols. 3 to 5).

It should be noted that nearly all of the countries studied use speeds, as shown in Table 13.12, which are related to design trucks and not to the driving behavior of passenger cars.

The beginning of an auxiliary lane corresponds to the point at which truck speeds fall below 50 to 70 km/h (col. 3). Switzerland is an exception because the 15th-percentile truck speed is dependent on the project speed of passenger cars, which means the Swiss procedure regards the driving behavior of trucks and passenger cars for the design of auxiliary lanes. Thus, this procedure appears to be a reliable one and was taken as the basis for "AL."

For comparative purposes, the speeds of approaching trucks at the foot of the grade are listed in col. 6 of Table 13.12. These approach speeds of about 80 km/h were taken from the truck speed evaluation backgrounds of the countries studied. Normally, these speeds do not coincide with the driving behavior of trucks at the 15th-percentile speed level. Based on actual speed measurements,[459] the approach speeds of trucks are assumed nowadays in the range from 60 to 65 km/h. The latter values for approach speeds correspond well to the speed assumptions shown in col. 3 of Table 13.12.

Table 13.13 shows the minimum lengths of auxiliary lanes, as applied in selected countries. Column 2 of Table 13.13 gives the minimum distances between two succeeding auxiliary lanes. If actual distances are shorter than those given in col. 2, then an uninterrupted auxiliary lane must be provided. As shown in cols. 1 and 2 of Table 13.13, minimum lengths of between 200 and 1,000 m and minimum interim distances of between 300 and 1,000 m are required for two-lane

TABLE 13.12   Speed Criteria (as Based on Design Trucks) for Establishing Auxiliary Lanes in the Upgrade Direction for Selected Countries (Normally Related to the 15th-Percentile Speed)

Country	Design criteria for auxiliary lanes		Criteria for the beginning and ending of auxiliary lanes				
	15th-percentile speed, $V15$, km/h (1)	15th-percentile speed reduction, $\Delta V15$, km/h (2)	Beginning		Ending		Approach speed, km/h (6)
			$V15$, km/h (3)	$\Delta V15$, km/h (4)	$V15$, km/h (5)		
Australia	<40						80
Austria			<70 for $V_d = 120 - 100$ <60 for $V_d = 100 - 80$ <50 for $V_d = 80 - 60$ <40 for $V_d = 60 - 40$		≥70 ≥60 ≥50 ≥40		80
Germany	>70 (no auxiliary lane) <20 (two-lane) <30 (multilane) Auxiliary lane is necessary in any case		<70		≥ 70		80
Hungary	<50		<50		≥50*		70
Ireland		≥25		≥15	≥50		80
Sweden			<65		≥60		90
Switzerland	<0.4 V_p† (multilane) <0.5 V_p† (two-lane)		<0.4 V_p† <0.5 V_p†		≥0.4 V_p† ≥0.5 V_p†		80
United States		≥16		≥16	≥ 72		88
AL	<0.5 $V85$(PC)		<0.5 $V85$(PC)		>0.5 $V85$(PC)		65

*Plus a 200-m safety section.

†V_p = project speed of passenger cars, km/h, based on Ref. 691.

Note: V_d = design speed, km/h.

rural roadways, depending on the country under study. According to Schulze,[638] the German values can be regarded as appropriate.

Tapered sections are designed with or without hatchings in different countries. In Switzerland, for instance, hatchings are provided at the beginning and at the end of auxiliary lanes. In Germany, the United Kingdom, and France, hatchings are present only at the end of auxiliary lanes. In the other countries under study, hatchings are not required or no information is given with respect to their use.

Table 13.14 gives the lengths of tapered sections in front of and behind an auxiliary lane without hatchings, and if a hatching is present at the end, the lengths correspond to cols. 2 and 3 (see also Fig. 13.1).

TABLE 13.13 Minimum Lengths and Interim Distances for Auxiliary Lanes in Selected Countries

	Minimum length, m (1)	Minimum interim distance, m (2)
Australia, two-lane	200	
Austria		
Multilane	1200 for $V_d = 120 - 100$	2500
Two-lane	1000 for $V_d = 100 - 80$	1000
	700 for $V_d \leq 80$	1000
Germany		
Two-lane	500	800
Multilane	1500	2500
Hungary, two-lane	—	300
Switzerland		
Two-lane	200	600
Multilane	1000	2000
AL		
Two-lane	500	800
Multilane	1000–1500	2000–2500

TABLE 13.14 Taper Section Lengths with and without Hatchings in Selected Countries

Country	Diverge taper length, Ld, m (1)	Merge taper length, Li, m (2)	Lateral displacement taper length, Lr, m (3)
Australia, two-lane	50 for $V_d = 50$	75*	
	60 for $V_d = 60$	90*	
	70 for $V_d = 70$	105*	
	80 for $V_d = 80$	120*	
	90 for $V_d = 90$	135*	
	100 for $V_d = 100$	150*	
Austria	120 for $V_d = 120 - 100$	250*	
Multilane	90 for $V_d = 100 - 80$	150*	
Two-lane	80 for $V_d \leq 80$	120*	
France, two-lane	130	230 (160)†	130
Germany			
Two-lane	≥ 150 for $V_d = 100$	90	≥ 150
	≥ 100 for $V_d \leq 80$	60	≥ 100
Multilane (new)	60	120	60
Multilane (existing)	≥ 200	120	≥ 200
Ireland	100	100*	
Sweden	150 for $V_d = 70$	200*	
	200 for $V_d = 90$	250*	
	250 for $V_d = 110$	300*	
United Kingdom			
Two-lane	100	100	100
Multilane	100	150	150
United States, two-lane	≥ 120 ft (45 m)	≥ 200 ft (60 m)*	
AL	130	230 (160)†	130

*Without hatching.

†Depending on $V85$ (see Fig. 25.17 of "CS").

TABLE 13.15   Auxiliary Lane Widths in Selected Countries

Country	Auxiliary lane width,* m
Australia, two-lane	3.00–3.50
France, two-lane	3.50
Germany	
Two-lane	3.25–3.75
Multilane	3.50–3.75
Hungary	
Major roads	3.50
Minor roads	3.00
Ireland, two-lane	3.00
Sweden, two-lane	3.00/3.50
United Kingdom, two-lane	3.20
United States, two-lane	10–12 ft (3.00–3.60 m)
AL	
Two-lane	3.25–3.75
Multilane	3.50–3.75

*Depending on the standard cross section.

As can be seen from Table 13.14, the diverge taper sections are generally shorter than the merge taper sections. This is quite evident in cases where hatchings are provided at the end, in which case the lengths of cols. 2 and 3 must be combined. Because France is quite experienced with two lanes in one direction and one lane in the other direction (2 + 1 intermediate cross section) and for safety reasons (deceleration maneuvers at the end of the auxiliary lane), the French values shown in Table 13.14 provide a sound solution.[700]

Table 13.15 gives a listing of auxiliary lane widths in selected countries. Normally, the width of the auxiliary lanes in these countries corresponds to the lane width of the standard cross sections.

Finally, Table 13.16 shows the power/weight ratios for the design trucks. As this table reveals, nearly all the countries under study use a design truck with 7.5 to 8 hp/ton. The only exception is Switzerland, which uses a stronger design truck that has a power/weight ratio of 10 hp/ton.

TABLE 13.16   Power/Weight Ratios of Heavy Design Vehicles in Selected Countries

Country	Design truck and power/weight ratios
Australia	8 hp/ton (semitrailer)
Austria	8 hp/ton*
Germany	Hypothetical truck
Hungary	7.8 hp/ton*
Ireland	7.9 hp/ton (semitrailer)
Sweden	7.2 hp/ton (semitrailer)
Switzerland	10.0 hp/ton*
United States	7.5 hp/ton*
AL	10.0 hp/ton (semitrailer/full trailer combination)

*No specific indication.

Note: 1 ton = 1000 kg = 10,000 N.

13.2.2.3 Design Truck. The engine performance of heavy trucks increased considerably in Europe during the last 10 to 15 years, as can be seen from the following list which shows the top motors of seven large truck manufacturers.

In 1984:

Scania: 420 hp

Iveco: 420 hp

Mercedes: 375 hp

Volvo: 370 hp

Renault: 365 hp

MAN: 360 hp

DAF: 330 hp

The corresponding permissible gross weight was 38 tons (Germany, as an example).

In 1995:

Mercedes: 530 hp

Renault: 530 hp

Volvo: 520 hp

Iveco: 515 hp

DAF: 507 hp

MAN: 500 hp

Scania: 500 hp

The corresponding permissible gross weight is 44 tons (EU standard).

Presently, MAN plans to go up to 600 hp, Renault to 560 hp, and Scania to 530 hp. Not only the engine performance of the top motorized trucks has increased but also the engine performance of the widely used types which has increased from a range of about 280 to 320 hp in 1984 to a range of about 370 to 420 hp in 1995.

These statements are supported by the data shown in Table 13.17 with respect to semitrailer motor distribution. This table shows that in every gross weight class, the percentage of the highest engine performance class increased, whereas the percentage of the other engine performance classes had decreased between 1987 and 1994. This is especially true for the gross weight class (>16 tons), where the percentage of the strongest engine performance class (>350 hp) had increased from 9.2 percent in 1987 to 49.9 percent in 1994.

Table 13.18 gives the power/weight ratios of single-unit trucks (col. 4), semitrailers, and full trailer combinations (col. 6) for three gross-weight classes. In order to select a reliable heavy design vehicle, two assumptions were made:

1. The design truck should correspond to a semitrailer or to a full trailer combination because, compared to a single-unit truck, its power/weight ratio is about one-half the latter (compare cols. 4 and 6), as a result of its much higher permissible gross weight (compare cols. 3 and 5).

2. The design truck should correspond to the largest gross weight class, that is, a gross weight of GW ≥ 16 tons according to Table 13.18.

With respect to the seven main truck manufacturers in Europe, the design truck should correspond to the power/weight ratios of typical truck types in Western Europe, which are shown in the bottom right corner box of Table 13.18. As can be seen from Table 13.18, the mean power/weight ratio is 9.86 hp/ton, which means a semitrailer/full trailer combination with a power/weight ratio

TABLE 13.17   Semitrailer Motor Distribution for the Years 1987, 1991, and 1994[31]

Engine performance, hp	Percentages		
	July 1, 1987	July 1, 1991	July 1, 1994
	6- to 11-ton permissible gross weight		
Up to 150	3.1	1.7	1.1
151–200	4.2	3.1	2.3
201–249	0.5	1.3	1.8
	12- to 15-ton permissible gross weight		
Up to 150	0.1	0.1	0.1
151–200	4.6	2.4	1.4
201–249	7.5	6.4	5.1
250–299	14.0	7.4	5.0
300–350	17.0	7.3	4.1
$\geq$351	5.0	5.3	7.2
	$\geq$16-ton permissible gross weight		
Up to 150	~0	0.1	0.1
151–200	0.1	0.1	0.1
201–249	1.6	1.6	1.3
250–299	13.8	12.1	8.6
300–350	19.2	15.8	12.1
$\geq$351	9.2	35.4	49.9

Note: 1 ton = 1000 kg = 10,000 N.

of 10 hp/ton appears to be appropriate as a heavy design vehicle for the design of auxiliary lanes in the upgrade direction.

It is interesting to note that the heavy design vehicle which corresponds to 10 hp/ton represents, to a large extent, today's composition of semitrailer/full trailer combinations (at least in western Europe). It has been selected as the design truck in Switzerland (see Table 13.16). Thus, it can be assumed that the truck speeds, which are determined from the Swiss evaluation background, could best represent the driving behavior on upgrade sections (see Figs. 13.2 and 13.6).

The other power/weight ratios of selected design trucks in different countries, as shown in Table 13.16, are too small to describe the actual truck speeds on auxiliary lanes in the upgrade direction, as will be shown later.

13.2.2.4   Actual Speed Measurements and Comparisons with Existing Speed Evaluation Backgrounds for Trucks.   In order to evaluate the driving behavior of trucks and passenger cars on upgrades (with auxiliary lanes) and on downgrades (without auxiliary lanes), eight test sections in the State of Baden-Württemberg in southwest Germany were selected for speed measurements. Table 13.19 gives an overview of the test sections with respect to lengths, maximum grades, minimum radii of curve, and maximum curvature change rates of the single curve. The total length of the test sections was 15.631 km. The maximum grades ranged from 5.1 to 7.8 percent. The sections were equipped with auxiliary lanes in the upgrade direction. As shown in cols. 5 and 6 of Table 13.19, five of the test sections (1 to 5) revealed somewhat generous horizontal alignments in the area of the Black Forest. Sections 6 to 8, on the other hand, had longitudinal grades which were superimposed by extremely narrow radii of curve and extremely high curvature change rates of the single curve of up to about 1450 gon/km.

Speed data collection, reduction, and analysis were made according to the procedures presented in "Developments in the United States" in Sec. 8.2.3.4.

TABLE 13.18 Power/Weight Ratios of Fully Loaded Common Truck Types (1995)

| Truck type (1) | Engine performance, hp (2) | Single-unit truck | | Semitrailer/full trailer combination | |
		Gross weight, tons (3)	Power/weight ratio, hp/ton (4)	Gross weight, tons (5)	Power/weight ratio, hp/ton (6)
6- to 11-ton gross weight					
DAF FA 45.150	147	7.5	19.6	17.5	8.4
Iveco 100 E 18 R	181	10	18.1	18	10.1
MAN 9.163	160	9	17.8	16.5	9.7
Mercedes 814	132	7.5	17.6	18	7.3
Renault S 180.09	175	9.5	18.4	21	8.3
Volvo FL 611/180	180	11	16.4	21	8.6
12- to 15-ton gross weight					
DAF FA 55.180	182	15	12.1	24	7.6
Iveco 120 E 23 R	227	12	18.9	26	8.7
MAN 14.232	230	14	16.4	28	8.2
Mercedes 1320	211	13	16.2	32.5	6.5
Renault M 210.13	205	13	15.8	28	7.3
Volvo FL 615/210	209	15	13.9	30	7.0
≥16-ton gross weight					
DAF FA 85.400	401	18	22.3	40	10.0
Iveco MT 180 E 27 R	267	18	14.8	32.5	8.2
MAN 19.463	460	18	25.6	40	11.5
Mercedes 1838	381	18	21.2	40	9.5
Renault AE 520	530	18	29.4	40	13.3
Scania R 113/320	320	18	17.8	40	8.0
Volvo FH 12/340	340	18	18.9	40	8.5
Average value >16-ton gross weight					9.86

TABLE 13.19 Design Elements of the Test Sections

No. (1)	Test section (2)	Length, m (3)	G_{max}, % (4)	R_{min}, m (5)	CCR_{Smax}, gon/km (6)
1	SR 10 Pforzheim	2100	6.5	600	95
2	SR 10 Enzweihingen	2540	6.55	475	97
3	SR 292 Obrigheim	3022	5.7	160	300
4	SR 292 Waibstadt	2438	5.3	310	135
5	SR 35 Knittlingen	1182	5.5	400	105
6	SR 314 Epfenhofen	2274	7.8	40	1255
7	SR 33 Nussbach	625	5.1	32	1477
8	SR 33 St. Georgen	1450	7.2	50	850
		Σ15,631			

Note: Maximum longitudinal grade = G_{max}, %; minimum radius of curve = R_{min}, m; and maximum curvature change rate of the single curve = CCR_{Smax}, gon/km.

In order to evaluate the driving behavior of heavy vehicles on upgrade and downgrade sections, the speeds of single-unit trucks, semitrailers, and full trailer combinations with permissible gross weights of at least 7.5 tons were measured under free-flow conditions (time gap of more than 6 s). Characteristics of the eight test sections are given in Table 13.19. Normally, three spot speed measurements were carried out at each of the test sections. The locations were set up in the following way: The first spot speed measurement was set up in the first part of the upgrade section, the second near the middle, and the third in the last part of the test section. Because of personal constraints (one to two persons per local speed measurement), the speed measurements of heavy vehicles on auxiliary lanes in the upgrade direction stood in the forefront. The second important speed class to investigate was the heavy vehicles in the downgrade direction. During the time periods when trucks were neither on the upgrade or downgrade directions, the speeds of free-flowing passenger cars were also collected. Normally, speeds of about 80 to 100 heavy vehicles were measured in the upgrade direction, during a time period of between 2 and 5 h.

In order to be able to follow the speed course of an individual truck on the auxiliary lane in the upgrade direction, the first two or three digits of the truck's license plate were written down along with the speed at every measurement spot. For trucks in the downgrade direction, and for passenger cars on both upgrades and downgrades, only the individual speeds were written down. Overall, 24 local speed measurements were conducted on the eight investigated roadway sections.

Characteristic speed distributions for trucks and passenger cars on upgrades and downgrades (6.4 percent) are presented in Fig. 13.19 where the longitudinal grade is superimposed with low CCR_S values and in Fig. 13.20 where the grade is superimposed with high CCR_S values. For both cases, with respect to truck speeds on auxiliary lanes in the upgrade direction, it can be seen that the overall speed distribution, including all heavy vehicles, is comparable to that which only includes semitrailers and full trailer combinations (see the upper parts of the figures). This result is true for nearly all of the 24 local speed measurements investigated, and proves the correctness of the assumption made in the previous chapter, which calls for selecting a semitrailer or full trailer combination with a power/weight ratio of 10 hp/ton as a heavy design vehicle.

Longitudinal Grades Superimposed by Horizontal Alignments with CCR_S Values <700 gon/km.   For two-lane rural roads in the upgrade direction, it was determined for grades of between 5 and 8 percent superimposed with relatively generous horizontal alignments (based on the results of Fig. 13.19 and on the other local speed measurements) that:

1. The 85th-percentile speed level of trucks (GW ≥ 7.5 tons) in the upgrade direction is between 55 and 65 km/h. The 15th-percentile level for the same direction of travel is about 30 km/h.

2. The 15th-percentile speed level of trucks in the downgrade direction is between 50 and 60 km/h, and is independent of the longitudinal grade (4 to 8 percent) and of the length of the downgrade section. The speed difference ($V85 - V15$) is in the range from 10 to 20 km/h only.

3. The 85th-percentile speed level of passenger cars in the upgrade direction is about 20 km/h higher than that in the downgrade direction.

4. For grades of more than 6 percent, the speed difference between the 15th-percentile speed of trucks and the 85th-percentile speed of passenger cars is quite large (about 80 km/h). For downgrades, this difference is only about 40 km/h.

Longitudinal Grades Superimposed by Horizontal Alignments with CCR_S Values >700 gon/km.   For two-lane rural roads in the upgrade direction, it was determined for grades of between 5 and 8 percent superimposed with high CCR_S values (based on the results of Fig. 13.20 and on the other local speed measurements) that:

1. The 85th-percentile speed level of trucks (GW ≥ 7.5 tons) in the upgrade direction is between 35 and 45 km. The 15th-percentile level for the same direction of travel is about 30 km/h.

2. The 15th-percentile speed level of trucks in the downgrade direction is between 30 and 35 km/h, and is independent of the longitudinal grade (4 to 8 percent) and of the length of the downgrade section. The speed difference ($V85 - V15$) is only about 10 km/h.

Test Section
SR 10 Enzweihingen - Measurement Spot No. 2
Longitudinal Grade: G = 6.4 %
Curvature Change Rate of the Single Curve: CCR_S = 0 gon/km

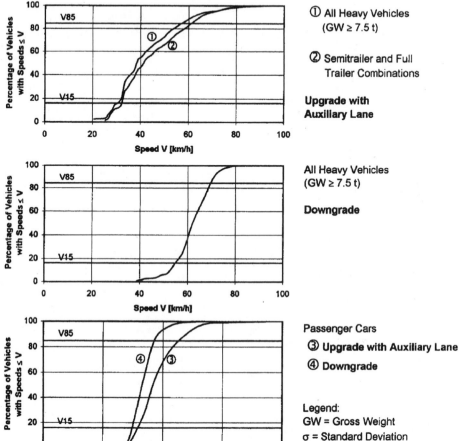

① All Heavy Vehicles
   (GW ≥ 7.5 t)

② Semitrailer and Full
   Trailer Combinations

Upgrade with
Auxiliary Lane

All Heavy Vehicles
(GW ≥ 7.5 t)

Downgrade

Passenger Cars

③ **Upgrade with Auxiliary Lane**

④ **Downgrade**

Legend:
GW = Gross Weight
σ = Standard Deviation

Measurement Spot No. 2						
	Number [veh.]	V15 [km/h]	V50 [km/h]	V85 [km/h]	V85 - V15 [km/h]	σ [km/h]	
All Heavy Vehicles	153	29.5	38.3	57.2	27.7	14.4	upgrade
Semi-/ Full Trailer	65	32	41.2	61.4	29.4	14.9	upgrade
All Heavy Vehicles	118	54.5	62.1	70	15.5	8.2	downgrade
Passenger Cars	126	78.4	92	114	35.6	19.8	upgrade
Passenger Cars	114	73.8	82.8	92.9	19.1	11.3	downgrade

FIGURE 13.19 Characteristic speed distributions for trucks and passenger cars on upgrades and downgrades with CCR_S values <700 gon/km.

Test Section
SR 314 Epfenhofen - Measurement Spot No. 2
Longitudinal Grade: G = 6.4 %
Curvature Change Rate of the Single Curve: CCR$_S$ ≈ 1100 gon/km

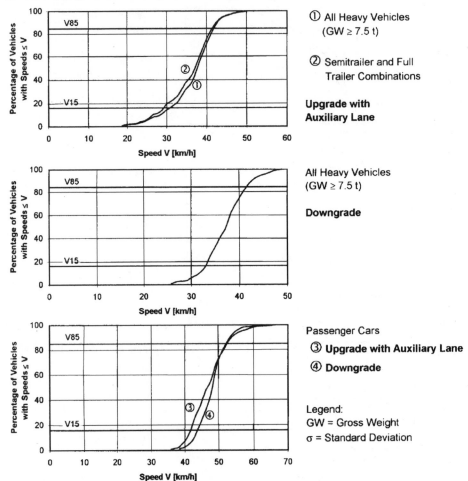

① All Heavy Vehicles
   (GW ≥ 7.5 t)

② Semitrailer and Full
   Trailer Combinations

Upgrade with
Auxiliary Lane

All Heavy Vehicles
(GW ≥ 7.5 t)

Downgrade

Passenger Cars

③ **Upgrade with Auxiliary Lane**

④ **Downgrade**

Legend:
GW = Gross Weight
σ = Standard Deviation

Measurement Spot No. 2						
	Number [veh.]	V15 [km/h]	V50 [km/h]	V85 [km/h]	V85 - V15 [km/h]	σ [km/h]	
All Heavy Vehicles	159	30.2	37.6	42.1	11.9	6.2	upgrade
Semi-/ Full Trailer	107	28.9	37	41.7	12.8	6.4	upgrade
All Heavy Vehicles	113	32.8	37.3	41.7	8.9	4.5	downgrade
Passenger Cars	132	41.4	46.4	52.5	11.1	5.6	upgrade
Passenger Cars	88	43.5	48.1	52.2	8.7	4.4	downgrade

FIGURE 13.20 Characteristic speed distributions for trucks and passenger cars on upgrades and downgrades with CCR$_S$ values >700 gon/km.

3. The speed distributions of passenger cars in the upgrade and downgrade directions are near-ly identical. The same is true for the speed distributions of trucks.

4. The speed differences between the 15th-percentile speed of trucks and the 85th-percentile speed of passenger cars for both driving directions are 20 and 25 km/h, respectively (see, for example, Fig. 13.20).

From the previous comparisons, it can be concluded that in the first design case, superimposition of longitudinal grades between 5 and 8 percent with low CCR_S values, the longitudinal grades dominate the speed behavior with partially great variations in speeds between trucks and passenger cars as well as between upgrade and downgrade directions.

For the design case, superimposition of longitudinal grades with high CCR_S values, it was determined that the design parameter curvature change rate of the single curve also plays an important role. For instance, independent of the longitudinal grades, it was found that with increasing CCR_S values (beginning with about 700 gon/km), this design parameter can influence the speed behavior of trucks (Fig. 13.5) and is always relevant for passenger cars (see Figs. 13.4 and 8.12 and Table 8.5). As CCR_S values increased, the variations in speed distributions within the individual vehicle classes or between the different vehicle classes (trucks and passenger cars) became smaller and smaller. That means the design parameter curvature change rate of the single curve not only influences the speed behavior of passenger cars but it also influences the speed behavior of trucks (at least for CCR_S values >700 gon/km).

So far, this fact has not been recognized or considered in the guidelines studied (see "Consequences of Existing Guidelines" in Sec. 13.2.2.2) with respect to designing auxiliary lanes (climbing lanes) on upgrade sections, with the exception of Germany. However, the influence of curvature change rate of the single curve was regarded in the new design procedure developed in "International Design Procedure with Case Studies" in Sec. 13.1.2.5.

On the basis of the in-depth review of the guidelines, presented in Sec. 13.2.2.2, it was decided to use the 15th-percentile speed as the appropriate truck speed for the design of auxiliary lanes in the upgrade direction. Furthermore, 24 spot speed measurements were conducted on eight roadway segments (see Table 13.19).[459] According to Schulze,[638] these speed measurements represent the actual driving behavior of trucks on auxiliary lanes in the upgrade direction in relation to grades and section lengths. It should be noted, however, that these 24 spot speed measurements are not sufficient to establish an evaluation background for trucks, like those in Figs. 13.2 and 13.18.

Thus, in order to prove the reliability, as well as the unreliability, of the spot speed measurements, a very work-intensive comparative procedure was used. This procedure encompassed the truck operating speed backgrounds of Australia, Austria, Germany, Ireland, Sweden, Switzerland, and the United States. For the gradients of each of the eight test sections, the 15th-percentile truck speeds were determined for the individual operating speed backgrounds of the seven countries under study and additionally at every speed measurement spot. Table 13.20 gives an overview of the results for test section no. 1 "SR 10—Pforzheim, Germany," in comparison to the actual measured 15th-percentile truck speeds at the local measurement spots M1, M2, and M3. As can be noted from Table 13.20, the truck speed evaluation background of Switzerland closely approximates the measured values. This procedure was repeated for the other seven test sections listed in Table 13.19.

Thus, the measured 15th-percentile speeds of trucks at the 24 different locations were compared with 168 truck speeds (24 × 7) which resulted from the individual speed evaluation backgrounds of the seven countries. Again, Switzerland represented best the measured values.

The following list compares the average deviations between the 24 truck speeds at the speed measurement spots, based on the individual evaluation background of the country under study, to the 24 measured 15th-percentile truck speeds at the test spots:

- *Switzerland:* +0.4 km/h
- *Germany:* −3.8 km/h
- *Sweden:* −4.7 km/h
- *United States:* −5.6 km/h

- *Australia:* −9.8 km/h
- *Austria:* −11.1 km/h
- *Ireland:* −13.9 km/h

It can be recognized that the truck speed evaluation background of Switzerland represents the lowest average deviation (+0.4 km/h), whereas the deviations of the other backgrounds range from −3.8 to −13.9 km/h. Thus, they represent a lower truck speed level than actually observed by the measurements on upgrade auxiliary lanes.

This result is to be expected because the speed evaluation background of Switzerland corresponds, unlike other backgrounds, to a design truck with a power/weight ratio of 10 hp/ton. The Swiss background represents today's best truck motorization and composition in western Europe, as previously discussed in "Design Truck" in Sec. 13.2.2.3.

These statements are confirmed by actual investigations in Canada in the Province of Quebec.[491] In this study, speed profiles of 760 heavy vehicles on six auxiliary lanes in the upgrade direction with longitudinal grades ranging between 1.5 to 10 percent and lengths between 600 to 2300 m were recorded and the 12.5th-percentile speed profile was compared to speed evaluation backgrounds, based on design trucks with a power/weight ratio of 11.2 and 7.5 hp/ton, which are used for the design of upgrade auxiliary lanes in Canada. Lupien, Baas, and Barber[491] found that the design truck with a power/weight ratio of 11.2 hp/ton adequately represented truck traffic on the majority of highways. However, it is recommended that the design of auxiliary lanes on upgrade sections should be based on the 7.5-hp/ton design truck for those highways where the traffic stream is composed of a significant number of heavily loaded trucks, for example, gravel or logging trucks.

13.2.3 Safety Facilities on Steep Downgrade Sections

13.2.3.1 *Emergency Escape Ramps.*
Basic information on the design of emergency escape ramps is given in Sec. 13.1.3. For further specific technical details, the reader should consult AASHTO.

TABLE 13.20 15th-Percentile Truck Speeds Based on Truck Operating Speed Backgrounds of Individual Countries in Comparison to Actual Speed Measurements for Test Section No. 1, "SR 10—Pforzheim, Germany"

Length of upgrade, m		110	290	260	180	170	70	260	160	10	210	200	180
Upgrades, %	B = 3.3	M1	6.5	6.2	6.4	M2	6.2	6.4	6.0	M3	5.2	1.7	E
Germany	60*	56	49	32	26	21	20	20	19	20	23	27	45
Switzerland	60*	57	50	39	34	30	29	28	27	28	30	34	47
United States	60*	56	48	28	25	24	23	23	23	24	26	28	40
Sweden	60*	56	49	28	25	24	24	24	24	25	27	30	43
Australia	60*	54	42	22	22	22	22	22	22	23	24	26	36
Ireland	60*	55	45	14	14	14	14	14	14	15	17	19	37
Austria (proposed)	60*	55	43	22	19	19	19	19	19	20	20	21	35
V15 (measured)†		56				31				28.5			

*Approach speed at the beginning of the test section.

†Semitrailer/full trailer combination.

Note: B = beginning of upgrade; M1 to M3 = local measurement spots; and E = end of upgrade.

A useful review of 59 facilities in the United States has been given by Williams, et al.[758] Details of the designs and testings of vehicles' arrestor beds in Australia have been made by Brown[83] and Cocks and Goodram.[117]

An alternate approach, described by Adam,[6] makes use of a half vehicle-width arrestor bed adjacent to the road shoulder along significant lengths of downgrade. Flexible guideposts separate the bed from the shoulder, and on the outer side a truck-strength concrete barrier is provided to control the direction of the runaway vehicle as it is brought to a halt. An advantage of this design is that arrestor beds can be made available to vehicles over a greater length of the grade than is possible using traditional methods.[37]

13.2.3.2 *Effective Speed Control by Automatic Radar Devices, Case Study: Elzer Mountain.* *

Regarding the relationship between speed and safety, Pigman, et al.[577] reported the following. Speed has been determined to be one of the most common contributing factors in vehicular accidents. The relationship between speed variance and safety has been investigated and it has been shown that, assuming equal exposure, the greater the variation in speeds, the higher the probability of an accident.[111,661] Another study examined speed variance and it was found that both slow drivers and fast drivers had accident rates that were approximately 6 times that of drivers operating close to the mean traffic speed.[750]

It has been documented that the greater the absolute speed, the greater the likelihood of increased accident severity.[214] The energy dissipated during a collision is directly proportional to the vehicle's weight and to the square of its speed. Therefore, increased speed results in more energy dissipation, which translates into greater damage to the vehicle and more injuries to the occupants.[577]

Experiences of the main author in Germany have indicated that the introduction of speed limits often only has a short-term effect on reducing speeds, unless police regularly enforce the speed limits. Posted speed limits alone will not guarantee compliance. Only when backed up by strict police enforcement can speed limits both reduce speed and alleviate accidents.

The presence of police enforcement has also been shown to have the effect of decreasing speeds in the United States.[269,599] The use of speed enforcement, a speed-check zone, or a parked patrol vehicle produce significant reductions in speeds in the vicinity of the enforcement unit.[133] Active police enforcement in conjunction with the use of radar units has been found in many situations to reduce speed.[577]

Thus, it is a documented fact in the United States and Europe that reasonable and systematic traffic surveillance can reduce the number and severity of accidents, especially on dangerous road sections. In many cases, the danger of a road section is directly attributable to improper use of speed by the driver. Besides alcohol abuse, not fastening seat belts, and inattention, most of the excessive speed errors occur with reference to road design, primarily by exceeding the critical speed for a curve or a downgrade section and thereby losing control.

Therefore, strong efforts should be made by police to enforce the speed limits, at least on dangerous road sections. The purpose of this section is to discuss the extent to which surveillance by police can produce an evident decrease in accidents. Experience gained in one of the most dangerous autobahn (interstate) sections in Germany, Elzer Mountain near the town of Wiesbaden, is used as the center of discussion.[189]

Speed and Accident Analysis.   Elzer Mountain lies in the route of Autobahn (Interstate) A 3 between Cologne and Frankfurt in hilly topography. It was built in the 1930s with two lanes in each direction plus a median and paved shoulders. Because of the high increase in traffic in the 1960s, a third traffic lane and an emergency lane were added in 1969 in both directions.

Figure 13.21 shows the major part of the horizontal and vertical alignment on Elzer Mountain. Note the generous horizontal alignment of this autobahn section, which consists of relatively safe radii of between 1500 and 2000 m.

---

*Elaborated based on Refs. 367, 395, and 397 and supported by J. Wacker, Ministry of Economy, Traffic and Regional Development, State of Hessen, Germany.

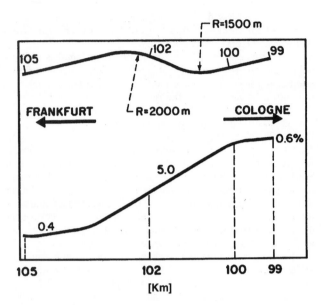

FIGURE 13.21   Horizontal and vertical alignment of Elzer Mountain.

The character of a generous, consistent alignment is further supported by three wide traffic lanes in each direction (lane width = 3.75 m) in addition to an emergency lane (Fig. 13.22). In this connection, the good horizontal alignment, superimposed by relatively high downgrade sections of up to 5 percent, poses a real danger which is extremely difficult to recognize by drivers. Drivers on such a downgrade tend to exceed the safe driving speeds, and thereby jeopardize their chances of stopping safely in an emergency.

For example, great variations in vehicle speeds existed between passenger cars and trucks. Speed measurements conducted in 1971 (see Fig. 13.23)—with no traffic control devices available at that time—indicated that the speed variations between the 85th-percentile speed of passenger cars in the left lane and the 15th-percentile speed of trucks in the right lane totaled 120 km/h.

In addition, the truck speeds in the right lane ranged from 15 to 110 km/h. The wide variations in vehicle speeds in the area of Elzer Mountain led to numerous hazardous driving situations.

Thus, despite adding a third lane—or because of adding a third lane—drivers were obviously unable to identify the steep downgrade sections of Elzer Mountain as dangerous, since about 200 accidents/y still occurred during 1970 and 1971 on a section length of 7.2 km (Table 13.21). That means that there were exactly 27 accidents/km/y on the downgrade direction. The data in Table 13.21 show the accident frequencies on Elzer Mountain from 1961 to 1996.[188,190,737] Note that until 1972 to 1973, or before the final installation of automatic radar devices, the downgrade direction from Cologne to Frankfurt was significantly more dangerous than the upgrade direction from Frankfurt to Cologne. Furthermore, following the installation of the traffic sign bridges (Fig. 13.22) with lane-related speed limits in 1972 and the installation and permanent operation of the radar devices in 1973, the total number of accidents and the number of fatalities and injuries were significantly reduced. Since the end of the 1970s, the numbers have stabilized and are now comparable to those in the upgrade direction (Table 13.21).

In order to be able to make appropriate proposals for improving the accident situation on Elzer Mountain, the Institute for Highway and Railroad Engineering, University of Karlsruhe, con-

FIGURE 13.22 Alignment of Elzer Mountain, downgrade direction.

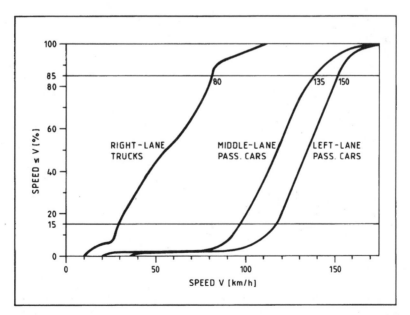

FIGURE 13.23 Characteristic distributions of speeds for passenger cars and trucks, downgrade direction (1971).

TABLE 13.21 Accident Development at Elzer Mountain, Long-Term Investigation (km 98.8 to km 106.0) (Elaborated based on Refs. 188, 190, and 737.)

Year (1)	Direction: Cologne–Frankfurt (downgrade)			Direction: Frankfurt–Cologne (upgrade)		
	All (2)	Fatalities (3)	Injuries (4)	All (5)	Fatalities (6)	Injuries (7)
1961	289	4	66	39	—	8
1962	232	2	67	42	1	9
1963	192	2	49	65	1	17
1964	250	5	79	64	—	13
1965	317	6	85	68	—	16
1966	325	5	90	91	—	26
1967	259	6	81	53	4	11
1968	257	8	85	66	—	14
1969	336	7	130	51	3	15
1969	Construction of a third lane in each direction					
1970	199	6	74	50	1	18
1971	199	8	83	66	1	20
1972	Installation of traffic sign bridges with lane-related speed limits					
1972	183	7	56	47	—	14
1973	Installation and permanent operation of automatic radar devices					
1973	84	2	29	34	1	9
1974	45	1	18	41	1	10
1975	56	3	18	28	—	8
1976	33	2	6	27	1	13
1977	30	—	10	26	—	5
1978	19	—	5	29	1	9
1979	30	1	6	36	—	17
1980	28	2	9	29	—	7
1981	29	—	6	31	—	8
1982	27	—	5	25	1	6
1983	21	—	10	38	—	10
1984	26	—	7	46	1	15
1985	44	1	5	41	—	8
1986	44	3	11	53	2	11
1987	44	—	12	47	—	12
1988	52	—	7	48	—	13
1989	55	1	7	62	1	14
1990	46	—	10	76	—	21

TABLE 13.21 Accident Development at Elzer Mountain, Long-Term Investigation (km 98.8 to km 106.0) (Elaborated based on Refs. 188, 190, and 737). (*Continued*)

Year (1)	Number of accidents					
	Direction: Cologne–Frankfurt (downgrade)			Direction: Frankfurt–Cologne (upgrade)		
	All (2)	Fatalities (3)	Injuries (4)	All (5)	Fatalities (6)	Injuries (7)
1991	70	—	13	53	1	11
1992	63	—	9	70	2	19
1993	50	—	12	60	—	7
1994	39	—	7	60	2	10
1995	66	—	15	67	1	13
1996	48	1	13	68	4	12

ducted a comprehensive analysis of speed distribution and the main causes of accidents.[367] The investigation period was from April 1, 1970 to March 31, 1972.

Figure 13.23 shows characteristic distributions of speeds for a group of passenger cars and trucks on the steep downgrade section on Elzer Mountain at km 102 (see Fig. 13.21). The 85th-percentile speed in the left lane was 150 km/h, in the middle lane it was 135 km/h for passenger cars, and in the right lane it was 80 km/h for trucks. That meant that 15 percent of both passenger cars and trucks still exceeded these speeds. The maximum design speed of 100 km/h, which had been permissible at that time in Germany for grades up to 5 percent, was exceeded by about 95 percent of the passenger cars in the left lane and by about 80 percent of the passenger cars in the middle lane. The general speed limit for trucks of 80 km/h in Germany was exceeded by 15 percent of vehicles of this type.

Figure 13.24 shows the accident rates for different accident types. For passenger cars, the accident type "congestion" was dominant, whereas for trucks the accident type "run-into" was more relevant. The accident type "run-into" is related only to two vehicles (for example, in a case when the driver of a following vehicle underestimates the speed of a vehicle in front and then normally causes a rear-end collision). The accident type "congestion," which includes by definition more than two vehicles, in many cases originates as a result of previous "run-into" accidents.

Because of the great differences in the driving speeds among trucks and between trucks and passenger cars on the downgrade section of Elzer Mountain (Fig. 13.23), the predominance of "run-into" accidents and, consequently, the high portion of "congestion" accidents could be expected. For example, 54 percent of the accident damage of about 3.1 million U.S. dollars, caused at Elzer Mountain during the "before" investigation period from April 1970 to March 1972, was produced by "run-into" accidents of trucks alone.

Therefore, to reduce the severity and number of accidents on Elzer Mountain, the traffic flow had to be made more uniform by narrowing the great variations in vehicle speeds.

Countermeasures.   To alleviate the accident situation at Elzer Mountain, in 1972 the Institute for Highway and Railroad Engineering, University of Karlsruhe, proposed lane-related speed limits that should be under surveillance by automatic radar devices.[367]

To correct the extremely heterogeneous distributions of the driving speeds (Fig. 13.23) and to reduce the accident risk (Fig. 13.24), the allowable maximum speed was limited to 100 km/h for passenger cars and to 40 km/h for trucks. In addition, the speed limit of 100 km/h for passenger cars was assigned to the left and middle lanes only, whereas the speed limit of 40 km/h was for the right lane only (Figs. 13.22, 13.25, and 13.26).

To prevent numerous passing and lane-changing maneuvers of passenger cars in the left and middle lanes, for both lanes the same speed limit of 100 km/h was selected, which corresponds to

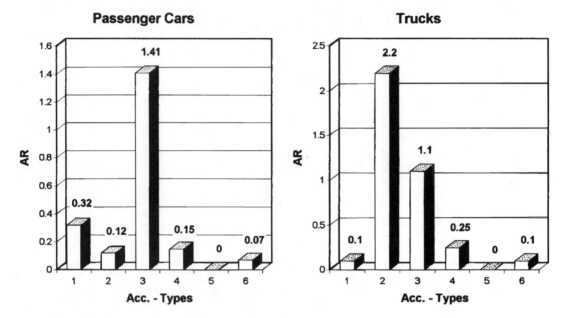

Legend:

AR = Accident Rate: Accidents per 10^6 vehicle-kilometers

Accident Types:

1. **Excessive Speed**
   (Single Vehicle Acc.)

2. **Run-Into**
   (Rear-End Collision, Only 2 Vehicles)

3. **Congestion**
   (More than 2 Vehicles)

4. **Changing Lanes and Passing**

5. **Intersection**

6. **Others**

FIGURE 13.24   Accident rates caused by different accident types, downgrade direction, investigation period: April 1, 1970 to March 31, 1972.

a typical maximum allowable design speed for a gradient of 5 percent in various countries (see Table 13.10). To prevent the middle lane from being occupied for passing maneuvers by slower vehicles, especially trucks, **DO NOT PASS** signs for trucks were also installed.

Figure 13.25 shows the traffic sign plan for Elzer Mountain in the downgrade direction. The speed limit signs were installed on four traffic sign bridges across the autobahn; the new speed and traffic regulations were introduced on April 1, 1972.

The experiences with general or local speed limits were often not satisfactory, especially in the case of rare surveillance by the police. Therefore, the second and the third traffic sign bridges were modified to use automatic radar devices (see Fig. 13.26).

Figure 13.27 shows the traffic sign bridge at km 101.3 with the automatic lane-related radar device for the right lane. The radar devices were installed for permanent operation in November 1973.

Furthermore, a psychologically striking warning sign (Fig. 13.28) was installed at the beginning of Elzer Mountain at km 98.8 before the first speed limits begin to inform the driver by a visual impression of the danger of the steep downgrade section ahead.

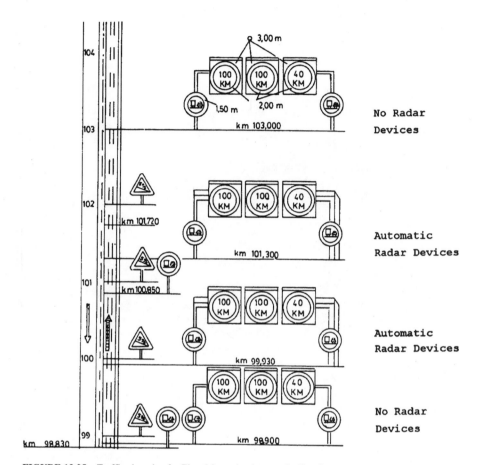

FIGURE 13.25   Traffic sign plan for Elzer Mountain, downgrade direction.

Long-Term Investigation: Speed and Accident Development.   Comparable speed measurements were conducted during the "before" investigation in 1971 after the introduction of the speed limits in 1972, after the installation of the automatic radar devices in 1974 and, in addition, in 1981 and 1983 at km 101.7. It should be noted that the observed differences between the 1974, 1981, and 1983 speed measurements were negligibly small and the 1981 speed distributions were representative for all speed measurements since the automatic radar devices were effective (see Figs. 13.29 and 13.30). New speed measurements in 1995 confirmed these results. That means that by the surveillance of the automatic radar devices, the speed distributions and the traffic flow revealed no evident changes over 20 y for passenger cars in the left and middle lanes in the downgrade direction at Elzer Mountain.

As can be seen from Fig. 13.29, the 85th-percentile speeds for passenger cars in the left lane were reduced by about 45 km/h (from 150 km/h in 1971 to about 105 km/h in 1981 to 1983). Corresponding speed reductions of about 35 km/h for passenger cars (from 135 km/h in 1971 to 100 km/h in 1981 to 1983, as in Fig. 13.30) were observed in the middle lane, and of between 40 and 45 km/h for trucks (from 80 km/h in 1971 to about 35 km/h in 1981 and 40 km/h in 1983 to 1995, as in Fig. 13.31) were observed in the right lane. The automatic radar devices were set to measure speeds exceeding 110 km/h in the middle and left lanes and exceeding 45 km/h in the

FIGURE 13.26 Traffic sign bridge for use of automatic radar devices.

FIGURE 13.27 Automatic lane-related radar device.

FIGURE 13.28   Warning sign at beginning of Elzer Mountain.

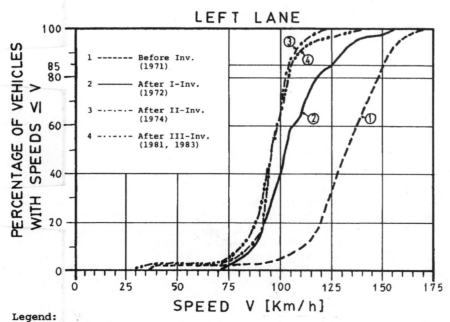

Legend:

Curves 3 and 4 were confirmed once more by speed measurements in 1995

FIGURE 13.29   Characteristic distributions of speeds for passenger cars in the left lane between 1971 and 1983 to 1995, downgrade direction at km 101.7.

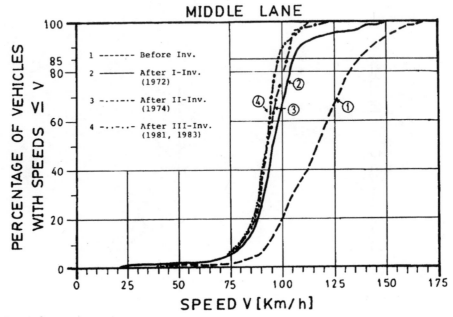

Legend:

Curves 3 and 4 were confirmed once more by speed measurements in 1995

FIGURE 13.30 Characteristic distributions of speeds for passenger cars in the middle lane between 1971 and 1983 to 1995, downgrade direction at km 101.7.

right lane. Furthermore, the steep increase in speed distributions in 1974 and 1981 to 1983 indicates that the traffic flow became decisively more uniform, which means the variability of speeds sharply decreased.

Long-term investigation of accidents revealed the following conclusions. The data in Table 13.21 shows the development of the accident frequency between 1961 and 1996 for the 7.2-km long section investigated at Elzer Mountain. As already mentioned, until 1972 to 1973, the downgrade direction from Cologne to Frankfurt was decisively more dangerous than the upgrade direction. But since 1972, with the introduction of the lane-related speed limits and especially since 1973 with the installation and the permanent operation of the automatic radar devices, the annual accident frequency has decreased from about 200 accidents in 1970 to 1971 to an average of 55 accidents in the 1990s. The number of fatal and personal-injury accidents went down too. From 70 to 80 injuries/y (6 to 8 fatalities included) in 1970 to 1971, this number decreased, on the average, to 11 injuries/y in the 1990s. It is interesting to note that since 1987, no fatal crash or only one fatal crash occurred in the downgrade direction.

The slight increase in the number of accidents and injuries between the 1980s and the 1990s is probably due to a sharp increase in volume of traffic following the unification of the two Germanies in 1988; this was also observed on the upgrade direction of Elzer Mountain. Furthermore, it should be noted that since the mid-1970s, the difference in accidents between the downgrade and upgrade directions has not been of major significance.

The decrease in accidents since 1973, combined with a low accident level in the downgrade direction since 1976, is not coincidental. It has to be attributed to the lane-related speed limits and to the strict surveillance by the police, which became possible for the first time in Germany

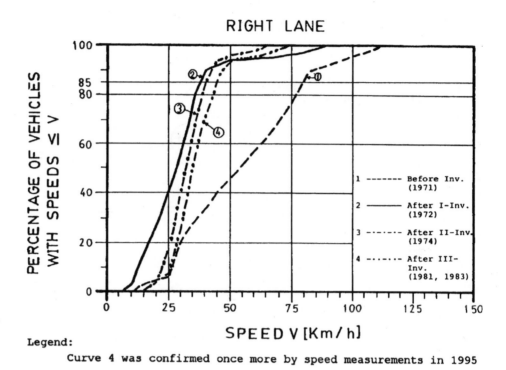

FIGURE 13.31 Characteristic distributions of speeds for trucks in the right lane between 1971 and 1983 to 1995, downgrade direction at km 101.7.

through the use of automatic radar devices. These statements are supported by the results shown in Fig. 13.32.

Figure 13.32 shows the accident rates for accidents with personal injuries and fatalities for the downgrade direction from Cologne to Frankfurt, for the upgrade direction from Frankfurt to Cologne at Elzer Mountain, and for the entire German autobahn (interstate) network. For the downgrade direction, the first success came at the end of the third-lane construction in 1970 and 1971. However, the accident rates for these 2 y were about 4.5 times higher than those corresponding to the upgrade direction at Elzer Mountain and to the average accident rates of the German autobahn network.

Following the introduction of lane-related speed limits in 1972 and surveillance by means of automatic radar devices in 1973, the accident rates for the downgrade direction decreased continuously until 1976. Since 1976, there have not been significant differences, based on the chi-square test for a 95 percent confidence level between the accident rates on the downgrade and upgrade driving directions at Elzer Mountain. The same is true for the average accident rates on the German autobahn network (see Fig. 13.32).

The experiences at Elzer Mountain have shown that reasonable speed regulations under surveillance through automatic radar devices can reduce the number and severity of accidents significantly. As related to personal injuries, the number of accidents and the accident rates were reduced by a ratio of about 9 to 1 between 1971 and the 1990s, whereas the average of 6.6 fatalities/y between 1965 and 1972 has decreased to "none" (respectively "one") fatality/y for the downgrade direction at Elzer Mountain during the last decade with the exception of the year 1986. Furthermore, the investigation period of more than 20 y appears to be long enough to verify that

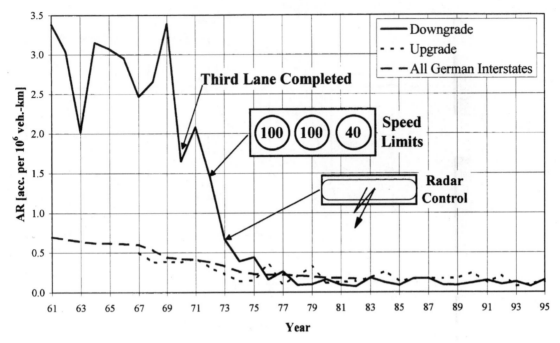

FIGURE 13.32   Accident rates for accidents with personal injuries at Elzer Mountain and for the German autobahn network (west).

the improvement of the accident situation is permanent and that the common impact of lane-related speed limits and surveillance by the police has reduced the number and severity of accidents to a level that can be indicated today as "normal."

In conclusion, it should be noted that similarly high absolute or relative accident decreases have not been experienced during the past two decades on any other autobahn section in Germany.

Police Procedures.   The automatic radar devices at Elzer Mountain have operated continuously since 1973. They are installed on traffic sign bridges across the autobahn and are lane-related (see Figs. 13.26 and 13.27). They are set to measure speeds exceeding 110 km/h for the left and the middle lanes and speeds exceeding 45 km/h for trucks in the right lane. If a vehicle exceeds these speeds, it is automatically measured and photographed. During nighttime, twilight, and rainy periods, additional strobe lights are used to illuminate the vehicles.

Figure 13.33 is a typical photograph which indicates the speed, site, date, time, and license plate of the speeding vehicle. At least once a day the rolls of film are changed and evaluated by the police. Traffic violation tickets are then delivered to the vehicle owners by mail, and they have to declare if they themselves drove the vehicle or identify the driver of the vehicle. On average, 63 percent of all vehicle owners or named drivers paid the fine at once, 27 percent had to be reminded a second time, and about 10 percent contested the traffic citation.

In addition to the continuous operation of the automatic radar devices, several times per year, especially during weekends and vacation time, the speed limits at Elzer Mountain are under direct surveillance by the police. During those periods, police officers control the driving speeds directly from the traffic sign bridges and inform their colleagues a few kilometers ahead by radio of speeding drivers. The drivers are stopped and have to pay the fine on the spot. This method also provides control of drivers from other European countries.[188,189]

FIGURE 13.33   Photograph made by an automatic radar device at Elzer Mountain.

Unfortunately, because of personnel constraints, this procedure has been more and more neglected by the police. This may explain the slight increase in personal injuries since the mid-1980s (see col. 4 in Table 13.21).

It should be noted that the main age group of drivers exceeding the speed limits at Elzer Mountain is between 35 and 44 years, followed by the age group of drivers between 25 and 34 years.

The radar devices at Elzer Mountain were frequently the topic of discussion in German newspapers, radios, and televisions, which aimed at informing the public that the radar devices were not radar traps but traffic safety devices. Nevertheless, a certain portion of drivers will always exist who will not obey reasonable traffic or speed regulations that are necessary for traffic safety, especially on extremely dangerous road sections like Elzer Mountain.

Therefore, the only way to keep the number and severity of accidents down is through strict surveillance by the police, even if such methods are not desired by traffic safety engineers. The authors would prefer that the individual drivers would be conscious of good driving habits and the safety of other highway users with a minimum of police enforcement but that will probably remain a dream.

13.2.4   Vertical Curves

The longitudinal profile of a road consists of a series of tangent grades and vertical curves. Vertical curves are applied to affect the transition between tangent grades. There are two types of vertical curves: convex vertical curves are known as crest curves and concave vertical curves are known as sag curves (see Fig. 13.12).

The major controls for the design of crest vertical curves are stopping sight distance or appearance requirements. At sag vertical curves, comfort associated with vertical acceleration, appearance, drainage, headlight performance or overhead restrictions to the line of sight may become relevant.

13.2.4.1 Application. All important equations and symbols for calculating vertical curves by applying the quadratic parabola are once again summarized in Fig. 13.34, in addition to the general statements about vertical curves in Fig. 13.12 and the more general equations given in Sec. 13.1.4.1.

13.2.4.2 Crest Vertical Curves

Important Relationships and Worldwide Comparison. An analysis of the driving behavior at crest vertical curves revealed that the motorists do not adjust their speeds to reduced sight distances. Based on measurements, observations, and accident investigations, a new design procedure was developed and standard and limiting values for crest and sag vertical curves were derived. For safety reasons the ranges for radii of crest vertical curves on two-lane roads (Table 13.6) are based on the consideration which avoids sight distances that are between one-half and full passing sight distances in order to lessen the number of passing maneuvers which are subject to insufficient passing sight distances.[157,159] The established limiting and standard values allow for safe driving and a good fit between the gradeline and the landscape.

The radii of crest vertical curves presented are based on new stopping sight distances derived in Chap. 15 from the overall regression equation [Eq. (10.40)] between the tangential friction factor and the design speed,[414,428] from new values for the aerodynamic drag force and vehicle mass,[451] and from new research on crest vertical curve design conducted by Durth and Levin.[157,477]

Table 13.22 gives a worldwide summary of minimum radii of crest (convex) vertical curves as a function of design speed. As demonstrated in Fig. 13.34, most countries use the quadratic parabola for vertical curve design. They apply minimum radii of crest vertical curves in order to satisfy stopping sight distance requirements. These radii correspond to K factors, or rates of vertical curvature:[361]

$$K = \frac{R_V}{100} \quad \text{m/\%} \tag{15.14}$$

A graphical presentation of the minimum K values in relation to design speed is given in Fig. 15.6 for selected countries.

As can be seen from Table 13.22 and Fig. 15.6, the minimum radii of crest vertical curves range from 1000 to 3000 m for a 60-km/h design speed, from 4100 to 12,500 m for a 100-km/h design speed, and from 10,500 to 20,200 m for a 120-km/h design speed.[361] These numbers show wide variations.

New research according to Table 13.22 shows that significant reductions of radii of crest vertical curves could be observed as compared to the still existing values in the guidelines of the specific country under study.[91,157,481] This conclusion expresses the opinion of many researchers, who argue that the design of minimum radii of crest vertical curves often has been too generous.

The developed values for "AL" follow the trend not to use too large radii of crest vertical curves, which means with respect to economic, environmental, and safety-related issues, balanced designs could be achieved.

The calculated minimum radii of crest vertical curves are related so far to:
Situation a

$$S \le L \qquad \text{Eq. (13.6)} \qquad\qquad \text{in Fig. 13.13}a$$

For minor changes in longitudinal grades:
Situation b

$$S > L \qquad \text{Eq. (13.7)} \qquad\qquad \text{in Fig. 13.13}b$$

may become relevant.

Driving Behavior and the Accident Situation. With respect to speed behavior in relation to vertical design, Levin[477] reported the following between measured speeds in the crest area (in

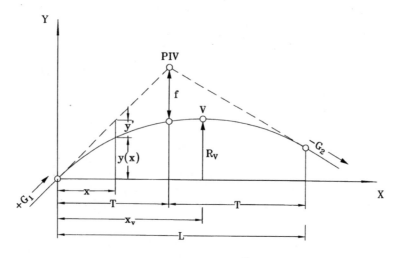

$$x_v = -\frac{G_1}{100} R_V \qquad\qquad (13.10)$$

$$G(x) = G_1 + \frac{x}{R_V} 100 \qquad\qquad (13.11)$$

$$y(x) = \frac{G_1}{100} x + \frac{x^2}{2 R_V} \qquad\qquad (13.12)$$

$$T = \frac{R_V}{2} \frac{|G_1 - G_2|}{100} \qquad\qquad (13.13)$$

$$f = \frac{T^2}{2 R_V} = \frac{T}{4} \frac{|G_1 - G_2|}{100}$$

$$= \frac{R_V}{8} \left(\frac{|G_1 - G_2|}{100} \right)^2 \qquad\qquad (13.14)$$

Sign rules:
   Upgrade: positive $(+G_1, +G_2)$
   Downgrade: negative $(-G_1, -G_2)$

in which PIV = point of intersection of tangent grades
   V = vertex
   R_V = radius of vertical curve, m
   T = tangent length, m
G_1, G_2 = tangent grades, %
   $G(x)$ = grade at arbitrary point of the vertical curve, %
   x = horizontal length in plan, m
   y' = tangent offset (vertical), m
   $y(x)$ = height difference from the beginning of the vertical curve, m
   x_v = horizontal length to vertex, m
   f = tangent offset at PIV, m
   L = length of vertical curve, m

FIGURE 13.34   Equations for calculating vertical curves by applying the quadratic parabola. (Elaborated based on Ref. 243.)

TABLE 13.22   Minimum Radii of Crest Vertical Curves as a Function of Design Speed in Various Countries (Worldwide) and AL*

Country	Design speed, km/h										
	40	50	60	70	80	90	100	110	120	130	140
	Minimum radius of crest vertical curve, m										
Austria, new research[481]	1,500	2,000	3,000	4,000	7,500	—	12,500	—	20,000	—	35,000
	—	—	2,356	2,974	3,308	4,093	5,170	—	7,824	—	—
Belgium	—	—	1,600	—	—	7,500	—	—	—	—	—
Denmark	—	—	—	—	3,500	—	6,000	—	15,000	—	—
Germany											
1984[243]	—	—	2,750	3,500	5,000	7,000	10,000	—	20,000	—	—
1995[246]	—	1,400	2,400	3,150	4,400	5,700	8,300	—	16,000	—	—
New research[157]	—	—	800	1,500	2,700	4,700	7,600	—	17,500	—	—
France	700	—	1,500	—	3,000	—	6,000	—	12,000	—	18,000
Italy	500	—	1,000	—	3,000	—	7,000	—	14,000	—	—
The Netherlands	—	—	—	—	1,800	—	4,100	—	12,400	—	—
Spain	—	—	—	—	3,500	—	6,000	—	12,000	—	—
Sweden	—	1,100	—	3,500	—	7,000	—	10,000	—	—	—
	—	(600)	—	(1,800)	—	(4,500)	—	(10,000)	—	—	—
Switzerland, new research[91]	1,500	2,100	3,000	4,200	6,000	8,500	12,500	20,000	20,000	—	—
	620	—	1,500	—	4,200	—	10,500	—	18,000	—	31,000
United Kingdom	—	1,100	1,900	3,300	5,900		10,500	—	18,500	—	—
Canada	400	700	1,500	2,200	3,500	5,500	7,000	8,500	10,500	12,000	—
United States	500	1,000	1,800	3,100	4,900	7,100	10,500	15,100	20,200	—	—
South Africa	600	—	2,000	—	5,000	—	10,000	—	20,000	—	—
Australia	—	540	920	1,570	2,400	4,200	6,300	9,500	13,500	19,500	—
Japan	—	800	1,400	—	3,000	—	6,500	—	11,000	—	—
AL + Greece	—	1,500	2,500	3,200	4,300	5,700	7,400	11,000	15,000	20,000	—

*Modified and completed based on Refs. 325, 361, and 363.

Note: ( ) = exceptional value.

front of the vertex) and radius of crest vertical curve (Fig. 13.35). According to Levin's study, a strong relationship was noted between the 50th-percentile speed, $V50$, as well as the 85th-percentile speed, $V85$, and the radius of the crest vertical curve. As the radius of the crest vertical curve increased, the speed also increased. There were wide variations between the 85th-percentile speeds and design speeds on crest vertical curves, as shown in Fig. 13.35. Only for very large radii did the gap between these two speeds become smaller. That means on small radii of crest vertical curves, motorists tend to select an 85th-percentile speed which is considerably higher than the design speed of the guidelines. This observation corresponds to that of the driving behavior on horizontal curves for which safety criterion I (Table 11.1) was developed in order to create an adjustment between the 85th-percentile speed and the design speed.

The fact that motorists were not so impressed with an upcoming crest vertical curve, which contradicts what has been assumed in the guidelines, is once again confirmed by the following research result: Independent of the radius of the crest vertical curve, the 85th-percentile speeds in the approach area of a crest, that is, in front of the vertex, were only about 2 km/h lower than the speeds on the remaining free roadway section.

Furthermore, Levin[477] reported the following about the accident situation, which corresponded to 498 accidents. The accident rate in the crest area of AR = 2.3 accidents/10^6 vehicle kilometers

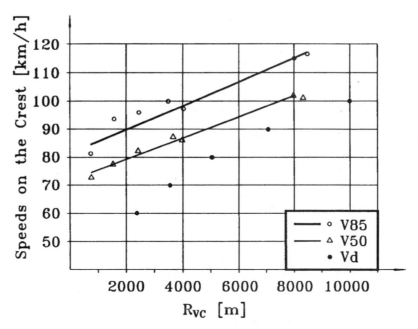

FIGURE 13.35   Relationship between radius of crest vertical curve and speed.[477]

was about 0.5 accidents/10^6 vehicle kilometers higher than that on level sections. Even though a statistically significant relationship could not be established between the radius of crest vertical curve and the increase in accident rates, there was a tendency for the accident rate on small radii of crest vertical curves to be considerably higher than that on generously designed ones.

Another accident investigation by Levin of 40,494 accidents outside built-up areas in Germany in 1988 yielded the following conclusions. The increase in accident rates is attributed to run-off-the-road and passing accidents. The first accident type accounted for 41 percent of accidents on crests; this was considerably higher than the average of 27 percent for all ROR accidents in the study. On the other hand, passing accidents accounted for 23 percent of accidents on crests; the average of all accidents for this accident type was 10.7 percent. With respect to severe accidents, the average accident severity on crests was about two-thirds higher than the average accident severity of the study.

McGee, et al.[503] indicated that crest vertical curves with large grade differentials (that is, greater than 6 percent) pose a higher risk in terms of accident potential to drivers than crest vertical curves with small grade differentials.

Furthermore, O'Cinnéide reported[553] the following. The TRB Special Report 214,[710] which includes an equation from which the accident frequency on a segment of roadway containing a single crest vertical curve and its tangent approaches can be estimated concluded that the geometry of vertical curves is not known to have a significant effect on accident severity. However, Srinivasan[669] stated that frequent changes in vertical alignment also result in a reduction in sight distance at the crest of vertical curves and these have been shown to be related to accidents, both with respect to frequency of occurrence and degree of sight obstructions. The combination of gradient and superelevation on curves has also been regarded as important.[553]

A study of accidents on two-lane rural roads in Texas by Urbanik, et al.[725] also indicated that limited sight distances, especially on crest vertical curves, could cause a marked increase in accident rates. An example would be a sharp horizontal curve hidden by a crest vertical curve (see Chap. 16).

For further information, see "Grade" in Sec. 9.2.1.3.

13.2.4.3   Sag Vertical Curves.   Sight distance on sag curves is not restricted by the vertical geometry in daylight conditions or at night with full roadway lighting unless overhead obstructions were present.

However, under night conditions on unlit roads, consideration has to be given to providing headlight sight distance on sags. Where horizontal curvature could cause the light beam to shine off the pavement (assuming 3° lateral spread each way), little is gained by flattening the sag curves.[139]

Furthermore, sag vertical curves should be designed to achieve the comfort control ($0.05g$ vertical acceleration).

Sight Distance and Comfort Control.   New research report by Durth and Levin,[157] "Specified Design of Vertical Curves," concluded the following. The recommended minimum radii of sag vertical curves (Table 13.23, range 1) guarantee the adherence of stopping sight distances at night for headlight conditions, a comfortable driving dynamic ride, and a good optical guidance. These values have been adopted by the "German Guidelines for the Design of Roads, Part Alignment."[159,246]

However, because these values represent minimum standards as compared to the other guidelines for design speeds of up to $V_d = 80$ km/h according to Table 13.24, the following considerations should be examined.

Table 13.23 (range 2) shows the minimum radii of sag curves, based on Eq. (13.8) for comfort control in relation to design speed. From ranges 1 and 2, it is obvious that the values recommended by Durth and Levin[157] (range 1), which include all relevant sag controls, exceed the comfort control values of range 2. Accordingly, nearly all guidelines regard "comfort" as having a less significant role.

However, these considerations are, so far, related only to the design speed. As previously discussed, especially in Chaps. 8 to 10, the design speed could, under real driving situations, be exceeded by substantial amounts. For this reason, safety criterion I was developed in order to coordinate the design speed, V_d, with the operating speed, $V85$, and to differentiate good designs from fair and poor designs, as shown, for example, in Table 11.1. Even though comfort control is

TABLE 13.23   Minimum Radii of Sag Vertical Curves Based on Different Approaches

Range 1 (Durth and Levin[157])		Range 2 (comfort: V_d)		Range 3 (good design: comfort)		Range 4 (fair design: comfort)	
V_d, km/h	R_{VSmin}, m	V_d, km/h	R_{VSmin}, m	$V85 = V_d + 10$, km/h	R_{VSmin}, m	$V85 = V_d + 20$, km/h	R_{VSmin}, m
50	500	50	400	60	550	70	750
60	750	60	550	70	750	80	1000
70	1000	70	750	80	1000	90	1250
80	1300	80	1000	90	1250	100	1550
90	2400	90	1250	100	1550	110	1850
100	3800	100	1550	110	1850	120	2200
110	—	110	1850	120	2200	(130)	2600
120	8800	120	2200	(130)	2600	—	—
(130)	—	(130)	2600	—	—	—	—

( ) = exceptional values.

Note: V_d = design speed, km/h; *V85* = 85th-percentile speed, km/h; and R_{VSmin} = minimum radius of sag vertical curve, m.

TABLE 13.24 Minimum Radii of Sag Vertical Curves as a Function of Design Speed in Various Countries (Worldwide) and AL*

	Design speed, km/h										
	40	50	60	70	80	90	100	110	120	130	140
Country	Minimum radii of sag vertical curves, m										
Austria	1,000	1,500	2,000	2,500	3,000	—	5,000	—	8,000	—	12,000
Belgium	—	—	550	—	—	1,250	—	—	2,200	—	—
France	—	—	1,500	—	2,200	—	3,000	—	4,200	—	—
Germany											
1984[243]	—	—	1,500	2,000	2,500	3,500	5,000	—	10,000	—	—
1995[246]	—	500	750	1,000	1,300	2,400	3,800	—	8,800	—	—
Greece	—	1,350	1,900	2,500	3,300	4,200	5,200	6,300	7,500	10,000	—
Italy	550	—	1,200	—	2,200	—	3,900	—	5,800	—	—
The Netherlands	—	—	550	—	1,000	—	1,500	—	—	—	—
Spain	—	—	—	—	2,500	—	3,500	—	5,000	—	—
Sweden	—	1,400	—	2,800	—	4,500	—	5,500	—	—	—
	—	(900)	—	(1,800)	—	(3,500)	—	(5,500)	—	—	—
Switzerland, new research[91]	800	1,200	1,600	2,500	3,500	4,500	6,000	8,000	8,000	—	—
	700	—	1,500	—	2,900	—	4,700	—	6,700	—	9,000
United Kingdom	900	1,300	2,000	2,000	2,200		2,600	—	3,700	—	—
Canada	700	1,100	2,000	2,500	3,000	4,000	5,000	5,500	6,000	6,500	—
United States	800	1,200	1,800	2,500	3,200	4,000	5,100	6,200	7,300	—	—
South Africa	750	—	1,600	—	3,000	—	5,000	—	7,000	—	—
Australia 1 (comfort)	300	—	600	—	1,000	—	1,600	—	2,300	—	—
Australia 2 (headlight)	—	1,700	2,800	4,800	7,400	13,100	15,000	—	—	—	—
Japan	—	700	1,000	—	2,000	—	3,000	—	4,000	—	—
AL	—	750	1,000	1,250	1,550	2,400	3,800	6,300	8,800	10,000	—

*Modified and completed based on Refs. 325, 361, and 363.

( ) = exceptional values.

related to driving dynamic forces in the vertical direction, it may be additionally superimposed by driving dynamic forces in the longitudinal and lateral directions.

Therefore, for safety reasons, the authors decided to relate comfort control to the 85th-percentile speed. According to Levin,[477] these assumptions point in the right direction (see Fig. 13.35); that is, significant differences exist between design speeds and 85th-percentile speeds (up to $\Delta V = V85 - V_d \approx 30$ km/h) on crest vertical curves as related, for example, to Germany. On sag vertical curves, these speed differences might be even higher.

On the basis of safety criterion I, the "comfort control" of ranges 3 and 4 in Table 13.23 was examined for good and fair design practices. As can be seen from this table, the minimum radii of sag vertical curves of range 3 (good design) reach or exceed the values of range 1 for design speeds of ≤60 km/h, whereas the calculated minimum radii for comfort control of range 4 (fair design) exceed the values of range 1 for design speeds of ≤80 km/h.

By incorporating the fair design range of safety criterion I into new recommendations for modern highway geometric design, it follows then that the values of range 4 up to design speeds of $V_d \leq 80$ km/h should form a basis for comfort control, whereas for design speeds $V_d \geq 90$ km/h the values of range 1 should represent reliable minimum radii of sag vertical curves for the other controls. On the basis of these considerations, Table 13.7 was developed.

Table 13.24 gives a worldwide view of minimum radii of sag (concave) vertical curves as a function of design speed. Sag vertical curves are generally considered less critical from a safety standpoint than crest vertical curves. Several different principles are applied as a basis for the design standards. Several countries base their values on headlight illumination distance to satisfy stopping sight distance requirements on unlit roadways at night. Other countries base their design values on driver comfort. For a 60-km/h design speed, for example, the lower values (550 to 750 m) correspond to a comfort criterion, and the higher values (1500 to 2000 m) are based upon headlight illumination.[361] In "AL," the comfort criterion is based on the assumptions of safety criterion I (fair design range) up to design speeds of $V_d \leq 80$ km/h (see Table 13.23).

A graphical presentation of Table 13.24, with respect to the K factor and design speed, is given in Fig. 15.7 for selected countries. As can be seen from Table 13.24 and Fig. 15.7, the minimum radii of sag vertical curves range from 1000 to 3500 m for an 80-km/h design speed, 1500 to 6000 m for a 100-km/h design speed, and 2200 to 10,000 m for a 120-km/h design speed. These values represent a wide variation.[361]

The newly proposed values of "AL" represent minimum levels for design speeds of ≤80 km/h, but they contain all relevant controls for sag vertical curve design. For higher design speeds of $V_d \geq 110$ km/h, relatively large values are required to compensate for excessive speeds on multilane highways. Overall, the proposed minimum radii of sag vertical curves lead to balanced designs in terms of economic, environmental, and safety-related issues.

13.2.4.4   Appearance Control.   The appearance control for crest and sag vertical curves, as recommended in Table 13.8, results from two interesting recommendations made in the guidelines of South Africa[663] and Germany.[243] Both guidelines state the following. For general appearance, minimum lengths of vertical curves for flat gradients are recognized to eliminate the impression of a kink in the alignment.

South Africa.   The length, in meters, of both sag and crest vertical curves should not generally be less than V_d, where V_d is the design speed, km/h. In terms of the minimum radius, the requirement is

$$R_{V\text{min}} = \frac{100 \, V_d}{|G_1 - G_2|} \tag{13.15}$$

Germany.   The tangent length, T (see Figs. 13.12 and 13.34), should be at least

$$T_{\text{min}} = V_d \tag{13.16}$$

According to Eq. (13.13) in Fig. 13.34, it follows that:

$$R_{V\text{min}} = \frac{200 \, V_d}{|G_1 - G_2|} \tag{13.17}$$

As can be seen from Eqs. (13.15) and (13.17), the Germans require for the appearance control criterion minimum radii of vertical curves that are twice as high as those of South Africa.

Very small changes of grade in relation to design speed were defined in the guidelines of the United Kingdom[139] according to the relationships presented in Table 13.8. Based on Eqs. (13.15) and (13.17), minimum radii of vertical curves were calculated, as shown in Table 13.25. The upper values shown in the bold-faced boxes represent the minimum radii of vertical curves recommended in South Africa, whereas the lower values represent the German recommendations for good appearance control. On the basis of Table 13.25, the authors regard the proposed ranges for lengths and radii of vertical curves for very small changes of grade in Table 13.8 as appropriate for a sound appearance control.

13.2.4.5   General Controls for Vertical Alignment.   In addition to the statements made in corresponding Sec. 13.1.4.5, the following information has been given in the South African Design

TABLE 13.25 Minimum Radii of Vertical Curves for Very Small Changes of Grade Based on the South African and German Guidelines

V_d		40	50	60	70	80	90	100	110	120	(130)
$\|G_1 - G_2\|$		R_{Vmin}, m, for minor changes in grades									
1.0	SAF	**4,000**	5,000	6,000	7,000	8,000	9,000	10,000	11,000	12,000	13,000
	GER	8,000	10,000	12,000	14,000	16,000	18,000	20,000	22,000	24,000	26,000
0.9	SAF	4,444	**5,556**	6,667	7,778	8,889	10,000	11,111	12,222	13,333	14,444
	GER	8,888	11,111	13,333	15,556	17,778	20,000	22,222	24,444	26,667	28,889
0.8	SAF	5,000	6,250	**7,500**	8,750	10,000	11,250	12,500	12,750	15,000	16,250
	GER	10,000	12,500	15,000	17,500	20,000	22,500	25,000	27,500	30,000	32,500
0.7	SAF	5,714	7,143	8,571	**10,000**	11,428	12,857	14,286	15,714	17,142	18,571
	GER	11,429	14,286	17,143	20,000	22,857	25,714	28,571	31,429	34,286	37,143
0.6	SAF	6,667	8,333	10,000	11,667	**13,333**	15,000	16,667	18,333	20,000	21,667
	GER	13,333	16,667	20,000	23,333	26,667	30,000	33,333	36,667	40,000	43,333
0.5	SAF	8,000	10,000	12,000	14,000	16,000	**18,000**	20,000	22,000	24,000	26,000
	GER	16,000	20,000	24,000	28,000	32,000	36,000	40,000	44,000	48,000	52,000
0.4	SAF	11,000	12,500	15,000	17,500	20,000	22,500	**25,000**	27,500	30,000	32,500
	GER	20,000	25,000	30,000	35,000	40,000	45,000	50,000	55,000	60,000	65,000
0.3	SAF	13,333	16,667	20,000	23,333	26,667	30,000	33,333	**36,667**	40,000	43,333
	GER	26,667	33,333	40,000	46,666	53,333	60,000	66,667	73,333	80,000	86,667
0.2	SAF	20,000	25,000	30,000	35,000	40,000	45,000	50,000	55,000	**60,000**	65,000
	GER	40,000	50,000	60,000	70,000	80,000	90,000	100,000	110,000	120,000	130,000

Note: SAF = values according to the South African guidelines; GER = values according to the German guidelines.

Standards[663] and in the French Highway Design Guide.[700] A road alignment is judged by its appearance in three dimensions. A good three-dimensional design will increase utility and safety, encourage a uniform speed, and improve appearance, almost always without additional costs.

Good optical guidance is important for safety and traffic operation. This is achieved, when the road alignment ahead has a continuous and flowing appearance and the course of the highway is clear and distinct, and can be readily perceived. The guidance is further enhanced by proper application of road markings and traffic signs, which should always be treated as integral parts of road design.[663]

A good coordination between the horizontal and the vertical alignment is necessary in order:

- To ensure good general conditions of visibility.
- To ensure, if possible, a certain visual comfort for new roads by avoiding an alignment that gives a broken or discontinuous impression; this will generally mean ensuring that the curves of the horizontal and vertical alignment coincide, and designing large vertical curves in comparison to those of the horizontal alignment.

However, for safety reasons, the beginning of horizontal curves (especially when their radii are less than 300 m) should not coincide with a high point on the vertical alignment (or be located in close proximity to it), which is likely to be highly detrimental to the perception of the curve.

Junctions and frontage access should not coincide with curves in the horizontal alignment nor with zones of reduced visibility.[700]

The appropriate coordination of the design elements of the vertical alignment with those of the horizontal alignment and cross section will be discussed in detail in Chap. 16, "Three-Dimensional Alignment."

CHAPTER 14
DESIGN ELEMENTS OF CROSS SECTION

14.1 RECOMMENDATIONS FOR PRACTICAL DESIGN TASKS*

14.1.1 General

As a vehicle traverses a circular curve, it is subject to a centripetal force which must be sufficient to balance the inertial forces associated with the circular path, as indicated in Chap. 10 and Sec. 12.1.2. For a given radius and speed, a set force is required to maintain the vehicle in the circular path. In road design, this is provided by the side friction which is developed between tire and pavement and by the superelevation.[37] Therefore, superelevation in horizontal curves is introduced in order to counteract the radial forces.

For normal values of superelevation, the following formula is used:

$$e + f_R = \frac{v^2}{g\,R} = \frac{V^2}{127\,R} \qquad (10.1;\ 14.1)$$

where e = superelevation which represents the inward tilt or transverse inclination given to the cross section of the roadway throughout the length of a horizontal curve to reduce the effects of centrifugal force on a moving vehicle. This is taken as positive, if the pavement slopes toward the center of the curve. Pavement superelevation is expressed as a percentage, %/100.

f_R = side friction factor or coefficient of the side frictional force which is developed between the vehicle tires and road pavement. It is taken as positive if the frictional force on the vehicle acts toward the center of the curve.

g = acceleration due to gravity = 9.81, m/s.^{2}

v = speed of the vehicle, m/s.

V = speed of the vehicle, km/h.

R = radius of curve, m.

The superelevation chosen is based primarily on safety and on other factors like comfort and appearance. The superelevation applied to a road should take into account:[37]

- The 85th-percentile speed of the curve
- The tendency of very slow-moving vehicles to track toward the center
- The stability of heavily loaded commercial vehicles
- The difference between inner and outer formation levels, especially in flat country
- The length available for the introduction of the necessary superelevation.[37]

---

*Elaborated based on Refs. 5, 35, 37, 241, 243, 246, 453, 663, 686, 692, and 700.

14.1.2   Superelevation

14.1.2.1   Superelevation on Tangents.   The superelevation on a roadway must be sufficient to avoid the accumulation of water on the road surface and to prevent any buildup of pools of water during rain. Therefore, the necessary superelevation rate of the pavement width has to be designed according to Fig. 14.1 for drainage requirements. The pavement width is considered to be the lane width(s), the edge strip(s), and the paved shoulder width(s) (like emergency lanes) taken together in order to facilitate road construction. [See Figs. 24.1 to 24.3 and 25.10 of Part 3, "Cross Sections (CS)."]

The minimum or standard superelevation rate of the pavement width on tangents is

$$e_{min} = 2.5\% \tag{14.2}$$

for all road category groups. The minimum superelevation rate of pavement crown sections is often called *cross slope* or *crossfall.*

For two-lane and three-lane roads, a one-sided superelevation is recommended. Three-lane roads are discussed in Secs. 25.1.2 and 25.2.2, "Intermediate Cross Sections," of "CS."

In many countries, however, a crown section is used on tangents as a normal design case. A crown section can also be economical and even desirable for redesigns and RRR projects of existing roads. In these cases, the pavement width is made up of two draining surfaces, with their intersection normally being at the road axis (see Fig. 14.1).

Through lanes on median-separated multilane highways are always designed with one-sided superelevation rates on tangents sloped toward the outside edge of the road (see also Fig. 14.1).

Left paved shoulders should have the same superelevation as the adjoining through lanes. Medians should be designed in such a way that they can drain water away from the through lanes.

For a stabilized hard shoulder (unpaved), a superelevation rate of 4 percent is recommended in the same direction as the adjoining pavement.

14.1.2.2   Superelevation on Circular Curves.   For driving dynamic reasons, superelevation on circular curves must normally be sloped toward the inside edge of the curve.

For a given speed, the maximum superelevation rate and the assumed value for maximum permissible side friction factor in combination determine the maximum curvature. The maximum rates of superelevation usable on highways are controlled by several factors: climate conditions, that is, frequency and amount of snow and ice, terrain conditions (flat or mountainous), area type (rural or urban), and frequency of slow-moving vehicles that would be subject to uncertain operation.[5] The joint consideration of these factors (see Table 10.1) leads to the following conclusions. The maximum superelevation rates in circular curves are limited:

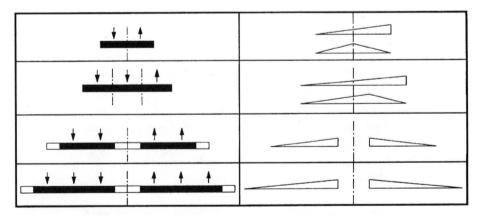

FIGURE 14.1   Superelevation types on tangents.[241,243,246]

- For roads of category group A (rural roads) in flat topography to

$$e_{max} = 8\ (9)\% \tag{14.3}$$

- For roads of category group A (rural roads) in hilly and mountainous topography to

$$e_{max} = 7\% \tag{14.4}$$

- For roads of category group B (suburban roads) to

$$e_{max} = 6\% \tag{14.5}$$

The value in parentheses for flat country is only valid in areas without ice and snow.

The following definitions apply to the different topography types.[663] A second definition is given in Subchapter 13.1. Both definitions are included in Part 2, "Alignment." Even though they generally yield similar conclusions, they provide somewhat differentiated interpretations:

Flat topography: Level or gently rolling country which offers few obstacles to the construction of a road having continuously unrestricted horizontal and vertical alignment (transverse terrain slope around 5 percent).

Hilly topography: Rolling, hilly, or foothill country where the slopes generally rise and fall moderately gently, and where occasional steep slopes may be encountered. It will offer some restrictions in horizontal and vertical alignment (transverse terrain slope around 20 percent).

Mountainous topography: Rugged, hilly, and mountainous country and river gorges impose definite restrictions on the standard of alignment obtainable and often involve long, steep grades and limited sight distances (transverse terrain slope up to 70 percent).[663]

By applying high values of superelevation rates, in combination with high longitudinal grades, the maximum relative slope of 10 percent should never be exceeded in order to avoid vehicle skidding under snow and ice conditions.[243]

For drainage reasons, the minimum superelevation rate in circular curves should be the same as in tangent sections:

$$e_{min} = 2.5\% \tag{14.6}$$

Based on the investigations and the assumptions made in Sec. 10.1.3 about maximum and minimum superelevation rates, as well as reliable utilization ratios of side friction factors for different category groups and topography classes, a coordination of radii of curve, superelevations, and 85th-percentile speeds could be taken from Figs. 14.2 to 14.4. The superelevations in these figures should be rounded to the next highest 0.5 percent.

The development of Figs. 14.2 to 14.4 is based on a linear interpolation of the results in Table 10.2 for maximum and minimum superelevation rates in relation to minimum radii of curve. In this way, the utilization ratios of side friction are also linearly interpolated and show a decrease with decreasing superelevation rates. To guarantee as much driving dynamic safety as possible, the superelevation rates in Figs. 14.2 to 14.4 are related to the actual driving behavior expressed by the 85th-percentile speeds and not to the design speed. Note that this is a very cost-effective way for improving safety by using higher superelevation rates. The 85th-percentile speeds can be determined for two-lane rural roads of category group A from the operating speed backgrounds of Fig. 8.12*a* or Table 8.5 for the country under study with respect to the curvature change rate of the single curve, CCR$_s$. For those countries for which operating speed backgrounds do not exist or cannot be established, the background of a country that has similar driving behavioral characteristics should be considered or the average of Fig. 8.12*b* should be selected.

For multilane median-separated rural roads of category group A, as well as for roads of category group B, recommendations were provided in Sec. 8.1.2 in order to establish reliable 85th-percentile speeds (see also Table 6.3).

For circular curves with very small deflection angles, a fully superelevated section has to be provided for a section length, in meters, which is passed in 2 s at the design speed level.[246]

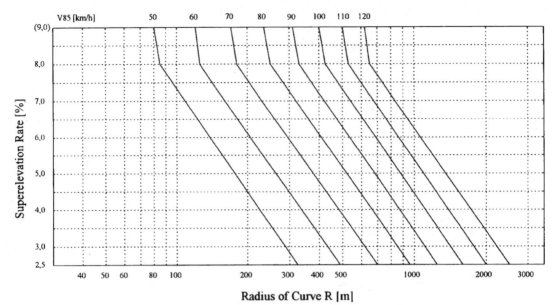

() exceptional case, applies only in areas without ice and snow

Legend:
Developed for
e_{max} = 8(9) % and n = 45 %,
e_{min} = 2.5 % and n = 10 %,
n = utilization ratio of side friction [%].

FIGURE 14.2 Superelevation rates in relation to 85th-percentile speeds and radii of curve of category group A (rural roads, flat topography) for new design.

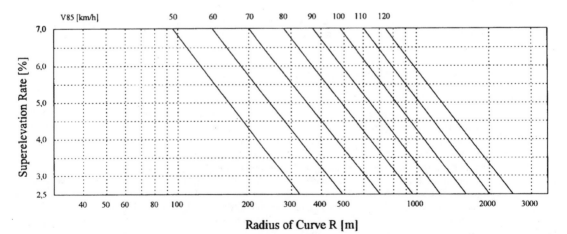

Legend: Developed for e_{max} = 7 % and n = 40 %,
e_{min} = 2.5 % and n = 10 %.

FIGURE 14.3 Superelevation rates in relation to 85th-percentile speeds and radii of curve of category group A (rural roads, hilly, and mountainous topography) for new design.

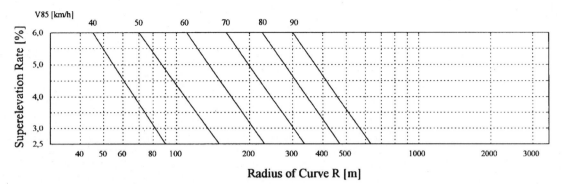

FIGURE 14.4 Superelevation rates in relation to 85th-percentile speeds and radii of curve of category group B (suburban roads, all topography classes) for new design.

Additional lanes, like emergency lanes, multiple purpose lanes, edge strips, and paved shoulders, should have the same superelevation rate and direction in the curve as the adjoining lanes.

14.1.2.3 *Negative Superelevation.*

Besides the most important design tasks, like the selection of an appropriate design speed, V_d, and the development of sound operating speeds, $V85$, as a basis for determining the limiting and standard design parameters of the horizontal and vertical alignments and of the cross section, the drainage of the road surface represents an additional important safety issue. In this connection, the design of weak drainage areas, for example, in the transition from a right-handed to a left-handed curve with low longitudinal grades, is often very critical. In addition, for large radii of curve, the minimum superelevation rate of $e_{min} = 2.5$ percent, which is used in tangent sections, is also recommended for application in these curves.

For solving the problem, two variants are available according to Redzovic.[289,596] The first variant is used for two- or three-lane, as well as for median-separated multilane, roads. In this case, the distortion of the road surface in the transition section leads to a superelevation rate which is directed toward the inside edge of the left-handed curve. It is called *positive superelevation* (+e). Variant I is presented in Fig. 14.5.

The second variant is only appropriate for median-separated multilane roads with extremely large radii of curve and insufficient longitudinal grades. In this case the minimum superelevation rate of $e_{min} = +2.5$ percent in the right-handed curve is maintained in the transition section and through the following left-handed curve. Accordingly, it is directed toward the outside edge of the left-handed curve. It is called *negative superelevation* (−e). Variant II is presented in Fig. 14.6.

Variant I.[596] In the case of the first variant (positive superelevation in right-hand and left-hand curves), high water-film depths can be expected in the area of the distortion section, since the superelevation rate goes through "zero," and possible hydroplaning effects may lead to dangerous traffic conditions.[321,336]

Based on driving dynamic considerations, the variant I solution is better, since the (positive) superelevation covers a portion of the centrifugal force of the vehicle passing through the curve and contributes to an increase in traffic safety (Fig. 10.15).

The basic driving dynamic formula for a positive superelevation is as follows:

$$\frac{v^2}{g\,R} \approx f_R(v) + e \qquad (10.29)$$

where v = speed, m/s
$\quad\quad g$ = acceleration of gravity, m/s^2
$\quad\quad R$ = radius of curve, m
$\quad f_R(v)$ = side friction factor
$\quad\quad e$ = superelevation rate, %/100

Variant II.[596]   The second variant, that is, directing the (negative) superelevation toward the outside edge of a left-hand curve after a right-hand curve (Fig. 14.6), is in contrast to the first variant of advantage for draining the road surface because the superelevation of the surface does not change in the course of the transition section. Consequently, no weak drainage areas exist.

However, from a driving dynamic point of view, this variant represents a less sound solution because the portion of the centrifugal force, which is normally covered by the positive superelevation, leads, in the case of a negative superelevation, to an additional utilization of the side friction factor, as can be seen from a comparison of Figs. 14.7 and 10.15. For instance, the resistant force, $Q \sin \alpha$, for positive superelevation (Fig. 10.15) acts in the opposite direction for negative superelevation (Fig. 14.7) and becomes a tractive force. Furthermore, it can be stated that in Fig. 10.15 the centrifugal force component, $F \sin \alpha$, supports, in the case of positive superelevation, the vehicle weight component, $Q \cos \alpha$, so the motor vehicle is additionally forced toward the road surface. This leads to an increase of the wheel loads. In contrast, the centrifugal force component, $F \sin \alpha$, acts, in the case of negative superelevation, in the opposite direction (Fig. 14.7) and reduces the vehicle weight component, $Q \cos \alpha$, and the wheel loads.[289,596]

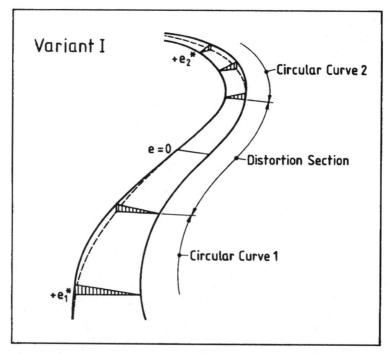

*   For large radii of curve: $e_1 = e_2 = + 2.5$ %

FIGURE 14.5   Transition from a right-hand to a left-hand curve with distortion section and positive superelevation directed toward the inside edge of the left-hand curve. (Elaborated based on Refs. 289 and 596.)

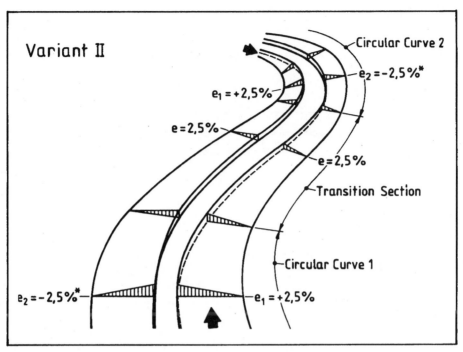

* For multilane, median-separated roads with large radii of
  curve and insufficient longitudinal slope: $e_2 = -2.5$ %

FIGURE 14.6   Transition from a right-hand to a left-hand curve and negative superelevation directed toward the outside edge of the left-hand curve. (Elaborated based on Refs. 289 and 596.)

By summing up the forces in the radial direction and by assuming equal side friction factors on all wheels, then, according to Fig. 14.7, the following equation for negative superelevation would result, if skidding is to be avoided:

$$f_R(v) \, (Q \cos \alpha - F \sin \alpha) \geq F \cos \alpha + Q \sin \alpha \tag{14.7}$$

[compare Eq. (10.27) for positive superelevation].

Thus, the basic driving dynamic formula for negative superelevation is as follows:

$$\frac{v^2}{g \, R} = f_R(v) + (-e) \tag{14.8}$$

A comparison of the basic driving dynamic formulas for positive and negative superelevation indicates, for the same speed and for the same radius of curve, that a higher side friction factor is utilized in the case of negative superelevation as compared to positive superelevation:

Positive superelevation:

$$f_R(v) = \frac{v^2}{g \, R} - e \tag{14.9}$$

Negative superelevation:

$$f_R(v) = \frac{v^2}{g \, R} + e \tag{14.10}$$

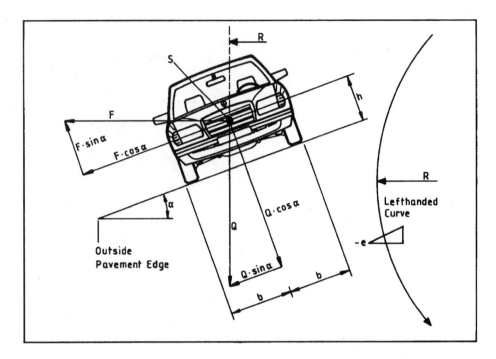

Legend (see also Figure 10-15):
F = centrifugal force [N],
Q = vehicle weight [N],
b = half of the vehicle width [m],
h = height of the center of gravity from the road surface [m],
α = angle of superelevation rate, tanα = -e [%/100],
S = center of gravity [-].

FIGURE 14.7   Motor vehicle driving through a left-hand curve with negative superelevation. (Elaborated based on Refs. 289 and 596.)

Thus, from a drainage point of view, negative superelevation should be regarded as favorable for safety issues. However, from a driving dynamic point of view, negative superelevation may contribute to unfavorable accident developments. Therefore, in order to keep the side friction demand as low as possible, negative superelevation rates should be used with extremely large radii of curve and only for insufficient longitudinal grades to minimize the previously discussed negative safety effects. Those design cases are typical for median-separated multilane roads in flat country.

According to Eq. (14.10), it follows then that the formula for the minimum radius of curve, in the case of negative superelevation, is as follows:

$$R_{min} = \frac{V85^2}{127\,[f_{Rperm} + (-e)]} \tag{14.11}$$

where R_{min} = minimum radius of curve, m

$V85$ = 85th-percentile speed, km/h, selected here for safety reasons

f_{Rperm} = maximum permissible friction factor in the radial (side) direction

n = utilization ratio of side friction

$(-e)$ = negative superelevation rate, %/100

Based on research evaluations, guideline comparisons, and new developments, as shown in Sec. 14.2.2.3, the following assumptions should be considered when determining the minimum radii of curve for negative superelevation:

$$e = 2.5\%$$

in agreement with the standard superelevation rate in tangents, and

$$n = 25\%$$

for safety reasons,

$$f_{R\text{perm}} = n\, 0.925\, f_{T\text{perm}} \tag{10.3}$$

$$f_{R\text{perm}} = 0.25 \times 0.925\, f_{T\text{perm}} \tag{14.12}$$

$$f_{T\text{perm}} = 0.59 - 4.85 \times 10^{-3}\, V + 1.51 \times 10^{-5}\, V^2 \tag{10.2, 10.40}$$

(The overall regression equation is based on Fig. 10.27.)

It follows that

$$f_{R\text{perm}} = 0.14 - 1.12 \times 10^{-3}\, V + 0.35 \times 10^{-5}\, V^2 \tag{14.13}$$

By introducing the preceding assumptions in Eq. (14.11), the minimum radii of curve shown in Table 14.1 could be determined. For negative superelevation, these values should not be reduced in order to guarantee the necessary side friction demand. Furthermore, it was decided to relate the minimum radii of curve for negative superelevation to the 85th-percentile speed and not to the design speed in order to achieve a higher safety margin, which means larger radii of curve. The relationship between design speed and 85th-percentile speed are given for median-separated multilane roads in Sec. 8.1.2 and in Table 6.3.

For a sequence of curves in the same direction of curvature, the direction of the superelevation rate (positive or negative) should not change at all.

14.1.3 Superelevation Runoff and Distortion

Superelevation runoff is the general term used to denote the length of highway needed to accomplish the change in superelevation between two adjoining circular curves with different superelevation rates or between a tangent section (one-sided superelevation or normal crown) to a fully superelevated circular curved section. In order to achieve the best solution for water runoff, the length of the superelevation runoff should be as short as possible; consideration of the limitations imposed by esthetics and rotation should be taken into consideration here.

TABLE 14.1 Minimum Radii of Curve in Relation to Negative Superelevation Rates Applied on Median-Separated Multilane Roads of Category Groups A and B

$V85$, km/h	R_{min}, m,* $e = -2.5\%$
70	700
80	1000
90	1500
100	2000
110	3000
120	3500
130	4500
140	5500

*Rounded.

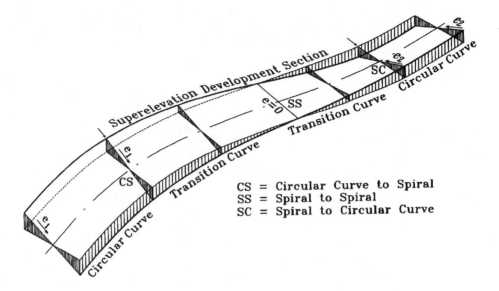

FIGURE 14.8   Three-dimensional presentation of a distortion section (reversing clothoid). (Elaborated based on Ref. 35.)

The careful design of the superelevation runoff increases the safety and comfort when driving through curves and improves the optical effect of curves. The drainage of the surface water should be guaranteed.

14.1.3.1   General.* A change of curvature in plan normally requires a change of superelevation, and it is accomplished by rotating the road surface around the centerline. The common case is the reversing clothoid where the superelevation rate goes through zero (Fig. 14.8).

For assessing the permissible change of superelevation (rate of rotation), the driving dynamic, optical, and drainage relevant conditions have to be reviewed.

Because of driving dynamic and optical considerations, one should strive for a long, smooth distortion; this is common in railroad construction. This request, however, often contradicts good drainage properties.

The change of superelevation rates takes place in the transition section, and it is called *superelevation development* or *distortion section*. In this section, the superelevation runoff is made and the traveled way is distorted.

[According to the definitions in Part 3, "Cross Sections (CS)," the term *traveled way* refers only to the travel lanes and edge strips, whereas the term *pavement* may also include the paved shoulders, like emergency or multiple purpose lanes, if present.]

The exact width of the traveled way is taken as a basis in most guidelines concerning the axes of rotation; as examples, see Figs. 24.1 to 24.3 and 25.10 of "CS."

The adopted position of the axis of rotation, that is, the point around which the traveled way is rotated in order to develop the superelevation depends on the type of the road facility, the cross section adopted, the terrain, and the location of the road. On a two-lane rural road, the superelevation is normally developed by rotating both halves of the traveled way around the centerline (axis of rotation) (see Fig. 14.9, case 1).

---

*Elaborated based on Refs. 35, 37, 243, 246, and 693.

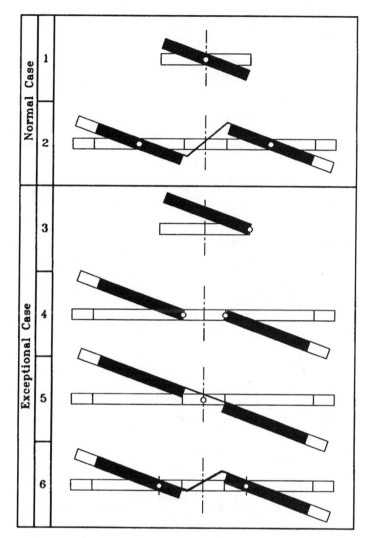

FIGURE 14.9 Examples for rotation axes of the traveled way. (Elaborated based on Refs. 243 and 246.)

On median-separated multilane rural roads, the rotation is normally performed around the centerlines of the traveled way for each direction of travel (see Fig. 14.9, case 2).

In exceptional cases, two-lane rural roads can be rotated around the edges of the traveled way (see Fig. 14.9, case 3).

Furthermore, on median-separated multilane highways where the median is relatively narrow, it may be desirable to adopt either the left (inside) edge of the traveled way in each direction of travel as the axis of rotation (Fig. 14.9, case 4) or the centerline of the median (Fig. 14.9, case 5).

For high-speed roads with design speeds $V_d = 100$ to 130 km/h, separated gradients for the traveled way in each direction of travel can represent economical solutions in the distortion section. In those cases, the rotation axis of the traveled way in each direction of travel does not have to be centered (see Fig. 14.9, case 6).[243,246]

If a transition curve is present, then the superelevation runoff should be performed within the transition curve, regardless of the type of applied superelevation (one-sided or normal crown) and the axis around which the traveled way is rotated. An encroachment of the superelevation runoff beyond the transition curve into a curved or tangent section should be avoided. If no transition curve is present, then the development of the superelevation runoff between the element sequence tangent–to–circular curve should be conducted on two-thirds of the tangent and on one-third of the circular curve.

14.1.3.2 Limiting Values.*

The length required to develop superelevation should be adequate to ensure a good appearance and give satisfactory riding qualities. The higher the design speed or wider the traveled way, the longer the superelevation development will need to meet the requirements of appearance and comfort. The criteria used to determine this length are:

- Rate of rotation of the traveled way, see Sec. 14.2.3.2
- Relative change of grade of the edges of the traveled way in relation to the grade of the axis of rotation, that is, "relative grade"

The relative grade can be determined from the following equation:

$$\Delta s = \frac{e_e - e_b}{L_e} d \qquad (14.14a)$$

where Δs = relative grade of the superelevation runoff, %
e_e = superelevation rate at the end of the superelevation runoff, %
e_b = superelevation rate at the beginning of the superelevation runoff, % (e_b is negative when e_b is opposite e_e)
L_e = length of the superelevation runoff, m
d = distance from the edge of the traveled way to the rotation axis, m

In order to avoid a too fast increase in the superelevation rate within the transition section, which may have unfavorable driving dynamic and optical effects, the maximum relative grade, max Δs, should not exceed the values shown in Table 14.2 for roads of category groups A and B.

*Elaborated based on Refs. 37, 124, 243, and 693.

TABLE 14.2 Limiting Values for Relative Grades*

V_d km/h	max Δs, %		min Δs,† %
	$d < 4.00$ m	$d \geq 4.00$	
(1)	(2)	(3)	(4)
50	0.5 d	2.0	
60	0.4 d	1.6	
70	0.35 d	1.4	
80	0.25 d	1.0	0.1 d
90	0.225 d	0.9	($\leq$ max Δs)
100	0.2 d	0.8	
110	0.2 d	0.8	
120	0.175 d	0.7	
130	0.15 d	0.6	

*Elaborated, modified, and completed based on Refs. 243 and 246.
†Only for areas of $e \leq 2.5\%$.

The minimum length of the superelevation runoff, $L_{e\,min}$, can be calculated from Eq. (14.14b) by regarding the maximum relative grade, max Δs, of Table 14.2 and the distance, d, from the edge of the traveled way to the rotation axis:

$$L_{e\,min} = \frac{e_e - e_b}{max\ \Delta s}\,d \qquad (14.14b)$$

where $L_{e\,min}$ = minimum length of superelevation runoff, m

max Δs = maximum relative grade of the superelevation runoff, %

Common examples of transitions between tangents and curves and between curves and curves are shown in Fig. 14.10 in addition to the resulting distortions of the traveled way and the corresponding superelevation runoffs.

A smoothing of the kinks which exist at the beginning and at the end of the distortion sections is not necessary, since the differences in heights are mostly within the ranges of structural tolerances.

14.1.3.3   Drainage Considerations.* In the distortion section, relative grades in the area of e_{min} = 2.5 percent through zero to e_{min} = −2.5 percent should not be allowed to fall short of the min Δs shown in Table 14.2.

$$min\ \Delta s = 0.10\,d \le max\ \Delta s \qquad (14.15)$$

where min Δs = minimum relative grade of the superelevation runoff, %

d = distance from the edge of the traveled way to the rotation axis, m

However, if such a case should come up, then a distortion in two stages must be arranged, that is, the area from e = +2.5 percent to e = −2.5 percent, should be sloped with min Δs. For the remaining section of the transition curve, the rest of the superelevation runoff should be accomplished until the necessary superelevation rate at the beginning of the circular curve is reached (see Fig. 14.11).

If no sufficient longitudinal grade can be ensured for the sequence circular curve–reversing clothoid–circular curve in the area of the superelevation zero point, this point can be displaced for a length of L = 0.1 A (where A = parameter of the clothoid) for roads of category group A and for L = 0.2 A for roads of category group B. This is also valid for the sequence tangent–clothoid–circular curve in order to achieve a better water runoff.

In addition, longitudinal grades and minimum relative grades of the superelevation runoff have to be coordinated using Eq. (13.1) in "Minimum Grades in Distortion Sections" in Sec. 13.1.1.2.

Further important information on drainage can be found in "Drainage" in Secs. 25.1.3.4 and 25.2.3.4 in Part 3, "Cross Sections."

14.1.4 Pavement Widening†

14.1.4.1   General. When vehicles traverse horizontal curves at normal speeds, they occupy larger widths of roadway compared to those on tangents. This is due to the fact that the rear wheels follow the front wheels on a narrower radius. The effect of this is an increase in the width of the vehicle in relation to the lane width of the roadway (Fig. 14.12). The amount of increase in occupation of the roadway depends on curve radius and on the vehicle's dimension. The curves made by the rear wheels are called *swept curves.*

---

*Elaborated based on Refs. 35, 124, and 243.

†Elaborated based on Refs. 5, 35, 243, 246, 563, and 607.

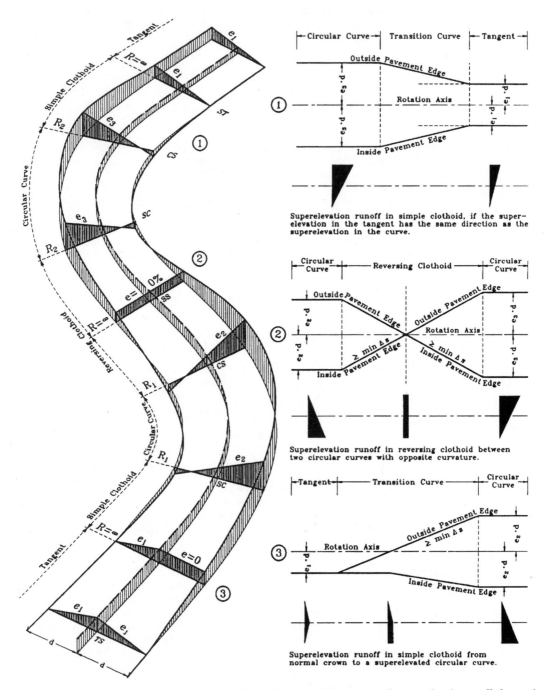

FIGURE 14.10 Examples of different distortions of the traveled way and the corresponding superelevation runoffs for transitions between tangents and curves and between curves and curves (see Fig. 12.8). (Elaborated, modified, and completed based on Ref. 693, SN 640, 195a.)

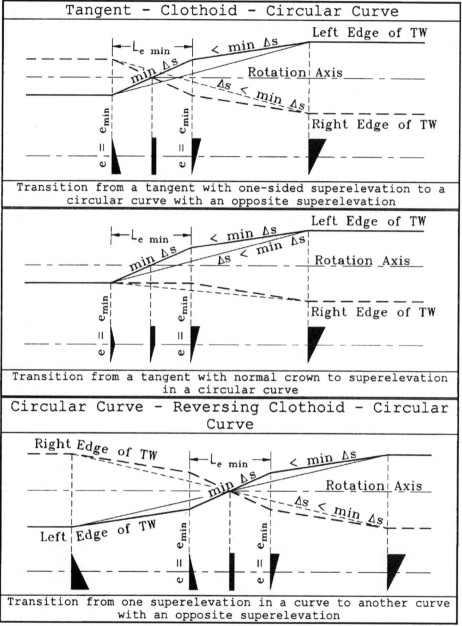

Legend: TW = Travelled Way
$L_{e\ min}$ represents here the exceptional case for min Δs and e_{min} = |2.5%|

FIGURE 14.11   Superelevation runoff development for relative grades Δs < min Δs for different transition cases. (Elaborated and modified based on Ref. 243.)

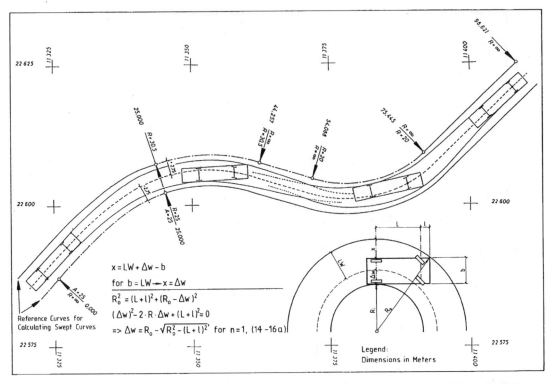

FIGURE 14.12   Roadway section showing the covered area by the swept curves for a two-axle design vehicle truck. (Elaborated based on Refs. 236 and 563.)

For the range of radii used on open highways, this additional increase in pavement width* can generally be ignored for passenger cars. For trucks or buses, however, this increase is rather significant and, therefore, should be considered in highway geometric design. Figure 14.12 shows the area covered by the swept curves of a two-axle design vehicle truck on a roadway section consisting of tangents and transition and circular curves, based on a CAD-program for swept curve examination.[236,563]

One formula that provides a reliable estimation of pavement widening at curved sites is expressed as follows[246] (see also Fig. 14.12):

$$\Delta w = n \left(R_o - \sqrt{R_o^2 - (L + l)^2} \right) \qquad (14.16a)$$

---

*The technical terms *pavement width* and *width of the traveled way* are the same for two-lane roads in this chapter. See Fig. 24.1 in Part 3, "Cross Sections."

where Δw = pavement widening, m
R_o = radius of the trajectory of the outer edge of the front overhang, m
R_i = radius of the trajectory of the inside rear wheel, m
R = radius of curve, m
A = parameter of the clothoid, m
L = length of the vehicle wheelbase, m
l = length of the front overhang, m
b = vehicle width, m
LW = lane width, m
n = number of lanes

For $R_o \approx R$, Eq. (14.16b) becomes relevant:[243,576]

$$\Delta w = n \left(R - \sqrt{R^2 - (L + l)^2} \right)$$ (14.16b)

For radii of curve of $R \geq 30$ m, Eq. (14.16b) can be replaced by the following approximated formula:[243,576]

$$\Delta w = n \frac{(L + l)^2}{2R}$$ (14.17)

For European design vehicles, the values given in Table 14.3 normally apply to the overall length, $L + l$ (length of the wheelbase plus length of the front overhang) (see Fig. 14.12).

The need for pavement widening at a curved site is dependent on one heavy design vehicle meeting another at that site, the frequency relative to heavy vehicle volume and distribution, curve radius, and speed. Failure to provide pavement widening at a curve will result in a higher concentration level as required by the driver and a possible reduction in speed. The need for pavement widening at a curve does not depend on the frequency of curves.[607]

14.1.4.2   Design Values

Two-Lane Roads.   On two-lane rural roads (pavement width of $w \geq 6.00$ m) where the number of heavy vehicles for both directions of travel does not exceed 15 vehicles/h, pavement widening is not required.[607] The same applies to pavement-widening increments of less than 0.50 m in order to avoid excessive construction costs with little gain in vehicle operation.

Figure 14.13 graphically shows the necessary widening values for curves with increasing radii for the representative design vehicles of Table 14.3 with respect to the opposing traffic. In general, widening should be realized on the inside edge of the curve and should be guaranteed over the whole length of the circular arc (Fig. 14.14).

Multilane Roads.   On multilane median-separated highways, vehicles may encounter other vehicles moving in the same direction. However, because of the normally generous alignments which

TABLE 14.3   Overall Lengths of Wheelbase and Front Overhang for Different Design Vehicles[246]

Vehicle type	Wheelbase plus front overhang ($L + l$), m
Passenger car	4.00
Single-unit truck	8.00
Full trailer combination (FTC)	10.00
Bus 1: single unit	8.50
Bus 2: articulated	9.00
Bus 3: megaliner	11.70

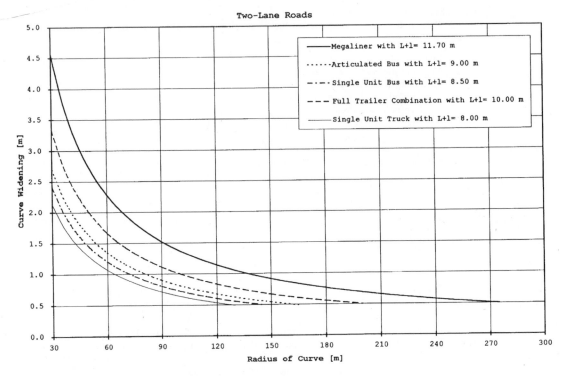

FIGURE 14.13   Curve widening for five representative design vehicles ($R > 30$ m).

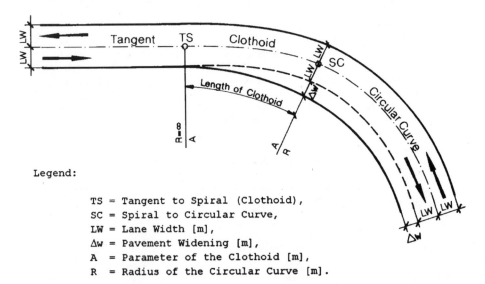

FIGURE 14.14   Widening attainment along the transition curve. (Elaborated, modified, and completed based on Ref. 35.)

correspond to these highways [in the majority of cases $R \geq 400$ m for design speeds of $V_d \geq 100$ km/h (Table 12.9)], the effects of vehicle off-tracking are small. Therefore, widening on multilane highways has to be considered only in very rare cases.

14.1.4.3   *Widening Attainment on Curves.*   The following rules apply for attaining widening on both ends of a curve (partially based on Ref. 5):

1. Widening should be done gradually and has to be realized on the inside edge of the pavement (Fig. 14.14).
2. In the case of a circular curve with transition curves on both ends, curve widening has to be attained along the full length of the transition curve (Fig. 14.14).
3. On highway curves without transition curves, widening should preferably be attained along the length of the superelevation runoff. A smooth-fitting alignment would result from attaining widening on one-half to two-thirds along the tangent and the remaining along the curve.
4. The pavement edge through the widening transition should be a smooth, graceful curve with rounded transition ends to avoid an angular break at the pavement edge.[5]
5. On widened curves, the final marked centerline should be placed midway between the edges of the widened pavement.

With the ever-increasing use of computer-aided design and drafting (CADD) for highway projects, the required pavement widening on curves can be analytically determined by calculating the swept curve paths of the different design vehicles moving along a roadway, as shown in Fig. 14.12.

For instance, in the swept curve program described in Refs. 277 and 563, a given horizontal curve will be traversed by a design vehicle at small intervals between which the differences in swept angles are additively computed in order to describe the changing position of the vehicle. The four coordinates of the relevant edges of the design vehicle, the stations, and distances to the centerline of the roadway are computed as well. The method allows the computation of the area covered by the swept curves for roadway sections with arbitrary sequences of design elements and their corresponding curvatures, as well as for the turning paths at intersections.

For pavement widenings in a hairpin curve design, it is reported by Pietzsch:[576]

> In mountainous topography, hairpin curves with high pavement widenings are necessary because of the narrow radii of curve. The hairpin curves are designed based on only driving geometric assumptions and not on driving dynamic demands. In order to inform the motorist in time about the change in road characteristics, hairpin curves should be introduced by a curve with opposing curvature. Furthermore, a sufficient sight distance space is necessary.

The minimum radii of curve should not fall short of $R_{min} = 12.50$ m for the road axis and $R_{min} = 5.30$ m for the inside pavement edge. The rules previously discussed for the transition from an unwidened to a widened cross section (Fig. 14.14) can only be used for hairpin curves down to a radius of $R = 30$ m. For smaller radii of curve down to a radius of $R = 12.50$ m, a swept curve design becomes necessary. If the relevant opposing case should also be possible within the hairpin curves because of the traffic relevance of the road, then pavement widenings have to be fulfilled individually for separate lanes of travel according to Fig. 14.15.[576]

14.1.4.4   *Lane Addition or Lane Drop.*   When the width of the traveled way has to be increased or decreased in order to add or to drop one or more lanes, the change in width of the traveled way should preferably be accomplished on curved sites by using a taper with an inclination of 1 to 40, as shown in Fig. 14.16. If this is not possible, then the change in width of the traveled way should be accomplished on tangent sections, but by using inclination values flatter than 1 to 40 for the taper. Both ends of the taper should be rounded by radii of $R = 1000$ m.[35]

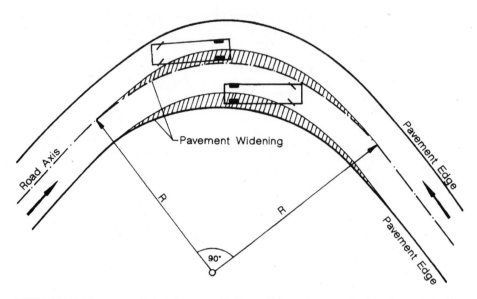

FIGURE 14.15   Swept curve design of narrow radii of curve in mountainous topography and pavement widenings of individual travel lanes.[576]

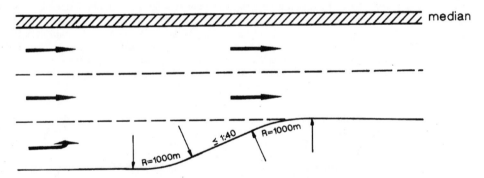

Legend: Taper inclination → 1:40 for curved sections

FIGURE 14.16   Change in width of the traveled way on a multilane highway (example: lane drop for a tangent section). (Elaborated, modified, and completed based on Ref. 35.)

The method described here works well and complies with various traffic demands, although it is conceptually simple. It has been applied successfully in Austria[35] and is also recommended here. Note that there are other existing methods which are more sophisticated from a geometric viewpoint, compare Sec. 14.2.4.

14.1.4.5 Preliminary Conclusion.   In proposing criteria and design principles for pavement widening and widening attainment on curves, the emphasis was to formulate a sound, easy-to-comprehend procedure, based on established methods.

Based on the accepted equation 14.17, the diagram in Fig. 14.13 was developed in order to determine curve widenings on two-lane roads, in relation to the "opposing" cases of the following representative design vehicles (Table 14.3):

1. Single-unit truck (wheelbase plus front overhang equal to 8.00 m)
2. Single-unit bus (wheelbase plus front overhang equal to 8.50 m)
3. Articulated bus (wheelbase plus front overhang equal to 9.00 m)
4. Full trailer combination (wheelbase plus front overhang equal to 10.00 m)
5. Megaliner bus (wheelbase plus front overhang equal to 11.70 m)

Figure 14.13 shows the curve-widening values in the range of radii of curve from $R \geq 30$ m to $R = 200$ m, which cover most of the design cases. For hairpin curve design ($R < 30$ m), other rules apply, which are mainly based on driving geometric assumptions (Fig. 14.15).

Based on the Austrian guidelines,[35] widening has to be applied gradually along the transition curve, as Fig. 14.14 reveals. This design is simple and also considers the Canadian and American recommendations in Sec. 14.2.4.

For a change in the width of the traveled way because of a lane addition or a lane drop, the Austrian approach[35] was also recommended. This approach is simple, practical, and requires no extra amount of work in the recommended design of the roundings at the taper ends according to Fig. 14.16.

14.2 GENERAL CONSIDERATIONS, RESEARCH EVALUATIONS, GUIDELINE COMPARISONS, AND NEW DEVELOPMENTS

14.2.1 General

According to AASHTO, the following general considerations are given about superelevation.[5] From accumulated research and experience, limiting values have been established for the superelevation rate, e, and the side friction factor, f_R (see Table 10.1). Using the e_{max} value with a conservative f_{Rperm} value in the basic equation 10.9 permits the determination of minimum curve radii for various design speeds (Table 10.2). For a given design speed, use of curves with radii larger than the minimum calls for balance in the factors involved in determining the desirable superelevation rates below the maximum.

There is a practical limit to the rate of superelevation. In areas subject to ice and snow, the rate of superelevation cannot be greater than that on which vehicles standing or traveling slowly would slide down the cross slope when the pavement is icy. At higher speeds, the phenomenon of partial hydroplaning can occur on curves with poor surface drainage that allows water buildup. Skidding occurs when the lubricating effect of the water film reduces available lateral friction below the demand being made by cornering. When traveling slowly around a curve with high superelevation, negative lateral forces develop and the car is held in the proper path only when the driver steers up the slope or against the direction of the horizontal curve. This direction of steering is an unnatural movement on the part of the driver and possibly explains the difficulty of driving on roads where the superelevation is in excess of that required for normal speeds. Such a high rate of superelevation is undesirable, especially for trucks with high centers of gravity.[5]

In highway curve design, it is necessary to determine superelevation rates which are applicable over the range of curvature for each design or operating speed. One extreme of this range is the maximum superelevation established by practical considerations and used to determine the maximum curvature for each speed. The maximum superelevation is different for different highway conditions. At the other extreme, minimum superelevation is needed for tangent highways or highways with extremely large radii of curve. For curvature between these extremes and for a given design speed, the superelevation should be distributed in such a manner that there is a logical relation between the side friction factor and the applied superelevation rate[5] as it is, for example, shown in Figs. 14.2 to 14.4. Furthermore, as the figures reveal, the superelevation rate should be determined based on the 85th-percentile speed instead of the design speed for a specific radius of curve. Because the 85th-percentile speed is normally higher than the design speed, a higher driving dynamic safety level can be provided by this assumption without any additional costs.

14.2.2   Superelevation

14.2.2.1   *Superelevation on Tangents.*   The main task of superelevation in tangents is to guarantee sufficient drainage of the pavement.

The minimum and standard superelevation rates in the tangent have been assessed to be e_{min} = 2.5 percent. The same minimum superelevation rate is valid for circular curves. According to Table 14.4, this value is also used in many other countries. Minimum superelevation rates normally range from 1.5 to 3.0 percent.

With respect to this range, two research investigations merit consideration. The first, conducted by Lamm, et al.,[404] concluded, based on empirical accident analyses on freeway sections with positive superelevation rates between 1.5 and 2.5 percent, that for lower minimum superelevation rates of e_{min} = 1.5 percent, no increase in accidents could be observed under wet surface conditions compared to the higher minimum superelevation rate of e_{min} = 2.5 percent; however, they indicated that a minimum superelevation rate of 2.5 percent is always substantiated by drainage requirements. On the other hand, they further indicated, based on driving dynamic investigations, that higher minimum superelevation rates were not more dangerous than lower minimum superelevation rates. Consequently, they concluded that the application of minimum superelevation rates between 2 and 2.5 percent did not yield significant differences in the accident situation.

The second investigation, an accident investigation conducted by Spacek[665] on 49 circular curves over a period of 5 years, yielded a minimum superelevation rate of 3 percent for all roads. This value was adopted in the Swiss norm SN 640 123a[692] because, theoretically, it can improve drainage conditions and reduce the negative effects of water pools.

14.2.2.2   *Superelevation on Circular Curves*

Background.   When driving through curves, a vehicle becomes subject to a centrifugal force. The extent of the centrifugal force depends on the speed of the vehicle: an increase in speed causes a considerable increase in the centrifugal force. A superelevation toward the inside edge of the roadway diminishes the effect of the centrifugal force.[686]

Furthermore, it is reported by AASHTO:[5] Where snow and ice are factors, tests and experience show that a superelevation rate of about 8 percent is a logical maximum to minimize slipping across a highway when stopped or attempting to slowly gain momentum from a stopped position. One series of tests[529] yielded coefficient of friction values for ice ranging from 0.050 to 0.200 depending on the condition of the ice, that is, wet, dry, clean, smooth, or rough. Other tests[253] corroborated these values.[5] These experiences are confirmed by the diagrams in Figs. 10.25 and 10.26.

Consequently, for roads of category group A (rural roads) in flat topography, a reliable maximum superelevation rate of

$$e_{max} = 8\% \tag{14.3}$$

was selected.

Furthermore, based on the assumption that the relative slope resulting from high superelevation rates and high longitudinal grades should not exceed 10 percent, a maximum superelevation rate for roads of category group A (rural roads) in hilly/mountainous topography of

$$e_{max} = 7\% \tag{14.4}$$

was considered reliable.

TABLE 14.4 Minimum and Maximum Superelevation Rates in Various Countries and AL*

Country	Minimum superelevation rate, %	Maximum superelevation rate, %
Australia	2.0–3.0	Flat terrain: 6–7 General maximum: 10 Mountainous terrain: 12
Austria	2.5	6–7
Belgium	2.5	5–6
Canada	2.0–1.5	6–8
Denmark	3.5–1.5	6
France	2.5	General maximum: 7 Mountainous terrain: 6
Germany	2.5	7
Greece	2.5	Flat topography: 8 Flat topography without ice and snow: 9 Hilly/mountainous topography: 7
Ireland	2.5	7
Italy	2.5	7
Japan	1.5–2.0	10
Luxembourg	2.0–2.5	5–6.5
Portugal	2.0	6 (8)†
South Africa	2.0–3.0	7
Spain	2.0	7 (10)†
Sweden	2.5–3.0	5.5
Switzerland	3.0 (2.5)†	7
The Netherlands	2.0 (2.5)†	5 (7)†
United Kingdom	2.5	5 (desirable maximum) 7 (absolute maximum)

United States	1.5–2.0	General maximum in areas:
		  with ice and snow: 8
		  with no ice and snow: 10
		  in exceptional cases: 12

AL‡	2.5	Flat topography: 8 Flat topography without ice and snow: 9 Hilly/mountainous topography: 7

*Elaborated, modified, and completed based on Refs. 153, 257, 361, and 363.

†( ) = exceptional cases.

‡AL = Part 2, "Alignment."

Where traffic congestion or extensive marginal development acts to curb top speeds, it is common practice to utilize a lower maximum rate of superelevation, usually 4 to 6 percent. Similarly, a lower maximum rate of superelevation is employed within important intersection areas or where there is a tendency to drive slowly because of turning and crossing movements, warning devices, and signals. In these areas, it is difficult to warp crossing pavements for drainage without negative superelevation for some turning movements.[5] However, a sufficient drainage always should be guaranteed, for example, by providing an adequate amount of longitudinal grade or relative slope.

Therefore, for roads of category group B (suburban roads), a reliable maximum superelevation rate of

$$e_{max} = 6\% \tag{14.5}$$

was selected.

Table 14.4 gives the minimum and maximum superelevation rate(s) for rural roadways in various countries.

With respect to minimum superelevation rate, it is interesting to note that there is a somewhat high degree of consistency between the countries. It appears that a 2.0- to 2.5-percent minimum superelevation rate is the most widely accepted value for design regardless of roadway class.[257]

With respect to maximum superelevation rates the following is reported:[361] Most countries have a single nationwide maximum rate, some of which are supplemented by a higher rate for exceptional cases. Maximum rates are limited by the risk of stationary vehicles sliding on icy or frozen pavement surfaces. In the largest countries (Australia, Canada, and the United States), which have wide ranges of climate, individual states or provinces may select their own maximum rates, and therefore a range of rates is indicated. For safety reasons, "AL" and Greece employ different maximum superelevation rates for flat and hilly/mountainous topography.

Most countries' normal maximum rates fall within the range of 6 to 8 percent. Sweden uses a maximum rate of 5.5 percent. A maximum rate of 10 percent is used in Japan and in rural areas not subject to ice and snow in some states in Australia and may be used under special circumstances in the United States.[361]

Safety Considerations

Approach of Part 2, "Alignment (AL)."   Adequate superelevation is an important design issue. Horizontal curves on high-speed highways are superelevated for safety and passenger comfort.[710] "AL" design policy specifies superelevation requirements based on radius of curve and 85th-percentile speed for *new designs* (see Figs. 14.2 to 14.4).

For a quantitative evaluation of the interplay between superelevation and side friction factor, safety criterion III was developed in order to distinguish between good, fair, and poor design practices (see Secs. 10.1.4 and 10.2.4). In this connection, improving superelevation plays an important role in increasing driving dynamic safety. For example, quantitative ranges are given in Table 11.1 for safety criterion III with respect to side friction assumed f_{RA} and side friction demanded f_{RD}.

Based on safety criterion III, the following case study shows how to examine *existing alignments* for sufficient or insufficient superelevation rates.

Case study (example: France): $R = 350$ m (no transition curves); $CCR_S = 182$ gon/km [Eq. (8.6)] or Table 8.4; $e = 3$ percent; and $V_d = 90$ km/h. For a CCR_S value of 182 gon/km, it follows from Fig. 8.12, related to the French operating speed background, or the corresponding regression model [Eq. (8.19)] given in Table 8.5 that the 85th-percentile speed is

$$V85 = 97 \text{ km/h}$$

The side friction assumed can be calculated from Eq. (10.10) in Sec. 10.1.4:

$$f_{RA} = 0.60 \times 0.925 f_{Tperm} \tag{10.10a}$$

with respect to f_{Tperm} [see Eqs. (10.2) or (10.40)]:

$$f_{RA} = 0.33 - 2.69 \times 10^{-3} V_d + 0.84 \times 10^{-5} (V_d)^2 \qquad (10.10b)$$

For $V_d = 90$ km/h, it follows that $f_{RA} = 0.156$.

The side friction demanded can be determined from Eq. (10.11) in Sec. 10.1.4:

$$f_{RD} = \frac{V85^2}{127 \, R} - e$$

$$f_{RD} = \frac{97^2}{127 \times 350} - 0.03 \qquad (10.11)$$

$$= 0.212 - 0.03$$

$$= 0.182$$

The difference between side friction assumed and side friction demanded of

$$f_{RA} - f_{RD} = 0.156 - 0.182$$

$$= -0.026$$

indicates that the observed curved roadway section falls into the fair range of safety criterion III where (see Table 11.1):

$$-0.04 \leq f_{RA} - f_{RD} < +0.01$$

or

$$-0.04 \leq -0.026 < +0.01$$

From the previous ranges for maximum superelevation rates, it becomes evident that an increase in superelevation of at least $\Delta e = 4$ percent, or better, $\Delta e = 5$ percent, would lead to a good design practice, as will be shown in the following.

For $\Delta e = 4$ percent, it follows then that the superelevation rate is now equal to $e = 3\% + 4\% = 7\%$. Consequently, the side friction demanded is as follows:

$$f_{RD} = \frac{97^2}{127 \times 350} - (e + \Delta e)$$

$$f_{RD} = 0.212 - (0.03 + 0.04)$$

$$= 0.142$$

Thus, the difference between the side friction assumed and that demanded is as follows:

$$f_{RA} - f_{RD} = 0.156 - 0.142 = +0.014 > 0.01$$

which falls into the good design range of safety criterion III (Table 11.1). Similarly, a superelevation increase of $\Delta e = 5$ percent (or $e = 8$ percent for the present case) leads to a difference of

$$f_{RA} - f_{RD} = 0.156 - 0.132 = +0.024 > 0.01$$

which definitely denotes a good design practice according to safety criterion III.

The previous example reveals that improving the superelevation on curves where superelevation is lower than that demanded according to safety criterion III, such as in the case of redesigns or as part of a RRR project, is a relatively simple and inexpensive way to improve driving dynamic safety by accounting for the centripetal force in Eq. (14.1).

Note, however, that the present example deals with an *existing alignment,* for which a utilization ratio of $n = 60$ percent for side friction assumed is regarded as justifiable [Eq. (10.10a)]. In contrast, for *new designs,* a utilization ratio of $n = 45$ percent is assumed for flat topography (Fig. 14.2) and $n = 40$ percent for hilly/mountainous topography (Fig. 14.3) in order to provide from the beginning as much driving dynamic safety as possible. Therefore, the previous results of the case study do not agree with the readings of Figs. 14.2 and 14.3. For an operating speed of $V85 = 97$ km/h ≈ 100 km/h, in combination with a superelevation rate of $e_{max} = 8$ percent, at least a minimum radius of curve of $R_{min} = 425$ m would be requested according to Fig. 14.2. From Fig. 14.3, where the maximum superelevation is limited to $e_{max} = 7$ percent, a minimum radius of $R_{min} = 480$ m is assumed for new designs in place of $R = 350$ m for the existing alignment, of the previous case study.

The results discussed do not represent contradictions but only attempt to use different driving dynamic safety levels for different application ranges. In doing so, for *existing alignments,* it is logical to recommend a relatively high utilization ratio of side friction when, at the same time, considering the already present and difficult to change economic and environmental impacts, since it is quite impossible to rebuild large portions of existing rural road networks.

However, in the case of new designs, it seems logical to provide a lower utilization ratio of side friction in flat country and the lowest value in hilly or mountainous topography where driving dynamic safety is additionally affected by the vertical alignment.

U.S. approach. *   A number of studies have attempted to link superelevation deficiencies to the cause or experience of accidents. For instance, a study by Zador, et al.[772] noted deficiencies in available superelevation at fatal accident sites compared with nearby control sites.

In a 1991 FHWA study by Zegeer, et al.,[781] a small but significant accident effect of too little superelevation was noted. The authors concluded that curve sites with a superelevation "deficiency" had significantly worse accident experience than curves with an adequate amount of superelevation. The *superelevation deficiency,* e_D, was defined as the difference between the recommended superelevation, e_R, according to AASHTO[5] and the actual superelevation, e_A, or $e_D = e_R - e_A$.

The percent reduction in the total number of accidents on curves due to improving superelevation is shown in Table 14.5. For example, assume the actual superelevation, e_A, on a curve is 4 percent and the AASHTO recommended superelevation, e_R, for a particular curve design is 6 percent. This corresponds to a superelevation deficiency, e_D, of 2 percent. According to Table 14.5, a 10 percent reduction in total curve accidents could be expected if adequate superelevation were provided to the horizontal curve. Such improvements to superelevation should be made to curves whenever the roadway section is redesigned or in cases of RRR projects.[781]

The 1991 study[781] also investigated the safety effect of too much superelevation. No adverse effects were found based on available data. Current design policy is implemented with an assumed upper limit on superelevation for areas with snow and ice. The presumption is that excess

---

*Elaborated based on Ref. 728.

TABLE 14.5   Accident Reduction Factors for Upgrading Superelevation on Existing Horizontal Curves[781]

e_D, %	Percent reduction in number of total accidents due to upgrading superelevation
1.0–1.9	5
≥2	10

Note: e_D = superelevation deficiency to be upgraded.

superelevation produces sliding down the curve under low-speed conditions, and hence increases accident potential. While this condition could theoretically occur at low-speed curve locations with sharp curvature and a high rate of superelevation, no evidence was found of any such significant adverse safety effects. Similar results for different approaches were revealed by Voigt and Krammes.[736]

Redesigns and RRR projects provide the opportunity to correct deficient superelevation at little or no extra costs. Correcting superelevation rates allows better drainage of the pavement's surface and improves vehicle control in wet weather.[710] Consequently, when applying the three proposed safety criteria (see previous case studies and those presented in Secs. 12.2.4.2 and 18.4) a careful attention given to necessary improvements of existing alignments, skid resistance, and superelevation will certainly lead to sound economic, environmental, and safer alignment designs. For instance, with respect to skid resistance, highway agencies should try achieving the skid resistance values that were previously recommended in Sec. 10.1.2.2.

It is furthermore required in the United States, with respect to resurfacing, that the normal pavement crown (cross slopes from the centerline on the tangent sections of two-lane roads that allow rainfall to drain on both sides of the roadway) be restored to generally match new construction requirements. Resurfacing projects provide highway agencies with the opportunity to correct deficient cross slopes at little or no additional cost. Even though the safety effects have yet to be evaluated in detail in future research, restoring cross slopes to match new construction standards is a good practice that highway agencies should routinely follow when resurfacing.[710,728] The same considerations are true for the normally recommended one-sided superelevated pavement in tangent sections of two- or three-lane rural roads (see Fig. 14.1).

14.2.2.3   Negative Superelevation. *   The use of negative (adverse) superelevation rate, that is, directed to the outside edge of left-turning curves of median-separated multilane highways, can be recommended as a sound solution for road surface drainage in terms of economics and environmental compatibility. Since in such a case distortion sections are no longer necessary, low-drainage zones in transition curves can be avoided. Furthermore, the expensive drainage in the median, which is very difficult to achieve at least in flat topography, is no longer relevant.

From a driving dynamic point of view, the negative superelevation causes a higher friction demand than the normal one and can have a negative effect on safety during critical driving maneuvers.

An increase in accidents at the end of the 1970s in curved areas with negative superelevation and at distortion sections on the autobahn system in Germany prompted fundamental investigations. A detailed discussion is given by Spacek in Ref. 665 and by Redzovic in Ref. 596.

Interstate sections with low longitudinal grades of $G \le 1.5$ percent, different radii of curve, and positive as well as negative superelevation rates between $e = |1.5|$ percent and $e = |2.5|$ percent were investigated with respect to

- Drainage aspects[321]
- Driving dynamic considerations[321,371]
- The accident situation[371,629]

Even though the results of these investigations differed with respect to the size of the radii of curve for negative superelevation, it was concluded, however, that radii of curve of $R \ge 5000$ m can be tolerated for a negative superelevation of $e = -2.5$ percent.

Furthermore, Krebs, Lamm, and Blumhofer[371] established a marginal unfavorable accident influence for the superelevation of $e = |2.5|$ percent as compared to $e = |1.5|$ percent and a tendency for a slight increase in the accident risk on sections with negative superelevation rates. Their evaluation of the accident situation also yielded the following conclusions:[371]

---

*Elaborated based on Refs. 289 and 596.

- For radii of curve ≥8000 m, a negative superelevation rate should always be applied.
- In the radii of curve range between 5000 and 8000 m, the application of negative superelevation rates is more favorable if the length of the circular curve and the succeeding transition curve fulfill the requirement $L \leq 2$ km. For longer lengths, distortion sections should be assigned.

However, according to Redzovic,[596] the driving dynamic investigations and accident analyses conducted so far do not allow reliable statements to be made on the friction potential and utilization when driving at different speeds through curves with different radii and negative superelevation rates. The same is true with respect to the assessment of limiting radii of curve for negative superelevation. To come to these conclusions, Redzovic used the test vehicle of the Institute for Highway and Railroad Engineering, University of Karlsruhe (Germany) and measured the driving dynamic forces relative to each of the wheels as well as to the center of gravity of the vehicle. His research illustrates the actual existing tangential and side friction demands for different test rides and driving maneuvers at curved sites with negative superelevation rate and varying radii of curve. Besides test rides in curves, driving maneuvers, which cause the largest side friction demands, were also investigated. These situations include passing, evasive, changing lane, and braking maneuvers in curves. Through the simultaneous utilization of tangential and side friction demands, Redzovic was able to show extreme impairments of the available friction.

Redzovic compared his friction utilization results to well-known and published maximum friction factors derived from different skid resistance measurements or backgrounds. His evaluation process encompassed both the tangential and radial directions. He used the following regression models for maximum permissible tangential friction factors in relation to speed; these models have been partially discussed in Sec. 10.2.2:

- Lamm and Herring[382] with respect to Eq. (10.38)
- Dames, et al.[130] with respect to Eq. (10.39)
- Litzka and Friedl[481] for humid condition, tire with high tread depth:

$$f_{T1 \, max} = 0.119 \left(\frac{V}{100}\right)^2 - 0.483 \left(\frac{V}{100}\right) + 0.623 \tag{14.18}$$

and for humid condition, tire with low tread depth:

$$f_{T2 \, max} = 0.204 \left(\frac{V}{100}\right)^2 - 0.537 \left(\frac{V}{100}\right) + 0.533 \tag{14.19}$$

- Lamm, et al.[420] with respect to Eq. (10.40) (overall regression equation for Part 2, "Alignment")

Figure 14.17 shows the resulting relationships between maximum permissible side friction factors and speed for a utilization ratio of $n = 30$ percent. This utilization ratio has been used in the German "Guidelines for the Design of Roads"[243,246] to calculate minimum radii of curve in the case of negative superelevation rates and was accepted by Redzovic for Fig. 14.17. In contrast, a utilization ratio of only 20 percent has been proposed by Spacek,[665] based on driving dynamic safety research for Switzerland.

In conclusion, the following statements were made in Ref. 596 with respect to Fig. 14.17:

- The calculated friction demands at the individual wheels, as well as the corresponding utilizations of available friction, revealed in comparison to the friction factors in the reviewed literature that for all test rides through curves (exception: $R \leq 1000$ m), large reserves are present in the tangential direction.
- For a utilization ratio of $n = 30$ percent, the calculated maximum permissible side friction factors, shown in Fig. 14.17, were reached during the test rides only in rare cases.

- When changing lanes at a constant speed, the existing friction demands at the individual wheels (except for the rear right wheel on left-handed curves), as well as related to the center of gravity of the vehicle, were smaller than the friction values adopted from the literature. That means driving dynamic safety is not impaired.

- Only for the extreme case, which refers to an evasive maneuver along a stationary obstacle, did the friction demands in the radial direction exceed the maximum allowable side friction factors taken from the literature (Fig. 14.17). However, note that this case represents a risky driving maneuver because of the very narrow radii of curve, which have to be passed by the test car, independent of the present superelevation rate.

- Based on the results, a negative superelevation rate of $e = -2.5$ percent can always be recommended in flat topography when designing with radii of curve of $R \geq 2000$ m ($V \geq 100$ km/h).

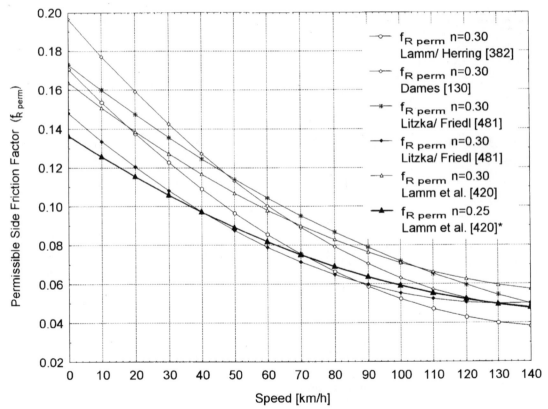

* Relationship used for determining minimum radii of curve in case of negative
  superelevation rates in AL.

FIGURE 14.17   Relationships between maximum permissible side friction factor and speed for a utilization ratio of $n = 30$ percent in the case of negative superelevation. (Elaborated and modified based on Refs. 289 and 596.)

TABLE 14.6 Minimum Radii of Curve for Negative Superelevation Used for Geometric Design in Several Countries and AL

Country	Negative superele- vation, %	Design or operating speed, km/h										
		40	50	60	70	80	90	100	110	120	130	140
Australia	3	—	—	—	900	1,250	1,500	2,000	3,000	4,000	5,000	—
Austria	2.5	2,000	2,000	2,000	2,000	2,000	2,000	2,000	—	3,000	—	4,000
Germany 1995	2.5	—	—	—	600	950	1,400	2,100	3,000	4,100	5,500	—
Germany 1984	2.5	—	300	500	800	1,250	2,000	3,000	4,500	6,500	9,000	—
Greece	2.5	—	240	400	600	900	1,300	1,700	2,300	3,000	3,900	—
Italy	2.5	500	800	1,100	1,500	2,000	—	2,000	—	3,000	4,500	4,500
France:[510]												
State roads	2.5	400	—	600	—	900	—	1,300	—	1,800	—	—
Interstates	2.0	—	—	—	—	2,000	—	3,000	—	4,000	—	5,000
Switzerland	2.5	1,500	1,500	1,500	2,500	2,500	3,200	4,000	5,000	7,500	—	10,250
United Kingdom	2.5	—	510	720	1,020	1,300	1,640	2,040	—	2,880	—	—
South Africa	2.5	1,200	1,200	1,200	1,500	2,000	3,000	4,000	5,000	7,000	—	—
AL	2.5	—	—	—	700	1,000	1,500	2,000	3,000	3,500	4,500	5,500

On the basis of these results, the following recommendations were made for multilane median-separated highways:[596]

1. The superelevation to the outside edge of a left-handed curve in the driving direction, that is, negative superelevation of the traveled way should be set equal to $e = -2.5$ percent because of drainage requirements.

2. Based on the empirically established friction factors, it can be stated that Eq. (14.11) in "Negative Superelevation" in Sec. 14.1.2.3 used to determine minimum radii of curve in case of negative superelevation, can be regarded as sound from a driving dynamic safety point of view. Equation (14.11) is confirmed empirically by the research of Redzovic[596] for a utilization ratio of permissible side friction of $n = 30$ percent (see Fig. 14.17).

For safety reasons, the utilization ratio was brought down to $n = 25$ percent here in Part 2, "Alignment (AL)" (see Fig. 14.17). On the basis of these assumptions, the minimum radii of curve for negative superelevation on median-separated multilane highways were calculated in Table 14.1 by introducing Eq. (14.13) into Eq. (14.11) in Sec. 14.1.2.3.

The minimum radii of curve for negative superelevation, used for geometric design in different countries, are presented in Table 14.6. (The values in Table 14.6 are first of all recommended for multilane median-separated highways.) As can be seen from the table, most countries use negative superelevation rates of $e = -2.5$ percent. The developed values for "AL" in Table 14.6 are, in comparison to those of other countries, somewhat lower than the average values. These values, however, are fully based on sound driving dynamic safety aspects, which means balanced designs can be expected with respect to economic and environmental issues.

14.2.3 Superelevation Runoff and Distortion

*14.2.3.1 General.** A vehicle is subjected to a rotation movement when moving on a road where there are changes in the superelevation. Rotation is a design variable used in the construction of the length of the superelevation runoff.

---

*Elaborated based on Refs. 35, 37, and 686.

The length of the superelevation runoff depends on the difference in grade between the edges of the traveled way and the axis of rotation, the speed, as well as the width of the traveled way. Related to the superelevation runoff, any sudden reaction by the driver with respect to the rotation of his or her motor vehicle around the longitudinal axis may become critical. Therefore, a limitation on the rate of rotation of the traveled way with respect to the design speed is necessary. Based on experiences gained in the 1970s, a rate of rotation from 3 to 6 percent per second is not detrimental for a comfortable, safe ride.[124]

According to Austroads,[37] the rate of rotation of the traveled way should generally not exceed 3.5 percent per second for design speeds up to 70 km/h and 2.5 percent per second for 80 km/h and over. Whereas the Austrian guidelines[35] consider a rotation rate of 4 percent per second as favorable, the Swedish standard specifications[686] propose varying rates of rotation from roughly 4 to 2 percent per second with increasing width of the traveled way (see Fig. 14.18).

However, based on modern vehicle fleet characteristics, a rotation rate of even 7 percent per second is regarded nowadays as acceptable and is used in several countries for establishing the relative grade, Δs, of the superelevation runoff, as shown in Table 14.7.

Like the other comfort and appearance criteria presented in "AL," the following values should be regarded as reasonable values and not as inherently correct.

14.2.3.2   Limiting Values.   The relative grade, Δs, that satisfies the rate of rotation, r, can be determined from the following formula:

$$\Delta s = \frac{3.6\, r\, d}{V_d} \tag{14.20}$$

where Δs = relative grade of the superelevation runoff, %
$\quad\quad r$ = rate of rotation of the traveled way, %/s
$\quad\quad d$ = distance from the edge of the traveled way to the rotation axis, m
$\quad\quad V_d$ = design speed, km/h

Maximum relative grades (max Δs) are given in Table 14.7 for several countries. For comparative reasons, a two-lane cross section was assumed here for a lane width of 3.75 m and an edge

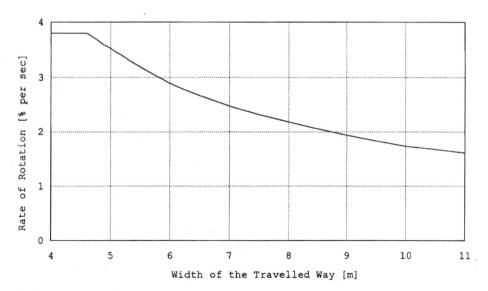

FIGURE 14.18   The design rotation for various widths of the traveled way.[686]

TABLE 14.7   Maximum Relative Grades (max Δs) of the Superelevation Runoff and Applied Maximum Rates of Rotation for Several Countries and AL (Example: Two-Lane Roads)

Country (1)	Applied rates of rotation, % (2)	50 (3)	60 (4)	70 (5)	80 (6)	90 (7)	100 (8)	110 (9)	120 (10)	130 (11)
		Maximum relative grades of the superelevation runoff: max Δs, %								
Australia	4–3	0.9	0.6	—	0.5	—	0.4	—	0.4	0.4
Austria	5–3	1.5	1.0	1.0	0.5	0.5	0.5	0.5	0.5	0.5
Germany[243]	7–6	2.0	1.6	1.6	1.0	1.0	0.8	0.8	0.8	—
Greece	8–6	2.0	1.6	1.6	1.0	1.0	0.9	0.9	0.9	0.9
Italy	12–7	2.0	2.0	2.0	2.0	2.0	1.0	1.0	1.0	1.0
Switzerland	7–4	1.0	1.0	1.0	1.0	0.8	0.8	0.8	0.8	0.8
Canada	4–2	0.5	0.5	0.5	0.5	0.5	0.5	0.5	0.5	0.5
United States	3–2	0.66	0.60	—	0.50	0.47	0.43	0.40	—	—
AL	7–6	2.0	1.6	1.4	1.0	0.9	0.8	0.8	0.7	0.6

strip of 0.25 m. In other words, the distance, d, from the edge of the traveled way to the rotation axis is 4.00 m. As can be seen from Table 14.7: (1) the maximum relative grade values (max Δs) decrease with increasing design speeds and (2) there are wide differences between the investigated countries at low design speeds. However, as design speeds increase, the differences in max Δs become smaller. Also, the calculated ranges of the applied rates of rotation, presented in col. 2 of Table 14.7, show wide variations between 2 and 12 percent per second.

The maximum relative grades of the superelevation runoff presented in "AL" on the basis of Eq. (14.20) are based on rates of rotation between 6 and 7 percent per second. Contrary to some guidelines, the proposed maximum relative grades decrease gradually with increasing design speeds in order to satisfy the increasing driving dynamic demands.

With respect to "AL," it can furthermore be stated that the applied maximum rates of rotation of 6 to 7 percent per second can be regarded as acceptable for a comfortable, safe ride along the superelevation runoff as related to modern vehicle fleet characteristics.

On the basis of these considerations, maximum relative grades of the superelevation runoff have been established in cols. 2 and 3 of Table 14.2 in Sec. 14.1.3.2.

Equation (14.21a) describes the relationship between the length of the superelevation runoff, the design speed, and the rate of rotation of the traveled way:

$$L_e = \frac{(e_e - e_b) \, V_d \, 100}{3.6 \, r} \qquad (14.21a)$$

where L_e = length of the superelevation runoff, m
e_e = superelevation rate at the end of the superelevation runoff, %
e_b = superelevation rate at the beginning of the superelevation runoff, % (e_b is negative when e_b is opposite to e_e)
V_d = design speed, km/h
r = rate of rotation of the traveled way, %/s

Consequently, the formula for the parameter of the clothoid, A, corresponds to

$$A = \sqrt{\frac{R\,(e_e - e_b)\,V_d}{3.6\,r}} \tag{14.21b}$$

on which, for example, the development of Fig. 12.6 is based for limiting values of the superelevation runoff.

14.2.3.3 Drainage Considerations.*

The arrangement of the minimum relative grade of the superelevation runoff (min Δs) is, for drainage reasons, of great importance in distortion sections where the superelevation runoff passes through zero and low longitudinal grades are present (see "Minimum Grades in Distortion Sections" in Sec. 13.1.1.2).

In order to consider that the water-film depth on the road surface increases with increasing width of the traveled way, the minimum relative grade is directly related to the width of the traveled way, as shown in the following equation (also see Table 14.2):

$$\min \Delta s = 0.10\,d \leq \max \Delta s \qquad \% \tag{14.15}$$

where d = distance from the edge of the traveled way to the rotation axis, m.

Equation (14.15) is also being used in the guidelines of Austria, Germany, Greece, Italy, and Switzerland.

This relationship guarantees that the length of the critical superelevation runoff section, where weak areas for drainage are most frequently present, does not increase with the increasing width of the traveled way. However, as already discussed in Sec. 14.1.3.3, the fast distortion is only necessary in areas where the minimum superelevation rate falls short of e_{min} = 2.5 percent (see Fig. 14.11). According to Eq. (14.14b) in Sec. 14.1.3.2 regarding the minimum relative grade, the critical central part of the superelevation runoff development has a constant length of 50 m for e_{min} = |2.5|% (Fig. 14.11). This means that the weak area for drainage is limited to a relatively short section.

14.2.4 Pavement Widening in Selected Countries

In General. When a vehicle is cornering, the right rear wheel follows a trajectory which might lie outside the edge of the travel lane, especially on sharp curves. In Fig. 14.19, the basic model used to calculate the trajectory of the right rear wheel is shown as curve r for the case when the trajectory of the right front wheel is known (curve f).

The natural equation used for the right rear wheel trajectory is

$$\kappa' = \frac{1}{R'} = \frac{\vartheta}{D} \tag{14.22}$$

where κ' = curvature of the right rear wheel trajectory, m^{-1}
  R' = radius of the right rear wheel trajectory, m
  ϑ = right front wheel angle, rad
  D = wheelbase, m

Note: "Rad" = radian, that is, the angle subtended at the center of a circle by an arc equal in length to the radius; 2π rad = 360° or 1° = 0.0174533 rad.

---

*Elaborated based on Ref. 35.

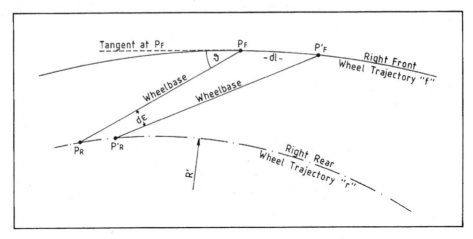

Legend:
```
        PF, P'F:   successive points of the right front wheel,
        PR, P'R:   successive points of the right rear wheel,
        dl      :  the differential length of the right front wheel trajectory,
        dε      :  the differential angle formed by two successive positions of
                   the longitudinal axis of the vehicle,
        ϑ       :  right front wheel angle.
```

FIGURE 14.19   Basic model used to determine the right rear wheel trajectory on a curve. (Elaborated based on Ref. 681.)

The angle, ϑ, of the right front wheel can be determined by solving the following differential equation:[681]

$$\kappa' = \frac{1}{R'(l)} = \frac{d\vartheta}{dl} + \frac{\sin\vartheta}{D} \tag{14.23}$$

The function, $\kappa'(l)$, denotes the natural equation of the right front wheel trajectory, which is assumed to be always known because it describes the traced radius.

For instance, when the front wheel is supposed to trace a circular curve with a radius of $R =$ constant, then the solution of the differential equation, Eq. (14.23), yields the following solution:[681]

$$\vartheta = \frac{D}{R'} - e^{-l/D}\left(\frac{D}{R'} - \vartheta_c\right) \tag{14.24}$$

where l = length traveled on the circular curve, m
$\quad\vartheta_c$ = initial wheel angle at the beginning of the circular curve, rad

By combining Eqs. (14.22) and (14.24), it follows then that the right rear wheel trajectory can be calculated from the natural equation as follows:

$$\frac{1}{R'} = \frac{1}{R} - e^{-l/D}\left(\frac{1}{R'} - \frac{\vartheta_c}{D}\right) \tag{14.25}$$

This natural equation represents the equation of the curvature as a function of the length (traveled on the circular curve). By applying the corresponding theory from differential geometry in a local coordinate system, the rear wheel path of the vehicle can be determined and by this the exact pavement widening (see Fig. 14.19).

In any case, the necessary calculations are not simple and normally require employment of numerical methods in order to perform the integration of the resulting trigonometric functions.

Based on the resulting rear wheel path, the necessary pavement widening can then be determined, in order to examine whether or not the vehicle is still on the travel lane and that it did not occupy other parts of the cross section, which are not provided for regular traffic, like for example shoulders, etc.

Even though the developed equation [Eq. (14.23)] delivers very accurate rear wheel paths in practice, either wheel templates or simpler equations [approximated equations with respect to Eq. (14.25)] are used to determine the rear wheel paths for different design vehicles and the corresponding necessary pavement widenings at curves. Such simple equations for pavement widenings are represented by Eqs. (14.16) or (14.17) in Sec. 14.1.4.1.

Design Values, Widening Attainment on Curves, Lane Addition, or Lane Drop in Selected Countries.   An international review of highway geometric design guidelines did not reveal much uniformity with respect to necessary pavement widenings in order to accommodate the rear wheel path and the trajectory of the outer edge of the front overhang. In the following subsections, some interesting solutions, which are found in the guidelines studied, are presented to determine the extra amount of pavement widening needed at (sharp) curves.

Austria.   According to the Austrian "Guidelines for the Alignment of Roads,"[35] pavement widening is only necessary for radii of curve less than 200 m. For these design cases, the necessary pavement widenings are given in Table 14.8 as a function of the radius of curve and the width of the traveled way.

Widening is performed linearly along the transition curve and is realized along the inside pavement edge, as shown in Fig. 14.14.

When lanes are added or dropped, the change in width of the traveled way is attained according to Fig. 14.16.

Canada.   According to the "Geometric Design Standards of Canada,"[607] the basis for determining the amount of pavement widening on curves is illustrated in Fig. 14.20 along with the corresponding equations [Eq. (14.26) to (14.28)].

For two-lane roads where the number of trucks in both directions is less than 15 vehicles/h, pavement widening is not required. Based on Fig. 14.20, pavement widening values for a single-unit vehicle are given, as an example, in Table 14.9. Additional tables exist for other design vehicles like for semitrailer or full trailer combinations.

On multilane highways pavement widening is not used.

Widening is performed over the length of the transition curve so that a smooth pavement edge is introduced.[607]

Germany.   According to the German "Guidelines for the Design of Roads" (Part 2, "Alignment"[243,246]), the necessary pavement widening at curves is determined from Eq. (14.17). Furthermore, the Germans recommend two opposing cases (FTC/FTC* or Bus 2/Bus 2) for pavement widening at curves which are dependent on the design vehicles shown in Table 14.3. The corresponding pavement-widening values are given in Table 14.10 with respect to the radius of the curve.

TABLE 14.8   Pavement Widening at Circular Curves (Dimension = cm), Austria[35]

	Radius of curve, m							
Width of the traveled way, m	60	80	100	120	140	160	180	200
≥7.00	120	80	60	40	30	20	0	0
<7.00	150	110	90	75	65	55	50	0

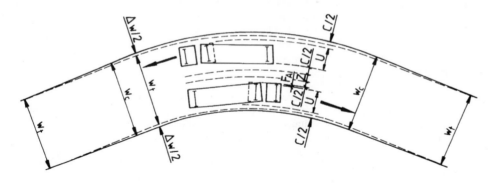

$$\Delta w = w_c - w_t \qquad (14.26)$$

where Δw = amount of widening
  w_c = pavement width on curve
  w_t = pavement width on tangent

$$w_c = 2(U + C) + F_A + Z \qquad (14.27)$$

where U = vehicle track width on curve, m
  C = nominal clearance between vehicles: 0.46 m for 6.0-m-wide pavements, 0.61 m for 6.6-m-wide pavements, 0.76 m for 7.0-m-wide pavements, and 0.91 m for 7.4-m-wide pavements
  F_A = front overhang, m
  Z = additional clearance, m, to compensate for difficulty of driving on curves where

$$Z = \frac{0.1046 \, V_d}{\sqrt{R}} \qquad (14.28)$$

  V_d = design speed of highway, km/h
  R = radius of curve, m

FIGURE 14.20   Basis for pavement widening at curves, Canada.[607]

For pavement widths of less than or equal to 6.0 m, no widening is used for widening amounts (Δw) less than 0.25 m, and for pavement widths of more than 6.0 m, no pavement widening is necessary for Δw less than 0.50 m.

For pavement widening at curves, the Germans use two quadratic parabolas with an interim tangent (see Fig. 14.21). The formula used for determining the length of the widening attainment is derived from geometric considerations for the element sequence tangent–clothoid–circular curve, as shown in the following:

$$L_Z = 2 (L + l) + \frac{L_A}{2} = 2 (L + l) + \frac{A^2}{2 R} \qquad (14.29)$$

where L_Z = length of widening attainment, m
  L = length of the vehicle wheelbase, m
  l = length of the front overhang, m
  A = parameter of the clothoid, m
  L_A = length of the clothoid, m
  R = radius of curve, m

TABLE 14.9 Pavement Widening Values on Curves (Example: Single-Unit (SU) Vehicles), Canada[607]

Design speed, km/h															
50		60		70		80		90		100		110		120	
R	Δw	R	Δw	R	Δw	R	Δw	R	Δw	R	Δw	R	Δw	R	Δw
Pavement width = 7.4 m															
50	1.2	50–55	1.3	55–65	1.3										
55–65	1.0	60–70	1.0	70–80	1.0										
70–85	0.8	75–95	0.8	85–110	0.7	90–125	0.8	130–140	0.7						
90–120	0.5	100–140	0.5	115–160	0.5	130–190	0.5	150–210	0.5	190–240	0.5	250–280	0.4		
125–200	0.2	150–240	0.2	170–280	0.3	200–320	0.3	220–380	0.3	250–450	0.3	300–500	0.2	340–550	0.2
Pavement width = 7.0 m															
50	1.5	50–55	1.5	55–60	1.5										
55–60	1.3	60–70	1.3	65–75	1.2										
65–80	1.0	75–90	1.0	80–105	1.0	90–115	1.0	130	0.9						
85–110	0.7	95–130	0.8	110–150	0.8	120–170	0.8	140–190	0.7	190–220	0.7	250	0.6		
115–170	0.5	140–200	0.5	160–240	0.5	180–280	0.5	200–320	0.6	230–380	0.5	280–420	0.5		
180–350	0.3	210–450	0.3	250–500	0.3	300–650	0.2	340–750	0.2	400–900	0.3	450–1050	0.2		
Pavement width = 6.6 m															
		50	1.7	55–60	1.7										
		55–65	1.5	65–75	1.4										
		70–85	1.2	80–95	1.2	90–110	1.3								
		90–115	1.0	100–130	1.0	115–150	1.0	130–170	0.9	190–200	0.9				
		120–180	0.8	140–210	0.8	160–250	0.7	180–280	0.7	210–320	0.7				
		190–350	0.5	220–420	0.5	280–500	0.5	300–600	0.5	340–700	0.5				
		380–1300	0.2	450–1700	0.2	525–2000	0.3	650–2500	0.2	750–2500	0.2				
Pavement width = 6.0 m															
		55	1.6	55–60	1.7										
		60–70	1.5	65–80	1.5	90	1.5	130	1.2						
		75–95	1.2	85–110	1.2	95–125	1.2	130–140	1.0						
		100–140	1.0	115–160	1.0	130–190	1.0	150–220	0.8						
		150–250	0.7	170–300	0.7	200–350	0.7	230–420	0.5						
		280–700	0.4	320–850	0.5	380–1100	0.5	450–1300	0.3						
		750–10000	0.2	900–10000	0.3	1150–10000	0.2	1400–10000	0.2						

Note: Widening values Δw, m, are based on SU design vehicles traveling at the design speed. R denotes centerline radius, m.

TABLE 14.10   Pavement Widening at Circular Curves, Two-Lane Roads, Germany[246]

Recommended opposing case	Pavement widening (for $n = 2$)		
	Δw	$w \leq 6.0$ m	$w > 6.0$ m
FTC/FTC*	$\dfrac{50\,n}{R}$	$30 < R \leq 400$	$30 < R \leq 200$
Bus 2/Bus 2	$\dfrac{40\,n}{R}$	$30 < R \leq 320$	$30 < R \leq 160$

Note: n = number of lanes; Δw = pavement widening, m; R = radius of curve, m; w = pavement width, m; *FTC = full trailer combination; and Bus 2 = articulated bus (Table 14.3).

Pavement widening is attained according to the rules shown in Fig. 14.21.

For changes in cross sections, like lane additions or lane drops, the Germans use two quadratic parabolas, but without an interim tangent (see Fig. 14.22). The required length needed to accomplish this is determined from the following formula for all road category groups:

$$L_Z = V_d \sqrt{\frac{\Delta w}{3}} \qquad (14.31)$$

where L_Z = length of widening attainment, m
$\quad V_d$ = design speed, km/h
$\quad \Delta w$ = pavement width increase, m, for example, the width of an auxiliary lane

Lane addition or lane drop are achieved by means of the rules given in Fig. 14.22.

As can be seen, the German procedure, used for curve widening, lane addition, or lane drop, is quite sophisticated but is somewhat difficult to handle. Further design cases for widenings in reversing or egg-shaped clothoids are also discussed.

United Kingdom.   The U.K. "Geometric Alignment Standards"[139] use a pavement widening of 0.3 m/lane for radii of curve between 90 and 150 m for standard widths of the traveled way:

7.3 m (two lanes)

11.0 m (three lanes)

14.6 m (four lanes)

For highways with widths less than the standard ones, pavement widenings in relation to radii of curve are applied as described in the following:

• 0.6 m/lane for radii between 90 and 150 m

• 0.5 m/lane for radii between 150 and 300 m

• 0.3 m/lane for radii between 300 and 400 m

The pavement widening is attained gradually along the transition curve. The total amount of widening should generally be realized on the inside edge of the circular curve (Fig. 14.14).

United States.   AASHTO policy[5] recognizes a minor importance of pavement widening on modern highways; however, pavement widening remains necessary for some conditions of speed, curvature, and width.

For these special cases, the basis for determining the necessary pavement widenings on curves is similar to that given in the Canadian guidelines[607] (see Fig. 14.20). While the Canadians offer a numerical listing for pavement-widening values on curves (Table 14.9), the Americans use

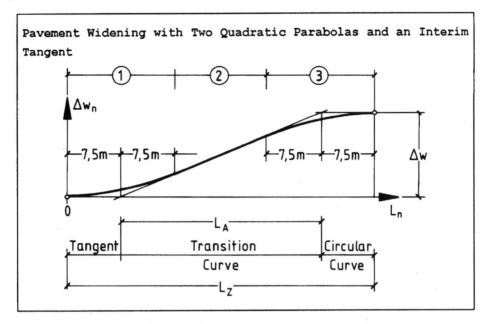

Pavement Widening with Two Quadratic Parabolas and an Interim Tangent

Note: See also the definitions based on Eq. (14.29).

$$\Delta w_n = \frac{\Delta w}{30 \text{ m} \times L_A} \cdot L_n^{\,2} \qquad (14.30a)$$

for

$$0 \leq L_n \leq 15 \text{ m}$$

$$\Delta w_n = \frac{\Delta w}{L_A}(L_n - 7.5 \text{ m}) \qquad (14.30b)$$

for

$$15 \text{ m} \leq L_n \leq (L_Z - 15 \text{ m})$$

$$\Delta w_n = \Delta w - \frac{\Delta w}{30 \text{ m} \times L_A} \cdot (L_Z - L_n)^2 \qquad (14.30c)$$

for

$$(L_Z - 15 \text{ m}) \leq L_n \leq L_Z$$

where Δw = pavement widening, m
  $L_Z = L_A + 15$ m
  Δw_n = widening at the observed point, m
  L_n = length of the widening attainment up to the observed point, m

FIGURE 14.21 Attainment of pavement widening for the element sequence: tangent–clothoid–circular curve, Germany.[243,246,576]

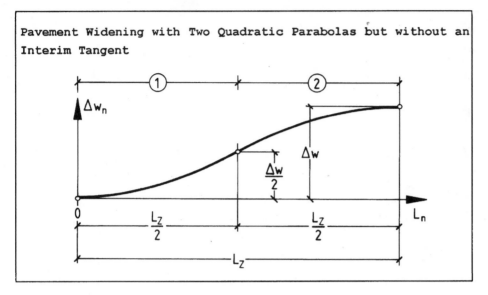

Note: For definitions, also see Fig. 14.21 and Eqs. (14.29) to (14.31).

$$\Delta w_n = \frac{2\,\Delta w\,L_n^{\,2}}{L_z^{\,2}} \qquad (14.32a)$$

for

$$0 \le L_n \le \frac{L_z}{2}$$

$$\Delta w_n = \Delta w - \frac{2\,\Delta w\,(L_z - L_n)^2}{L_z^{\,2}} \qquad (14.32b)$$

for

$$L_z/2 \le L_n \le L_z$$

FIGURE 14.22   Attainment of pavement widening for lane addition or lane drop, Germany.[243,246,576]

nomograms to describe the relationships shown in Fig. 14.20 [see AASHTO 1990,[5] Section III ("Elements of Design"), Fig. III-25, p. 215].

A minimum widening amount of 0.60 m is accepted; lower values are disregarded.

In contrast to the other guidelines studied, for multilane highways AASHTO proposes that widening on a two-lane, one-way pavement of a divided highway should be the same as that on a two-lane, two-way pavement.

Preliminary Conclusion.   A review of the previously discussed methods of pavement widening and widening attainment in different countries indicates that despite the fact that the practice of pavement widening is widely accepted, little or no uniformity exists between the different studied guidelines.

Therefore the recommendations for practical design tasks were developed in Subchapter 14.1 to present sound solutions for the design of pavement widenings, widening attainment on curves, as well as for lane addition or lane drop by an easy-to-comprehend procedure.

According to most of the guidelines studied, pavement widening is not required on multilane highways because of increased construction costs and questionable benefits.

CHAPTER 15
SIGHT DISTANCE

15.1 RECOMMENDATIONS FOR PRACTICAL DESIGN TASKS

15.1.1 General Considerations

Traffic safety and quality of traffic flow require certain minimum sight distances for stopping in time (*stopping sight distance*) or for safe overtaking (*passing sight distance*).[243]

According to AASHTO,[5] the path and speed of motor vehicles on highways and streets are subject to the control of drivers whose ability, training, and experience are quite different. With respect to safety on the highway, the designer must provide a sight distance of sufficient length that would allow drivers to control the operation of their vehicles and avoid striking an unexpected object on the traveled way. Certain two-lane highways should also have sufficient sight distances to enable drivers to occupy the opposing traffic lane for overtaking other vehicles without hazard. Two-lane rural highways, generally, should provide such passing sight distances at frequent intervals and for substantial portions of their length.[5]

The required stopping sight distance is important for the evaluation of sight conditions on two-lane and multilane roads of all category groups. In addition, for evaluating sight conditions on two-lane rural roads with traffic in both directions (category group A), the required passing sight distance must be determined. For roads of category group B, the passing sight distances are not as important.[243]

15.1.2 Stopping Sight Distance

Stopping sight distance (SSD) is the most fundamental of the sight distance considerations in highway geometric design, since adequate stopping sight distance is required at every point along the roadway. Stopping sight distance is the distance that a driver must be able to see ahead along the roadway in order to identify hazards in the roadway and bring his or her vehicle safely to a stop where necessary. SSD can be limited by both horizontal and vertical curves (see Figs. 15.3 and 15.4). Thus, horizontal and vertical curves on roadways must be designed with SSD in mind.[268]

The required stopping sight distance, SSD, is the distance, which a driver needs in order to stop his or her vehicle before reaching an unexpected obstacle on the road when riding at the 85th-percentile speed.

Stopping sight distance is the sum of two distances: the distance traversed by the vehicle from the instant the driver sights an obstacle necessitating a stop to the instant the brakes are applied (perception reaction distance) and the distance required to stop the vehicle from the instant brake application begins (braking distance).[5]

The SSD model, which follows from the basic principles of physics, is

$$SSD = \text{reaction distance} + \text{braking distance}$$

The following procedure, which determines sound stopping sight distances, was first developed by Lamm.[385] It has been applied in Germany since 1973,[241,243,246] in Austria since 1981,[35] and in Greece since 1995[453] using different assumptions.

The SSD model incorporates the effect of a speed-dependent friction factor, as well as the aerodynamic drag force on the decelerating vehicle. Furthermore, the 85th-percentile speed, $V85$, is regarded as the operating speed. Operating speed backgrounds for different countries are presented in Fig. 8.12 and Table 8.5.

With respect to perception-reaction time, a value of $t_R = 2.0$ s is used.

The stopping sight distance is calculated from the following equations:[385]

$$\text{SSD} = S_1 + S_2 \tag{15.1}$$

with

$$S_1 = \frac{V_0}{3.6} \cdot t_R \tag{15.2}$$

and

$$S_2 = \frac{1}{3.6^2 g} \int_{V_1}^{V_0} V f_T(V) + (G/100) + (F_L/Q) \ dV \tag{15.3}$$

where SSD = stopping sight distance, m
$\quad S_1$ = perception reaction distance, m
$\quad S_2$ = braking distance, m
$\quad V_1$ = speed at the end of the braking maneuver, km/h
$\quad V_0$ = speed at the beginning of the braking maneuver, km/h
$\quad t_R$ = driver perception reaction time, s ($t_R = 2.0$ s), for roads of category groups
$\quad\quad$ A and B

$$f_T(V) = 0.59 - 4.85 \times 10^{-3}\, V + 1.51 \times 10^{-5}\,(V)^2 \tag{10.2, 10.40}$$

$\quad g$ = acceleration of gravity, m/s^2 (9.81 m/s^2)
$f_T(V)$ = tangential friction factor
$\quad G$ = grade, % (+ for upgrade, − for downgrade)
$\quad F_L$ = aerodynamic drag force, N

$$F_L = 0.5\, \gamma\, C_W\, \text{FA} \left(\frac{V}{3.6}\right)^2 \tag{15.4}$$

where γ = air density, kg/m^3 = 1.15 (for 15°C, 722 torr)[91,159]
$\quad C_W$ = aerodynamic drag coefficient = 0.35[91,159]
$\quad$ FA = projected frontal area of passenger cars, m^2 (FA ≈ 0.9 × track × height[62,]
$\quad\quad$ FA = 2.08 m[291,159])
$\quad Q$ = weight of a passenger car, N
$\quad Q$ = m × g, m = mass of vehicle, kg, m = 1304 kg[159]

On the basis of the newest investigated vehicle fleet,[159] the expression, F_L/Q, in the braking distance formula is

$$\frac{F_L}{Q} = \frac{0.5 \times 1.15 \times 0.35 \times 2.08}{1304 \times 9.81} \left(\frac{V}{3.6}\right)^2 = 0.327 \times 10^{-4} \left(\frac{V}{3.6}\right)^2 \tag{15.5}$$

Because most countries do not have skid resistance backgrounds for evaluating sound tangential friction factors in relation to speed, the overall regression equation [Eq. (10.2), (10.40)] is recommended for modern highway geometric design. This model was developed from the tangential friction factors which are in use in the highway geometric design guidelines of France, Germany,

Sweden, Switzerland, and the United States (see Fig. 10.27). Because of good agreement with actual pavement friction inventories in the United States and in Germany (see Figs. 10.28 and 10.29), Eq. (10.2) can be regarded as reliable,[414,420,428,447,448] and will be used hereafter to derive the stopping sight distances (Table 15.1).

Based on the equations presented in Table 15.1, stopping sight distances were calculated for speeds of $V85 = 60, 70, 80, 90, 100, 110, 120$, and (130) km/h, and grades between ± 12 percent. The results are shown in Fig. 15.1.

The results of Fig. 15.1 are recommended for application in most countries because: (1) they include a generally valid relationship between tangential friction factor and speed $f_T(V)$, as represented by Eq. (10.2); (2) they consider the effect of the aerodynamic drag force; and (3) they represent modern vehicle fleet characteristics, as will be discussed in detail in Sec. 15.2.2.3.

For two-lane rural roads of category group A, the 85th-percentile speeds shown in Fig. 15.1 or in Table 15.1 can be determined for independent tangents and curved roadway sections on the basis of the operating speed background of the country under study (Fig. 8.12.2a or Table 8.5). For those countries that do not yet have operating speed backgrounds, then the speed background of a country that has similar driving behavior characteristics should be considered or the average operating speed background of Fig. 8.12b may be used.

So far, the 85th-percentile speeds have been related to longitudinal grades of up to 5 to 6 percent. For higher longitudinal grades of $G > 6$ percent and curvature change rates of the single curve of $CCR_S > 600$ gon/km, an additional operating speed background in difficult topography was developed in Sec. 18.5.1.[164]

For multilane rural roads of category group A, as well as for roads of category group B, recommendations relative to 85th-percentile speeds are given in Sec. 8.1.2 (also see Table 6.3).

It was decided to use the same stopping sight distances for straight and curved roadway sections. This assumption for curved sites is not totally correct. However, in the Austrian guidelines,[35] it has been shown that when individual stopping sight distances were additionally calculated for curves, the error between the SSD values on tangents and curves (even with minimum radii for the respective design speed) was insignificant and did not exceed 3.2 percent. Since the assumptions made in the Austrian guidelines are comparable to those of the present procedure

TABLE 15.1   Derivation of a Reasonable Stopping Sight Distance Formula[451, 453]

$$SSD = S_1 + S_2 \tag{15.1}$$

$$S_1 = \frac{V_0}{3.6} t_R = \frac{V_0}{3.6} 2 = \frac{V_0}{1.8} = \frac{V85}{1.8} \tag{15.2}$$

(Valid for roads of category groups A and B.)

The integral in Eq. (15.3) takes into account the braking distance from $V_0 = V85$ to $V_1 = 0$, with $f_T = f_T(V)$ according to Eq. (10.2), and F_L/Q according to Eq. (15.5). The solution of the integral is represented by

$$S_2 = 223.2 \ln\left(\frac{1.76 \times 10^{-5} \, V85^2 - 4.85 \times 10^{-3} \, V85 + G/100 + 0.590}{0.590 + G/100}\right)$$

$$+ \frac{2.16}{\sqrt{1.81 \times 10^{-5} + 7.05 \times 10^{-5} \, G/100}}$$

$$\cdot \arctan \frac{V85 \sqrt{1.81 \times 10^{-5} + 7.05 \times 10^{-5} \, G/100}}{1.18 + 2 \, G/100 - 4.85 \times 10^{-3} \, V85} \tag{15.6}$$

where $V_0 = V85 = $ 85th-percentile speed, km/h

$\quad\quad G = $ average grade over the braking distance, % (negative for downgrades)

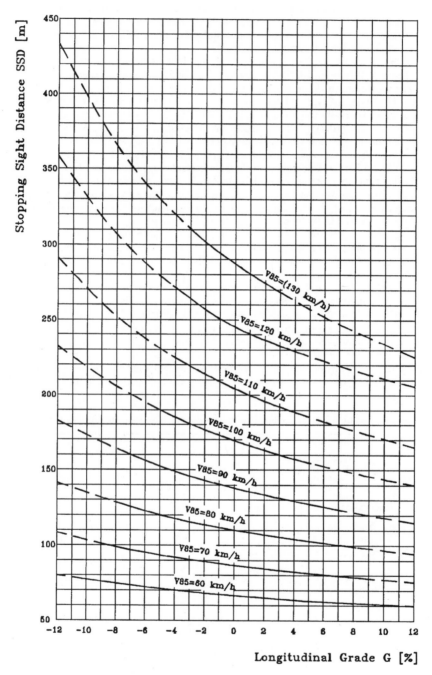

FIGURE 15.1   Required stopping sight distances.[451,453]

with respect to the speed-dependent tangential friction factor and vehicle fleet characteristics, it can be presumed that similar results could be expected. Thus, in order to simplify the procedure somewhat, the SSD values in tangents (Fig. 15.1) will also be used for curved roadway sections.

Opposing (Meeting) Sight Distance.   In addition to the previous discussions of stopping sight distances, Austrian and South African guidelines[35,663] consider an additional sight distance model, which could close the gap between stopping and passing sight distances. The *opposing sight distance* (OSD) corresponds to the roadway section which is needed to allow two opposing vehicles to stop in time and avoid a collision. It is equal to the sum of the stopping sight distances for the two vehicles. For roads with traffic in opposite directions, the opposing sight distance should be available on the entire roadway section.

The opposing sight distance represents a lower limiting value that guarantees safety when starting passing maneuvers. Therefore, the opposing sight distance is considered a *critical passing sight distance*. It is the minimum section length, which has to be clearly observed, in order to allow a timely stop between the overtaking vehicle and the oncoming one.

If the opposing sight distance cannot be ensured because of economic and environmental constraints, then a double line between the two opposing traffic lanes must be provided. Doing so prevents a collision in the same driving lane.

For safety reasons, an examination of OSD is always recommended, especially for determining "no passing zones."

15.1.3 Passing Sight Distance

According to AASHTO,[5] most roads are considered to qualify as two-lane rural highways on which faster moving vehicles frequently overtake slower ones, the passing of which must be accomplished on lanes regularly used by opposing traffic. If passing is to be accomplished safely, the driver should be able to see a sufficient distance ahead, clear of traffic, to complete the passing maneuver without cutting off the passed vehicle in advance of meeting an opposing vehicle appearing during the maneuver.[5]

The *required passing sight distance,* PSD, is the distance necessary for the safe performance of a passing maneuver. It is the summation of the following distances: the distance traversed by the passing vehicle, the distance traversed by the opposing vehicle during the time period of the passing maneuver, as well as the safety distance between both vehicles at the end of the passing maneuver.[241,243] For safety reasons, since passing maneuvers are fast, PSD should depend on the 85th-percentile speed. The 85th-percentile speeds can be determined from the operating speed backgrounds shown in Fig. 8.12*a* and Table 8.5 for the country under study. Those countries that do not yet have those backgrounds should consider the speed background of a country that has similar driving behavior characteristics or should use the average operating speed background of Fig. 8.12*b*.

The passing sight distance model is presented in Fig. 15.2. The required passing sight distances are listed in Table 15.2 in relation to the 85th-percentile speeds.

15.1.4 Sight Distance Recommendations and Controls*

The actual present stopping/opposing and passing sight distances are the result of horizontal and vertical alignment, cross section, and road environment.

In order to determine existing sight distances, the following rules apply:

- The determination must be performed with regard to the horizontal alignment (Fig. 15.3) and the vertical alignment (Fig. 15.4), since both form the three-dimensional space of the road. Furthermore, all road features, as well as the present and intended future planting, must be considered.

---

*Elaborated based on Refs. 35, 241, 243, and 246.

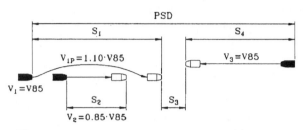

    ◼◼ Vehicle at the beginning of the passing maneuver
    ▭◻ Vehicle at the end of the passing maneuver
    PSD [m] = Required passing sight distance

```
PSD = S₁ + S₃ + S₄
V₁   [km/h] = speed at the beginning of the passing maneuver
V₁ₚ  [km/h] = speed during the passing maneuver
V₂   [km/h] = speed of the passed vehicle
V₃   [km/h] = speed of the opposing vehicle
S₁   [m]    = traversed distance of the passing vehicle
S₂   [m]    = traversed distance of the passed vehicle
S₃   [m]    = safety distance
S₄   [m]    = traversed distance of the opposing vehicle
```

FIGURE 15.2   Passing sight distance model. (Elaborated based on Refs. 152, 246, 249, and 250.)

TABLE 15.2   Required Passing Sight Distances for Two-Lane
Rural Roads of Category Group A[152,246,249,250]

V85, km/h	60	70	80	90	100
PSD, m	475	500	525	575	625

- The determination has to be performed for each sight distance category and each driving direction.
- For the determination of the sight distance in plan, the eye of the driver is assumed to be in the center of the right lane as seen in the driving direction. The object for the stopping sight distance is also assumed to lie in the center of the right lane and for the opposing and passing sight distances in the center of the opposing lane (Fig. 15.3). For crest vertical curve design, the height of the driver's eye (Fig. 15.4) is assumed to be 1.00 m above the road surface (Table 15.3). The same is true for the height of the object of the opposing/passing sight distances. The height of the object for the stopping sight distance depends on the 85th-percentile speed (see cols. 5 and 6 of Table 15.3).
- For safety reasons, the stopping sight distance must be present over the whole length of the roadway for all road categories.
- The same should be true for the opposing sight distance even though this is not always possible because of economic and/or environmental constraints.
- In addition, the passing sight distance should be present for a sufficient portion of the roadway on all two-lane roads outside built-up areas (rural) with opposing traffic. As a standard value for average conditions, at least 20 to 25 percent of the observed roadway section should guar-

a) Stopping Sight Distance

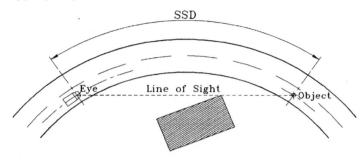

b) Passing or Opposing Sight Distance

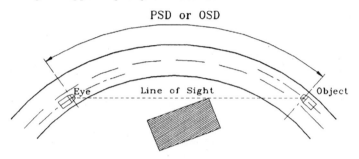

FIGURE 15.3   Sight distance in plan. (Elaborated and completed based on Ref. 689.)

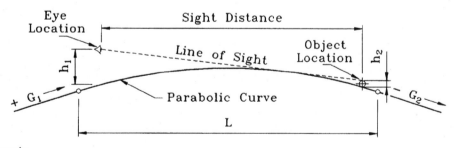

Legend:

L = length of curve,
G_1 = percent grade of approach tangent,
G_2 = percent grade of exit tangent,

h_1 = height of driver's eye, and
h_2 = height of object.
h_1 = h_2 for PSD and OSD

FIGURE 15.4   Sight distance in profile. (Elaborated and completed based on Ref. 689.)

TABLE 15.3 Basic Values for the Determination of Existing Sight Distances in Plan and Profile*

| (1) | Eye | | Object | | |
	Location (2)	Height, h_1, m (3)	Location (4)	V85, km/h (5)	Height, h_2, m (6)
Stopping sight distance (SSD)	In the center-line of the driving lane	1.0	In the center-line of the driving lane	40	0.00
				50	0.00
				60	0.00
				70	0.05
				80	0.15
				90	0.25
				100	0.35
				110	0.40
				120	0.45
				130	0.45
Passing sight distance (PSD)		1.0	In the center-line of the opposing lane		1.0
Opposing sight distance (OSD)		1.0	In the center-line of the opposing lane		1.0

*Elaborated and completed based on Refs. 35, 241, and 243.

antee passing maneuvers. The distribution of sections with passing possibilities should be uniform. It should be noted that a large portion of the roadway cannot be used for passing maneuvers. Those sections have to be marked with **No Passing** signs. Sufficient passing sight distances are not needed as much on roads inside built-up areas (suburban or urban roads) and sometimes they are even undesirable. If the existing portion of a roadway with passing possibilities is smaller than the percentage given previously, for example, in the case of protecting the landscape or for economic reasons, and a change of the alignment is difficult to achieve, then sufficient passing possibilities can be created, for example, by providing passing lanes according to the intermediate cross-sectional type $b2 + 1$. (See Part 3, "Cross Sections," Sec. 25.1.2.) Road sections of this kind have to be included in the portion of sections with sufficient passing sight distances.

- Changes in sight distances (increases or decreases) should be well balanced, and a decrease in sight distance should be gradual.

Within the sight field, which has to be kept free from obstructions, all obstacles which could be detrimental to the sight distance (for example, slopes, walls, parked vehicles, etc.) should be avoided or forbidden in order to improve the line of sight. Rows of trees in loose groupings, as well as single trees and bushes, can remain within the sight field if they do not create an immediate hazard or if they serve as an optical guidance.

The existing sight distance has to be compared with the required sight distance by using, for example, sight distance profiles for both directions of travel.[243]

15.2 GENERAL CONSIDERATIONS, RESEARCH EVALUATIONS, GUIDELINE COMPARISONS, AND NEW DEVELOPMENTS

15.2.1 General Considerations

The ability to see ahead and observe potentially conflicting traffic is critical for safe, effective highway operation. *Sight distance,* an important element in the geometric design of highways, is the length of roadway over which a driver has an unobstructed view.

15.2.2 Stopping Sight Distance*

15.2.2.1 Simplified SSD Model and Comparisons.   A review has been conducted of the SSD criteria used in Australia, Britain, Canada, France, Germany, Greece, South Africa, Sweden, Switzerland, and the United States. This review found that most countries' SSD criteria are based on the same model, but that assumptions made by different countries concerning the parameters used in that model vary.

SSD is generally defined as the sum of two components: perception reaction distance and braking distance. The SSD design situation assumes that there is a hazard in the roadway, such as an object, and that the driver of an approaching vehicle must first detect the object's presence and then brake to a stop. The perception reaction distance is the distance traveled by the vehicle from the instant the object comes into view to the instant at which the driver applies the brakes. The braking distance is the distance traveled by the vehicle from the instant the brakes are applied until the instant the vehicle comes to a complete stop. In contrast to the SSD model developed in Sec. 15.1.2, many countries use a simplified SSD model that disregards the effect of a speed-dependent tangential friction factor, $f_T(V)$, or the aerodynamic drag force on the decelerating vehicle.

This simplified SSD model is expressed by

$$\text{SSD} = 0.278 \, V_0 \, t_R + \frac{V_0^2}{254 f_T} \tag{15.7}$$

where SSD = stopping sight distance, m
V_0 = initial speed, expressed by the design speed or 85th-percentile speed, km/h
t_R = driver perception reaction time, s
f_T = coefficient of braking friction between the tires and the pavement surface (also known as the *tangential friction factor*)

Stopping sight distance is also affected by roadway grade, that is, stopping distances decrease on upgrades and increase on downgrades. Specifically, grade effects on stopping sight distance can be expressed by

$$\text{SSD} = 0.278 \, V_0 \, t_R + \frac{V_0^2}{254 \, (f_T \pm G)} \tag{15.8}$$

where G = percent grade/100 (+ for upgrades, − for downgrades).

Table 15.4 and Fig. 15.5 compare the minimum required SSD design values for the countries reviewed. It can be seen that the U.S. design values are near the upper end of the range, whereas the French, Italian, and Swiss values are near the lower end of the range.[268] Furthermore, it can be seen that the developed stopping sight distances in Part 2, "Alignment (AL)," are somewhat lower than the average values and can be regarded from economic and environmental points of view as

---

*Elaborated based on Ref. 268. (The major portion of the work presented hereafter was conducted by D.W. Harwood, Midwest Research Institute, Kansas City, Missouri.)

TABLE 15.4   Minimum Required Stopping Sight Distances on Level Terrain for Selected Countries and AL*

Country	t_R, s	20	30	40	50	60	70	80	90	100	110	120	130	140
		\ Design or operating speed, km/h												
		\ Stopping sight distance, m												
Australia	2.5	—	—	—	—	—	—	115	140	170	210	250	300	—
	2.0	—	—	—	45	65	85	105	130	—	—	—	—	—
Austria	2.0	—	—	35	50	70	90	120	—	185	—	275	—	380
Canada	2.5	—	—	45	65	85	110	140	170	200	220	240	260	—
France	2.0	15	25	35	50	65	85	105	130	160	—	—	—	—
Germany	2.0	—	—	—	—	65	85	110	140	170	210	255	305	—
Greece	2.0	—	—	—	50	65	85	110	140	170	205	245	290	—
Italy	1.2†	—	—	30	—	60	—	105	—	160	—	230	—	315
Japan	2.5	20	30	40	55	75	—	110	—	160	—	210	—	290
South Africa	2.5	—	—	50	65	80	95	115	135	155	180	210	—	—
Sweden	2.0	—	35	—	70	—	165	—	—	—	195	—	—	—
Switzerland:														
High-performance roads	2.0	—	—	—	—	60	75	90	110	135	160	190	—	—
Other roads	2.0	—	25	35	50	70	90	110	—	—	—	—	—	—
United Kingdom	2.0	—	—	—	70	90	120	—	—	215	—	295	—	—
United States	2.5	—	30	44	63	85	111	139	169	205	246	286	—	—
AL	2.0	—	—	—	50	65	85	110	140	170	205	245	290	—

*Elaborated based on Ref. 268.

†Derived from SSD calculation.

reasonable, since they contain additional driving dynamic safety considerations, as discussed in Sec. 15.1.2.

15.2.2.2   Vertical Curve Design.   Stopping sight distances on vertical curves can be based on the average grade, G, over the deceleration distance (also see Secs. 13.1.4 and 13.2.4). The minimum lengths of vertical curves are controlled by required stopping sight distance, driver eye height, and object height. The required length of curve is such that, at a minimum, the stopping sight distance calculated from Eq. (15.8) is available at all points along the vertical curve.

Note that, contrary to the European approach presented in Secs. 13.1.4 and 13.2.4, which is directed toward determining the required radius of the vertical curve, the U.S. approach, which is discussed in the following, calculates the required length of the vertical curve. The simple conversion formula is

$$R_V = 100 \frac{L}{AD} \tag{15.9}$$

where R_V = required radius of the vertical curve, m
$\quad\quad L$ = required length of the vertical curve, m
$\quad\quad AD$ = algebraic difference in grade, % = $|G_1 - G_2|$ [upgrade = positive $(+G_1, +G_2)$ and downgrade = negative $(-G_1, -G_2)$]

The following formulas are used to determine the required lengths of crest and sag vertical curves by taking into consideration adjacent grades, object, and eye heights.

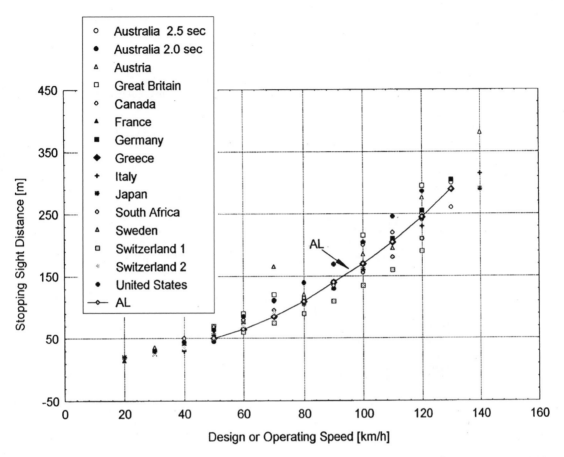

FIGURE 15.5   Minimum required stopping sight distances on level terrain for selected countries and AL. (Elaborated based on Ref. 268.)

For crest vertical curves:

$$L = \frac{AD\, S^2}{200\, (\sqrt{h_1} + \sqrt{h_2})^2} \qquad \text{when } S < L \qquad (15.10)$$

and

$$L = \frac{2\, S + 200\, (\sqrt{h_1} + \sqrt{h_2})^2}{AD} \qquad \text{when } S > L \qquad (15.11)$$

where S = sight distance, m

h_1 = eye height above the roadway surface, m

h_2 = object height above the roadway surface which is hidden from the driver's view, m

For sag vertical curves:

$$L = \frac{AD\, S^2}{2\, (h_3 + S \tan \phi)} \qquad \text{when } S < L \qquad (15.12)$$

and

$$L = 2S - \frac{(2h_3 + S\tan\phi)}{\text{AD}} \qquad \text{when } S > L \qquad (15.13)$$

where h_3 = height of vehicle headlights above the roadway surface, m
ϕ = upper divergence angle of headlight beam, degrees (most countries use 1°; some other countries use 0°)

The curvature of crest and sag vertical curves is often characterized by the K factor, which is defined as the length of the vertical curve divided by its algebraic difference in grade. In other words,

$$K = \frac{L}{\text{AD}} = \frac{R_V}{100} \qquad (15.14)$$

Minimum K values for crest vertical curves which have been taken from the guidelines of various countries[268,361] are presented in Fig. 15.6. The minimum K values are based on the required SSD as well as eye and object heights. Many countries specify parabolic vertical curves; most European countries specify circular vertical curves but for convenience lay them out in the field

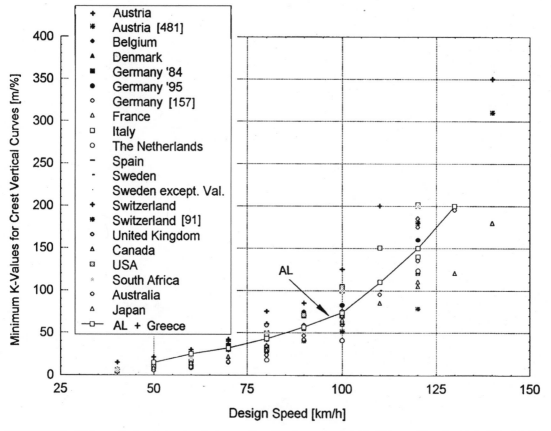

FIGURE 15.6   Minimum K values for crest vertical curves for selected countries and AL. (Elaborated based on Refs. 268 and 361.)

as parabolic curves. For a circular vertical curve, the K value represents the radius of the vertical curve. However, it should be recognized that for a given K value, the alignment of parabolic and circular vertical curves only differ by a few centimeters.

Minimum K values for sag vertical curves are presented in Fig. 15.7. Some of the countries under study, including the United States, use sag vertical curve criteria which are based on headlight height; other countries consider sag vertical curves as less critical with respect to safety, and they base sag vertical curve design on comfort and appearance.[268]

In "AL," minimum radii of crest vertical curves are based on SSD and PSD criteria (Table 13.6). For sag vertical curves, they are based on headlight sight distance and comfort control (Table 13.7).

The developed values in "AL" for crest vertical curves (Fig. 15.6) represent approximately the average values in comparison to the countries under study. That means balanced designs from economic, environmental, and safety points of view can be expected. Compare also Table 13.22.

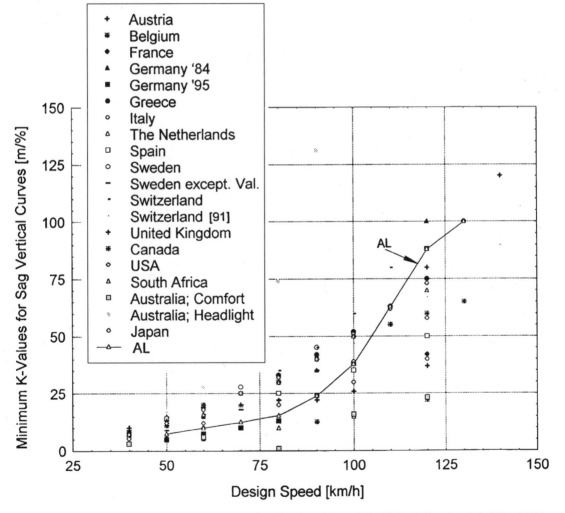

FIGURE 15.7   Minimum K values for sag vertical curves for selected countries and AL. (Elaborated based on Refs. 268 and 361.)

The proposed values in "AL" for sag vertical curves (Fig. 15.7) represent minimum standards for design speeds of at most 80 km/h. They contain, however, all controls needed for sound sag vertical design. For higher design speeds, $V_d \geq 110$ km/h, relatively large values are required to compensate for excessive speeds on multilane highways. Compare also Table 13.24.

15.2.2.3   Parameters Used in Sight Distance Models and Vertical Curve Design

Driver Perception Reaction Time. *   Most of the investigated countries use perception reaction times (PRT) of 2.0 s for rural roads, with the exception of Australia, which uses this PRT only for higher speeds. Canada, Japan, South Africa, and the United States use 2.5 s (see Table 15.4).[268] Both values were found to be adequate according to recent research investigations. Takoa,[696] for instance, concluded the following. It appears that the AASHTO design value of 2.5 s may correspond to the response time of the 95th-percentile driver. Therefore, the stopping sight distance design driver assumption is satisfactory at the present time.

An interesting Evaluation of Driver/Vehicle Accident Reaction Times was conducted by Wilson, Sinclair, and Bisson from the University of New Brunswick.[759] The aim of the research project was to study a driver's perception and reaction times in collision avoidance situations.

The authors reported the following:

> Perception time is the time which elapses from the instant an obstacle or dangerous situation appears in the path of a driver's vehicle to the instant where the driver has recognized the conflict and decided to undertake braking or other evasive actions. The driver reaction time is the time required by a driver to apply braking, turn the steering wheel, or complete other movements after deciding upon a course of action.

The time components needed to describe perception and brake reaction times are given in Table 15.5.

The values for driver perception and reaction times are derived from actual roadway driving conditions. Table 15.6 gives a summary of the study findings for different male and female groups of different ages, as well as the total survey sample for driver perception and reaction times for braking maneuvers. The vehicle braking response time has been included in the calculations to determine a total value.

A significant difference did not exist between the mean values for perception and reaction times of the male and female subjects. The perception time at the 95 percent level of subjects in the under 35-year age group was found to be marginally shorter than that of the over 35 age group.

According to Table 15.6 (last line), the average driver perception time was determined to be 0.56 s with the 99th-percentile value being 0.90 s. Reaction time for the braking maneuver was

---

*Elaborated based on Ref. 443.

TABLE 15.5   Breakdown of Collision Avoidance Time Components[759]

Component	Description
Perception:	
Latency	Time needed for the eyes to begin to move to the stimulus
Eye movement	Time needed for the eyes to move to the stimulus
Recognition	Time needed for the brain to recognize the image
Decision	Time needed to select procedure
Brake reaction:	
Movement time	Time needed to move the foot from the accelerator to the brake
Vehicle response time	Time lag between braking and wheel lock-up or measurable deceleration

TABLE 15.6 Comparison of Perception and Reaction Times for Different Male and Female Groups of Different Ages[759]

Age group, y	Driver perception time				Driver reaction time				Brake response time					
	Mean, s	SD, s	Variance	99th, s	Mean, s	SD, s	Variance	99th, s	Mean, s	SD, s	Variance	Average total, s	99th, s
Males over 35	0.57	0.12	0.01	—	0.21	0.03	0.01	—	0.12	—	—	0.90	—
Males under 35	0.54	0.20	0.04	—	0.30	0.10	0.01	—	0.12	—	—	0.96	—
Females over 35	0.65	0.13	0.02	—	0.26	0.08	0.01	—	0.12	—	—	1.03	—
Females under 35	0.52	0.15	0.02	—	0.30	0.10	0.01	—	0.12	—	—	0.94	—
Total survey sample	0.56	0.16	0.03	0.90	0.28	0.10	0.01	0.58	0.12	0.03	—	0.96	1.60

Note: SD = standard deviation and 99th = 99th-percentile value.

calculated as 0.28 s with the 99th-percentile value at 0.58 s. When the vehicle brake response time of 0.12 s is added to the reaction time, the total time was found to be 0.70 s.

The average total perception, reaction, and vehicle response time for the study sample was found to be 0.96 s. At the 99th-percentile level, this value increased to 1.6 s.[759]

The current design standards of Australia, Canada, Japan, South Africa, and the United States use a perception and reaction time of 2.5 s. Comparing this value with the study findings of Ref. 759 at the 99th-percentile level (1.6 s) indicates that the previously mentioned design standards are conservative and that the normally used values of 2.0 s in most of the European countries appear to be adequate (Table 15.4).

Reaction time is defined by Koerner[351] as the time needed by an unalerted driver until the braking system of his or her vehicle is activated. The reaction time can be subdivided into different time intervals, which result from the human reaction mechanism as well as from the vehicle braking system (Table 15.7).

In order to be able to correctly determine the reaction time needed for estimating SSD, all phases of the emergency braking process have to be examined. These phases are:

- Visibility of the rigid obstacle
- Visual perception of the obstacle
- Recognition of the reaction need
- Accelerator release to brake contact
- Touching the brake
- Beginning of the increase in brake pressure
- Reaching the end pressure for a full braking maneuver
- Stopping of the vehicle immediately in front of the obstacle

In the evaluation from the different phases of the braking process, one must consider two cases. The first case handles the development of the reaction time outside builtup areas, whereas

TABLE 15.7 Reaction Phases of the Emergency Braking Process[351]

Reaction phases in dangerous situations to start a braking maneuver (reaction and perception time)	Human reaction phases (driver)	Perceive recognize conceive act
	Mechanical reaction phases (motor vehicle)	Response phase of the braking Half swell time system

the second case refers to urban streets. This is necessary because research has shown that driving monotony and fatigue have a significant influence on the alertness of the driver, especially on interstates (freeways or autobahnen) where it was found that driving for about 2 h resulted in an increase in the reaction time by about 0.5 s.[483]

The time intervals of the various phases, which are needed during an emergency braking process, are described in detail in Ref. 351. The following percentile values were determined:

Reaction time (50th-percentile value) = 1.260 s

Reaction time (85th-percentile value) = 1.665 s

Reaction time (99th-percentile value) = 2.280 s

These reaction time values refer to an alert driver as is the case of drivers in urban areas. It can be concluded that a reaction time of 1.5 s in urban areas will be exceeded by about 30 percent of the drivers.

For an unalert driver, as is the case in rural areas, the brake reaction time must be increased by approximately 0.5 s. Thus, adopting a reaction time of 2.0 s on rural highways seems quite reasonable.

Even though the preceding values of reaction time, that is,

$$t_R = 1.5 \text{ s} \qquad \text{for urban streets}$$

and

$$t_R = 2.0 \text{ s} \qquad \text{for rural highways}$$

do not completely cover 100 percent of the driver population, it appears that higher values would result into longer stopping sight distances. This leads to larger radii of crest vertical curves and to uneconomical alignment designs.[351]

A recent in-depth literature investigation of perception reaction time was conducted by Buehlmann, Lindenmann, Spacek,[91] which is partially based on Ref. 150. It is concluded that the total perception reaction time falls in the range from 1.485 to 2.030 s.

Generally, because increased attention is expected inside builtup areas, a driver perception reaction time of 1.5 s can be assumed on urban roads.

On roads outside builtup areas (rural), a perception reaction time of 2 s, as used in most European countries, is regarded as adequate.[91] By and large, these results are confirmed by the overall review process presented in this section.

*Tangential Friction Factors Used in Different Countries**

Table 15.8 and Fig. 15.8 show the assumed tangential friction factors used to determine SSD. With reference to Table 15.8 and Fig. 15.8, it should be kept in mind that, during the deceleration process, the tangential friction factors of Austria, Germany, Greece, and of Part 2, "Alignment (AL)," are represented by a speed-dependent relationship, $f_T(V)$, in the stopping sight distance model [Eq. (15.3)]. On the contrary, the tangential friction factors of the other countries represent a constant (average) rate over the entire deceleration process [Eq. (15.8)].

The friction values of "AL," Eq. (10.2) or (10.40), are already in use in the new Greek guidelines.[453] The speed-dependent relationships of the tangential friction factors for Austria and Germany are given here: the Austrian guidelines[35] use the following equation to describe the tangential friction factor at any speed during the deceleration process (Fig. 10.25). This equation was first developed by Lamm and Herring in 1970.[382,383] It has also been applied in the German guidelines, editions 1973 and 1984:[241,243]

$$f_T(V) = 0.214 \left(\frac{V}{100}\right)^2 - 0.640 \left(\frac{V}{100}\right) + 0.615 \qquad (15.15, 10.38)$$

---

*Elaborated based on Refs. 268 and 451.

TABLE 15.8 Criteria for Tangential Friction Factors Used for Stopping Sight Distance Design for Selected Countries and AL*

Country	Design or operating speed, km/h									
	30	40	50	60	70	80	90	100	110	120
Australia	—	—	0.52	0.48	0.45	0.43	0.41	0.39	0.37	0.35
Austria	0.44	0.39	0.35	0.31	0.27	0.24	0.21	0.19	0.17	0.16
France	—	0.37	—	0.37	—	0.33	—	0.30	—	0.27
Germany	0.51	0.46	0.41	0.36	0.32	0.29	0.25	0.23	0.21	0.19
Greece	0.46	0.42	0.39	0.35	0.32	0.30	0.28	0.26	0.24	0.23
South Africa										
Passenger cars	0.42	0.38	0.35	0.32	—	0.30	—	0.29	—	0.28
Heavy vehicles	0.28	0.25	0.23	0.21	—	—	—	—	—	—
Sweden	0.46	0.45	0.42	0.40	0.37	0.35	0.33	0.32	0.30	—
Switzerland:										
High-performance roads	—	—	—	0.49	—	0.44	—	0.40	—	0.36
Other rural roads	—	0.43	—	0.35	—	0.30	—	—	—	—
United States	0.40	0.38	0.35	0.33	0.31	0.30	0.30	0.29	0.28	0.28
AL	0.46	0.42	0.39	0.35	0.32	0.30	0.28	0.26	0.24	0.23

*Elaborated based on Ref. 268.

Note: The tangential friction factors of Austria, Germany, Greece, and "AL" are assumed to increase with decreasing speed during the deceleration process. The tangential friction factors for the other countries represent a constant (average) rate over the entire deceleration process.

Based on the new skid resistance background for Germany, which was developed by Dames in 1992,[131] the German guidelines[246] use the following equation to describe the tangential friction factor at any speed during the deceleration process (Fig. 10.26):

$$f_T(V) = 0.241 \left(\frac{V}{100}\right)^2 - 0.721 \left(\frac{V}{100}\right) + 0.708 \qquad (15.16, 10.39)$$

As can be seen from Fig. 15.8, the tangential friction factors of "AL" are somewhat lower than the average values, and can be regarded as adequate from economic and environmental points of view. Furthermore, they contain all the driving dynamic safety considerations which were discussed in Chap. 10.

*Modern Vehicle Fleet Characteristics.** Since the mid-1970s, passenger cars have developed essentially better aerodynamic drag coefficients. In general, the passenger cars of today have aerodynamic drag coefficients C_w in the range of 0.30 to 0.45. For the existing vehicle fleet in western Europe, the average aerodynamic drag coefficient, C_w, is 0.38. As the trend continues toward vehicle shapes with more favorable aerodynamic drags, the introduction of a lower C_w value for evaluating SSD is regarded as reasonable.

Consequently, for the elaboration of the new German guidelines,[246] as well as of the Swiss norm,[689] an average C_w value of 0.35 was recommended.[91,159] This value was also introduced in the SSD model in Sec. 15.1.2 and in the new Greek guidelines.[453]

Besides the aerodynamic drag coefficient, C_w, the aerodynamic drag force, F_L, is also influenced by the projected frontal area of the vehicle, FA [see Eq. (15.4)]. In addition to the previous vehicle parameters, the mass of the vehicle is also of importance for the SSD model discussed in Sec. 15.1.2.

Research studies by Buehlmann, et al.,[91] who investigated the passenger car fleet in Switzerland during 1985, and by Durth, et al.,[159] who conducted a similar investigation in

---

*Elaborated based on Refs. 91, 159, and 451.

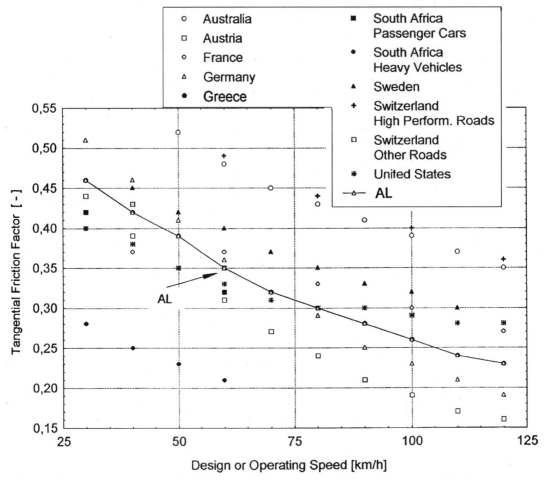

FIGURE 15.8   Criteria for tangential friction factors used for stopping sight distance design for selected countries and AL. (Elaborated based on Ref. 268.)

Germany during 1991, concluded the following with respect to the projected frontal area of passenger cars:

FA = 2.08 m² for Germany, 1991, and "AL" (see Table 15.9)
FA = 2.10 m² for Switzerland, 1985

With respect to the weighted mean of vehicle mass, the vehicles were assumed to be loaded with three-quarters of the difference between the gross mass and the net mass added to the net mass. Consequently, the following values were obtained:

m = 0.75 (1428 − 934) + 934 = 1304 kg for Germany, 1991, and AL (Table 15.9)
m = 1250 kg for Switzerland, 1985

For comparative purposes, the 1981 Austrian guidelines,[35] for instance, made use of the following assumptions to determine the aerodynamic drag force:

TABLE 15.9 Determination of Specific Vehicle Parameters Based on the Existing Vehicle Fleet in Germany, 1991 (Federal Motorvehicle Administration)[159]

Make (1)	Model (2)	Year (3)	Number of vehicles (1991) (4)	Track, cm (5)	Height, cm (6)	Frontal area,* m² (7)	Net mass (unloaded), kg (8)	Gross mass, kg (9)
Audi	80/90	1990	577,114	169.5	139.7	2.13	1080	1540
BMW	3 Series	1990	735,875	164.5	138.0	2.04	1125	1585
Ford	Fiesta	1990	487,603	158.5	137.6	1.96	810	1250
Ford	Escort	1990	485,353	164.0	140.5	2.07	855	1300
Ford	Sierra	1990	485,779	169.8	140.7	2.15	1095	1600
Mercedes	190	1990	713,283	167.8	139.0	2.10	1140	1640
Opel	Corsa	1990	436,276	153.2	136.5	1.88	757	1245
Opel	Kadett	1990	1,115,651	166.3	140.0	2.10	897	1385
Opel	Vectra	1990	366,327	170.0	140.0	2.14	1070	1590
Opel	Omega	1990	286,850	177.2	153.0	2.44	1229	1745
VW	Polo	1990	1,114,064	158.0	135.5	1.93	750	1230
VW	Golf	1990	2,581,581	166.5	141.5	2.12	870	1400
		Σ =	9,385,756					
		Mean		165.4	140.2	2.09	973	1459
		Weighted mean				2.08	934	1428

*FA ≈ 9.3 × Track × Height, m².

Drag coefficient: $C_W = 0.46$

Projected vehicle frontal area: FA = 2.21 m²

Vehicle mass: m = 1175 kg

In conclusion, it can be stated that the aerodynamic drag coefficient and the projected frontal area of passenger cars revealed a decreasing trend between 1981 and 1991, whereas the vehicle mass was increasing, at least in Europe.

For the sight distance model in Sec. 15.1.2 and for the Greek guidelines,[453] the newer German values were adopted.

Criteria for Driver Eye Height and Object Height.   Table 15.10 summarizes the differences between driver eye height and object height for determining vertical curve length in selected countries. All of the assumed driver eye heights are in the range of 1.00 to 1.15 m for a passenger car driver. Object height assumptions are more varied. Australia, Britain, Sweden, Switzerland, and the United States each assumes a small object with a height in the range of 0.15 to 0.26 m. Canada and France use an object height based on vehicle taillight height in the range of 0.35 to 0.38 m. Germany, Greece, and "AL" use a value of object height that varies with operating speed from 0 m at low speeds to 0.45 m at high speeds (Table 15.3). A unique feature of the Swedish guidelines is that they specify a minimum portion of the object (1 minute of arc) that must be visible.[268]

Recent investigation in the field of eye and vehicle height developments have been conducted by Durth and Levin in relation to the German passenger car fleet.[156,157,477]

The most important results of their investigation are as follows:

1. In the last few years, a further decrease of the eye and vehicle height in the passenger car groups could be observed. However, the trend to lower heights has evidently diminished.

2. The measured eye heights of the motorists in the traffic flow are between 87 and 123 cm, with an average of 111 cm.

3. The vehicle heights of the observed passenger car sample lie between 108 and 158 cm, with an average of 135 cm.

4. Obviously, no relevant reductions of the eye and vehicle heights have to be expected in the future. A further decrease of the present height distribution of the passenger car fleet should not be expected.

The results of the investigation indicate that the eye height value of 1.00 m is still considered safe because only 2.0 percent of the drivers fall short of this value.[156]

With respect to object height, Kahl and Fambro[318] analyzed a representative sample of accident data in order to evaluate the types of objects encountered on the roadway that could affect the stopping sight distance. [It should be noted here that the SSD model of AASHTO,[5] for instance, uses 0.15 m as the critical object height (see Table 15.10)].

Kahl and Fambro conclude the following:[318]

1. Two percent of all accidents involved objects or animals on the roadway, and only 0.07 percent of the accidents involved objects or animals less than 0.15 m high. Therefore, small objects and animals were not struck often enough to justify their use as the critical encounter in the stopping sight distance model.
2. The roadway alignment was not a major contributory factor in the object and animal-related accidents because over 90 percent of the accidents occurred on straight, level roads. Thus, the driver's visibility was not limited by the geometry of the roadway.
3. Most of the object and animal-related accidents occurred at night; thus, longer SSD and curve lengths would not necessarily increase the driver's visibility in these situations.
4. Most accidents with objects and animals did not often involve severe occupant injuries; therefore, a small object is not the critical, hazardous encounter in the SSD situation.

These findings do not support the use of the small critical object heights which are currently applied in several guidelines, as shown in Table 15.10. As recommended in Ref. 318, it would be appropriate to use an object height which is greater than 0.15 m. For instance, the taillight of a vehicle might be a better object height because it is a hazard that the driver frequently encounters.

As shown in Table 15.10, most of the countries under study and "AL" make use of these findings by applying higher object heights, at least for higher design speed levels.

TABLE 15.10   Criteria for Driver Eye Height and Object Height Used in Vertical Curve Design for Selected Countries and AL[268]

Country	Driver eye height, m		Object height, m
	Passenger car	Truck	
Australia	1.15	1.80	0.20
Austria	1.00	—	0.00–0.19*
Canada	1.05	—	0.38
France	1.00	—	0.35
Germany	1.00	2.50	0.00–0.45*
Greece	1.00	—	0.00–0.45*
Italy	1.10	—	0.15
Japan	1.20	(1.50)†	0.10 (0.75)†
South Africa	1.05	1.8	0.15–0.60*
Sweden	1.10	—	0.20
Switzerland	1.00	2.50	0.15
United Kingdom	1.05	—	0.26
United States	1.07	—	0.15
AL	1.00	—	0.00–0.45*

*Depends on speed.

†Crossings (structures).

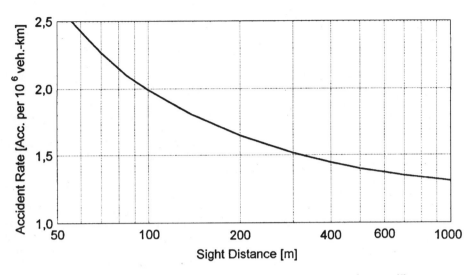

FIGURE 15.9 Accident rate as function of sight distance on two-lane rural roads ($R^2 = 0.91$).[368]

Safety Considerations. As already discussed in "Sight Distances" in Sec. 9.2.1.3, accident studies on two-lane rural roads yielded the following conclusions (see also Fig. 15.9):

1. As sight distance increases, the accident risk decreases.

2. High accident rates are associated with sight distances less than 100 m.

3. Between 100 and 200 m, the accident rates are about 25 percent lower than those associated with sight distances less than 100 m.

4. For sight distances greater than 200 m, no major improvements in accident rates are noted.

A study by Hiersche,[285] which was based on a number of sight distance investigations indicated that: (1) 44 percent of accidents caused by the alignment occur as a result of lack of sight distances, (2) an increase in sight distance leads to a decrease in accident frequency, and (3) the number of sight obstructions has a considerable impact on the accident situation. With respect to the latter finding, Hiersche noted that the accident risk increases with an increase in the number of obstructions up to a certain point at which it begins to decrease again. A possible explanation for this follows: an increase in the frequency of sight restrictions, that is, continuous short sight distances, affects the speed considerably and also the accident risk. Therefore, consistency in sight distance should be provided along the roadway, as well as consistency in horizontal alignment, according to the developed safety criteria in order to achieve an adequate driving behavior.

15.2.3 Passing Sight Distance Criteria in Different Countries*

Passing sight distance is needed where passing is permitted on two-lane highways to ensure that passing drivers who use the lane, which is normally reserved for traffic in the opposite direction

---

*Elaborated based on Ref. 268.

have a sufficiently clear view ahead to minimize the possibility of collision with a vehicle travel-ing in the opposite direction. Passing sight distance (PSD) is considered in the geometric design of a highway to ensure that the completed highway will operate efficiently.

Figure 15.10 shows the various components of the passing maneuver used for explaining and comparing the policies of various countries. The figure shows the position of the passing, passed, and oncoming vehicles at various points in time. At point *A*, the passing vehicle (vehicle 1) starts from a position trailing the passed vehicle (vehicle 2). The passing vehicle accelerates and, at point *B*, begins to enter the lane of traffic in the opposite direction. At point *C*, the passing vehi-cle reaches the "critical position" or "point of no return" at which the sight distance required to abort the pass is equal to the sight distance required to complete the pass. Beyond point *C*, the driver of the passing vehicle is committed to completing the pass because more sight distance would be required to abort the pass than to complete it. At point *D*, the passing vehicle completes the passing maneuver and returns to its normal traffic lane.

According to Fig. 15.10, it is assumed that the most critical vehicle traveling in the opposite direction (vehicle 3) that would still result in acceptable operations would move from point *H* to point *G* in the time that the passing vehicle moves from point *A* to point *B*; then, the opposing vehicle would move from point *G* to point *F* in the time the passing vehicle moves from point *B* to point *C*, and the opposing vehicle moves from point *F* to point *E* in the time the passing vehi-cle moves from point *C* to point *D*. This results in a clearance margin equal to the distance from point *D* to point *E* at the end of the passing maneuver.

The PSD criteria used in geometric design in different countries are based on varying assump-tions about which of the distances shown in Fig. 15.10 should be included in PSD and on varying assumptions about the speeds, accelerations, decelerations, and clearance margins that will be used by the passing, passed, and oncoming vehicles.

The design values for passing sight distances, as used in different countries, are presented in Table 15.11.[268]

The values of "AL" in Table 15.11 correspond to those of Germany and Greece and are con-sidered reasonable because they are based on reliable safety research work in this field.[152,249,250]

Table 15.12 summarizes the values of driver eye height and object height which are assumed for measuring the passing sight distance.

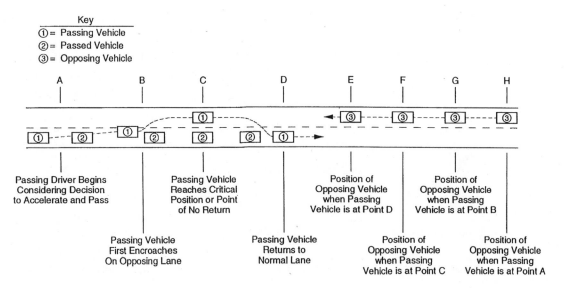

FIGURE 15.10   Components of the passing maneuver used in passing sight distance criteria in various countries.[268]

TABLE 15.11 Passing Sight Distance Criteria Used in Geometric Design in Several Countries and AL*

Country	Based on distance shown in Fig. 5.10	Design or operating speed, km/h											
		30	40	50	60	70	80	85	90	100	110	120	130
Australia	A to H	—	—	330	420	520	640	—	770	920	1100	1300	1500
Austria	B to G	—	—	—	400	—	525	—	—	650	—	—	—
Canada	A to F	—	—	340	420	480	560	—	620	680	740	800	—
Germany	B to G	—	—	—	475	500	525	—	575	625	—	—	—
Greece	B to G	—	—	—	475	500	525	—	575	625	—	—	—
Italy	—	—	200	—	300	—	400	—	—	500	—	600	—
South Africa	A to F	—	—	340	420	490	560	—	620	680	740	800	—
Switzerland	B to G	—	—	—	450	500	550	—	575	625	—	—	—
United Kingdom	B to G	—	—	290	345	410	—	490	—	580	—	—	—
United States	A to F	217	285	345	407	482	541	—	605	670	728	792	—
AL	B to G	—	—	—	475	500	525	—	575	625	—	—	—

*Elaborated and modified based on Ref. 268.

TABLE 15.12 Criteria for Driver Eye Height and Object Height Used for Measuring Passing Sight Distance in Several Countries and AL*

Country	Driver eye height, m	Object height, m
Australia	1.15	1.15
Austria	1.00	1.00
Canada	1.05	1.30
France	1.00	1.00
Germany	1.00	1.00
Greece	1.00	1.00
Japan	1.20	1.20
South Africa	1.05	1.30
Sweden	1.10	1.35
United Kingdom	1.05	—
United States	1.07	1.30
AL	1.00	1.00

*Elaborated based on 250 and 268.

Note: All values in the table are related to passenger cars; none of the countries under study considers trucks in their PSD criteria.

CHAPTER 16
THREE-DIMENSIONAL ALIGNMENT

16.1 RECOMMENDATIONS FOR PRACTICAL DESIGN TASKS

The essential form of highways expresses their function, which is to move people and goods safely and rapidly from one place to another. Highways should have a pleasing appearance, and they should fit gracefully into their surroundings and become acceptable components of the landscape as viewed from outside the highway.[21] The coordination or proper fitting together of the horizontal and vertical alignments and the cross section is an important technique for achieving an esthetically pleasing highway design.

In many countries,[5,37,328] highway esthetics is generally considered a desirable goal in design because anything worth doing is worth doing well. The safety benefits of esthetically pleasing highways have not been well-quantified; nonetheless, in *Practical Highway Esthetics,*[21] it is stated that there is a subtle interrelationship between highway esthetics and highway safety. Measures, such as a smooth continuous alignment, wide recovery areas, broad rounded ditches, flat slopes, and erosion control, all of which make a highway beautiful, also make it safer for traffic. Not only is the highway actually safer but it also appears safer to the driver and passengers, which is important to their enjoyment of the roadside areas and the scenery. *Practical Highway Esthetics* also makes the case for "safety in variety"; that is, monotony is the enemy of both good esthetics and safe operation and it dulls the enjoyment of the visual experience and diminishes the alertness that is essential for safe driving.[657]

The U.S. approach,[5] for instance, promotes the concept of alignment coordination principally for its esthetic value. In addition, excellence in design owing to the coordination of vertical and horizontal alignments increases usefulness and safety, encourages uniform speed, and improves road appearance, and these things are almost always accomplished without additional cost.[5] The German approach concentrates more directly on the three-dimensional alignment design—from a structural standpoint—and differentiates directly between individual good and poor design cases.[240,246]

16.1.1 Design Approach*

16.1.1.1 General Discussion. The foregoing chapters have concentrated primarily on the physical attributes of proper road design with respect to the horizontal and vertical alignments and the cross section to satisfy the requirements of safety and performance.

It has been indicated that, while these needs are of prime importance, some compromise may be necessary in the interests of convenience and economy, especially with respect to three-dimen-

---

*Produced under licence from Austroads Incorporated, Australia, and modified based on Ref. 37.

sional alignments. Ultrasafe roads are of little value if access is too inconvenient or if the additional cost reduces the number of projects that can be undertaken.

Such issues are relatively tangible and amenable to some form of assessment. There remains, however, one area for consideration which is far from tangible or quantifiable but which is nevertheless of increasing importance, and that is the issue of amenity. Considerations of *amenity* are those which concern the effect that a road and its traffic have upon the environmental and esthetic senses of users and of those others who are affected by its construction and operation. The pleasing coordination of alignment and grading, the fitting of the road to the natural contours of the landscape, and the preservation or enhancement of the natural vegetation are all involved.

This chapter considers those less tangible factors which, while not subject to rigid analytical treatment, are nonetheless important in the determination of the total efficacy of a road.

The traditional method of designing roads has been based on the limitations of manual techniques in which the problem is considered, separately, in three views: plan, longitudinal profile, and cross section. Such an approach is a result of the semigraphical techniques usually employed, and can clearly produce satisfactory results if carried out by an experienced designer.

At the same time, it can produce poor results if the designer only considers each view independently of the other two, without proper consideration of the interrelation between the various views.

Even conscientious adherence to the appropriate tabulations and charts included in the previous sections will not guarantee a satisfactory result if the three traditional views are treated independently. The road user sees the road as a constantly changing three-dimensional continuum, and unless designers take full cognizance of this fact, they may not appreciate how the finished design will appear to the road user. It is the appearance of the road to the driver that determines the driver's behavior, and unless the road appears to the driver as the designer intended, the design will have failed in one of the most important attributes, that of satisfying the needs of the user.

The road, therefore, must be considered *at all stages* of design as a three-dimensional structure which should be not only safe, functional, and economical but also esthetically pleasing.[37]

16.1.1.2   *The Driver's View.*   The driver sees a foreshortened and, thus, distorted view of the road, and unfavorable combinations of horizontal and vertical curves can result in apparent discontinuities in the alignment, even though the horizontal and vertical designs each comply separately with the provisions of previous chapters. Such combinations can mask from the driver a change in horizontal alignment or even a sag curve deep enough to conceal a significant hazard such as a hidden dip.

Not only is the driver's view constantly changing, but the duration of his or her view of successive elements of a road also varies. Features situated in a long, low sag remain in view for a considerable length of time, whereas other features at or near an abrupt crest or on a tight curve are in view only fleetingly. It follows then that important features, such as intersections, are best located on long sag curves.

Visual clues to the driver from peripheral areas must be treated with adequate attention. While the designer views the whole road layout at once and is aware of all changes in alignment, the driver sees much less at any one time. The driver's inherently restricted view can be further limited at night or in other times of poor visibility. The designer must, therefore, provide the driver with as many clues as possible as to what lies ahead, but must make sure that the roadside conditions do not convey messages which are ambiguous or misleading.[37] The visual clues presented to the driver by the view of the road surface are especially important on sections with sharp vertical (crest) or horizontal curves or both combined.

16.1.2   Design of Driving Space*

16.1.2.1   *Elements of Three-Dimensional Alignment.*   Sections 16.1.2 to 16.1.4 are based, to a large extent, on the German "Guidelines for the Design of Rural Roads," Part: Alignment,

---

*Elaborated based on Refs. 240, 246, and 657.

Section 2: Three-Dimensional Alignment (RAL-L-2), 1970,[240] the new German "Guidelines for the Design of Roads," Part: Alignment, 1995,[246] and the most recent knowledge in this field.[160,433,657]

Based on an in-depth review by the authors of many guidelines and published works in various countries on different continents, it can be concluded that the German approach to three-dimensional alignment can be regarded today as one of the most reliable and effective methods with respect to practical design tasks. It has been referred to in a number of references, including Refs. 328, 432, 690, and 730.

By applying perspective methods, the view to the road can be shown in a single drawing. In these recommendations, only the perspective view of the driver should be considered. Other perspective views should not be used for the three-dimensional evaluation of the road. For example, a perspective view from a "bird's eye" may show a sharp curve (Fig. 16.1a), which in reality is not critical because of driving dynamics or optical deficiencies (Figs. 16.1b and 16.1c).

The goal of the following recommendations is to produce the best alignment that provides optimum safety and traffic quality. Well-balanced road sections, in which each single design element contributes to a good road characteristic, should be created. Well-balanced sections eliminate unsafe feelings and driver discomfort. With the use of these recommendations, the designer will be able to recognize and evaluate preliminary road designs that result from superimposing selected horizontal and vertical design elements. In this way, the designer can create three-dimensional design elements that achieve perceptible, sound road characteristics.

Although the design of a road may consist of individual elements, the combination of the horizontal alignment (plan) and the vertical alignment (profile) results in a spatial or three-dimensional creation. The resulting driving space can be described in its sequence with the concept of road characteristics. It includes all of the structural elements and determines the driving behavior of the motorist. Road characteristics should not be greatly changed over short roadway sections. A consistent sequence of images of the driving space should be balanced in relation to the design parameters among themselves (relation design). Different roadway sections should be connected by gradual transitions.[160,444]

Thus, in the design of highway alignments, horizontal and vertical design elements are necessarily superimposed. Combining the cross section of the road, which includes shoulders, pavement width, pavement lane, and edge markings, with horizontal and vertical design elements results in a three-dimensional design element. The design of any road is made up of a series or a sequence of three-dimensional design elements. Typical three-dimensional design elements are shown in Fig. 16.2.

The creation of a good view of the road (optical guidance by the roadway) requires a coordinated design of the roadway edge (surface guidance) and of the driving space (spatial guidance) with regard to the function of the road. It can be positively influenced by the sensible selection and use of all given possibilities (three-dimensional design elements of the roadway, pavement markings, slopes, embankments, plantings, engineering structures, traffic signing, and directional signing).[160]

A good view of the road (optical guidance) is important for safety and traffic flow along a roadway section. Usually, this can be achieved if the view of the road appears to blend into the surroundings and if the direction of the road is readily apparent.

Furthermore, the optical guidance by the road is created by the perspective view of the road. For example, the direction of the road becomes more obvious as pavement edges and lane lines are marked more distinctly (Fig. 16.1d). Pavement markings are of special significance in superelevated sections and where lanes are widened (surface guidance). Detailed information, including numerous examples that show three-dimensional alignments of highways (spatial guidance), is given in Figs. 16.3 and 16.4.

16.1.2.2 *Horizontal Design Elements*

Tangents.   Long tangent sections of highways are monotonous and fatiguing. They can mislead the driver into traveling at excessive speeds and increase the danger from headlight glare at night. Therefore, long tangents with constant grades must be avoided, and the maximum length

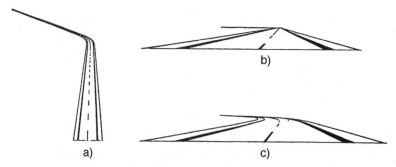

a) Picture of Road from the Bird's-Eye View,
b) same as a), however from Driver's View,
c) same as b), however at the beginning of the circular curve

The Road from Different Perspective Views

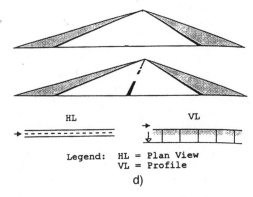

Legend: HL = Plan View
VL = Profile

d)

Improvement in Optical Guidance by Pavement Markings

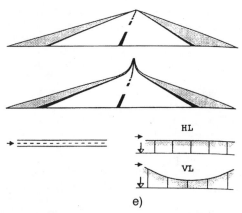

e)

Tangent in Plain and in Sag Vertical Curve

FIGURE 16.1 Examples of different perspective views. (Elaborated based on Ref. 240.)

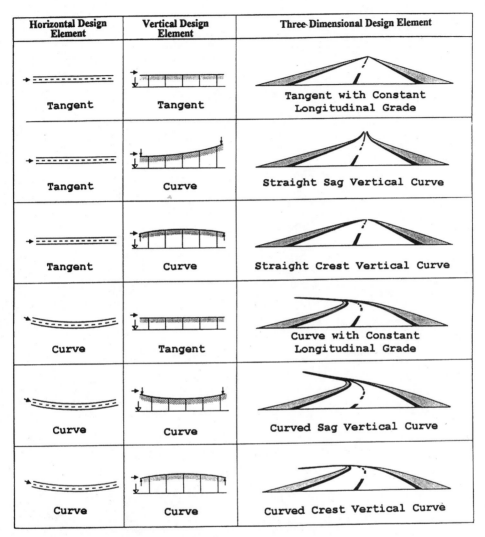

Horizontal Design Element	Vertical Design Element	Three-Dimensional Design Element
Tangent	Tangent	Tangent with Constant Longitudinal Grade
Tangent	Curve	Straight Sag Vertical Curve
Tangent	Curve	Straight Crest Vertical Curve
Curve	Tangent	Curve with Constant Longitudinal Grade
Curve	Curve	Curved Sag Vertical Curve
Curve	Curve	Curved Crest Vertical Curve

FIGURE 16.2   Three-dimensional design elements created by superimposing tangents and curves. (Elaborated based on Refs. 240, 246, and 657.)

should be limited to a numerical value, in meters, of approximately 20 times the design speed, in kilometers per hour (see Sec. 12.1.1.2).

Tangents should be provided to allow passing necessities in cases where they can be well adapted to the topography.

The unfavorable impression caused by long tangents in hilly topography can be reduced by the use of a sag vertical curve with a long length and large radius (Fig. 16.1e).

Short tangent segments between two horizontal curves in the same direction should be avoided (Fig. 16.3a). If such designs cannot be eliminated, it is important that a minimum length be used between the two curves. The minimum length of the tangent segment should correspond to

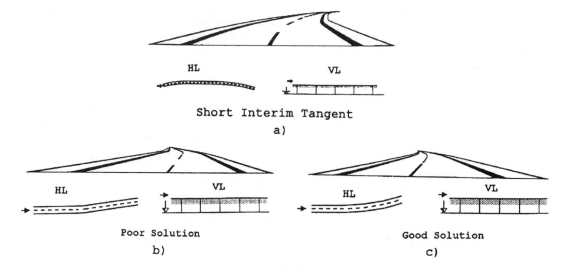

HL   VL

Short Interim Tangent
a)

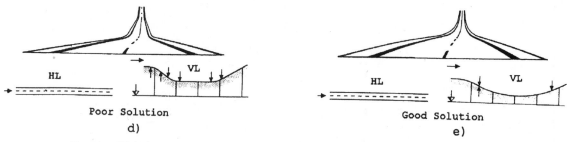

HL   VL   HL   VL

Poor Solution   Good Solution
b)   c)

Perspective View With and Without a Visual Break

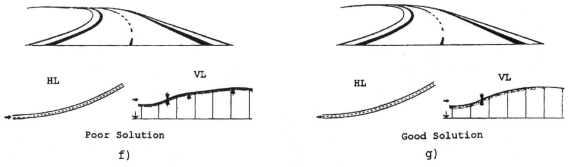

HL   VL   HL   VL

Poor Solution   Good Solution
d)   e)

Short Tangent between Two Succeeding Sag Vertical Curves

HL   VL   HL   VL

Poor Solution   Good Solution
f)   g)

Short Tangent between Two Succeeding Crest Vertical Curves

Legend: HL = Plan View;   VL = Profile

FIGURE 16.3   Examples of poor and good solutions. (Elaborated based on Refs. 240, 246, and 657.

a numerical value, in meters, of approximately 6 times the design speed, in kilometers per hour, to maintain consistency of the optical guidance (see Sec. 12.1.1.2).

Further discussions of tangent lengths, nonindependent tangents, and independent tangents are provided in "Evaluation of Tangents in the Design Process" in Sec. 12.1.1.3.

Curves.   Short circular curves between tangents appear as optical breaks (Fig. 16.3*b*) from the perspective view of the driver. Such an optical break can be avoided (Fig. 16.3*c*) by connecting the two tangents with a long horizontal curve (see also Refs. 444 and 690).

16.1.2.3   *Vertical Design Elements*

Tangents.   The tangent in profile is a segment with constant grade. With respect to the three-dimensional alignment, this part of the alignment is not critical.

A short tangent used between two succeeding sag vertical curves can give the impression of a crest vertical curve (Fig. 16.3*d*) and should be avoided. A better solution, using a long sag vertical curve, is shown in Fig. 16.3*e*. The same is true for a short tangent between two succeeding crest vertical curves. Such a design may give the impression of a sag vertical curve (Fig. 16.3*f*) and should be avoided. A better solution, using a long crest vertical curve, is shown in Fig. 16.3*g*. In addition, the greater the distance a motorist can see ahead on the road, the longer a sag vertical curve should be to eliminate visual breaks.

Sag Vertical Curves.   The sag vertical curve is the three-dimensional design element with the best visual qualities and optical guidance (Fig. 16.1*e*). However, there is one exception: the use of a short sag vertical curve between long sections with constant grades should be avoided. In this case, it does not matter whether the horizontal alignment is on a tangent section or on a curve (Figs. 16.4*a* and 16.4*b*, respectively). In both cases, a visual break in the perspective view occurs. The length of sag vertical curves on embankments usually can be increased considerably without a large increase in earthwork costs.

Crest Vertical Curves.   The crest vertical curve represents the *most* critical design element when considering good visual qualities. The influence of a crest vertical curve is especially critical with short lengths that cause insufficient sight distances (Fig. 16.4*c*). Crest vertical curves with minimum stopping sight distances should be avoided on the mainline roadway if at all possible. The main consideration in using longer lengths is earthwork costs. With the availability of user-friendly earthwork programs and the perspective plot capabilities with the programs, it is now easy to design and test many alternative profiles.

Consequences.   On main roadway sections, visual breaks that are the result of short horizontal and vertical curves or their combination should be avoided. Instead, strive to use longer design elements. Short curves lead to inconsistencies at the roadway's edge. These statements are better understood by comparing Figs. 16.3*a* and 16.3*b* (poor solution) with Fig. 16.3*c* (good solution) and Figs. 16.3*d* and 16.3*f* (poor solution) with Figs. 16.3*e* and 16.3*g* (good solution). In addition, Figs. 16.4*a*, 16.4*b*, and 16.4*c* also show designs that should be avoided.

The following should also be avoided:

- *Diving:* The partial disappearance of the road from the driver's view with reappearance in the extension of the just-passed roadway section (Fig. 16.4*d*).

- *Jumping:* Similar to diving but with displaced reappearance (Fig. 16.4*e*).

- *Fluttering:* Multiple diving or a rapidly rolling profile (Fig. 16.4*f*).

- *Broken-back vertical curve:* A short tangent section between two sag vertical curves (Fig. 16.3*d*).

Most of these designs may lead to critical driving maneuvers because of visual misconceptions, since portions of the alignments become hidden from the driver's view, which may, in turn, mislead the driver about the course of the roadway and opposing traffic. These visual misconceptions are especially dangerous in the case of passing maneuvers.

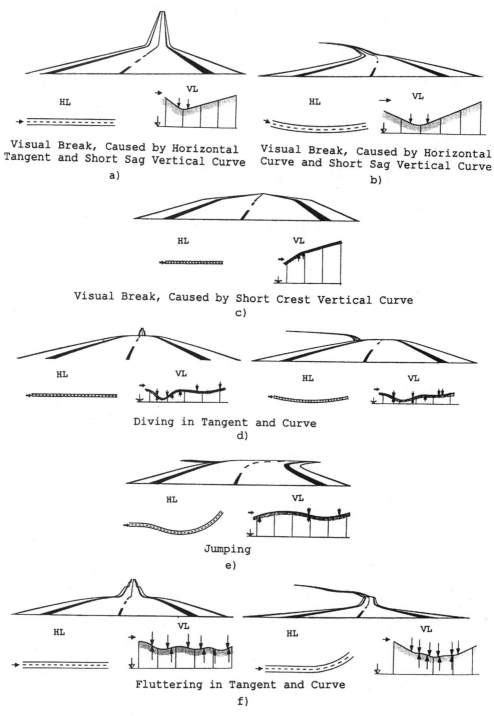

FIGURE 16.4   Design cases to be avoided. (Elaborated based on Refs. 240, 246, and 657.)

16.1.3 Sequence of Design Elements and Superimposition of Elements*

16.1.3.1 Horizontal Alignment.   The size of succeeding design elements in plan can be determined from the radii relations shown in Figs. 9.1 and 9.36 to 9.40 for selected countries.

The safety of a motorist is not potentially impaired by the use of a series of smaller-radius curves for a winding alignment. Despite the sharp curvature, a more or less consistent alignment does exist (Fig. 16.5a). However, isolated sharp curves in the course of a gentle alignment are nonconsistent and should be avoided (Fig. 16.5b).

Therefore, especially for small and average radii of curve, the relationship between preceding and succeeding radii of curve sequences with respect to the previously mentioned relation design backgrounds, should be strictly adhered to.

16.1.3.2 Vertical Alignment.   For sequences of design elements in the vertical plane, the following must be considered (Figs. 16.5c and 16.5d):

1. In hilly/mountainous topography, the radii of crest vertical curves should be larger than the radii of sag vertical curves. This concept provides a longer sight distance for the crest vertical curve (Fig. 16.5c). Significantly longer sight distances provide a greater feeling of safety to the driver.

2. For smaller differences in the elevation of a roadway (up to 10 m) and on a roadway with a flat topography, the radii of sag vertical curves should be larger than those of crest vertical curves. This concept takes into consideration that a motorist can see the road for a longer distance in flat terrain and therefore provides the motorist with a smoother, more visually satisfying view of the course of the road (Fig. 16.5d). It follows from considerations 1 and 2 that:

Hilly topography:

$$R_{crest} > R_{sag}$$ (16.1)

Flat topography:

$$R_{sag} > R_{crest}$$ (16.2)

3. Quick sequences of short crest and sag vertical curves should be avoided.

16.1.3.3 Superimposition of Elements.   With respect to the superimposition of horizontal and vertical alignments, the ratio between radii of horizontal curves R and radii of sag vertical curves R_S cannot be arbitrarily selected, and these radii must be related or tuned to each other. To achieve a satisfactory three-dimensional solution, experience shows that the ratio, R/R_S, should be as small as possible. The ratio should be in the range of 1/5 to 1/10. The main reason for this is that horizontal curves superimposed by sag vertical curves may mislead the motorist in terms of his or her perspective view by presenting an alignment that appears more generous than it actually is.[393,562]

If these values are exceeded, a perspective analysis of the roadway section is recommended. This can be accomplished easily by using modern computer systems and user-friendly earthwork programs now available. Such programs are usually part of a computer-aided drafting and design (CAD) system and also involve the use of perspective plot programs.

The flatter the topography, the larger the radii of crest, and sag vertical curves with respect to the radii of horizontal curves should be selected.

A favorable alignment of the road is guaranteed, based on visual, drainage, and driving dynamic considerations if the reversing points of both the horizontal and vertical alignments are set to coincide approximately. This is shown in Fig. 16.5e, and it can be accomplished if the curves in

---

*Elaborated based on Refs. 240, 246, and 657.

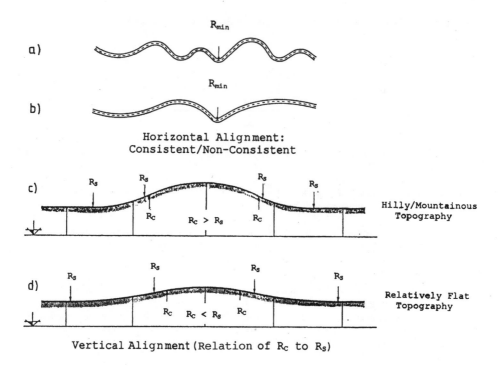

Horizontal Alignment:
Consistent/Non-Consistent

Vertical Alignment (Relation of R_C to R_S)

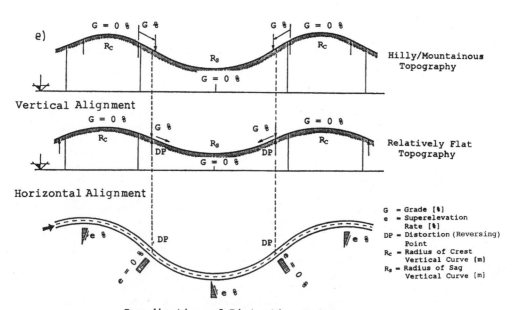

Coordination of Distortion Points
in Horizontal and Vertical Alignments

FIGURE 16.5 Element sequences and superimposition. (Elaborated based on Refs. 240, 246, and 657.)

both horizontal and vertical alignments are placed at approximately the same location and have about the same length. In this way, the reversing (distortion) points then lie at about the same spot and a sufficient longitudinal slope for drainage is guaranteed at the zero points of the superelevation rate, in plan and a sufficient lateral slope at the zero points of the grade in profile.

With such coordination, the number of reversing points in the horizontal and vertical planes should normally be the same. In addition, by designing areas with low superelevation rates in this way, sufficient longitudinal grades for drainage are attained, and, in areas with low longitudinal grades, sufficient superelevation is available.

In hilly or mountainous topography with steeper longitudinal grades, it may be desirable to select a segment of constant grade between the ends of consecutive crest and sag vertical curves. (See the upper vertical alignment of Fig. 16.5e.) In this case, the distortion point of the horizontal alignment should be set nearer to the beginning of the sag vertical curve. This type of design enables the driver to recognize, in advance, the distortion point of the horizontal alignment.

If the local topography does not allow the designer to fit together the reversing points as shown in Fig. 16.5e, then any directional change in the alignment should be clearly visible within the present sight distance. In areas of crest vertical curves, horizontal curves should never be hidden. For example, the views of the road shown in Figs. 16.4d and 16.4e should be avoided.

16.1.4 Intersections and Bridges*

For reasons of recognition and visual perception, intersections should be located in sag vertical curves if possible (Fig. 16.6). The recognition of intersections can be enhanced by appropriate measures such as planting, delineation, traffic devices, etc. The necessary sight distances have to be considered.

Engineering structures, such as bridges, must be coordinated with the alignment (Fig. 16.7b, for example). Flat bridges, such as that shown in Fig. 16.7a, should be avoided if possible.

Large bridges should be recognizable by the driver in time to adjust his or her speed to prevailing conditions, including side wind effects, icy pavements, and so on.

Visually unfavorable engineering structures are those that block the view of an upcoming curve (Fig. 16.7c, for example). Therefore, within the engineering structure, the alignment should already be curved to be in tune with the succeeding course of the roadway (Fig. 16.7d, for example).

---

*Elaborated based on Refs. 240 and 246.

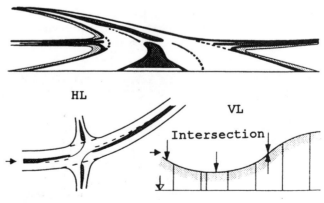

FIGURE 16.6   Intersection in sag vertical curve.[240,246]

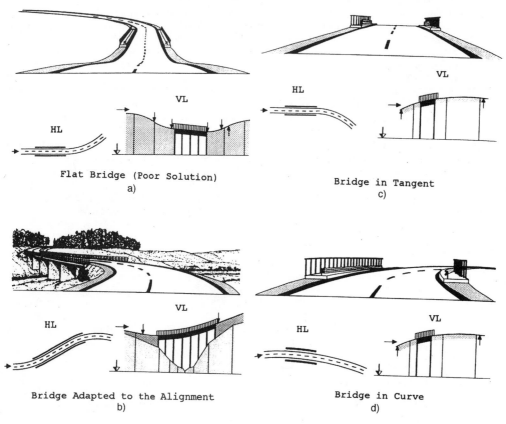

FIGURE 16.7 Examples of engineering structures from a three-dimensional perspective.[240,246]

In conclusion, it can be stated that many recommendations and rules for achieving a good visual, three-dimensional alignment (mostly on the basis of practical experience) exist. However, sound quantifiable design criteria with special emphasis on traffic safety could not be found so far. Three-dimensional alignment has to be regarded today as the most complex component in the highway geometric design process. It still represents the weakest link in the overall design of roads.

16.2 GENERAL CONSIDERATIONS, RESEARCH EVALUATIONS, GUIDELINE COMPARISONS, AND NEW DEVELOPMENTS

A highway, characterized by its horizontal and vertical alignments, as well as by its cross section, is a three-dimensional surface. This surface is mathematically referred to as a *ruled surface* and is examined and described by differential geometry.

The mathematical definition of a highway configuration in three-dimensional space requires the definition of the motion of triplet mutually orthogonal unit vectors.[585] However, transformation of the functional and configuration aspects, as well as the design criteria of a highway into motion equations of a triplet of three-dimensional vectors, is a very difficult task that requires an overall revision of the classical engineering design concept on three separated levels (plan, profile, and cross section).

Furthermore, a three-dimensional highway alignment design would also impose a number of practical problems. In this case, for instance, the individual design elements, like tangents, circular and transition curves, make no sense anymore and must be replaced by new three-dimensional design elements. Figure 16.2 shows those elements schematically but without any mathematical descriptions. For example, Psarianos[585] indicated that the introduction of the classic design elements in a three-dimensional space would, at least theoretically, result in a highway design geometry with up to 720 combinations of various design element variations. It is obvious that such a complicated geometry could hardly be reviewed by a highway engineer.

As a result, a three-dimensional alignment nowadays is especially limited to esthetical approaches and examinations; the mathematical and physical consequences in this case remain unconsidered, with few exceptions. Therefore, the following discussion is related primarily to highway esthetics.

In this context, the vertical and horizontal design of a highway should, when combined, have a pleasing appearance. They should also fit gracefully into their surroundings and become acceptable components of the landscape as viewed from outside the highway. The coordination or proper fitting together of the horizontal and vertical alignments is an important technique for achieving an esthetically pleasing highway alignment design. Even though the safety benefits of esthetically pleasing highways have not been well documented in the past, the literature contains statements about the subtle interrelationship between highway esthetics and highway safety; that is, those things that make a highway beautiful can also make it safer for traffic. In addition, an esthetically pleasing highway also appears to be safer to the users, which is important for their enjoyment of that highway.[657]

16.2.1 Design Approach

16.2.1.1 General Discussion.   According to Neuzil,[547] principles and guidelines underlying esthetic highway design are by no means recent developments in highway design technology but produced in response to current widespread public concern over the impact of engineered works on physical and social environments. A small, but historically significant, number of segments of esthetically designed parkways and turnpikes were built in the eastern United States and Europe long before the interstate highway system was begun.[659,719] In this respect, the formal study of the esthetics of high-speed road alignments began in Germany in the 1930s with the work of Lorenz [486] and others. The German engineers went to considerable trouble and expense to eliminate or modify combinations of vertical and horizontal curvature which looked awkward when viewed in perspective from a low angle. However, at least 25 years have passed since AASHO[2] and the German guidelines[240] summarized the most important rules for producing a rural highway design that would be pleasing to the eye as well as functional and economical. Indeed, both guidelines have indicated that esthetic design recommendations can often be satisfied without conflicting with economic and environmental requirements in highway construction and operation, and careful attention to design esthetics can help to ensure a safe, smooth flow of traffic over the completed facility.[290,487,546,568]

In addition, Neuzil[547] reported the following:

> The pleasurable experience of driving on an attractively designed highway may help reduce driving strain and fatigue factors that bear significantly on highway safety. More directly, one must consider the perceived safety or perceived hazard and the driver's response to them. For example, the minimum curve for a given design speed physically provides the driver with an adequate level of operational safety. As he approaches the curve, however, he may perceive the curve as somewhat unsafe in appearance (even at his reasonable approach speed), particularly if other geometric design features prevail over the segment of highway—vertical alignment conditions, presence or absence of a transition curve, etc. If the driver's perception and judgment of safety of the curve ahead cause him to slow down unnecessarily, traffic friction, accident potential, and motoring strain and discomfort may be unnecessarily increased for the motorist and possibly for nearby drivers on the road as well. Each mile of travel provides many such potential experiences. In this regard, com-

pare driving on a highway with graceful flowing alignment with driving on a highway with design-minimum horizontal and vertical curves and uncoordinated long tangent–short curve alignment and profile.

In their classic treatise on esthetics of highway design, Tunnard and Pushkarev[719] cite accident fatality rates on 13 parkways and turnpikes, which appear to indicate a rather strong relationship between esthetic quality and safety. Highways with a monotonous road characteristic were shown to have fatality rates generally twice that of more attractive highways.[547] Studies which presented particular ways in which the highway designer can be assisted in his or her task of designing a visually stimulating highway include, to name a few, Refs. 89, 205, 540, 582, 626, and 656.

16.2.1.2   *The Driver's View.*

In the field of esthetic design practice, the area of effective vision is of prime importance. In this connection, Kawczynski[328] reported the following on design practice and highway esthetics.

The fundamental principle of road safety engineering is to ensure that road user perceptions of the highway environment and its inherent risks are at least equal to the actual standards used. Actual standards may be less than desirable but it is the road user's perception of the limitations imposed by these standards which is most important. It is not enough that the surfaces and lines which form the road alignment meet the minimum requirements with regard to horizontal and vertical radii, as these may be distorted by the way they are perceived from the driver's perspective. Their actual dimensions and layout may be modified by the creation of optical distortion or illusion such as inflection, discontinuity, and concealment which may have an adverse impact on driver behavior and hence be a source for accidents.[332] From the driver's perspective, the driver's anticipation space is important (see Fig. 16.8).

Figure 16.8 shows the reduction of the zone in the driver's field of vision on which his or her attention is concentrated when his or her speed increases. Anticipation space is determined by the area of effective vision and focusing distance, which depend on speed, as shown in Table 16.1. The majority of esthetic principles are based on the desire to fit the road within the anticipation space.[328]

For every driver there is an optimal density of objects for confidently driving a vehicle. The optimal density of objects is usually achieved when the driver's view is limited to his or her focusing distance. Both an insufficient viewing distance and an extended viewing distance lead to an increase in driver's psychological stress (see also Chap. 19).

An extended viewing distance, for example, a long straight section in monotonous open country with a low intensity of traffic, leads to excessive speed and to what is frequently referred to in literature as a peculiar half-drowsy state bordering on sleep, which can lead to accidents. This state is facilitated by the hypnotizing action on drivers of the regular vibration of the vehicle and

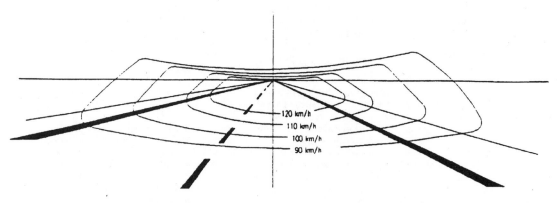

FIGURE 16.8   Driver's anticipation space.[328]

TABLE 16.1   Relationship between Speed, Area of Effective Vision, and Focusing Distance[328]

Speed, km/h	90	100	110	120
Angle of effective vision, degrees	25	20	16	12
Focusing distance, m	500	600	700	750

of the view of the bright pattern of road pavement.[328] According to Babkov,[40] 3 percent of all drivers do not feel it, 23 percent are subjected to it to a great extent, and 74 percent feel it but easily withstand or resist it. Fighting the monotony is what underlies the recommendations in several design standards on the subject of the curvilinear alignment, at least for multilane roads in hilly topography.

Statistical data from Germany, as reported in Ref. 328, show that on the comparatively winding Interstate Ulm—Karlsruhe, for which the parameter S is 60 percent, the accident rate connected with fatigue of a driver was 0.325 accidents/million vehicle kilometers. On the almost straight Interstate Karlsruhe—Mannheim, for which the parameter S is 20 percent, the accident rate with a similar volume of traffic was 0.882 accidents per million vehicle kilometers—three times greater. (S represents the ratio of the arc length of curved roadway sections to the total length of the alignment as a first indication of the smoothness of the alignment.)

Related to the findings presented here, Kawczynski[328] stated that, in a number of European countries, roads had begun to be aligned in the form of a combination of circular and transition curves without straight sections. Note, however, in this connection the new statements about curvilinear alignment made in Secs. 9.1.3.2 and 9.2.3.2, where it is expressed that relation design today means more than curvilinear alignment. The issue of curvilinear alignment, for example, in hilly topography, stands side by side with the issue of sound transitions between independent tangents and curves, for example, in flat topography.

In conclusion, the driver's anticipation space, especially focusing distance, should be considered in the design of horizontal and vertical alignments and the combination of both.

16.2.2 Recommended Design Practices in Various Countries

In the following, the design practices for three-dimensional alignment in the United States, Switzerland, and Australia are studied.

*16.2.2.1 United States.** In this section, important recommended practices in the United States are presented and are compared to the practices recommended in Germany, which are described in Sec. 16.1.

The following are some guidelines for the satisfactory three-dimensional appearance of highway alignment. They are taken from *Practical Highway Esthetics,*[21] the Green Book,[5] Pushkarev's "The Paved Ribbon,"[590] and Cron.[129]

- *Curvature in the horizontal plane should be accompanied by comparable curvature in the vertical plane, and vice versa:*[21,590] Thus, the gradeline for a long, flat horizontal curve should be smooth and flowing and not interrupted by short dips and humps. Figure 16.9a shows an unpleasant view and Fig. 16.9b shows a more pleasing view.

     Comment: The earlier discussion of vertical design elements (sags) in Figs. 16.4a and 16.4b and in Fig. 16.4f show what can happen if this recommended practice is not followed.

- *Awkward combinations of curves and tangents in both the horizontal and vertical planes should be avoided:*[5] The most prominent of these combinations is the broken back gradeline, which is two sag curves in the same direction connected by a short tangent (Fig. 16.9c).

---

*Elaborated based on Ref. 657.

Vertical broken backs are visibly prominent only when short vertical curves are used. The broken back appearance can be corrected by using longer vertical curves at each end of the short grade tangent or by eliminating the short grade tangent. The remedy for horizontal broken back curves is to replace the tangent with a flat curve or to use at least 500 m of tangent section between the two horizontal curves in the same direction.

- *Horizontal and vertical curvature should be coordinated to avoid combinations that appear awkward when viewed from a low angle:*[5,21,129,590] Ideally, the vertices of horizontal and vertical curves should coincide (Fig. 16.9*d*). This statement corresponds to the earlier discussion of the superimposition of elements (Fig. 16.5*e*); however, this is not always possible. A reasonably satisfactory appearance will result, however, if the vertices of the horizontal and vertical curves are kept apart by not more than one-quarter phase. Skipping a phase in the plan while keeping the profile vertices in phase will result in reasonably good coordination and appearance (Fig. 16.9*e*). A shift of one-half phase will result in poor coordination and appearance, as shown in Fig. 16.9*f*.

- In the Green Book,[5] the following is noted:

> Sharp horizontal curvature should not be introduced at or near the top of a pronounced crest vertical curve. This condition is undesirable in that the driver cannot perceive the horizontal change in alignment, especially at night when the headlight beams go straight ahead into space. The difficulty of this arrangement is avoided if the horizontal curvature leads the vertical curvature, that is, the horizontal curve is made longer than the vertical curve. A suitable design can also be achieved by using design values well above the minimums for the design speed.

Comment: This agrees with the earlier discussion on crests and the superimposition of elements. See Figs. 16.4*d* and 16.4*e*, which show designs that are to be avoided.

- The Green Book[5] continues:

> Somewhat allied to the above, sharp horizontal curvature should not be introduced at or near the low point of a pronounced sag vertical curve. Because the road ahead is foreshortened, anything but flat horizontal curvature gives an undesirable distorted appearance. Further, vehicular speeds, particularly of trucks, often are high at the bottom of grades, and erratic operation may result, especially at night.

Comment: This statement addresses the same concerns as described in "An Interesting Phenomenon" in Subchapter 16.2.3.2. Therefore, the authors suggest the following review. Using a CAD system, prepare a series of perspective drawings from the driver's viewpoint. The objective should be to determine if sag vertical curves superimposed with horizontal curves result in perspective views that make the horizontal curve appear flatter than in reality (see Fig. 16.20).

A properly designed study, as hypothesized in Sec. 16.2.3.2, could likely determine the visual effects of varying the lengths and of overlapping the curves. If the hypothesis is true, then the situation is one in which a driver's expectancy is violated. This could lead to improper actions by drivers and, possibly, increased safety problems.

- *The length of highway that can be seen at one time by the motorist should be limited, but adequate sight distance should be preserved:*[21] There should not be any more than two course changes in horizontal alignment (Fig. 16.10*a*) or three breaks in the vertical gradeline in the view of a driver at any point (Fig. 16.10*b*). In particular, a disjointed appearance should be avoided. This may occur when the beginning of a horizontal curve is hidden from the driver by an intervening summit while the continuation of the curve is visible in the distance beyond[5,21] (see Figs. 16.4*d* and 16.4*e*). It may also occur when long tangents are laid in rolling terrain such that the road appears as a series of segments of diminishing size as it passes over successive hilltops ahead (see Fig. 16.4*f*).

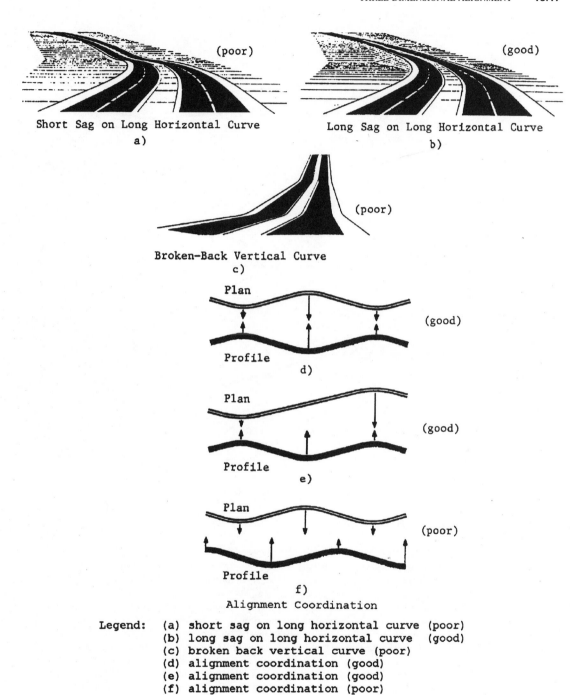

FIGURE 16.9 Examples of poor and good solutions, United States.[590]

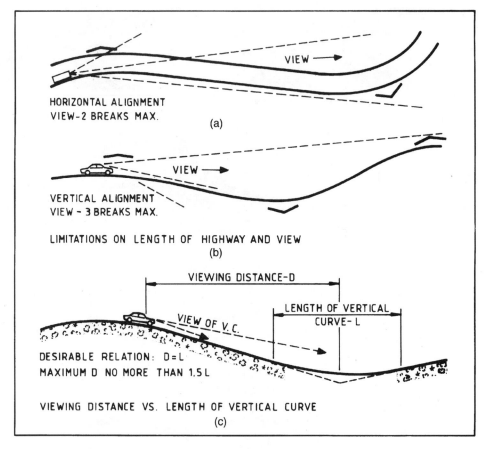

Legend: (a) horizontal alignment, view: two breaks maximum
           (b) vertical alignment, view: three breaks maximum
           (c) viewing distance vs. length of vertical curve

FIGURE 16.10   Fundamental issues for optical (visual) guidance, United States.[5,21]

- For satisfactory appearance, curves, both horizontal and vertical, should usually be considerably longer than the minimum design standards would require, based on safety and operational ease. This is particularly true for sag vertical curves which appear to the driver as sharp angles or kinks when seen from a distance. To avoid the appearance of a kink, the length of a sag vertical curve should be about the same as the viewing distance from which the curve is first perceived by the driver or, as a minimum, at least 0.6 of the viewing distance (Fig. 16.10c). The length of the horizontal curves, in meters, should be at least 3 times the design speed, in kilometers per hour, and preferably twice that length. This statement seems to contradict the minimum lengths of circular curves presented in Table 12.7 in Sec. 12.1.2.2. However, Table 12.7 indicates the basic minimum curve lengths required to produce satisfactory results for performance and safety, whereas appearance criteria may require considerably higher standards in some circumstances.

   Comment: The discussions in "Vertical Alignment" in Sec. 16.1.3.2 appear to support this viewpoint.

- Pushkarev[590] correctly claims that crest vertical curves do not pose the esthetics problems that sag curves do if they are so high that the road terminates visually on the horizon line near the crest. He makes an interesting observation that the minimum stopping sight distances (190 to 260 m at about 110 km/h[5]) beyond which the driver does not see while going over the crest "Visually often appears quite precarious, even though it is functionally safe." He implies that: (1) much larger radius crest curves (generally longer crest curves) should increase the driver's feeling of safety and security and (2) a driver's feeling of safety derived from the view of the road is very important.

    Comment: To some extent, this agrees with the statements made in "Vertical Alignment" in Sec. 16.1.3.2. Perhaps someday we will be able to quantify: (1) how to provide the driver with a feeling of safety and security and (2) the importance, safety-wise, of such driver feelings.[657]

16.2.2.2 *Switzerland.** The Swiss norm largely supports the German guidelines and the U.S. approach, but it also contains some interesting variations, in addition to in-depth graphic interpretations.

The basis for a balanced alignment is a harmonious tuning of design elements and element sequences in order to guarantee a well-perceived alignment and not an abruptly changing road characteristic.

Changes in vertical curvature should be limited as far as possible, especially in the case of a strong winding alignment. The curvatures in the vertical plane should agree with the curvatures in the horizontal plane (Fig. 16.11a). This arrangement facilitates the drainage of the road surface. A good esthetic effect results wherever the length of the horizontal curve corresponds approximately to the length of the vertical curve. Similar statements have been made with regard to Figs. 16.5e and 16.9d and 16.9e. It is interesting to note, however, that in contrast to the German approach, Switzerland attempts from the beginning to limit the changes in vertical curvature.

Small directional changes in the plan (Fig. 16.11b) and small gradient changes in the profile (Fig. 16.11c) should be improved by curves with large radii, as, for example, shown in Figs. 16.3c (plan) and 16.3e (profile).

A short interim tangent between two horizontal curves in the same direction (Fig. 16.11d) should be replaced by a large horizontal curve. Short sections with constant grades between two sag vertical curves (Fig. 16.11e) should be replaced by a long sag vertical curve [see also Figs. 16.3c (plan) and 16.3e (profile)].

Whenever directional changes are hidden by a crest vertical curve (for example, beginning of a horizontal curve or distortion point of a reversing clothoid beyond a crest vertical curve), those situations should be avoided to improve safety (see Fig. 16.12a).

In general, the combinations of elements of the horizontal and vertical alignments presented in Fig. 16.12 may lead to an erroneous perception of the road space, and should be avoided as far as possible for safety and economic reasons. If in doubt, an examination by perspective views is recommended.

If a sag vertical curve follows a crest vertical curve, then the course of the roadway may disappear from the sight of the motorist, but it reappears later after a certain distance (Figs. 16.13a and 16.13b). This sight loss is caused by diving and has an irritating effect on the driver, especially for distances $\overline{AB}$ in the shaded area of Fig. 16.13c.

By all means, sight losses must be avoided whenever they mask dangerous spots such as intersections or unexpected directional changes in the alignment.[690]

16.2.2.3 *Australia.†* As in the German, U.S., and Swiss guidelines, the Australian guidelines provide a discussion of the most important results of the current state of knowledge. However, while the previous guidelines analyzed and evaluated, more or less, individual three-dimensional

---

*Elaborated based on Ref. 690.

†Produced under licence from Austroads Incorporated, Australia, and modified based on Ref. 37

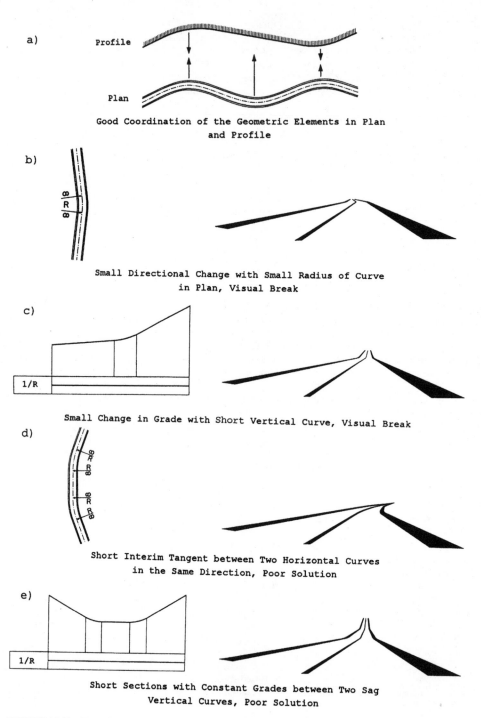

FIGURE 16.11   Examples of: (*a*) good and (*b* to *e*) poor design combinations, Switzerland. (Elaborated based on Ref. 690.)

No.	Combinations	Deficiencies	Improvements
a	Curve beginning or distortion point of reversing clothoid lie beyond crest	Hidden change of direction (Safety)	Coordination of curves to make visible the beginning of horizontal curve in front of crest 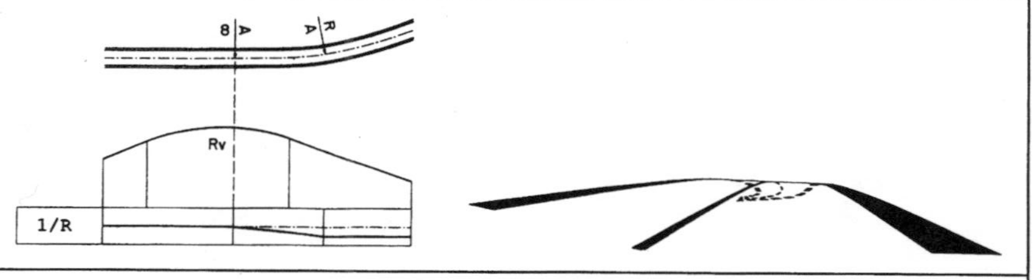
b	Curve beginning immediately after sag	Visual break, extreme case: simulation of false curve direction	Rv / R ≥ 6 Shift elements in such a way that they agree in plan and profile 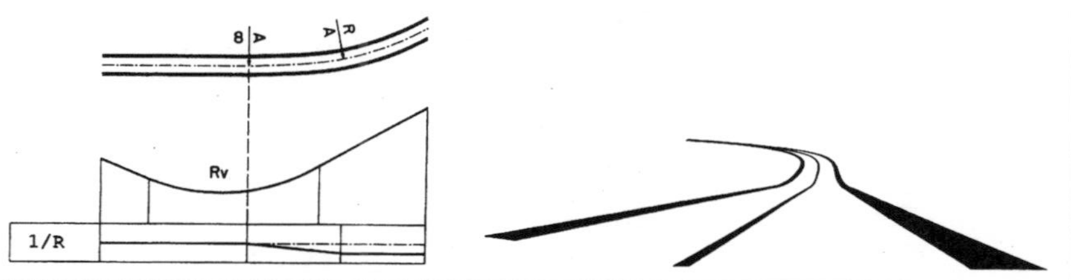
c	Short crest in long horizontal curve	Visual break, extreme case: simulation of false curve direction	Select Rv / R as large as possible 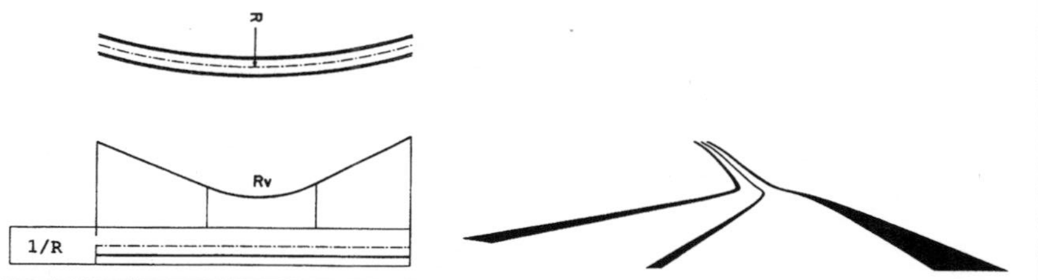

FIGURE 16.12 Erroneous coordination of elements on horizontal and vertical alignments, Switzerland.[690]

No.	Combinations	Deficiencies	Improvements
d	Sag immediately after horizontal curve	Visual break, extreme case: simulation of false curve direction	Shift elements in such a way that they agree in plan and profile 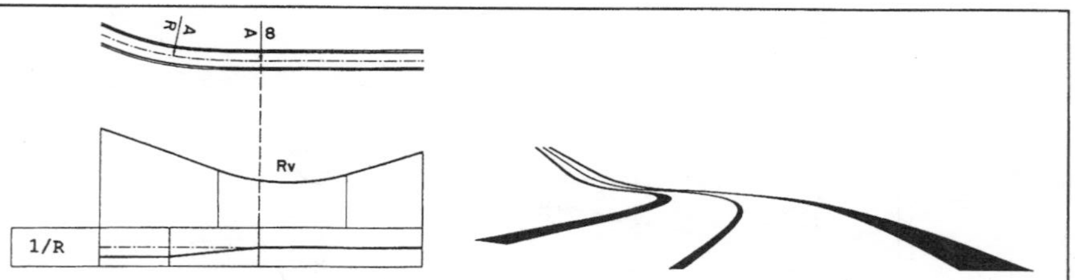
e	Distortion point of reversing clothoid coincides with lowest point of sag	Narrowing of visual perception of the road, drainage problems	Shift elements in such a way that sag corresponds to one horizontal curve 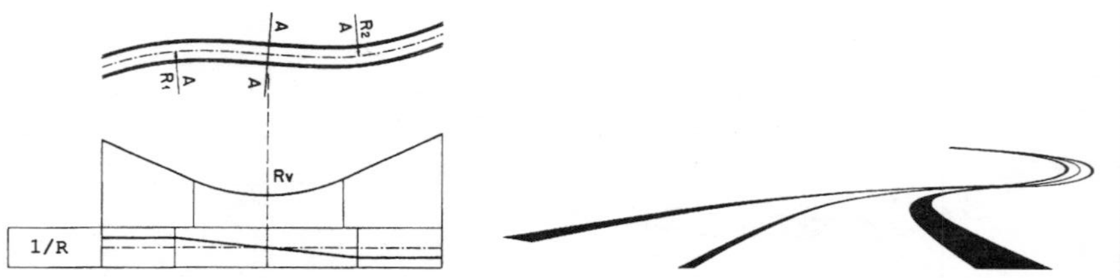

FIGURE 16.12 (*Continued*) Erroneous coordination of elements on horizontal and vertical alignments, Switzerland.[690]

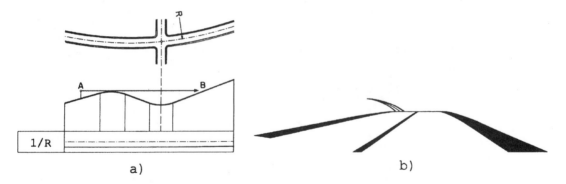

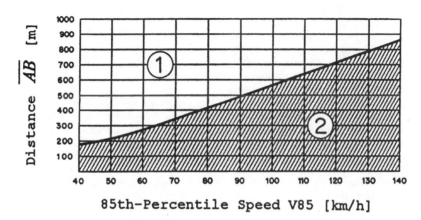

Sight Loss Caused by Diving in the Course of the Roadway

Distance $\overline{AB}$ between the Motorist and the Point
where the Course of the Roadway Reappears

Legend:

1) Sight losses are not critical
2) Sight losses are critical

FIGURE 16.13   Sight losses, Switzerland.[690]

design cases, as well as the superimposition of element sequences on horizontal and vertical alignments, the Australian guidelines dealt with the overall subject of fitting the road to the terrain and combining the horizontal and vertical curvature from an esthetic point of view. It is interesting to note that the Australian guidelines consider relation design issues that support the findings of the authors elaborated in Secs. 9.1.3.2 and 9.2.3.2 with respect to relation design of curvilinear alignments and sound transitions between independent tangents and curves in relation to topography and other issues alluded to previously.

General Recommendations.   The following presentation provides important elaborated excerpts from the Australian guidelines:[37]

> It has been shown that the speed conditions of a road are partly influenced by the nature
> of the terrain, and partly by the horizontal alignment. It follows, therefore, that if the indications of these two factors are similar, the road will provide the best level of consistency

in driver expectancy, and thus safety. Further, a road having both horizontal and vertical curvature carefully designed to conform with the terrain will result in a structure having the desirable esthetic quality of being in harmony with the landform.

In flat open terrain, long straight road sections are common, but generally there is advantage in avoiding excessive lengths of straight road. A gentle curvilinear design especially in hilly terrain always helps to keep the operating conditions "under control" and at the same time, affords scope for far more sympathetic fitting of the road to the terrain. The increased flexibility of this approach enables more pleasing designs to be produced at no extra cost; economies in earth works can often be achieved by fitting the road more closely to the terrain. In addition, safety is enhanced by making the driver more aware of his speed, by allowing him to make better assessments of the distances and speeds of other vehicles, by reducing headlight or sun glare in appropriate circumstances, and by reducing boredom and fatigue. Even in flat country curvilinear designs can be used, at least partially.

Curvilinear design is most readily applicable to divided roads with their less stringent sight distance requirements but the principles are just as relevant to two-lane rural roads provided that care is taken to ensure adequate passing opportunities. In this connection, the relation design background for Australia in Fig. 9.36 applies for curvilinear design, whereas Tables 12.4 and 12.8 provide the basis for sound transitions between independent tangents and curves with and without transition curves.

If the topography is such that "natural" curvature precludes the provision of overtaking sight distance, then artificially introduced long tangents could be lacking in harmony with the terrain and may be subject to inhibiting vertical curvature or require large amounts of earthwork. In such cases, a harmonious curving alignment with the provision of overtaking zones may produce an economical as well as an esthetic solution.

Figure 16.14 illustrates an example of the method and the benefits of proper fitting of the road to the terrain and of proper coordination of horizontal and vertical elements. In addition, there are some examples of poor design form with the appropriate remedial measures indicated. These latter examples are typical of the likely results if the designer does not consider the vertical and horizontal views simultaneously, particularly if a "minimum" vertical standard is superimposed on a relatively unrestricted horizontal regime.

The graphs are not intended to be comprehensive, but serve merely to demonstrate the general concepts that should (or should not) be followed. In all cases, recognition of the deficiency is sufficient to indicate the appropriate remedy, and recognition of the deficiency requires only that the designer takes a three-dimensional, rather than a two-dimensional, view of the problem.[37]

Horizontal Alignment.[37]   The dynamic and safety aspects of horizontal alignment have already been discussed, but further considerations must be given if the resulting recommendations are not to be used mechanically as minima which can sometimes produce esthetically poor results. While Secs. 12.1.2, 12.2.2, 12.1.3, and 12.2.3 indicate the basic minimum curve and transition lengths required to produce satisfactory results for performance and safety, appearance criteria may require considerably higher standards in some circumstances. The following considerations assume the greatest importance when the geometric elements of the road are visible for some distance, that is, generally in flat terrain. In mountainous areas, while appearance criteria will have some relevance, considerations of safety and economy may assume greater importance.

Difficult terrain imposes sufficient, obvious restraints that little problem is encountered in providing road geometry in harmony with the terrain, and the driver can readily appreciate the reasons for any restriction. Flat terrain, however, displays no such obvious restraints and the designer must, therefore, ensure that the geometry does not appear forced or unnatural. For this reason, curve radii considerably greater than the minimum specified in Table 12.5 may be justified in flat terrain. Arc lengths of at least 500 m may be required to ensure that curves with small deflection angles do not appear as kinks.

Independent tangents may be acceptable in flat terrain, as the artificial introduction of curves may result in the appearance of a forced alignment. Where possible, isolated curves should be avoided in flat terrain, even if a fully curvilinear alignment is not employed (see Fig. 16.5b). Several curves in succession, from time to time, add welcome change to a journey in otherwise featureless terrain. Rarely is the terrain so flat that some legitimate curvature cannot be introduced. In this respect, the recommendations about relation design provided in Secs. 9.1.3.2 and

a)

The ideal combination. A smooth flowing appearance results when vertical and horizontal curves coincide. Ideally, horizontal curves should slightly overlap the vertical.

b)

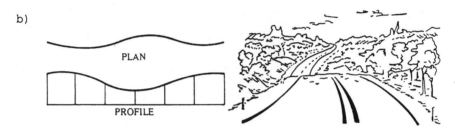

The summit vertical curve restricts the drivers view of the start of the horizontal curve and can produce a dangerous situation.

c)

This example has visually poor alignment with unrelated horizontal and vertical curves, and a broken backed horizontal curve.

d)

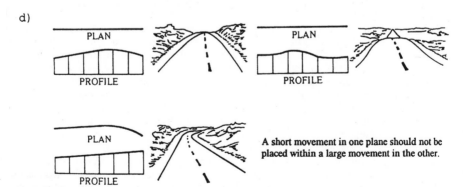

A short movement in one plane should not be placed within a large movement in the other.

FIGURE 16.14   Examples of: (*a*) good and (*b* to *d*) poor design combinations, Australia. (Elaborated based on Ref. 37.)

9.2.3.2 should always be considered for curvilinear alignments, as well as for sound transitions, between independent tangents and curves.

Generally, compound curves should be avoided unless the radii are large because of the difficulty the driver has in detecting the change of curvature. However, compound curves may be considered where:

- The topography, or some other control, for example, a bridge, makes the use of a single-radius curve impracticable.

- The compound curve replaces an otherwise broken back horizontal curve.

The provisions of "Compound Circular Curves" in Secs. 12.1.2.4 and 12.2.2.4, regarding safety of compound curves, are important.

A free-flowing form which naturally fits the terrain usually incorporates reversed horizontal curvature. Such an alignment frequently looks better than any other, particularly in rolling terrain. The designer, apart from satisfying the appropriate dynamic requirements, should ensure that the curves are of correctly related radii and that they are sufficiently well separated to allow pleasing warping of the superelevation. The recommendations of Sec. 14.1.3 set the requirements.

Horizontal curves in the same direction joined by short tangents should be avoided. Such a design leads to visual breaks (see, for example, Fig. 16.3a), since it is difficult to produce a pleasing grading of pavement edges. Furthermore, it is virtually impossible to provide the correct amount of superelevation throughout.

Vertical Alignment.[37]   Much of the philosophy of horizontal alignment is equally valid for vertical alignment. Even though the vertical scale is invariably much smaller than the horizontal, injudicious selection of vertical components can be equally hazardous and as unsightly as poor horizontal design. Indeed, some aspects of vertical design may warrant even closer attention to detail than similar aspects of horizontal design. Research has indicated that a driver's behavior is influenced mainly by horizontal alignment and terrain; drivers do not seem to respond to variations in vertical geometry. Vertical geometry cannot, therefore, be determined in isolation but must be related to the corresponding horizontal geometric features.

Vertical grading composed of long tangents and generous vertical curves should be adopted in preference to sections with numerous grade changes between short tangents. That means hidden dips will be avoided (see Figs. 16.4d, 16.4e, 16.12a, and 16.14b to 16.14d).

Road profiles should not include minor humps or hollows, which curtail sight lines, when the horizontal alignment is adequate to provide overtaking sight distance (see Fig. 16.4f).

Broken back profiles consisting of two or more vertical curves in the same direction separated by short tangents should be avoided. Such profiles are disjointed and look unsightly, especially in sag situations, and should be replaced by single, longer vertical curves (compare Figs. 16.3d, 16.9c, and 16.11e). The difference in levels, and hence earthwork quantities, between the two situations is rarely significant.

Reversed vertical curves can be designed to have common vertical tangent points, and such practice produces pleasing, flowing gradelines which are more likely to be in harmony with the natural landform. The rate of change of vertical acceleration at the tangent point must be considered, however, if either of the component curves is designed for near-limiting values of vertical acceleration. The algebraic sum of the adjacent rates of vertical acceleration should remain within the acceptable limit, usually $0.1g$ [m/s^2] if comfort criteria are to be maintained.[37]

Combined Horizontal and Vertical Curvature.[37]   The most pleasing three-dimensional result is achieved when the horizontal and vertical curvatures are kept in phase, as this relates most closely to naturally occurring forms. Where possible, the vertical curves should be contained within the horizontal curves (see Figs. 16.5e, 16.9d, 16.9e, and 16.11a). This enhances the appearance in sag curves by reducing the three-dimensional rate of change of direction and improves the safety of crest curves by indicating the direction of curvature before the road disappears over the crest. Thus, the best appearance requires the scale of the vertical and horizontal movements to be comparable: a small movement in one direction should not be combined with a large movement in the other.

Horizontal curves combined with crests have less influence on the appearance of a road than those combined with sags. Nevertheless, the effect on safety can be much greater, as the crest can obscure the direction and severity of the horizontal curve (see Fig. 16.12a). Minimum radius horizontal curves, therefore, should not be combined with crest vertical curves.[37] However, observe also the interesting phenomenon, discussed in Sec. 16.2.3.2, when superimposing horizontal curves combined with sags.

Design Aids.[37]   The foregoing discussion has emphasized the importance of the total three-dimensional scope of a road at the design stage if safety and appearance are to be optimized. Traditional design methods are based on two-dimensional concepts and drawings, and thus require a designer with considerable experience if fully satisfactory results are to be achieved.

Various design aids have been developed over the years to assist the designer in his or her task. Conservative aids for examining three-dimensional aspects of proposed designs consist of building models such as scale models, design models, and display models. However, such models are relatively time consuming to build. Typical examples of three-dimensional models are shown in Fig. 16.15. However, in this computer age they are fading away.

An adjunct to computer techniques is the graphics facility afforded by large computer packages. Modern computer-aided design (CAD) systems offer two main areas of interest to the road designer:

- Interactive graphics can be used to examine and amend trial designs without the need for intermediate printouts or plots.

- Perspective views can be used to examine proposed designs as they will appear to the driver. Such views are invaluable aids to assessing esthetic, as well as safety, aspects of proposals but would be prohibitively laborious to produce using manual methods. The speed of the computer, however, renders the production of such plots very practicable and even offers the opportunity of ultimate animation of a series of successive views into a pseudomotion picture view of the three-dimensional alignment of the road.[37] In this respect, the Institute for Highway and Railroad Engineering, University of Karlsruhe, Germany, years ago developed such motion pictures to examine three-dimensional alignments in relation to different driving speed levels.[540,626] Figures 16.16 and 16.17 reveal typical examples of this development stage.

Another interesting possibility for application, developed by Kupke,[379] is driving through a test section with the help of a simulator, for example, by a sample of test persons (see Fig. 16.18). In this connection, many research possibilities are opening up for investigating problems of many different kinds. Unfortunately, technical and financial limitations interrupted these promising developments.[540]

Much of the tedium of producing normal repetitive drawings of road projects can be eliminated by the use of computer-compatible plotting facilities. High-quality drawings can be produced accurately for different demands (see Fig. 16.19) and quickly, leaving the designer free to concentrate on the more creative aspects of the problem.[37]

16.2.3   Safety Aspects

16.2.3.1   General Statements.   The horizontal and vertical alignments can restrict the driver's speed, sight distance, and passing opportunities. It is difficult to separate the safety effects of the different alignment elements.[553]

Although each element may be designed separately, the effect that vertical and horizontal alignments have on each other in combination should be carefully considered in highway design. It is desirable that they increase safety and encourage uniform speed along a highway section. Poorly designed vertical and horizontal alignment combinations can detract from the desirable features and aggravate the deficiencies of each. Instead, vertical and horizontal components should complement each other[728] (see also Fig. 9.32).

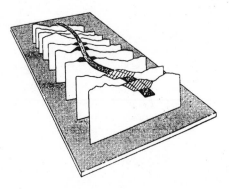

Area Model in Rib Construction
a)

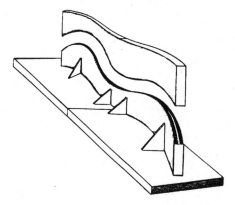

Gradient Model Adjusted to Horizontal Alignment
b)

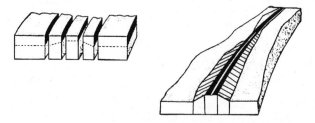

Block Model Manufacture and Setting
c)

FIGURE 16.15   Examples of three-dimensional models.[206,626]

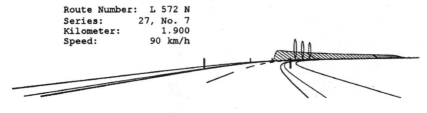

Route Number: L 572 N
Series: 27, No. 7
Kilometer: 1.900
Speed: 90 km/h

RESULTS BY HN-PROGSYST

Signal Effect of Surrounding Elements

a)

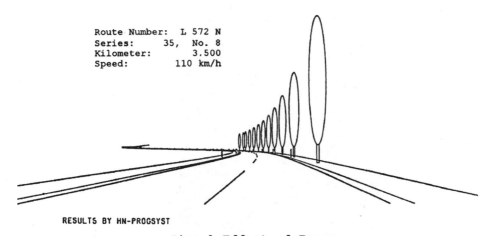

Route Number: L 572 N
Series: 35, No. 8
Kilometer: 3.500
Speed: 110 km/h

RESULTS BY HN-PROGSYST

Signal Effect of Trees

b)

FIGURE 16.16 Examples of motion picture views.[540]

Two important considerations in highway design are design speed and sight distance. These, in turn, vary depending on the horizontal and vertical alignment characteristics of the highway. Favorable horizontal and vertical alignments allow for higher design speed and extended sight distance, and are generally expected to result in increased motorist safety. However, little information exists on accident effects of the three-dimensional alignment. Also, while certain combinations of unfavorable horizontal and vertical alignment are expected to increase accident frequency and severity, not much is known about the quantitative combinations of such "high-accident" alignments.[657]

The impact of individual design elements (such as radius of curve, lane width, grade, sight distance, etc.), their superimposition, and their interactions on the safety aspects of a highway were analyzed (based on international investigations) in "Influence of Design and Operational Parameters on the Accident Situation" in Sec. 9.2.1.3. With respect to horizontal alignment, it was found that radii of curve greater than 400 m have relatively little effect on the accident situation. With respect to the vertical alignment, it was noted that: (1) grades below 5 percent also have relatively little effect on the accident rate and (2) a sharp increase in the accident rate was noted on grades of more than 6 percent.

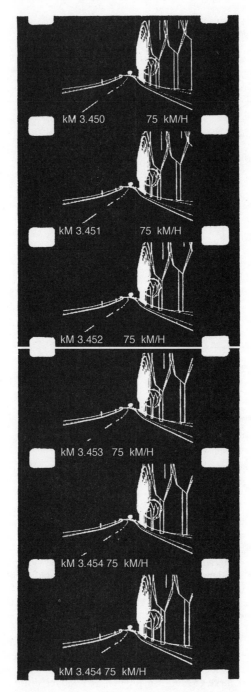

FIGURE 16.17 Perspective motion picture cut.[94,626]

FIGURE 16.18 Syntheticly induced simulator image.[379]

Furthermore, it was found that the new design parameter curvature change rate of the single curve including transition curves CCR_S accounted for the largest variability in operating speeds and accident rates on two-lane rural highways. In addition, a classification system, based on accident investigations of large databases in the United States and Germany for grades less than 6 percent and AADT values less than 12,000 vehicles/day, was developed to distinguish good, fair (tolerable), and poor design practices on the basis of CCR_S. In this respect, and as revealed in Tables 9.10, 10.12, and 18.14, a CCR_S value of less than or equal to 180 gon/km ($R > 350$ m) represents good design. This range formed the basis for developing the corresponding good design ranges of safety criteria I to III. It follows then that any superimposition of horizontal and vertical alignments which meets the quantitative good design ranges of safety criteria I to III (Table 11.1) and gradients less than 5 percent would lead to sound spatial alignment designs. Of course, the previously discussed specific qualitative rules with respect to three-dimensional alignment also have to be considered here.

Note that roadway sections with horizontal alignments up to 180 gon/km and grades less than 5 percent normally represent the largest part of the rural road network in most of the countries investigated.

With respect to the fair design or even poor design ranges of the three safety criteria in connection with grades even greater than 6 percent, new research work has recently started at the Institute for Highway and Railroad Engineering, University of Karlsruhe, Germany, in order to pinpoint more adequately deficiencies in three-dimensional alignment of existing roadways.

The previously mentioned statements about the good design ranges of safety criteria I to III are more or less supported by the results of a study conducted by Zador, et al., that evaluated the effects of vertical and horizontal alignment on safety by studying accident histories at several sites in New Mexico and Georgia.[773] The results of the study in both states indicated that sharp left-hand curves and sharp downgrades are considerably more common at crash sites than at any of the other site types. In New Mexico, the 10th-percentiles of curvature and grade at crash sites

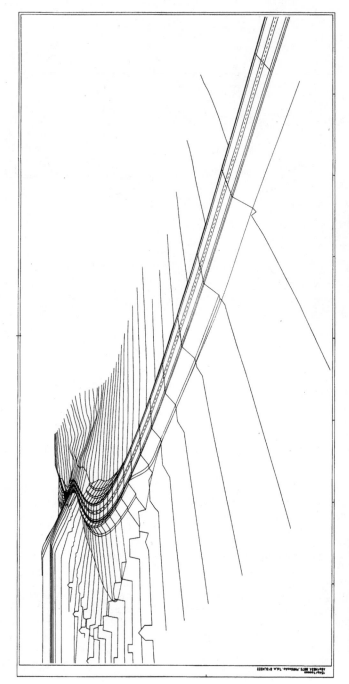

FIGURE 16.19 Terrain modeling by cross-sectional profiles in the perspective view. (*Source:* Highway Administration, Baden-Wuerttemberg, Germany.[626])

were 180 gon/km and -4 percent, respectively. Sections that exceeded both of these values had fatal rollover crashes about 15 times more frequently per volume of travel as did average sections. This combination accounts for approximately 3.5 percent of all fatal rollover accidents in the state. In Georgia, the 10th-percentiles of curvature and grade were 230 gon/km left and -3.3 percent, respectively. Sections that exceeded this combination accounted for 4.6 percent of all fatal rollover accidents in the state. In both states, these sites accounted for approximately 0.25 percent of all travel volume and less than 1 percent of all roadway kilometers, thus indicating a very high over-involvement of crashes at these sites.

Note that the dangerous road sections reported by Zador[773] fall clearly into the fair and poor design ranges of safety criteria I to III combined with grades >5 percent.

Similar conclusions can be drawn with respect to run-off-the-road accidents, although not as strongly as for rollover accidents.

16.2.3.2 *An Interesting Phenomenon.** The problems of visual reception when horizontal and vertical curves are superimposed was recognized very early in Germany by Osterloh,[561,562] who indicated that perspective views of roads are very difficult to evaluate. In this respect, repeated tests by experts to estimate the design elements of horizontal and vertical alignments from the driver's seat failed time and time again. In the case of superimposed horizontal and vertical curvatures, the estimations for instance were widespread and deviated by large margins from the actual values. These under- or overestimations may be understood from the following issues:

1. According to Gregory,[235] a person can see clearly three-dimensionally only up to a distance of 6 m. For longer distances, a person relies, to a large extent, on experience about size relations of objects and their locations to recognize the corresponding distances from his or her eyes.

2. When constructing a perspective view, the horizontal and the vertical curve lead together to one specific image curve. However, going backwards and taking this image curve as a basis, and then modifying the horizontal curve a little, would result in another vertical curve. This procedure, repeated as many times with varying horizontal curves, would always lead to different vertical curves. The same holds true when modifying vertical curves for the same image curve. In other words, for every image curve there exists an infinite number of paired curves consisting of different horizontal and vertical curves.

By considering both of the previously mentioned issues, the following statements can be made: The relevant distances for decision making by the driver fall into the range of between 50 and 150 m for which a real three-dimensional view is no longer possible. Because the viewed image on the retina allows an infinite number of combinations of horizontal and vertical curves, a clear recognition or a proper estimation of curve elements becomes practically impossible for the driver.

In other words, regardless of the many experiences a driver might have, he or she will be *misled* about the true elements of the roadway section ahead. A curve, for instance, might look sharper than what it actually is, causing deceleration; this consequently may lower the accident risk. In other instances, a curve might look flatter than what it actually is, thus causing the driver to accelerate, thereby increasing the accident risk.

As can be seen from the upper part of Fig. 16.20,[561,562] superimposing a horizontal tangent with a crest vertical curve would decrease the perspective view (Fig. 16.20b), whereas a superimposition with a sag vertical curve would increase the perspective view (Fig. 16.20c). On the other hand, the perspective views in the lower part of Fig. 16.20 indicate that the horizontal curve with superimposed crest vertical curve (Fig. 16.20e) is sharper than the horizontal curve shown in Fig. 16.20d. On the contrary, Fig. 16.20f shows, in comparison with Fig. 16.20e, the stretched effect of superimposing a horizontal curve with a sag vertical curve.

This leads to the hypothesis that sag vertical curves superimposed with horizontal curves may result in perspective views which make the horizontal curve appear flatter than in reality.

---

*Elaborated based on Refs. 369, 393, and 657.

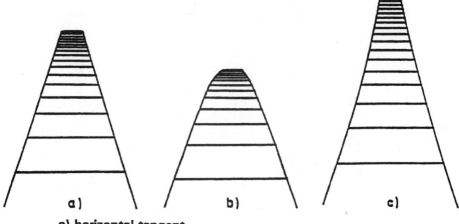

a) horizontal tangent

b) horizontal tangent with superimposed
crest vertical curve

c) horizontal tangent with superimposed
sag vertical curve

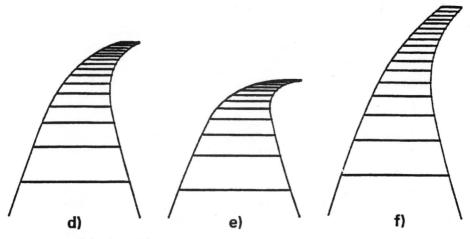

d) horizontal curve

e) horizontal curve with superimposed
crest vertical curve

f) horizontal curve with superimposed
sag vertical curve

FIGURE 16.20 Different three-dimensional views obtained by superimposing vertical and horizontal curves.[393,561,562,657]

Therefore, in those situations, higher operating speeds, in relation to design speeds on which driving dynamic calculations are based, might be expected on horizontal curves, thus increasing the accident risk. A superimposition of a crest vertical curve with a horizontal curve should yield the opposite results.[394,657]

These statements are supported through the findings of the accident studies which were conducted on four two-lane rural roads in Germany with a total length of 40 km and an accident history of 654 accidents during a 3-y period.[369,393] Three of the investigated sites consisted of horizontal curves combined with sag vertical curves. The fourth consisted of a horizontal curve combined with a crest vertical curve (see Fig. 16.21). This figure shows the geometrics, accident locations, and accident causes for horizontal/sag vertical curve superimposition (Figs. 16.21a to 16.21c) and horizontal/crest vertical curve superimposition (Fig. 16.21d). The investigated section length where superimposition largely took place was approximated at 500 m for each of the

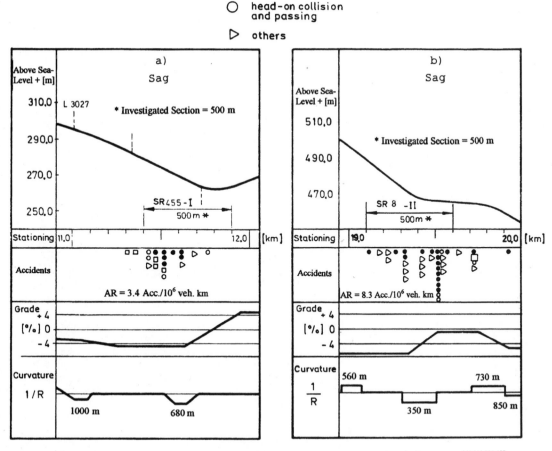

FIGURE 16.21   Influence on the accident situation by the superimposition of horizontal and vertical curvature.[369,393,394,657]

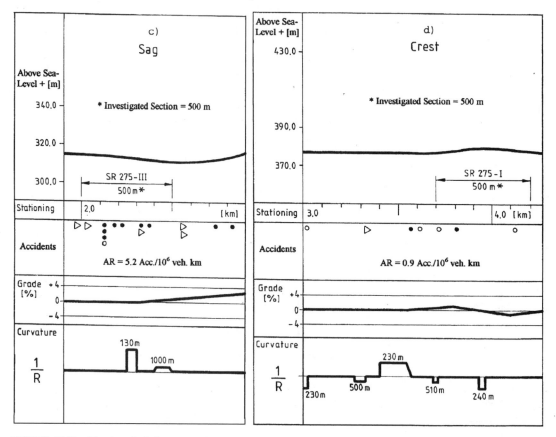

FIGURE 16.21 (*Continued*) Influence on the accident situation by the superimposition of horizontal and vertical curvature.[369,393,394,657]

four cases. From the graphical layout in Fig. 16.21, it can be seen that there is a high concentration of accidents on the three cases for sag curves as opposed to a low concentration on the crest curve case superimposed by horizontal curves. The accidents in the areas of sags often resulted in injuries or even fatalities. Excessive speed was cited as the most frequent cause of these accidents. The accident rates corresponding to the three sag vertical curve cases were 8.3, 5.2, and 3.4 accidents/million vehicle kilometers (see Figs. 16.21a to 16.21c). With respect to the crest vertical curve case, the accident rate was 0.9 accidents/million vehicle kilometers (see Fig. 16.21d). The average accident rate was 2.15 accidents/million vehicle kilometers for the four roadway sections investigated with an overall length of 40 km.

These results support the hypothesis that superimposed sag vertical curves create a potential hazard by making the horizontal curve appear flatter than it really is. In order to avoid the negative effect of a sag vertical curve, it is recommended that the beginnings of horizontal and sag vertical curves be placed somewhat distant from each other. Either the beginning of the sag must be shifted far enough into the horizontal curve to negate its sharpness effect or the end of the sag vertical curve must be shifted to the point where the horizontal curve begins.[657] However, these statements contradict, to a certain extent, the common experience based on drainage considerations and expressed by Figs. 16.5e and 16.9d. Therefore, new research is urgently needed to solve this problem.

16.2.4 Preliminary Conclusion

The purpose of the present chapter, "Three-Dimensional Alignment," is to assist designers in avoiding horizontal and vertical designs that, in subtle negative ways, may diminish the driver's feeling of comfort, certainty, and safety and that, at times, may violate the driver's expectations. The proposed recommendations should help to create highways that are much freer from operating speed inconsistencies and should reduce high-risk driving maneuvers because of errors in driver judgment. Modern geometric highway designs that correspond to the recommendations outlined previously in Secs. 9.1.3.2 and 9.2.3.2 and in the present chapter, "Three-Dimensional Alignment," would thus include more safety-related issues. However, it should be noted that most of the safety assumptions are evaluated only in a qualitative manner.

Overall, it can be concluded that many recommendations and rules for achieving good visual three-dimensional alignments are available, such as the good design ranges of safety criteria I to III which were examined for superimposing horizontal curvatures up to 180 gon/km with grades less than 5 percent, and the practical experiences presented for many individual critical cases in the previous sections. However, it also should be noted that sound quantifiable design criteria with special emphasis on traffic safety could so far not be developed for highways where grades are greater than 6 percent. No statistically sound accident analysis research work has been conducted for such grades with respect to the superimposition of horizontal and vertical alignment, either in Europe or in the United States. Three-dimensional alignment, a very complex component in the highway geometric design process, still represents the weakest link in the overall design of highways.

CHAPTER 17
LIMITING VALUES
OF DESIGN ELEMENTS

In Table 17.1, the limiting values from the most important design elements are listed, separated into the design components:

- Horizontal alignment (plan)
- Vertical alignment (profile)
- Cross section
- Sight distance

In col. 1, a short description of the individual design element is given together with the technical term and the corresponding dimension. In contrast to other guidelines,[246,453] minimum radii of curve, maximum grades, and maximum superelevation rates are consciously directed to different topography classes, based on safety considerations.

For a comprehensive understanding, the relevant chapter or table numbers are given in col. 2. In this way the interested reader can easily find the original data and the corresponding relationships.

Column 3 provides information about the road category group. Group A represents first of all rural roads, while group B mainly encompasses suburban roads (see Table 6.2).

The relevant speed, differentiated by design speed V_d or 85th-percentile speed $V85$, is given in col. 4. This column shows which speed has been taken as the basis for determining the individual design element.

Columns 5 to 13 cover the speed range between 50 and 130 km/h. For steps of 10 km/h, the limiting values for the design parameters were calculated according to the assumptions developed in Chaps. 12 to 15. In this connection, it is interesting to note that the design of two-lane rural roads of category group A is normally laid out up to speeds of 100 km/h, while multilane median-separated motorways of group A represent, in most cases, the speed range between 100 and 130 km/h. Suburban roads are overwhelmingly designed for speeds up to a maximum of 80 to 90 km/h.

TABLE 17.1 Limiting Values of Design Elements

Design elements	See chapter or table no.	Roads of category group	Relevant speed, km/h	Limiting values for V, km/h, according to col. 4								
				50	60	70	80	90	100	110	120	(130)
(1)	(2)	(3)	(4)	(5)	(6)	(7)	(8)	(9)	(10)	(11)	(12)	(13)
Plan												
Maximum tangent length, L_{max}, m	Chap. 12.1.1.2	A	V_d	20 V_d								
Desirable tangent length, L_{der}, m	Chap. 12.1.1.2	A	—	600 < L_{der} < 1000								
Short tangent length, TL_s, m	Table 12.1 or 12.2†	A	V85	110	125	145	165	180	200	—	—	—
Minimum tangent length between curves in the same direction of curvature, L_{min}, m	Chap. 12.1.1.2	A	V_d	6 V_d								
Minimum radius of curve: Flat topography, R_{min}, m	Table 12.5	A	V_d	85	125	180	250	330	425	530	650	(790)
Hilly and mountainous topography, R_{min}, m	Table 12.5	A	V_d	95	140	200	280	370	(480)	(600)	(740)	—
Suburban roads, R_{min}, m	Table 12.6	B	V_d	70	110	160	225	300	—	—	—	—
Minimum length of circular curve, L_{min}, m	Table 12.7	A, B	V_d	30	35	40	45	50	55	60	65	(75)
Minimum radius of curve for the element sequence independent tangent to curve: With transition curve, R_{min}, m	Table 12.4	A	V85$_T$	700	700	≥400			425 (500)	530	650	(790)
Without transition curve, R_{min}, m	Table 12.8	A	V85	≥800 (1000)								
Minimum radius of curve for negative superelevation rate (e = −2.5%), R_{min}, m	Table 14.1	A, B	V85	700	700	700	1,000	1,500	2,000	3,000	3,500	4,500
Parameter of the clothoid, A_{min}, A_{max}, m	Table 12.1.3 and 12.2.3	A, B	V_d	$A_{min} = R/3$; $A_{max} = R$; in addition see Fig. 12.6 for superelevation runoff requirements								

Profile

Parameter	Reference												
Maximum grade:													
Flat topography, G_{max}, %	Table 13.1	A	V_d	—	—	6(8)	5(7)	4(6)	4(5)	3(5)	3(5)	3(5)	3(4)
Hilly topography, G_{max}, %	Table 13.1	A	V_d	—	—	7(9)	6(8)	5(7)	5(6)	4(6)	4(6)	4(6)	—
Mountainous topography, G_{max}, %	Table 13.1	A	V_d	—	—	9(10)	8(9)	7(9)	7(8)	6(8)	—	—	—
Suburban roads, G_{max}, %	Table 13.1	B	V_d	8(12)	7(10)	7(10)	6(8)	5(7)	4(6)	—	—	—	—
Minimum radius of crest vertical curve, R_{VCmin}, m	Table 13.6 or Table 13.22	A, B	V_d		1,500	2,500	3,200	4,300	5,700	7,400	11,000	15,000	20,000
Minimum radius of sag vertical curve, R_{VSmin}, m	Table 13.7 or Table 13.24	A, B	V_d		750	1,000	1,250	1,550	2,400	3,800	6,300	8,800	10,000
Minimum grade in distortion section, G_{min}, %	Chap. 13.1.1.2	A, B	—	0.7 or [$G - \Delta s \geq 0.2\%$ (without curb)]									

Cross section

Parameter	Reference												
Minimum superelevation rate, e_{min}, %	Chap. 14.1.2.1	A,B	—	2.5									
Maximum superelevation rate in curves:													
Flat topography, e_{max}, %	Chap. 14.1.2.2	A	—	8 (9)*									
Hilly and mountainous topography, e_{max}, %	Chap. 14.1.2.2	A	—	7									
Suburban roads, e_{max}, %	Chap. 14.1.2.2	B	—	6									
Maximum relative grade of the superelevation runoff ($d \geq 4$ m), max Δs, %	Table 14.2	A, B	V_d	0.5 d / 2.0	0.4 d / 1.6	0.35 d / 1.4	0.25 d / 1.0	0.225 d / 0.9	0.2 d / 0.8	0.2 d / 0.8	0.2 d / 0.8	0.175 d / 0.6	0.15 d / 0.6
Minimum relative grade of the superelevation runoff, %	Table 14.2	A, B	V_d	0.1 d d [m]: distance from the edge of the traveled way to the rotation axis									

Sight distance

Parameter	Reference												
Minimum stopping sight distance for $G = 0$ [%], SSD_{min}, m	Fig. 15.1, Table 15.4	A, B	V85	50	65	85	110	140	170	205	245	290	
Minimum passing sight distance, PSD_{min}, m	Table 15.2	A	V85	—	475	500	525	575	625				
Required minimum roadway portion with passing sight distance, %	Chap. 15.1.4	A	—	20 to 25								Normally multilane highways	

( ) = exceptional value.

† = short tangent lengths correspond to the maximum allowable lengths of nonindependent tangents (Tables 12.1 and 12.2). For recent developments, see Sec. 18.5.2.

17.3

CHAPTER 18
SAFETY EVALUATION PROCESSES FOR TWO-LANE RURAL ROADS

Road traffic accidents have many different causes arising from the complex relationships between the driver, the motor vehicle, and the road. Nevertheless, if accidents are evaluated, based on statistically significant numbers, they can be shown to depend on definable influencing factors.

For the statistician, a serious accident is, as a single case, a "rare occurrence"; however, as a mass occurrence, traffic accidents become susceptible to a rational analysis and consequently can be forecast. The same is true for the three safety criteria presented in this book, which are based on accident research with respect to road characteristics in order to achieve design, operating speed, and driving dynamic consistency in highway geometric design.

Consideration of "motor vehicle safety" good results has been achieved during the last two decades by the decisive improvement in active and passive vehicle safety,[387,633] as will be discussed in detail in Sec. 20.3.

The driver, as the "most important regulator," is indirectly included in all accident investigations but is difficult to comprehend or to analyze. Therefore, Chap. 19 was elaborated to clarify—at least partially—the influences of human factors on accident situations.

As complex as the generation of traffic accidents is, so are the existing qualitative countermeasures, which are mostly based on the examination of already generated accident "black spots."

Therefore, the elaboration of procedures for the arrangement and evaluation of roadway sections, according to their danger potential in road networks, has appeared to be very important since the 1970s. In particular, highway departments have requested procedures which would allow the evaluation of the accident situation based on objective assumptions in order to conduct redesigns or RRR strategies on those sections of the network which are particularly dangerous. Such procedures are presented in the next two subsections.

18.1 ARRANGEMENT AND EVALUATION OF DANGEROUS ROAD SECTIONS IN NETWORKS BY RELATIVE ACCIDENT NUMBERS

The first procedure for arranging and evaluating dangerous road sections in networks was presented by Lamm and Kloeckner in 1976.[388,389] The procedure is based on four relative accident numbers, which were already discussed and defined in "Relative Accident Numbers" in Sec. 9.2.1.3, as follows:

AR = accident rate, accidents per 10^6 vehicle kilometers, Eq. (9.1)

ACR = accident cost rate, monetary unit per 100 vehicle kilometers, Eq. (9.2)

AD = accident density, accidents per kilometer per year, Eq. (9.3)

ACD = accident cost density, monetary unit per kilometer per year, Eq. (9.4)

The relative accident numbers *accident rate* and *accident cost rate* are related to vehicle kilometers traveled and, consequently, represent the danger, which is the individual risk in terms of accident number and severity on the observed roadway section, whereas *accident density* and *accident cost density* provide economic information about the accident number and costs per kilometer and year. The monetary unit used in the following case study is German marks (DM) and the personal damages were calculated at the time of the investigation according to Willeke, et al.[757]

Comparing the cost estimates of Willeke for fatalities and serious and light injuries for 1967 with those of the German Federal Highway Research Institute (BASt) for 1996 in Table 9.3, one recognizes that a cost explosion of more than 30 times could be observed with respect to fatalities and serious injuries during the last three decades. Therefore, note that the calculated accident cost rates and accident cost densities in the following case study correspond to the 1970s and not to the 1990s. However, this does not change, in any way, the systematic representation or outcome of the proposed procedure.

For the case study, a typical network of a rural county in northern Germany "County Leer" was selected. A database of detailed information from police reports for a time period of 6 years (1968 to 1973) was available for federal and state roads and county routes. The investigation was based on a road length of 546 km; the portion of federal roads (F) was 21 percent, the portion of state roads (S) was 23 percent, and county routes (C) represented 56 percent of the rural road network of the county "Leer." The network to be investigated was subdivided into 137 individual roadway sections.

Overall, 4511 accidents were recorded on the 137 sections from 1968 to 1973. The number of fatalities was 255, the number of seriously injured persons was 1723, and the number of slightly injured persons was 1954. That corresponds to 655 fatalities or injuries per year. The recorded property damage by the police amounted to about 15 million German marks, or 2.5 million German marks/y. According to the previous discussions, this total would nowadays be many times higher. With respect to important accident types, the run-off-the-road type, which is directly related to erroneous speed estimations, represented 1197 accidents (27 percent). In addition, 1594 intersection accidents (35 percent), 217 pedestrian accidents (5 percent), and 770 passing or rear-end accidents (17 percent) occurred.

In addition to the comparison of absolute accident numbers, the accident situation of the County Leer and of each of the 137 roadway sections was evaluated based on the analysis of relative accident numbers.

For the road classes federal road (F), state road (S), and county route (C), the absolute and relative accident numbers are compiled in Table 18.1. The relative accident numbers were calculated according to Eqs. (9.1) to (9.4), defined in "Relative Accident Numbers" in Sec. 9.2.1.3. As can be seen from Table 18.1, with decreasing importance of the road classes (expressed, for example, by traffic volumes), the accident rate and also, to a certain extent, the accident cost rate increase, whereas accident density and accident cost density decrease. That means, as previously discussed, for low-volume roads, the accident risk stands in the forefront, while for average/high-volume roads, the evaluation of the endangerment, expressed by economic values like accident numbers and costs per kilometer, becomes relevant. In spite of the fact that during the last 25 years traffic volumes have significantly increased, the preceding fundamental statement is true in general.

To confirm these findings, Table 18.2 was developed for typical traffic volume classes in County Leer. It shows clearly that, with increasing traffic volume, the accident cost density significantly increases; however, the accident risk, as expressed by the accident rate, significantly decreases.

TABLE 18.1 Absolute and Relative Accident Numbers for Different Road Classes[388]

	Federal road	State road	County route
Accidents	2,311	870	1,330
AADT, vehicles/day	4,985	1,116	634
Length, km	113	128	305
Accident rate, accidents/10^6 vehicle kilometers	1.87	2.77	3.14
Accident cost rate, DM*/100 vehicle kilometers	1.43	2.02	1.88
Accident density, accidents/km/y	3.40	1.13	0.72
Accident cost density, DM*/km/y	25,973	8,270	4,353
Average overall damage per accident, DM†	7,620	7,301	5,990

  *DM = German marks.

  †Including personal and property damages.

TABLE 18.2 Relative Accident Numbers for Different Traffic Volume Classes[388]

Traffic volume, vehicles/day	Length portion, %	Vehicle kilometers traveled, %	Accident rate, accidents/10^6 vehicle kilometers	Accident cost density, DM/km/y
Very low (0–2,000)	77.6	33.5	3.10	4,985
Low (2,000–5,000)	11.4	21.5	2.31	18,360
Average (5,000–10,000)	11.0	44.9	1.72	39,442

Thus, on the one hand, a safety evaluation process, based only on accident rate and/or accident cost rate, would group low-volume roads (normally county routes) as relatively high with respect to an objective endangerment level. On the other hand, a safety evaluation process, based only on accident density and/or accident cost density, would lead to results which classify average or high-volume roads (normally state or federal roads) as relatively high with regard to an objective endangerment level. To avoid this, the following procedure attempts to combine the interactive relationships between these four relative accident numbers in such a way that they commonly define the endangerment level of individual roadway sections within the overall framework of the road network.

In Table 18.3, the four relative accident numbers are shown for a selection of the 137 roadway sections in County Leer. Additionally, the section number and length, the absolute number of accidents, the average annual daily traffic, and the average overall damage per accident are listed. For example, one can clearly recognize that a low-volume county road can express, by high accident rates (AR) and accident cost rates (ACR), a high accident risk for the traffic participant, while, in contrast, accident densities (AD) and accident cost densities (ACD) may be relatively low on the same section. (Example: C 6300 with AR = 10.13 and ACR = 4.01 as well as AD = 2.22 and ACD = 8814). At the other end, an average or high-volume road can be considered, with respect to accident risk, as favorable for the traffic participant because of low accident rates and accident cost rates; however, accident densities and accident cost densities may be high (Example: F 7006 with AR = 1.35 and ACR = 1.27 as well as AD = 4.79 and ACD = 44,964).

Therefore, an exact analysis is necessary so that the available funds are used most effectively and in an optimum way with respect to safety improvements. In many countries, the funds are normally spent for accident reduction measures on average or high-volume roads. Risk reduction on very low- or low-volume roads is usually not considered, since the tangible benefit could be easily rejected as uneconomic. However, it should not be forgotten that those roadway sections represent the largest portion of most rural road networks (see, for example, Table 18.1). Therefore,

TABLE 18.3 Accident Numbers for Individual Roadway Sections (Selection)[388]

	SN	Length, m	Number of accidents	AADT	AOD	AR	ACR	AD	ACD
F	4381	3.8	92	4,033	6,032	2.74	1.65	4.04	24,352
F	4382	5.4	70	3,862	7,284	1.53	1.12	2.16	15,728
F	4383	3.5	109	4,232	4,607	3.36	1.55	5.19	23,933
F	7002	3.0	68	7,159	9,274	1.44	1.34	3.77	34,928
F	7003	9.0	245	6,288	9,199	1.98	1.82	4.55	41,842
F	7004	2.5	74	6,707	6,141	2.01	1.24	4.93	30,245
F	7005	1.5	59	7,633	7,336	2.35	1.72	6.55	48,074
F	7006	3.8	109	9,707	9,392	1.35	1.27	4.79	44,964
F	7007	1.2	35	6,744	8,699	1.97	1.72	4.85	42,222
F	7009	5.0	97	4,611	12,627	1.92	2.42	3.23	40,840
F	7010	2.9	49	3,982	9,721	1.94	1.88	2.82	27,435
F	7201	2.4	34	6,078	5,448	1.06	0.58	2.35	12,823
F	7202	7.9	78	1,671	7,006	2.70	1.89	1.65	11,548
F	7203	3.4	49	2,905	7,349	2.26	1.66	2.40	17,627
F	7501	4.1	75	7,782	9,760	1.07	1.05	3.04	29,690
F	7502	2.3	119	7,782	6,447	3.03	1.96	8.61	55,537
F	7503	2.5	63	5,362	8,806	2.14	1.89	4.19	36,915
F	7504	2.8	115	4,550	4,436	4.12	1.83	6.85	30,380
S	0101	2.4	14	905	8,275	2.94	2.43	0.97	8,044
S	0102	1.4	15	1,107	5,082	4.42	2.24	1.79	9,084
S	1200	5.0	28	1,258	12,582	2.03	2.56	0.93	11,739
S	1401	3.4	43	2,200	9,101	2.62	2.39	2.11	19,165
S	1402	1.6	30	4,314	7,369	1.98	1.46	3.12	22,996
S	1501	9.0	49	723	6,257	3.43	2.15	0.91	5,669
C	6300	1.5	20	601	3,963	10.13	4.01	2.22	8,814
C	6400	3.8	19	801	5,511	2.85	1.57	0.83	4,596
C	6500	5.8	7	364	3,863	1.51	0.58	0.20	776
C	6601	2.1	3	210	5,759	3.11	1.79	0.24	1,374
C	6602	5.0	3	300	8,361	0.91	0.76	0.10	834
C	6800	4.1	9	300	4,854	3.34	1.62	0.37	1,777
C	6900	2.6	1	200	300	0.88	0.03	0.06	19

Note: C = county route, S = state road, and F = federal road; SN = section number; AADT = average annual daily traffic, vehicles/day; AOD = average overall damage per accident, DM; AR = accident rate, accidents/10^6 vehicle kilometers; ACR = accident cost rate, DM/100 vehicle kilometers; AD = accident density, accidents/km/y; and ACD = accident cost density, DM/km/y.

these sections also have to be included in an objective procedure for evaluating the endangerment levels of individual roadway sections in the network.

To evaluate the endangerment of a certain section, the relative accident numbers: accident rate, accident cost rate, accident density, and accident cost density are used. Each of these four relative accident numbers characterizes the endangerment of the section according to a specific criterion. The individual section can be identified by one or another relative accident number as dangerous, while the remaining ones do not indicate this. The simplest case for the selection of dangerous road sections would be if all four accident numbers confirmed the endangerment of specific sections; however, this only happens in exceptional cases.

To date, acknowledged statistical procedures for the evaluation of the endangerment of roadway sections in the network are so far unknown to the authors. Note that the three quantitative safety criteria presented in this book are strictly related to alignment design; intersections are not included. However, the following procedure[388] attempts to encompass all accidents, alignment-related ones, as well as intersection-related ones, and others such as, for example, pedestrian accidents.

Step 1.   In a first step the individual roadway sections were numbered in sequences with regard to the endangerment level expressed by each individual relative accident number. Tables 18.4 to 18.7 present the sequences of the endangerment levels for the accident rate, accident cost rate, accident density, and accident cost density. The first positions represent the respective road sections, which are especially endangered by a certain criterion (for example, in Table 18.4 according to the accident rate). The respective average values for County Leer are also given in the tables. The average values for the accident rate and accident cost rate have the 70th and the 69th position, respectively. For the accident density and accident cost density, they have the 52nd and the 45th position.

For the evaluation of statistical relationships, the 15th- and 85th-percentile levels are normally used for the examination of occurrences with respect to one individual parameter.[104,365,406,408] The 15th-percentile level characterizes the conditions in the inferior range and the 85th-percentile level in the upper range. Between the 15th- and the 85th-percentile level lie 70 percent of the

TABLE 18.4   Sequence of Endangerment Levels with Respect to Accident Rates[388]

Position	Section	AR	Position	Section	AR
1	C 6300	10.13	70	Average	2.29
2	C 2901	9.89	.		
3	C 2902	8.13	.		
4	C 2500	7.60	119	F 7501	1.07
5	C 0700	7.37	120	F 7201	1.06
6	C 5100	6.94	121	C 6602	0.91
7	C 6000	6.69	122	C 6900	0.88
8	C 4903	6.38	123	C 1800	0.86
9	S 3101	6.17	124	C 2800	0.71
10	C 5800	5.52	125	C 4400	0.69
.			126	F 7008	0.52
.			127	C 4801	0.22
.			.		
20	85th-percentile	4.23	.		

Note: C = county route, S = state road, F = federal road; AR = accident rate, accidents/10^6 vehicle kilometers; and 85th-percentile = 85th-percentile limit.

TABLE 18.5   Sequence of Endangerment Levels with Respect to Accident Cost Rates[388]

Position	Section	ACR	Position	Section	ACR
1	C 5100	7.93	69	Average	1.62
2	C 2500	5.55	.		
3	C 2901	5.45	.		
4	C 0700	4.91	119	S 2100	0.54
5	C 2902	4.56	120	F 7008	0.46
6	C 4701	4.35	121	C 2201	0.41
7	C 6000	4.16	122	C 5200	0.37
8	S 1801	4.93	123	C 2301	0.25
9	C 4903	4.09	124	C 1800	0.17
10	F 6300	4.01	125	C 2800	0.05
.			126	C 6900	0.03
.			127	C 4801	0.01
.			.		
20	85th-percentile	2.90	.		

Note: ACR = accident cost rate, DM/100 vehicle kilometers.

TABLE 18.6   Sequence of Endangerment Levels with Respect to Accident Densities[388]

Position	Section	AD	Position	Section	AD
1	F 7506	11.25	52	Average	1.38
2	F 7510	9.54	·		
3	F 7502	8.61	·		
4	F 7504	6.85	119	C 4000	0.14
5	F 7005	6.55	120	C 3900	0.13
6	F 4383	5.19	121	C 4602	0.13
7	S 3101	5.00	122	C 6602	0.10
8	F 7004	4.93	123	C 1800	0.10
9	F 7007	4.85	124	C 4400	0.08
10	F 7006	4.79	125	C 4801	0.06
·			126	C 6900	0.06
·			127	C 2800	0.05
·			·		
20	85th-percentile	3.13	·		

Note: AD = accident density, accidents/km/y.

TABLE 18.7   Sequence of Endangerment Levels with Respect to Accident Cost Densities[388]

Position	Section	ACD	Position	Section	ACD
1	F 7506	71698	45	Average	9747
2	F 7502	55537	·		
3	F 7510	49347	·		
4	F 7005	48074	119	C 4000	678
5	F 7006	44964	120	C 1200	645
6	F 7007	42222	121	C 4602	574
7	F 7003	41842	122	C 2201	447
8	F 7009	40840	123	C 2301	279
9	F 7503	36915	124	C 1800	193
10	F 7002	34928	125	C 2800	40
·			126	C 4801	28
·			127	C 6900	19
·			·		
20	85th-percentile	23562	·		

Note: ACD = accident cost density, DM/km/y.

occurrences, and one can assume that all occurrences in this range can be regarded as not especially predominant. That means that all roadway sections in Tables 18.4 to 18.7 which lie above the 85th-percentile level are predominant. Therefore, all relative accident numbers which lie above the 85th-percentile limits, listed in Tables 18.4 to 18.7, classify the concerned roadway sections as somehow endangered.

It can again be seen from Tables 18.4 to 18.7 that accident rate and accident cost rate generally classify county routes as especially endangered, which means low-volume roads with high individual risks. In contrast, accident densities and accident cost densities evaluate mainly federal roads as endangered, on which average or high traffic volumes lead to high accident numbers and high overall damages. State roads lie somewhere in between.

Thus, of the 137 roadway sections in County Leer, Tables 18.4 to 18.7 identify, in each case, 20 road sections which are characterized as endangered by at least one of the investigated four relative accident numbers, based on the 85th-percentile criterion. This gives 80 identifications of road sections which are endangered. A particular section could be identified one, two, three, or four times.

TABLE 18.8 Evaluation Scheme for Characterizing Dangerous Roadway Sections Based on the 85th-Percentile Criterion for Four Relative Accident Numbers[388]

Section		AD	ACD	AR	ACR	Section		AD	ACD	AR	ACR
F!	7506	+	+	0	0	C!	6300	0	−	+	+
F!	7510	+	+	0	0	C!	2902	0	0	+	+
F!	7502	+	+	0	0	C	2500	−	−	+	+
F!	7504	+	+	0	0	C	0700	−	−	+	+
F!	7005	+	+	0	0	C	5100	−	−	+	+
F!	4383	+	+	0	0	C!	5800	0	0	+	+
S!	3101	+	0	+	−	C	2600	0	−	+	−
F	7004	+	+	−	−	S!	1801	0	0	+	+
F!	7007	+	+	−	0	C	0801	0	0	+	0
F	7006	+	+	−	−	C	3201	0	−	+	0
F!	7003	+	+	−	0	S	1601	0	−	+	0
C!	4903	+	+	+	+	S!	3102	0	0	+	+
F!	7503	+	+	−	0	S	0102	0	−	+	0
F	7507	+	+	−	−	C	4901	−	−	+	0
F!	4381	+	+	0	0	C!	2401	0	0	+	+
F	7002	+	+	−	−	C	4701	0	−	0	+
C!	5700	+	0	+	+	C!	6000	0	−	+	+
C!	2901	+	0	+	+	C	5303	−	−	0	+
F!	7009	+	+	−	0	S	1502	0	0	0	+
C	0802	+	0	0	0	C	1700	−	−	0	+
F	7501	0	+	−	−	C	1002	−	−	0	+
F	7010	0	+	−	0	C	3300	−	−	0	+
F	7511	0	+	−	−						
F	7509	0	+	−	−						

Note: (+) = above the 85th-percentile limit, (0) = between the average and the 85th-percentile value, and (−) = below the average value.

Step 2. For a general characterization of endangered road sections in networks, an evaluation scheme was developed, as presented in Table 18.8. In this table, the sections which represent the 85th-percentile limit and above, and are thus defined as somehow endangered, have been marked by the symbol (+). A designation by the symbol (0) was used if the corresponding section lies between the average and the 85th-percentile value. Relative accident numbers below the average values were designated by the symbol (−).

In the following, a roadway section is interpreted as generally dangerous if it contains the symbol (+) at least twice and the symbol (0) once. These sections are marked by the symbol (!) in Table 18.8.

As can be seen from Table 18.8, only one section (C 4903) is represented by the symbol (+) four times. Two sections (C 5700 and C 2901) are characterized by the symbol (+) three times and by the symbol (0) once. The case where the symbols (+) and (0) are both present twice occurred for 11 sections. Finally, eight sections revealed the symbol (+) twice and the symbol (0) once. Thus, overall, 22 sections could be characterized as being especially endangered according to the preceding qualitative definitions, as shown in Table 18.8 by the symbol (!). Accident analysis confirmed that the selected 22 road sections are indeed very dangerous ones among the 137 investigated roadway sections in County Leer.

A qualitative evaluation scheme for characterizing dangerous roadway sections, based on statistical numbers, like the 85th-percentile limits and the average values of the four relative accident numbers, has thus been developed. However, no guidance has yet been given on how a sequential weighting process, according to the strength of different endangerment levels, could be reached in order to give the responsible authorities an idea of where the funds should be invested.

Step 3.   In Tables 18.4 to 18.7, the roadway sections in the County Leer were numbered in sequence according to their respective endangerment levels, expressed by one relative accident number. The position numbers in the tables can be understood as weighting indices for the endangerment levels. For the four relative accident numbers in Tables 18.4 to 18.7, there are four different weighting indices, expressed by the position numbers for each roadway section. For example, the roadway section C 4903 corresponds to the weighting index of "8" based on the accident rate (Table 18.4), to the weighting index of "9" based on the accident cost rate (Table 18.5), and to the weighting indices of "12" and "16" based on the accident density and accident cost density, respectively. After summing up these four weighting indices and establishing the average value, the new overall weighting index enables construction of a new sequence for describing the overall endangerment levels of the roadway sections. The results of this procedure are presented in Table 18.9. For example, the previously mentioned roadway section C 4903 now obtains the overall weighting index of "11" and, thus, represents the position number "1", which means this roadway section has to be regarded as the *most dangerous* one in the County Leer (Table 18.9).

Comparing Tables 18.8 and 18.9, one recognizes that the new overall endangerment sequence in Table 18.9 agrees very well with the assessment of dangerous roadway sections according to the 85th-percentile criterion in Table 18.8. The 22 sections classified in Table 18.8 as dangerous by the symbol (!) show overall weighting indices which group them all into the range of the 30 most dangerous sections in the County Leer according to Table 18.9. The rest of the sections, which were not indicated as dangerous in Table 18.8, normally have the symbol (+) once and the

TABLE 18.9   Evaluation Scheme for a Sequential Order of Roadway Sections According to their Overall Endangerment Indices[388]

Position	Section	OWI	Position	Section	OWI
1	C 4903	11	25	S 1601	38
2	C 2901	12	26	F 7503	38
3	C 5700	15	27	F 7003	39
4	C 2902/1	20	28	F 4381/1	40
5	C 2902/2	20	29	F 4381/2	40
6	S 3102	20	30*	F 7007	41
7	F 7504	23			
8	F 7510	24	31	C 4701	41
9	C 6300	24	32	C 5100	42
10	S 1801	25	33	S 1002	43
11	C 2401	25	34	F 7010	45
12	F 7502	26	35	C 4901	46
13	C 5800	26	36	C 5303	48
14	S 1502	27	37	S 3300	48
15	C 6000	28	38	F 7004	48
16	C 0801	29	39	S 3103	51
17	C 2500	30	40	F 7006	53
18	F 7506	32	41	F 7002	53
19	S 3101	33	42	F 7507	56
20	C 0700	34	43	F 7509	56
21	F 7005	35	44	C 1700	56
22	F 4383	35	45	F 7511	59
23	S 0102	36	46	C 2600	59
24	F 7009	37	47	F 7501	63

*Last of the 22 sections which were identified according to Table 18.8 as especially endangered.

Note: OWI = overall weighting index.

symbol (0) two or three times. However, since they are classified by the overall weighting process in Table 18.9 as dangerous, they will also be considered here.

The comparative evaluation for especially endangered roadway sections according to Tables 18.8 and 18.9 selected 30 roadway sections of the 137 sections in County Leer. These sections were analyzed in detail for the possible introduction of accident reduction measures such as traffic warning devices, guardrails, lower speed limits, stationary radar devices, RRR measures or even redesigns, etc. Based on the 30 sections identified as dangerous, such a safety improvement process was conducted in County Leer in the years 1977 to 1979 in the sequence of the identified endangerment levels. Beginning with the year 1980, a significant decrease in the overall accident development in County Leer could be observed and one of the formerly most dangerous counties in Germany could be brought down to a level of average accident situations. The proposed procedure was used in the 1980s by several other highway agencies, especially in southwestern Germany, with great success.

In conclusion, based on the two developed evaluation schemes, the dangerous roadway sections could be grouped in networks and put into a sequential order with respect to their overall endangerment levels. To do this, four relative accident numbers, which describe, on the one hand, the individual risk of the traffic participant and, on the other hand, the general risk with respect to economic aspects, were used. Their ranking, according to Tables 18.8 and 18.9, has led to reliable and understanding results including low-volume, as well as average, or high-volume roads. Each procedure complements the other, and using both leads to a control of the results. The resulting ranking of the investigated roadway sections, based on overall endangerment levels (Table 18.9), provides, in combination with the characterization of dangerous sections (Table 18.8), objective criteria for the responsible authorities to decide where the available funds are best invested from the viewpoint of traffic safety improvements.

The characterization and ranking of dangerous road sections is not confined to a single road class, but encompass, as Tables 18.8 and 18.9 clarify, both low-volume county routes, as well as high-volume federal roads, classified by their overall endangerment as derived from the four relative accident numbers. The ability to include low- and even very low-volume roads into a general safety evaluation scheme should be recognized as especially important.

In spite of the qualitative assumptions, for example, with respect to the definition of dangerous road sections in Table 18.8 or the weighting process in Table 18.9, the results confirm that the presented procedure is able to detect dangerous road sections in networks, as was confirmed numerous times by the police departments involved. Of course, it is left to the discretion of the interested reader to use modified assumptions or definitions.

The presented procedure should only be considered as a first step in the complex field of safety evaluation processes in road networks and, for future research in this area, should be interpreted as an impact. Such research is urgently needed. More scientifically based and still better quantifiable solutions are desirable. In this connection, Committee 3.8, "Traffic Accidents," of the German Road and Transportation Research Association is attempting to develop a more sophisticated procedure in order to incorporate important safety requirements in road network design. This is necessary to compensate for the fact that the Part "Network" is, so far, overwhelmingly directed only to the functional classification issue. However, since the new procedure is still in the development stage, it could not be incorporated in this book.

A new procedure to quantitatively evaluate safety issues in road networks with respect to alignment design is, for the first time, presented in the next section.

18.2 DEVELOPMENT OF A SAFETY MODULE FOR ROAD NETWORKS*

Three quantitative safety criteria for evaluating alignment design have been developed in order to address this important target area for reducing accident frequency and severity. These safety criteria are:

---

*Elaborated based on Refs. 449, 595, 672, and 673.

Safety criterion I: Achieving design consistency (Secs. 9.1.2 and 9.2.2)

Safety criterion II: Achieving operating speed consistency (Secs. 9.1.3 and 9.2.3)

Safety criterion III: Achieving driving dynamic consistency (Secs. 10.1.4 and 10.2.4)

The quantitative ranges for safety criteria I to III are presented in Table 11.1. A methodical procedure for the safety evaluation process, consisting of 13 hierarchical steps, is given in Sec. 18.4.1. For a better understanding of the sometimes complex calculation and evaluation processes, a number of case studies are presented in Secs. 12.2.4.2 and 18.4.2.

Based on the classification ranges of Table 11.1, the results of safety criteria I to III indicate the existence of roadway sections that exhibit different design safety levels with respect to the individual safety criterion. The reason for this is that each of the safety criteria represents a separate safety aspect in highway geometric design. It may happen, for example, that according to safety criterion II, the transition section between an independent tangent and a curve would correspond to poor design, whereas safety criterion I with respect to the design speed or safety criterion III with respect to side friction assumed, or both, are well in order for the observed curved site or vice versa.

As an overall safety evaluation procedure, the previously discussed three safety criteria are combined here in an overall safety module. Table 18.10 shows the classification system of the safety module, as based on criteria I to III, for good, fair, and poor design levels. All three criteria are weighted equally. With one exception, at least two of the three criteria must correspond in the decision process to assess the design safety level. The procedure developed represents the current state of knowledge. Changes or improvements concerning the boundaries of the safety module will be examined by new research in the near future (see Table 18.11).

Fundamentally, the evaluation process for the safety module corresponds to the flowchart in Fig. 12.31. The necessary design input data (such as radii of curve, parameters of the clothoid, lengths of elements, and superelevation rates) were provided for the following case study by the road data bank of the Ministry for Environment and Traffic, State of Baden-Wuerttemberg, Germany. The case study is based on the road network of Ehingen County in southwest Germany.

All pertinent equations for calculating 85th-percentile speeds, design speeds, side friction assumed/demanded, and evaluating tangent lengths, as well as the ranges for the three safety criteria in Table 11.1, are contained in the safety computation program presented in Fig. 12.31. The presented procedure, consisting of 13 hierarchical steps, is discussed in detail in Sec. 18.4.1. The accident data for selected roads in the network of Ehingen County has also been provided by the previously mentioned Ministry for Environment and Traffic. The present databases contain road

TABLE 18.10   Classification of the Safety Module for Good, Fair, and Poor Design Levels

Classification				
By criteria I to III (1)				For the safety module (2)
3×	good			Good design
2×	good/1×	fair		
2×	good/1×	poor		Fair design
3×	fair			
2×	fair/1×	good		
2×	fair/1×	poor		
1×	good/1×	fair/1×	poor	
2×	poor/1×	good		
3×	poor			Poor design
2×	poor/1×	fair		

sections with gradients up to 5 to 6 percent and traffic volumes between 500 and 10,000 vehicles/day.

The following case study is based on the German operating speed background, expressed in Table 8.5 by Eq. (8.18).

Since, for this study, geographical information on road sections in the network of Ehingen County is available, such as design elements, operating speeds, accidents, etc., a GIS appears to be the most suitable tool for solving the complex relationships and superimpositions for the safety module proposed here. A Canadian program known as SPANS MAP[22,352,534,672] was used for the analysis of the various safety criteria which make up the core of the overall safety module. The benefit of a GIS is that data of different formats and from different origins can be entered, analyzed, and displayed together. In this study, the display is made on a digitized map (see Fig. 18.1) on which are superimposed the results of the individual safety criteria.

It should be noted that the safety evaluation processes can be done manually. However, this is very time consuming when the three safety criteria need to be combined in the safety module. Therefore, it is more efficient to use a GIS.

The results of safety criterion II, "achieving operating speed consistency," are shown in Fig. 18.1 as an example. Originally, discriminating colors were used to provide easy recognition of good, fair, and poor designs. Since colors could not be presented here, identifying symbols had to be applied. Each symbol represents one curved site with the corresponding transitions, and the symbol describes the safety design level. The results shown in Fig. 18.1* are related to safety criterion II; the results shown in Fig. 18.2* are related to the overall safety module. Accordingly, the results of safety criteria I and III can also be established.

The roadway sections in Figs. 18.1 to 18.3 which are not graphically interpreted by graphical symbols were not included in the investigation because design and/or accident data were not completely available.

Figure 18.2 shows the results of the overall safety module for the road network of Ehingen County in southwest Germany. The highway engineer can immediately recognize different safety levels, which, in this figure, represent the combined impact of the three safety criteria which are equally weighted in accordance with Table 18.10.

The procedure indicates that the process of evaluation of roadway sections or networks by an overall safety module is possible and that this safety module is able to differentiate between good design (sound alignments), fair design (endangered sections), and poor design (dangerous sections).

To determine the degree of agreement between the developed safety module and actual accident rates on observed roadway sections, a 3-year case study was conducted. The results are shown in Fig. 18.3. As can be observed from this figure, the circular symbol, which represents full agreement, and the triangular symbol, which represents a lower accident rate than the safety module would predict, predominate. Thus, it can be concluded that, in the majority of investigated road sections, the actual accident rate corresponds well with the developed safety module or the results are at least on the safe side. The actual accident rate is higher than the predicted one only where the square symbol is shown, which is rare.

Additional investigations, not fully presented in Figs. 18.1 to 18.3, are now based on 440 curved roadway sections in Ehingen County.[673] The results revealed that the case, safety module "poor"—actual accident rate "uncritical," occurred in 11.1 percent of all cases, whereas the case, safety module "good"—actual accident rate "critical," was found only to represent 1.1 percent of the investigated curved segments. In all the other cases (87.8 percent), the comparison revealed that the results of the safety module indicating "good, fair, and poor design practices," corresponded well to the observed actual accident rates. This indicates that the safety module at the current stage of development leads to only minor misinterpretations of the actual endangerment.

The figures presented so far show the graphical layouts of the results for individual safety criteria (for example, Fig. 18.1) and for the overall safety module (Fig. 18.2). However, the GIS used

---

Source: ISE/AKG Software. (ISE = Institute for Highway and Railroad Engineering, University of Karlsruhe, Germany. AKG = AKG Software Consulting GmbH, Ballrechten-Dottingen, Germany.)

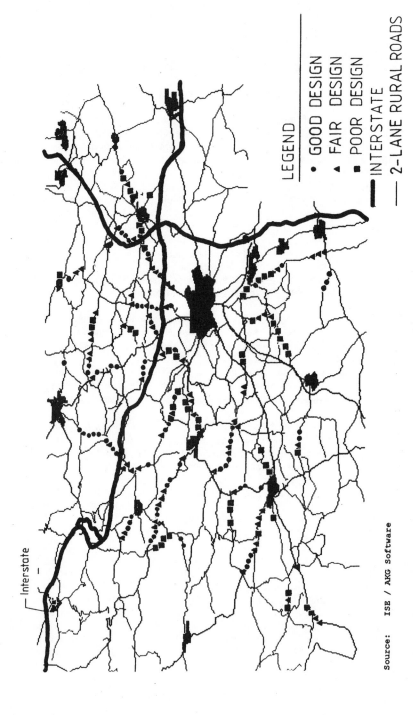

LEGEND

• GOOD DESIGN
▲ FAIR DESIGN
■ POOR DESIGN

━━ INTERSTATE
── 2-LANE RURAL ROADS

Source: ISE / AKG Software

FIGURE 18.1 Digitized map of the road network of Ehingen County in Germany superimposed with the results of safety criterion II.

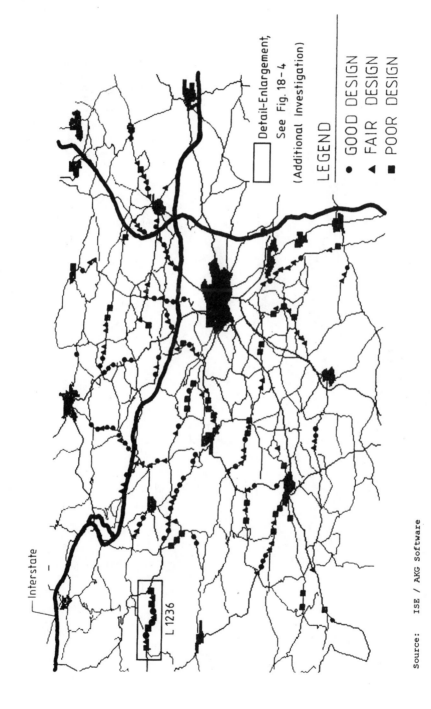

Source: ISE / AKG Software

FIGURE 18.2 Results of the overall safety module for Ehingen County, Germany.

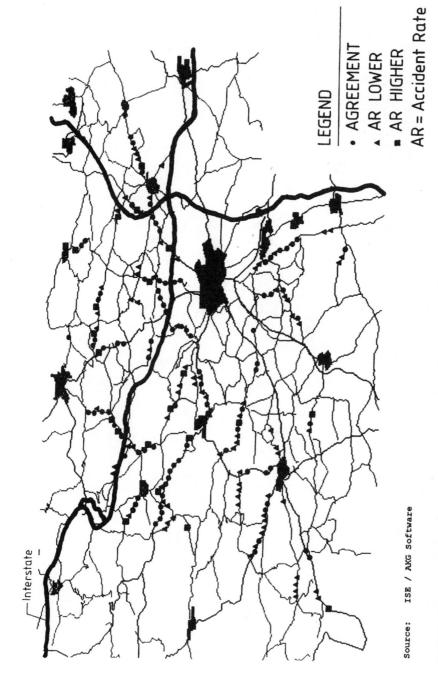

LEGEND

· AGREEMENT
▲ AR LOWER
■ AR HIGHER
AR = Accident Rate

Source: ISE / AKG Software

FIGURE 18.3 Level of agreement between safety module and actual accident rate for Ehingen County, Germany.

to develop the safety module also provides further communications and operational functions such as data reading and processing, printing, zooming, etc. Furthermore, in addition to the graphical layout, a text area is incorporated. Figure 18.4 shows the detailed enlargement of State Route L 1236 in Fig. 18.2. An arrow is directed to the actual element to be evaluated in depth. All relevant information for this element (774) and the following ones (775 to 780…) are presented in the text area of Fig. 18.4 as well as in the graphical layout.

The text area indicates the information directly on the screen about the results of the safety module as well as about the individual safety criteria I to III for each element. Further information is given about the identity of the roadway section (L 1236), the road designator (from net node to net node), and geometric and traffic parameters (radius of curve, grade, curvature change rate of the single curve, lane width, superelevation rate, AADT). The last parameters are not visible on the present screen layout of Fig. 18.4.

With the cursor, any element in the graphical layout can be selected and the available information is displayed in the text area of the screen. The program can also be used to search for a certain road section or element in the text area, and its position is then directly shown by the arrow, for example, in the graphical layout of Fig. 18.2. With this program, it is possible to analyze and evaluate, relatively easily and quickly, entire road networks or individual roadway sections with respect to quantitative safety performances. At the same time, all relevant data of the individual investigated design element are provided.

In the present case of State Road 1236 between net node 752 4011 and net node 752 5024 in Ehingen County, four curved segments including the transitions are evaluated as "poor" by the safety module according to Fig. 18.4. In these segments, critical driving maneuvers and a corresponding increased accident risk can be expected. Those areas should undergo a critical analysis by the responsible officials concerning more stringent traffic warning devices, speed limits, more intensive police enforcement (for example, by radar devices), or RRR strategies or even redesigns. However, it may occur, by all means, that sections of "poor design" are not especially endangered, since in the field of accident research the relationships are very often complex, and accident occurrences are often the result of the interplay of many factors besides the three safety criteria which are primarily design-related. It is also possible that the accident situation of roadway sections that were originally poorly designed has already been improved by appropriate countermeasures. In this case, the safety module can be considered as an additional control concept for the success of safety measures already implemented.

Note that the data processing and CAD backgrounds developed to date are mainly usable by experts for scientific applications. Future development for everyday use is intended.

The results of the overall safety module, which include for the first time, three quantitative safety criteria in geometric highway design, appear to be pointing in the right direction for evaluating roadway sections or networks with respect to design, redesign, rehabilitation, and restoration strategies.

Note that safety strategies, such as the ones developed here, have been known for decades in other civil engineering fields such as structural engineering, water resources engineering, and so forth.

So far, for the general evaluation process the three safety criteria were combined (equally weighted) in the overall safety module according to Table 18.10. But further research is needed into the possibility of assigning individual weights to the three safety criteria for combining them in a safety module.

Based on new accident research, it appears that safety criterion II is more important than safety criterion III, and safety criterion III may be more important than safety criterion I. Therefore, the proposal for the weighting coefficients in Table 18.11 may be more appropriate than equal weights. Unfortunately, this idea could not be pursued to date because of financial constraints. It is unquestionable that further accident research is needed to establish reliable weighting coefficients with respect to the overall safety module. However, a new and interesting approach was developed by Eberhard[164] in Table 18.15.

Safety evaluation processes, based on the three individual safety criteria (combined or not combined in a safety module) discussed here, should be a substantial part of modern highway geometric design.

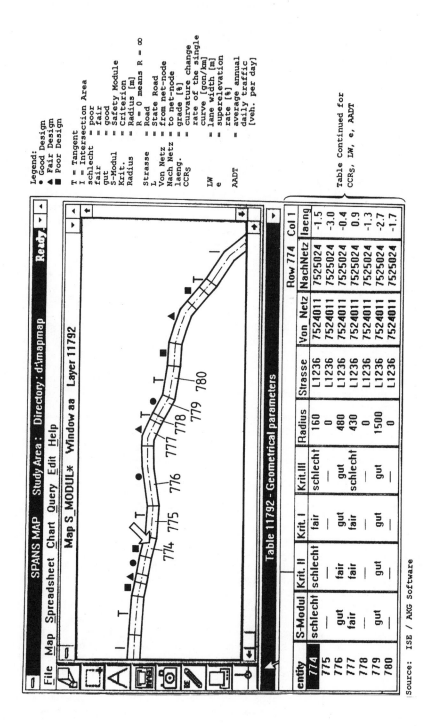

FIGURE 18.4 In-depth safety evaluation process of an individual roadway segment in a network. (Detailed enlargement of State Route 1236 from Fig. 18.2.)

Source: ISE / AKG Software

TABLE 18.11 Proposal for Future Weighting Coefficients with Respect to the Overall Safety Module

Safety criterion	Weighting coefficients for design levels		
	Good	Fair	Poor
I	0	0.5	1.0
II	0	1.0	2.0
III	0	0.75	1.5

18.3 CONCLUSIVE REMARKS ABOUT THE CLASSIFICATION SYSTEM FOR GOOD, FAIR, AND POOR DESIGN PRACTICES BASED ON ACCIDENT RESEARCH*

18.3.1 Databases and Discussion of Accident Rate and Accident Cost Rate

The classification system of Table 11.1 for different design safety levels in coordination with the three safety criteria has been based on the following three accident databases for two-lane rural roads.

Accident database 1: Germany, Table 9.10,[449, 672, 673] 204 curved roadway sections, some of which have tangents separating them. Accident type: run-off-the-road accidents only.

Accident database 2: United States, Table 9.10,[406, 408, 412] 261 curved roadway sections, some of which have tangents separating them. All accidents included.

Accident data base 3: Germany, Table 10.12,[632] 657 curved roadway sections. Accident type: run-off-the-road accidents only.

In order to examine the design (CCR$_S$) classes and the corresponding ranges of the three safety criteria with respect to good, fair, and poor design practices according to Table 11.1, Schmidt[628] conducted an in-depth accident investigation which is related not only to the accident rate but also to the accident cost rate.

The formulas for the accident rate (AR) and the accident cost rate (ACR) are presented by Eqs. (9.1) and (9.2) in "Relative Accident Numbers" in Sec. 9.2.1.3. Since these equations have been refined by Schmidt[628] in the following for different length-related assumptions in calculating AR and ACR, the general form of the equations is repeated here:

$$AR = \frac{accidents \times 10^6}{AADT\ 365\ T\ L} \qquad accidents/10^6\ vehicle\ kilometers \qquad (9.1)$$

where AR = accident rate
AADT = average annual daily traffic, vehicles/24 h
 L = length of the investigated section, km
 T = length of the investigated time period, y
 365 = number of days/y

$$ACR = \frac{S\ 100}{AADT\ 365\ T\ L} \qquad monetary\ unit/100\ vehicle\ kilometers \qquad (9.2)$$

where S = sum of property and personal damages in the time period T observed (monetary unit of the country under study).

---

*Elaborated with the support of G. Schmidt based on Ref. 628.

The accident rate, AR, evaluates all accidents equally and does not differentiate between them by accident severity. The investigations to date, as related to the assessment of sound design (CCR_s) classes in Tables 9.10, 10.12, and 11.1, are related so far solely to accident rates.

In contrast, the accident cost rate considers the sum of all property and personal damages and quantifies the accident situation in a monetary scale.

Both relative accident numbers are related to vehicle kilometers traveled and, consequently, represent the danger, which means the individual risk of the motorist in terms of accident number (AR) and accident severity (ACR) on the observed roadway section.

The alignment database for the following investigations encompassed 30 two-lane rural roadway sections with an overall length of about 90 km in the State of Baden-Wuerttemberg in southwestern Germany. The roadway sections consisted of 515 curved or independent tangent segments. The curvature change rate values of the single curve, CCR_s, were calculated for the individual segments according to Eq. (8.6) in combination with Figs. 8.1 and 8.2. The database contained roadway sections with grades up to 5 to 6 percent, and AADT values between 1000 and 10,000 vehicles/day. The roadway sections investigated consisted of about 30 percent new alignments and about 70 percent old alignments. About two-thirds of them were located in flat topography and about one-third in hilly topography. The 30 roadway sections, therefore, represented several different kinds of alignment design.

The accident database was established based on the actual traffic accident reports of the police for the years 1991 to 1993. As discussed numerous times in Chaps. 9 and 10, the three quantitative safety criteria developed are directed primarily to the driving behavior in curves and the corresponding transitions. Therefore, run-off-the-road accidents, caused mainly by excessive speed, and deer accidents were considered as the relevant accident types. With respect to deer accidents, it should be mentioned that they are also closely related to speed, at least indirectly, since they cause strong braking maneuvers, which often lead to uncontrolled driving situations, for example, run-off-the-road. The accident rate and accident cost rate were calculated for the length of the individual curved or straight segment. However, if an accident occurred up to 150 m behind the preceding curve in the tangent section, it was attributed to the curve, since it can normally be assumed in those cases that the accident occurred while traveling through the curve. For the evaluation of tangent section accidents caused by passing maneuvers, which play less of a role in curves, were also considered. In this way, a total of 464 accidents for the time period from 1991 to 1993 were identified as being relevant for the investigation.

18.3.2 Assumptions for Calculating Accident Cost Rates

For determining the accident cost rate, the same variables as for the accident rate are, in general, valid [see Eqs. (9.1) and (9.2)]. However, instead of the absolute accident frequencies used for calculating accident rates, the sum of all personal and property damages is used for calculating accident cost rates. For the determination of property damages, the damage amount could directly be taken from the traffic accident reports of the police. If no information was given, a minimum value of 6000 German marks/accident (about 3000 U.S. dollars) was taken as a basis.

For personal damages, the average values of the German Federal Highway Research Institute (BASt) for the year 1991 were used (Table 18.12). For the individual personal damages in this table, the following definitions are valid.

Slight Injury: A person is slightly injured if he or she has to undergo an out-patient treatment and/or is unfit for work for a time period of 1 to 2 days.

Serious Injury: A person is seriously injured if he or she has to undergo an in-patient treatment and/or is unfit for work for a time period of more than 2 days.

Fatality: The death of a traffic participant is attributed to the respective accident if the death occurs up to 30 days after the accident.

TABLE 18.12 Average Personal Damages through Road Traffic Accidents, in German Marks

Damage category (1)	Number of persons (1991) (2)	Costs per person, DM (1991) (3)	Costs per person, DM (1996) (4)
Fatality	11,300	1,350,000	1,600,000
Serious injury	131,093	61,000	74,000
Slight injury	374,442	5,400	7,200

Since the accident investigations are related to individual curved or tangent segments of a road, it should be noted[385] that, for the typical low accident numbers with fatalities, one single death can extremely influence the accident cost rate because of the high average personal damage (Table 18.12). However, in the case of such an accident, it is pure chance whether a participant is seriously or fatally injured. Therefore, it appears to be justified to combine the two categories "serious injury" and "fatality" to a combined average personal damage amount.

According to the German conditions in Table 18.12, the following combined average amount for seriously and fatally injured persons was calculated:

$$C_{F/SI} = \frac{11{,}300 \times 1{,}350{,}000 + 131{,}093 \times 61{,}000}{11{,}300 + 131{,}093}$$

$$= 163{,}292 \text{ German marks (about 80,000 U.S. dollars), 1991}$$

where $C_{F/SI}$ = combined average personal damage amount for fatally and seriously injured persons (German marks).

Accordingly, the sum of all personal and property damages, S, in the formula for the accident cost rate [Eq. (9.2)] can be calculated for the observed time period and the individual road segment, as follows:

$$S = n_1\, C_{F/SI} + n_2\, C_{LI} + n_3\, C_{PD} \tag{18.1}$$

where n_1, n_2, n_3 = number of fatally/seriously or slightly injured persons and number of property damages

C_{LI} = average personal damage amount for slightly injured persons (5400 German marks or about 2700 U.S. dollars according to Table 18.12), 1991

C_{PD} = property damage costs according to the traffic accident reports of the police or a minimum amount of 6000 German marks (about 3000 U.S. dollars), 1991

For comparison, the costs for the personal damage categories of the year 1996 are also given in col. 4 of Table 18.12. One can recognize from this table the strong increase in personal damage costs over 6 y (from 1991–1996), at least for Germany (see also Table 9.3).

18.3.3 Calculation of Weighted Mean Accident Rates and Accident Cost Rates

Based mainly on the three accident databases previously discussed in Tables 9.10 and 10.12, a classification system for good, fair, and poor design practices was established which contains the following design (CCR_S) classes:

$$0\text{–}35 \quad \text{gon/km} \rightarrow \text{tangent sections}$$
$$>35\text{–}180 \quad \text{gon/km} \rightarrow \text{good design}$$
$$>180\text{–}360 \quad \text{gon/km} \rightarrow \text{fair design}$$
$$>360 \quad \text{gon/km} \rightarrow \text{poor design}$$

The quantitative ranges of safety criteria I to III were coordinated in Table 11.1 to this classification system and, again, differentiated for good, fair, and poor design levels.

It is the task of the present in-depth accident investigation conducted by Schmidt[628] to confirm or not to confirm the classification system used to date, including the accident cost rate. In order to be able to address the actual accident situation in the best way, it was decided to calculate the so-called weighted mean accident rates and accident cost rates for the four individual CCR_S classes. This means that the accident rate and the accident cost rate of each of the 515 observed curved or tangent segments have to be multiplied with the respective length of the partial segment and then the individual values have to be summarized for each of the four CCR_S classes. Note that in the first step only those segments on which at least *one* accident occurred were considered. Furthermore, for the calculation of the weighted mean accident rate or cost rate, the exact length of a curved segment is not always relevant, but rather a length based on where the farthest accident behind the curve occurred. This additional length for a curved section can amount to a maximum of 150 m in each driving direction. Thus, the weighted mean accident rate or cost rate can be calculated for each individual CCR_S class according to the formula:

Weighted mean accident rate:

$$\text{AR}_{W,M} = \frac{\text{AR}_1 L_1 + \text{AR}_2 L_2 + \cdots + \text{AR}_n L_n}{L_1 + L_2 + \cdots L_n} \quad \text{accidents/}10^6 \text{ vehicle kilometers} \quad (18.2)$$

where $\text{AR}_{W,M}$ = weighted mean accident rate for the respective CCR_S class, accidents/10^6 vehicle kilometers
$\text{AR}_{1,...,n}$ = accident rates of the partial segments $1,...,n$ in the observed CCR_S class, accidents/10^6 vehicle kilometers (with at least one accident present)
$L_{1,...,n}$ = lengths of the partial segments $1,...,n$ in the observed CCR_S class, m

Weighted mean accident cost rate:

$$\text{ACR}_{W,M} = \frac{\text{ACR}_1 L_1 + \text{ACR}_2 L_2 + \cdots + \text{ACR}_n L_n}{L_1 + L_2 + \cdots L_n} \quad \text{monetary unit/100 vehicle kilometers}$$
$$(18.3)$$

where $\text{ACR}_{W,M}$ = weighted mean accident cost rate for the respective CCR_S class, German marks/100 vehicle kilometers
$\text{ACR}_{1,...,n}$ = accident cost rates of the partial segments $1,...,n$ in the observed CCR_S class, German marks/100 vehicle kilometers (with at least one accident present)

The results of Eqs. (18.2) and (18.3) are presented in Table 18.13.
Based on the results of Table 18.13, the following statements can be made:

- Weighted mean accident rates and accident cost rates increase with increasing CCR_S classes.
- CCR_S ranges with a CCR_S value up to 180 gon/km reveal a low accident risk (expressed by the accident rate) and a low accident severity (expressed by the accident cost rate).

TABLE 18.13 Weighted Mean Accident Rates and Accident Cost Rates with Respect to Individual Design Classes[628]

Design/CCR_S classes, gon/km	$AR_{W,M'}$ accidents/10^6 vehicle kilometers	$ACR_{W,M'}$ German marks/100 vehicle kilometers	Evaluation
0–35 (tangent)	0.64	1.57	Good
≥35–180	1.28	4.79	Good
≥180–360	2.54	13.60	Fair
≥360	4.29	26.08	Poor

Note: Including run-off-the-road and deer accidents.

- The accident risk in curved sections with a CCR_S value between 180 and 360 gon/km is already twice as high and the accident severity about 3 times as high as in curves with a CCR_S value of up to 180 gon/km.

- The accident risk in curves with a CCR_S value greater than 360 gon/km is nearly 3.5 times higher and the accident severity is about 5.5 times higher than in curves with CCR_S values of up to 180 gon/km.

The information presented here fully confirms the results of the former accident investigations presented in Tables 9.10 and 10.12 with respect to the accident rate. The results of the accident cost rate, which is included here for the first time, strongly support the arrangement of the individual design (CCR_S) classes in Table 11.1, which are coordinated with the quantitative ranges of safety criteria I to III for examining good, fair, and poor design practices.

Note that the length-related considerations for the establishment of Table 18.13 were not included in the development of Tables 9.10 and 10.12. Despite this, the agreement of the results is convincing, although the in-depth investigation of the present chapter has to be evaluated as more reliable from a scientific point of view.

18.3.4 Calculation of Relative Weighted Mean Accident Rates and Accident Cost Rates

Tables 9.10, 10.12, and 18.13 were established exclusively for curved or tangent road segments, on which at least one accident occurred. It can, therefore, be speculated that the mean accident rates and cost rates calculated so far may be biased, since the length-related influences are not analyzed sufficiently. Therefore, the following calculations also include those roadway segments where no accident occurred. The relative accident numbers calculated this way are called *relative weighted mean accident rates and accident cost rates*. The equations for each CCR_S class are as follows:

Relative weighted mean accident rate:

$$AR_{R,W,M} = \frac{AR_{W,M} \, L_A}{L_O} \qquad \text{accidents/}10^6 \text{ vehicle kilometers} \qquad (18.4)$$

where $AR_{R,W,M}$ = relative weighted mean accident rate for the respective CCR_S class, accidents/10^6 vehicle kilometers

$AR_{W,M}$ = weighted mean accident rate for the respective CCR_S class from Table 18.13, accidents/10^6 vehicle kilometers

L_A = overall length of partial segments, where at least one accident occurred in the observed CCR_S class, m

L_O = overall length of *all* partial segments in the observed CCR_S class, m

Relative weighted mean accident cost rate:

$$\text{ACR}_{R,W,M} = \frac{\text{ACR}_{W,M}\, L_A}{L_O} \qquad \text{monetary unit/100 vehicle kilometers} \qquad (18.5)$$

where $\text{ACR}_{R,W,M}$ = relative weighted mean accident cost rate for the respective CCR_S class, German marks/100 vehicle kilometers.

The other technical designations are the same as those in Eqs. (18.3) and (18.4).

Table 18.14 reveals the results for the relative weighted mean accident rates and cost rates.

As could be expected, based on the additional segment lengths included in Table 18.14, the absolute values of Table 18.14 are lower than those of Table 18.13. However, it is interesting to note that the accident risk and the accident severity relationships between the individual CCR_S classes are nearly the same.

Again, the results of Table 18.14 indicate that:

- Accident rates and accident cost rates increase with increasing CCR_S classes.
- CCR_S ranges up to 180 gon/km experienced a low accident risk and a low accident severity, classified as "good design."
- The accident risk in curves with CCR_S values between 180 and 360 gon/km was at least twice as high and the accident severity nearly 3 times as high as that on curves with CCR_S values up to 180 gon/km, classified as "fair design."
- The accident risk in curves with CCR_S values greater than 360 gon/km exceeds the corresponding value for good design by about 4.5 times and the accident severity by about 7.5 times.

The results of Tables 18.13 and 18.14 confirm the classification system in Table 11.1 in a convincing manner. Note that this classification system was already established, based on three older accident databases, as discussed at the beginning of this chapter and presented in Tables 9.10 and 10.12.

Thus, the developed classification system presented in Table 11.1 already allows *rough* evaluations of good and fair design strategies and the detection of poor design practices with respect to the assessed design (CCR_S) classes. Note that the quantitative ranges for the three safety criteria were directed strictly to these design classes, as presented in detail in Table 9.11 for safety criterion I, Table 9.13 for safety criterion II, and Table 10.11 for safety criterion III. Based on these assumptions, the safety criteria allow *in-depth* insights into the detailed requirements of alignment design tasks concerning:

- Sound or unsound differences between design speed and operating speed (achieving design consistency: safety criterion I)
- Sound or unsound operating speed differences for the transitions between independent tangents and curves, or curves to curves (achieving operating speed consistency: safety criterion II)
- Sound or unsound differences between side friction assumed and demanded (achieving driving dynamic consistency: safety criterion III)

In contrast to the rough evaluations according to the CCR_S classes, the safety criteria allow differentiated, quantitative, and specific statements concerning the previously mentioned important alignment goals with respect to good, fair, and poor design levels.

In conclusion, the developed CCR_S classes, based on accident research, form the basis for the successful application of the three safety criteria and the combined safety module.

With respect to the 30 roadway sections investigated, separated into 515 curved or independent tangent segments, Schmidt[628] reported that, if the safety criteria revealed "good design," low accident risks and severities would normally be observed, while if they indicated "poor design," significantly higher accident risks and severities would have to be expected.

Comparable results could be obtained through the case studies in Secs. 12.2.4.2 and by the safety module presented in Sec. 18.2.

TABLE 18.14 Relative Weighted Mean Accident Rates and Accident Cost Rates with Respect to Individual Design Classes[628]

Design/CCR$_S$ classes, gon/km	AR$_{R,W,M}$ accidents/10^6 vehicle kilometers	ACR$_{R,W,M}$ German marks/100 vehicle kilometers	Evaluation
0–35 (tangent)	0.14	0.35	Good
≥35–180	0.79	2.94	Good
≥180–360	1.60	8.57	Fair
≥360	3.63	22.04	Poor

Note: Including run-off-the-road and deer accidents.

The alignments of seven of the 30 two-lane rural roadway sections investigated in this chapter were taken as the basis for the case studies in Sec. 18.4.2.

18.4 EXAMINATION OF EXISTING ALIGNMENTS BASED ON THE THREE SAFETY CRITERIA AND THE SAFETY MODULE FOR INTERNATIONAL OPERATING SPEED BACKGROUNDS

The predominant portion of two-lane rural road networks (worldwide) consists of alignments which do not fulfill modern highway design requirements from a traffic safety point of view. Therefore, one main task of the responsible authorities in many countries is to improve the safety level of current dangerous roadway sections, since it is impossible to rebuild major parts of the existing road networks. Up to now, the safety evaluation of those sections has been conducted more or less qualitatively, for example, based on the analysis of accident black spots. Consequently, a procedure to detect, analyze, and improve safety deficiencies quantitatively, before accidents occur, is urgently needed. Therefore, one of the main topics of this book is the development of a reliable procedure to achieve this goal.

Besides the safety evaluation for new designs, the procedure presented is directed toward examining existing (old) alignment designs with respect to good, fair (tolerable), and poor design levels as a basis for future redesigns or RRR projects. The procedure enables the highway engineer to evaluate alignment design up to grades of 6 percent and AADT values up to 12,000 vehicles/day by means of three quantitative safety criteria, which have been discussed thoroughly in Chaps. 9 and 10. Quantitative ranges for the three safety criteria, in order to achieve "good designs," to classify "fair designs," and to detect "poor designs," based on individual curved or tangent segments, as well as the transitions in between, are given in Table 11.1.

Furthermore, the effects of the three safety criteria are at least partially incorporated in Chaps. 12 to 15 for the assessment of limiting and standard values for reliable design elements with respect to horizontal alignment, vertical alignment, cross section, and sight distance.

18.4.1 Methodical Procedure and Comments*

The safety procedure presented consists of 13 hierarchical steps for two-lane rural roads:

1. Assess the road section to determine where safety examinations of the existing alignment should be conducted.

2. Determine the kind of design elements (circular curves, clothoids, tangents) present in the roadway section, the corresponding geometric parameters (R, A), and lengths (L), as well as the superelevation rates (e) at curved sites.

---

*Elaborated based on Refs. 460 and 461.

3. Differentiate between curves and tangents and between independent tangents and nonindependent tangents according to the formulas in Table 12.3 or according to the limiting ranges in Tables 12.1 or 12.2. Only independent tangents are considered further in the evaluation process in Sec. 12.1.1.3.

4. Calculate the curvature change rate of the single curve with transition curves, CCR_S, if present. CCR_S is calculated using Eqs. (8.6a) to (8.6f) in Fig. 8.1, Sec. 8.1.3. For independent tangents, $CCR_S = 0$ gon/km.

5. Determine the 85th-percentile speed, $V85$, for each curved site and independent tangent with respect to the design parameter CCR_S on the basis of the operating speed background for the country under study as given in Table 8.5 or Fig. 8.12, Sec. 8.2.3.7.

6. Assess the design speed, V_d, for the examined roadway section. Note that, for existing alignments, the design speed is often not known. A <u>sound</u> design speed can be estimated in the following way: Calculate the length-related average $\overline{CCR_S}$ value based on all the curves in the observed roadway section according to <u>Eq. (9.12)</u> in Sec. 9.2.2.1. Tangent sections should not be <u>included</u>. Determine for this average $\overline{CCR_S}$ value the corresponding average 85th-percentile speed, $\overline{V}85$, from the operating speed background of the country under study (Table 8.5 or Fig. 8.12, Sec. 8.2.3.7). This average 85th-percentile speed represents a good estimate for the assumed design speed, V_d, regarding future redesigns or RRR projects of the existing alignment under investigation.

7. Determine side friction assumed f_{RA} for a utilization ratio of $n = 60$ percent, Eq. (10.10b), Sec. 10.1.4. Determine side friction demanded f_{RD} according to Eq. (10.11).

8. Evaluate safety criterion I: Calculate the difference between $V85_i$ and V_d according to the classification system in Tables 9.1 and 11.1 for good, fair (tolerable), and poor design levels for each individual design element.

9. Evaluate safety criterion II: Calculate the difference in 85th-percentile speeds between successive design segments (independent tangent i to curve $i + 1$ or curve i to curve $i + 1$) according to the classification system in Tables 9.2 and 11.1 for good, fair (tolerable), and poor design levels.

10. Evaluate safety criterion III: Calculate the difference between side friction assumed f_{RA} and side friction demanded f_{RD} according to the classification system in Tables 10.3 and 11.1 for good, fair (tolerable), and poor design levels for each individual design element.

11. Analyze the results of safety criteria I to III with respect to good, fair, and poor design practices throughout the length of the existing alignment being examined. In the case of good design, no adaptations or corrections are necessary. In substantiated individual cases, the fair design level for one or more safety criteria might be sufficient. However, it should be noted that, in those cases, the expected accident rate and accident cost rate will be about 2 or 3 times as high as those for good design practice, as has been proven by large accident databases, according to Tables 9.10, 10.12, and 18.14.

12. In cases of poor design, for one or more safety criteria, critical accident frequencies and severities may be expected and may lead to an uneconomic, unsafe operation. For example, Table 18.14 reveals that, for poor design, the accident risk in curves and the corresponding transition sections exceeds the value for good design by about 4.5 times and the accident severity by about 7.5 times.

13. For a comprehensive overview, combine the three safety criteria into an overall safety module according to the classification system in Table 18.15.

Note that any safety improvement has to be examined by the repetition of steps 1 to 13 in order to confirm its intended safety effectiveness. Later on, a "before" and "after" investigation is recommended to prove the safety benefit by statistical and economic procedures, as presented, for example, in Refs. 211 and 225.

Comments.   With respect to the examinations of existing alignments, for example, presented in Tables 18.16 to 18.22, the following comments have to be considered for an in-depth comprehension:

General. Proposals for safety improvements are very complex and depend strongly on local conditions, since economic, environmental, and administrative aspects often play an important role in addition to safety considerations. Therefore, when evaluating existing alignments with respect to safety evaluation processes, not all questions can be answered up to the last detail, since highway engineers are not able to redesign or reconstruct large portions of their existing networks, mainly because of administrative, financial, or environmental constraints. In this connection, all we can do is concentrate on the most dangerous roadway sections.

Statistical Viewpoints. The limiting ranges for the design classes developed and the corresponding safety criteria in Table 11.1 have been established based on the results of large accident databases (Tables 9.10, 10.12, and 18.14). Therefore, if a roadway section of an existing alignment reveals "poor design" for one or more safety criteria or even for the overall safety module, that does not automatically mean that an accident black spot there has to exist. In this connection, "poor design" means that the probability for the creation of accident frequencies and/or severities is high, while for "good design," it is low. Since accidents are rare occurrences, sound statistical statements about the real accident situation depend on large numbers. Thus, only large accident databases are able to classify for the majority of critical or uncritical design cases concerning the actual accident situation; with respect to an individual case, the results may be erroneous. However, it should be kept in mind that the first case study in this field (discussed in Sec. 18.2) revealed that the results of the safety module, indicating good, fair, and poor design practices, corresponded to about 88 percent with the observed actual accident rates on 440 investigated curved segments.

Safety Criterion I. Note that the assumed design speed, V_d, for safety criterion I in step 8 represents the average 85th-percentile speed ($\overline{V}85 \rightarrow V_d$) of the curved segments in the observed roadway section according to "Existing (Old) Alignments" in Sec. 9.2.2.1. In this way, the assumed design speed, V_d, coordinates, in an optimum way, the existing operating speeds, $V85_i$, and, at the same time, considers environmental and economic issues for upgrading existing highway alignments or full-blown RRR projects. Because of this coordination between $V85_i$ and V_d, safety criterion I normally shows good results according to col. 11 in Tables 18.16 to 18.22. However, note that if the original design speed had been known, other consequences may have resulted, based on safety criterion I, since large discrepancies in selected design speeds could often be observed when studying one and the same existing (old) roadway section. Therefore, the authors consider the search for the original design speed(s) unnecessary and recommend from the beginning the procedure described in step 6 and developed in Sec. 9.2.2.1. They even recommend the use of the procedure for new designs in order to ensure that the design speed is coordinated well from the beginning with the proposed alignment.

Safety Criterion III. The following considerations are related only to the poor design level of safety criterion III in step 10, if redesigns are not intended or impossible to conduct and safety criteria I and II as well as the safety module that do not reveal poor design levels.

Side friction assumed f_{RA} in safety criterion III is related to the overall regression curve for maximum permissible tangential friction with respect to the design speed [Eqs. (10.2) and (10.40)]. Furthermore, side friction assumed f_{RA} depends on the utilization ratio, n, which was selected for new designs to

$n = 40$ percent (hilly/mountainous topography) according to Table 10.1, Eq. (10.5)

$n = 45$ percent (flat topography) according to Table 10.1, Eq. (10.4)

For these assumptions, the minimum radii of curve were determined in Table 12.5 for maximum superelevation rates. The selected side friction factors for new designs and the corresponding minimum radii of curve provide as much driving dynamic safety as possible, as was discussed numerous times in Chap. 10, and represent skid resistance values which cover about 80 to 90 percent of the road surfaces of the so far known sound skid resistance backgrounds (Figs. 10.28 and

10.29). Furthermore, the assessed minimum radii of curve agree well with the worldwide comparison in Table 12.9.

In order not to be too conservative, the utilization ratio for redesigns or RRR projects of existing alignments was increased to $n = 60$ percent [Eq. (10b) in Sec. 10.1.4], which leads to an average reduction of about 20 percent with respect to the minimum radii of curve in Table 12.5 (flat topography).

However, when examining existing (old) alignments, every experienced designer knows that frequently narrow curves are present with radii considerably lower than those presented in Table 12.5, even if a 20 percent radii reduction is already considered. Therefore, it is logical that, in those cases, the comparison of friction assumed f_{RA} and friction demanded f_{RD}, for which the safety evaluation process of safety criterion III is laid out in Sec. 10.2.4.3, often reveals poor design. This fact can be observed relatively frequently for safety criterion III in col. 13 of Tables 18.16 to 18.22 when conducting safety evaluations of existing alignments.

Therefore, the following recommendations about countermeasures at curved sites are only valid for existing alignments, where only safety criterion III reveals poor design and where no redesigns are intended.

In this connection the following two countermeasures at curved sites may be appropriate in order to avoid too many costly redesigns:

- A cost-effective safety improvement can be achieved by reconstructing the pavement and increasing the superelevation rate. In the case of resurfacing, great emphasis should be placed on skid resistance. Maximum superelevation rates should be used.
- Speeds must be reduced at curved sites for radii which are more than 20 percent shorter than those which are recommended as minimum radii for the respective design speed in Table 12.5.

Skid resistance, which is often ignored when establishing safety improvements for highway alignment design, has to be considered more important in the future.

In this connection, many researchers require the achievement of high skid resistance values, especially at curved sites, since skid resistance has to be considered as a major safety issue when resurfacing roadways.[81,217,496] Therefore, a skid-resistant pavement surface is a *key* element in prevention of skidding accidents on wet pavements.

As already discussed in Sec. 10.2.2.1, it is interesting to note that Europeans and Americans always require good skid-resistance values for resurfacing projects (especially in curves), since, in recent decades, on both continents, insufficient skid resistance was often provided. This fact led to numerous accident black spots, especially in curved roadway sections, since the accident risk strongly increases as skid resistance decreases.

The threshold values for skid resistance were assessed in Secs. 10.1.2.2 and 10.2.2.1, as follows:

0.42 for $V = 40$ km/h

0.35 for $V = 60$ km/h

0.30 for $V = 80$ km/h

These skid-resistance values cover from 80 to 90 percent of the road surfaces, which means a wide variety of pavements. With respect to good skid-resistant pavement, which can be achieved today by sound resurfacing projects, practitioners are convinced that the margin of safety may reach a factor of 1.5 for wet pavements. That would correspond to a skid-resistance value of 0.45 for $V = 80$ km/h. Comparing this value with the new skid-resistance background of Germany in Fig. 10.26, it can be seen that a value of 0.45 still covers about 30 percent of the road surfaces for a speed of $V = 80$ km/h. Note that this value is assumed in Germany as a "target value," and it should be possible to reach this skid-resistance range in other countries as well, at least in critical curves, where safety criterion III reveals poor design practice.

Taking this value as an example, side friction assumed could be estimated according to Eq. (10.10a) as follows:

$$f_{RA} = 0.60 \times 0.925 \times 0.45 = 0.25 \qquad \text{for } V_d = 80 \text{ km/h}$$

In comparison, the threshold value for side friction assumed would be, according to Table 10.6

$$f_{RA} = 0.60 \times 0.925 \times 0.30 = 0.17 \qquad \text{for } V_d = 80 \text{ km/h}$$

Thus, the structural increase of skid resistance represents one possibility for increasing safety in narrow curves. However, practical experience with respect to the case studies in Tables 18.16 to 18.22 revealed that, for tight curves of ($R \leq 150$ m), this procedure alone is not sufficient to achieve good or even fair design levels according to safety criterion III.

Therefore, the speed behavior at those critical curve sites also has to be reduced. Besides speed limits, which normally have no significant influence on the driving speed of the motorist, questions about sound and perceivable visual guidance are of primary importance. In this connection, correct signing and pavement markings, the installation of appropriate traffic warning devices (such as arrows and chevrons), and guardrails may lead to improved visual guidance. For example, incrementally reduced speed limits from 80 to 60 km/h or even to 40 km/h or from 70 to 50 km/h may support the perception process of the upcoming curve. Good visual guidance is meant primarily to reduce the mental workload of the driver, so that he or she is not surprised by sudden alignment or operational changes, but can adapt step-by-step his or her driving behavior based on driving experience (see Chap. 19).

Unfortunately, estimates about possible speed reductions in curves with sound hardware for visual guidance, whether supported or not by speed regulations, do not so far exist, although a corresponding research project has just started at the Institute for Highway and Railroad Engineering, University of Karlsruhe, Germany. First tendencies may support estimates that the 85th-percentile speeds in curves with radii between $R = 175$ m and $R > 100$ m are obviously 5 to 10 km/h lower and between $R \leq 100$ m and $R = 50$ m about 10 to 15 km/h lower, as compared to the operating speed backgrounds in Fig. 8.12 and Table 8.5. In order not to have to depend on these perhaps inaccurate speed estimates, local speed measurements can, of course, be conducted at critical curve sites.

Note once more that only in the case of safety examinations for existing alignments, where redesigns are excluded for whatever reasons, the two possibilities discussed—increase in skid resistance and/or slowing down operating speeds—are recommended to alleviate the poor design level of safety criterion III.

A typical example for the previous design case is represented by the case study of France in Table 18.18. Safety criteria I (col. 11) and II (col. 12) show, in the majority of cases, good or fair design practices, while safety criterion III (col. 13) reveals exclusively poor design.

For a better understanding, let us as an example concentrate on element 7 with a radius of curve of $R = 100$ m in Table 18.18. Safety criterion I indicates fair design. The operating speed transitions to element 7, expressed by safety criterion II, represent good or fair design. (Elements 6 and 8 are, as nonindependent tangents (NIT), not considered in the safety evaluation process). Only safety criterion III reveals poor design. The overall safety module in col. 14 also shows fair design for both driving directions, which means redesigns are not necessarily required.

Therefore, in order to improve the poor design level of safety criterion III, the present side friction factor assumed $f_{RA} = 0.15$ in col. 9 of Table 18.18 can be increased by good structural skid-resistance measures to, for example, $f_{RA} = 0.25$, based on the previous considerations. Furthermore, the present side friction factor demanded $f_{RD} = 0.38$ in col. 10 could be reduced to $f_{RD} = 0.26$ if an 85th-percentile speed reduction of 10 km/h is assumed through the installation of appropriate control devices for reducing the operating speed and, at the same time, increasing the existing superelevation rate from 7 to 8 percent (step 7). Thus, the new evaluation range of safety criterion III in col. 13 of Table 18.18 would correspond to fair design according to Table 11.1:

$$f_{RA} - f_{RD} = 0.25 - 0.26 = -0.01$$

This would result in a significant safety increase for skidding accidents (run-off-the-road), assuming that the two combined actions (skid resistance increase and operating speed decrease) would be successful.

Note that this procedure is recommended to be used only if, besides safety criterion III, neither of the two other safety criteria reveal poor design, the overall safety module represents at least fair design and no realignments are planned.

In the case where at least two safety criteria express poor design practice, redesigns or stationary radar devices (to strictly control the allowable operating speeds) are normally considered the only way to alleviate the probability of high accident risks and severities in critical curves and the corresponding transitions. However, new research of the Institute for Highway and Railroad Engineering, University of Karlsruhe, Germany, revealed, that very stringent traffic control devices, such as speed limits combined with chevrons and guardrails, may sometimes serve as surrogate measures.

Safety Criterion II.   The proposed safety evaluation process for safety criterion II in step 9 is valid for new designs, redesigns, RRR projects, as well as for the examination of existing alignments. No further comment is necessary.

Safety Module.   For a fast, comprehensive overview of existing alignments, safety criteria I to III are combined into an overall safety module (step 13).

In this connection, Eberhard[164] developed a simple calculation scheme in Table 18.15 for evaluating the safety module according to good, fair, and poor design levels. As in Table 18.10, the three safety criteria are equally weighted: However, this time a specific weighting factor is assigned to each design level. Summing up the weighting factors for the safety criteria, as is shown based on different calculation examples in Table 18.15, the calculated average value represents, in combination with the given limiting ranges, an evaluation scheme for the safety module. Table 18.15 corresponds, to a large extent, to Table 18.10, but is now mathematically provable, which is important for the use of data processing systems.

The results of the safety module with respect to seven case studies are presented in col. 14 of Tables 18.16 to 18.22. The safety module has to be examined for both driving directions.

If the safety module reveals good design, no adaptations or corrections of the existing alignment are necessary.

If the safety module expresses fair design, the installation of sound traffic control devices for good visual guidance, whether or not combined with appropriate speed regulations, is normally sufficient to alleviate existing safety deficiencies. Sometimes, a cost-effective safety improvement can be achieved by reconstructing the pavement for increasing skid resistance and superelevation rate according to the detailed discussions in comment under "Safety Criterion III," given previously. However, note that in the fair design range, higher accident rates and cost rates are to be expected than in the good design range (see Tables 9.10, 10.12, and 18.14).

If the safety module indicates poor design, upgrading of the existing alignments or full-blown RRR projects are normally recommended. According to Tables 18.10 and 18.15, this most critical case includes with respect to the individual safety criteria:

- Poor design according to two criteria and fair design according to one
- Poor design according to all three criteria

After the global assessment of poor design through the safety module, different solutions for realignments or RRR strategies may be directed to the individual outcomes of the three safety criteria, as partially discussed, based on the results of the following case studies in Tables 18.16 to 18.22. However, it should be understood that the authors can only give general recommendations, since not every critical case or combination thereof in the complex field of existing alignment design can be dealt with. The safety module and the safety criteria have to be understood to give valuable indications about critical and noncritical roadway segments from a safety point of view. Based on this information, the responsible highway engineer has to make the decisions about whether safety improvements are necessary.

TABLE 18.15 New Evaluation Scheme for the Safety Module Based on the Three Safety Criteria

Weighting Factors for the Three Safety Criteria with Respect to the Different Design Levels

good	→	+ 1.0
fair	→	0.0
poor	→	− 1.0

Limiting Ranges for the Average Value of the Safety Module

$x \geq 0.5 \rightarrow$ good

$- 0.5 < x < 0.5 \rightarrow$ fair

$x \leq - 0.5 \rightarrow$ poor

Calculation Examples

Safety Criterion I	Safety Criterion II	Safety Criterion III	Sum Σ	Safety Module Average Value	Classification
+ 1.0	− 1.0	− 1.0	− 1.0	− 0.33	fair
0.0	− 1.0	− 1.0	− 2.0	− 0.67	poor
0.0	+ 1.0	IT*	+ 1.0	+ 0.50	good
+ 1.0	+ 1.0	+ 1.0	+ 3.0	+ 1.00	good

Legend:

IT*: The symbol describes an independent tangent for which Safety Criterion III is not relevant, since no centrifugal force exists. Therefore, Safety Criterion III will not be considered in the evaluation scheme for independent tangents.

Finally, note that the developed safety evaluation processes may be used down to minimum radii of $R = 50$ m, since for radii of $R < 50$ m, operating speed and driving dynamic issues are more frequently replaced through driving geometric considerations.

18.4.2 Case Studies

In this section, the outcome with respect to the three safety criteria and the safety module is tested based on international operating speed backgrounds presented in Fig. 8.12 and Table 8.5 for existing alignments. Originally, 30 existing alignments were investigated by Schmidt in his master's thesis[628] for German conditions. The exact design data for the existing, mostly old, alignments were provided by the Regional Offices of the Ministry for Environment and Traffic, State of Baden-Wuerttemberg, Germany. The authors decided to relate the following case studies to this database, since it can be assumed that, especially for old alignments, similar roadway sections (design element sequences) also exist in other countries.

The evaluation tables (Tables 18.16 to 18.22) are generally organized according to the sequential elaboration steps 1 to 13 of the proposed methodical procedure in Sec. 18.4.1. These tables contain the necessary design input data (cols. 3, 4, and 8), the calculation of the curvature change rate of the single curve, CCR_S (col. 5), the determination of the 85th-percentile speed, $V85$ (col. 6) for the operating speed background of the country under study, the estimate of a sound design speed, V_d (col. 7), and the calculation of side friction assumed, f_{RA} (col. 9) as well as side friction demanded, f_{RD} (col. 10). In this way, all relevant input data for the evaluation processes with respect to safety criteria I to III (cols. 11 to 13) and the safety module (col. 14) are presented.

The following case studies are located in flat or hilly topography with maximum grades of up to 6 percent and AADT values between 1000 and 12,000 vehicles/day.[628]

18.4.2.1 Australia.   The Australian operating speed background was developed by McLean.[505] The functional relationship between the 85th-percentile speed and the curvature change rate of the single curve is transformed to the following formula (see Table 8.5 and Fig. 8.12):

$$V85 = 101.2 - 0.043 \ CCR_S \qquad (8.20)$$

Table 18.16 shows a relatively generous alignment of an old two-lane rural road, consisting of independent tangents and curves without transition curves. The safety module reveals overwhelmingly good design for both driving directions. Only for element 5, including the transition between elements 4 and 5, does the safety module indicate fair design in the corresponding driving direction. Traffic warning devices and the increase of the superelevation rate from 2.4 to 8 percent appear to be sufficient solutions. In case of a redesign, a radius of curve of $R = 400$ m for element 5, including a transition curve between elements 4 and 5, is recommended according to Table 12.4.

18.4.2.2 Canada.   The Canadian operating speed background was provided by Morrall, et al.[508,527] and transformed to the following equation (see also Table 8.5 and Fig. 8.12):

$$V85 = e^{(4.561 - 5.27 \times 10^{-4} \times CCR_S)} \qquad (8.22)$$

Table 18.17 reveals an alignment of an old two-lane rural road without transition curves, which is, for long segments, generously designed with the exception of one extremely small radius of curve. Correspondingly, the safety module reveals good design in most cases. However, the safety criteria and the safety module indicate poor design at all levels for element 7, including the transitions between elements 5 to 7 and elements 8 to 7, seen in the respective critical driving direction. (Element 6, as a nonindependent tangent, is not considered in the safety evaluation process.) As a result, an upgrading of the alignment between elements 5 and 8 or even 9 is recommended. Since the alignment consists in this part of curves in the same direction of curvature with an NIT and an IT in between, a good idea may be to replace the whole section, including ele-

TABLE 18.16 Case Study for the Examination of an Existing Alignment According to the Operating Speed Background of Australia

Element no. (1)	Station, km (2)	Parameters, m (3)		L_r, m (4)	CCR_S, gon/km (5)	$V85_r$, km/h (6)	V_d^*, km/h (7)	e, % (8)	f_{RA} (9)	f_{RD} (10)	Criterion I, $|V85_i - V_d|$, km/h (11)		Criterion II, $|V85_i - V85_{i+1}|$, km/h (12)		Criterion III, $f_{RA} - f_{RD}$ (13)		Safety module $1\rightarrow5$ (14)	$5\rightarrow1$
1	0.050 / 0.330	R	−450	280	142	95	95	4.0	0.15	0.12	0	+			0.03	+	→	+
2	0.330 / 1.410	IT		1080		101	95				6	+	6	+			+	+
3	1.410 / 1.505	R	450	95	142	95	95	4.0	0.15	0.12	0	+	6	+	0.03	+	+	+
4	1.505 / 2.005	IT		500		101	95				6	+	6	+			+	+
5	2.005 / 2.115	R	200	110	319	88	95	2.4	0.15	0.28	7	+	13	o	−0.13	−	o	←

Legend for the Case Studies in Tables 18.16 to 18.22: + = good design, o = fair design, and − = poor design. V_d^* = design speed is unknown (V_d was determined according to step 6 and the comment under "Safety Criterion I" in Sec. 18.4.1). NIT = nonindependent tangents, which are not considered in the safety evaluation process. IT = independent tangents, considered in the safety evaluation process (see Sec. 12.1.1.3). ↑↓ = the safety module has to be examined for both driving directions (col. 14). The calculation process according to Table 18.15 always starts with the transition (SC II, col. 12) in front of the succeeding curve and then the other two criteria (SC I, col. 11, and SC III, col. 13) follow. With respect to an independent tangent, only SC II and SC I have to be considered in the evaluation scheme for the safety module since no centrifugal force exists. SC = safety criterion, based on Table 11.1.

TABLE 18.17 Case Study for the Examination of an Existing Alignment According to the Operating Speed Background of Canada

Element no. (1)	Station, km (2)	Parameters, m (3)	L_r, m (4)	CCR_s, gon/km (5)	$V85_r$, km/h (6)	V_d^*, km/h (7)	e, % (8)	f_{RA} (9)	f_{RD} (10)	Criterion I, $|V85_i - V_d|$, km/h (11)	Criterion II, $|V85_i - V85_{i+1}|$, km/h (12)	Criterion III, $f_{RA} - f_{RD}$ (13)	Safety module 1→12 (14)	Safety module 12→1 (14)
1	0.070 0.200	R −250	130	255	84	85	4.3	0.16	0.18	1 +		−0.02 o	→	o
2	0.200 0.260	NIT	60								2 +		+	+
3	0.260 0.325	R 300	65	212	86	85	2.7	0.16	0.17	1 +	1 +	−0.01 o	+	+
4	0.325 0.455	R −350	130	182	87	85	3.9	0.16	0.13	2 +	1 +	0.03 +	+	+
5	0.455 0.650	R 400	195	159	88	85	3.1	0.16	0.12	3 +		0.04 +	+	+
6	0.650 0.715	NIT	65								39 −		−	−
7	0.715 0.795	R 50	80	1274	49	85	4.0	0.16	0.33	36 −	38 −	−0.17 −	o	−
8	0.795 1.035	IT	240		87	85				2 +	0 +		+	+
9	1.035 1.125	R 350	90	182	87	85	3.3	0.16	0.14	2 +	0 +	0.02 +	+	+
10	1.125 1.215	R −350	90	182	87	85	2.2	0.16	0.15	2 +		0.01 +	+	+
11	1.215 1.315	NIT	100								6 +		+	+
12	1.315 1.590	R 1000	275	64	93	85	3.2	0.16	0.04	8 +		0.12 +	+	←

Note: For the legend, see Table 18.16.

18.32

ments 5 and 9, by one curve with a radius of 350 or 400 m with transition curves to additionally avoid kinks in the alignment.

The fair design cases of safety criterion III for elements 1 and 3 can be improved to good by simply increasing the superelevation rates.

18.4.2.3   France.   The French relationship between $V85$ and CCR_S was taken from the Highway Design Guide of France.[700] The transformed equation reads (see also Table 8.5 and Fig. 8.12):

$$V85 = \frac{102}{1 + 346\,(CCR_S/63,700)^{1.5}}\tag{8.19}$$

Table 18.18 reveals an existing alignment of an old two-lane rural road with radii of curve ranging from $R = 80$ to 700 m. Transition curves exist only at the beginning of the observed roadway section.

It is interesting to note that, despite the small radii of curve, the transitions between successive curves according to safety criterion II reveal fair or even good design levels. The same is true for safety criterion I and for the overall safety module with the one exception of poor design in each case.

That means the most critical safety deficiencies are expressed through safety criterion III for this case study, which reveals poor design levels for all investigated curved sites. Safety evaluations, such as presented in Table 18.18, can frequently be found when examining old alignments. Since it is impossible to rebuild all these sections, the comment under "Safety Criterion III" in Sec. 18.4.1 was elaborated, in particular, for possible safety improvements through an increase in skid resistance and a decrease in operating speed instead of realignment.

In combination with the reconstruction of the pavement, a slight flattening of the very short and narrow curve element 5 by including the nonindependent tangent 6 is recommended to better coordinate the present poor design level of safety criterion I and the corresponding one of the safety module.

However, if upgrading of the existing alignment can be considered, the redesign should be related to at least the minimum radius of Table 12.5 for a design speed of $V_d = 90$ km/h, considering at the same time a 20 percent reduction.

In such a case, transition curves should be used, maximum superelevation rates should be applied, and the nonindependent tangents should be avoided in order to achieve a good curvilinear alignment.

18.4.2.4   Germany.   The German operating speed background from the "Guidelines for the Design of Roads, Part: Alignment," editions 1973 and 1984[241,243] was taken as the basis for the following case study. The resulting equation is, according to Table 8.5:

$$V85 = 60 + 39.70 * e^{(-3.98*10^{-3}CCR_S)}\tag{8.18}$$

The case study in Table 18.19 describes an existing alignment of a two-lane rural road with large changes in radii of curve, including clothoids. The radii vary between 70 and 1,000 m. With respect to safety criterion II, the transitions between elements 2 and 3, as well as between elements 5 and 6, reveal poor design practices. This is fully confirmed by the safety module for the respective critical driving direction. As the appropriate way to alleviate the safety problems here, redesigns have to be recommended, at least for the previously mentioned element sequences. However, because of the large radii of curve of elements 2 and 6, the minimum radius of curve of Table 12.5, depending on V_d, should be used without considering any reduction normally allowed for redesigns. It is a matter of course that the introduction of transition curves and of appropriate superelevation rates also has to be considered. If a major upgrading of the alignment is taken into consideration, the radius of curve of element 1 should be flattened accordingly.

TABLE 18.18 Case Study for the Examination of an Existing Alignment According to the Operating Speed Background of France

Element no. (1)	Station, km (2)	Parameters, m (3)		L_r, m (4)	CCR_S, gon/km (5)	$V85_r$, km/h (6)	V_d^*, km/h (7)	e, % (8)	f_{RA} (9)	f_{RD} (10)	Criterion I, $|V85_i - V_d|$, km/h (11)		Criterion II, $|V85_i - V85_{i+1}|$, km/h (12)		Criterion III, $f_{RA} - f_{RD}$ (13)		Safety module $1\rightarrow9$ (14)	$9\rightarrow1$
1	2.069 2.314	A	130	245	220	95	90	4.5	0.15	0.27	5	+			−0.12	−	→	○
		R	−225															
		A	120															
2	2.314 2.859	A	120	545	297	92	90	7.0	0.15	0.35	2	+	3	+	−0.20	−	○	○
		R	200															
		R	160															
		R	700															
		R	160															
3	2.859 2.934	NIT		75													○	○
4	2.934 3.049	R	−120	115	531	81	90	7.0	0.15	0.36	9	+	11	○	−0.21	−	○	○
5	3.049 3.084	R	80	35	796	69	90	7.0	0.15	0.40	21	−	12	○	−0.25	−	−	○
6	3.084 3.134	NIT		50														○
7	3.134 3.224	R	−100	90	637	76	90	7.0	0.15	0.38	14	○	7	+	−0.23	−	○	○
8	3.224 3.369	NIT		145									20	○			○	○
9	3.369 3.619	R	−300	250	212	96	90	3.5	0.15	0.21	6	+			−0.06	−	○	←

Note: For the legend, see Table 18.16.

TABLE 18.19 Case Study for the Examination of an Existing Alignment According to the Operating Speed Background of Germany

Element no. (1)	Station, km (2)	Parameters, m (3)		L_r, m (4)	CCR_S, gon/km (5)	$V85_r$, km/h (6)	V_d^*, km/h (7)	e, % (8)	f_{RA} (9)	f_{RD} (10)	Criterion I, $|V85i - V_d|$, km/h (11)		Criterion II, $|V85i - V85_{i+1}|$, km/h (12)		Criterion III, $f_{RA} - f_{RD}$ (13)		Safety module $1\rightarrow 7$ (14)	$7\rightarrow 1$
1	4.175 4.245	A	70	70	258	74	75	3.5	0.17	0.25	1	+			−0.08	−	→	o
		R	−150															
		A	70										17	o				
2	4.245 4.325	R	−1000	80	64	91	75	2.5	0.17	0.04	16	o			0.13	+	o	o
													27	−				
3	4.325 4.490	A	70	165	593	64	75	5.5	0.17	0.40	11	o			−0.23	−	−	o
		R	−70															
		A	65										12	o				
4	4.490 4.680	A	100	190	228	76	75	2.5	0.17	0.16	1	+			0.01	+	+	+
		R	250										13	o				
5	4.680 4.855	A	70	175	611	63	75	5.5	0.17	0.40	12	o			−0.23	−	o	−
		R	70															
		A	70										25	−				
6	4.855 5.005	R	500	150	91	88	75	2.5	0.17	0.10	13	o			0.07	+	o	+
		A	200										8	+				
7	5.005 5.245	A	200	240	170	80	75	2.5	0.17	0.18	5	+			−0.01	o	+	←
		R	−250															

Note: For the legend, see Table 18.16.

18.35

The poor design levels for safety criterion III resolve themselves by a realignment or by increased superelevation rates (for example, element 7).

18.4.2.5   Greece.   In connection with the development of the Greek "Guidelines for the Design of Highway Facilities, Part: Alignment,"[453] the following operating speed background was established by Psarianos[586] (see also Table 8.5 and Fig. 8.12):

$$V85 = \frac{10^6}{10,150.1 + 8.529\,\mathrm{CCR}_S} \tag{8.16}$$

The roadway section of Table 18.20 represents a new alignment. The two-lane rural highway corresponds to the road category A III in Table 6.2 and primarily has to satisfy connector functions between two municipalities. Accordingly, a design speed of $V_d = 90$ km/h was selected from Table 6.2. The radii of curve are well aligned and vary between $R = 300$ m and 560 m, including transition curves. Besides the curved elements, the alignment includes an independent tangent (element 5).

As Table 18.20 reveals, all three safety criteria and the safety module express good design levels and, thus, they confirm a good alignment without any safety deficiencies. The present example confirms the practical relation design conceptions of this handbook, which means curvilinear alignment sections can be interspersed with independent tangents with sound transitions between them, for example, to improve passing possibilities.

18.4.2.6   Lebanon.   The operating speed background for Lebanon was developed by Choueiri and Lamm in Table 8.5:[108]

$$V85 = 91.03 - 0.056\,\mathrm{CCR}_S \tag{8.21}$$

The alignment to be examined in Table 18.21 represents an old two-lane rural road section with radii ranges between $R = 150$ and 1500 m and no transition curves between them. Despite the large differences in radii of curve, Table 18.21 overwhelmingly shows good design. The fair design level of safety criterion II, for the transitions between elements 4 to 3 and elements 7 to 8, can probably be handled by the proper application of traffic warning devices perhaps combined with a speed limit to warn motorists of the upcoming slight discontinuity of the road. The present case study represents a good example, that is, that many old alignments with minor safety deficiencies also exist. Therefore, the advantage of the safety evaluation process presented is not only to detect poor design but also to confirm good or fair design.

18.4.2.7   United States.   For the United States, there exist two operating speed backgrounds (Table 8.5), developed by Lamm and Choueiri for the State of New York[406,408] and by Ottesen and Krammes developed for several states in the United States.[565] It was decided to evaluate the following alignment by the latter one. The transformed equation reads:

$$V85 = 103.04 - 0.053\,\mathrm{CCR}_S \tag{8.14}$$

The alignment examined in Table 18.22 once again represents an old alignment of a two-lane rural road without considering transition curves and radii ranging from $R = 100$ to 600 m. As often happens with old alignments, the design levels of the safety criteria are widely scattered, and often indicate fair design with respect to the safety module. However, the transition between elements 5 and 3 is very critical, as indicated by safety criterion II and by the safety module for the critical driving direction. Therefore, at least as a partial solution, a realignment by flattening the radius of curve of element 3 to an appropriate level with transition curves and including the nonindependent tangents (elements 2 and 4) is recommended. For alleviating the poor design levels of safety criterion III for elements 1, 8, and 10 under comment "Safety Criterion III" in Sec. 18.4.1

TABLE 18.20 Case Study for the Examination of an Existing Alignment According to the Operating Speed Background of Greece

Element no. (1)	Station, km (2)	Parameters, m (3)		L_t, m (4)	CCR_S, gon/km (5)	$V85_t$, km/h (6)	V_d^*, km/h (7)	e, % (8)	f_{RA} (9)	f_{RD} (10)	Criterion I, $\|V85_i - V_d\|$, km/h (11)		Criterion II, $\|V85_i - V85_{i+1}\|$, km/h (12)		Criterion III, $f_{RA} - f_{RD}$ (13)		Safety module (14)	
																	1→6	6→1
1	1.975 2.140	A	120	165	148	88	90	6.0	0.15	0.14	2	+			0.01	+	→	+
		R	300															
		A	120										1	+				
2	2.140 2.365	A	120	225	156	87	90	6.0	0.15	0.14	3	+			0.01	+	+	+
		R	−300															
		A	150										4	+				
3	2.365 2.675	A	150	310	104	91	90	5.5	0.15	0.11	1	+			0.04	+	+	+
		R	400															
		A	250										2	+				
4	2.675 3.195	A	275	520	68	93	90	4.5	0.15	0.08	3	+			0.07	+	+	+
		R	−560															
		A	400										6	+				
5	3.195 3.635	IT		440		99	90				9	+					+	+
													6	+				
6	3.635 3.770	A	175	135	71	93	90	6.0	0.15	0.08	3	+			0.07	+	+	←
		R	500															
		A	175															

Note: For the legend, see Table 18.16.

TABLE 18.21 Case Study for the Examination of an Existing Alignment According to the Operating Speed Background of Lebanon

Element no. (1)	Station, km (2)	Parameters, m (3)	L_t, m (4)	CCR_s, gon/km (5)	$V85$, km/h (6)	V_d^*, km/h (7)	e, % (8)	f_{RA} (9)	f_{RD} (10)	Criterion I, $|V85_i - V_d|$, km/h (11)	Criterion II, $|V85_i - V85_{i+1}|$, km/h (12)	Criterion III, $f_{RA} - f_{RD}$ (13)	Safety module 1→10 (14)	Safety module 10→1 (14)
1	0.090 0.155	R −250	65	255	77	80	4.5	0.17	0.14	3 +		0.03 +	→	+
2	0.155 0.245	NIT	90								4 +			
3	0.245 0.300	R −200	55	319	73	80	5.0	0.17	0.16	7 +	16 o	0.01 +	+	+
4	0.300 0.535	R 1500	235	42	89	80	2.5	0.17	0.02	9 +	10 +	0.15 +	+	+
5	0.535 0.630	R −300	95	212	79	80	4.0	0.17	0.12	1 +	2 +	0.05 +	+	+
6	0.630 0.720	R 250	90	255	77	80	4.5	0.17	0.14	3 +	5 +	0.03 +	+	+
7	0.720 0.815	R −400	95	159	82	80	3.5	0.17	0.10	2 +	15 o	0.07 +	+	+
8	0.815 0.880	R 150	65	425	67	80	6.0	0.17	0.18	13 0		−0.01 o	o	o
9	0.880 0.925	NIT	45								6 +			
10	0.925 1.035	R −200	110	319	73	80	5.0	0.17	0.16	7 +		0.01 +	+	←

Note: For the legend, see Table 18.16.

18.38

TABLE 18.22 Case Study for the Examination of an Existing Alignment According to the Operating Speed Background of the United States

Element no. (1)	Station, km (2)		Parameters, m (3)		L_t, m (4)	CCR_S, gon/km (5)	$V85_t$, km/h (6)	V_d^*, km/h (7)	e, % (8)	f_{RA} (9)	f_{RD} (10)	Criterion I, $|V85_i - V_d|$, km/h (11)		Criterion II, $|V85_i - V85_{i+1}|$, km/h (12)		Criterion III, $f_{RA} - f_{RD}$ (13)		Safety module 1→10 (14)	10→1 (14)
1	0.000	0.137	R	120	137	395	82	85	2.8	0.16	0.41	3	+			−0.25	−	→	○
			R	260															
2	0.137	0.176	NIT		39									13	○				
3	0.176	0.312	R	−100	136	637	69	85	3.2	0.16	0.34	16	○			−0.18	−	○	−
4	0.312	0.380	NIT		68									28	−				
5	0.380	0.444	R	600	64	106	97	85	1.4	0.16	0.11	12	○			0.05	+	○	+
6	0.444	0.662	R	−500	218	127	96	85	2.4	0.16	0.12	11	○	1	+	0.04	+	+	○
7	0.662	0.800	NIT		138														
8	0.800	1.035	R	170	235	375	83	85	6.2	0.16	0.26	2	+	13	○	−0.10	−	○	○
9	1.035	1.110	NIT		75														
10	1.110	1.254	R	−150	144	425	81	85	2.9	0.16	0.32	4	+	2	+	−0.16	−	○	←

Note: For the legend, see Table 18.16.

should be referred to (increase in skid resistance and superelevation rates as well as decrease of operating speeds by a proper application of traffic warning devices).

This example reveals very well the complexity which the highway engineer has to face with respect to sound decisions for reducing accident risk and severity.

18.4.3 Preliminary Conclusion

Based on Tables 18.16 to 18.22, the responsible safety authorities are provided with quantitative safety information about sound, risky, and dangerous roadway sections. Although important comments and typical recommendations for possible safety improvements were given by the authors in the previous chapters, any definite decision about the use of countermeasures, for example, traffic warning devices, speed regulations, increase of superelevation rates, skid-resistance improvements, sight distance adaptations, etc., or even redesigns, rehabilitation strategies, and stationary radar devices, are consciously left to the discretion of the responsible authorities because of the unknown complex interrelationships between local, administrative, environmental, and economic conditions at the investigated locations.

The procedure presented is only intended to recognize sound, endangered, and dangerous design practices, while the highway design or traffic safety engineer basically has to decide which countermeasures are appropriate for the specific roadway location, segment, or section.

18.5   RECENT DEVELOPMENTS*

18.5.1 Operating Speed Background for Two-Lane Rural Roads in Mountainous Topography

The operating speed backgrounds developed so far for two-lane rural roads are valid up to longitudinal grades of 5 percent (maximum 6 percent) and are graphically presented for curvature change rates up to 600 gon/km in Fig. 8.12 (see also Table 8.5). These operating speed backgrounds are the basis for the practical application of the three quantitative safety criteria. It was the aim and purpose of the study to clarify the relationship between the driving behavior of passenger cars and structural parameters on roadway sections with high longitudinal grades superimposed by large curvature change rate values in order to examine the extent to which the developed safety criteria may be applicable in difficult terrain.

To evaluate the relevant 85th-percentile speeds by regression analyses, a broad database was established, consisting of about 100 roadway segments with different longitudinal grades (ranging from 6 to 14 percent), curvature change rates of the single curve (ranging from 0 to 4,500 gon/km), pavement widths (ranging from 5.50 to 7.50 m), and various superelevation rates in the Black Forest area in southwestern Germany. Local speed measurements were conducted for both driving directions only for passenger cars under free-flow conditions with a minimum time gap of about 6 s. The test segments consisted of about one-fourth new and three-fourths old alignment designs.

Various regression analyses plus statistical test procedures confirmed that the curvature change rate of the single curve, CCR_S, revealed a significant impact on operating speed and strongly superimposed the much lesser influences of the other investigated design parameters. The impact of the independent longitudinal grade G, for example, shows only marginal significance for upgrade sections with small curvature change rates ($CCR_S < 130$ gon/km). For larger CCR_S values on upgrade sections, as well as for the whole database with respect to downgrade sections, no significant influence of the grade on the driving behavior could be tested. The same is true for the independent variables: pavement width and superelevation rate.

---

*Elaborated by O. Eberhard based on Ref. 164.

Thus, also in mountainous topography ($G > 6$ percent), the design parameter CCR_S has by far the strongest impact on operating speed, while the longitudinal grade plays a less important role. The same conclusions resulted from Sec. 8.2.3 for flat and hilly topography for longitudinal grades of $G \leq 5$ to 6 percent.

Since the functional graphs $V85 = f (CCR_S)$ for various regression analyses revealed maximum differences of ± 2.5 km/h between upgrade and downgrade direction, the two data sets were combined into one database. Figure 18.5 shows the regression equation and the graphic layout, including the data points between $V85$ and CCR_S valid for both driving directions in mountainous topography. The high coefficient of determination of $R^2 = 0.88$ confirms that the relationship is a strong one.

As already mentioned at the end of Sec. 18.4.1, it does not make much sense to use the safety evaluation processes presented for radii of curve of $R < 50$ m. Therefore, the range of application of safety criteria I and II should be limited to a maximum value of $CCR_S \approx 1250$ gon/km, as illustrated in Fig. 18.5.

For radii of curve smaller than the recommended minimums in Table 12.5 (-20 percent), safety criterion III will normally reveal poor design. For a better understanding and to counteract this fact, the comment "Safety Criterion III" in Sec. 18.4.1 was elaborated with respect to existing (old) alignments.

Furthermore, it has to be considered that safety criteria I to III were first developed with respect to sound relation design issues between successive curves and between independent tangents and curves while at the same time considering the recommendations for minimum radii of curve. Those design cases normally are appropriate for flat and hilly topography, whereas they are rarely applied in mountainous topography, especially not on old alignments.

Since safety criteria I to III are developed to evaluate design, operating speed, and driving dynamic consistencies and to detect corresponding inconsistencies, it is logical that the case "poor design" most frequently occurs in difficult terrain. In order to examine the extent to which the safety criteria are also appropriate for the safety evaluation of existing alignments in mountainous topography, 10 roadway sections with an overall length of about 35 km were investigated on the basis of the functional relationship presented in Fig. 18.5. The most important statements are discussed briefly in the following:

- Safety criterion I accounted for nearly 90 percent of all investigated roadway segments (curved sites and independent tangents) with good or at least fair design practices. Only sections with very narrow or even hairpin curves were generally evaluated as "poor design," which had to be expected. In exceptional cases, this was also true for independent tangents, where relatively low design speeds V_d in mountainous topography, calculated according to step 6 in Sec. 18.4.1 for the operating speed background shown in Fig. 18.5, were exceeded by the actual operating speeds for more than

$$|V85_i - V_d| > 20 \text{ km/h} \qquad \text{(see Table 11.1)}$$

- Safety criterion II testifies for 80 percent of all investigated transitions between independent tangents and curves as well as between curves and curves good or at least fair design. Realizing the functional relationship of Fig. 18.5, it becomes clear that the poor design level of safety criterion II

$$|V85_i - V85_{i+1}| > 20 \text{ km/h} \qquad \text{in Table 11.1}$$

can only occur if one of the curved segments is characterized by radii of curve of $R < 100$ m, this means that for safety criterion II, especially ($CCR_S > 600$ gon/km) poorly coordinated transitions to narrow or even hairpin curves are conspicuous.

- The most unsatisfying results are provided by safety criterion III. Based on the safety assumptions for this criterion, radii of curve smaller than the minimum ones in Table 12.5 (-20 percent) normally reveal side friction deficiencies. Since radii of curve in this low range are often used in mountainous topography, it is not surprising that safety criterion III confirmed "poor design" with respect to side friction assumed and side friction demanded

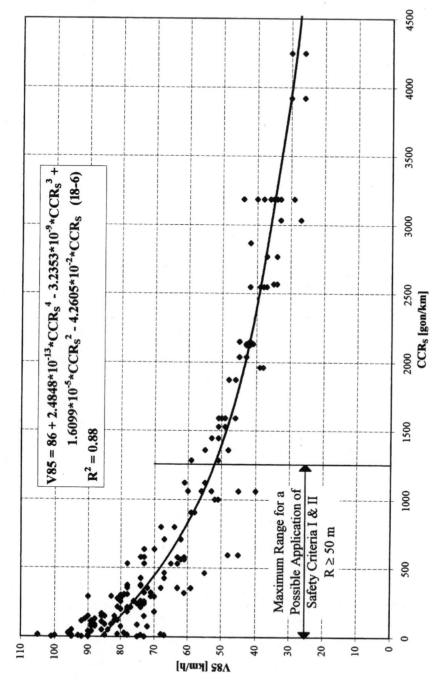

FIGURE 18.5 Operating speed background for mountainous topography ($G > 6$ percent).[164] (*Source: Institute for Highway and Railroad Engineering, University of Karlsruhe, Germany.*)

The equation shown in the figure:

$$V85 = 86 + 2.4848 \times 10^{-13} \times CCR_S^4 - 3.2353 \times 10^{-9} \times CCR_S^3 + 1.6099 \times 10^{-5} \times CCR_S^2 - 4.2605 \times 10^{-2} \times CCR_S \quad (18–6)$$

$$R^2 = 0.88$$

Maximum Range for a Possible Application of Safety Criteria I & II $R \geq 50$ m

$$f_{RA} - f_{RD} < -0.04 \qquad \text{in Table 11.1}$$

for about half of the investigated curved sites.

The examination of the developed safety evaluation processes revealed that, for mountainous topography with respect to existing alignments of old two-lane rural roads, by far more safety deficiencies were detected than for flat or hilly topography with longitudinal grades of $G \leq 5$ to 6 percent. Assuming that corresponding alignment configurations are unavoidable because of spatial constraints in mountainous topography, redesign measures are often not practicable, for example, in narrow or even in hairpin curves. Therefore, for existing (old) alignments, it often has to be satisfactory to provide a sound and perceivable visual guidance in order to recognize inconsistencies in alignment design in time. As already mentioned in the comment, "Safety Criterion III," in Sec. 18.4.1, correct signing and pavement markings, incrementally reduced speed limits, the installation of appropriate traffic warning devices (such as arrows and chevrons), and guardrails may lead to an improved visual guidance. Reducing operating speeds is the most important issue to improve poor design levels, designated by the three safety criteria in mountainous topography. In this connection, good visual guidance means adaptation of the driver's mental workload in such a way that he or she is able to adjust step-by-step driving behavior for upcoming alignment or operational changes.

One main issue of the developed safety procedure was to detect poor design levels for possible future redesigns or RRR strategies. The procedure is based on balanced safety considerations, which encompass not only the size of limiting values, such as minimum radii of curve, but also the size of related values, such as relation design issues. However, the fact is that the safety aspect for roadway design in mountainous topography, especially with respect to old alignments, often played a less important role than the construction costs. Therefore, the characteristic of those roads is, first of all, determined by the presence of specific constraints. Thus, a safety evaluation procedure, as developed in this book, has to document design deficiencies already there, where correct design rules are not sufficiently considered. As the previous statements confirm, this is especially true for old alignments in mountainous topography. Therefore, safety criteria I to III, if revealing poor design levels, only have limited influence for old alignments in mountainous topography with respect to redesigns or RRR strategies. However, they pinpoint those risky locations or roadway segments, where an optimum visual guidance is necessary to sufficiently adapt actual driving behavior to difficulties in the existing terrain.

In this connection, the following statements may be of interest:

- The Australian "Guide to the Geometric Design of Rural Roads"[37] indicates that difficult terrain imposes sufficient obvious restraints, and the driver can readily appreciate the reasons for any speed restriction. Terrain that has no difficulties, however, displays no such obvious restraints and the designer must therefore ensure that the geometry does not appear forced or unnatural, and sound operating speeds and driving dynamic assumptions have to be guaranteed.

- The German "Guidelines for the Design of Roads"[246] note that in mountainous topography, where to overcome large height differences, the alignment consists of narrow or hairpin curves, the driver does not expect a road which allows a comfortable drive but counts on the road ahead to have a stretch of rich curves that are difficult to maneuver. So far, as inconsistencies in alignment design are made clear to the driver, that means a good visual guidance, and relation design does not have to be applied.

Thus, both guidelines indirectly assume, but do not prove, that difficult terrain automatically influences the drivers in such a way that they reduce their operating speeds. This consideration is true, to a certain extent, by comparing the operating speed background for mountainous topography in Fig. 18.5 with the operating speed backgrounds of different countries in Fig. 8.12 for flat or hilly topography. However, these speed reductions are obviously not large enough, since the actual operating speeds of Fig. 18.5 often lead to poor design levels with respect to the three safety criteria, at least for existing alignments of old two-lane rural roads, as the previous investigations revealed. Obviously, the only solution is the enforcement of a good visual guidance, for example, by appropriate traffic control devices and/or speed regulations, which also is recommended in the German guidelines.[246]

However, note that the recommended safety considerations should also be consciously regarded and obeyed in mountainous terrain for new designs with respect to safety criteria I and II, at least for radii of curve of $R \geq 100$ m (if possible, down to $R > 50$ m). If it is not possible to satisfy safety criterion III for all curved sites, the comment "Safety Criterion III" in Sec. 18.4.1 may give valuable insights. As had to be expected, the study[164] proved that the three safety criteria revealed no poor design levels for roadway sections constructed in conformity with modern design rules, as presented in this handbook for mountainous topography. Thus, for new designs, the typical excuses "terrain, economical and/or ecological constraints" should only be accepted in exceptional cases.

As mentioned numerous times before, the three safety criteria are mainly related to design and operating speed considerations as well as to driving dynamic impacts. Since those assumptions are only valid down to a certain radius of curve, where driving dynamic influences are less important than driving geometric maneuvers, the application range of the safety criteria is limited. This cross-point in radii of curve lies, and in particular cases depends, besides road design, on various influencing factors (human, vehicle, environment, weather condition, etc.) and their interdependence. According to the experience of the authors, based on hundreds of local speed measurements, it is recommended that the application range should not fall short of radii of curve of $R < 50$ m, which means curvature change rates of the single curve of about $\mathrm{CCR}_s > 1250$ gon/km. Of course, this consideration is only related to the application range for the three safety criteria, which also includes poor design levels, and does not mean that sound alignments can be expected for those limiting values.

18.5.2 Modification of the Procedure for Determining Tangent Speeds and Lengths in the Safety Evaluation Process

Eberhard[164] modified and simplified the procedure for the evaluation of tangents in the design process presented in "Evaluation of Tangents in the Design Process" in Secs. 12.1.1.3 and 12.2.1.3. So far, an attainable speed difference of $\Delta V85 \leq 20$ km/h or $\Delta V85 > 20$ km/h was introduced for distinguishing nonindependent tangents from independent tangents. The modified version does not include such a limiting value, since the desired control of the design quality for the transition between curved and tangent segments is automatically guaranteed through the subsequent application of safety criterion II.

The modified procedure is graphically presented in Fig. 18.6. For the classification of tangents in the design process, Eqs. (18.7) to (18.9) were developed for the three design cases in Fig. 18.6, based on the fundamental equation [Eq. (12.1)] in Sec. 12.1.1.3.

Case 1: For $\mathrm{TL} \leq \mathrm{TL}_{min} \rightarrow$ nonindependent tangent (Fig. 18.6):

$$\mathrm{TL}_{min} = \frac{|(V85_1)^2 - (V85_2)^2|}{2 \times 3.6^2\, a} \tag{18.7}$$

$$\mathrm{TL}_{min} = \frac{|(V85_1)^2 - (V85_2)^2|}{22.03} \tag{18.7a}$$

where a = acceleration/deceleration rate, m/s^2 = 0.85 m/s^2 (see Secs. 12.1.1.3 and 12.2.1.3). For other symbols, see Fig. 18.6.

In Eqs. (18.7) and (18.7a) $\mathrm{TL} \leq \mathrm{TL}_{min}$ means that the tangent segment is the maximum length which is necessary for adapting the operating speeds between curves 1 and 2. Thus, no additional length for accelerating and decelerating maneuvers in tangent segments is available. In this case, the element sequence curve-to-curve, and not the interim (nonindependent) tangent, controls the evaluation process according to safety criterion II for differentiating between good, fair, and poor design practices (see Table 11.1).

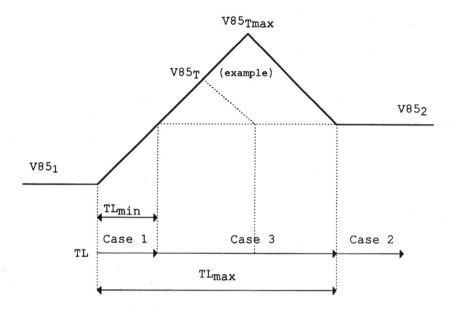

Legend:

$V85_{1,2}$ = 85th-percentile speeds in curves 1 and 2 [km/h],

$V85_{Tmax}$ = Maximum operating speed in tangents [km/h]
               (depending on the regression equations in Table 8-5
               for different countries or for mountainous topograhy
               from Figure 18-5 for CCR_S = 0 gon/km),

$V85_T$ = Operating speed in tangents [km/h]
               ($V85_T$ can maximum reach $V85_{Tmax}$),

TL   = Existing tangent length between two successive
               curves [m],

TL_{min} = Necessary acceleration/deceleration length between
               curves 1 and 2 [m],

TL_{max} = Necessary acceleration/deceleration length to reach
               $V85_{Tmax}$ between curves 1 and 2 [m].

Case 1: TL ≤ TL_{min}
         (non-independent tangent, not considered
         in the safety evaluation process).

Case 2: TL ≥ TL_{max}
         (independent tangent, considered in the
         safety evaluation process).

Case 3: TL_{min} < TL < TL_{max}
         (independent tangent, considered in the
         safety evaluation process).

FIGURE 18.6 Systematic sketches for determining tangent speeds and lengths in the safety evaluation process.[164]

Case 2: For TL ≥ TL_{max} → independent tangent (Fig. 18.6):

$$TL_{max} = \frac{(V85_{Tmax})^2 - (V85_1)^2}{2 \times 3.6^2\, a} + \frac{(V85_{Tmax})^2 - (V85_2)^2}{2 \times 3.6^2\, a} \tag{18.8}$$

$$TL_{max} = \frac{2\,(V85_{Tmax})^2 - (V85_1)^2 - (V85_2)^2}{22.03} \tag{18.8a}$$

for a definition of the symbols, see Fig. 18.6 and Eq. (18.7).

In Eqs. (18.8) and (18.8a) TL ≥ TL_{max} means that the existing tangent segment is long enough to allow an acceleration and deceleration maneuver up to the maximum operating speed ($V85_{Tmax}$) on tangents (see Fig. 18.6). In this case, the element sequences independent tangent-to-curve or curve-to-independent tangent become relevant for the evaluation of safety criterion II in Table 11.1.

Case 3: For TL_{min} < TL < TL_{max} → independent tangent (Fig. 18.6):

$$\frac{TL - TL_{min}}{2} = \frac{(V85_T)^2 - (V85_1)^2}{22.03} \qquad \text{for } V85_1 > V85_2 \tag{18.9}$$

$$\Leftrightarrow \quad V85_T = \sqrt{11.016\,(TL - TL_{min}) + (V85_1)^2} \tag{18.9a}$$

For a definition of the symbols, see Fig. 18.6. Always use the larger value of $V85_{1,2}$.

The existing tangent length lies somewhere between TL_{min} and TL_{max}. Although the tangent segment does not allow accelerations up to the highest operating speed ($V85_{Tmax}$), additional acceleration and deceleration maneuvers are possible (see Fig. 18.6). In this case, the realizable tangent speed ($V85_T$) has to be calculated according to Eq. (18.9a) for the evaluation of safety criterion II.

For a better understanding with respect to the previously discussed three design cases, the following examples are given in Table 18.23.

Accordingly, safety criterion I has to be controlled in tangents with respect to Table 11.1 for the differences between the design speed, V_d, and the 85th-percentile speeds in independent tangents, $V85_T$ or $V85_{Tmax}$, for design cases 2 and 3. Corresponding to safety criterion II, nonindependent tangents (case 1) are too short to be considered in connection with safety criterion I.

Tangent considerations have no influence on safety criterion III.

In comparison to Secs. 12.1.1.3 and 12.2.1.3, it is recommended to use the modified version of the tangent procedure for those roadway sections, where operating speed differences of more than $\Delta V85_i = 20$ km/h frequently exist between two curves enclosing a tangent segment. This is especially true for roads in mountainous topography, where narrow or hairpin curves are unavoidable in overcoming large height differences.

TABLE 18.23 Example Applications for Controlling Safety Criterion II

Example no.	$V85_1$, km/h	$V85_2$, km/h	$V85_{Tmax}$, km/h	TL, m	TL_{min}, m	TL_{max}, m	Case	$V85_T$, km/h
①	70*	50*	86*	100	109	336	1	—
②	95*	90*	100*	150	42	131	2	100
③	85*	60*	102*	330	165	453	3	95

*Calculated based on the corresponding regression equation in Table 8.5 or Fig. 18.5 for the individual CCR_S-value of the observed curved or tangent site.

Safety evaluation for relevant transitions according to criterion II in Table 11.1:

① Germany, mountainous, operating speed background according to Fig. 18.5 and Eq. (18.6)[164]

$TL \leq TL_{min} \rightarrow$ case 1 $\rightarrow$ nonindependent tangent

Curve 1 $\rightarrow$ curve 2: $|V85_1 - V85_2| = |70 - 50| = 20$ km/h $\Rightarrow$ "fair design"

② Germany, old, operating speed background according to Table 8.5 and Eq. (8.18)[241,243]

$TL \geq TL_{max} \rightarrow$ case 2 $\rightarrow$ independent tangent

Curve 1 $\rightarrow$ tangent: $|V85_1 - V85_{Tmax}| = |95 - 100| = 5$ km/h $\Rightarrow$ "good design"

Tangent $\rightarrow$ curve 2: $|V85_{Tmax} - V85_2| = |100 - 90| = 10$ km/h $\Rightarrow$ "good design"

③ France, operating speed background according to Table 8.5 and Eq. (8.19)[643,700]

$TL_{min} < TL < TL_{max} \rightarrow$ case 3 $\rightarrow$ independent tangent

Curve 1 $\rightarrow$ tangent: $|V85_1 - V85_T| = |85 - 95| = 10$ km/h $\Rightarrow$ "good design"

Tangent $\rightarrow$ curve 2: $|V85_T - V85_2| = |95 - 60| = 35$ km/h $\Rightarrow$ "poor design"

CHAPTER 19
HUMAN FACTORS

19.1 INTRODUCTION

Safety is a complex problem. It involves driver behavior, vehicle characteristics, roadway features, and driving conditions. Safety must be approached both as a question of vehicle and roadway design and as a question of use, driving habits, traffic regulation, law enforcement, and risk management. Safety is not just an individual concern. The industry has an important role to play. All levels of the government—local, state, and federal—are involved. There is no single, simple solution to the problem of highway safety, and the public policy questions relating to the safety of the automobile transportation system are among the thorniest that must be faced.[557]

Chapters 19 ("Human Factors") and 20 ("Road Safety Worldwide") are intended to:

1. Present, from a civil engineering standpoint, the safety-related aspects and criteria related to Part 2, "Alignment (AL)," and Part 3, "Cross Sections (CS)."

2. Discuss and analyze, from different points of view and not just from a road engineering point of view, a number of issues in the wide field of *traffic safety*.

Good coordination between the human being, the vehicle, and the road allows a better evaluation of the relationships and interrelationships of traffic safety and provides ways to improve the accident situation at present and in the future.

Last but not least, Chaps. 19 and 20 aim to inform civil engineers about the state of the art of current safety practices and research, in addition to the special field of *roadway engineering,* and the state of techniques and sciences applied in other related disciplines. These chapters were added here in order to provide civil engineers, in particular, with a broad overview of the current status of road safety worldwide.

19.2 GENERAL COMMENTS

As the main controlling element, the driver plays a major role in determining the success or failure of a highway system. More than 90 percent of all accidents are attributed, to a large extent, to human error or improper human behavior.[489] Only proper driving performance results in safe, efficient traffic operations.

For optimum highway design and traffic operation, an understanding of human factors is essential. However, engineers need more than just a rudimentary understanding of driving behavior and task performance, since they are responsible for road design, traffic operation, and regulation in order to avoid driver errors, wherever possible.

To date, geometric and/or driving dynamic safety evaluations have been used in a more or less general way. However, site-specific evaluation of highway safety can be achieved only by means of a combined procedure, based on driving geometrics, driving dynamics, and driving psychology in order to examine, for example, every curve site and the corresponding transitions according to good, fair, or poor design practice. In this connection, the quantitative safety criteria previously developed present a first important step for problem diagnosis and for possible safety improvements.

According to Lunenfeld and Alexander,[489] a safety evaluation process should be based on the premise that competent drivers can be given enough appropriate information about hazards and inefficiencies to avoid errors. This process combines highway and traffic engineering, as well as human performance methods and procedures, to produce a highway information system that matches the driving attributes and demands of the situation. Errors are seldom caused by one feature. Therefore, every problem location must also be examined using driver workload- and perception-based procedures to identify site-specific problems which contribute to driver error and/or improper driving task performance.[489]

19.3   DRIVER-VEHICLE-ROADWAY/ENVIRONMENT

The highway system consists of an array of different subsystems and elements. The system is dynamic, with subsystems and elements often interacting in a transitory way.[524]

19.3.1   The Driving Task

As reported by Sievert,[648] the terms *active and passive safety* have been used in the field of transportation since the 1960s. Because passive safety could be improved more easily and at much lower costs than active safety, the main focus since that time has been to improve passive safety. Today's existing high-performance components in automobile engineering and electronic devices like deformation rigidity of the passenger cell, restraint systems, and air bags have improved safety considerably. For detailed information, see Sec. 20.3. Therefore, further research on passive safety may have less effect on the accident situation than research on active safety.

Consequently, active safety is becoming a key issue more and more in improving traffic safety. *Active safety* includes, for example,

- Driving dynamic safety
- Driving behavior
- Perception/visibility
- Ergonomics
- Conditions (for example, air conditioning, etc)

with driving dynamic behavior being the most important source for improving active safety and preventing accidents.[648]

Inappropriate behavior of the driver is, besides vehicle and environmental influences, the main cause of accidents (see also Fig. 20.10). Inappropriate behavior results from deficiencies in human-vehicle interaction and/or from a misunderstanding of upcoming driving situations with respect to the roadway or environment.

In this connection, the most effective ways to improve traffic safety are by aiding the driver in his or her interaction with the vehicle and by designing highways in such a way that features which are relevant to the driving task are perceptible, understandable, and soundly designed, based on driving dynamic, driving behavior, and psychological points of view. The basic relationship can be seen in Fig. 19.1 according to Donges.[145]

The driving task is influenced mainly by the information received and used. This information is compared with the information already processed by the driver. Decisions are made and actions performed. This task encompasses a number of discrete and interrelated activities. Driving consists of three performance levels, referred to as *navigation, guidance,* and *control.* The model

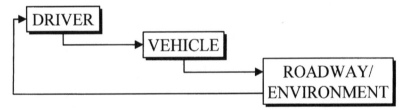

FIGURE 19.1   Interaction between driver-vehicle-roadway/environment.[274]

developed by Donges[144,145] in Fig. 19.2 indicates that the levels present a hierarchy which exhibits a decreasing complexity regarding information handling as we move from navigation to control. Regarding safety, the hierarchy exhibits decreasing importance as we move from control to navigation.

As discussed in Refs. 5, 145, 489, and 648, the performance levels can be defined, as follows:

Control: Control includes all the activities involved in the driver's interaction with the vehicle, that is, steering and speed control and its interfaces, that is, controls and displays. Task performance ranges from relatively undemanding (passenger car) to relatively demanding tasks (tractor trailer with multiple gears and clutches).

Information is received from the "feel" of the vehicle, from its displays, and from the roadway. Drivers continually make minute adjustments and use the feedback to maintain control. Most control activities, once mastered, are performed "automatically," with little conscious effort.

Guidance: At the guidance level, the driver's main activities involve the maintenance of a safe speed and a proper path relative to roadway and traffic elements, for example, following the road and maintaining a safe path in response to road and traffic conditions. Guidance activities are characterized by judgment, estimation, and prediction within a dynamic, constantly changing environment. Guidance level decisions are translated into speed and path maneuvers in response to alignment, grade, hazards, traffic, and the environment. Information is gathered from the highway and its components, traffic, and the highway's information system.

Navigation: The navigation level consists of a *pretrip phase,* where trips are planned and routes selected, and on an *in-trip phase,* where the trip plan is followed. Pretrip information sources include maps and verbal instructions. In-trip information sources include landmarks and route guidance signs. Navigational activities are generally cognitive and verbal in nature.[145,489,648]

The three performance levels are shown graphically in Fig. 19.3.

Depending on alignment, traffic, and environment, the driving task may be complex and demanding. Several individual activities need to be performed simultaneously. This requires smooth, efficient handling and integration of information.[5]

Driving often takes place under certain demanding circumstances such as high speeds, time pressures, unfamiliar locations, and rapidly changing environmental conditions. On the other hand, driving may be undemanding and monotonous, so that drivers become inattentive and may fall asleep. The key to safe, efficient performance is, as far as possible, error-free information handling.[5]

According to Braun, et al.,[70] it is obvious that every "participation" in traffic automatically includes a certain amount of risk of being involved in an accident (see Fig. 19.4).

With respect to Fig. 19.4, critical driving situations cannot be described exactly within the context of one simple situation of inadequate behavior. This is only possible within the complex relationships between the interaction of driver-vehicle-roadway/environment.[648]

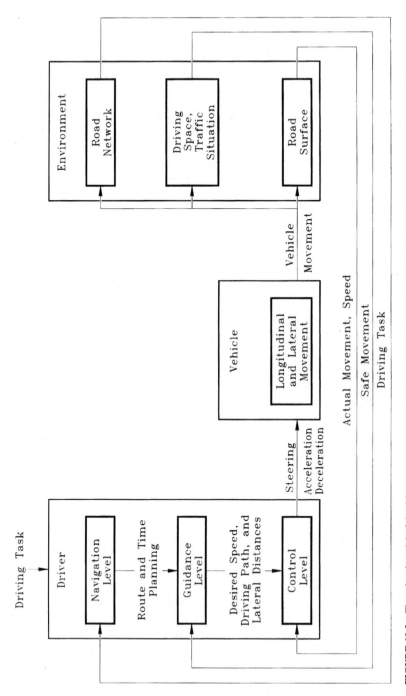

FIGURE 19.2 Three-level model of the driver-vehicle-roadway/environment system. (Elaborated based on Refs. 144 and 145.)

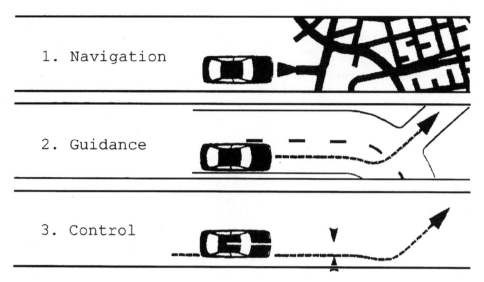

FIGURE 19.3   Performance levels of the driving task. (Elaborated based on Ref. 648.)

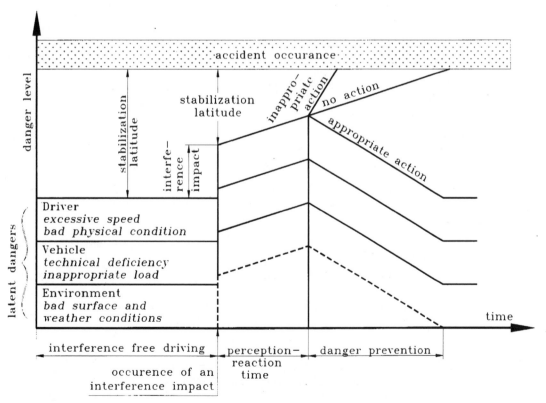

FIGURE 19.4   Flowchart of a critical driving maneuver. (Elaborated and modified based on Refs. 70 and 648.)

Latent dangers can already exist in all three components even during normal unimpaired driving. The level of the latent dangers depends on the arrangement of the elements in the system. It increases if the behavior of at least one of the components is not normal. If impairments occur in that part of the system, the driver will attempt to stabilize the system through anticipated or immediate actions.[648]

Thus, success in preventing an accident depends, to a great extent, on the level of the latent dangers and the correct intervention of the driver.[648]

19.3.2 Information Gathering

The key to a successful task performance is efficient information gathering and processing. This is often relatively easy when drivers only have to perform one activity at a time. However, the driving task usually requires performance of a number of activities across all performance levels. At any instant, drivers generally gather information from more than one source, establish priorities relative to what information to attend to and process, make decisions, and perform control actions often under time pressures. Drivers use their sensory input channels to gather and/or receive information. However, while most senses are used in driving, as much as 90 percent of all information is gathered and received visually.[489,524,648,752]

According to Ritchie, McCoy, and Welde,[606] one of the important skills in driving is the perception of the lateral force which results from a change in the direction of motion. The extent to which this kinesthetic perception is related to visual perception is not clearly established. The ability to drive safely at high speeds may depend strongly on the interaction between these two senses. Drivers habitually choose the speed at which they will negotiate a curve before entering the curve. If the curve is unfamiliar, this choice must be based on visual perception. Therefore, when the driver chooses a speed, he or she is predicting the lateral force which will be felt in the curve, and, depending on the predicted lateral force, the chosen speed will be an inverse function. The results show clearly a strong inverse relationship between speed and lateral acceleration for speeds above 20 mi/h (32 km/h).

The lateral force developed in the curve is the primary criterion for the choice of speed in curves. The kinesthetic perception of lateral force is the primary mode by which the driver obtains this information for use in the driving task. There is an interaction between this kinesthetic perception and visual perception which develops quickly during the acquisition of driving skill and which may be degraded by such agents as drugs or alcohol or changes due to age. The observed decrease in available side friction as speed increases reflects the driver's estimate of increasing danger.[606] Similar conclusions can be found in Refs. 220, 282, and 545.

19.3.2.1 *Perception and Reception of Visual Information*   Although it is not possible to identify and describe all pertinent visual reception factors, the following are important vision-related considerations:[381,489]

1. The visual channel is selective. For visual information to be received, it must be looked at and attended to.
2. The visual information source must be within the driver's field of view and "cone of clear vision" at operating speeds.
3. Drivers must have the capability (for example, visual acuity and color vision) to receive the information.

On aspects of visual perception in road traffic, it is reported by Cohen:[118–120]

> In order to be able to steer a vehicle safely in traffic, the features of the environment have to be permanently perceived and evaluated in order to make corresponding decisions. The fundamental assumption for this is that the driver has first to discover these features in order to analyze them properly.

The limitations of the usable field of vision first of all depend on stimulating features and on the capabilities of the retina. It is proven that an increasing load impairs the capability of the retina perceiving a stimulation. This phenomenon is known as narrowing the usable field of vision.

For an average visual load it could be observed that the maximum extent of the visual field can be assumed to be 9 degrees around the focus of vision. This case may exist when following another traffic participant on a wide straight section. However, for unfavorable influencing factors (for example opposing traffic on a narrow road) the narrowing of the usable field of vision could be proven. On the other hand the usable field of vision can also decrease for very favorable road conditions, for example driving unhindered on a wide straight section.

The results reveal compensational occurrences, since obviously the driver changes his attention according to the visual load. For a low visual load the level of attention decreases the narrowing of the visual field is the consequence. If the visual load is increasing, the test person increases his attention and can even overcompensate the impairing effect. However, when the impairment exceeds the limitations of the mental capacity, for example if two or three impeding influencing factors occur simultaneously, then the test person can no longer compensate the visual load completely, which means important visual information is no longer perceived. In conclusion, it is interesting to note that the highest performance can be reached for average visual loads.[118–120,278]

With respect to motion and speed, which represent two important issues of this book, the previous observations can be interpreted as follows.

According to Bubb,[90] a selected speed that is too high for a specific situation is the main cause of accidents with considerable property damages and personal injuries. In general, one cannot assume that the driver rushes blindly into danger by selecting a speed that is too high. Rather, based on a subjective evaluation of the exterior situation, the driver evaluates the endangerment level and selects a desired speed which seems appropriate. In the ideal case, the driver would be able to control the actual vehicle speed by the speedometer to compare this speed with the desired speed and to coordinate both speeds.

However, in reality, the validity of the desired speed, based on risk evaluation by the driver, has to be doubted. For example, desired speeds depend strongly on subjective influence factors, for example, the need to hurry, psychological conditions, effect of drugs, etc. On the other hand, the actual speed is controlled to an essentially lesser extent by means of the speedometer as assumed.[120,524] That means the driver obtains his or her feeling for the speed mainly through available senses. In this connection, the visual sense is of major importance, while the kinesthetic (motion) and haptic (touch) senses play an additional, but minor, role.

Visual Sense   During the driving process, the image on the driver's retina is a constantly changing picture, which is called *perspective in motion*.

Two main characteristic features of perspective are used as sources for speed estimation. The *perspective gradients* of a horizontal surface are used for estimates of distance and speed (Fig. 19.5), while the pattern of the *radial vector field* on the retina is responsible for the perception of speed direction (Fig. 19.6).

The driver perceives the environment in perspective; objects far away appear smaller than those nearby. Therefore, the density of equally distributed objects in the distance seems to be greater than those close to the observer. The extent of increasing density with respect to distance from the observer is called the gradient (Fig. 19.5). In this connection, it is easy to imagine that objects close by appear to move greater distances (greater speed vectors) than objects far away (smaller speed vectors) during the motion of the driver (Fig. 19.6). The *speed vector* is defined as the visual motion of an object or a characteristic feature during a specific time interval.

In contrast to speed perception, accelerations are visually perceptible only indirectly. The peripheral visual field is very sensitive to the motion of objects. As shown in Fig. 19.6, it can be easily explained that objects in the visual periphery appear to move greater distances than those perceptible by central vision (see Fig. 19.6).

Kinesthetic Sense (Sense of Motion)   The *human motion sensitivity* located in the ear allows the evaluation of translatory and rotational accelerations because of its physical capability. Speed differences can be estimated only indirectly. For the perception of accelerations, threshold values

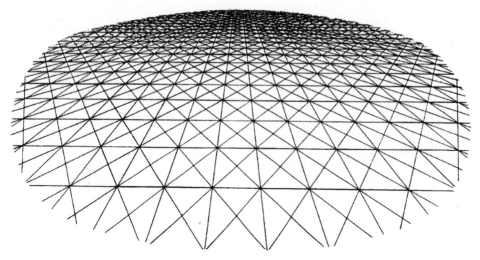

FIGURE 19.5   Perspective gradients of a horizontal surface.[213]

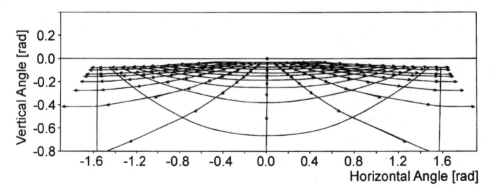

FIGURE 19.6   Radial vector field on the retina of the driver's eye in motion.[227]

between 0.01 and 0.02g are valid.[610] Only if these threshold values are exceeded can a perception of a speed change be expected based on the kinesthetic sense.

Haptic Sense (Sense of Touch)   Haptic senses include all sensitivities related to recognizing vibrations, steering forces, and others.

The driver integrates the impressions from the different information channels into a unique perception for his or her location and motion in space with respect to local objects and other moving traffic participants. Part of this perception is also related to speed.[90]

In conclusion, it can be noted with respect to perception and estimation of speeds, according to Cohen:[118–120]

- The misinterpretation of the desired speed by the driver is not the exception but the rule. Low speeds will be overestimated; high speeds will be underestimated.

- The driver gets used to high speeds. If the driver reduces speed only a little, after driving at a high speed over a longer time period, the driver feels that he or she is driving especially slowly, for example, after leaving an intestate. This effect begins to fade away after about 4 min.

- The speedometer is hardly ever used to correct erroneous speed estimations, especially not if a correction is urgently needed in a complex traffic situation.

- The main cause for the misinterpretation of driving speeds is that the speed perception originates as a consequence of the integrated interplay of several senses.[118–120]

Existing knowledge indicates that the normally limited capability of the driver does not lead to adequate speed perceptions or estimations. By interpretation of this, it can be assumed that inappropriate driving speed, which frequently leads to collisions and serious consequences, is, to a large extent, not caused on purpose by the driver.[121,122]

19.3.2.2 Information Handling and Processing* Driving is an "information-decision-action" task. Drivers gather information from a number of sources and use it with stored information (for example, knowledge, skills, and expectancies) to make decisions and perform actions. Drivers may have to handle too many things at the same time. They often have overlapping information needs associated with various activities. To satisfy these needs, they scan the environment, detect information, receive and process it, make decisions, and perform control actions in a continual feedback process.

Humans are serial processors in that they can only handle one source of visual information at a time. Given the need to parallel process while driving, they compensate by "juggling" several information sources. Drivers integrate various activities and maintain an appreciation of a dynamic, changing environment by sampling information in short glances and shifting attention from source to source. They rely on judgment, estimation, prediction, and memory to fill in the gaps, to share tasks, and to eliminate lower-priority information.

Drivers may become overloaded when they have to process too many sources of information or when an information source has too much information content. *Overloaded* drivers may become confused or miss important information sources due to high processing demands. Drivers may become *underloaded* when they have virtually no information to process. An underload causes drivers to miss important new information sources due to lack of paying attention.[489]

Therefore, the highway system should be designed to provide a steady pace of information processing demand. There should not be any "peaks," where too much information must be handled, "valleys," where there is too little information to process, or "spikes," where there is a sudden surge of high processing demand following a plateau of low processing load.[489] Such a case would be, for example, the transition from an independent (long) tangent to an isolated narrow curve.

19.3.3 Memory and Cognitive Capabilities

Incoming information is preprocessed by the driver. Potentially relevant sources are transferred to the driver's short-term memory for storage and rapid access, retrieval, and processing. If information in short-term memory is not relevant, reinforced, repeated, or retrieved and processed, it is usually forgotten. The short-term memory lasts from 30 s to 1 or 2 min with only a few information sources. Important information sources may be transferred to the long-term memory, which has no limitations on the amount of information it can store or on the time frame for retrieval.[5,251,489]

19.3.4 Reaction Time†

An important issue in mental capacity is reaction time. *Reaction time* is the time between the receipt of information, its being processed, and an action taken. It varies from driver to driver, and is a function of alertness, complexity, and expectancy. Complex decisions take longer to process

---

*Elaborated based on Lunenfeld and Alexander.[489]

†Elaborated based on Refs. 5 and 489.

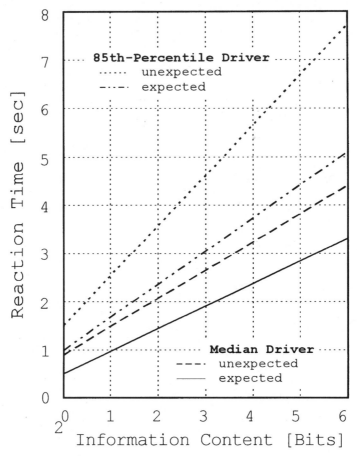

FIGURE 19.7 Reaction time of the median and the 85th-percentile driver. (Elaborated and modified based on Refs. 5 and 489.)

than simpler ones. Reaction time and driver error show similar behavior. The longer the reaction time, the greater the chance for error.[5] Compare "Driver Perception Reaction Time" in Sec. 15.2.2.3.

The relationship between reaction time, in seconds, and complexity, in bits (plotted to the base 2), for a median and 85th-percentile driver is presented in Fig. 19.7. *One bit* equals the amount of information needed to resolve uncertainty between two equally probable responses. Figure 19.7 shows that even a zero-bit decision (one alternative) takes time to process and that reaction time increases exponentially. Few drivers can process more than a 3- or 4-bit (8 or 16 alternatives) decision in transit. In addition, processing a complex decision takes time and attention away from other needed information sources. Thus, in high processing demand situations, it is usually easier and faster to make several simple decisions than a highly complex one.[489] That means the design of the road should be as simple as possible without complex element sequences and intersections. All relevant interrelationships with respect to reaction time and sight distances are discussed in detail from the design point of view in Chap. 15.

19.4 DRIVER ERROR

19.4.1 General Comments*

In general, the causes of accidents will be discussed in Sec. 20.2.8. The present section focuses only on the direct errors of drivers as the principal contributing factor in most accidents. Some driver errors are caused by fatigue due to nondriving-related stress, sensory-motor deficiencies, or alcohol or drug abuse. These are normally beyond the scope of highway safety procedures for improving highway design.

Highway safety procedures can assess errors due to:

- Excessive task demands
- Unusual maneuvers of task requirements
- Poor forward sight distance
- Expectancy violation
- Too high processing demand
- Too little processing demand
- Deficient, ambiguous, confusing, or missing information
- Misplaced, blocked, or obscured information

In the case where errors occur because of the nature of the task, the demands of the situation, the lack of visibility of hazards, expectancy violations, and/or deficiencies in road layout, error sources should be eliminated through application of highway safety evaluation procedures,[145,359,489] for example, through the application of the three safety criteria.

19.4.2 Accidents Attributed to Driver Error†

In traffic psychology, numerous investigations have been conducted with the primary goal of determining the psychological factors contributing to human error. For example, Table 19.1 shows the corresponding percentages for different factors, as developed by Spoerer[667] and Undeutsch.[721] As can be seen from this table, the percentile distributions and rankings are comparable with few exceptions.

A common characteristic of many high-accident locations is that they place large or unusual demands on the information processing capabilities of drivers. Inefficient operation and accidents usually occur where the chances for information handling are high. At locations where design is deficient, the possibility of error and inappropriate driver performance increases.

According to Table 19.1, it is interesting to note that the primary factors contributing to accidents up to rank 5 can be attributed to missing information and erroneous mental processing of perceived information. Both investigations reveal that physical deficiencies of the driver are relatively low with respect to the accident situation (also see Sec. 19.7).

The secondary contributing factors have an additional effect on the accident situation. Unfamiliarity with the traffic situation and time pressure, which usually lead to high workload values, rank first and second according to Ref. 667.

It can be concluded from Table 19.1 that designs of traffic facilities which lead to misinterpretations, with respect to present and upcoming traffic situations, violate expectations and fail in directing attention, for example, to poor curve design, insufficient sight distance, improperly coordinated three-dimensional alignments or misplaced optical guidance by traffic warning devices. Therefore, those design and operational deficiencies should be avoided for new designs, redesigns, or RRR strategies and corrected for existing alignments, as far as possible (also see Sec. 19.9).

---

*Elaborated based on Refs. 5, 70, and 489.

†Elaborated based on Refs. 5, 25, 46, 177, 251, 566, 603, 631, 667, 721, and 740.

TABLE 19.1   Primary and Secondary Factors Contributing to Road Accidents Caused by Human Error*

	Spoerer[667]†		Undeutsch[721]	
	Percentage	Rank	Percentage	Rank
Primary factors contributing to road accidents:				
Misinterpretation of the traffic situation	37.0	1	22.0	1
Inattention	36.4	2	16.5	3
Routine	25.3	3	15.0	4
Violated expectancy	17.9	4	14.5	5
Misinterpretation of subsequent traffic situations	12.3	5	18.0	2
Situational demands contrary to general human behavior	8.0	6	4.0	7
Physical deficiencies	1.9	7	10.0	6
Secondary factors contributing to road accidents:				
Unfamiliar with the traffic situation	37.0	1	—‡	7
Time pressure	34.6	2	23.1	1
Emotional condition	16.7	3	22.1	2
Lack of driving experience	16.0	4	—‡	3
Temporal impairments (for example, uncertainty and slow reaction)	16.0	5	—‡	4
Desire to show off or aggressiveness	5.5	6	—‡	8
Lack of driving skills	4.3	7	—‡	5
Insufficient knowledge of traffic rules	4.3	8	—‡	6
Permanent impairments (for example, boredom and fatigue)	0.6	9	1.9	9

*Elaborated and modified based on Ref. 667 and 721.

†Summarized percentages are not equal to 100 percent due to multiple occurrences.

‡Percentages not available.

Another interesting approach is the concept of local accident investigations, which was initi-ated by the Federal Highway Research Institute (BASt).[46] The goal of these investigations is to get as much information as possible about an accident. That means, in particular, analysis of the causes, the sequence of development of an accident and the psychological, technical, and medical effect of an accident. Detailed questions about misleading human behavior and the corresponding reasons have to be considered. Therefore, the acquisition of as much accident data as possible, such as damages, characteristics of the vehicles and the road, that could give answers about col-lision speed, decelerations, injuries, and much more, was important. The accident causes, as well as the physical and psychological condition of the drivers, were investigated. Furthermore, inter-views on the subjective view of the accident participants were conducted. Table 19.2 summarizes the results of the previously cited research[46] with respect to human errors.

Table 19.2 reveals, in more detail than Table 19.1, the relationship between traffic accidents and psychological processes. It is known that the probability of human error is high for mental overload as well as for mental underload conditions. Overload is characterized by high demands on information processing speed, information complexity, and workload. Large sets of informa-tion sources can confuse drivers during the evaluation process. On the other hand, mental under-load, caused by a monotonous environment and low workload demands, leads to inattention and driver fatigue. Due to low traffic demands, drivers tend to neglect requirements with respect to their driving tasks, since they entertain themselves, for example, by tuning the radio, etc. This may cause problems when suddenly increasing demands arise. The following subsections present a more detailed evaluation of the results shown in Tables 19.1 and 19.2.

19.4.2.1   *Error Due to Deficient, Ambiguous, Confusing, or Missing Information**   As Table 19.1 reveals, the highest frequency of driver error is due to misinterpretations of traffic situations,

---

*Elaborated based on Refs. 524, 645, and 646.

TABLE 19.2   Percentile Distribution of Human Errors with Respect to Road Accidents[740]

	Percentage
Causes of mental overload:	
High-speed demand of information processing	36.4
Highly complex information sources	9.7
High workload demands	6.3
Large set of information sources	3.8
	Σ 56.2
Causes of mental underload:	
Low workload demands	24.7
External interference	8.2
Internal interference	6.8
Inattention	3.1
	Σ 42.8

Note: The missing 1 percent is related to the environment and the vehicle.

followed by inattention. These are mainly errors of information sampling, reception, and classification. Insufficient attention and signal detection play a major role in the development of driver errors.[524] It may seem trivial that, if a well-motivated, skilled, and knowledgeable driver acquires less than satisfactory information on which he has to base decisions, those decisions and the resulting behavior are likely to be in error. However, when information is inherently ambiguous or incomplete, even an optimal decision maker and action taker cannot be expected to produce what an outside observer would define as an optimal behavior.[524] Such a situation arises for certain misleading design solutions which may impair or confuse the driver's decision-making process based on perspective views while in motion. With respect to the accident situation, this phenomenon is especially critical in curves and curve transitions, and it is called the *illusive curve.*

According to Shinar,[645,646] an *illusive curve* is a curve which, from a distance, does not appear to be as sharp as it really is. Thus, the speed reduction demands of the illusive curve are typically underestimated. This is manifested in sharp braking actions in the curve or, in the extreme case, an accident. On a more subjective level, the driver recognizes that he or she is negotiating the curve too fast. It is, therefore, hypothesized that, in the case of the illusive curve, a divergence exists between the physical parameters of curvature and their psychophysical correlates. In such a case, accidents result from large discrepancies between physical and perceived curvature, the latter affecting the entry into the curve.[645] Compare, for example, "An Interesting Phenomenon" in Sec. 16.2.3.2.

Thus, on the road, regardless of the actual curvature, a partially obscured curve may seem to be shallower than a more visible curve of similar radius. It is, therefore, possible that, from a critical distance prior to the curve, the high-accident curve is the one which is less visible than a geometrically similar low-accident curve.[645]

Therefore, the question arises, what are the determinants for the driver's decision to slow down before a curve and what are the perceptual factors associated with accidents in a curve? In this connection, the perceived sharpness of the curve seems to be the major determinant for drivers to slow down, followed by the perceived width. Therefore, drivers would slow down more for curves that appear sharper, narrower, and closer. But high-accident curves do not appear sharper or more dangerous. In contrast, they are misperceived as closer and more visible than a low-accident curve. Additionally, the driver's position in relation to the curve has to be taken into account. Obviously, the driver does not possess all the necessary information to evaluate a curve's true length and angle. The driver sees only a portion of the curve and even that portion is viewed in perspective.[524,645]

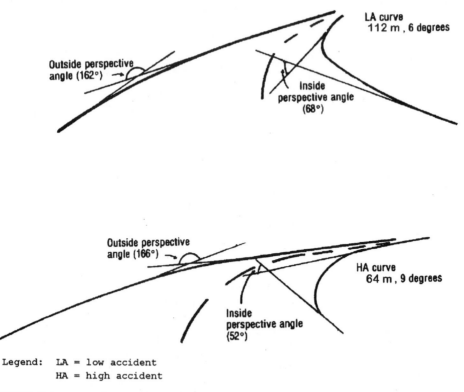

Legend: LA = low accident
HA = high accident

FIGURE 19.8 Edgeline tracing of a low-accident and a high-accident curve.[645]

Figure 19.8 traces the edge lines and centerlines for a high-accident and a low-accident curve. The sharper the curve is, the smaller the inside perspective angle.

As the true angle of a curve increases (that is, as the curve becomes sharper), the inside and outside perspective angles decrease. However, for a given horizontal curvature, the perspective angle is also influenced by the vertical curvature and degree of superelevation. In general, a sag vertical curve increases the perspective angle and a crest vertical curve decreases it, making it appear sharper (see Fig. 16.20). The effects of superelevation are limited to the outside perspective angle: positive superelevation decreases the perspective angle, making the curve appear sharper.[645]

In general, the drivers' subjective judgment in terms of relative visibility is erroneous and might be based on irrelevant cues. The discrepancy between the physical and the psychophysical measures suggests that drivers overestimate the perceived length of high-accident curves relative to low-accident curves.

19.4.2.2 *Error Due to Situation Demands** Drivers often make errors when they have to perform several highly complex tasks at the same time under extreme time pressure. This type of error can occur in both urban and highly complex rural areas. Resultant information processing demands beyond the driver's capabilities may cause confused or inadequate understanding or overloaded thought processes (see Table 19.2, upper part).

Other locations present the opposite situations. Some rural locations have widely spaced decision points, sparse land use, smooth alignment, and light traffic. In this case, drivers' thought

---

*Elaborated based on Ref. 5.

processes are more likely to be underloaded. Errors are caused by decreased vigilance, during which time drivers fail to detect, recognize, or respond to new, infrequently encountered, or unexpected elements of information sources[5] (see Table 19.2, lower part). Accident causes, such as inattention and routine as shown in Table 19.1, reveal that drivers often fail in those situations.

Thus, inappropriate motivation of the driver seems to be one main accident cause. But the potential dangers of the road, especially at high speeds, make it clear that care and alertness are required. Unless we assume that people are highly irrational, the failure of motivational approaches to reduce accidents suggest that the causes of accidents may be such that even well-motivated people often fail to prevent themselves from making errors.[524]

This basic finding suggests that there must be other cognitive mechanisms that are so strong they may override motivation as a dominant cause factor in error genesis. A more detailed evaluation of this relationship is shown in Secs. 19.4.3, 19.7, 19.9, and 19.10.

19.4.2.3   Error Due to Deficient Driver Capability   Many errors are caused by deficiencies in drivers' capabilities in connection with an inadequate design or difficult situation.[25] Table 19.1 reveals that deficient driver capabilities contribute, especially as "secondary factors," to the occurrence of accidents. For example, insufficient experience and training often contribute to the inability to recover from a skidding situation.[5] Similarly, inappropriate perceptual behavior[123] and risk taking[121] may lead to driver errors (see Sec. 19.7).

But, in terms of Table 19.1, physical deficiencies show only a limited influence on driver error and accident situation. This can be explained by the fact that tests of physical capabilities, for example, vision tests during a driver's licensing process, eliminate a considerable portion of the accidents possible due to physical deficiencies. On the other hand, drivers with physical impairments, for example, older drivers, may compensate for this with more careful driving behavior. For more details, see Sec. 19.7.

19.4.3   Accident Models for Driver Error*

One often reads, in publications related to accident research, the admission that a general theory about accident occurrences does not exist.

The main advantage if such a theory exists, would be that it would provide a framework on which existing knowledge could be added and from which derivations and explanations could be made which would serve as a better foundation for further questions and research and would also allow the elaboration of coordinated safety measures.

If one speculates about why no all-embracing theoretical concept about accident occurrences has been developed yet, one must consider the various aspects of the lack of coordination in the efforts so far. One of the main reasons for that is that attempts at interdisciplinary cooperation between engineering and behavioral sciences has not progressed beyond a modest start.

However, it is true that the human factors relevant in the occurrence of traffic accidents can basically be explained by the knowledge that was developed in general and perception-oriented psychology in relation to other living and working situations.

For a long time, the so-called accident tendency model, which means that accidents occur primarily through human failure, has been both used and criticized. As a consequence, two fundamental types of countermeasures have been developed in the application of this theory:

1. Those persons who obviously cannot cope with the dangerous situations that may occur (elimination through psychological tests of people who are obviously incapable, for example, because of alcohol or drug abuse or people with an extremely high number of traffic violation points) should be kept away from traffic participation.

2. Influence and educate people in such a way that they become capable of coping with dangerous situations.

Elaborated based on Refs. 155, 279, and 533.

In contrast, in the technical field, attempts have been made to apply to human beings the methods that have been used to overcome technical defects. The problem with this, however, is that technical systems are evaluated based on their performance and on how they are supposed to perform. One can trace mistakes back to elements that do not fulfill the functions for which they were intended. Applied to "human elements," this means the noncompletion of tasks within the time in which they should have been completed.

However, this consideration is problematic when applied to people. People have numerous functions and, because of that, can make many and very different kinds of errors. In addition, they have their own understanding of tasks as well as their own goals outside of the defined task. However, errors based on this consideration are defined exclusively by the defined task and not by the person who carries out the task.

An additional problem with the transfer of technical reliability concepts to people lies in the meaning of "error" for behavior. Technical reliability rests on the idea that errors are bad and that the goal is, therefore, to prevent their occurrence. When technical errors are compared with human mistakes, this analogy collapses. People are erring, as well as error-correcting, beings. There are no people who do not make mistakes. They are highly sensible and have control mechanisms that comprehend deviations from the proposed course. Acknowledged deviations are the stimulus for the entire human decision and learning process and cannot be excluded from consideration.

In the development of accident models, the human being is considered an element in a mutual system of driver-vehicle-roadway/environment. From this viewpoint, human behavior is understood as a response to the other system elements.

The following models provide information about different approaches for accident genesis and the interactions of humans in these situations. These models indicate where derivations and damages can occur within a system in order to reveal the levels on which humans are able to intervene for possible changes.

Based on a normal condition, the accident model according to McDonald[501] in Fig. 19.9 analyzes possibilities of deviations and their control. Interference impacts can affect the system and cause destabilization, which means the system moves farther and farther from its normal condition and, finally, cannot be brought back to its normal condition.

As a metastable condition, the phase from the occurrence of an interference impact or a loss of control is understood until a detrimental occurrence or an accident occurs. The duration of this phase varies and depends on the dynamics of the system; for example, swerving of a vehicle can last for either a short or long time. During this phase, a person is able to introduce measures to avoid damage, as, for example, intercepting actions supported by automatic active safety measures, as explained in detail in Sec. 20.3.2. If this is unsuccessful, the system enters the instable condition and a certain amount of damage is unavoidable. Introduced measures are only able to diminish the extent of damage from the accident. After the accident occurrence, the sole safety measures for damage (especially injury) reduction are of the passive kind such as passenger cell stability, restraint systems, airbags, etc. For details, see Sec. 20.3.3.

In the model shown in Fig. 19.9, the person is not considered a "system" but rather as a controlling element for the system vehicle-roadway/environment. Within this process, the person can play a double role. On the one hand, he or she may cause deviations from the normal condition; on the other hand, he or she may intercept those deviations. With respect to the accident genesis, it can be noted that the later the measures for stabilization are introduced, the less time is available to complete them successfully. Thus, with increasing instability of the term "condition" in Fig. 19.9, the complexity and difficulty of necessary actions for a system stabilization increases, while the available time for the realization of these actions decreases. Both lead to an increase in the physical and mental load on the driver.[501]

In the previous model, the person was only considered a controlling element. The person as a system of inclinations, based on experiences, cognitive processes, actions, etc., for coping with demands, was so far not explicitly investigated.

The model according to Hale and Hale,[256] shown in Fig. 19.10, reveals the interior processes of the person and the corresponding information processing that occurs during coping with a traffic situation. In this model, it is presumed that an accident occurs if the driver is unable to handle a situation successfully. For example, this can happen when either the information available to the

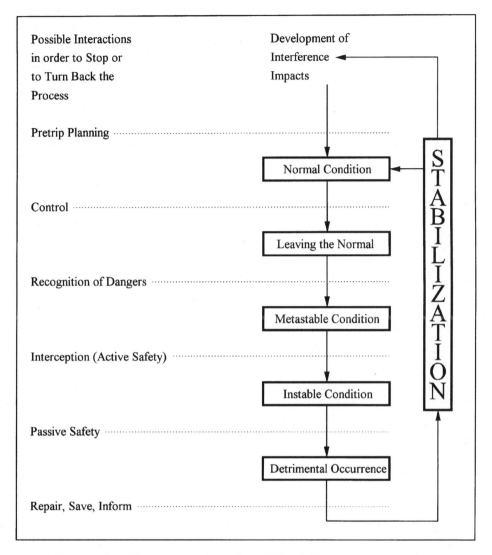

FIGURE 19.9 Flowchart of the accident model according to McDonald.[501]

driver is incomplete or incompatible with what is expected, or the driver's goals, preferred risk, and benefit, as well as available skills and experience, cause decisions which may lead to inappropriate actions in driving behavior.

The model in Fig. 19.10 shows a relationship in the form of a closed loop and describes the interplay with the information which is perceived and processed by the human being. At each step of the information perception and decision-making process, several possibilities exist to influence, infere with, or even impair this process and to corrupt the sampled data.[256]

Step 1 (Fig. 19.10).   Before a driver is able to make appropriate decisions and actions, information is needed about the particular traffic situation. The highway and environment serve as the primary information source. Secondary information can be drawn from traffic signs and other driver information systems. But the information which is available to the driver can be incorrect or incomplete.

The other possible information source may come from the driver's expectations about the subsequent development of the traffic situation and other available information sources which are based on the driver's experience and stereotype. Normally, this anticipatory behavior is used for preprocessing a set of possible follow-up actions. In the case where expectancies are violated, these considerations can distract the driver's perception in such a way that certain aspects of a situation are missed or overemphasized.

Step 2. The perceived information comes from two sources: the available information which exists outside the person and the expected information which is present internally. Available information can be limited or distracted by physical deficiencies such as reduced vision or hearing. On the other hand, fatigue, stress, and drugs or alcohol can have such an influence that drivers believe they see objects which do not exist or do not see objects which should have been visible (see Fig. 19.10, Step 2).

Step 3. After perception and recognition of the situation, the driver has to decide which actions should be performed and in what time sequence. Figure 19.10, Step 3, shows a predecision or preprocessing task during which the driver selects appropriate actions according to goals, plans for achievement, and a repertoire of available skills. Furthermore, the selection process is influenced by basic rules, training, experience, and innate capabilities which guarantee a high probability of success by performing this particular action in an appropriate way.

However, this complex process also shows clearly that, at this step, many different factors can lead to errors. On the one hand, errors can arise from the selected method for coping with the situation. On the other hand, the available capabilities and skills of the individual may limit the scope of the decision. The capabilities and skills of a human can be critical during the last seconds of a driving process for the difference between injury (fatality) and recovery, for example, from skidding or an impending collision.

Step 4. If the alternatives for action are clear, the decision about the actual handling are influenced additionally by cost-benefit considerations. In this connection, the potential risk is set versus the potential success (see Fig. 19.10, Step 4).

The subjective balance of risk and benefit has a major influence on determining driving behavior. Assuming nearly constant physical capabilities of the drivers, different subjective levels of risk taking and assumed benefit can cause differences in driving behavior between drivers traveling at medium speeds and those taking high risks, for example, in passing or traveling at speeds that are too high in curves (also see Sec. 19.7).

Motivation, experience, and innate behavior determine, for a specific situation, which level of risk is appropriate for the individual driver with respect to the assumed benefit. The higher the benefit, the higher the risk taking.

Step 5. The quality and effectiveness of the action performed depend strongly on the correctness and completeness of the previously processed information and the decision making of each step with respect to a specific situation. Errors in perception and reception can result in inappropriate decisions, which lead to unfavorable actions causing dangerous driving behavior or even an accident. Normally, the occurrence or failures with respect to missing an information source can be attributed to both the human being, for example, the inexperienced driver, and the situation, for example, deficiencies in highway design or traffic operation.[256]

The model of accident genesis shows that human decision making is very complex and depends on a variety of influencing factors. Most possible errors are cognitive in nature. To prevent those errors, the other components of the highway system have to be coordinated in such a way that they match the expectancies of the driver and are easy to perceive and are applicable for standardized actions with high cost-benefit ratios. This can be achieved by consistent, sound alignment design, for example, according to safety criteria I to III.

The model of accident genesis shown in Fig. 19.10[256] considers, to a far larger extent, the human factors rather than the accident model according to McDonald[501] in Fig. 19.9. However, it

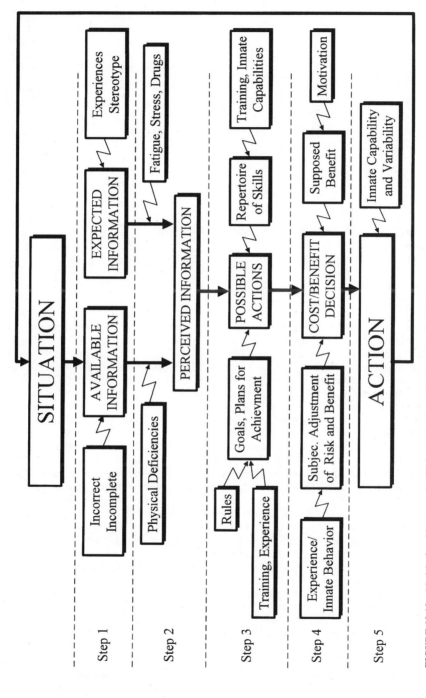

FIGURE 19.10 Model of accident genesis. (Elaborated and modified based on Ref. 256.)

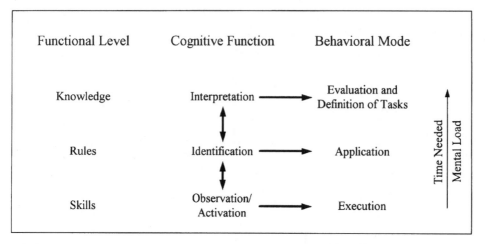

FIGURE 19.11   Model of functional levels. (Elaborated and modified based on Ref. 592.)

does not consider that some behavioral sequences are based on conscious actions, while others are handled more or less automatically and unconsciously.

Rasmussen[592] developed (Fig. 19.11) a model of three different functional levels according to the automation of actions. At the skill-based level, incoming information results directly in an automated behavioral answer which is executed without conscious control.

If no automated behavioral mode is present or if a person has to select between several possible behavioral modes, the process is transferred to the rule-based level, where an appropriate behavioral mode can be selected from the "memory" and executed.

If no appropriate rule exists or if the driver does not want to use available rules, the process has to be transferred to the knowledge-based level, where in-depth considerations about the solution of the problem take place.

The extent of thought operations, the necessary time, and the mental load increase with increasing functional level, which can be decisive in the case of an accident.

The driver acquires capabilities and skills during driving practice which can be executed automatically on the skill-based level. The assumption is that this is a learning effect which occurs only after multiple experiences with the corresponding sequences of events. Coping with unknown demands can only be conducted on the knowledge-based level (Fig. 19.11), which means the action is correspondingly slow because of increased mental load. According to Sievert,[648] the driver is only able to act on a rule or skill-based level in the area of driving experience, as presented in Fig. 19.12.

The area of experience is presented in Fig. 19.12 with respect to lateral and longitudinal vehicle accelerations.

Outside of this area, uncommon physical behavioral modes exist with respect to vehicle-handling characteristics such as accelerations, forces, torques, etc. According to McDonald,[501] the vehicle moves, in such cases, from the metastable to the instable condition (Fig. 19.9). These changed conditions can be handled only on the knowledge-based level according to Fig. 19.11. If the mental workload becomes too great or exceeds the performance capabilities of the driver, it leads to a collapse of the system and, consequently, to an accident.

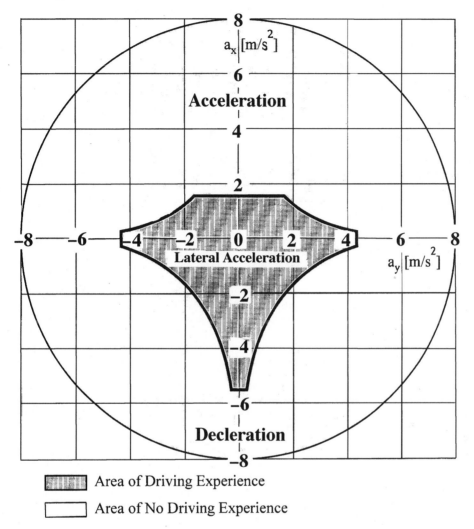

Area of Driving Experience

Area of No Driving Experience

FIGURE 19.12   Area of experience and inexperience for the normal driver. (Elaborated based on Ref. 648.)

19.5 *EXPECTANCY**

Expectancies are formed by the driver's experience and training. Situations that generally occur in the same way, and successful response to these situations, are incorporated into the driver's store of knowledge.[5] *Expectancy* relates to a driver's readiness to respond to situations, events, and information in predictable and successful ways. It influences the speed and accuracy of driver information processing and is a major factor in design, operation, and traffic control. Aspects

*Elaborated based on Refs. 5, 359, and 489.

of the highway system that agree with commonly held expectancies facilitate the driver's task. Expectancies that are violated, on the other hand, lead to longer reaction time, confusion, inappropriate responses, and errors.

Expectancy affects all task levels. *Control expectancies* pertain to vehicle handling and response. *Guidance expectancies* involve highway design, traffic operations, hazards, and traffic control devices. *Navigation expectancies* affect pretrip and in-transit phases and relate to routes, service, and guide signs.[489]

Firm expectancies help drivers to respond rapidly and correctly. Unusual, unique, or uncommon situations that violate expectancies may cause longer reaction times, higher workload values, inappropriate responses, or errors. The design features of the highway are sufficiently similar to form expectancies with respect to common geometric, operational, and road characteristics. Designing in accordance with predominant expectancies is one of the *most* important ways to raise performance. Unusual or nonstandard design should be avoided. Special care should be taken in transitions from one design element to another. (This requirement can be completed with the appropriate application of the previously developed safety criterion II.) When drivers perceive the information, they predominantly expect from the highway their performance will be error-free. If their expectancy is violated, that means drivers do not get what they expect *or* get what they do not expect, it can result in erroneous behavior.[5] Safety criteria I to III were developed to avoid expectancy violations in highway alignment design in the future in order to achieve design and operating speed consistency as well as driving dynamic consistency.

A Priori Expectancy[359,489]   *A priori expectancies* are widely held expectancies which most drivers bring into the driving task from a lifetime of experience and habit. Unusual geometric features, for example, one-lane bridges, features with unusual dimensions, for example, very long and/or very sharp horizontal curves, and features combined in an unusual way, for example, an intersection hidden beyond a crest vertical curve, may violate a priori expectancies.

Ad Hoc Expectancies[359,489]   *Ad hoc expectancies* are short-term location and site-specific expectations that drivers formulate based on site-specific practices and situations encountered while traveling in a given geographic area. Geometric features whose dimensions differ significantly from upstream features, for example, a horizontal curve significantly sharper than preceding curves, may violate ad hoc expectancies.

Design and Operation Considerations   Both a priori and ad hoc expectancies should be considered in design and operations. Appropriate expectancies should be reinforced and expectancy violations eliminated through the use of consistent, standardized designs and appropriate uniform traffic control devices.[489] However, consider in this connection the classification of good, fair, and poor design practices for safety criteria I to III, based on accident research according to Tables 9.10, 10.12, and 18.14. That means that traffic control or warning devices do not automatically guarantee safety improvements, especially if poor design practices exist.

Consistency should be maintained within and between jurisdictions, and it should be recognized that upstream practices affect downstream expectancies.

19.6 PRIMACY*

Primacy addresses the relative importance of competing information both between levels of performance and within a given level when information competes for a driver's attention or when performance of one activity distracts attention from information for another.

Primacy is assigned on the basis of the consequence of not, or not exclusively, paying attention to an information source. When two or more information sources or information needs have to be satisfied, the one(s) with the highest primacy should be processed first.

---

*Elaborated based on Refs. 5, 122 and 489.

For example, among the different levels of Fig. 19.2, control assumes the highest primacy, followed by guidance and navigation. The primacy concept suggests that information on the control level must be processed completely before processing at the guidance level can begin. Further, this level must be satisfied before navigational information will be processed. With respect to accident investigation, it can be reported that "loss of control" contributes most to accidents, followed by "failure to track the road," and "getting lost" the least contributing factor. Similarly, when two or more activities at the same level have to be satisfied, the activity with either the greatest threat and/or the closest proximity assumes the higher primacy. Accordingly, the design should focus the driver's attention on the safety critical elements or high-priority information sources.

The importance of this concept in safety design is that maintaining a high level of lane identification and demarcation, as well as preventing optical illusions and providing clear sightlines and good visual quality, reduces the demand for the amount of time to process information on the control level, and, thereby, increases the time available to process guidance and navigational information. This, in turn, reduces the potential for erratic maneuvers and the accident potential.[5,122,489]

19.7   DRIVER ATTRIBUTES AND POPULATIONS

19.7.1   General Comments

Attributes[489]   The driver population in each country encompasses a broad spectrum of attributes. For example, there are somewhat more male than female drivers. They range in age from 15 to over 80, in education from less than high school to college, and in training and experience from novices to professionals.

Capabilities[489]   Sensory-motor capabilities also vary, with a considerable range in vision, hearing, and reaction time. Most drivers need glasses to correct visual acuity, approximately 8 percent of the male population is color weak, and the majority of the older drivers (>65) experiences some degree of visual, hearing, and processing impairments that worsen with age.

Populations[489]   Drivers often belong to a unique segment of the overall population by virtue of their vehicles (for example, trucks or motorcycles), their age (for example, young or old), or their language fluency (for example, English or Spanish in the United States). These populations' special attributes can affect a facility's safety and operational efficiency, particularly when they represent a large portion of the traffic stream and when they require special treatment or information.

19.7.2   Novice Drivers*

The highest accident risk and severity is represented by the age group 15 to 24 years (see Sec. 20.2.5 and Figs. 20.7 and 20.8). Inexperience and the willingness of younger drivers to take high risks are found to be the main causes for this fact.

Cohen[121] conducted an investigation into accident risks and accident causes of novice drivers. In this connection, it is common knowledge that training improves skills and capabilities and, thus, accident risks decrease up to a certain level. For example, Fig. 19.13 shows a typical relationship between years of employment and accident frequency.

Similar to other activities, driving is a learnable sensory-motor activity. Therefore, with increasing experience, drivers should tend to be more reliable (Fig. 19.13). But with respect to road traffic, these basic findings are only partially true. Generally, it can be stated that with increasing experience over the years the accident frequency will decrease but in road traffic, an increasing accident frequency exists during the first 2 years of driving experience.

---

*Elaborated based on Ref. 121.

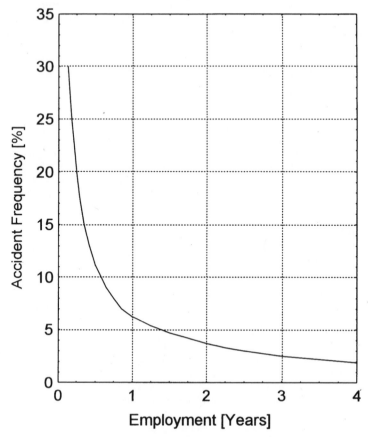

FIGURE 19.13   Work accident risk. (Elaborated based on Ref. 121.)

Cohen[121] designated four phases of risk exposition in road traffic (Fig. 19.14):

1. Novice drivers
2. Practiced drivers
3. Experienced drivers
4. Routine drivers

As in other fields of practice, in the first phase (up to 1 year of driving experience), the novice driver shows a high accident risk due to the lack of driving experience. In contrast to the trend shown in Fig. 19.13, it should be noted that, in road traffic, an increase in accident frequency can be expected for up to 2 years of driving experience (Fig. 19.14). This time period characterizes the second phase of risk exposition in road traffic. As already mentioned, the accident frequency with 2 years of driving experience is higher than a few months after receiving a driver's license.

With the beginning of the third phase, after 2 years of driving experience, a decrease can be seen in accident risk and after 6 to 10 years, a fourth phase starts, characterized by constantly low accident risk, which continues up to the senior age.

The preceding considerations reveal that accident risk in road traffic depends not only on learning but also on sensory-motor deficiencies which decrease slowly with increasing age. That

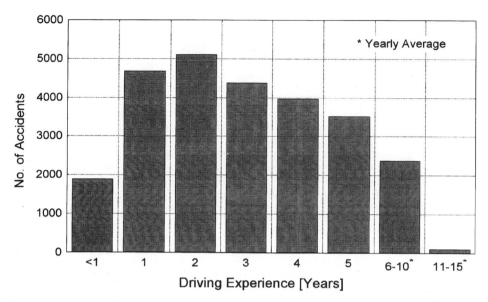

FIGURE 19.14   Accidents in road traffic with respect to driving experience. (Elaborated based on Ref. 121.)

means decreasing physical and mental skills and capabilities starting at an age of approximately 25 years. Both findings contradict the observed accident behavior.

The previous research revealed that no simple solutions exist because accident statistics reflect the result of an integrated effect of many influences which may be superimposed by stochastic events, for example, driver error in an unfavorable traffic situation.

If human reliability depends, to a large extent, on learning and sensory-motor capabilities, other contrary stimulating effects must exist, which lead to the observed behavior. Increased risk taking, decreased risk cognition, and personality-related issues, such as emotions, etc., cannot be used to explain the increasing accident frequency in the second phase because they are effective in the first phase too.

This phenomenon can be explained only in the context of perceived risk and estimated capabilities. Cohen[121] designated three spheres of risk exposition through which the endangerment of the novice driver and the related high accident frequency can be explained.

These three spheres of risk are

- Self-assurance
- Sensory-motor functions
- Visual orientation

Self-Assurance   Driving behavior is an optimizing process of balancing opposing demands. On the one hand, there is safety, which is demanded from all drivers, and, on the other hand, there are time, speed, comfort, and others.

In the optimizing process, the driver takes some risks into account in order to have more comfort, speed, and available free time. According to the risk compensation theory (Wilde[753]), every driver tends to maintain a constant, tolerable, individual level of risk.

With increasing driving skills, the driver notices decreasing risk and attempts to take advantages of this through faster and more risky driving to find his or her own limitations. Increased safety resulting from better driving skills will be compensated and overcompensated for by riskier and faster driving. The driver's subjective estimation of the safety before and after may be the same because the objective risk results from the actual endangerment and the individual ability to cope.

Inexperience and increased risk taking lead to high endangerment for younger drivers. In the second phase, self-assurance increases this trend, which results in a high accident frequency (see Fig. 19.14, 2 years of driving experience). Thus, the achieved safety profit is compensated or overcompensated for by increased risk taking as a result of self-assurance. The learning process lasts for a long period and continues to have an effect on driving behavior. The reason may be delayed feedback from dangerous situations because the lesson which each driver learns can be different. Assuming success in coping with such situations, the driver may believe that his or her success results from better driving skills which will further increase and lead to additionally increased performance in the future. This behavior is erroneous in that no mechanism for preventing endangerment and decreasing risk can be considered in advance.

In this context, the achieved safety improvement by increased driving skills is considerably lower than the increased risk by self-assurance (see also Secs. 19.3 and 19.9).

Beginning with the third phase, the feedback process takes effect and, supported by increased driving skills, the driving behavior changes from coping to preventing critical situations in advance.

Sensory-Motor Functions   Beginning at the age of 25 years, the sensory-motor performance will slowly decrease with increasing age, but, as several research studies reveal, the reliability of human actions increases up to the senior age. This fact can only be explained if a mechanism exists which compensates for this negative process regarding human performance. Sensory-motor functions, in other words, human orientation and action, are mentally connected via the brain. Attention directs which information is selected by the visual sense, how the input is processed, which consequences based on previous experiences result, and which actions are performed to reach the best results. It is common knowledge that, in certain situations, this process can be such a difficult task that the novice driver is not prepared.

Therefore, anticipated actions lead to better results than quickly executed actions without previous thinking. Only in very rare emergency situations are quickly executed and also exactly performed actions required.

Visual Orientation   In comparison to experienced drivers, novice drivers only have limited experience in efficient information selection. They have problems in recognizing and assessing the perceived information. Perception in traffic and efficient viewing have to be learned.

Perceptual learning increases the performance of the driver in three ways. Increasing perceptual experience:

1. Leads to a more effective localization of relevant information
2. Increases the driver's ability to select the appropriate driving behavior based on a clue (also see Ref. 524)
3. Develops a system of codes, which allows the driver to perform a highly efficient assessment of relevant features of the roadway and the environment

With increasing driving experience, knowledge will be increased, where possible information is located in the traffic space, and which relevance with respect to the driving task is assigned to each information source. Furthermore, the driver will be able to integrate this information more effectively in his or her internal representation of the environment, which leads to more appropriate actions and extends the number of alternative actions. After years of driving experience, the perceptual behavior of the novice driver changes to the task-related behavior of the experienced driver.

In conclusion, Cohen[121] revealed the following characteristic differences between novice and experienced drivers:

1. Novice drivers often fix on locations near their own vehicle rather than those further away. They give most of their attention to the control task because novice drivers incorrectly estimate

the exact position of their vehicle on the road. This results in neglecting other important tasks, such as recognizing dangers in advance, and leads to higher time pressure in handling nonanticipated critical driving situations.

2. Novice drivers use large eye movements and fix on objects with information of little relevance, for example, the road surface. This is because of their limited ability to select appropriate information. The determination of relevant information depends on individual cognition. Only with sufficient experience can the driver distinguish between relevant information and repetition and concentrate attention on relevant sources.

3. Inexperienced drivers do not properly adapt their visual search on the distribution of relevant information sources.

4. Novices do not anticipate future workload, for example, a sharp and difficult perceptible curve from a long distance. They process all relevant information when already negotiating the curve or shortly in front of the curve. This results in a high mental workload in the curve and a decreased capacity for unexpected traffic events. In contrast, decreased mental work leads to an increasing usable field of view, which gives the driver the ability to better recognize relevant objects and decreases the mental effort for the same task performance.[121]

The complexity of the information sampling process and its mental assessment makes it clear that novice drivers have problems in adjusting their visual behavior to the requirements of the road and traffic. This fact has to be considered in designing roadways and the adjacent environment. In this connection and with respect to older drivers, the exclusive use of the "normal driver" as the design criterion in most guidelines may lead to unexpected results in following the trend to reduce important design parameters with respect to modern highway design.

19.7.3 Older Drivers[489]

There is no accepted age for "old," as people age differently. However, "old" is usually considered to be 65+. Currently, 15 percent of the Americans and Europeans are 65+. Most older drivers retain their licenses and drive daily, although generally not to work and often not at night.

Age-Related Sensory-Motor Impairments   All drivers ultimately experience age-related sensory-motor impairments that vary from driver to driver. These include a gradual loss of vision, hearing, and memory. Common problems include poor night vision and glare recovery, decreased visual acuity, hearing loss, increased reaction time, loss of short-term memory, and poor attention span.

Compensation of Age-Related Impairments   Older motorists compensate by driving slower, avoiding stressful situations, and relying on experience. However, they have a higher than average accident rate, and are often involved in multivehicle collisions at merges, unprotected left turns, and intersections (see also Sec. 20.2.5 and Figs. 20.7 and 20.8).

Measurements in Design to Aid Older Drivers   Methods to aid older drivers include improved sight distance, enhanced signs and markings, better maintenance of traffic control devices, lower speeds, and alternative transportation. When the percentage of older drivers in the traffic stream is greater than 15 percent, their diminished capabilities should be taken into account by designing for the older, rather than for the average, driver.[489]

19.8 HAZARDS

According to Lunenfeld,[489] a *hazard* is any object, condition, or situation which, when a driver fails to respond successfully, tends to produce a highway system failure. Noncatastrophic failure includes any unintended system inefficiency that does not result in an accident (for example, lost or confused drivers, erratic maneuvers, or traffic conflicts). Catastrophic system failures result in accidents.

Hazard Classification   Object hazards can be fixed or moving. *Fixed objects* include trees, light poles, and sign supports. *Moving objects* include pedestrians, animals, and vehicles. *Highway condition hazards* include horizontal and vertical curves, potholes, and lane drops. *Situation hazards* are combinations of object and condition hazards that may be transitory, such as a stopped queue over a crest vertical curve, and may include environmental elements such as rain or fog (for example, a slippery-when-wet curve in the rain).[489]

Hazard Amelioration   The most effective way to ameliorate the adverse effects of hazards is through their elimination. When this is not feasible, drivers should either be protected from the hazard or it should be made "forgiving." When elimination, protection, or forgiveness of hazards cannot be achieved in the short term, then drivers should be informed of their presence and threat so they can be avoided. A basic premise is that competent drivers, provided with appropriate information, can select or be given a proper hazard avoidance speed, path, and/or direction.[489]

Hazard Avoidance   Hazard avoidance is dependent on a driver's ability to search for hazards, detect their existence, recognize and identify them, determine their hazard level, decide on (or be given) an appropriate hazard avoidance strategy, and perform the requisite maneuver as part of a continual feedback process.

Similarly, appropriate hazard avoidance maneuvers can vary from stopping to avoid a pedestrian to steering around a disabled vehicle.[489]

Hazard detection and recognition varies with hazard type, visibility, consciousness, primacy, expectancy, and visual complexity. Common hazards, such as trees and vehicles, are relatively easy to detect and recognize. Others, such as potholes and inoperative traffic signals, may be very difficult to detect.

19.9 RISK COMPENSATION—DRIVER ADAPTATION*

19.9.1 General Comments

With the introduction of safety improvements on highways and streets, it often happens that after a considerable period of normal traffic operation, the safety benefit is smaller than expected or no longer exists because drivers adapt their driving behavior to the new conditions of the roadway and the environment.

This phenomenon can be described as a secondary behavioral adaptation of traffic participants to the primary effect of previously introduced safety measures. *Driver adaptation* is defined as an unintended behavior of drivers which may occur after introducing changes in the roadway system (Fig. 19.15).

Every change in one of the components of the driver-vehicle-roadway/environment system will influence at least one of the other components (see Secs. 19.3 and 19.4). If a driver perceives changes on the roadway or in the environment which may determine subsequent driving behavior, he or she will act according to this changed situation with optimized effort-benefit ratio (see Fig. 19.10). Therefore, behavioral adaptation can be understood as the driver's reaction to changes in the traffic system, which may either increase or decrease safety.[573]

Driver adaptation can be separated into three kinds (Fig. 19.15):

- Ad hoc adaptation
- Delayed adaptation
- Adaptations due to changes in risk exposure

---

*Elaborated based on Pfafferott, et al.[573]

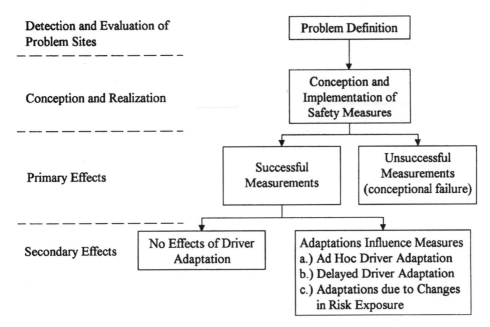

FIGURE 19.15 Conception, realization, and effects of safety measures.[573]

Ad hoc adaptations usually occur in situations which are already known to the driver. Therefore, an appropriate, already mentally preprocessed action can be chosen from the driver's individual behavioral repertoire (see the model of Hale and Hale in Fig. 19.10).

Example: Because of safety deficiencies, a speed limit sign has been installed at an interstate exit. However, the driving speeds are higher than the proposed speed limit, since many drivers know from previous experience that those speed limitations are normally assigned very conservatively.[573]

Delayed adaptations usually happen in situations which are not well known to the driver. The driver must first become familiar with the situation based on subjective risk estimations. This can only be achieved through a long learning process.

Example: After installing an automated speed warning device at an accident black spot, which shows the driver's actual speed, in case the speed limit is exceeded, a considerable safety effect was reached in the first year. But in the following years the safety effect diminished more and more because the automated speed warning device was not combined with law enforcement actions. Instead of, or in addition to, the intended effect, an unintended effect occurs. Thus, the drivers returned to their previous behavior.[573]

Adaptation as a result of changes in risk exposure occur, for example, if parents of small children, who undergo traffic education at school, allow them to go to school alone without supervision by their parents. In this case, the probability of having a traffic accident is much higher than if they were escorted by their parents.

19.9.2 Stimulating Effects of Highway Safety Measures Due to Driver Adaptation

The mental process which leads to driver adaptation after introducing safety measures has not yet been fully explored. However, based on evaluation of empirical data, effects on highway safety measures can be determined which may stimulate driver adaptation.

In general, it can be stated that the following features have a stimulating effect on the occurrence of driver adaptations:[573]

1. *Possibility of interaction with the effects of safety measures:* The perception of the effects of safety measures is required for adaptation of drivers on changed roadway/environment conditions. If the driver is able to influence the outcome of safety effects, then she or he is able to interact with the measure. In this case, the drivers have the possibility to adjust their behavior for the lowest effort and highest benefit.

2. *Immediate feedback experience:* This effect is especially likely in the case where the feedback of the drivers' behavior occurs immediately. For example, if a driver negotiates a curve with a higher speed than posted and nothing happens to advise him or her that this behavior is critical or inappropriate, the driver will try this again. Thus, immediate feedback reduces the delay in driver adaptation.[573]

3. *Possibility of extending their own limits:* If changes in the system driver-vehicle-roadway environment give drivers the ability to extend their own limits, for example, higher speeds or reducing uncomfortable speed changes, the driver will usually accept a higher risk. In this case, the probable driver adaptation will influence safety negatively.

4. *Raising the perceived subjective safety level:* Behavioral adaptations with decreasing safety effects are more likely in cases where the driver perceives an increased subjective safety level, while the objective safety level remains at the previous level or does not increase as much as is perceived. The driver convinces himself that higher performance capabilities provide the ability to cope with more critical situations (see Sec. 19.7).

5. *Experience of performance and fun:* Driver behavior depends on drivers' motivation, driving skills, or actual experiences. The more safety improvements allow performance and "fun" in driving, the higher the probability of decreasing safety effects due to driver adaptation.[573]

19.9.3 Behavioral Models to Explain Driver Adaptation

Driver behavior with respect to driver adaptation can be explained in the context of several driving behavioral and driver accident models. It is necessary that those models consider the safety effect of driving behavior and driver adaptation.

The Model of Subjective and Objective Safety[335]   Adaptation can be interpreted from the psychological contradiction between the cognition of the objective and the subjective risk. Klebelsberg[335] differentiates between objective and subjective safety. Objective safety, for example, can be evaluated by the three developed quantitative safety criteria, while subjective safety is understood as the safety experienced.

It is assumed with respect to the model in Fig. 19.16:

1. Objective and subjective safety put one another into perspective in the actual situation. Therefore, a change in one of the two components could lead to a change in the balance of both components.

2. The situation in appropriate traffic behavior depends on the fact that objective safety would at least be greater or equal to subjective safety because, in this case, the corresponding physical margins are not surpassed.

According to Klebelsberg,[335] risk behavior depends on performance and safety aspects which are personified more through the current situation than personality. In the model of subjective and objective safety, it is not assumed that for an individual person a specific balance exists between the two components but from their own situational relation to one another.

For example, the improvement of the visibility at an intersection (higher safety) could lead to a higher speed at this intersection (through the increased subjective safety). The result could mean less safety on the behavioral level.[573]

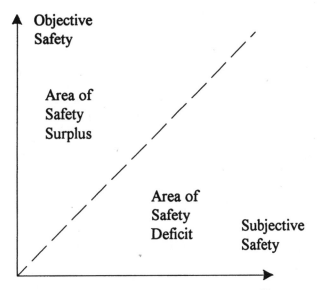

FIGURE 19.16 Model of objective and subjective safety.[335]

Theory of Risk Homeostasis[753]   Wilde[753] asks the question "Why do drivers accept a certain level of risk?" He hypothesized a risk homeostasis to try to explain this with the following issues (see Fig. 19.17):

1. Drivers constantly compare the level of subjective risk with the level of accepted risk, which corresponds to the personal attention or workload level.
2. If there are discrepancies between subjective and accepted risk, the individual tends to reduce or eliminate this gap.
3. The risk increases through the adaptation process.
4. The risk acceptance is assumed as an independent variable which determines the accident rate.
5. The summarization of the objective risk for all traffic participants for a certain time period (1 to 3 years) shows the accident rate.
6. The accident rate is assumed to be constant over the observed time period.

According to Wilde,[753] the cognition of the risk depends on the perceptibility of the risk.

In contrast to the previous model in Fig. 19.16, Wilde[753] predicts, precisely, that if measures are taken which reduce the objective risk, this would not increase the safety as long as the accepted risk is not also reduced at the same time.

The law of risk homeostasis was not produced for the individual but for the social system of the driver population. And through this the theory reaches the limits of the possibility to disprove it because general assumptions cannot be precisely formulated.[573]

Theory of Risk Behavior[533]   In the "zero-risk theory of driver behavior," it is assumed by Naatanen and Summala that the driver can control the risk through a simple pattern of situations in road traffic and can generally avoid behavior which calls for fear or for the anticipation of fear. Furthermore, it is assumed that the drivers try to satisfy their mobility needs, for example, driving fast, as long as there is no apparent accident risk.

In this connection, special relationships between stimulating and inhibiting factors of the driver behavior have to be differentiated. The most important factor of the inhibitory motives is the subjective risk (Fig. 19.18).

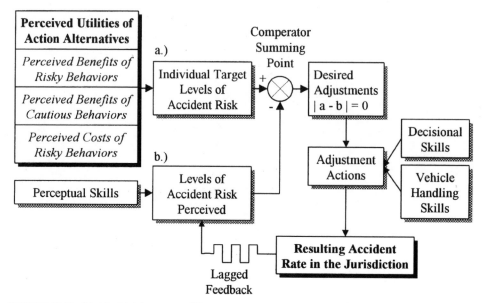

FIGURE 19.17   Model of risk homeostasis.[753]

Naatanen and Summala hypothesized that:

1. The subjective risk as part of the cognition of danger is an important motivational factor in the behavior of the driver

2. The subjective risk is not sufficiently pronounced

According to Fig. 19.18, the control loop starts at the stimulus situation.[573]

Perception is an active, selective process, which is guided by the driver's motives and experience. The perception initiates expectancies and decisions are made with respect to behavioral changes, as far as they are necessary. In this model, it is assumed that accidents happen because the objective risk is too underestimated. The reason for this is the driver's overestimation of abilities. In addition, perception also plays a role: The estimation of speed is often erroneous, and depending on the situation, often too low and the physical forces which occur in the case of an impact are often underestimated. This tendency is supported through the subjective, seemingly easy driving task. Thus, the main thesis is to weed out the tendency of drivers to select higher speeds, since the lack of risk cognition leads to risky driving behavior.[573]

19.9.4   Preliminary Conclusion

It was intended to reveal general mechanisms from a theoretical view regarding behavioral adaptations of drivers with respect to safety measures. A comprehensive theory to describe the adaptation process could, under specific circumstances, be the basis for a risk homeostasis model. However, for this, the problem of determining when the balance is reached would have to be solved. The theory should not be defined on the basis of the accident situation; in contrast,

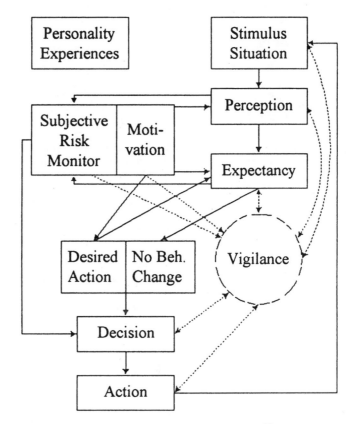

FIGURE 19.18   Flowchart for decision making by drivers.[533]

it should predict the accident level. From a theoretical viewpoint, we can make the following conclusions:

1. Adaptation mechanisms are phenomena which are observable for the driver and which have a positive or negative effect on traffic safety.

2. Adaptation mechanisms can be observed for different hierarchical steps of driving tasks and develop a feedback process as a function of time.

3. At the center of the theories which the adaptation mechanisms explain lies the concept of risk, and this has an objective and a subjective component.

4. Subjective risk lies in an interrelationship between the motivational components of the driver and the perception of the situation which the driver is in in the moment.

5. Adaptation mechanisms are aided through:

   Overestimation of one's own capabilities, as function of the perception of a situation
   Decisions to accept high risks
   The incapacity to evaluate the risk of a certain situation appropriately.[573]

19.10 DRIVER WORKLOAD AND MENTAL STRESS CONCEPT

19.10.1 General Comments

It is reported by Krammes, et al.[359] that workload is an important concept in the design of systems controlled by humans. Successful operation of such systems demands that the mental workload imposed on operators does not exceed their processing capabilities. On the other hand, it is also important that mental workload levels do not fall below levels where attention-to-task requirements become difficult. Systems should be developed that efficiently use the operator's workload capabilities. A model of workload is of greatest importance in the design phase of a system in order to predict which configurations will maximize performance efficiency and still leave operators some residual capacity to meet unexpected task demands.[769]

The Yerkes-Dodson law, illustrated in Fig. 19.19, describes the relationship between human mental workload and performance.

At low workload levels, the level of human performance is also low because information is missed due to inattention. As the workload of performing a task increases, the level of human performance increases until a maximum level is reached. The area around this maximum performance is the optimal workload for the given task.[359] Beyond this level, additional mental workload demands cause a dramatic decrease in performance because there is too much information to process[754] (see Fig. 19.19).

Since error rates generally increase as performance decreases, it can be expected that accident rates would be highest in those areas associated with either very high workloads or extremely low workloads. Furthermore, it is assumed that sudden increases in workload would also be associat-

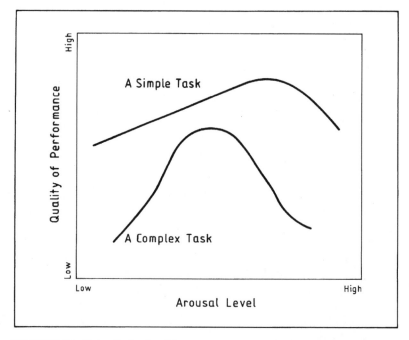

FIGURE 19.19   Yerkes-Dodson law.[789]

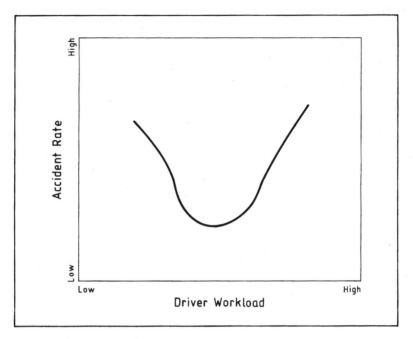

FIGURE 19.20   Hypothesized relationship between workload and accident rate.[789]

ed with increased accident risk. The general shape of the hypothesized relationship between work-load and accident rate is illustrated in Fig. 19.20.[789]

Definition of Workload   The driving task imposes work on the driver; this work varies greatly in task difficulty and task frequency. The level of this workload and its effects on driver performance would seem to be greatly affected by driver expectations and driver capabilities (see Secs. 19.5 and 19.7). Roadways with inconsistencies in their designs would be expected to violate driver expectancies and impose higher workloads on drivers. An understanding of the basis of workload and its impact on performance is desirable to analyze design consistency.

Workload has been defined by Senders[788] as a measure of the effort expected by a human operator while performing a task, independently of the performance of the task itself. Venturino[734] defined *mental workload* as an expenditure of mental capacity required to perform a task or combination of tasks.[789]

Another definition of *workload* was given by Knowles[787] as consisting of the answer to two questions: How much attention is required? and How well will the operator be able to perform additional tasks? Jex[313] defined this with other words as the operators' evaluation of the attentional load margin between their motivated capacity and the current task demands while achieving adequate task performance in a mission relevant context.

The definition presented by Knowles and Jex seems very appropriate to the driving environment, since it consists of many overlapping tasks, each requiring a portion of driver's attention. A method for examining the workload demands placed on the driver would appear to be a way of directly arriving at the capabilities of the driver as he or she negotiates a given roadway.

19.10.2 Measurement of Driver Workload*

Most techniques for measuring human mental workload assume that there is an upper limit on the human operator's ability to process information and to select appropriate responses.[284] Thus, systems should be developed that keep operators at a level of workload that forces them to pay attention to the given task while allowing enough excess mental capacity to handle unexpected emergencies.[359]

When selecting a technique for measuring mental workload, two properties of the techniques must be examined: (1) the sensitivity of the measure to changes in workload and (2) the intrusiveness of the measure of the primary task. A measure of workload must be able to accurately and consistently measure changes in workload. Otherwise, it cannot be used with any credibility. If results between subjects are inconsistent, then it is difficult to draw conclusions about the task in question. If a measurement technique intrudes upon a primary task, that is, diverts mental processing time from the primary task, then the results may reflect the combined workload of both the measurement technique and the primary task rather than the workload of the task in question.[359]

Subjective measures attempt to measure workload by having subjects rate their perceived workload on a given scale. The ratings represent the subject's conscious judgment regarding the difficulties encountered in the performance of the evaluated task. These measures are easy to obtain and have high face validity.[226] Some researchers argue that if a person says that he or she is loaded, then that person is loaded no matter what the behavioral and performance measures indicate. Different rating scales have been developed by many different agencies. The overall outcome of these efforts is confusing and, to some extent, disappointing. Human subjects have no difficulty in rating their experiences but the experimenter has the burden of selecting the appropriate scale. The most serious concern about subjective measures is the recurrent finding of a general dissociation between subjective estimates and objective indices of task performance.[226,359]

Another technique for measuring mental workload is the direct measurement of the primary task. If someone wants to know the workload associated with a certain task, that person should be able to directly measure the performance of someone performing that task. Gartner and Murphy[204] argued, however, that the sensitivity of primary task measures might not be adequate. A driver might show the same level of performance on two different curves but his or her actual workload on one curve might be considerably higher. Another problem with these types of measures is that they are difficult to perform in the field. A third limitation of the direct measurement of the primary task is that it can only discriminate between overload and nonoverload conditions; it cannot provide relative measures of workload in the nonoverload condition.[284] An advantage of the direct measurement technique is that the measure is nonintrusive. Only the behavior under study is measured. Another positive aspect of these types of measures stems from a study by Wierwille and Gutmann.[751] Their study indicated that by measuring five primary tasks instead of only one, the results were far superior in their reliability. This approach is case sensitive. Different measures must be developed for each experimental situation examined. The difficulty in generalizing different tasks must be considered when choosing methods for evaluating mental workload.[284,359]

According to Helander,[280] another possible way of obtaining knowledge about a driver's information processing is to use indirect evidence of nervous adaptation. The concept of activation level is used as a suitable index of behavioral intensity. A low level of activation implies a low performance level, a moderate level is reflected in a high quality of performance, and a high level of activation, due to an overflow of the information processing system, implies a collapse of the integrative functions which determine driver behavior.

A high level of activation or arousal may imply that the probability of obtaining a performance quality not adequate with the traffic situation will increase.

---

*Elaborated based on Krammes, et al.[359]

In the proposed theory, the properties of the environment are inherent in a concept called *environmental complexity*. The perceptual complexity is the driver's interpretation of the environmental complexity. It varies from driver to driver and depends on the driver's mental and physical state. Thus, the concept describes the total influence of the environment on the driver.

The level of perceptual complexity determines the activation level in the driver which, in turn, has implications for both the perceptual processes and the driver's decision-making process. The latter is manifested through control operations influencing the movement of the vehicle, which, in turn, influences the environmental complexity (see Secs. 19.3, 19.4, and 19.7).

In this theoretical system, some of the concepts can be quantified. As suggested, the level of activation can be assessed using correlates as skin conductance reaction (SCR), heart rate, muscle tension, and eye blink rate. As indicators of the driver's control operations, measurements of the brake pressure applied and the movements of the steering wheel can be used. Furthermore, the vehicle's behavior is quantified using indicators such as velocity and acceleration.[280]

But successful physiological measurement of operator workload has, unfortunately, been much easier to conceptualize than to achieve.[760] These types of measurements are unintrusive and can be taken as the operator is performing a task. A new method of measuring workload by physiological means is through the use of cardiac arrhythmia. Measurements of cardiac arrhythmia are calculated by measuring the duration of interbeat intervals and converting each into an estimate of instantaneous heart rate. The variance of these instantaneous heart rates is a quantification of cardiac arrhythmia. Results from tests using cardiac arrhythmia have shown a marked improvement over other physiological tests.[284,359]

19.10.3 Driver Workload Homeostasis

Driver workload homeostasis can be described as the unintended behavior of drivers to keep their workload in an optimum range. Based on driving experiences, the driver knows that overload, as well as underload, increases the probability of error. If behavior tends to be error-free, an optimal workload level has to be maintained. It can be reached by a consistent, sound alignment flow without sudden changes in driver expectancy. To satisfy such a requirement, the good design level of safety criteria I and II has to be fulfilled.

Zeitlin[784] reported, based on a 4-year study of driver workload on rural and urban roads in New York State, that indications of workload homeostasis were evident as drivers appeared to modify their performance to keep workload within a comfortable range. There is strong interference of a form of workload homeostasis operating in the operator controlled, self-paced driving task. Thus, the driver is able to control enough parameters of the driving task to hold the workload at an acceptable level. The driver may vary speed, modify attention to information sources (road signs or the radio), or modify participation in nonessential activities (conversation or smoking). Drivers were observed trying to increase their overall workload under easy driving conditions (speeding) nearly as often as they tried to decrease their workload under difficult conditions. Similar results could be found by Godthelp and Van Winsum.[221,731]

With respect to highway safety, it can be concluded that after the occurrence of a highway feature which imposes a higher workload on the driver, as the driver feels comfortable, he or she will compensate for this effect by reducing speed and driving more carefully.

If this premise is true, the first critical feature of a sequence will impose a higher workload on the driver than subsequent equally critical features. For example, if a sharp curve follows a gently designed alignment, this sharp curve will have an effect on the inconsistency. High workload on drivers and increased accident values may result. But if the successive curves have the same sharpness, the driver already decelerates before negotiating these curves, since their sharpness could be expected. Accordingly, lower workload values will occur. Accident experience and speed consistency measures reveal the same general tendencies. If the separation distance between two subsequent inconsistencies is too short, cumulative or carryover effects of workload can occur.

19.10.4 Driver Workload Based on Highway Safety Evaluation Procedures*

According to Wooldridge,[789] roadway designers are faced with many choices in the design or rehabilitation of a roadway. The designer must meet or exceed the requirements placed by engineering guidelines and standards, in some cases choosing which requirements will be met. The designer must then request design exceptions for those requirements that are not met. One of the requirements placed on the designer is to meet drivers' expectancies.[5] Given the vague guidance provided in highway design, most of the designer's attention is usually directed toward the clear-cut requirements for discrete elements of the design, neglecting an overall examination of the driving environment. When today's designer attempts to reconstruct segments of old routes as needs dictate (accident frequency and severity) and money becomes available, attention must be placed on the issue of driver expectancy so that inconsistencies are not built into the highway system.[789]

Highway designers are vitally concerned with building the most efficient, most cost effective, and safest highways possible. However, to increase the safety of a highway, the engineer must know which portions of the roadway merit improvement. With respect to these considerations, safety criteria I to III were developed in Chaps. 9 and 10 to evaluate good and fair design practices and to detect poor ones, quantitatively.

Little need would exist for accident study and analysis if drivers could readily assess the risks that they encounter as they drive. However, it is common knowledge that wide discrepancies exist between the driver's subjective perception of risk that they encounter as they drive and the objective risk level (also see Sec. 19.4). In some specific situations (see Sec. 19.3.2), drivers do not appear to be capable of accurately assessing the risks they encounter in driving along the roadway, and they are not able to adequately modify their driving behaviors accordingly.[789]

A method for examining the workload demands placed on the driver would appear to be a way of evaluating the capabilities of the driver as he or she negotiates a given roadway.

Methodology for Evaluating Geometric Design Consistency[512]   A method for evaluating driver workload was presented by Messer.[512] By gathering empirical evidence about driver expectations of roadway features and relating violations of those expectancies to workload, a model was formed. The model is based on the presumption that the roadway itself provides most of the information that the driver uses to control his or her vehicle; hence, the roadway imposes a workload on the driver. This workload is higher during encounters with complex geometric features and can be dramatically higher when drivers are surprised by encounters with unexpected or unusual combinations and sequences of geometric features.[789]

This methodology is applicable to all rural nonfreeway facilities. Because of reliance on subjective ratings, expert opinion, and limited empirical evidence, the procedure can be used as a methodology for evaluating the geometric consistency of rural highways. The Messer driver workload procedure[512] quantifies design consistency by computing a value for driver workload WL_n. The technique relies on a set of assigned ratings for various roadway elements.[789] In the following, different factors and impacts are discussed to calculate driver workload values according to Eq. (19.2).

Workload Potential Ratings R_f[512]   The probability that a particular geometric feature may be inconsistent in a particular situation depends on numerous factors. The more influential factors that relate to the feature itself include the following:

1. Type
2. Relative frequency of occurrence
3. Basic operational complexity and criticality in the driving task
4. Overall accident experience in general

---

*Elaborated based on Messer.[512]

Other important design variables that will affect the apparent criticality of a feature include:

1. Time available
2. Sight distance to the feature
3. Separation distance between the features
4. Operating speed
5. Prior roadway design features

Driver familiarity, traffic, topography, and roadside environmental effects will also influence the resulting criticality of the individual feature.

Average criticality ratings R_f were developed for nine basic geometric features by using a seven-point rating scale for identification of hazardous locations based on driver expectancy considerations. Within this rating scale, 0 is no problem and 6 is a critical problem.[512] The results are presented in Table 19.3.

Sight Distance Factor S[512]   In order to provide feature visibility in proportion to criticality of workload rating, the designer should seek to provide as much sight distance as is practicable on roadways that approach geometric features. The greater the workload rating, the greater the sight

TABLE 19.3 Geometric Feature Ratings R_f for Average Conditions on Rural Highways*

Geometric feature	Two lanes	Four lanes
Bridge		
Narrow width, no shoulder	5.4	5.4
Full width, no shoulder	2.5	2.5
Full width, with shoulders†	1.0	1.0
Divided highway transition		
Four lane to two lane		4.0
Four lane to four lane		1.8
Intersection		
Unchannelized	3.7	2.4
Channelized	3.3	2.1
Railroad grade crossing	3.7	3.7
Shoulder width change		
Full drop	3.2	2.1
Reduction	1.6	1.0
Alignment		
Reverse horizontal curve	3.1	2.0
Horizontal curve	2.3	1.5
Crest vertical curve	1.9	1.2
Lane width reduction	3.1	2.0
Crossroad overpass	1.3	0.8
Level tangent section†	0.0	0.0

*Elaborated and modified based on Ref. 512.

†Assumed.

Note: Four-lane divided highways are usually assumed to equal 0.65 of two-lane highway design highway ratings. Value system: 0 = no problem, 6 = big problem.

distance needed. Adequate sight distance and feature visibility provide the amount of time unfamiliar drivers will need to correct false expectations, decide on the appropriate speed and path, and make the required traffic maneuver.

The maximum sight distance to each feature should be estimated by using the measurement criteria for stopping sight distance (see, for example, Figs. 15.3a and 15.4). A motorist can be assumed to look through features to see other features downstream if they are visible. The midpoint of the curve or the obviously most critical location of the feature should be used as the target point, including the determination of separation distances. After estimating the 85th-percentile speed (see, for example, Fig. 8.12, Table 8.5, or Fig. 18.5) and the measured maximum sight distance, the sight-distance adjustment factor S can be determined from Fig. 19.21. This factor adjusts rating values from average speed and sight distance levels to specific site conditions.[512]

Carryover Factor C[512] Generally, it is necessary to avoid creating compound features. A compound geometric feature contains two or more of the basic geometric features listed in Table 19.3 at the same location or with a separation distance between the centers of the adjacent features of 457 m (1500 ft) or less. The distance can be reduced to 305 m (1000 ft) where 85th-percentile speeds are less than 80 km/h or the compound feature is composed of two of the lower-valued basic features (<2.0). Tangent sections are excluded from compound feature analysis.

Compound features should be eliminated by an appropriate separation distance to provide the unfamiliar motorist with enough time to recover from the experience of driving into the first unex-

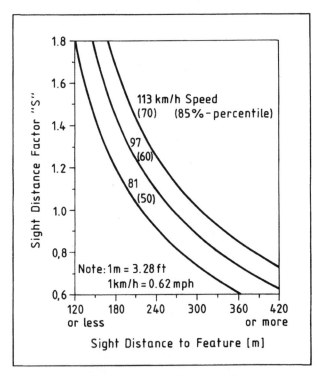

() = 85th-percentile speeds in mph

FIGURE 19.21 Sight distance factor S due to sight distance to feature as related to the 85th-percentile speed.[512]

$(CCR_s \approx 110$ gon/km) greater than the preceding horizontal curve may be considered not similar to the previous curve. Flatter curves are always similar when immediately following a sharper curve.[512]

Driver Unfamiliarity Factor U[512]   The higher the percentage of motorists' unfamiliarity with the highway, U, the higher the probability of drivers being surprised by relatively unusual geometric features. It is assumed that the percentage of unfamiliar drivers on higher classified roadways is higher than on low classified roadways, as shown in Table 19.4.

TABLE 19.4   Determination of Driver Unfamiliarity Factor*

Classification system (rural)	Road category†	Driver unfamiliarity factor U
Principal arterial	AI	1.0
Minor arterial	AII	0.8
Collector road	AIII	0.6
Local road	A IV/V	0.4

*Elaborated and modified based on Ref. 512.

†Adapted to the classification system of Table 6.2.

Calculation of Driver Workload Value WL_n and Level of Consistency[512]   The evaluation of the potential for the geometric feature to be inconsistent by using this procedure is based on the calculated driver workload value, WL_n. WL_n refers to the workload value being calculated for the actual feature, whereas WL_1 refers to the workload value for the previous feature. The workload value, WL_n, is determined from the following equation, which uses the previously described factors and the workload value calculated for the last feature, WL_1:

$$WL_n = U \times E \times S \times R_f + C \times WL_1 \qquad (19.2)$$

where WL_n = driver workload value of the actual feature
  WL_1 = driver workload value of the previous feature
  U = driver unfamiliarity factor
  E = feature expectation factor
  S = sight distance factor
  R_f = workload potential rating for average conditions
  C = carryover factor

The previous factors have been combined to provide information that could indicate which unexpected geometric features are creating a problem. But at what value of workload this conclusion can be drawn is actually a subjective decision. However, in an effort to standardize the process and to allow relative comparisons, the reasonable criteria are presented in Table 19.5.

As revealed in Table 19.5, it can be concluded that a $WL_n > 6$ is defined as an apparent geometric inconsistency. This table can be used to estimate the level of consistency, LOC_n, assigned to the calculated workload value.

In highway design the procedure can be used to minimize both the absolute level of workload and also the jump in workload between features.[512]

With respect to alignment consistency and the developed design classes, as well as the safety criteria in this handbook (Table 11.1), the levels of consistency A, B would correspond to "good design" levels, C to E to "fair (tolerable) design," and level F to "poor design" practice.

pected feature before cognition of a subsequent surprise featur
needs to recover, perceive, and react to a subsequent unexpe
from 5 to 10 s. Assuming a viewing and maneuvering distan
range, the overall separation time of 10 to 20 s is desired.

If the separation distance from the last feature and the 85th-
example, Fig. 8.12, Table 8.5, or Fig. 18.5), the workload carr
from Fig. 19.22. This factor adjusts conditions from isolated
stances. It takes into consideration driver memory loss, decisio
average viewing distances (not sight distances) used in the dri

Feature Expectation Factor E[512]   As revealed in Sec. 19.5,
according to what the drivers expect. Drivers tend to build up
ing roadway will be like based on their previous driving experi
basically unexpected at any location because of their limited
design. Other features, such as horizontal curves and interse
design attributes or operational demans that are unexpected.

The feature expectation factor, *E,* adjusts for the potential
where the prior feature is similar to the current feature. If the
ture, the formula

$$E = 1 - C$$

for calculating the feature expectation factor should be used.
the last one, *E* should be equal to 1. Horizontal curves that ha

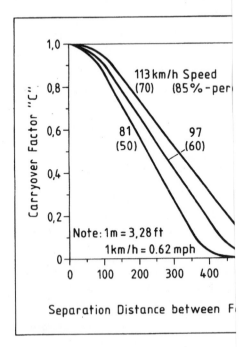

() = 85th-percentile speeds in

FIGURE 19.22   Carryover factor *C* due to separat
features as related to the 85th-percentile speed.[512]

TABLE 19.5 Driver Expectation Related to Workload Value WL_n[512]

Driver expectation	Level of consistency (LOC_n)	Workload value WL_n
No problem expected	A	≤ 1
	B	≤ 2
Small surprises possible	C	≤ 3
	D	≤ 4
	E	≤ 6
Big problem possible	F	> 6

19.10.5 Comparison of a Driver Workload Procedure with Speed Consistency Analyses and Accident Data*

Several different speed consistency and workload measures have been compared in this chapter. Lamm, et al.[405] presented a hypothetical alignment (Fig. 19.23) for which three different speed consistency analyses have been performed, based on Leish,[470] Swiss,[691] and German[241] assumptions. McLean[790] used the same hypothetical alignment to contrast an Australian methodology with the three consistency analyses presented by Lamm, et al.[405] A fifth measure of consistency, based on workload, has been added through the application of the Messer procedure.[512] The Leisch, German, and Swiss procedure evaluate curve 1 and curve 3 as unacceptable because of the speed consistency criterion. The same conclusion comes from McLean, based on speed consistency or side friction criteria. Curves 1 and 3 are also evaluated as unacceptable by Messer because of high workload values. This comparison reveals that all five procedures assess from different viewpoints the same curves as critical. The same is true with respect to safety criterion II, developed in this book for various operating speed backgrounds according to Table 8.5.

In addition to curves 1 and 3 in Fig. 19.23, the driver workload procedure predicts that problems may also exist for curve 2 and, to a lesser extent, for curves 4 and 5.

The driver workload procedure[512] is highly sensitive to severe horizontal curvature. This sensitivity probably accounts for the differences in results. The speed profile procedures assume that once a driver has slowed down for one curve, another similar curve is of little consequence. Messer,[512] however, assumes that combining high-workload features in close proximity results in a higher workload for the second and the following features.[789] With respect to the driver workload homeostasis, the drivers would decelerate in the case of too high workload and the expected cumulative effect, suggested in the Messer procedure,[512] may be compensated for.

On the other hand, the influence of design elements, such as intersections, narrow bridges, lane width changes, and changes in the number of lanes, can only be examined by the driver workload procedure.[512,789]

Furthermore, Wooldridge[789] performed an evaluation associating driver workload with accident rates. The driver workloads were determined, based on the method of Messer,[512] for individual portions of roadway as well as the accident rates on those portions of roadway. The evaluation was accomplished by using two individual variables. In an attempt to determine the influence of a priori expectancy, a grouping was made of those segments with workloads within the six levels of consistency according to Table 19.5. Ad hoc expectancy was examined through the determination of the yaw of the workload on individual features of the roadway.

Driver workload yaw was defined as the difference between the moving average workload and a specific feature's workload. In this way, the yaw provided a measure of the change in workload

---

*Elaborated based on Wooldridge.[789]

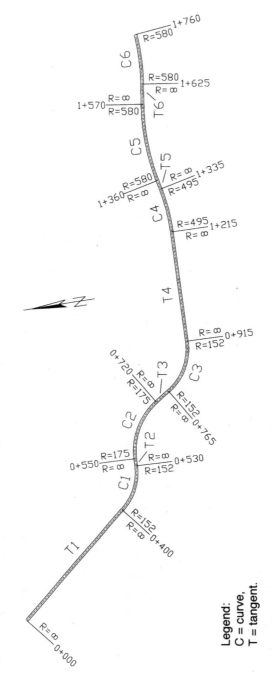

FIGURE 19.23 Hypothetical alignment for comparison of speed consistency and workload measures.[405]

Legend:
C = curve,
T = tangent.

based on the design element level. Driver workload yaw was calculated according to the following equation:

$$\text{Driver workload yaw} = \text{workload rating of the feature (WL}_n) - \text{moving average workload}$$

The extent of roadway length considered in the determination of the moving average was 305 m (1000 ft). It was decided to examine each feature's associated workload in relation to the workload in the previous 305 m (1000 ft). Both the workload ratings and the yaws were grouped into ranges and the numbers and cumulative lengths of features with those characteristics were determined. The number of accidents in those grouped features was then determined.[789]

The roadway features were grouped into two different schemes. The level of consistency according to Messer,[512] which ranges from A to F, was chosen as the first scheme. The number of accidents per 10^8 vehicle miles was then calculated for the cell groupings. The results are given in Fig. 19.24. As revealed in Fig. 19.24, accident rates dramatically increased for those features with effective workloads greater than 6 (LOC F, see Table 19.5).

In a second grouping scheme, roadway segments belonging to various categories of yaw were grouped. As shown in Fig. 19.25, the accident rate was much higher for the segments with yaws greater than or equal to 4.

Those roadway segments that had high effective workloads and those that had effective workloads much higher than the moving average had much higher accident rates than those segments that had lower effective workloads and those segments that had low yaw values. However, an elevated accident rate for areas with extremely low workloads and yaws was not found. A possible cause for this might have been that the driver workloads on the study roadways might not have been low enough in absolute terms, since roadways were chosen by criteria that included the requirement that all of the roadways pass through rolling terrain.[789]

The evaluation of the study on roadways showed that large changes in workload over a short distance were strongly associated with high accident rates. When feature workloads were compared with the average workload in the previous 305 m (1000 ft), it was found that roadway segments exhibiting a large positive change in workload experienced a greatly increased accident rate when compared with those on other segments of the studied roadways. This finding would seem to indicate that when ad hoc driver expectancies are not met, accident risk increases.[789]

Although conclusive statistical evidence has not been provided, the hypothesis that accident rates would be highest in areas associated with high workloads could be confirmed. Features with high workload values can be expected to have "big problems." Similar conclusions could be drawn from the t-test results of mean accident rates for different CCR_S classes in Tables 9.10, 10.12, and 18.14 when differentiating according to good and poor design practices. Poor design levels, which are certainly associated with high workload values, reveal significantly higher accident rates and cost rates than those designated as good design. Since the safety criteria I to III are correspondingly coordinated, similar results have to be expected between the workload procedure and the safety evaluation processes presented in this handbook.

19.10.6 The Relationship between Curve Characteristic, Operating Speed, and Driver Workload*

The different conclusions from the wide-ranging research with respect to the driving behavior in curves shows that a final conclusion to clarify the reasons for the frequently misleading behavior has only partially been explored.

A high driver workload can be a factor in driver misbehavior but it is also an indicator of a higher level of driving complexity. To describe the relationships between road design, speed, and driver workload, the following parameters were selected:

---

*Elaborated based on Refs. 274–276 and 604.

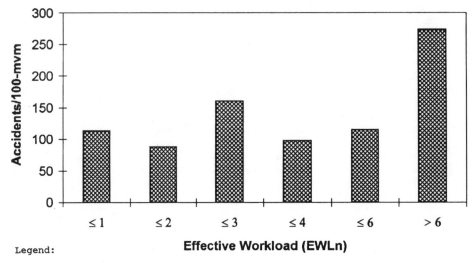

Legend:

mvm = million vehicle-miles.

FIGURE 19.24   Evaluation of effective workload.[789]

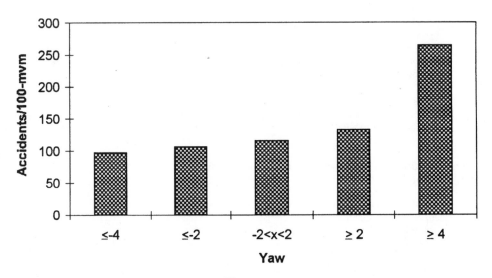

FIGURE 19.25   Evaluation of driver workload yaw.[789]

Curve characteristic: Curvature change rate of a single curve, CCR_S

Operating speed: 85th-percentile speed $V85$

Driver workload: Average values of the z-transformed skin conductance reaction, calculated for each driver and day separately (SCR_T)

The driver workload is considered in Fig. 19.26 as the dependent variable, while the curvature change rate and the 85th-percentile speed are taken as independent variables. The radius of curve has significant effects on the accident situation as discussed in detail in "Radius of Curve" in Sec. 9.2.1.3. However, the *radius of curve,* defined as the distance between the center and the periphery of a circular curve, cannot be perceived by the driver. In contrast, the driver perceives changes

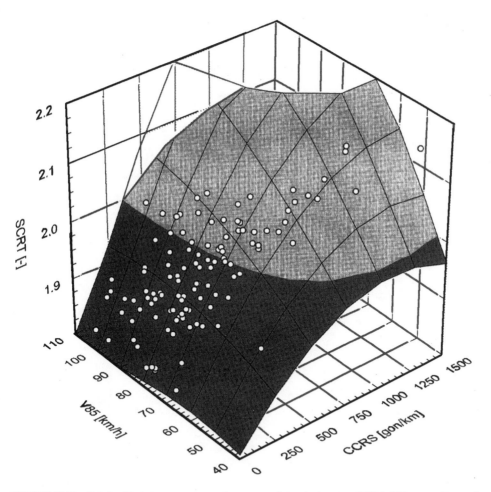

FIGURE 19.26   Relationship between curvature change rate of the single curve (CCR_S), 85th-percentile speed ($V85$), and driver workload (SCR_T). (Elaborated and modified based on Refs. 274, 276, and 604.)

in the curvature of the roadway while driving through the curve. The dynamically and perspectively perceived information affects the driver's behavior. For this reason, the curvature change rate of the single curve was chosen which best describes the road characteristics, as revealed in Sec. 8.2.3.6.

The speed behavior of passenger cars under free-flow conditions is most appropriately described by the 85th-percentile speed measured around the vertex of the curve.[274,276,604]

Recent research[604] has revealed that the skin conductance reaction is an appropriate measure for describing workload.

The relationships between road characteristics (expressed by CCR_S), operating speed (expressed by $V85$), and workload (expressed by SCR_T) are presented by the three-dimensional surface plot in Fig. 19.26. As can be seen from this figure, with increasing CCR_S and/or $V85$, the workload, SCR_S, is increasing. Furthermore, it can be stated that, for good design levels of CCR_S ≤ 180 gon/km ($R \geq 350$ m) according to Table 11.1, no increase in driver workload is noticeable in the studied operating speed range between 50 and 110 km/h. However, for poor design levels of $CCR_S > 360$ gon/km ($R < 175$ m), with increasing CCR_S values, strong increases in workload can be observed.

Thus, it is evident according to Fig. 19.26 that complex relationships exist between workload, road characteristics, and operating speed. Further research is urgently needed to also consider workload issues in future highway geometric design standards.

With respect to safety evaluation processes, it is interesting to note that Richter, et al. state, in their book *Driving Behavior and Driver Mental Workload as Criteria of Highway Geometric Design Quality,* the following: "The Safety Module according to Lamm was found to be the most sensitive and adequate one. All tested road sections were rated according to the three Safety Criteria (Good, Fair and Poor Design)."[604]

CHAPTER 20
ROAD SAFETY WORLDWIDE

Every year, worldwide at least 500,000 people are killed in road accidents. This chapter provides an overview of the present state of affairs in the area of road safety.

The anticipated growth of (motorized) mobility on a global scale will, without effective road safety management, lead to an increase in the number of fatal accidents and injured persons. From experience we know that the problem of road accidents is not *unassailable,* but it is not to be expected that one single all-embracing measure can be found to solve the problem. It is rather more the necessity to carry out existing measures and activities in a better way based on synergy and permanent implementation. Furthermore, preventive and explicit road safety considerations will have to include decisions concerning the planning and investment in road infrastructure (design, construction, and maintenance) in order to prevent road safety problems rather than solving them with *hindsight* ("when the steed is stolen, the stable door is locked").[746]

Road safety is a worldwide problem that has not been solved to satisfaction anywhere. It is a complex problem that involves driver behavior, vehicle characteristics, roadway features, and driving conditions. Generally, it must be addressed both as a question of vehicle and roadway design and as a question of use such as driving habits, traffic regulations, law enforcement, risk management, etc.

Road safety is not just an individual concern. In this respect,

1. The industry has an important role to play here, in addition to the different levels of government (local, state, and federal agencies).

2. Targeted and integrated actions should be taken as soon as possible in order to reduce drastically the human grief and the high economic costs incurred by road traffic accidents.

3. Road safety is a question of civilized behavior to our fellow citizens. We believe we have to talk more about traffic morals, which does not necessarily mean we should moralize on the subject. We must find the courage to change our habits. After all, the only way of improving the situation, with regards to road safety, is by personal commitment, taking responsibility for our actions, and setting a good example. Those of us who are responsible for road safety must set the trend in these respects.[181]

Road safety worldwide and promising targeted and integrated actions are alluded to in the following.

*20.1   ROAD ACCIDENTS: WORLDWIDE A PROBLEM THAT CAN BE TACKLED SUCCESSFULLY**

20.1.1   Road Accidents

Road safety tends to be regarded by most countries as a social problem, albeit a high level of social apathy can be noted. Generally speaking, on a day-to-day basis, the aim of reducing the number of road accident victims does not receive high priority.

Many different ways to shed some light on road hazard problems have been tried so that "eyes have been opened." For example, instead of presenting the annual (small) probability of dying in an accident, an attempt has been made to draw attention to the problem by calculating the probability of an individual becoming involved in an accident some time during his or her life, or by estimating the economic consequences of road accidents (1 to 2 percent of a country's gross national product), or by pointing out that young people often become the victim of a road accident (the principal cause of death among people between the ages of 15 and 45), and, therefore, that road hazard leads to a high number of "lost years of life," or by showing how many people will have a fatal accident in the next 20 years "if policy remains unaltered" (for example, 400,000 in all countries of central and eastern Europe), or by reporting that there is no technological system in existence today (such as transport by rail, air, energy production, etc.) that has led to so many fatalities—and will continue to do so—as the current road transport system. But, to date, all of these descriptions have not seemed to have been successful in drawing the attention of the general public and the politicians to the problem of road hazard.

There are a number of explanations for this. The large majority of all traffic accidents hardly even makes an impression on the general public or on the press, in contrast, for example, to airplane disasters or shipping disasters. In addition, both the nature of the road hazard problem and the approach to be adopted is complex. There are many possible causes while the relationship between the causes and the many actors or stakeholders involved in the approach adopted is extremely intricate. The finger cannot be pointed at any one source.

Another explanation is that many people believe that road accidents are unavoidably associated with road transport systems.

Finally, the fact should be mentioned that road safety measures are rarely popular—neither among road users nor politicians—because they cost money and frequently restrict individual freedom, while the anticipated positive result of these measures can rarely be determined.

Perhaps attention can be drawn to the safety problem in an entirely different way: by demonstrating that it can be tackled successfully. Improving road safety has proven not to be an impossible task. In highly motorized countries, there has been no question of a considerable reduction in the number of road accident victims, even though mobility has constantly been on the rise. However, growth in mobility has not led to a parallel growth in the number of road accident victims. More striking still is the fact that a growth in mobility in many countries has even been accompanied by a drop in the number of road accident victims.

Could both these approaches (profiling road hazard as a real social problem and a problem that can be tackled successfully) lead to the support of politicians and policy makers, of course, supported by a more high-profile (and professionally conducted) lobby from the public and social organizations?

After all, to date there is no country anywhere in the world where people are totally satisfied with the way in which the problem is tackled or the results of this approach. There is no country in the world where the high level of road hazard is accepted as inevitable, even in those countries where a high level of mobility is linked to a relatively low number of road accident victims. What

*Elaborated by F. Wegman.[746] With contributions from Peter Holló, KTI, Institute for Transport Sciences, Budapest, Hungary; Stein Lundebye, The World Bank, Washington, United States; Grant Smith, Transport Canada, Ottawa, Canada; Luc Werring, European Commission, Brussels, Belgium; and with the kind help of members of the PIARC Committee on Road Safety. Stichting Wetenschappelijk Onderzoek Verkeersveiligheid (SWOV), Leidschendam, The Netherlands.

is the perspective for road safety if road transport as we know it today does not undergo a fundamental change in the coming decades (roads in a historically and geographically determined area, various modes of transport that use the same physical space, variations in speed limits between 5 to 150 km/h, the individual as an independent, decision-making being, and roads where few restrictions are imposed)? Can the number of road accident victims be (further) reduced by 10, 50, or even by 90 percent? And how can the various countries in the world learn from each other?

20.1.2   Recent Developments in Road-Safety Trends

Two indicators are regularly used as a yardstick to compare road safety in one country with that in another: traffic safety and personal safety (Trinca, et al.[717]). *Traffic safety*—sometimes indicated in terms of fatality rate or casualty rate—is a measure of how safely the road transport function is performed. It is commonly measured in terms of deaths per 10,000 registered motor vehicles or per 100 million vehicle kilometers traveled. The other—*personal safety*—indicates the degree to which traffic accidents affect the safety of the population. It can be considered a public health indicator: the number of traffic fatalities per 100,000 population (mortality).

Data on traffic safety and personal safety are given in Tables 20.1*a* and 20.1*b*, prepared by the U.K. Transport Research Laboratory using data from the International Road and Traffic Accident Database (OECD and German Federal Highway Research Institute, BASt).

The following conclusions can be drawn from the data on personal safety and traffic safety in highly motorized countries according to Wegman, 1995:[746]

- Generally speaking, the personal safety rate is improving (in the last 7 years, between 10 and 40 percent). However, the reduction rate differs; the reduction rate was higher in the 1970s than in the 1980s.

- Generally speaking, the traffic safety rate is improving and the reduction rate is higher than that of the personal safety rate (in the last 7 years, between 15 and 60 percent).

- The reduction rate for traffic safety differs in each country so there is no constant per country over time; nowadays the decline in the fatality rate is lower than it used to be.

- A good road safety record (low traffic safety rate) corresponds to a short half lifetime of traffic safety rates; no evidence of a law of diminishing returns has been found. Examples are the United Kingdom, Norway, and the Netherlands with relatively low safety rates and high reduction rates, also in recent years.

It is impossible to come to a consistent conclusion for developing countries. This is understandable in view of the completely different situations in terms of road transport and motorization in that part of the world. During the last 10 years, the number of vehicles per person multiplied by 10 in some countries (for example, South Korea and Thailand), while in others the growth was more modest (South America). In many countries in Africa the motorization remains very low, between 1 and 100 vehicles/1000 inhabitants). The traffic safety record for many developing countries (a quality indicator for the safety of road transport) remains very high compared to highly motorized countries (decuple). Any conclusion about developments in time is obscured by the poor quality of available data.

There is no doubt that the situation worsened at the end of the 1980s and the beginning of the 1990s in countries in central and eastern Europe. In various papers, Oppe and Koornstra[560] have successfully modeled the development of road fatalities based on long-term developments in traffic growth (motorized kilometers) and in fatality rates (road deaths per distance of travel). The so-called logistic function, which is an *S*-shaped curve, fits the long-term trend of traffic growth for many highly motorized countries.

In the long term, the growth of motorization in many countries is accompanied by an exponentially decreasing curve for fatality rates, that is, a reduction in annual road fatalities per kilometer driven with a constant percentage (log-linear trend), although this percentage differs from one year

TABLE 20.1a   Road Accident Fatalities and Fatality Rates by Country[746]

Country	Year	Source	Fatalities	Fatalities/10^4 vehicles	Fatalities/10^5 persons
Europe					
Austria	1993	IRTAD*	1,283	3.1	16.2
Belgium	1993	IRTAD	1,660	3.4	16.5
Bulgaria	1992	IRF†	1,299	6.5	15.3
Cyprus	1993	IRF	115	3.3	18.4
Denmark	1993	IRTAD	559	2.7	10.8
Finland	1993	IRTAD	484	2.1	9.6
France	1993	IRTAD	9,568	3.4	16.6
Germany	1993	IRTAD	9,949	2.2	12.3
Great Britain	1993	IRTAD	3,814	1.5	6.7
Greece	1993	IRTAD	2,104	6.6	20.3
Hungary	1993	IRTAD	1,678	N/A‡	16.3
Iceland	1993	IRTAD	17	1.2	6.4
Ireland	1993	IRTAD	431	3.7	12.1
Italy	1993	IRTAD	7,177	2.0	12.6
Lithuania	1993	IRF	892	10.2	24.0
Luxembourg	1993	IRTAD	76	3.1	19.2
The Netherlands	1993	IRTAD	1,252	1.9	8.2
Norway	1993	IRTAD	281	1.3	6.5
Poland	1993	IRF	6,341	6.9	16.5
Portugal	1993	IRF	2,171	6.0	22.0
Rumania	1993	IRF	2,826	11.5	12.4
Spain	1993	IRTAD	6,378	3.6	16.3
Sweden	1993	IRTAD	632	1.5	7.3
Switzerland	1993	IRTAD	723	1.8	10.5
Turkey	1993	IRTAD	8,078	N/A	14.1

*JRTAD = International Road and Traffic Accident Data Base

†JRF = International Road Federation

‡N/A = not available

TABLE 20.1b Road Accident Fatalities and Fatality Rates by Country[746]

Country	Year	Source	Fatalities	Four wheelers ×10³	Two wheelers ×10³	Total vehicles ×10³	Population ×10³	Fatalities/10⁴ vehicles	Fatalities/10⁵ persons
					Africa				
Botswana	1993	TRL	379			94	1,425	40.3	26.6
Cameroon	1987	IRF	1,034			180	10,700	57.4	9.7
Central African Republic	1993	IRF	30	1	1	2	3,041	150.0	1.0
Ethiopia	1991	IRF	1,169	59	2	61	52,000	191.6	2.2
Ivory Coast*	1989	Sweden	606			340	13,000	17.8	4.7
Liberia	1987	IRF	80			11	2,221	72.7	3.6
Mauritius	1991	IRF	163	67	69	136	1,078	12.0	15.1
Morocco	1990	IRF	1,921			749	25,091	25.6	7.7
Niger	1987	IRF	148	41	0	41	7,439	36.1	2.0
Rwanda	1990	IRF	331	10	8	18	7,200	183.9	4.6
Seychelles	1992	TRL	8			7	72	11.4	11.1
South Africa	1992	IRF	10,142	5,108	285	5,393	31,917	18.8	31.8
Swaziland	1990	IRF	193	51	3	54	726	35.7	26.6
Tanzania*	1989	Sweden	1,116			250	25,000	44.6	4.5
Tunisia	1989	IRF	1,172	494	12	506	7,910	23.2	14.8
Uganda*	1989	Sweden	1,271			200	17,400	63.6	7.3
					North and South America				
Barbados*	1991	TRL	23		N/A	65	250	3.5	9.2
Brazil	1993	IRF/EYB	5,500	13,469	N/A	13,469	156,578	4.1	3.5
Canada	1992	IRF	3,485	17,010	339	17,349	27,297	2.0	12.8
Chile*	1992	TRL	1,700			1,361	13,813	12.5	12.3
Colombia	1989/90	IRF	2,564	1,329	245	1,574	32,317	16.3	7.9
Costa Rica	1993	IRF	235	335	46	381	3,167	6.2	7.4
Ecuador	1991	IRF	1,057	384	N/A	384	10,502	27.5	10.1
Honduras	1990	IRF	400	107	8	115	4,800	34.8	8.3
Jamaica*	1989	Sweden	400			150	2,500	26.7	16.0
Mexico*	1989	Sweden	7,401			8,000	86,000	9.3	8.6
Suriname	1993	IRF/WHO	55	58	N/A	58	446	9.5	12.3
United States	1992	IRF	39,235	190,362	4,065	194,427	255,078	2.0	15.4

(*Continued*)

TABLE 20.1b Road Accident Fatalities and Fatality Rates by Country[746] *(Continued)*

Country	Year	Source	Fatalities	Four wheelers × 10³	Two wheelers × 10³	Total vehicles × 10³	Population × 10³	Fatalities/10⁴ vehicles	Fatalities/10⁵ persons
Asia and Middle East									
Bahrein	1993	IRF	56	141	1	142	538	3.9	10.4
Brunei	1991	IRF	47	128	4	132	261	3.6	18.0
China*	1990	CARS	49,271			10,095	1,100,000	48.8	4.5
Hong Kong	1993	IRF	351	462	28	490	6,020	7.2	5.8
India	1991	IRF	56,525	4,025	4	4,029	844,000	140.3	6.7
Indonesia	1990/92	IRF/EYB	10,887	3,256	6,987	10,243	179,379	10.6	6.1
Iraq	1989	IRF	4,625	1,041	N/A	1,041	17,785	44.4	26.0
Japan	1993	IRF	10,942	63,266	16,395	79,661	123,788	1.4	8.8
Korea South	1993	IRF	10,402	6,274	1,936	8,210	44,056	12.7	23.6
Kuwait	1989	IRF	301	610	5	615	2,095	4.9	14.4
Malaysia	1993	PDM	4,666	3,229	3,483	6,712	19,050	7.0	24.5
Nepal	1992	TRL/RSP	530			68	18,916	77.9	2.8
Pakistan	1993	IRF	6,299	774	1,166	1,940	122,801	32.5	5.1
Philippines	1990	IRF/CARS/EYB	1,099	1,219	383	1,602	61,480	6.9	1.8
Saudi Arabia	1991	IRF	3,719	5,103	13	5,116	11,861	7.3	31.4
Singapore	1993	IRF	258	441	118	559	2,873	4.6	9.0
Sri Lanka	1992	IRF	1,795	381	516	897	17,894	20.0	10.0
Taiwan	1993	IRF	2,349	4,208	10,948	15,156	20,944	1.5	11.2
Thailand	1993	IRF	9,496	4,136	7,106	11,242	58,336	8.4	16.3
Yemen Arab Republic	1990	IRF	1,334	484	N/A	484	14,020	27.6	9.5
Oceania									
Australia	1992	FORS/IRF	1,977	9,954	292	10,246	17,483	1.9	11.3
Fiji	1994	TRL	88	53	1	54	761	16.3	11.6
New Zealand	1993	IRF	600	1,900	61	1,961	3,500	3.1	17.1
PNG	1992	TRL	290	50	N/A	50	3,920	58.0	7.4

*Source other than International Road Federation (IRF), for example, World Year Book 1993 (WYB), World Health Organization Stats. 1993 (WHO), Road Accidents Great Britain (1994), Road Traffic Accidents in Europe and North America 1995, United Nations (UN).

Notes: N/A = not available. TRL = Transport and Road Research Laboratory, Crowthorne, United Kingdom.

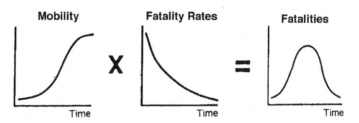

FIGURE 20.1   Trend in fatalities as a result of trends in mobility and fatality rates.[560]

to the next. The percent decline per year differs in each country. Based on empirical data, Oppe and Koornstra concluded that the cyclic modifications should be added to the long-term macroscopic trend of mobility growth and fatality rates, although some room for discussion remains. Simply by combining both developments as a product (fatalities = fatalities/kilometer × kilometer), the development of fatalities could be described (see Fig. 20.1).

This leads to the conclusion that a reduction in the number of fatalities ought to be the result of a higher decrease in the fatality rate rather than an increase in mobility growth. Should the growth in mobility accelerate, for example, due to high economic growth, extra attention should then be devoted to (road-safety) measures with the aim of further decreasing the road traffic risk or else an immediate increase in fatalities will result. The model used by SWOV has been calibrated with statistics from many countries around the world.

To illustrate this approach, an example for central and eastern Europe can be used. This example was presented to a Policy Seminar on "Road Safety for Central and Eastern Europe" in Budapest in 1994.[301] The SWOV model used available data from both the past and present. Making a conservative estimation of growth in vehicle ownership and combining this development with a rather optimistic (reduction rate of 5.7 percent) and pessimistic development (reduction rate of 4.4 percent) in fatality rates, the resulting estimates of fatalities are as shown in Fig. 20.2, with the cumulative difference during the period 1995–2010 amounting to some 400,000 lives saved.

However, in no sense can a correlation between the growth of mobility and the reduction of fatality rates be the result of natural law or spontaneous development. We might consider this correlation as a collective influence to adapt society to increasing traffic, which requires an increased, renewed, improved, and well-maintained road traffic system. This increase in traffic growth and the corresponding adaptation results in better and newer roads, a higher level of average driver experience, newer and safer vehicles, and appropriate traffic regulations and enforcement. All highly-motorized countries went through this adaptation to mass-motorization, however, at different speeds.

Research indicates two interesting results of these modeling activities, which are of great importance for policy making. First, the more explosive the motorization growth is, the larger the annual decrease in the fatality rate will be. Secondly, a lagged correlation exists between traffic growth and fatality rate reduction. Hence, after some years high traffic growth leads to higher reductions in fatality rates. This could be understood as the time lag needed to implement effective countermeasures for risk reduction. A long time lag could be considered a poor answer to traffic growth. A road-safety policy that anticipates traffic growth is the effective answer to this time lag.

20.1.3   Road Safety Around the World

The next question that should be answered is: how can a reduction in the fatality rate be achieved? What factors can explain an established reduction and to what degree has road-safety policy been

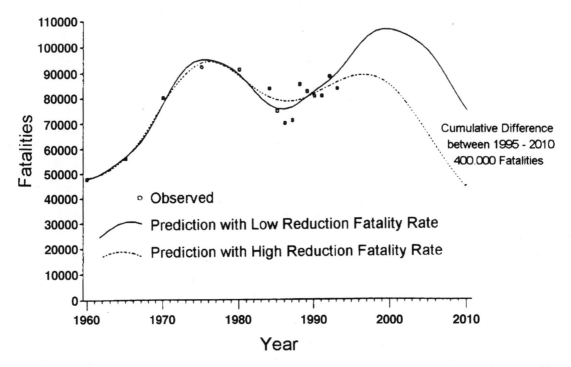

FIGURE 20.2 Fit and prognosis of road fatalities for central and eastern European countries, Russian federation, and countries of the former Soviet Union.[301]

a contributing factor? In considering these questions, it is, of course, impossible to treat all parts of the world in the same way when considering these questions.

20.1.3.1 European Union. Each year, accidents are the cause of about 50,000 deaths and more than 1.5 million injuries on the roads of the European Union. Since the Treaty of Rome (1957) was signed, almost 2 million people have been killed in the 12 countries which were Community Members (until 1995) and almost 40 million have been injured. But road accidents do not only have dramatic consequences in human terms; the economic cost is also substantial.

At the Community level, the principal actions taken so far in the area of road safety have been concerned with the harmonization of rules relating to motor vehicles through the adoption of over 100 directives; for example, the maximum number of hours for a driver, minimum tire tread depth for private cars, the periodic inspection of vehicles, including harmonized standards for the testing of brakes, general standards for obtaining a Community model driver's license, mandatory wearing of seat belts, including restraints for children, and speed limiters for heavy vehicles. In addition, two important draft directives are on the Council table: harmonization of speed limits for commercial vehicles and maximum blood alcohol concentration permitted.

Nevertheless, even with the advances that have been made in technical and behavioral standards, the road-safety record of the member states varies significantly, as can be seen in Table 20.1*a*. The fatal accident rate (expressed per kilometer of travel) differs more than sevenfold between the most advanced member states and those with the worst figures. Moreover, the trend also varies considerably, with some states improving their position much more than others while in a few countries, the situation is actually deteriorating. Using the same basis for measurement, the average risk on Community roads is nearly twice that in the United States. If it were possible to attain the same level as in the United States, more than 20,000 deaths could be avoided every year. The scope for improvement is clearly considerable.

The European Commission decided to set up a Community program on road safety with qualitative (not quantitative!) targets and the identification of priorities. The Commission decided that the process of harmonization through legislation and the development and application of common research projects should be considered the main type of Community action applied in the fields of user behavior, vehicles, and infrastructure. Unfortunately, nothing has been heard of "specialized Community body on road safety or road-safety agency" since.

Nowadays, two organizations are active in the field of European road-safety policy: the European Transport Safety Council (ETSC) and the European Road-Safety Federation (ERSF) in which different nongovernmental associations are joining forces. Both, together with the Commission, a high-level group of road-safety directors of the member states, and other interested parties in road safety [for example, the Federation of European Road Safety Research Institutes (FERSI)], are now looking for an effective organization, procedures, priorities, etc., sailing between Scylla and Charybdis in the European Union; that is, member state autonomy and European Union subsidiarity.

Of course, the emphasis of activities in the European Union lies with the member states themselves. It is not within the scope of this chapter to provide a description of the policy plan for each member state, but the following references include interesting publications about safety plans and programs from the various countries (see, for example, Refs. 140, 166, 186, 355, 517, 619, and 695) and about a European policy for road safety.[212]

It is hardly possible to summarize all European road-safety plans in a few lines. However, some trends can be recognized:

- Creating more awareness of road-safety problems among politicians, intermediate players, or stakeholders (regional and local authorities, private organizations, the private sector) and the general public/road users
- Targeted road-safety programs comprised of the most cost-effective measures and monitoring aids and, if necessary, revised programs
- Paying more attention to the various options to implement certain measures
- Involving other participants because central governments should not be the only participant in developing road safety strategies and in implementing those strategies
- Looking for funds to finance road-safety measures other than the regular budgets spent by the central government

20.1.3.2 *Developing Countries.*

According to estimates by the World Bank, 350,000 people are killed in automobile accidents in the developing world every year. Two-thirds of the accidents involve pedestrians, most of whom are children. Two-thirds of the accidents occur in cities or in the surrounding areas. As a result of the surge in urban areas and vehicle ownership, the figures are mounting. In fact, the numbers have reached such levels that among people in the 5- to 44-year age group, road accidents are the second-most frequent cause of death. As a result, accidents cost developing countries a whopping $1.4 to $2 billion or an estimated 1 to 2 percent of their GNP. To make matters worse, much of these costs must be paid for in foreign currency, since many vehicles, spare parts, and medicines are imported. In Asia, for example, car ownership jumps by 12 to 18 percent each year, yet streets, highways, and safety measures have not kept pace. Thus, the rate and number of accidents have increased dramatically.

Much of the problem is related to a shortage of funds, both for owners of vehicles and for the governments. Many private and commercial vehicles are old and in poor condition. Streets have deteriorated. Different kinds of traffic—pedestrians, nonmotorized transport, buses, trucks, and cars—all share the same streets. Sidewalks are often nonexistent, particularly in outlying areas. Streets and vehicles are poorly lit, increasing the risk at night.

The improvement of road safety in developing countries requires the development of sustainable project components to match public awareness and government commitment. Countries exhibit different degrees of readiness to implement road-safety measures, depending on the government's sensitivity to the problem and its importance on the political agenda. The lower the awareness level, the less likelihood there is of government interest and ability to absorb safety components in projects sponsored by the World Bank or other donors. World Bank consultants in the field of road safety have pinpointed three levels of awareness.

1. *Awareness level 1:* In these countries, there is little safety awareness. Accident data may or may not be collected and any data system will be primitive. Little is known about trends or road users at risk. No one is working specifically on safety matters. General interest by the government is low.

2. *Awareness level 2:* The government is aware of the road-safety problem but has given it little priority. Accident data are sparse. Occasionally, there may be road-safety pressure groups. The media may be beginning to press for action. Some university research may be under way.

3. *Awareness level 3:* The government has recognized the need for assistance. An improved data system has been established and staff trained in safety operations. Analysis of black spots is undertaken and identification of the road user groups most at risk. A National Road Safety Council (NRSC) coordinates a national road-safety program. Efforts are made to improve driving tests and vehicle examinations, develop children's traffic education programs, and improve legislation. A core of people exist who specialize in safety and who are keen to tackle the problem but lack resources. Road-safety research is being undertaken and the media is active in pushing for action.

The nature of the World Bank loans is related to the level of awareness. Of course, major differences between countries can be seen with respect to awareness, measures, and policy. A general impression is that developing countries only devote attention to road hazards in a very modest fashion and have other priorities, but developments in highly motorized countries that have already taken place clearly show the negative consequences of such an attitude. The attention paid by the World Bank to road safety is, therefore (unfortunately), scant but indications are that there is a growing level of World Bank interest in road safety (see, for example, the major efforts of the Bank in central and eastern Europe).

In a contribution prepared by the World Bank, examples have been given of recent advances in a number of developing countries, where, with the support of the World Bank, activities are being developed in the field of road safety. Examples are cited in Ref. 746.

Although positive intervention by the World Bank to improve road safety in developing countries could be pointed out, the conclusion seems to be inevitable that these efforts are too incidental to result in a reduction in the number of accidents and casualties. Road-safety programs have to be launched on a more massive scale and all parties involved (developing countries, international banking organizations, and donor agencies) are urged to commit themselves to such programs.

The most recent information about road safety in developing countries, which confirms the previous statements, was reported by Jacobs and Kirk at the XIIIth International Road Federation World Meeting in Toronto, Canada, in June 1997. The authors concluded the following.[311]

Independent studies by both the Transport Research Laboratory, United Kingdom, and the World Bank have estimated that about 500,000 people lose their lives each year as a result of road accidents and over 15 million suffer injuries. The majority of these, about 70 percent (350,000 people), occur in those countries which the World Bank classifies as low or middle income. While the road accident situation is improving in the high-income countries (Table 20.3), most developing countries face a worsening situation. For example, in those African countries where data were available, there has been a 300 percent increase and in Asian countries something like a 200 percent increase. As infectious diseases are increasingly brought under control, road deaths and injury rise in relative importance. In Thailand, for example, more years of potential life are lost through road accidents than from tuberculosis and malaria combined. The authors believe that, while there can be no such phenomenon as a completely "safe" road system (due in no small part to human behavior), very high road accident death rates are not the inevitable price that has to be paid by these countries for the mobility of people and goods and a great deal can be done to improve their road safety situation while developing into a more industrialized society. Apart from the humanitarian aspects, it should also be remembered that road accidents cost any country, be it developing or developed, a considerable sum of money each year. The most recent estimates suggest that in all countries of the developing world combined, road accidents cost approximately U.S. $36 billion each year.[311] This number is by far higher than the one estimated by Wegman[746] at the beginning of this chapter.

20.1.3.3 *Central and Eastern Europe.*

Recent estimates indicate a road accident toll of about 75,000 road deaths every year in central and eastern European countries. Since 1986, a dramatic increase in accidents and deaths has occurred in these countries. Of course, such an increase is closely related to the recent political, economic, and social changes. It was reported in many CEECs that almost all absolute and relative indices of accidents have been increasing.

However, the models used by SWOV, as described in Sec. 20.1.2, led to the conclusion that if, between 1995 and 2010, appropriate road-safety policies were put in place and implemented, at least 400,000 lives could be saved (see Fig. 20.2).

The political and economic changes at the end of the 1980s also seem to be expressed in terms of growth in the annual number of road accident fatalities. Of course, reasons can be given to explain this phenomenon:[743,744] a rapid growth in the number of vehicles, many new and inexperienced drivers on the road, many new and second-hand cars from the west driving at relatively high speeds on inadequate and insufficiently maintained roads, much driving under the influence of alcohol or drugs in situations where there is little police enforcement and a poorly equipped police force, etc. These are possible explanations, but scientifically supported evidence is not available.

In the meantime, several CEE countries are active in preparing and implementing road-safety plans. Hungary is used here to illustrate these developments (Fig. 20.3).

Figure 20.3 shows the changes in the number of road vehicles, personal injury accidents, and fatalities between 1976 and 1994 in Hungary. Since the peak in the number of accidents in 1990, both the number of accidents and persons injured and killed in road accidents have shown a decrease. In 1994, the number of fatalities diminished below the level experienced in 1976. This is a considerable change because since then the number of road vehicles in Hungary has increased by over a million. The decreasing tendency also continued during the first part of 1995.

Such a reduction in the number of accidents is, of course, the result of numerous factors: partly generated by social and economic changes and by carefully selected measures to improve traffic safety. Probably, the positive change has been the result of the following factors:

- A gradual cessation of the negative "side effects" due to rapid and significant social change.

- A decrease in vehicle kilometers due to the increase of the cost of vehicle operation and maintenance (7 to 8 percent between 1990 and 1991, increased by 2 to 3 percent between 1991 and 1992, and practically unchanged since then); however, the recent reductions in the number of casualties cannot be explained by decreased road traffic volumes.

- Modification of the highway code (introduced March 1, 1993): introduction of speed limits of 50 km/h in built-up areas, the compulsory use of daytime running lights (DRL) on main roads outside urban areas and on motor roads, and the compulsory use of seat belts on rear seats outside built-up areas. The mandatory use of DRL outside built-up areas was introduced June 1, 1994.

- Increased and improved enforcement activities combined with publicity campaigns and higher penalties.

In spite of these efforts, a significant improvement in the severity of accidents has not been achieved. In order to achieve significant improvements in this respect, the strictness and consistency of enforcement, further increase of seat-belt usage, enhancement of the passive safety of vehicles and roadway furniture, improvement of first-aid and rescue services are needed in Hungary and the other countries of central and eastern Europe.

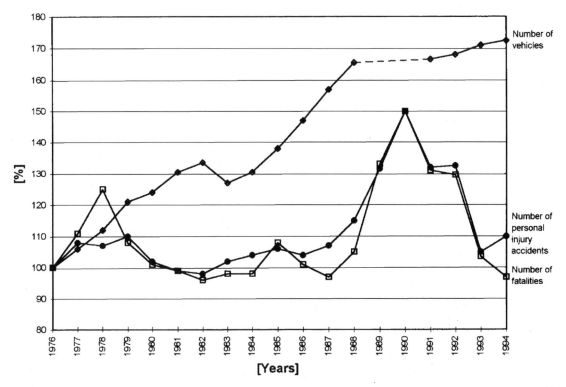

FIGURE 20.3   Changes in the number of vehicles, personal injury accidents, and fatalities in Hungary between 1976 and 1994.[746] (*Source:* KTI)

20.1.3.4   Australia.   Australia deserves the international reputation it has gained for successfully reducing the number of road accidents. It was one of the first to introduce many well-known road-safety measures: compulsory seat-belt use, helmets for motorcyclists and bicyclists, and random breath testing. A national road safety strategy was drawn up 1992.[34] It was the first national approach by federal, state, and local governments, as well as industry and community groups, to reduce road accidents. Australia is aiming for a level of road trauma below current levels despite an expected 18 percent increase in the population and a 25 percent increase in road travel by the year 2001.

Australian strategy defines three overall goals, four specific goals, and eight key priorities. The eight "key priorities" are: alcohol and drug abuse, speeding, protection of vehicle occupants, driver fatigue, road hazards, heavy vehicles, novice drivers and riders, and improved trauma management. Policy objective will be achieved through: the coordination among, and involvement of, all agencies in making the best use of their resources, participant commitment through partnership and agreeing on ways to measure the success of the strategy, cost-effective measures to set priorities and coordinate research and development.

The State of Victoria (4 million inhabitants in a rather urbanized area, including the capital Melbourne) has been acknowledged as one of the states making the best progress. An average annual change in fatal crashes between 1989 and 1993 of -13.5 percent has been registered; for all of Australia this percentage was -7.9. Together with New South Wales (-9.9 percent), Victoria is in the lead.

A number of studies have been carried out to evaluate which countermeasures and other factors have contributed to the substantial and unprecedented reduction in road accident fatalities and serious injuries in Victoria since 1989. The following have been identified (Cameron, et al.[96]): increased random breath testing; speed cameras; compulsory bicycle helmets; lowering the 110-km/h freeway speed limit; improvements to the road system, especially through treatment of accident black spots; reduced economic activity; and reduced alcohol sales. The road-safety programs are estimated to have contributed to reductions in serious casualty crashes of 26 to 29 percent during 1990/1992.

The bulk of the reductions can be reasonably attributed to the two major road-safety programs: the random breath-testing program and the speed camera program. These two programs in Victoria can be considered as extremely interesting for other jurisdictions because of the strategic use of road-safety television advertising combined with police enforcement leading to general deterrence (drinking and driving) and specific deterrence (speeding). The research results indicate clear links between levels of publicity supporting the speed and alcohol enforcement programs and reductions in casualty rates. Less positive results can be reported from a publicity campaign ("Concentrate or kill") not directly linked to enforcement.

20.1.3.5 Canada.   During 1993, 3601 road users were killed in reportable traffic accidents in Canada, an increase of 2.9 percent over traffic deaths in 1992, but still 3.2 percent fewer than the average number of road users killed during the previous 3 years. The 1993 figure represents the first increase in traffic deaths in Canada since 1989. For the third successive year, the number of road users killed in traffic accidents has remained at approximately the same level (3500 to 3700 persons). It would appear that this level has replaced that of the 1982–1990 period, when the number of annual traffic deaths ranged from 4100 to 4350 persons. This achievement has largely been the result of ongoing safety programs and initiatives conducted at all levels of government, safety, and enforcement organizations. The most successful continues to be the National Occupant Restraint Program (NORP), which targeted a 95 percent seat-belt usage by 1995.

Nationally, seat-belt usage among passenger-car drivers is currently 90.1 percent. No doubt, this level of use has largely contributed toward keeping the number of fatally injured motor vehicle occupants at the same level during the past 3 years, despite increases of 4.7 percent and 6 percent in the number of motor vehicles registered and licensed drivers, respectively.

Furthermore, the following activities are presented as the most prominent ones:[518] introduction of a graduated licensing system, special attention to the problems of the older driver, high-risk groups, introduction of high-tech tools for police enforcement, and increasing seat-belt usage. (For detailed information about high-risk target and vulnerable road user groups, see also Sec. 20.2.)

20.1.3.6 Japan.   In recent years, about 11,000 road accident fatalities have been recorded in Japan annually (Japanese statistic: victims who die within 24 h). In a population of 125 million, this represents a mortality rate of 8.8 (30 days: 11), indicating that Japan is one of the relatively safer countries in the world with regard to road traffic. The number of road accident victims/10,000 vehicles is similarly 1.4.

It is significant, however, that while the annual number of road accident victims was halved during the 1970s, it has since risen from 8500 up to 11,000 (Fig. 20.4).

This means that Japan forms the exception with respect to other highly motorized countries. Halving the number of road accident victims in the 1970s was mainly the result of technical and engineering traffic measures that led to a better structure of the road network and improved design of the roads and verges.

According to Koshi it can be stated:[356]

> In the 1980s however, those measures that were effective in the decade before approached their saturation levels in terms of both the number of installations and the extent to which they could prevent accidents; hence they became less and less effective. The Japanese failure to maintain the decreasing tendency of accidents in the 1980s was also due to the fact that Japan continued to implement conventional measures along the same lines as in the 1970s. Stress should now be placed on a new policy to improve driver quality (especially education of young drivers), traffic regulations and enforcement.

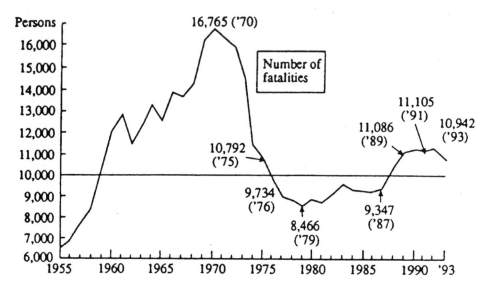

FIGURE 20.4   Trend of road accident fatalities in Japan.[298,299]

The experience gained in Japan demonstrates that a sound implementation of known traffic measures is effective (1970s), but also that an increase in the number of road accident victims in association with a growth in mobility is not unrealistic. Furthermore, road-safety measures should be maintained; an effect, once achieved, does not offer any guarantee for the future!

In more recent Japanese publications, the following items are mentioned for improving the situation: countermeasures for the elderly, for women, for nighttime accidents, for accidents during holidays and vacations, accidents on expressways and in residential areas, accidents involving pedestrians, motorcyclists and mopeds, and accidents related to on-street parking.[497]

20.1.4   Effective Recent Initiatives

Based on contributions from the various authors and on a literature review, the following subjects have been included in this chapter. The chosen subjects can offer a significant contribution to the scope of road hazard for which possible realized improvements have been proven:

- Drinking and driving under the influence of alcohol (Zaal[771])
- Excessive and inappropriate driving speeds [European Transport Safety Council (ETSC)[175] and Zaal[771]]
- Assistance given to road accident victims [Trinca, et al.[717] and National Road Trauma Advisory Council (NRTAC)[537]]
- Restraint systems, that is, seat belts, child restraints, etc. (SWOV[355])
- Vehicle safety (ETSC[174])

Many examples can be given of programs which have proved successful in promoting road safety. This overview will describe a number of the better known worldwide programs. Section 20.1.5 deals with improvements to the road infrastructure which are of major importance for improving road safety.

20.1.4.1 Drinking and Driving. In 1988, Canada convened a National Road Safety Symposium for the purpose of identifying priorities and targets for the 5-year period from 1990 to 1995. In all, 12 major issues were identified which were then forwarded to appropriate agencies for review and action. After extensive review, the government chose to concentrate on two major issues: to increase the use of occupant-restraint systems and to decrease impaired driving.

In 1991, 48 percent of all drivers killed in road crashes had consumed some alcohol prior to the crash. The largest group of fatally injured drinking drivers—62 percent—comprised those with blood alcohol levels over 150 mg%.

Strategy to Reduce Impaired Driving (STRID). STRID's objective was to reduce roadway collision deaths attributed to alcohol by 20 percent by the end of 1995. Each province has introduced its own programs designed to reduce deaths related to impaired driving. Programs vary from good host/hostess programs aimed at the person who organizes a social event to programs aimed at potential passengers in vehicles operated by an impaired driver, to television advertisements aimed directly at vehicle operators. On a national level, a 6.4-percent decrease in the problem (as measured by the ratio of fatally injured impaired drivers to fatally injured nondrivers) has been achieved.

20.1.4.2 Speed. Speed is a core issue in road safety although the relationship between speed and accidents is a very complex one. Two "laws of physics" are considered relevant (Maycock[498]). The first law is that the stopping sight distance (in an emergency) is proportional to the square of the speed [compare the simplified SSD model, Eq. (15.7) in Sec. 15.2.2.1]. It is reasonable to state that higher speeds result in a higher probability of becoming involved in an accident. The second relevant law is that the kinetic energy is proportional to the square of the vehicle's speed, which means that a high-speed accident will involve more damage and more serious injuries [compare Eq. (10.28) for the radial direction in Sec. 10.2.1.2].

Based on empirical data, Transport and Road Research Laboratory (TRL)[718] concludes that if mean speeds can be reduced by 1 km/h, then, on average, injury and accidents will be reduced by about 3 percent (Finch, et al.[194]). An increase of 1 km/h leads to an increase of 3 percent. More severe accidents (fatal) will be reduced by a greater amount according to the "power laws" of Nilsson.[551] This means that only minor changes in speed will lead to significant effects.

Nowadays, the problem is that in many countries, on many roads and under many circumstances, many drivers in almost all types of vehicles drive at excessive speed with no regard for the prevailing conditions (although some drivers drive too slow and cause dangerous situations for other road users!). This makes speed management one of the most complex problems. A wide range of possible interventions are available or will become available in the future. But the complex nature of the problem, the pleasant and rewarding consequences for a driver driving at high speed, the social acceptability nowadays of speeding, and the actual high performance of modern cars (top speeds of 160 km/h are not exceptions any more) demand that policy makers, who have the responsibility for being ahead of public opinion, move just far enough ahead to maximize the acceleration of change without alienating the public (Allsop[17]). We should be amazed that we accept the present levels of road traffic speed!

Most of the time speed management means the limitation of personal freedom. This has to become acceptable to the majority of drivers by proper road design, proper speed limits, and education and information while the minority of irresponsible drivers have to be confronted with legal sanctions.

Authors' comment: Since "speed" has the greatest impact on traffic safety, the main goal in developing this book has been the establishment of three safety criteria to evaluate new designs, redesigns, and RRR projects, respectively, to examine existing alignments with respect to good, fair (tolerable), or poor design practices. Since all three criteria are directly and quantitatively related to speed (design speed/operating speed), it can be expected that their application will lead to sound highway geometric design in the future and to a reduction in accident frequency and severity. Of course proper road design does not include all safety impacts of the driver-vehicle-road system but a very important one. Therefore, providing quantitative and, where so far not possible, at least qualitative rules for the relationship speed-design-accident situation, as discussed numerous times in Chaps. 8 to 19, means a significant step forward with respect to successful traffic safety work.

20.1.4.3   *Assistance Given to Road Accident Victims.* Road deaths occur either within a few minutes of the crash at the scene of the accident, within a few hours during transport to the hospital, or within 30 days. In the Netherlands, 57 percent die at the scene of the accident, 22 percent the same day, and 21 percent within 30 days. Very little can be done for the first group. The lives of the second group depend on adequate management, that is, first aid and emergency calls as soon as possible, a correct initial diagnosis by the telephone operator, the quick response by qualified medical personnel to the accident spot, correct diagnosis at the scene and optimal stabilization of the patient by properly trained and equipped personnel, fast (and safe) transport to a hospital and proper care during that transport, and treatment in a hospital with a specialized trauma unit. This is a chain, and a chain is only as strong as its weakest link. It requires an integrated approach: the education and training of the public, the emergency services such as the police and firefighters, and the emergency medical services such as ambulance attendants. A second important element is communications: public and car telephones equipped with emergency number access—with only one emergency number that is well known to the public. Additional communications might be useful: for example, emergency telephones on highways, radios in heavy vehicles, etc. The organization behind the telephone has to be a fast and professional one. Victims can be transported by ambulance or by helicopter if they are multitrauma patients. Well-managed (general) hospitals and specialized trauma hospitals are needed. Time lags have to be as short as possible to allow proper treatment in 'the Golden Hour'.

The positive effects of this chain approach have been well-documented, leading to conclusions about preventable deaths and decreasing financial costs of medical consumption and (social) insurance costs, e.g. Draaisma,[147] NRTAC,[537] and Trinca, et al.[717]

20.1.4.4   *Seat Belts.** (Proper) seat-belt usage has been recognized as a major road-safety measure and is part of road-safety policy in almost all countries of the world. Based on careful analysis, the effects of seat-belt usage in front seats appears to be about 40 percent, which means that if seat-belt usage increases from 0 to 100 percent, the number of people killed in a car crash will decrease by 40 percent. The effects of the use of rear seat belts have not yet been studied extensively but are expected to be somewhat less than for front seat belts. Air bags (without the use of seat belts) reduce driver fatalities by 20 percent. Combining an airbag with seat belts results in another 25 percent effect or about 45 percent (Evans[179]).

One of the most effective means for improving rates of seat-belt use is to make it compulsory by law (Hagenzieker[252]): compulsory for front-seat usage seems to be very common and also child restraints. A lot of progress is being made for rear seats. Rates of seat-belt use of over 90 percent have been measured in a few countries: Germany, United Kingdom, and Finland. The high compliance in the United Kingdom was not achieved by legislation, but, in fact, was the end of a long process to raise seat-belt use rates by publicity.

20.1.4.5   *Vehicle Safety.* Any agenda dealing with accident avoidance (precrash measures or primary/active safety) and injury minimization (crash measures or secondary/passive safety) should include the following items:

- Speed control, that is, maximum speed and performance limits using intelligent speed control technology in which a car communicates with the surroundings and where speed limits are set depending on the type of road or risk-influencing circumstances (for example, bad weather). Special attention has to be given to public acceptance, fail-safe technology, legal problems, and the introduction of such facilities.
- Improvement of vision and conspicuousness, that is, daytime running lights.

---

*See also Sec. 20.3.3.3 and 20.3.3.4.

- Improvement of protection from frontal impact (offset impact), side impact, compatibility between cars and underride protection, protection of pedestrians, and cyclists and motorcyclists.
- Seat belts, air bags, and head restraints.

More detailed information about vehicular safety is given in Sec. 20.3.

20.1.5 Better Roads Improve Road Safety

Investment to expand and improve road infrastructure is a major factor in improving road safety. Many countries can illustrate this statement by their own experiences (for example, Japan). The principles of safe network design and road design are rather well known, as demonstrated by Part 1, "Network," Part 2, "Alignment," and Part 3, "Cross Sections" in this book and the important references, for example, Refs. 103, 244, 246, 247, 308, 442, 453, 460, 461, 559, 621, 710, 718, and 728. However, practical conditions, criteria other than safety, limited funds, and environmental demands lead rather often to suboptimal road design.

A comparison of the fatality rates for various types of roads reveals that the traditional roads (main roads in and outside built-up areas with no traffic restrictions) are among the most hazardous (Fig. 20.5).

The fact that the proportion of safe types of roads (traffic calming "type 7" and motorways "type 1") along the total length of the infrastructure has increased and that the proportion of mobility on these roads and streets has increased even more sharply has certainly contributed to the drop in the fatality rate. However, some eroding safety effects can be observed as a result of the deterioration of our infrastructure.

Authors' comment: Note, in many countries, the injury and especially the fatality rates of road types 3 and 4 (outside built-up areas) are significantly higher than for road types 5 and 6 (inside built-up areas) (compare, for example, Fig. 9.2).

Roads are built with one major function in mind:[442] to enable people and goods to travel from one place to another. However, it is necessary to differentiate traffic functions because the character of the travel process differs (long distance, allowing for access, etc.), and these differences require different road design. An explanation for the high accident rates on "all-purpose roads" might be that these roads are multifunctional (flow/connector function, distributor/collector function, and access/local function) and allow different types of road users in the same space and, at the same time, at relatively high speeds and big speed differences. Design guidelines and road design standards are available in many countries (Ruyters[621]). But, to date, these existing national standards only rarely contain information on the safety effects of the road design that is recommended or even prescribed. To enable the design of safer roads, more clarity is needed about the relationship between the layout and the safety aspects of the infrastructure elements. Several approaches could be used to improve the safety quality of a road network. The rather traditional approach is to analyze accidents in accident-prone locations (black spots, black routes, and black areas) by finding similarities between the types of accidents. Improvements in road design then have to be found in order to eliminate these accidents (Delmarcelle[136]).

This so-called black-spot approach is effective in reducing fatalities, injuries, and accidents; effects of up to 50 percent have been reported. This approach could be considered as the first generation of road-safety measures and the start of road-safety policy dealing with "the road." It can be strongly recommended, especially when low-cost engineering measures are proposed (Slop[652]).

Starting with road-user behavior and expectations, a second generation of road-safety measures can be defined. The design philosophy of this approach could be characterized by consistency and uniformity in road design, homogeneity in traffic streams (speed!), and predictability for the road user. With proper road design, the road user's expectations can be induced and have to be met, always and everywhere. These principles find concrete shape in road classification or categorization and well-founded design standards. Experiences from different countries indicate the positive safety effect of applying modern design standards compared with those from some

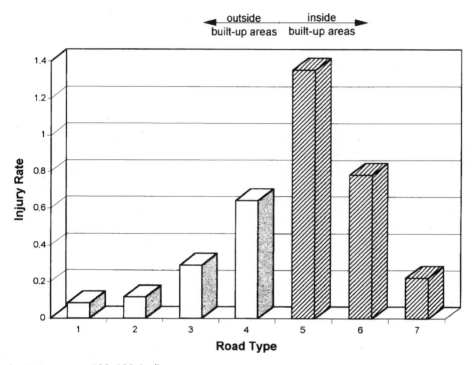

1. Motorway: 100-120 km/h
2. Motorroad: 100 km/h
3. Arterial rural road: 80 km/h
4. Local rural road: 80 km/h

The most elementary classification for the built-up area is subdivided into:

5. Arterial roads with a speed limit of 50 km/h (sometimes 70 km/h)
6. Residential streets with a speed limit of 50 km/h
7. "Woonerf" and residential street (approx. 8 km/h to 30 km/h)

Legend: Compare also the road classification in Table 6-2.

FIGURE 20.5 Injury rates in The Netherlands (1986) per million motor vehicle kilometers.[746]

decades ago (tens of percents).[286] The harmonization of standards, at least in terms of the assumptions underlying these standards, must play an important role, and future research results will have to decide which assumptions and which scientific evidence should form the real core of our knowledge. More attention has to be paid to the human operator of the road system (Chap. 19) and how the road user responds to changes in the road system (Evans[179]).

Authors' comment: It is the goal of this book not only to fulfill this second generation of road-safety measures but to extend it in order to offer reliable recommendations for practical design tasks with special emphasis on traffic safety in a quantitative, and not just qualitative, manner. On the basis of speed and accident research, the following important issues, relationships, and interrelationships should be considered and incorporated in a future overall framework for safer road design:

- Design consistency (safety criterion I)
- Operating speed consistency (safety criterion II)
- Driving dynamic consistency (safety criterion III)
- Relation design, consistency of curvilinear alignments, and sound transitions between independent tangents and curved sections
- Tuning of design elements of horizontal/vertical alignment, cross section, and sight distance

The third generation of road-safety measures is described under "Sustainable Safe Traffic" in Sec. 20.1.6.3. The inspiration behind this concept is that we should no longer accept the fact that we are handing over a road traffic system to the next generation which will lead to hundreds of thousands of fatalities and millions of injured every year. We should try to reduce drastically the probability of accidents in advance, primarily by means of the infrastructure design and by accepting fewer compromises than we do today.

20.1.6 Promising New Developments

Of course, different countries with different cultures, different levels of motorization, and different histories of road-safety policy have different means for improving road safety. Some promising new developments are presented here.

20.1.6.1 Road-Safety Targets. An outline of good practice for targeted road-safety programs has been presented in a recent OECD report[555] (see also Read[593]). A targeted road-safety program is based on a clear target and consists of a set of countermeasures designed to reach the target.

The following steps can be distinguished:

- *Analyzing accident trends and road-safety problems:* It is useful to provide a description of road-safety trends and the forces underlying them, both in order to understand past changes and to gain an idea of how road safety may develop in the future. A systematic description of road-safety problems can help in identifying vulnerable road users or high-risk groups that need special attention in a road-safety program (see Sec. 20.2).

- *Assessing the potential of countermeasures:* Assessing the potential of the various countermeasures requires knowledge both of their effectiveness and the target group. Although this knowledge will either be lacking or highly uncertain, it is considered good practice to estimate their potential. The theoretical safety potential can rarely be fully realized. Practical, political, and economic obstacles are always present and realistic road-safety programs should address the implications of these obstacles.

- *Assessing the effects of confounding factors:* Exogenous factors affecting safety are always present and their implication for road-safety policy making should be addressed as well.

- *Setting the targets:* Clearly formulated road-safety targets can guide policy making in a better way than less clearly formulated road-safety targets. Highly ambitious quantified road-safety targets can help policy making but there is no guarantee that the desired results will be obtained.

- *Formulating alternative plans of action:* In order to find the least costly alternative, alternative plans of action should be examined systematically. The phases involved should generate different alternatives, estimate cost-effectiveness or a cost-benefit ratio for each countermeasure, and construct a cost-effective program.

- *Monitoring and feedback:* The careful monitoring of targeted road-safety programs is needed in order to explain policy performance and, if necessary, to revise the targets and/or the plans and programs. Monitoring is crucial in a targeted program, and leads to the need to improve the quantity and quality of road-safety data; the lack of relevant, accurate, accessible, timely, standardized, and integrated data hampers the development, delivery, and evaluation of road-safety countermeasures.

20.1.6.2   Safety Auditing.   Road safety is a quality issue of road traffic and has to be balanced with other issues, such as level of service, access to destinations, environmental impact, costs, etc., when it comes to deciding the infrastructural projects for investment. In making decisions on infrastructural projects, road-safety issues already have been considered, as explicitly as possible, during the planning phase.

On a strategic level, the safety consequences of changes (redistributions) in traffic over a road network due to infrastructural projects (new roads or new layout of roads) have to be assessed by using a *scenario technique.* This technique uses the fact that different categories of roads (with different road and traffic characteristics) turn out to have different road-safety records that depend on traffic volume. By modeling the road type, the values of relevant safety indicators and traffic volumes, road-safety impacts of different alternatives can be calculated. Secondly, on a project level, an audit technique can be used to make, as explicit as possible, the safety consequences of certain choices in the detailed planning and the design process and to optimize a road design. The primary objective of using an audit technique is to ensure that road safety is optimally incorporated during the design and realization phase of infrastructure projects in order to arrive at a road design which is simple and recognizable for future road users, thereby minimizing potential error.

This new tool for accident prevention is in operation in the United Kingdom,[137,138,302] the United States,[309] Australia,[38] and New Zealand. In France, two audit experiments were undertaken on French toll motorways in 1995.

The road safety impact assessment (including the audit technique) is a useful tool for evaluating those issues relevant to road safety at an early stage and during all the subsequent phases of road design (for new and also for existing roads). To obtain optimum benefit from this concept, the following recommendations were formulated by Wegman, et al.:[745]

- Use independent auditors
- Publish the auditors' reports
- Commence a safety audit with strategic scenario results
- Publish audit reports after completion of the preliminary design, after completion of the detailed design, and just before the road is opened
- Use the audit reports to advise the initiators of a road project (the initiator, though, remains fully responsible)
- Develop and use "best practice" checklists

Of course, to carry out road-safety audits, teams of well-trained experts are needed.

Although the concept of auditing looks very promising because of its preventive and explicit nature and the fact that audits are used successfully in other fields, so far no research is available in which these positive expectations are confirmed in terms of accident reduction.

20.1.6.3   Sustainable Safe Traffic.   A scientifically supported, long-term concept for the implementation of an essentially safer road traffic system can best be achieved by tackling the causes underlying accidents, removing areas of conflict, or making these controllable by road users. Where accidents still occur, the risk of serious injury should be virtually excluded.

A sustainable safe traffic system has:

- An infrastructure whose proper road design is adapted to the limitations of human capacity. (Corresponding proposals are elaborated in Chaps. 8 to 19 and in Part 3, "Cross Sections.")
- Vehicles designed with ways to simplify the task of the road user and constructed to protect human beings as effectively as possible (Sec. 20.3).
- A road user who is adequately educated, informed, and, where necessary, controlled. (First efforts are presented in Chap. 19.)

As to the infrastructure, the key to arriving at sustainable safety lies in the systematic, consistent application of three safety principles that reduce, in advance, the liability of encounters with implicit risk. The three safety principles are:

- The functional use of the road network by preventing unintended use of each road
- The homogeneous use by preventing large differences in vehicle speed, mass, and direction of movement
- The predictable use, thus preventing uncertainty among road users by enhancing the predictability of the road's course and the behavior of fellow road users

This approach could be characterized as preventive, systematic, and consistent, which is in contrast to the curative, incidental, and the compromise approach of many road-safety policies of today.

20.1.6.4 Improved Police Enforcement.   For many years, the police have played a prominent role in improving road safety. It is generally accepted that the success of police enforcement in changing human behavior depends on the ability to create a general and a specific deterrence. *General deterrence* relies on the perception of the road user that traffic laws are enforced and violators are prosecuted and punished. *Specific deterrence* deals with actual experiences of violators who are detected, prosecuted, and punished. Both deterrences have to be part of a far broader road-safety strategy to prevent an impossible enforcement task.

The general principles of effective police enforcement are rather well documented.[224,742,771] These principles have been applied to major offenses (drinking and driving, speeding, red-light violation, and seat-belt enforcement) in many countries and much practical information is available. The most important principles are as follows:

- Combine police enforcement with publicity before and during enforcement activities; it is essential that road users actually observe increased levels of enforcement.
- Influence positive social acceptance of enforcement by publicity, which could lead to better safety effects.
- Combine continuity with the short-term intensive enforcement (for example, Secs. 13.1.3.2 and 13.2.3.2) and blitzes of high-risk behavior and locations.
- Combine a selective mix of visible and less visible controls.
- Apply other legal sanctions than just fines, such as license suspension and revocation of licenses, curfew, point demerit schemes, etc., and ensure that the chance of violations being detected are high.

This leads to the conclusion that enough knowledge is available and that improved and successful police enforcement is based on better management, better education and information, better motivation, a better build-up of a chain with strong links, and better use of available knowledge and expertise. It now comes down to implementation with the better use of existing forces rather than more manpower and more equipment. However, our knowledge about the cost-effectiveness of different strategies is still poor and it is recommended that a meta-analysis be made in this field and that research projects comparing different strategies be carried out. It is evident that in central and eastern Europe and in developing countries, the nature of the police enforcement problem and possible solutions will differ but the general principles remain the same.

20.1.6.5 Telematics.   Telematics (Advanced Transport Telematics ATT) is increasingly being considered as a means to improve traffic and transport management as well as road safety. Road-safety arguments turn out to be a good sales argument. High expectations are created around telematic applications and their expected positive effect on road safety, that is, expectations which are not completely fulfilled. In addition, the developments in this area are rarely guided by relevant social and policy-making developments but rather by a "technology push." From the point of view of road safety, the following questions are of importance in an assessment of telematic applications:

- Controlled traffic growth (that is, route planners and fleet management with GPS)
- Optimal distribution of traffic over time and space (actual radio traffic information)
- Management of traffic streams (that is, homogenizing driving speed)
- Reducing risks (that is, warnings of extreme road and weather conditions, improvement of visibility, and intelligent speed limiters)
- Restriction of the negative effects of accidents (accident alarm system and coordination of assistance)

There are also developments afoot which could have a negative influence on the driving behavior and driving performance (that is, the car used as office, including telephone and fax). It is necessary that safety assessments on such developments be made, which could perhaps lead to regulations. Over and above this, there should at least be a code of practice, a checklist such as exists in England whereby possible negative consequences can be established. Furthermore, from the point of view of the user, the acceptance of all these developments is of importance and also the comprehensibility (man/machine interaction), especially in an aging society.

Governments and representatives of road users should closely and more carefully follow telematic developments and applications (individual and collective systems as well), make assessments at management level, and, if necessary, try to make corrections. Of course this should be done on an international basis.

20.1.7 Preliminary Conclusions and Recommendations

1. It can be reasonably expected that the annual number of road accident victims is more likely to increase than to fall, particularly in developing countries and probably also in countries in central and eastern Europe. However, in highly motorized countries, where a considerable drop was registered in the 1970s, a less significant drop has been seen in recent years, and sometimes even a "rebound" effect. A further drop in road accident statistics may be achieved through an effective road-safety policy.

2. Although an increase in mobility can be linked to a drop in the number of road accident victims, as has been shown in many highly motorized countries, it must be feared that a high growth in mobility will not be associated with a parallel reduction in the road accident statistics.

3. There is no question about the effectiveness of some remedial measures in preventing accidents and their severity: prevention of drinking and driving, use of seat belts and crash helmets, improvements to road infrastructure and car design, the maintenance of vehicles, and proper assistance for road accident victims. Sufficient research results are available on the principles of these measures and of "best practices" as well. Although remedial measures could not simply be copied, adaptation to prevailing conditions are possible. Emphasis should be put on systematic and long-term sustained implementation.

4. No options seem to be available for simple, large-scale, new, effective measures which can further promote road safety. On the contrary, we should focus on the problems and measures that have already been recognized and adopted. Better implementation of well-known measures, leading to larger and longer-lasting effects at less expense, is more appropriate than searching for a new "universal remedy." A large number of countries are currently focusing on three areas of priority: driving under the influence of alcohol or drugs, speeding, and insufficient use of seat belts and helmets. Furthermore, attempts are being made to improve road safety for high-risk groups (young drivers) and vulnerable road users (pedestrians, children, and the elderly).

5. The international community has been unable—even in a world where a multitude of technology is available to simplify communications over long distances (fax and internet)—to organize the transfer of knowledge effectively. It should be an important mission for international organizations and of individual countries, together with the professional and research community, to make an effort to realize such a transfer of knowledge.

The authors thank Mr. F. Wegman for his permission to use his paper of the PIARC Conference in Montreal 1995[746] for Sec. 20.1.

20.2 WAYS TO IMPROVE TRAFFIC SAFETY THROUGH ACCIDENT ANALYSIS AND EVALUATION: UNITED STATES VERSUS WESTERN EUROPE

20.2.1 Introduction

Most countries compare their own state of affairs or achievements with other countries. The countries chosen for comparison are usually those considered to be in the same "league" as the comparing country as far as the subject to be compared is concerned. Highly developed, industrialized countries usually compare themselves with other such countries in the belief that differences will not be great. Most countries go even further by concentrating on comparisons with neighboring highly developed, industrialized countries in the belief that, since they are geographically close, these countries will have even fewer differences than countries further away. Countries compare their economies and health and welfare systems with each other as well as many other aspects of life. There are generally two aspects which fall under the comparison:

1. The present level

2. The historical development, which often leads to the expected future development[694]

Countries are pleased if they appear to be doing better than other countries and disappointed if the reverse is the case. In the latter case, questions are asked about a country's poor performance and it looks to those countries performing better in the hope of learning from them. In road-safety policy and research, the situation is the same—there is more interest in those countries that appear to be safer. More detailed comparisons are made (their laws and traffic behavior) with these countries with the purpose of increasing one country's own safety.[694]

In 1949 Smeed published the first of his papers on the international comparison of road safety,[654] which contained the 1938 data for 20 highly developed countries and expressed a statistical relationship between the number of road deaths per inhabitant (mortality) and the number of motor vehicles per inhabitant (motorization). This relationship is one of a declining number of deaths per vehicle as motorization increases. He later extended his work to more detailed comparisons (modal split) and to lesser-developed countries all over the world (see, for example, Smeed[655] and Adams[7]).

Since then, many researchers, including the authors, have made international comparisons. Two subdivisions can be made: first, comparisons between two (or a few) countries and multinational comparisons, and second, the use of aggregated data and in-depth disaggregated data.

Because of the amount of work involved and the availability of data, the multinational comparisons limit themselves to the use of aggregated data, such as death rates, whereas those projects using in-depth data generally limit themselves to two countries.

Examples of the first type (multinational, aggregated data) are comparisons made by the authors between the United States and the countries of Europe. These comparisons have appeared in numerous papers and technical reports and in the proceedings of international conferences over the past two decades (see Refs. 105–107, 110, 340, 398, 400, 403, 413, 419, and 425–427). An example of the second type (binational, in-depth data) is a comparison between the Federal Republic of Germany and Great Britain.[254]

20.2.2 Background and Objectives

Worldwide, road accidents cause a great number of lost lives. In developed countries, the number of people killed in road accidents amounts to approximately 200,000/y. In developing countries, this figure is around 350,000/y.[180] One forecast has suggested that, by the year 2000, fatalities could reach 1 million.[511]

Traffic casualties can never be totally reduced by countermeasures to the design of roads and vehicles. Nevertheless, every effort has to be made to try to reduce human suffering and the cost to society.

The objective of this chapter, which addresses traffic safety in developed countries, is to discuss the road accident situation in the United States and the following six European countries: Austria, Denmark, France, Germany, The Netherlands, and Switzerland. These countries were selected on the basis of available accident data, especially with respect to fatality rates per billion driven vehicle kilometers, and *somewhat* comparable vehicle ownership ratios, that is, number of motor vehicles/1000 population (see NMVR in Table 20.2). Selected reference values for the countries considered are also given in Table 20.2.

Some of the data used in this paper have been derived from the "International Road Traffic and Accident Database" (IRTAD)[184] and "Statistics of Road Traffic Accidents in Europe and North America" (SRTAENA).[165] Although primary sources are not listed individually in this paper, IRTAD and SRTAENA contain complete listings, relying heavily on data provided by national statistical offices throughout the world.

20.2.3   International Comparisons

Clearly, the absolute number of road fatalities alone does not give an objective norm for comparison. This can be attributed to differences between the countries with respect to, for example, surface area, population, culture, topography, traffic laws and regulations, infrastructure, facilities provided for road users, and so forth (see Table 20.2). For this reason, the number of traffic fatalities per billion vehicle kilometers (mileage fatality rate)* and the number of road fatalities per 100,000 population (population fatality rate) were also taken into consideration. The first criterion (fatalities per billion vehicle kilometers) is usually considered to be the most reliable source for objective comparison.

Specifically, the objectives of this section are to:

1. Show quantitatively the changes in fatalities in the European countries mentioned earlier and in the United States during the time period 1980 to 1995.

2. Show quantitatively the changes in fatality rates during the time period 1980 to 1995.

3. Determine whether there were statistically significant changes in the traffic accident characteristics studied.

4. Identify those age groups that were most frequently involved in fatal accidents.

5. Identify those road users that were most frequently involved in fatal accidents.

6. Formulate mathematical relationships between fatalities/fatality rates and time.

---

*Fatality rate = fatalities per 10^g vehicle kilometers.

TABLE 20.2   Selected Reference Values for the Year 1995

	Country						
	Austria	Denmark	France	Germany	The Netherlands	Switzerland	United States
POP	8,040	5,216	58,027	81,538	15,425	7,019	262,755
LPR	106,397	71,255	954,700	640,000	113,419	71,027	6,285,629
LM	1,589	786	8,030	11,143	2,207	1,187	73,343
AR	83,850	43,069	551,208	357,039	41,526	41,293	9,363,353
NMV	4,404	2,104	28,515	47,487	6,598	4,121	192,337
NTW	532	49	2,768	3,981	760	704	3,932
NPC	3,480	1,611	24,210	40,405	6,129	3,229	121,997
NMVR	548	407	491	582	428	587	732

Note: POP = Home population, in 1000s; LPR = total length of all public roads, in km; LM = network length of all motorways, in km; AR = area of country, in km²; NMV = number of motor vehicles, except moped/mofa, in 1000s; NTW = number of motorized two-wheelers, in 1000s; NPC = number of passenger cars and station wagons, in 1000s; and NMVR = number of motor vehicles/1000 people.

TABLE 20.3   Traffic Fatalities, 1980 to 1995

Year	Country						
	Austria	Denmark	France	Germany	The Netherlands	Switzerland	United States
1980	2,003	690	13,672	15,050	1,996	1,209	51,091
1981	1,949	662	13,547	13,635	1,807	1,130	49,301
1982	1,933	658	13,527	13,450	1,709	1,156	43,945
1983	1,967	669	13,021	13,553	1,757	1,124	42,589
1984	1,814	665	12,737	12,041	1,615	1,068	44,257
1985	1,524	772	11,387	10,070	1,438	881	43,825
1986	1,495	723	11,947	10,620	1,527	1,003	46,056
1987	1,469	698	10,742	9,498	1,485	923	46,390
1988	1,620	713	11,497	9,862	1,366	917	47,087
1989	1,570	670	11,476	9,779	1,456	897	45,582
1990	1,558	634	11,215	11,046	1,376	925	44,529
1991	1,551	606	10,483	11,300	1,281	834	41,462
1992	1,403	577	9,900	10,631	1,285	834	39,235
1993	1,283	559	9,568	9,949	1,252	723	40,150
1994	1,338	546	9,019	9,814	1,298	679	40,716
1995	1,210	582	8,891	9,454	1,334	692	41,798
Percent change, 1995/1980	−39.6	−15.7	−35.0	−37.2	−33.2	−42.8	−18.2

20.2.3.1   Traffic Fatalities.   Table 20.3 provides the absolute number of traffic fatalities. It can be seen that, although the number of registered motor vehicles in the countries considered have increased since 1980, the number of traffic fatalities has decreased. For instance, Switzerland experienced the largest decrease (42.8 percent), followed by Austria (39.6 percent), Germany (37.2 percent), and France (35.0 percent). The United States and Denmark experienced decreases of 18.2 percent and 15.7 percent, respectively.

The findings for the United States are rather impressive considering the fact that Americans spend nearly twice as much time on the road as European drivers. Americans use motor vehicles for 82 percent of their trips compared with 48 percent for Germans, 47 percent for the French, and 42 percent for the Danes. An average U.S. motorist now drives the equivalent of two round trips a year between Los Angeles and New York City. And there are at least 23 million more vehicles than licensed drivers in the United States.[583]

Generally, the favorable fatality developments in the western European countries and in the United States are due, in large part, to increased public concern about highway safety; improved highway design and construction; improved crashworthiness; increased availability of air bags and automatic safety belts; improved emergency medical service capacity to treat crash victims quickly and skillfully; new emphasis on motorcycle licensing, training programs, and helmet use; and transfer from more dangerous to less dangerous modes of travel.

20.2.3.2   Fatality Rates per Billion Driven Vehicle Kilometers.   For years, traffic engineers and researchers have made great effort to identify and analyze traffic accidents and, subsequently, to recommend and apply countermeasures. Their efforts have been successful in the countries considered, as can be seen from the declining trends in fatality rates per billion vehicle kilometers shown in Table 20.4. Note that the improvement in active and passive vehicle safety also plays an important role here (compare Sec. 20.3).

Examination of Table 20.4 reveals the following:

- All the countries considered experienced more or less similar decreasing trends in fatality rates per billion vehicle kilometers, ranging from 67.0 percent in Austria to 45.6 percent in Denmark.

TABLE 20.4   Fatality Rates per Billion Vehicle Kilometers, 1980 to 1995

Year	Country						
	Austria	Denmark	France	Germany	The Netherlands	Switzerland	United States
1980	56.27	24.98	44.10	39.81	26.69	30.19	20.91
1981	54.75	24.63	42.65	37.34	24.45	26.97	19.84
1982	53.08	25.02	41.57	35.75	22.59	26.96	17.22
1983	52.13	24.50	39.06	34.95	22.39	25.61	16.10
1984	46.46	23.10	37.29	30.16	19.91	24.05	16.08
1985	38.23	25.55	32.53	25.05	17.78	19.41	15.44
1986	35.08	22.56	33.14	24.80	17.99	21.05	15.69
1987	29.72	20.71	28.05	21.03	16.76	18.50	15.09
1988	32.48	20.37	27.25	20.74	14.60	17.64	14.53
1989	29.43	18.82	26.50	19.98	15.13	16.58	13.59
1990	27.89	17.32	25.72	19.96	14.27	16.68	12.91
1991	26.96	16.17	23.40	19.68	12.90	14.65	11.93
1992	23.33	14.78	21.43	18.02	12.47	14.36	10.85
1993	20.93	14.22	20.31	16.65	12.06	12.33	10.86
1994	20.97	13.65	18.52	16.61	12.02	11.30	10.72
1995	18.59	13.60	17.93	15.67	12.18	11.39	10.70
Percent change, 1995/1980	−67.0	−45.6	−59.3	−60.6	−54.4	−62.3	−48.8

- The United States has the lowest fatality rates.
- Of the European countries, Switzerland has the lowest fatality rate whereas Austria had the highest in 1995.
- The fatality rate of Austria (18.59 fatalities per billion vehicle kilometers) and France (17.93) in 1995 are higher than the fatality rate of 17.22 already reached by the United States in 1982.
- The 1995 fatality rate for Austria is about 1.7 times that of the United States, 1.6 times that of Switzerland, 1.5 times that of The Netherlands, 1.4 times that of Denmark, 1.2 times that of Germany, and slightly higher than that of France.

Because fatality rates describe the fatality risk of being involved in a fatal accident, it follows then that the numbers shown in Table 20.4 should be considered multipliers of the chance of becoming involved in a fatal accident. For instance, compared to the United States, the fatality risk in Austria and France is about 1.7 times as high, in Germany about 1.5 times as high, in Denmark about 1.3 times as high, and in The Netherlands and Switzerland about 1.1 times as high relative to 1995.

In other words, the fatality rates of some European countries are becoming more closely comparable to that of the United States. The rates for other European countries are still decisively higher than those of the United States or other European countries.

Furthermore, it is interesting to note that, despite major differences in the 1980 fatality rates between the countries under study (Table 20.4), the percentile changes between 1980 and 1995 varied within relatively narrow margins. That means in those countries where a high safety standard was already present relative to 1980, a continued quality of good safety engineering has led to a further significant decrease in fatality rates during the observed time period (compare, for example, Denmark and the United States).

20.2.3.3   *Fatality Rates per 100,000 People.*   Yearly fatality rates per 100,000 people between 1980 and 1995 are shown in Table 20.5, from which the following observations can be made:

TABLE 20.5   Fatality Rates per 100,000 People, 1980 to 1995

Year	Country						
	Austria	Denmark	France	Germany	The Netherlands	Switzerland	United States
1980	26.55	13.47	25.44	19.25	14.17	19.17	22.48
1981	25.81	12.92	25.07	17.39	12.72	17.84	21.47
1982	25.48	12.85	24.90	17.15	11.96	18.14	18.94
1983	26.03	13.08	23.84	17.32	12.25	17.54	18.18
1984	24.03	13.01	23.23	15.44	11.22	16.61	18.72
1985	20.17	15.10	20.68	12.96	9.95	13.64	18.36
1986	19.78	14.13	21.61	13.67	10.51	15.47	19.10
1987	19.41	13.62	19.35	12.21	10.16	14.16	19.06
1988	21.35	13.90	20.62	12.66	9.28	13.96	19.16
1989	20.66	13.06	20.49	12.47	9.83	13.55	18.36
1990	20.26	12.35	19.92	13.96	9.24	13.87	17.85
1991	19.97	11.78	18.54	14.17	8.53	12.36	16.44
1992	17.83	11.18	17.31	13.24	8.49	12.20	15.38
1993	16.11	10.79	16.63	12.29	8.22	10.47	15.57
1994	16.69	10.51	15.60	12.07	8.46	9.74	15.64
1995	15.05	11.16	15.32	11.59	8.65	9.86	15.91
Percent change, 1995/1980	−43.3	−17.2	−39.8	−39.8	−39.0	−48.6	−29.2

- In 1995, France had the highest fatality rate (15.32 fatalities per 100,000 people in Europe), but rates of 15.91 for the United States and 15.05 for Austria are similar. The lowest rate of 8.65 is in The Netherlands, followed by 9.86 for Switzerland, 11.16 for Denmark, and 11.59 for Germany.

- All the countries experienced decreases in fatality rates per 100,000 people, ranging from 48.6 percent in Switzerland to 17.2 percent in Denmark.

Despite the somewhat confusing numbers shown in Tables 20.4 and 20.5, both tables indicate that Austria and France have the highest fatality rates, whereas The Netherlands, Switzerland, and Denmark have the lowest; Germany falls somewhere in between. That means both fatality rates yield similar conclusions with respect to the fatality situations in the investigated European countries.

This statement, however, does not apply to the United States. For instance, with respect to fatality rates per billion vehicle kilometers, the United States has the lowest rate of all the countries under study (see Table 20.4). However, with respect to fatality rates per 100,000 people (see Table 20.5), the United States has the highest.

This can probably be attributed to the significantly higher vehicle mileage driven in the United States compared to that in Europe. The latter could certainly have an effect on the outcome shown in Table 20.4 but not on the outcome of Table 20.5, which is only related to population figures.

Despite the biased results for the United States, the criterion (fatalities per billion vehicle kilometers) is regarded by many authors and, as stated earlier, as the most reliable source for an objective comparison because the vehicle mileage indicator expresses in a direct way the actual traffic situation which exists in a country, while the population figure can only provide us with indirect information about the traffic situation. However, because the vehicle mileages driven cannot be subdivided according to different age, road user, or other traffic participant groups, the fatality rate per 100,000 people, therefore, could be used as the only criterion for making comparisons, for example, with respect to age or road-user groups.

20.2.4 Statistical Significance

To determine if there were statistically significant changes in the accident characteristics studied during the period 1980 to 1995—with respect to the total number of fatalities, fatality rates per 10^9 vehicle kilometers, and fatality rates per 100,000 people—the statistical analysis t test[84] (small samples, unequal variances) was used in this study for comparing fatalities and fatality rates. The test statistic is the t' value defined as

$$t' = \frac{X_1 - X_2}{[s_1^2/n_1 + s_2^2/n_2]^{1/2}} \tag{20.1}$$

where X_1 and X_2 = sample means of the two populations
s_1 and s_2 = population standard deviations
n_1 and n_2 = population sample sizes

The t' value is assumed to follow a t distribution with the degrees of freedom given by

$$df = \frac{f(s,n)}{g(s,n)} \tag{20.2}$$

where $f(s,n) = [s_1^2/n_1 + s_2^2/n_2]^2$
$g(s,n) = [(s_1^2/n_1)^2/(n_1 - 1) + (s_2^2/n_2)^2/(n_2 - 1)]$

In the null hypothesis, H_0, the sample means of the two populations do not differ. However, for the alternative hypothesis, H_1, the sample means of the two populations do differ. Because H_1 does not state the direction of the predicted differences, the region of rejection is two-fold.

The following time periods were considered for analysis in each of the subject countries:

- Time period I includes the years 1980 to 1984 to describe the fatality situation at the beginning of the 1980s.
- Time period II includes the years 1985 to 1989 to describe the fatality situation in the second half of the 1980s.
- Time period III includes the years 1990 to 1995 to describe the fatality situation at the beginning of the 1990s.

Table 20.6 shows the t-test results for traffic fatalities in the countries studied for different time periods. The table indicates the following:

- Between time periods I and II, the western European countries, with the exception of Denmark, experienced significant improvement in safety for the total number of fatalities and for the fatality rates in question. The improvements in the United States were significant for the fatality rate per 10^9 vehicle kilometers and insignificant (marginal) for the total number of fatalities and the fatality rate per 100,000 people.
- Between time periods II and III, the United States experienced significant improvement in safety for the total number of fatalities and for the fatality rates in question. The European countries, on the other hand, experienced significant improvement for the fatality rate per 10^9 vehicle kilometers as well as significant and marginal improvements for the total number of fatalities and for the fatality rate per 100,000 people. Germany was the only country to experience an insignificant deterioration for the total number of fatalities and for the fatality rate per 100,000 people. This may be due to the reunification of the two Germanies in 1989, as can be deduced, for example, from Fig. 20.6 with respect to fatalities per 100,000 people.[88]
- Between time periods I and III, the countries under study experienced improvement in safety which were statistically significant.

TABLE 20.6 *t*-Test-Results for Different Fatality Categories and for Different Time Periods

Country	Time periods I to II (1980–1989) t test			Time periods II to III (1985–1995) t test			Time periods I to III (1980–1995) t test		
	$t_{calc.}$	$t_{crit.}$	S	$t_{calc.}$	$t_{crit.}$	S	$t_{calc.}$	$t_{crit.}$	S
Traffic fatalities									
Austria	9.50 >	2.31	x	2.27 <	2.36	o	8.19 >	2.31	x
Denmark	−2.62 <	−2.57	+	6.17 >	2.31	x	5.97 >	2.36	x
France	7.17 >	2.31	x	3.80 >	2.36	x	8.53 >	2.36	x
Germany	6.99 >	2.57	x	−1.13 >	−2.31	—	5.64 >	2.36	x
The Netherlands	4.69 >	2.57	x	4.66 >	2.36	x	7.81 >	2.57	x
Switzerland	6.85 >	2.31	x	3.16 >	2.36	x	7.72 >	2.31	x
United States	0.26 <	2.57	o	4.84 >	2.26	x	2.70 >	2.45	x
Fatalities per 10^9 vehicle kilometers									
Austria	8.27 >	2.31	x	4.41 >	2.26	x	13.08 >	2.26	x
Denmark	2.36 <	2.57	o	5.10 >	2.45	x	13.48 >	2.31	x
France	6.17 >	2.31	x	4.49 >	2.31	x	11.43 >	2.26	x
Germany	6.90 >	2.36	x	3.52 >	2.36	x	10.21 >	2.45	x
The Netherlands	5.11 >	2.36	x	4.94 >	2.45	x	8.91 >	2.57	x
Switzerland	6.40 >	2.36	x	4.47 >	2.26	x	9.95 >	2.31	x
United States	2.98 >	2.57	x	6.74 >	2.26	x	6.32 >	2.57	x
Fatalities per 100,000 people									
Austria	9.76 >	2.31	x	2.83 >	2.45	x	8.25 >	2.36	x
Denmark	−2.54 >	−2.57	—	6.15 >	2.31	x	6.01 >	2.45	x
France	7.20 >	2.31	x	4.13 >	2.36	x	8.75 >	2.31	x
Germany	6.91 >	2.57	x	−0.19 >	−2.31	—	5.95 >	2.31	x
The Netherlands	4.74 >	2.57	x	5.46 >	2.36	x	7.57 >	2.57	x
Switzerland	6.84 >	2.31	x	3.61 >	2.36	x	8.13 >	2.31	x
United States	1.33 <	2.78	o	6.41 >	2.36	x	4.13 >	2.45	x

Note: S = statistical significance; x = significant improvement in safety; o = insignificant (marginal) improvement in safety; — = insignificant (marginal) deterioration in safety; and + = significant deterioration in safety.

Overall, the safety research in the six western European countries under study and in the United States should be regarded as impressive during the observed three time periods. As can be seen from Table 20.6, the symbol "x," which means significant improvement in safety during the time period considered for the country in question, applies to the majority of the cases under investigation. The symbol "+," which has been used by the authors in other studies to denote a significant deterioration in safety, is only partially available for Denmark between time periods I and II. In this connection, Fig. 20.6, which was prepared by Bruehning,[88] is of interest because it proves the positive development in fatality rates per 100,000 people in some of the countries under study.

The importance of the results presented in Table 20.6 is that they provide traffic safety officials with a tool to analyze the development of different traffic fatality categories. For instance, in previous studies by the authors,[106,425] the traffic safety situations in western Europe (WE) [western Europe included: Austria, Belgium, Denmark, Germany (West), France, Great Britain, Italy, The Netherlands, Norway, Sweden, and Switzerland] and in the United States were compared by *t*-test results, including additionally the involvement of different age groups and road

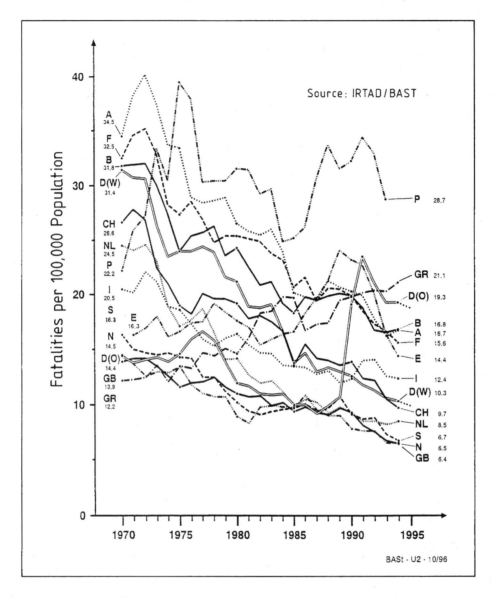

Legend: A = Austria, B = Belgium, CH = Switzerland, D(W) = Germany (West),
D(O) = Germany (East), GB = Great Britain, GR = Greece, E = Spain,
F = France, I = Italy, N = Norway, NL = The Netherlands, P = Portugal,
S = Sweden.
IRTAD = International Road and Traffic Accident Data Base,
BASt = German Federal Highway Research Institute.

FIGURE 20.6 Fatalities per 100,000 people for selected European countries.[88,184]

TABLE 20.7 Summary of Findings (*t* Tests) for Different Fatality Categories and Different Time Periods, Western Europe Versus United States[106]

	Time periods I to II, 1970–1977		Time periods II to III, 1975–1980		Time periods, III to IV, 1978–1983		Time periods, IV to V, 1981–1987	
	Western Europe	United States	Western Europe	United States	Western Europe	United States	Western Europe	United States
Total number of fatalities	x	x	x	+	x	o	x	o
Fatalities per 10^9 vehicle kilometers	x	x	x	—	x	o	x	o
Fatalities per 10^5 population	x	x	x	+	x	o	x	o
Age group:								
0–14	x	x	x	x	x	x	x	o
15–24	o	o	—	—	o	o	x	o
25–64	x	x	x	+	x	o	x	—
Over 64	x	x	o	—	o	o	x	+
Pedestrians	x	x	x	+	o	o	x	o
Bicyclists	x	+	x	o	o	o	x	—
Motorcyclists	o	—	x	+	o	o	x	o
Passenger car occupants	x	x	x	+	x	o	x	o
Truck and bus occupants	x	—	—	+	o	o	x	—

Note for symbol definitions, see Table 20.6.

users. A typical example of such studies is given in Table 20.7 for the time period between the early 1970s and mid-1980s. Based on this table, significant "+" and marginal "—" deteriorations in safety could be seen for the United States up to the mid-1980s. In contrast, western Europe experienced, for the most part, significant improvement in safety between the time periods compared for the different fatality categories. The favorable results shown in Table 20.6 for the United States are probably due to the positive accident development since the middle of the 1980s.

It should be noted, however, that the example shown in Table 20.7 clearly indicates where improvements in traffic safety are warranted, especially for those fatality categories which have experienced significant safety deterioration. Using results like those presented in Tables 20.6 and 20.7, traffic safety engineers can see the positive and negative trends, as well as developments during given time periods. They can, for instance, compare the results relative to their countries with those of other countries (making international comparisons) or compare on a national level the results relative to different states, provinces, counties, etc. In doing so, reliable decisions or urgently needed improvements with respect to individual traffic safety work could be made, evaluated, and followed up on.

20.2.5 Fatalities by Age Groups

By all means, traffic accidents are a plague to modern society—the main cause of death in certain age groups—but ranging high for all age groups.[315] An example is given in Fig. 20.7, which reveals the distribution in percent of traffic fatalities by age. The age groups used to develop the figure were: 0–14, 15–24, 25–64, and over 64. Unfortunately, a finer breakdown of traffic fatalities for persons between the ages of 25 and 64 was not available in the United Nations' publication, "Statistics of Road Traffic Accidents in Europe."[723]

Examination of Fig. 20.7 reveals the following:

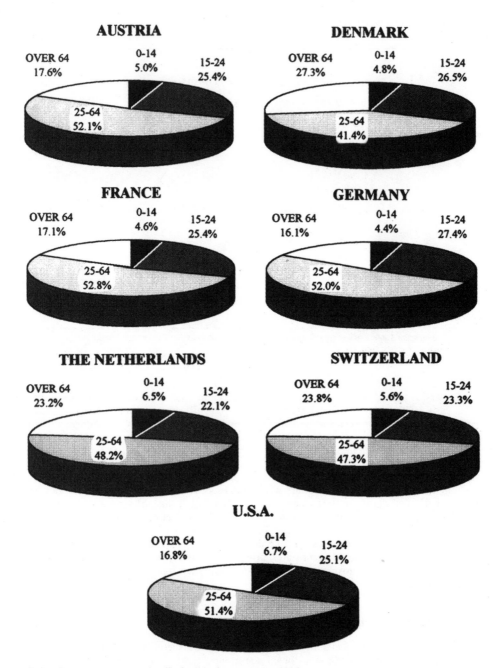

FIGURE 20.7   Percentage of traffic fatalities by age groups, 1995.

- Between 41 and 53 percent of persons killed in traffic accidents are in the 25- to 64-year age group, or on average about 10 to 13 percent, for persons between 25 to 34, 35 to 44, 45 to 54, and 55 to 64.

- Between 22 and 27 percent of persons killed in traffic accidents are in the 15- to 24-year age group. The average value for western Europe is 26.4 percent.[110]

- In Denmark, The Netherlands, and Switzerland, one-fourth of persons killed in traffic accidents are in the *over* 64-year age group. In the other countries considered, the percentage share is between 16 and 17 percent. The average value for western Europe is 19.7 percent.[110]

- Between 4 and 7 percent of persons killed in traffic accidents are in the 0- to 14-year age group. The average value for western Europe is 5.2 percent.[110]

The average values for western Europe referred to here are based on a 1989 study by the authors[110] which included additionally the countries Belgium, Great Britain, Italy, Norway, and Sweden.

In conclusion, on the basis of Fig. 20.7, the following statements can be made:

- In 1995, young adults (15 to 24 years of age) accounted, on average, for 25 percent of traffic fatalities in both western Europe and the United States; yet people in this age group made up only 15 to 16 percent of the populations of western Europe and the United States.[110] That means the proportion of traffic fatalities for young adults is approximately 1.7 times their proportion of the population. This result emphasizes the serious need for action to effectively tackle the problem facing this age group in both western Europe and the United States.

- The percentage of fatalities for persons 65 years and over is also higher than the percentage this age group makes up of the total population (15 percent in western Europe and 12.3 percent in the United States[110]). Comparing the population percentage with the fatality percentages for persons in this age group, as shown in Fig. 20.7, also indicates that immediate action should be taken to lower the accident involvement of these persons.

Similar results were reported in "Statistical Viewpoints" in Sec. 9.2.1.1 (see Fig. 9.3) in order to support the development of safety criteria I and II, whereas the aim of the present section is to analyze and evaluate the overall accident situation and its development on two continents by regarding as many viewpoints as possible.

The slight differences in percentages for the different age groups, as shown in Figs. 9.3 and 20.7, are due to the different years of investigation—1991 and 1995—and to uncertainties related to different data sources[165,184,723] used to develop both figures. Despite different assumptions, the very good agreement between Figs. 9.3 and 20.7 indicates that the outcomes of both investigations are not biased.

The results of Figs. 9.3 and 20.7 will be further analyzed by studying additionally the fatality rates per 100,000 people shown in Fig. 20.8.

Examination of Fig. 20.8 reveals the following:

- Austria, Denmark, France, Germany, Switzerland, and the United States show the highest fatality rates for persons in the 15- to 24-year age group, whereas The Netherlands show the highest fatality rate for persons in the over 64-year age group.

- With respect to persons in the 15- to 24-year age group, Austria shows the highest fatality rate (29.7 fatalities/100,000 people), followed by the United States (29.2), France and Germany (27.9), Denmark (21.9), Switzerland (18.9), and The Netherlands (14.3). The average value for western Europe is 25.3.[110]

- With respect to persons in the over 64-year age group, the United States shows the highest fatality rate (20.8 fatalities/100,000 people), followed by Denmark (19.9). The lowest rate of 12.1 was in Germany. The average value for western Europe is 17.9.[110]

In conclusion, it can be stated, based on Fig. 20.8, that the age groups 15 to 24 and over 64 are the most endangered of all age groups in the countries under study. Which of the age groups is

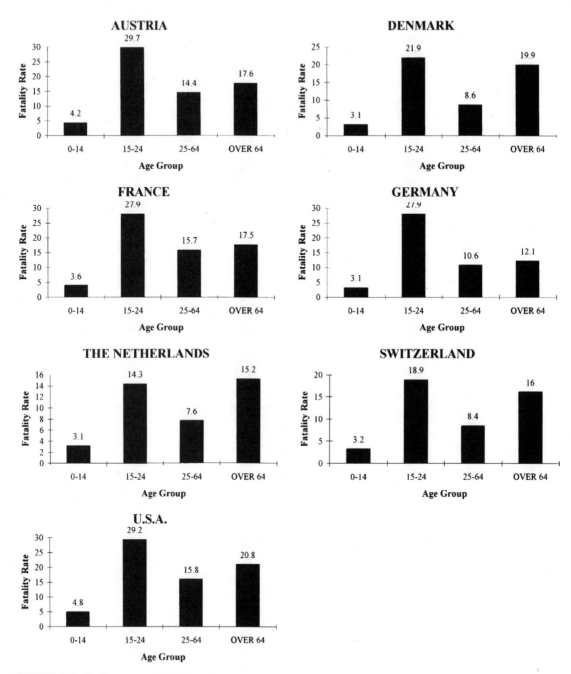

FIGURE 20.8 Fatality rates per 100,000 people by age groups, 1995.

more endangered depends on the individual country, as discussed earlier, and is probably due to different lifestyle patterns and village structures, that is, more pedestrians and bicyclists are on the streets, or on the size of the country, network functions, that is, long and short distances of travel, etc.

These statements confirm the previous results, and, consequently, the traffic safety work of the investigated countries, including the ones shown in Fig. 9.3, which should be directed to one or, better, to both of the identified age groups.

Finally, it should be noted that the fatality rates per 100,000 people by age groups shown in Table 20.8 and Fig. 20.8 are once again quite low in The Netherlands and Switzerland, but are somewhat high in Austria, France, and the United States.

The authors do not wish to speculate on the reasons behind these differences, but would merely like to identify the high-risk target groups, which have been shown in this section to be the age groups 15 to 24 and over 64.

The absolute numbers of traffic fatalities by age groups for the years 1980 and 1995 are given in Table 20.9, from which the following observations can be made:

- All the countries experienced significant decreases in traffic fatalities for persons in the 0- to 14-year age group, which ranged from 63.9 percent in Germany to 25.4 percent in the United States.

- All the countries show a significant decrease in fatalities for persons in the 15- to 24-year age group, ranging from 30 to 50 percent. That means even though this age group still represents, according to Fig. 20.8, a high-risk group, a successful safety work has been done during the observed time period from 1980 to 1995.

- There were wide variations in the decreases in traffic fatalities for persons in the 25- to 64-year age group. For instance, while Switzerland shows the largest decrease (36.3 percent), Denmark shows a decrease of only 3.2 percent.

- While the European countries show a decrease in traffic fatalities for persons in the over 64-year age group, which ranged from 52.4 percent in Germany to 7.0 percent in Denmark, the United States shows an increase of more than 30 percent. That means the results of Table 20.9 clearly indicate that the United States, in particular, should undertake measures to alleviate the fatality situation for persons 65 years and over, which is confirmed by the symbol "+" representing a significant safety deterioration in Table 20.7 for this age group already in the 1980s.

20.2.6 Fatalities by Road-User Groups

Road-user behavior is of paramount importance to road safety. Drunken driving, speeding, driving through red lights, and not wearing seat belts are examples of road-user behavior which increases the accident risk.[59] It has been estimated that about a 30 to 50 percent reduction in traffic casualty accidents could be achieved if speed limits could be realistically enforced. A similar picture emerges from studies investigating seat-belt use, compliance with traffic signals, violation of priority rules, driving under the influence of alcohol or drugs, and compliance with the regulations for professional drivers with respect to driving, working, and resting hours and maximum vehicle load.[616]

TABLE 20.8   Fatality Rates per 100,000 People by Age Groups, 1995

Age group	Country						
	Austria	Denmark	France	Germany	The Netherlands	Switzerland	United States
0–14	4.2	3.1	3.6	3.1	3.1	3.2	4.8
15–24	29.7	21.9	27.9	27.9	14.3	18.9	29.2
25–64	14.4	8.6	15.7	10.6	7.6	8.4	15.8
Over 64	17.6	19.9	17.5	12.1	15.2	16.0	20.8

TABLE 20.9 Traffic Fatalities by Age Groups for 1980 and 1995, the Endpoints of the Timespan under Study

Country	1980	1995	Percent change
	0 to 14 years		
Austria	132	60	−54.6
Denmark	50	28	−44.0
France	909	413	−54.6
Germany	1,159	418	−63.9
The Netherlands	203	87	−57.1
Switzerland	76	39	−48.7
United States	3,747	2,794	−25.4
	15 to 24 years		
Austria	591	307	−48.1
Denmark	220	154	−30.0
France	4,289	2,259	−47.3
Germany	4,917	2,593	−47.3
The Netherlands	606	295	−51.3
Switzerland	327	161	−50.8
United States	18,459	10,486	−43.2
	25 to 64 years		
Austria	891	630	−29.3
Denmark	249	241	−3.2
France	6,177	4,689	−24.1
Germany	5,761	4,916	−14.7
The Netherlands	726	643	−11.4
Switzerland	513	327	−36.3
United States	23,215	21,423	−7.7
	65 years and over		
Austria	386	213	−44.8
Denmark	171	159	−7.0
France	2,123	1,522	−28.3
Germany	3,196	1,521	−52.4
The Netherlands	461	309	−33.0
Switzerland	293	165	−43.7
United States	5,341	6,991	+30.9

An overall view of any situation must always begin with the individual. We must not be blinded by technology and systems so that we forget that, first of all, it is the driver or the pedestrian and his or her behavior which determines the degree of traffic safety.[572]

The percentage share of traffic fatalities by road-user groups for the year 1995 is shown in Fig. 20.9. With respect to endangerment relative to each of the countries under study, the figure reveals the following:

- In Denmark, the percentage of pedestrians in the total number killed is much higher than that of the other countries considered (20.3 percent as against 10.6 percent in The Netherlands, for instance). The average value for western Europe is 18.3 percent.[110]

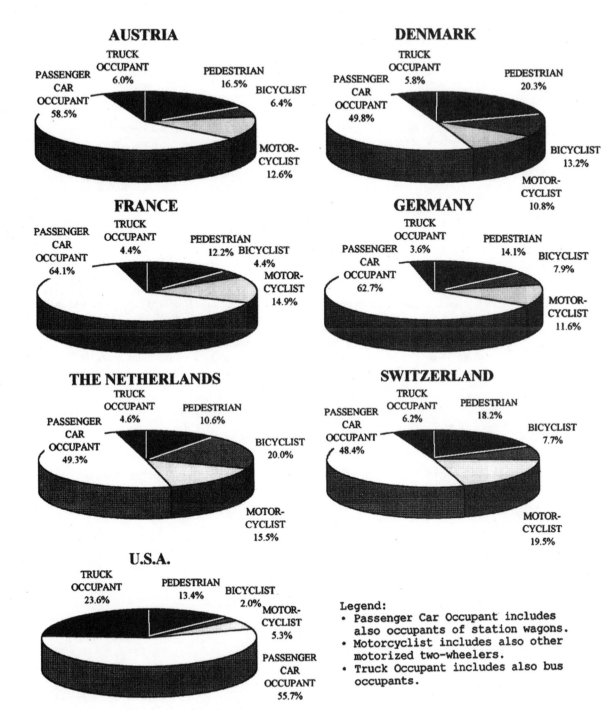

FIGURE 20.9 Percentage of traffic fatalities by road-user groups, 1995.

- The percentage of bicyclists in the total number killed in The Netherlands is much higher than that of the other European countries considered and 10 times that of the United States (20.0 percent as against 2.0 percent). The average value for western Europe is 7.3 percent.[110]

- In Switzerland, the percentage of motorized two-wheelers in the total number killed is much higher than that of the other European countries considered and about 4 times that of the United States (19.5 percent as against 5.3 percent). The average value for western Europe is 14.9 percent.[110]

- The percentage of occupants of passenger cars and station wagons (64.1 percent) in the total number killed in France is higher than that of the other countries considered. The average value for western Europe is 54.7 percent.[110]

- In the United States, the percentage of truck and bus occupants in the total number killed is 23.6 percent. In the European countries considered, this percentage ranges from 6.2 percent in Switzerland to 3.6 percent in Germany. The average value for western Europe is 4.2 percent.[110]

Based on Fig. 20.9, it may be concluded that pedestrians, bicyclists, and motorcyclists represent high-risk target groups in western Europe, whereas truckers are a high-risk target group in the United States.

It might be possible that the low percentage of bicycle and motorcycle fatalities in the United States is due to the fact that, contrary to western Europe, in the United States there are fewer bicyclists and motorcyclists as a share of the driving population. On the other hand, the difference in truck and bus occupant fatalities between western Europe and the United States may be related to the fact that, contrary to western Europe where speed limits of 60 to 80 km/h are in effect for trucks, truckers in the United States normally drive between 100 and 120 km/h.

Figure 20.9 shows additionally that the road-user group passenger car occupants represents—with about 55 percent of fatalities—the largest fatality category on both continents. Improved crashworthiness, active and passive safety systems, such as air bags combined with lap and shoulder safety belts, could, by and large, help reduce the risk of death or serious injury to this road-user group. It has already been estimated that, since 1977, by equipping passenger cars with air bags and lap belts the number of fatalities could be reduced by 50 percent and the number of injuries by about 25 percent.[557]

The absolute numbers of traffic fatalities by road-user groups for the years 1980 and 1995 are given in Table 20.10, from which the following observations can be made:

- All the countries experienced decreases in traffic fatalities for pedestrians, with Denmark having the smallest decrease (14.5 percent).

- All the countries experienced decreases in traffic fatalities for bicyclists, again with Denmark experiencing the lowest decrease (8.3 percent).

- All the countries experienced decreases in traffic fatalities for occupants of motorized two-wheelers, which ranged from 58.4 percent in Germany to 35.5 percent in The Netherlands.

- There were wide variations in the decreases of traffic fatalities for occupants of passenger cars and station wagons, with Switzerland having the largest (41.9 percent) and Denmark the smallest (3.3 percent).

Again, traffic safety officials can directly deduce from comparisons like those shown in Table 20.10, where safety deteriorations exist, and consequently propose appropriate measures to combat them.

To gain a better perspective of the involvement of different age groups as road users in motor vehicle accidents, Table 20.11 was developed. This table, which shows fatality rates per 100,000 people for different age groups and road users indicates that the most endangered age groups are clearly the age groups 15 to 24 (as motorcyclists and passenger car occupants) and over 64 (as pedestrians, bicyclists—especially in western Europe—and passenger car occupants). For the age group 15 to 24 (as motorcyclists and passenger car occupants), the accident cause is most likely excessive speed, risk-taking, inexperience, immaturity, alcohol, etc.[391,392,499]

TABLE 20.10 Traffic Fatalities by Road-User Groups for 1980 and 1995, the Endpoints of the Timespan under Study

Country	1980	1995	Percent change
Pedestrians			
Austria	459	200	−56.4
Denmark	138	118	−14.5
France	2,482	1,086	−56.2
Germany	3,720	1,336	−64.1
The Netherlands	295	142	−51.9
Switzerland	254	126	−50.4
United States	8,070	5,585	−30.8
Bicyclists			
Austria	86	77	−10.5
Denmark	84	77	−8.3
France	708	395	−44.2
Germany	1,338	751	−43.9
The Netherlands	425	267	−37.2
Switzerland	75	53	−29.3
United States	965	830	−14.0
Occupants of motorized two-wheelers			
Austria	316	152	−51.9
Denmark	131	63	−51.9
France	2,556	1,322	−48.3
Germany	2,631	1,095	−58.4
The Netherlands	321	207	−35.5
Switzerland	265	135	−49.1
United States	5,144	2,221	−56.8
Occupants of passenger cars and station wagons			
Austria	1,066	708	−33.6
Denmark	300	290	−3.3
France	7,267	5,696	−21.6
Germany	6.915	5,929	−14.3
The Netherlands	910	657	−27.8
Switzerland	577	335	−41.9
United States	27,455	23,290	−15.2

The high death rates among persons older than 64 years of age, especially as pedestrians, reflect their greater susceptibility to complications when injured and their substantially poorer prognosis.[41] This effect is greatest for the least severe injuries. Minor injuries that are rarely fatal in younger people result in significant mortality rates at ages 65 years and over, whereas for the most severe injuries, age plays a less important role in survival. Also, it points to their lack of abil-

TABLE 20.11   Road Users Killed per 100,000 People by Age Groups, 1995

Country	0 to 14	15 to 24	25 to 64	Over 64
Pedestrians				
Austria	1.1	1.0	1.8	7.7
Denmark	0.8	1.6	1.5	7.2
France	0.9	1.2	1.4	5.3
Germany	0.9	1.0	1.1	4.8
The Netherlands	0.6	0.5	0.5	3.5
Switzerland	1.5	0.8	0.6	7.1
United States	1.3	1.7	2.1	3.7
Bicyclists				
Austria	0.7	0.6	0.9	1.6
Denmark	0.8	1.5	1.2	3.0
France	0.7	0.6	0.6	1.0
Germany	0.7	0.7	0.7	2.1
The Netherlands	1.9	1.3	1.0	5.1
Switzerland	0.6	0.7	0.5	1.7
United States	0.5	0.4	0.3	0.2
Occupants of motorized two-wheelers				
Austria	0.1	5.9	1.7	1.3
Denmark	0.0	3.4	1.1	1.0
France	0.3	7.0	2.2	0.7
Germany	0.1	4.5	1.3	0.4
The Netherlands	0.1	3.3	1.2	1.3
Switzerland	0.1	4.1	1.8	2.4
United States	0.0	2.0	1.1	0.1
Occupants of passenger cars and station wagons				
Austria	1.5	23.1	8.4	6.0
Denmark	1.3	10.2	4.9	8.4
France	1.8	18.3	10.1	11.0
Germany	1.2	19.6	7.2	4.5
The Netherlands	0.7	7.0	4.4	5.0
Switzerland	0.9	9.7	4.7	5.3
United States	2.2	18.8	7.7	13.9

ity to perceive and respond to the movement of vehicles. With regard to pedestrians, Table 20.11 indicates that the fatality rate per 100,000 people for pedestrians over 64 years of age in western Europe is significantly higher than that of the same age group in the United States. In contrast, the fatality rate for the same age group, but as passenger car occupants, is significantly higher in the United States than in Europe. One could draw the conclusion that older Americans are more likely to die while in a car, whereas older Europeans are more likely to die while on the street.

In conclusion, it should be noted that it is relatively easy for countries which have not been included in the present comparative analyses to conduct investigations like the ones presented in Secs. 20.2.4 to 20.2.6 in order to pinpoint any significant or even marginal safety deterioration. In doing so—that is, by conducting accident analyses and evaluations—timely countermeasures could be developed to improve the safety of the most endangered target groups.

20.2.7 Discussions Relevant to Sec. 20.2.6

Analyses of road traffic death statistics show that, on average, about 40 percent of all traffic-related deaths in western Europe are pedestrians, bicyclists, and motorcyclists (see, for example, Fig. 20.9). It is common knowledge that, in terms of vehicle mileage, lack of vehicle protection, etc., pedestrians, bicyclists, and motorcyclists are more vulnerable than, for example, passenger car occupants and truck and bus occupants. Because of the high fatality involvement relative to all traffic-related deaths by the previously mentioned three road-user groups, they will therefore be the subject of discussion in the following sections.

20.2.7.1 Pedestrians.   Generally speaking, pedestrian collisions are predominantly an urban phenomenon. Five out of six injuries to pedestrians and two out of three deaths occur in urban areas. However, for pedestrian collisions in rural areas, the ratio of deaths to injuries is about 3 times as high as in urban areas, reflecting the generally higher impact speeds in rural areas.[41]

Among all categories of people injured by motor vehicles, pedestrians have the highest ratio of deaths to injuries. Studies show that the number of deaths per 1000 injuries reported by the police for pedestrians is about twice that for motorcyclists and over 4 times that for motor vehicle occupants.[41]

One-fifth of all collisions that are fatal to adult pedestrians occur at intersections. Among children, nearly 80 percent who are killed as pedestrians are in the roadway but not at intersections.[41,536]

Vehicle size and design are major factors in pedestrian injuries (see Sec. 20.3). In Great Britain, for example, the death rate among pedestrians struck by various types of vehicles was found to range from 4/1000 struck by bicycles to 106/1000 struck by heavy trucks. Of people struck by passenger cars, 31/1000 were killed.[29]

Percent changes in the absolute numbers of pedestrian fatalities between 1980 and 1995 are shown in Table 20.10. Examination of this table reveals that pedestrian deaths in the countries under study decreased significantly between 1980 and 1995. Overall, it can be stated that the pedestrian fatality development during the last decade was favorable in the majority of the European countries under study, with reductions of more than 50 percent during the investigated time period from 1980 to 1995 (see Table 20.10).

However, it should not be forgotten that, according to Table 20.11, persons over 64 years of age are the most vulnerable of all age groups, especially as pedestrians.

The favorable pedestrian fatality development in Europe, in particular (see Table 20.10), is due to the following factors:[187]

- Europeans have given more consideration to accommodating pedestrians in business districts and residential areas. Most European cities have extensive pedestrian zones. These zones are a variety of street closings combined with malls.

- In Europe, considerable concern has been given to the planning and design of residential areas. Much emphasis has been placed on returning residential areas to residents and restricting the through movement of automobiles in these areas.

- Pedestrian separations are provided in Europe in areas of intense pedestrian activity and/or complicated urban intersections. Often, pedestrian tunnels or bridges are provided.

- Enforcement of regulations prohibiting pedestrians on freeways is extensive.[187]

20.2.7.2   Bicyclists.   The bicycle is a means of transport for which no license is required. For children who are not yet allowed to drive a moped or car and for older people who are no longer able to drive a car, the use of a bicycle is often the only way of covering relatively long distances independently.[494]

The use of bicycles for general transportation is on the increase in the United States and, in time, may approach the level of major European cities. One reason for this increase in bicycle usage, especially by the older segments of the population, is the availability of the lightweight multigear bicycle. However, as the general use of bicycles for transportation has increased, the pattern of accident involvement has undergone a change. The major result has been an increase in the number of accidents and injuries.[612] The most frequently injured parts of the body in bike accidents are the head and face, arms and hands, and legs and feet. The most common types of injury are lacerations, contusions, and abrasions, followed by fractures, strains and sprains, and concussions.

It is generally recognized that human error plays an important role in most traffic accidents. The human factor in bicycle accidents can be related to the degree of knowledge of the traffic regulations, attitudes towards traffic safety and rule compliance, and the degree of motor and cognitive abilities. Since the foregoing characteristics are, in large part, age-dependent, they could at least be held partly responsible for the high accident involvement of specific age groups.[494]

The major causes of bicycle accidents are:[651]

1. *Rider error (a conservative estimate)*: 75 percent
2. *Vehicle driver error (bicyclist could not escape)*: 10 percent
3. *Poor bicycle maintenance:* 5 percent
4. *Defective bicycle trail design or dangerous road condition:* 3 percent
5. *Defective bicycle owner's manual (inadequate instructions)*: 3 percent
6. *Defective bicycle assembly at retail level:* 2 percent.
7. *Defective bicycle (assembly or design of frame or component)*: 2 percent

The favorable bicycle fatality development in Europe, in particular (see Table 20.10), is due to the European philosophy which supports the separation, wherever possible, of bicycles from motor vehicles.[187] To a lesser extent, Europeans believe in separation from pedestrians although the degree of separation is sometimes minimal due to space constraints. In addition, at trouble spots, such as intersections, bridges etc., separation is generally viewed as necessary. In many cases, bicycle-pedestrian separation is indicated by a painted line delineating areas designated for each mode. Whenever possible, separation is effected by curbing and reinforced by different surface types. Signing for bicycle-pedestrian separation coincides with curbing and/or lane marking. These signs are generally blue and circular in design with a bicycle and/or pedestrian symbol superimposed on it. Bicycle signals in Europe are present at locations where conflicting motor vehicle and bicycle traffic, especially right-turning vehicles, is heavy.[187]

United States.   Generally speaking, research studies show that bicycle safety could be drastically improved by requiring bicyclists to wear helmets. Certainly, helmets are not the sole solution to the problem since many bicyclists do not follow traffic rules but they can be a very effective means of easing the suffering. Statistics concerning bicycle safety in the United States[493,662,685] show the following:

- About 50,000 cyclists suffer serious *head injuries* each year.
- An estimated three-fourths of the more than 800 bicycle-related accident fatalities involve *head injuries.*
- One out of 7 children under age 15 suffers *head injuries* in bicycle-related accidents.
- Each day a bicyclist under the age of 15 is killed. More than 75 percent of those killed die as a result of *head injuries.*

Findings from the Harborview Injury Prevention and Research Center in Seattle, Washington, indicate that 80 percent of bicycle fatalities are directly related to *head injuries*; this result coincides with the figures mentioned previously. Yet only 5 to 16 percent of all bicyclists wear helmets in the United States. *Bicycle helmets reduce the risk of serious head injuries by 85 percent.*[482,685] Similar results were reported in Refs. 41, 127, 146, and 193. These studies indicated the following:

- The head or neck is the most seriously injured part of the body in 5 out of 6 fatally injured bicyclists. Typically, bicyclists who die from serious head injuries do not have other life-threatening or potentially disabling injuries. Thus, if bicyclists used helmets, many fatalities and serious head injuries would not occur.
- Despite evidence that wearing helmets can reduce the risk of head injuries dramatically, protective headgear has not caught on with many bicyclists in Europe and the United States.

Australian studies cited in Ref. 146 reported the following:

- Head injuries were reported for 68 percent of bicyclists involved in road crashes to which an ambulance was called in Adelaide, South Australia.
- About 48 percent of injured bicyclists admitted to hospitals in western Australia had suffered head injuries.

Trying to convince people to wear protective headgear has become the main thrust of helmet advocacy groups and manufacturers in many countries. For example, the League of American Wheelmen, a 22,000-member bicycle organization based in Baltimore, Maryland, approved a resolution that bicyclists participating in its national and regional rallies must wear helmets.[482] In this connection, it is interesting to note that the Chairman of the German Social Democratic Party received a serious head injury in 1996 during a bicycle excursion, in which he did not wear a helmet. This example shows that there is still more to do in order to convince people all over the world—even high-ranking politicians—to wear helmets for their own protection.

One of the biggest misconceptions among bicyclists is that lower speeds put them at a low risk. Research studies, however, show that head injuries occur not because of the bicycle's speed, but because of the vertical distance—how far the head travels to hit the pavement. According to these studies:[685]

- It takes about ½ s for the head to go from a bike to the pavement; blink once and that is how long one has—not enough time—to protect the head in a fall.
- The average height of a person sitting on a bike is 5.3 ft (or 1.6 m). At that distance, the head hits the ground at 12.6 mi/h (or 20 km/h), which is the threshold of irreversible injury to the brain. Other studies have indicated that a fall from only 3 ft 10 in (or 1.2 m) at about 11 mi/h (or 17.7 km/h) can cause serious brain damage.[685]

Finally, the energy of falling objects is measured in terms of g's, the same term used by astronauts to describe the force their rocket needs to escape the earth's gravitational field. When a falling object impacts a surface, its sudden deceleration is also given a g value. Research studies[644] show that a bicyclist traveling at 15 to 20 mi/h (or 24 to 32 km/h), who takes a typical fall directly over the handlebars into a fixed object, will decelerate or impact his or her head at an unbelievable $1000g$; compare this value to the g force of only 3 that space shuttle astronauts experience during launch. The bicyclist would surely die under such an impact.[644]

Europe.[161]   The frequency and severity of a conflict depends, first of all, on the driving behavior of the bicyclist. Especially dangerous is

- Crossing a right-of-way road
- Making a left-turn from a right-of-way road
- Making a left-turn from a road with no right-of-way into a road with a right-of-way

The following irregularities, by order of succession, are especially rich with conflicts:

- Cutting off the right-of-way of a motor vehicle
- Using the traffic lanes instead of the bicycle lane
- Using the bicycle lane in the opposite direction
- Getting on in the wrong lane
- Passing motor vehicles on the right side
- Driving through red lights

It follows that typical false driving behavioral attitudes at intersections should be regarded as a main cause of accidents involving bicyclists.

While on the road, the bicyclist is exposed to a higher accident risk than the motorist or the pedestrian. Relative accident numbers show that it is about 8 times more dangerous to ride a bicycle than to drive a car. Related to the exposure vehicle/bicycle kilometer, the risk of being killed in a traffic accident is 7 to 10 times higher than that of a passenger car occupant.[20] Also, comparing the accident situation of bicyclists to that of pedestrians shows that the accident risk for bicyclists is more unfavorable. Related to the exposure bicycle/pedestrian kilometer, the figure is about twice as high. Related to the exposure duration of traffic participation, it is even 5 times higher than that for pedestrians.

The accident risk for bicyclists can also be differentiated by specific age and sex groups. As previously noted, it is estimated that the accident risk for children under 15 years of age and for older persons between 65 and 74 years of age is about 2 to 2.5 times higher than that of average age groups. For senior citizens, 75 years of age or more, the risk is even 5 times higher than that of average age groups.

The lower-risk behavior for female bicyclists seems to be responsible for the fact that girls or younger women show an overall lower accident risk than men of the same age group. This is especially more pronounced for the problem groups, including children under 15 years of age and senior citizens. The accident risk for female bicyclists is about 20 to 70 percent lower in this case than that for male bicyclists.[20]

Several investigations which have been conducted over the last decade have come to a surprising conclusion in terms of separating bicycles from motor vehicles. In this connection, it was found, with respect to the accident rate (AR) on road networks with bicycle lanes, that the safety level of the latter was not much different from that of road networks without bicycle lanes.

Outside built-up areas, Kloeckner[341] calculated the following accident rates (accidents per 10^6 bicycle kilometers):

Roads without bicycle lanes, AR = 1.92

Roads with bicycle lanes, AR = 1.77

Within built-up areas, Knoche[342] conducted comparative safety investigations between roads with and without bicycle lanes. He established the following accident rates:

Roads without bicycle lanes, AR = 3.2

Roads with bicycle lanes, AR = 2.9

The following reasons, among others, provide some sense with respect to the low safety effect of bicycle lanes:[341]

- The bicyclist shows up unexpectedly in front of the motorist on the bicycle lane at intersection areas. Consequently, the motorist very often misinterprets the directional intentions of the bicyclist.
- Many accidents occur in close proximity to property entrances (41 percent) because the motorist gets onto the street from a private property (often driving backward) without paying adequate attention to the traffic on the bicycle lane; this often leads to collisions with bicyclists.[19]
- Collisions occur with motor vehicles which are parked on the bicycle lane or with passengers who suddenly open car doors on the side of the bicycle lane. Ten percent of bicycle accidents occur in this way.
- Seventeen percent of accidents occur when bicyclists turn into traffic lanes to avoid vehicles which are parked on the bicycle lane.

Thus, no proof has been given so far on significant reductions in the accident risk on roads with bicycle lanes on either one or both sides.

With respect to unreported accidents with bicyclists, it should be noted that a good number of them are not usually reported to the police. Consequently, these are not included in the accident statistics.

Therefore, a conclusion which can be made is that official accident statistics are incomplete and provide only a partial picture. In reality, considerably more accidents do occur and, essentially, more persons are injured.[341]

Based on German experiences, the number of injured bicyclists and accident consequences can be broken down by main age groups as follows:

- *Children* (*under 15 years of age*): This age group accounts for about one-fourth of all fatally, seriously, and slightly injured bicyclists. The main reasons for this include a high traffic participation by this group as bicyclists and limited experience in dealing with road traffic safety situations which may occur.[20]
- *Teenagers and adults* (*between 15 and 65 years of age*): This group is not a problem group.
- *Senior citizens* (*older than 65 years of age*): This age group alone accounts for about one-third of all fatally injured bicyclists in road traffic; yet this group accounts for only 9 percent of overall injured bicyclists. Thus, one can conclude that with increasing age the severity of accident consequences also increases.

With respect to classifying the number of injured persons by sex, community size, and road site, the following statements can be made:

- The accident consequences on rural roads are considerably heavier than those on suburban roads because of the higher speeds on the former. Also, the number of bicycle accidents with heavy personal injuries in communities with populations under 10,000 inhabitants is twice as high as that in larger communities.[20]
- Males are more frequently involved in accidents than females. Within communities the proportion is 3 to 2; outside communities the proportion is 2 to 1.[20]
- Two-thirds of all victims as children are male. The main reason for this is a higher risk-taking driving behavior.[87]

The classification of traffic accidents by accident types is intended to describe the traffic-related conflict situation that led to the accident. With reference to the frequency distribution by accidents types concerning bicycles, Kloeckner[341] indicated the following. About two-thirds of bicycle accidents are attributed to left-turning and right-turning and crossing movements.

Accident in the longitudinal direction is a result of conflicts with the moving traffic in the same or in opposite directions of travel. This type accounts for about 10 percent of all accidents. Finally, accidents with the standing traffic represents about 5 percent of bicycle accidents.

The number of reported bicycle accidents varies according to the accident type and nature of collision. Collisions with passenger cars and trucks are included more often in the statistics than those with other traffic participants. The highest nonreported portion is single bicycle accidents for which only 3 to 18 percent are reported. Thus, a comparatively high number of unreported accidents can be assumed for juvenile bicyclists and children because the latter often get involved in accidents while alone.

The different degrees of reported or unreported accidents lead to obvious deviations from actual accident statistics. Thus, accidents with passenger cars/trucks are overrepresented in the official accident statistics, whereas in reality single bicycle accidents probably dominate.

With reference to accident sites, it has been shown that inside built-up areas bicycle accidents occur at intersection areas, whereas outside built-up areas the accident site is usually between intersections.

As with pedestrians, persons over 64 years of age can be regarded as the most vulnerable group of all bicyclists (see Table 20.11).

The previous statistical figures clearly show the urgent need to require bicyclists to wear protective headgear and to push forward appropriate legal measures. From the viewpoint of traffic safety, helmet laws and their enforcement by the police should be valued much higher than all of the previously discussed aspects concerning traffic-related behavior and the bicycle accident situation.

Worldwide, the studies mentioned earlier show, unfortunately, that the use of helmets by cyclists is not stressed in many countries.[185]

20.2.7.3 *Motorcyclists.*

Motorcycles are unique vehicles. They travel at highway speeds like cars and trucks but they are less stable, harder to see, and offer less protection for riders in an accident. It is, therefore, not surprising that the death rate per 100 million person kilometers of travel is more than 15 times the rate for cars. The ratio of deaths to reported injuries is twice as great for motorcyclists as for occupants of passenger vehicles.[41,105] In the case of a motorcycle–passenger car crash, 45 times more fatalities are suffered by motorcyclists than by passenger car occupants.[602]

Percent changes of the absolute number of motorcycle fatalities are shown in Table 20.10. Examination of this table reveals that the countries under study experienced significant decreases in motorcycle fatalities between 1980 and 1995.

A previous study by the authors[427] indicated that 50 percent of the motorcycle accident deaths in 1989 were persons in the age group 15 to 24 years. Yet persons in this age group made up only 15.3 percent of the population of western Europe. These statistics indicate that the high-risk age group 15 to 24 should be of major concern to the authorities that are involved in traffic safety research and decision making.

The age distributions of persons killed in motorcycle accidents may become more meaningful if they were compared with the age distributions of the general population. Table 20.11 gives the motorcycle fatality rates per 100,000 inhabitants by age for the year 1995. A review of this table demonstrates the following:

- Motorcycle fatality rates vary tremendously by age. As can be expected, they are highest for persons in the age group 15 to 24 years.

- In 1995, the fatality rates per 100,000 people for persons in the age group 15 to 24 years in western Europe fell between 3.3 in The Netherlands and 7.0 in France (Table 20.11).

The differences in fatality rates may be explained, for example, by differences in the composition of the total number of motorcycles, differences in road use (traffic intensity and available road length), differences in driving experience, etc.[98]

In view of these findings, it can be concluded that persons in the age group 15 to 24 years are the most vulnerable of all age groups as motorcyclists.

The favorable motorcycle fatality developments in Europe and the United States (see Table 20.10), may be due in part to improved traffic education for riders, licensing programs, use of helmets, and enforcement of traffic laws by the police.

Helmet-wearing rates by motorized two-wheel riders have been relatively stable in most western European countries. However, an increasing number of motorized two-wheel riders in Europe do not use appropriate protective clothings.[185]

While stringent laws for wearing helmets exist in most western European countries, they do not, however, exist in a great number of countries around the world.

The following statements, which are related to U.S. experiences, could shed light on the importance with respect to traffic safety of wearing protective headgear. This importance, for instance, was addressed in a 1990 report by Cable News Network (CNN),[95] from which the following observations can be made:

- Motorcyclists are 4 times more likely to die in California than in Georgia where helmet-use laws are strongly enforced.

- South Carolina experienced a 184 percent increase in motorcycle fatality rates following the relaxation of helmet-use laws.

- Wyoming experienced a 73 percent increase in motorcycle fatality rates following the relaxation of helmet-use laws.

- Partial helmet-use laws are no more effective than no laws at all.

A 1988 study,[178] which used Fatal Accident Reporting System (FARS) data for 1975 to 1986, concluded the following: helmet use reduces the fatality risk to motorcycle riders by 28 percent. Other studies of helmet use[231,261,684] reported increases of between 22 and 30 percent in motorcycle fatalities following the repeal of helmet-use laws.

Studies conducted in 1981 by Berkowitz[48,49] and the National Highway Traffic Safety Administration[727] suggested very large increases in fatalities were associated with the repeal of helmet-use laws. Similarly, a 1980 study by Watson, et al.[741] indicated that states that revoked their helmet-use laws after a 1976 change in federal requirements experienced a substantial drop in helmet use and an increase of about 40 percent in fatalities.

Other studies have indicated that motorcycle safety could also be improved by requiring motorcyclists to leave their headlights on during the daytime. According to the New York State Department of Motor Vehicles,[548] a motorcycle with its lights off is twice as likely to go unnoticed by other road users. A 1981 study by Hurt, et al.[297] indicated that motorcyclists with headlights on have about one-fourth the risk (0.266) of daytime multiple-vehicle collisions than motorcyclists without their headlights on. A 1977 study by Vaughan, et al.[733] reported that the relative risk of accident involvement is about 3 times higher among motorcyclists not using their headlights. Finally, a 1977 study by Waller and Griffin[739] and a 1976 study by Robertson[609] reported that about one-quarter of daytime multiple-vehicle collisions could be prevented by headlight-use laws.

20.2.8 Accident Causes*

The Institute for Research in Public Safety at Indiana State University (Bloomington, Indiana)[716] published, in 1979, a "TRI-Level Study on the Causes of Traffic Accidents," which can still be regarded today as reliable.

The tri-level methodology involved baseline data collection on level A, on-site investigations by technicians on level B, and in-depth investigations by a multidisciplinary team on level C. The results for level C, as an example, are presented in Fig. 20.10. According to level C, human factors (92.6 percent) were more frequently cited as accident causes than either environmental factors (33.8 percent) or vehicular factors (12.7 percent).

---

*Elaborated based on Refs. 391, 392, and 716.

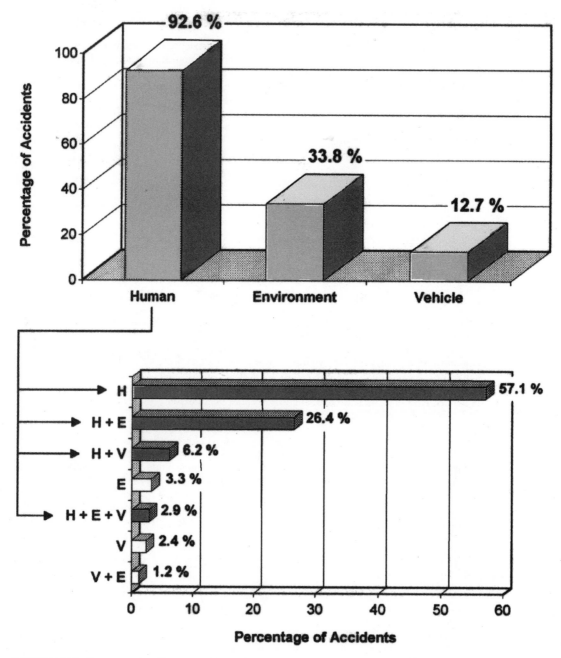

FIGURE 20.10 Percentage of accidents caused by human, vehicular, and environmental factors[716] (level C results).

However, it is important to note, according to Fig. 20.10, that human factors alone were identified as probable level C causes of only 57.1 percent of all cases, while human and environmental factors were identified as probable causes in 26.4 percent, human and vehicular factors in 6.2 percent, and human, vehicle, and environmental factors in 2.9 percent. All these factors together, including environmental and vehicle factors, then make up the high percentage of more than 90 percent of accidents which are normally referred to in national accident statistics as human-related accidents. The environmental factors refer, for example, to weather-related conditions, etc. Therefore, a reconsideration should finally take place because it is by far too simple and misleading to separate accident causes by the three categories human, environment, and vehicle, while disregarding, at the same time, the interrelationships that exist between them as shown, for example, in the lower part of Fig. 20.10. (The more than 100 percent for level C, as shown in Fig. 20.10, is understandable considering the fact that often, for a given accident, more than one accident cause is responsible.) In Germany, for example, the average ratio is 1.6 accident causes/accident.

Of the specific direct human causes, improper lookout, excessive speed, and inattention were most frequently cited (Fig. 20.11).

Surprisingly, the accident cause improper lookout accounts for nearly one-fourth of the accident history and even exceeds the accident cause excessive speed. This result emphasizes the importance of sight distance requirements in highway geometric design, as discussed in Chap. 15, as well as vehicle safety considerations with respect to sight conditions, as shown in Sec. 20.3. Improper lookout accidents resulted when drivers changed lanes, passed, or pulled out from an alley, street, or driveway without looking carefully enough for oncoming traffic. Half of the drivers failed to make any surveillance effort; the remainder had looked, but failed to see, the oncoming traffic, which should have been visible. For the drivers who looked but failed to see, approximately 40 percent experienced a view obstruction, which added to the difficulty of their surveillance task, even though it was determined that this difficulty could, and should, have been

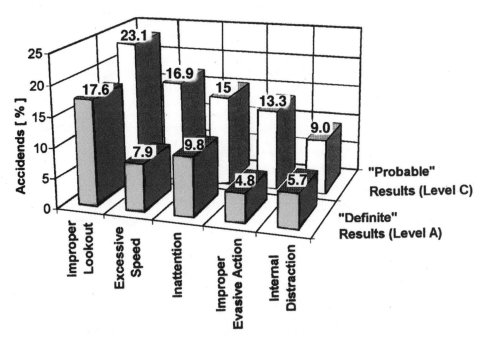

FIGURE 20.11 Percentage of accidents caused by specific direct human causes (example U.S.A.).[716] (The percentages in Europe may differ.)

easily overcome. Also of significance was the overinvolvement of drivers 65 years of age and over who committed improper lookout mistakes. Of drivers over 65 who caused accidents, approximately half had made errors of this kind.[716]

Thus, further research is needed to identify the behavioral components and level of attention which comprise proper lookout, so that adequate training, licensing, and enforcement measures can be devised. Future research should try to identify the relevant mechanism (for example, mechanical difficulties in turning the head, reduction in the visual field or other visual skills, or changes in field dependence) in order to suggest appropriate countermeasures such as specialized training programs for older traffic participants.

With 16.9 percent, the accident cause excessive speed represents the next most dangerous level (Fig. 20.11). Particularly relevant in considering countermeasures for the excessive speed category is the overrepresentation of males and females in the age group 15 to 24 years. This fact explains, once more, the distribution of traffic fatalities by age in Figs. 9.3 and 20.7.

Most of the excessive speed errors occurred with reference to "road design," primarily in the sense of exceeding the critical speed for a curve and thereby losing control, as discussed a number of times in the previous chapters. Therefore, strong efforts must be made by the police in monitoring the driving behavior of motorists on dangerous sections of two-lane rural roads where most of the excessive speed-related accidents occur (see Fig. 9.2). In case of new designs, redesigns, or RRR projects, the sound application of the developed safety criteria I to III in Chaps. 9 and 10 will certainly lead to considerable safety improvement in highway engineering (see case studies in Secs. 12.2.4.2 and 18.4.2).

The motivation underlying risk-taking behavior among young drivers (particularly males), as well as their skills in vehicle handling and judgment of roadway requirements, may require a closer examination and possibly a reevaluation of present driver training programs.

In addition, the driving behavior of U.S. truckers, from the standpoint of European drivers, must be noted here. In the United States, in contrast to Europe, trucks have the same speed limit as passenger cars. The result is that trucks again and again are passing passenger cars, which potentially is quite a dangerous act. A reasonable recommendation would be to establish appropriate speed limits for trucks in the United States. In this respect, speed limits for trucks have had a convincing success in Europe for about 30 y, as expressed by Fig. 20.9, when comparing the percentages of truck occupant fatalities between western European countries and the United States.

The third specific human accident cause was inattention with 15 percent (Fig. 20.11). Inattention most frequently involved a delay in determining that the traffic ahead had either stopped or decelerated, and less frequently it involved a failure to observe critical road signs and signals.

Aside from informing drivers (through public information and driver education programs) of the importance of attentiveness on the driving task, possible areas of improvement include:

- Changes in the size or placement of road signs and signals
- Environmental changes in order to reduce the incidence of sudden stops
- Further improvements of in-vehicle communications systems
- Actuation systems to avoid contact in rear-end situations
- Installation of improved brake lights (for example, with possible changes in intensity, color, or pulse characteristics)[716]

Several of these previously mentioned devices have been developed or installed or are being tested.

20.2.9 Statistical Modeling

The study described in this chapter also had the objective of developing statistical models which describe the relationships between traffic fatalities/fatality rates and time. To achieve this, two approaches were undertaken. The first viewed the potential relationship between the fatality rate per billion vehicle kilometers and time, whereas the second attempted to account for the relationship between the *cumulative* number of traffic fatalities and time.

The model forms calibrated are as follows:

Linear:

$$y = a + bx \tag{20.3}$$

Exponential:

$$y = \exp(a + bx)$$
$$= \exp(a)\exp(bx)$$
$$= A\exp(bx) \tag{20.4}$$

where $y =$ is the fatality rate per billion vehicle kilometers (FRK) or the cumulative number of traffic fatalities (CF)

$x =$ the time

$a, A,$ and $b =$ constants

The initial step of the statistical modeling process was to plot the variables in order to determine the presence or absence of patterns and to locate erroneous data points.

With respect to the fatality rate per billion vehicle kilometers, the data from Table 20.4 appeared to lie close to a curve that could be exponential. The functional relationship, however, would be more convincing if a plot of the logarithm of the fatality rate versus time turned out to be linear. The data obtained nearly determined a line as shown in Fig. 20.12. Consequently, a

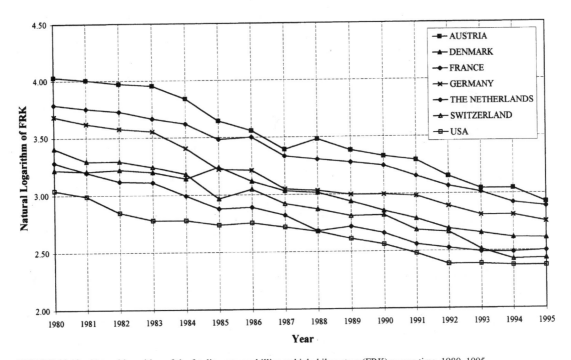

FIGURE 20.12   Natural logarithm of the fatality rate per billion vehicle kilometers (FRK) versus time, 1980–1995.

computer-assisted analysis was performed to determine the best regression equation, in terms of statistical significance and overall form. Comparing the R^2 values among models, it was found that an exponential curve is a reasonable fit for the fatality rate per billion vehicle kilometers. Thus, the resulting equations between the fatality rate (FRK) and time are as follows:

Austria:	FRK = 59.1336 exp [−0.0770 (year−1980)]	$R^2 = 0.9730$
Denmark:	FRK = 24.2776 exp [−0.0644 (year−1985)]	$R^2 = 0.9760$
France:	FRK = 46.0533 exp [−0.0629 (year−1980)]	$R^2 = 0.9887$
Germany:	FRK = 38.2063 exp [−0.0642 (year−1980)]	$R^2 = 0.9466$
The Netherlands:	FRK = 25.1485 exp [−0.0558 (year−1980)]	$R^2 = 0.9638$
Switzerland:	FRK = 30.0692 exp [−0.0658 (year−1980)]	$R^2 = 0.9791$
United States:	FRK = 19.8182 exp [−0.0443 (year−1980)]	$R^2 = 0.9590$

With respect to the cumulative number of traffic fatalities (CF), the data were found to lie in a perfect linear trend, as shown in Fig. 20.13. Consequently, in terms of statistical significance and overall form, the least-squares regression technique produced the following linear models between the cumulative number of traffic fatalities (CF) and time:

Austria:	CF = 2913.90 + 1569.97 (year−1980)	$R^2 = 0.9965$
Denmark:	CF = 783.21 + 660.16 (year−1980)	$R^2 = 0.9981$
France:	CF = 19,357.35 + 11,269.88 (year−1980)	$R^2 = 0.9968$
Germany:	CF = 20,997.21 + 10,769.84 (year−1980)	$R^2 = 0.9976$
The Netherlands:	CF = 2784.32 + 1447.49 (year−1980)	$R^2 = 0.9970$
Switzerland:	CF = 1778.79 + 919.69 (year−1980)	$R^2 = 0.9951$
United States:	CF = 56,911.69 + 43,978.12 (year−1980)	$R^2 = 0.9995$

The value for the coefficient of determination, R^2, which is used to indicate how effectively the regression equation fits the data, was definitely strong. This value, as indicated in the preceding equations for both FRK and CF, was extremely high, which is quite exceptional for fatality data.

If current trends continue unhindered, it may be estimated from the preceding regression models that in the year 2010 the number of fatalities per 10^9 driven vehicle kilometers in the countries under study will be as follows:

5.9 in Austria

4.9 in Denmark

7.0 in France

5.6 in Germany

4.7 in The Netherlands

4.2 in Switzerland

5.3 in the United States

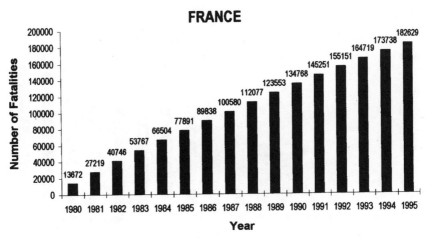

FIGURE 20.13   Cumulative number of traffic fatalities versus time, 1980–1995.

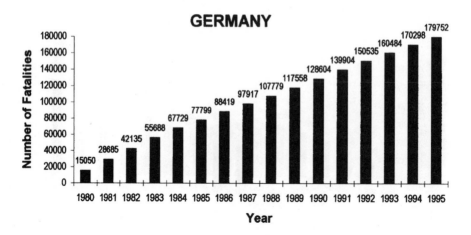

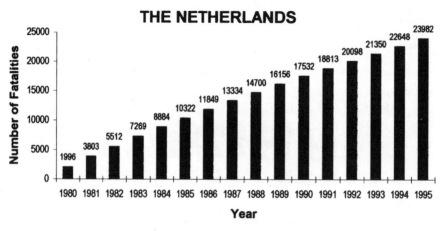

FIGURE 20.13 (*Continued*) Cumulative number of traffic fatalities versus time, 1980–1995.

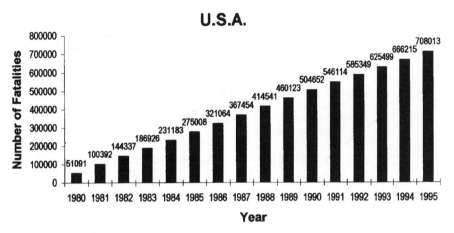

FIGURE 20.13   (*Continued*)   Cumulative number of traffic fatalities versus time, 1980–1995.

Similarly, with respect to the cumulative number of traffic fatalities, it may be estimated from the previous regression models that around

    51,000 persons in Austria

    21,000 persons in Denmark

   360,000 persons in France

   345,000 persons in Germany

    47,000 persons in The Netherlands

    30,000 persons in Switzerland

1,380,000 persons in the United States

would have lost their lives in automobile accidents between 1980 and 2010.

20.2.10 Preliminary Conclusion

Even though the number of registered motor vehicles continues to increase, this section has shown that traffic fatalities and fatality rates in the countries considered—Austria, Denmark, France, Germany, The Netherlands, Switzerland, and the United States—are falling.

With respect to traffic fatalities for different age groups, all countries experienced decreases in traffic fatalities for persons in the 0- to 14-year age group, 15- to 24-year age group, and 25 to 64-year age group. While the European countries experienced decreases in the number of fatalities for persons in the over 64-year age group, the United States experienced an increase of more than 30 percent during the time period from 1980 to 1995 (see Table 20.9).

Looking at fatality rates per 100,000 people for different age groups, the highest fatality rates were for persons in the 15- to 24-year age group in Austria, Denmark, France, Germany, Switzerland, and the United States. In The Netherlands, however, the highest fatality rate was for persons in the over 64-year age group (see Fig. 20.8).

With respect to traffic fatalities for different road-user groups, it was shown that (see Table 20.10) all countries under study experienced decreases in traffic fatalities for pedestrians, bicyclists, motorcyclists, and occupants of passenger cars and station wagons.

With respect to traffic fatalities for different road users per 100,000 people by age groups, it was shown that (see Table 20.11):

- The highest pedestrian fatality rates were for persons in the over 64-year age group.
- The highest bicycle fatality rates were for persons in the over 64-year age group in Europe and for persons in the 0- to 14-year age group in the United States.
- The highest fatality rates for occupants of motorized two-wheelers and occupants of passenger cars and station wagons were for persons in the 15 to 24-year age group.

Generally speaking, the age groups 15 to 24 and over 64 could be regarded as the most vulnerable age groups, as related to fatalities per 100,000 people.

With respect to statistical modeling, it has been shown, at least for the countries considered and the time period observed, that:

- An exponential form is an appropriate mathematical model to describe the relationship between the fatality rate per billion vehicle kilometers and time.
- A linear form is an appropriate mathematical model to describe the relationship between the cumulative number of traffic fatalities and time.

The developed models could also be used for prediction purposes, assuming that current fatality trends continue without any significant change.

In closing, the following points can be made:

- Specifically, the study identified changes in fatalities and fatality rates for different road-user and age groups, determined whether there were statistically significant changes in the fatal traffic accident characteristics studied, and formulated mathematical relationships between fatality rates and time.
- The strength of this study is in the opportunity it provides to compare trends in fatalities across populations that are similar enough to be comparable, yet different enough to justify making the comparison. The information provided may both stimulate and guide efforts to discover the sources of the differences revealed and thereby point the way to changes that will lower fatality rates where they are highest. The results are important in that they pinpoint the problem areas in traffic safety and, hopefully, cause the authorities responsible for traffic safety to concentrate more on the troubled areas.
- Perhaps the flatter experience of the U.S. fatality rates compared with European statistics is, in part, due to the fact that the United States may be approaching a safety "limit." Very simply, there may be more opportunities for improvements in Europe than in the United States; therefore, a comparison of trends will show more marked changes for European countries than the United States.
- The fatality development for persons older than 64 could become especially critical in both the United States and Europe. Generally speaking, statistics show that older people are the fastest growing section of the populations of many countries. In the year 2000, one in six persons will be older than 65, especially in highly developed nations. If current fertility rates continue, by 2030, one in four people will be older than 65. As a group, older adults' health and financial circumstances have improved to such an extent that they continue to participate fully in life for many years after their retirement.[471] Judged on face value alone, these are disturbing trends— more vehicles being driven on the same roads by drivers with special needs can only mean increased hazards.
- This study is just a small step toward a better understanding of what we can learn from a comparison of accident characteristics and trends in different countries and continents. Future studies should look at developments in, for example, other continents, including developed and developing countries.

20.3 VEHICLE SAFETY*

20.3.1 Introduction

In the year 1896 there were only two motor vehicles in the State of Ohio. Roughly 4 weeks after the purchase of the second motor vehicle, the two vehicles collided head-on on a highway. This was the first reported accident which was caused by motor vehicles in the State of Ohio.[392]

Following the development of quantitative safety evaluation processes for roadway engineering and the discussions of human involvement in traffic matters, the issue of vehicle safety becomes the main topic of this section. Braess[65] reported that the efforts with respect to road traffic safety are as old as road traffic itself. At the beginning of the automobile era the most important goal was to raise the operational safety of motor vehicles and the standards of roads in order to allow the driver to reach his or her destination within a reasonable time interval.

The increase in traffic densities and operating speeds prompted manufacturers, especially after World War I, to improve driving and braking behavior.

In spite of improvements, at least related to certain issues of traffic safety, accident figures significantly increased in all industrialized nations after World War II. Consequently, a systematic investigation of accident causes, accident types, and accident consequences has taken place during the last 40 years.[65]

With respect to the automobile itself, at a relatively early period, methods for predicting the behavior of the automobile in critical situations[27] were being sought. In this connection, studies of the crash behavior of motor vehicles were especially important. Beginning with the biomechanical stress limits of the human being up to the optimal structural design of the motor vehicle with respect to its active and passive safety, these studies cover a wide, multidisciplinary, and complex field.

Therefore, the aim of a vehicle designer must be to build motor vehicles which, in case of an accident, are as harmless as possible to their occupants (passive safety) and prevent accidents, as far as possible, from happening (active safety). Thus, motor vehicle safety should be considered in general, and differentiated by active and passive safety.[312] Figure 20.14 shows a detailed representation of vehicle safety.

There are numerous impacts on traffic safety. A frequent cause of accidents is the failure of the human being as vehicle operator. There are several reasons for this, including an overestimation of one's driving skills, lack of driving experience, insufficient physical condition, use of alcohol, medicine, drugs, etc.

Surprisingly however, also minor issues, like comfort, may exhibit a strong influence on safety. Examples include adjustable seats in the horizontal and vertical directions, adjustable steering wheels, and switches which are used simply to identify and set things in motion. Bagatelles often also have an important meaning. For example, for an ergonomically sound solution, a driver should not change his or her seating position to adjust a mirror. It is quite dangerous for the driver to lean forward in order to adjust the mirror, since at this moment his or her head and eyes are no longer focusing on the driving conditions ahead. In order to examine the driver's field of view, the so-called eyellipse templates have been used to describe the positions of the eyes of different drivers in different seating positions. Based on these eyellipse templates, the driver's field of view through the windshields, the sight conditions by means of the inside and outside rear mirrors, and sight obstructions from the so-called body pillars have been defined. How important those at the first blink minor investigations are shown in Fig. 20.11. As this figure reveals with respect to accident causes, improper look-out exceeds, for example, excessive speed, as a cause of accidents, at least in the United States.

The examples just mentioned contribute to accident avoidance. Nevertheless, human failure must be considered and cannot be excluded—not even by great technical display measures—and thus accident prevention measures often become ineffective.

---

*Elaborated by M. Schoch based on Ref. 633.

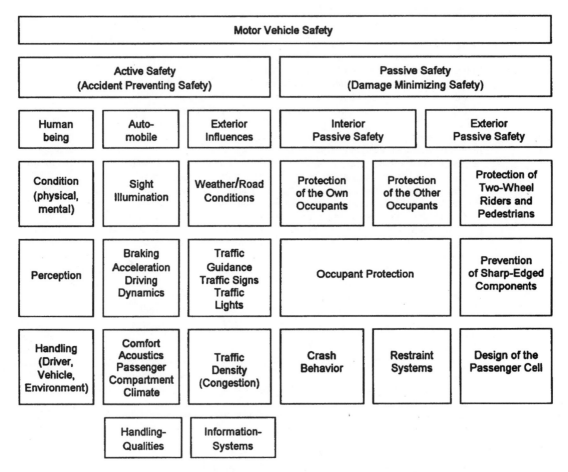

FIGURE 20.14   Motor vehicle safety subdivided into active safety and passive safety. (Elaborated based on Ref. 642.)

Reviewing, again, Fig. 20.14, the following sections describe current technical developments in the field of vehicle safety in relation to facilities which promote *active safety* (that is, contribute to accident prevention) and to facilities which benefit *passive safety* (that is, help reduce accident severity).[312,756]

20.3.2 Active Safety

Active safety covers the wide and complex field of enhancing safety by means of improvement in accident avoidance.[467] In order to develop accident prevention systems, the conditions and causes of accidents must be known. Therefore, accidents have to be reconstructed, simulated, and examined. Accident evaluation has led to the conclusion that the normally closed-loop system driver–motor vehicle environment becomes disturbed when an accident occurs (Fig. 10.4).[632]

As already discussed in Sec. 10.2.1, a driver action causes a vehicle reaction, which leads to a new driver action, and so forth. Both the driver and the vehicle are affected by different disturbances and environmental factors on the road, such as roadway design, traffic mix and composition, road surface condition, weather, time of day, etc., which influence the driver's action and the movement of the vehicle. In order to be able to regard all these influencing factors, intelligent systems have to be considered when dealing with active safety.

Several of these systems will be discussed in brief in the following. They have been developed, to achieve good accelerating or decelerating capabilities in cases of extreme conditions. Some of these systems have been combined for better solutions.

ABS = antiblock system

The primary objective of the ABS is to determine when the wheels are about to lock and to act to prevent it. The system constantly extrapolates an estimated vehicle speed from the speed of the wheels. This estimated speed and the wheel speed information from the sensors are used to control the brakes. When the brakes are applied, the tires start to slip. The system determines when lock will occur from the degree of slip and reduces hydraulic pressure to prevent wheel lock. Thus, because the wheels continue to roll, albeit at a slower speed than the body of the vehicle, directional stability and steerability—and thus driver control of the vehicle—are maintained (see Fig. 20.15).[706]

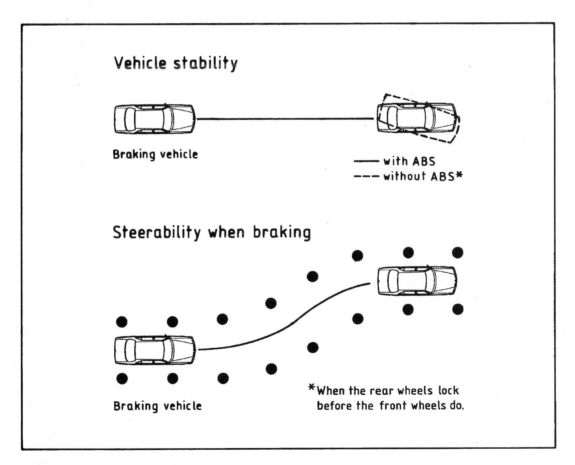

FIGURE 20.15   ABS effectiveness.[706]

ASR (TSC) (ASC):	Antrieb - Schlupf - Regelung (German) (traction slip control/anti-slip control). The system prevents the wheels from spinning when accelerating and consequently the swerving motion of the vehicle.
ASC:	Automatic stability control. This system prevents one drive wheel from spinning. Electronic control of the throttle valves reduces the propulsion power.
ASC + T:	T for traction. The preceding effect is intensified by an additional brake control on the spinning wheel.
EBD:	Electronic brake force distribution. The EBD system distributes, in conjunction with ABS, the braking forces between front and rear wheels in an optimum way, independent of loading and pedal pressure.
ESP:	Electronic stability program. A microcomputer coordinates the driving behavior of the vehicle in case of swerving movements through intervention in the motor management and an automatic brake intervention at the individual wheels.
FDR (DDC):	Fahr-Dynamik-Regelung (German) (driving dynamic control). This system reduces, in cases of extreme steering maneuvers, the danger of swerving and enables control of the vehicle during critical driving situations.

The previous examples represent only some of the systems which contribute to accident prevention within the scope of active safety.

In the following sections the main aspects of passive safety, especially the passive safety of motor vehicle bodies, will be considered and discussed.

20.3.3 Passive Safety

20.3.3.1 General.   It does not make too much sense to build an air bag into an automobile where the vehicle body does not provide sufficient survival space when involved in an accident. Likewise, a stiff passenger cell is a basic necessity for interior safety; it is, however, of little advantage if restraint systems do not provide an adequate effect. The preceding examples indicate that the following safety components should work together properly in order to ensure a high degree of passive safety:

- Rigid passenger cell
- Deformation zones in the front and back
- Lateral protection (side impact)
- Improvements in the vehicle for the protection of pedestrians and two-wheel riders
- Safety door locks
- Divided steering column and safety steering wheel
- Lap-shoulder seat-belt system (pretensioning device, belt-force limiter)
- Front and side air bags

With respect to vehicular safety, it should be kept in mind that the protection of one traffic participant does not necessarily create a safety risk for another traffic participant. However, vehicle body designs which do well in crash tests because of high-compression resistance may, in reality, be harmful to pedestrians and two-wheel riders in actual accident situations. Only a safety partnership can open the perspectives that are necessary to further reduce the risk of accidents and injuries in road traffic.

Vehicle manufacturers always aim to build, as far as possible, safe vehicles because this is the best sales argument and gives customers an incentive to buy that car make, especially following an accident.

Based on experiences gained from accident research investigations which resulted in a number of safety measures, it appears that the passive safety of motor vehicles could be drastically improved. However, this results frequently in a heavier total weight of the vehicle (see "Modern Vehicle Fleet Characteristics" in Sec. 15.2.2.3). This weight spiral could be brought down by using lightweight design or new materials.

Therefore, safety measures should also be transformed to smaller, lighter vehicles. This can be realized! However, the feasibility is narrowed by the price the customer is willing to pay. The customer must be able to afford safety, and the cost of these vehicles should not be within the reach of only a few wealthy people. Furthermore, it is not only the interior passive safety of a vehicle which contributes to sale decisions; there is also the exterior passive safety which concerns pedestrians, two-wheel riders, and occupants of other vehicles. Commercial vehicles, especially, present an increased danger for other road users. Aiming for more commercial vehicle safety requires that special attention be devoted to "compatibility," that is, the partner's protection.

20.3.3.2 Compatibility. *Compatibility* is the mutual behavior of motor vehicles which take part in an accident. For example, in the case of accidents between passenger cars and trucks, problems which are difficult to solve could result in terms of the hazard potential of motor vehicles. One has to distinguish between mass aggressiveness and form aggressiveness.

Mass aggressiveness results from the weight of the vehicle and could only be reduced for trucks if the vehicles were lighter or could drive slower. A truck trailer unit with a weight of 40 tons has a kinetic energy of 9.9 MJ at a speed of 80 km/h. In order for a fully loaded passenger car with a weight of 1700 kg to reach the same kinetic energy, it would have to approach a speed of 400 km/h. The formula for kinetic energy is as follows:

$$E_{kin} = \frac{1}{2} \, mass \times speed^2 \qquad MJ \qquad (20.5)$$

With respect to *form aggressiveness,* for a passenger car–truck accident involved in a head-on collision, the hood of the passenger car would come under the bumper of a heavy truck. Thus, the deformation zone of the passenger car can hardly produce any effect. The truck's bumper forces its way into the passenger compartment with little resistance because the bumper has a rigid front end. In this respect, Ricci reported in 1979[602] in a study of passenger car–truck accidents that the number of fatalities suffered by the occupants of passenger cars is 24 times that of trucks. The form aggressiveness of trucks could be reduced in case a deformation zone is present such as by means of an energy-absorbing front underride protection.

A major portion of the fatalities in vehicle-vehicle collisions has been the occupants of small cars. The risk of being killed in a motor vehicle crash increases with a decrease in the vehicle mass. This trend can be reduced by means of design. A description of the physical aspects which explain the handicaps of smaller cars with respect to occupants follows.

When a vehicle collides with a rigid wall, then a clear relationship exists between deceleration and the available deformation distance. For the same deformation distances, the same decelerations and, consequently, the same occupant strains would be reached, irrespective of the vehicle mass.

When two vehicles with different masses collide with each other, the speed changes and, consequently, the decelerations would be inversely proportional to the masses of the vehicles in question. In reality, however, there are many aspects which still have to be added and which complicate the state of affairs. These aspects, for instance, include the strength of the passenger compartments of the vehicles in question, the lengths and rigidities of the deformation zones, the collision speeds, and the tangling ups of vehicles. Therefore, a high structural rigidity, as well as a particular restraint system design, are required for small, light motor vehicles.

Consequently, the compatibility of a lightweight vehicle demands, in the case of collisions with heavier vehicles, that the front of the lightweight vehicle deform later and at greater forces as compared to the heavier vehicle because deformation zones are barely present in lightweight vehicles. That means that the heavier vehicle, with respect to the mass of the more aggressive vehicle, should absorb most of the kinetic energy.

Furthermore it is important to note that the deformations on the vehicles should occur in a fully plastic (absorbable) manner because in case of an elastic rebound, the speed changes are even higher.

Based on these considerations, Schwant and Stocklose[641] have developed the following main propositions for passive safety:

---

1. Main Proposition for Passive Safety

For a vehicle speed change caused by an impact, the stresses on occupants are inversely proportional to the distance traversed by the occupants using a restraint system. Furthermore, the relative motion of the occupants as compared to the vehicle has to be regarded.

Therefore, the basic parameters that can be influenced by the developer in order to reduce the stress on vehicle occupants, are the length of deformation of the front part of the vehicle and the available forward bending possibilities for the occupants of the vehicle.

---

2. Main Proposition for Passive Safety

For a vehicle speed change caused by an impact, the stresses on vehicle occupants become lower and lower as the speed change is more uniform along the stopping distance.

Therefore, when designing motor vehicles, a major consideration must be given to the fact that the speed should decrease in case of an accident in a uniform manner (without speed peaks). This can be achieved by using design measures that can be examined by computer simulations.

These two main propositions summarize what is essential for passive safety. The details of these propositions are presented in the following considerations which are found to be important.

The realization of long deformation lengths, which reduce energy, and stable passenger cells are the best life insurance for vehicle occupants. In the deformation zones, barriers are built in, so that by means of a properly coordinated deformation not every minor accident would result in penetration into the passenger compartment. The final, and biggest, resistance must be provided by the passenger cell itself. In order to reach the desired uniform speed change, the thicknesses of body sidewalls should be designed differently and must be reinforced.

Achieving a constant vehicle deceleration is made possible by up-to-date design procedures. The difficulty lies in the realization that a vehicle body shell produces high decelerations at the beginning and then low deceleration levels afterwards. This can be achieved by using, for example, a combination of framework structure, which exhibits a high initial stability before it buckles, and up-to-date frame structures.

In order to develop small cars with small deformation lengths, special attention must be given to the deformation characteristics of the vehicle.

20.3.3.3   Success of Restraint Systems in (West) Germany.   West Germany refers to the federal states which existed before the unification of the two Germanies—East and West.

In spite of increased safety awareness, there are always traffic participants who jeopardize the results achieved so far. Therefore, the use of seat belts by passenger car occupants is an effective measure for protection against injury in case of an accident. Figure 20.16 shows official accident statistics related to seat belts. Seat-belt use rates in passenger cars are shown in Fig. 20.17. Interesting information can be drawn from these figures with respect to accident development in relation to seat-belt use rates by passenger car occupants.

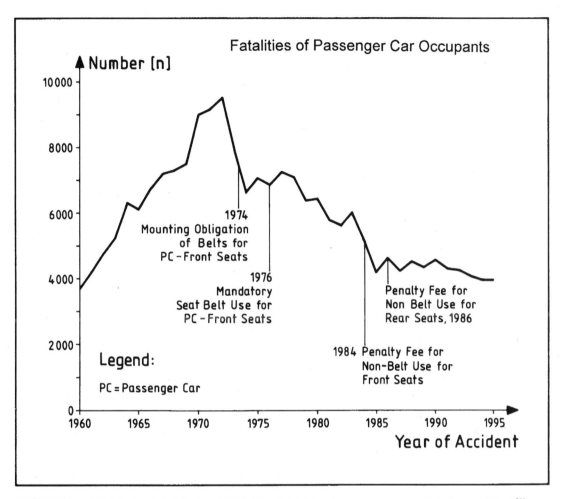

FIGURE 20.16 Official accident statistics from 1960 to 1995 for fatalities of passenger car occupants in Germany (West).[564]

The first mandatory installation of front seat belts in passenger cars was introduced in Germany in 1974. This was followed by a mandatory seat-belt use law in 1976. Between 1976 and 1984, as Fig. 20.16 shows, a more or less continuous decrease in fatalities could be noticed; this was attributed, to a certain extent, to the mandatory seat-belt use laws. This assumption is confirmed by the results of Fig. 20.17 which reveal that, in the time span from 1975 to 1984, seat-belt use rates increased from 40 to about 60 percent for drivers and front-seat passengers, which means that a decrease in fatalities was noted with an increase in seat-belt use.

In 1984, a penalty fee was introduced for not wearing seat belts in front seats. As Fig. 20.17 shows, this action resulted in a sudden increase of seat-belt use by more than 90 percent, which also led to a further decrease in fatalities (see Fig. 20.16). With the introduction of a penalty fee for not wearing seat belts in rear seats in 1986, a significant increase in seat-belt use rates, from about 10 percent in 1985 to about 70 percent in 1993, was noted (see Fig. 20.17).

Based on the previously discussed traffic laws, seat-belt use rates by car occupants are still quite high in Germany on all types of roadways. According to the Federal Highway Research Institute (BASt), the following seat-belt use rates for adults were observed in 1995:[185]

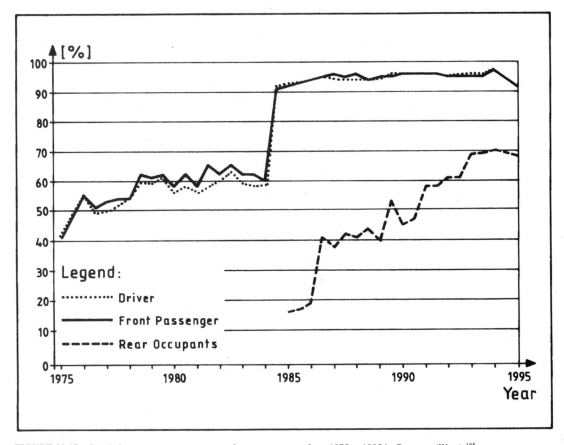

FIGURE 20.17 Seat-belt use rates, as a percentage, for passenger cars from 1975 to 1995 in Germany (West).[185]

Inside built-up areas: 86 percent

Two-lane rural roads: 93 percent

Interstates (autobahnen): 96 percent

The German example may also provide other countries with interesting insights and could provide the groundwork for the question: What can be expected from the introduction of different seat-belt use laws in relation to seat-belt use rates and accident (fatality) developments?

Since 1993, a mandatory law requiring children under 14 years of age to use seat belts and/or child seats has been in effect. This led to a considerable reduction in child fatalities and injuries. The use rates of these safety devices, individually or combined, have also been given by the Federal Research Institute (BASt) for 1995 for different roadway types:[185]

Inside built-up areas: 82 percent

Two-lane rural roads: 89 percent

Interstates (autobahnen): 96 percent

Since January 1992, seat belts have also been installed in new trucks, semi-trailers, and full trailer combinations, but without any applicable mandatory seat-belt use laws. The introduction of mandatory seat-belt use in buses could also lessen the number of injuries.

20.3.3.4 Interior Passive Safety

Passenger Cars and Station Wagons. Achieving a high level of protection for a vehicle's occupant is one of the most important tasks that helps in reducing fatalities and injuries. In this connection, the interior passive safety encompasses all motor vehicle–related measures which are aimed at keeping the forces and decelerations down on occupants in case of an accident, to secure a survival space, and to provide possibilities for rescue operations of the occupants, if necessary. That means a highly stable passenger cell combined with a sufficient energy absorption of the vehicle's front and rear sections as well as doors.

Seat belts are not always sufficient for preventing serious accidents. In these cases, impact cushions (and/or air bags) have to absorb the rest of the energy. It is important to know exactly what consistency the cushion materials are because they should not shatter and must withstand high temperatures. The steering wheel also represents a critical safety-related component, which can seriously injure a driver when an intrusion takes place into the passenger compartment.

Another safety aspect, which should not be underestimated, concerns the stability of head rests. According to the "European News about Traffic Safety,"[173] injuries to the neck area occur more and more, and make up more than 50 percent of the physical damages to accident victims. However, this kind of injury is difficult to diagnose. Victims often suffer for quite a long time after an accident from the painful consequences of neck injuries. Indeed, many motor vehicles are equipped today with head rests; these, however, are often not adjusted correctly. The head of the driver or passenger should be located at the same height as that of the highest point of the head rest. In addition to improper adjustment, it should be noted that many head rests were designed based on 20-y-old legislation.

In order to counteract the negative safety trend of head rests, the European Commission has assessed the overall height of head rests to be 800 mm. In this respect, for instance, Volvo offers a head rest which is able to reduce neck injuries by considerable amounts or levels. With its newest development, Saab will soon equip its models with an active head rest which moves forward and upward in case of a crash in order to catch the neck (see Fig. 20.18). These are two examples which, hopefully, will be followed by other car manufacturers.[173]

A further safety aspect is the stability of the rear seat backrests, such as in station wagons. In case of accidents, the anchorage of the backrests is often insufficient to keep the load in the luggage compartment; for the rear occupants, this could lead to fatal consequences. A simple way would be to fasten the load. However, in the majority of cases hooks for securing loads are often not present. In case of rollover accidents, it is not unusual for vehicle occupants to be killed by their own luggage. Especially dangerous also are unsecured fire extinguishers or gasoline tanks. For example, the "Saab 900" has a strong transversal bar at the backrest, which is anchored safely in the construction of the vehicle's body, that serves to attach three lap-shoulder belts. Together with the divided folding backrest, this transversal bar forms a safe barrier against luggage. In addition, fastening brackets are in place to secure heavy luggage.

There is no doubt that air bags, in combination with restraint systems, have significantly increased passive safety. In 1977, the U.S. Department of Transportation published the estimations shown in Table 20.12 for fatality and injury reductions after 1980.[557,726] As can be seen from the table, enormous decreases in the fatality numbers were expected at that time with respect to unbuckled passengers, concerning a level II vehicle crashworthiness combined with air-cushioned restraint systems. Under level II, sustainable crashworthiness represents a frontal impact of up to 40 mi/h (64 km/h) and a side impact of up to 20 mi/h (32 km/h). As can be seen from Table 20.12, the estimated percent reductions in fatalities ranged from 43 to 65 percent. Meanwhile, many of these expectations have already been fulfilled but much is still to be done. Today, we may be somewhere in the middle of this progress.

One important issue concerned with passive safety is that air bags and fastening seat belts belong together. An opinion survey, initiated by the *Journal of the General German Automobile Club* (ADAC-Motorworld), revealed that 2.2 million German motorists risk their lives because they believe that they can disregard the belt if their car is equipped with an air bag.[173]

In order to correct this dangerous error, ADAC conducted crash tests with two identical passenger cars equipped with air bags at a speed of 50 km/h. The dummies were only buckled up in one car. In a real situation, while the occupants protected with seat belts and air bags would have

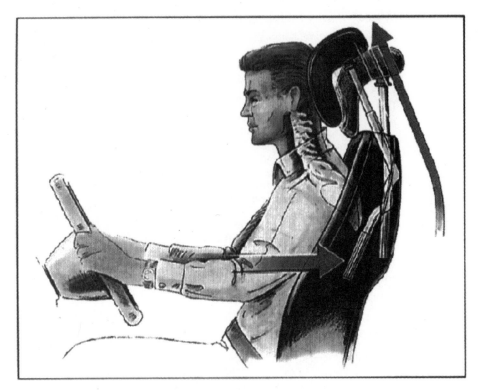

FIGURE 20.18   Saab's active head rest.[173]

TABLE 20.12   Estimated Effectiveness for a Level II Crashworthiness Combined with Air-Cushioned Restraint Systems, 1977[557,726]

Safety precaution(s)	Reduction of fatalities, %	Reduction of injuries, %
Air cushion	43	20
Air cushion and lap belt	50	24
Level II and air cushion	59	23
Level II, air cushion, and lap belt	65	27

Note: Front, side, and rear-end collisions included. Rollover accidents not included.

sustained only bruises and light injuries, the unbelted passengers would have suffered serious injuries like fractures to the legs and thorax, deep flesh wounds, a hip joint fracture, and a fractured knee cap. As a rule, it can be stated that, despite air bags, the knees, hip joints, and pelvis exclusively are protected by the belt. In addition, only the seat belt prevents ejection from the car, which has to be regarded as the most serious accident consequence.

In this connection, the latest German research findings, as reported by the Minister of Transportation Bruederle of the Federal State of Rhineland Palatinate,[86] indicated the following:

> It is wrong to believe that an air bag provides sufficient protection in case of an accident. An airbag is only a supplement to the safety belt and not a replacement. Studies of severe accidents indicate that at least 50 percent of the fatal injuries and 87 percent of the other serious injuries could have been prevented by simply buckling up. The belt is the "Number One" life-saver!

Furthermore, rear occupants also have to be buckled up or else the driver and front passenger(s) will be additionally endangered by the possible impact energy from the rear-seat passenger.[173]

The most recent research results by U.S. automobile manufacturers, like General Motors, Ford and Chrysler, have warned in an identical statement against the danger of air bags, especially with respect to children as front-seat passengers.[749] They stated that air bags and seat belts protect life; for children, however, air bags can be fatal. According to the report, front passenger air bags caused the death of at least 28 children and 19 adults, most of whom were small women. The fatal cause was as follows: The front passenger air bags inflate with too high an energy level. Most of the fatally injured persons were not or were incorrectly buckled up, or were seated too close to the air bags, which inflate in fractions of a second at a speed of about 250 km/h.

Meanwhile, 22 million car owners have been warned of the dangers associated with air bags by the three U.S. manufacturers. In the future, these manufacturers will develop "intelligent" air bags. The energy and speed of these air bags will depend, when inflating, on the impact energy, the size of the vehicle's occupants, and the seat distance to the air bag. Also, in Germany, a number of accident casualties could be traced back to front passenger air bags.[749]

In this connection it should be noted that scientists of Boston University in the United States have already developed an intelligent air bag system. The new technique is hidden in the seat's upholstery and consists of a foil from the space research "smart skin," which transforms the pressure on the seat into on electric charge, and a computer calculates the position and weight of the passenger. In case of an accident, the system knows within 5 ms the momentum with which the air bag has to be inflated in order to catch the occupant without injury.[63]

Associated indirectly with interior passive safety are two other future devices, on which the French automobile manufacturer Renault is working. One device represents the first prototype for an in-the-automobile integrated breathing alcohol test. With this new system, the motorist can easily and reliably examine his or her alcohol level in order to determine if the alcohol level, which is mandated by law, has been exceeded or not. In this respect, alcohol consumption is detrimental to the driver's sight conditions, it impairs evaluation of risks, and it reduces reflexes. This useful accessory is to be welcomed. It will be available for production prior to the year 2000.[173]

Furthermore, Renault is working in another general safety project. It is a system which prevents a driver from falling asleep at the steering wheel.

This is accomplished by a small camera mounted on the dashboard facing the driver. It continuously surveys the driver's eyelids (Fig. 20.19). If the eyes close, that is, the driver falls asleep, then an alarm would sound to wake the driver up.

According to Renault, the antifatigue system will become standard equipment in the year 2000. It will be installed first of all in buses and trucks and later in passenger cars.[173]

Specific Aspects of Interior Passive Safety for Commercial Vehicles.   The passive safety used to reduce accident consequences for commercial vehicles can only be increased to a certain extent because these vehicles have a higher total weight and higher overall dimensions as compared to passenger cars. When designing the superstructures of commercial vehicles, occupant protection must be an important element. In this connection, two aspects have a considerable influence: preserving the survival space and limiting acceleration. On the basis of repeated crash tests, front structures could be further optimized in order to provide a better balance between the surviving space and low acceleration. In this connection, it should be noted that a commercial vehicle with a small hood has an advantage over the so-called cab-over-engine truck because the hood could be used as a crush zone. For heavy trucks and buses, effective deformation zones with bumpers and corresponding attachments are realized. Because the driver and the occupants are seated relatively high, they face relatively low risk in case of a collision with a passenger car.

A front-end guard (front underride protection) below the driver's cab serves as a partner protection for passenger cars (exterior passive safety). Basically, it is important, for limiting accelerations

1. To provide a long deformation distance for the commercial vehicle
2. To ensure, within this distance or during the corresponding collision time interval, a possible constant force with plastic energy absorption without a rebound

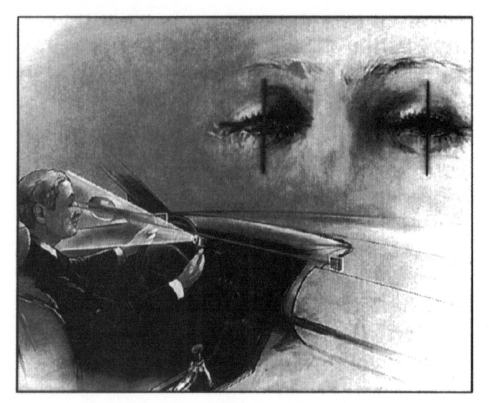

FIGURE 20.19   Renault's antifatigue system.[173]

A further safety problem with commercial vehicles is load. In this respect, in case of an inappropriate load and attachments, commercial vehicles could quickly become rolling time bombs. Time pressure and the driver's lack of competence are often cited as the reason for the improper attachment of the load.

Specific Aspects of Interior Passive Safety for Buses.   Of special importance is passive safety in buses because many occupants may be involved in case of a traffic accident. For instance, in case of a rollover accident, the bus turns on its roof pillar and serious consequences could result from the buckling of the superstructure and the impairment of the surviving space in the head area of occupants. This must be prevented through the development of strong rollover-protection superstructures. Furthermore, a frequent bus accident type is that where it collides with a truck from behind, which represents the most dangerous accident situation for the bus driver. Frequently, the bus driver is jammed and is seriously or fatally injured. Also, the occupants are exposed to a certain injury risk in this accident type; the number of fatalities, however, is considerably lower in this accident type as compared to a rollover.

In contrast to passenger cars, with respect to buses, legislators are far behind the state of knowledge with their standards, at least in Europe. Individual standards like Economic Commission of Europe Regulation (ECE) R66, which regulates the stability of the superstructure, and ECE R80, which determines if the stability of seat brackets is basically correct, do not often have any legal force.

The following statement is the opinion of the principal author and not the opinion of Schoch, who elaborated on Sec. 20.3.

With respect to the safety standards of school buses, Canada and the United States have meanwhile achieved a safety level (Fig. 20.20) which is far beyond that of Europe regarding vehicle

**Typical School Bus in the U.S.A.
(Safety Tank)**

**Rear Wall with Rear Lighting System
and Emergency Door**

FIGURE 20.20   The school bus—a safety vehicle in Canada and the United States.[391]
(*Source:* Lamm)

Examination of the Roof Stability according to
Safety Standard No. 220 for the Case of Rollover
Accidents (Vehicle Research and Test Center,
NHTSA; East Liberty, Ohio)

Traffic Flashing Light with Speed Limit,
as Indicator for a School-District

FIGURE 20.20 (*Continued*) The school bus—a safety vehicle in Canada and the United States.[391] (*Source:* Lamm)

and legislative developments. So far, it has not been understood fully in many countries outside North America that school buses *must* be designed for the transport of children and not adults. Of course, it should not be forgotten that the design of school buses, as shown in Fig. 20.20, results in an aggressive accident opponent to other road users.

20.3.3.5 Exterior Passive Safety.   *Exterior passive safety* involves all vehicle-related measures which are meant to minimize the severity of injuries outside the vehicle for all traffic participants. It reflects also the capability of a vehicle to provide care for nonoccupants as well.

The forward structure of a vehicle should be designed in such a way that pedestrians and two-wheel riders can survive an accident by minimizing the severity of injuries. According to Fig. 20.9, the percentage of traffic fatalities in Europe for pedestrians, bicyclists, and motorcyclists is somewhere between 35 and 45 percent; many of these injuries result from collisions with passenger cars and commercial vehicles. A *partner protection* means that the hood is, to a certain extent, deformably (softly) designed, the wiper linkage is not critical, the windshield and the exterior rear-view mirror frames exhibit only a small injuring design. In addition, the forward structure must be designed in such a way that in case of a frontal crash with a smaller and lighter vehicle, the occupants will not suffer a catastrophy.

Specific Aspects of the Exterior Passive Safety for Commercial Vehicles

Rear underride protection.   Nowadays, nearly all trucks and trailers are equipped with an underride protection. Exceptions to this rule are, for example, semitrailer tractors, two-wheel trailers which transport long materials, and construction site trucks. The rear underride protection is attached to the rear end of the vehicle at a height of 550 mm above the road surface. This underride protection prevents, in case of a rear-end impact, an underriding under the platform or the superstructure of the truck. For all-terrain vehicles, foldable designs are used because the rear underride protection reduces ground clearance.

Lateral underride protection.   Since 1992 (in Germany), new trucks, semitrailers (without the tractors), or trailers have to be equipped with lateral underride protection on both sides. Beginning in 1994, older vehicles also had to have lateral underride protection.

The lateral underride protection has been required by accident researchers for quite a long time. It prevents, in case of an accident, pedestrians, two-wheel riders, and passenger cars from getting into the free spaces between the road surface and the lower edge of the vehicle frame in between the areas of the vehicle's axles and are consequently rolled over. A further advantage of lateral covering is to improve the sight conditions for passing vehicles on wet road surfaces because a major portion of the water is led by an air draft which is adjacent to the vehicle contour.

Front underride protection.   Accident analyses indicate that head-on collisions between trucks and passenger cars result in a high degree of accident severities. This fact is explained by the relatively high collision speeds and by the differences in mass, form, and rigidity between trucks and passenger cars. In addition, such accidents frequently involve an underriding of the truck front by a passenger car and, thus, the intrusion of the truck's forward structure into the passenger compartment because the passenger car is unable to activate its front safety structure. Figure 20.21 shows the distribution of the principal points of impact in case of truck–passenger car accidents. This figure clearly shows that the majority of accidents occur in the front area of the truck. Thus, a front underride protection on trucks is quite important.

The front underride protection on a truck absorbs the energy in case of an accident. The best solution is to use front protection with an energy absorber because of a more effective safety gain and because collision forces are better reduced by an energy-absorbing underride protection than by a simple rigid bumper.

20.3.4 Crash Kinds and Types

The aim of crash tests is to demonstrate the safety level of a vehicle to the vehicle manufacturers, the legislators, the general public, and the customers. The results of these crash tests aid the construction of body design. They serve as admissions to the market and enable comparative evalu-

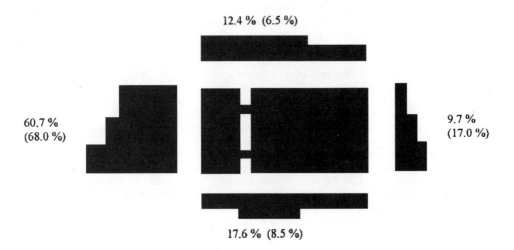

12.4 % (6.5 %)

60.7 %
(68.0 %)

9.7 %
(17.0 %)

17.6 % (8.5 %)

Legend:
The percentages in parantheses represent impacts resulting in fatally injured
traffic participants.

FIGURE 20.21   Distribution of principal points of impact in case of truck–passenger car accidents (as a percentage).[633]

ations. An important assumption for meaningful crash tests is that they should closely represent
an accident situation. In considering the various individual accident types, it is understandable
that the automobile designer cannot possibly assign every individual accident to a simulation test,
with respect to possible accident opponents and expected speed changes.

For example, Fig. 20.22 shows the state of knowledge regarding the frequency of different
accident collision types for Volkswagen for the year 1991.[319]

According to Fig. 20.22, the following list shows the frequency, in percent, of accident collision types in decreasing order:

1. Front (21 percent)

2. Oblique from forward (19 percent)

3. Oblique lateral into the side of the door (17.2 percent)

4. Oblique frontal against the exterior vehicle edge (14.6 percent)

5. Offset (12.3 percent)

6. From the side, right angle (7.2 percent)

7. Oblique from rearward into the rear door (3.6 percent)

8. Rollover (2.8 percent)

9. Rear end (2.3 percent)

To find out how the most endangered person groups, that is, the drivers and right-front passengers, die in a crash, the principal points of passenger car impacts together with the percentages
of fatalities for drivers and right-front passengers have been identified in Fig. 20.23. This figure
shows that approximately 50 percent were killed in front collisions and 30 percent from left or
right side impact.[182,391,392]

Even though the percentages are not directly comparable because Fig. 20.22 shows accident
collision types and Fig. 20.23 shows accident fatalities, the results, to a certain extent, agree with
each other.

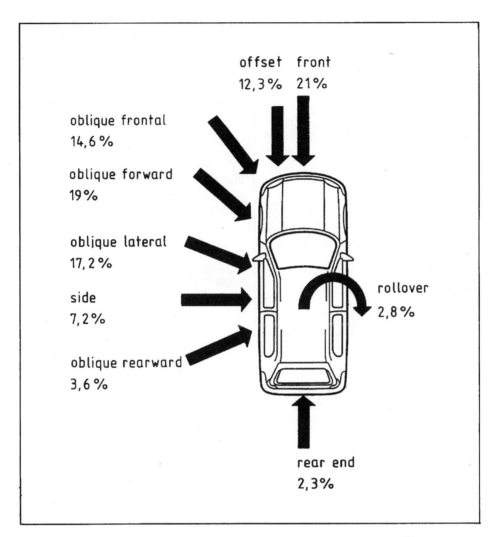

offset front
12,3% 21%

oblique frontal
14,6%

oblique forward
19%

oblique lateral
17,2%

side
7,2%

oblique rearward
3,6%

rollover
2,8%

rear end
2,3%

FIGURE 20.22 Percentile frequency of different accident collision types, Volkswagen, 1991.[319]

20.3.4.1 Front Impact Tests. The front impact against a rigid barrier with 100 percent over-lapping causes the strongest vehicle deceleration and, correspondingly, a high stress on the buck-led-up occupants. However, the structural stress for the passenger compartment is relatively low because the kinetic energy, which must be absorbed, is distributed over the whole vehicle width. That means a reduction in the survival space is relatively small. The front crash with 100 percent overlapping against a rigid wall at a speed of 50 km/h is, so far, the only crash test regarded in Europe as mandatory. Cases 1 to 3 of Fig. 20.24 show this kind of crash configuration.

Another impact test is the 30° inclined barrier, with respect to the vertical axis of the vehicle either to the left or to the right. The test vehicles slide off the barrier in the lateral direction. The force component, lateral to the driving direction, causes a twist of the vehicle (see case 4 in Fig. 20.24). However, this case cannot simulate real accident situations. For example, two colliding vehicles with 50 percent overlap become tangled up with each other and perform a slight twist. From this tangling up, other vehicle deformations and occupant stresses originate, as the barrier,

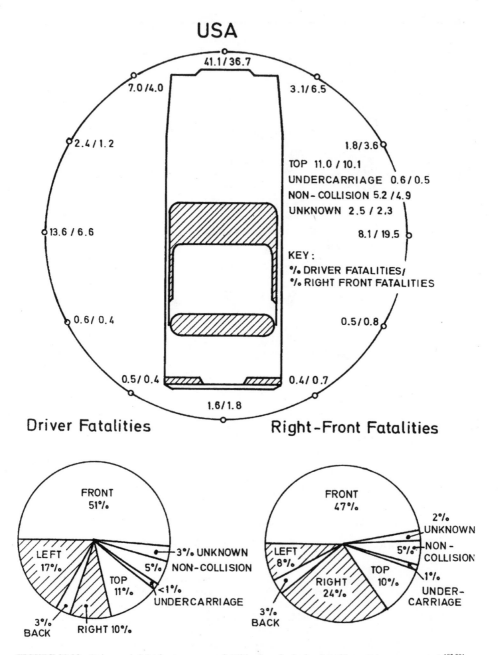

FIGURE 20.23 Driver and right-front passenger fatalities by principal point of impact (passenger cars).[182,391]

Case	Crash Kind	Configuration	Speed [km/h]	Vehicle Load	Test Reason
1	0°	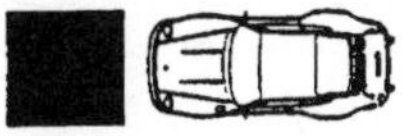	4-8	Curb Weight Test Weight	bumper
2	0°	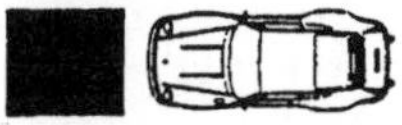	15-28	2 Occupants and Additional Load	sensor design, interior stress, structural bevior
3	0°	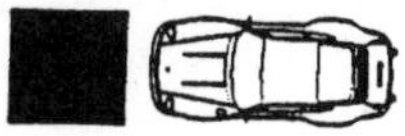	48.3-56.4 (30-35 mph)	Curb Weight up to 5 Occupants and Additional Load	occupant stress, vehicle structure, steering wheel displacement, gas tank tightness, passenger compartment, behavior of extremely small or big occupants, rescue behavior
4	30° left / right	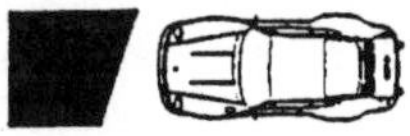	25-53	2 Occupants and Additional Load	sensor design, occupant stress, vehicle structure, gas tank tightness, passenger compartment
5	30° left / right Anti-Side Device	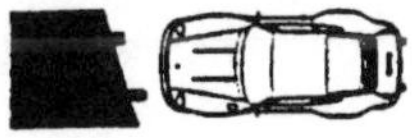	50	2 Occupants and Additional Load	occupant stress, vehicle structure, gas tank tightness, passenger compartment, sensor design
6	40 % Overlapping	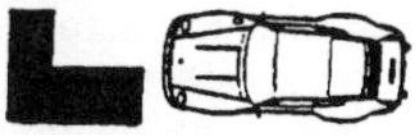	15	1 Occupant	repair costs
7	40 % Overlapping	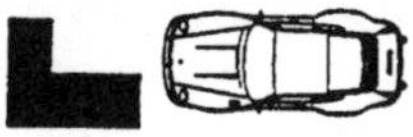	50	2-3 Occupants and Additional Load	occupant stress, vehicle structure, gas tank tightness, passenger compartment, sensor design
8	40 % Deformable Barrier	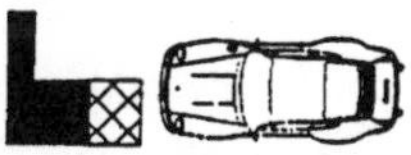	56	2 Occupants and Additional Load	occupant stress, vehicle structure, gas tank tightness, passenger compartment, sensor design
9	Pole, Central / Offset	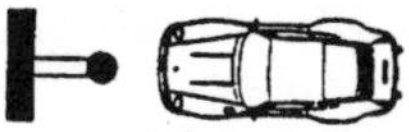	50	2 Occupants and Additional Load	occupant stress, vehicle structure, gas tank tightness, passenger compartment, sensor design
10	Vehicle / Vehicle		50-80	2 Occupants and Additional Load	occupant stress, vehicle structure, gas tank tightness, passenger compartment, sensor design

FIGURE 20.24 Selection of front crash arrangements of the Porsche safety development.[633]

according to case 4 in Fig. 20.24, is able to show. Also, equipping the barrier with the so-called antislide devices (case 5, Fig. 20.24) cannot prevent a total slide-off.

In order to model an actual accident situation in a better way, since 1990 front impact tests have been conducted against rigid barriers with 40 percent overlapping (offset) (see cases 6 and 7 in Fig. 20.24). According to the degree of overlapping, the tests reveal different results, depending, for example, on how strongly the engine block is impacted by the barrier and how it contributes to the absorption of the kinetic energy. However, in order to guarantee high passenger compartment stability, the 40 percent overlapping test gives valuable insights on the vehicle design in addition to the 100 percent wall crash.

In the actual accident situation, the deformations depend on the local rigidity of the forward structure of vehicles. However, this consideration is neglected for crash tests against rigid barriers. Only an appropriate deformation characteristic of the barrier would enable, in reality, a realistic vehicle deceleration.

20.3.4.2   The Offset-Crash against a Deformable Barrier.   A rigid, plane barrier exhibits the same rigidity along the whole impact area. This causes constant development of deformation on the vehicle, independent of its specific rigidity. Therefore, the impact against a rigid barrier corresponds to a front collision of two vehicles of the same weight, with the same frontal structure, and with the same impact speed. It is only valid in case of an impact with a full and exact overlapping, where structural areas with equal rigidity hit each other.

However, because the forward structure is made of different structural zones in terms of the width, height, and depth of the vehicle, in real offset crashes different structural areas collide with each other. Therefore, contrary to the rigid barrier, the vehicle deformation depends, in the case of a deformable barrier, on the specific rigidity of the impacted area, such as during actual vehicle-vehicle impacts (see case 8 in Fig. 20.24).

Therefore, the compatibility aspects are better considered by deformable barrier tests for front impact procedures. In case of an impact against a rigid barrier, the impact energy has to be absorbed exclusively by the examined vehicle; in case of a deformable barrier, a distribution of the deformation energy takes place with respect to the different structural zones of the vehicle. The size of barrier deformation can provide valuable information on the compatibility of the examined vehicle.

For instance, the General German Automobile Club (ADAC) conducted crash tests in 1995 on the new "VW-Polo" against a rigid barrier and a deformable barrier. The crash test against the rigid barrier indicated that the "VW-Polo" provides safety for its occupants. However, the crash against the deformable barrier indicated that this car represents an aggressive accident opponent, that is, its fork-like frame side rails penetrate like spears into the crush zone of the opponent.[73]

20.3.4.3   Lateral Collisions.   Because occupants are becoming more and more protected in the most frequently occurring front collisions by well-designed seat-belt systems and air bags, the proportion of injured people is on the increase in lateral collisions. Therefore, the research work of the last few years has been concentrating, more and more, on occupant protection systems for lateral collisions. The lateral protection should already be outlined as an integrated component during the development stage of the automobile in order to enable an appropriate design. Essential for the lateral crash is the stability of the body-hinges-doors-locks combination.

Several automobile manufacturers have developed lateral air bags. BMW offers a double air bag for head and chest, Volvo has integrated the air bag into the seat, and Mercedes-Benz has installed the side bag above the arm rest in the front doors for the E-class. Meanwhile, most manufacturers offer lateral air bags, often as standard equipment.

U.S. and European lateral impact procedures are discussed in detail in the following:

- The impact of a rigid 4000-lb barrier (about 1818 kg) of 90° against the side of a test vehicle is the legal procedure in the United States (see case 1, Fig. 20.25). The test has to be conducted on both sides, with an impact speed of 20 mi/h (32 km/h). The relatively heavy barrier has a noncontoured (smooth) surface.

Case	Crash Kind	Configuration	Speed [km/h]	Vehicle Load	Test Reason
1 U.S.A.	90° left / right 1800 kg	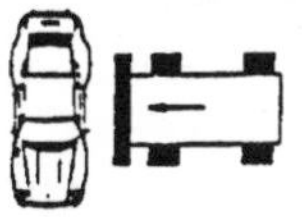	32-35	1 Occupant and Additional Load	gas tank tightness, occupant stress, vehicle structure, rescue behavior
2 U.S.A.	27° 1365 kg	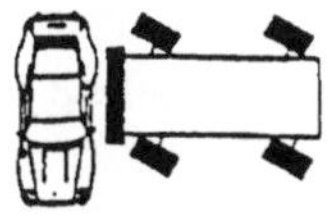	54	1-2 Occupants and Additional Load	occupant stress, vehicle structure, gas tank tightness, passenger compartment , rescue behavior
3 Europe	90° left / right 950 kg	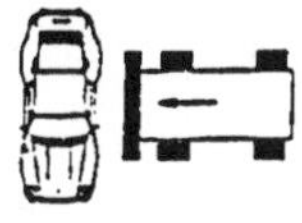	50	1 Occupant and Additional Load	occupant stress, vehicle structure, gas tank tightness, passenger compartment, rescue behavior

FIGURE 20.25   Selection of lateral crash arrangements of the Porsche safety development.[633]

- Since 1993, an additional lateral impact test has been introduced in the United States, according to the Federal Motor Vehicle Safety Standards (FMVSS 214). This dynamic lateral impact procedure was used beginning with 1994 vehicles. For this test, a barrier of 1365 kg is driven at a speed of 54 km/h at an angle of 27° into the side of the vehicle to be tested (see case 2 in Fig. 20.25). With this so-called crabbed barrier, the relative speed between two collision vehicles is simulated. The barrier is deformable with aluminum honeycombs.

European accident researchers believe that the U.S. deformation element is too strong and too wide (66 in = 165 cm). The U.S. barrier corresponds, by means of its dimensions and rigidity, to the forward structure of a U.S. truck. The deformation encompasses a wide area and extends from the A pillar (front pillar) to the rear end of a passenger car. In contrast, the European proposal is to use a smaller, lighter-weight barrier. The deformation part (likewise) with aluminum honeycombs) is softer and corresponds to about the foreward structure of a European passenger car (see case 3 in Fig. 20.25). The impact angle of 90°, which is not ideal, has to be reviewed because it does not consider the relative speed of the opponents. The impact speed is 50 km/h, and the mass of the deformable barrier is 950 kg.

It is interesting to note that smaller, shorter vehicles mostly do better in lateral crashes as compared to larger, longer vehicles. The reason is that the small rigid zones of the passenger compartment are commonly impacted. In addition to that, small cars have a lower steady-state condition and, consequently, they are easily pushed away. Automobiles with longer passenger compartments mainly counteract the impact energy by the B pillar (center pillar), and withstand longer, before being pushed away, because of the higher mass. In general, this means that shorter vehicles tend to have smaller deformations in case of lateral crashes. Because of this, the disadvantage is nearly compensated for, that is, the occupant is seated a smaller distance from the outer skin of the door and less space exists for padding in the door.

20.3.4.4 *Rear-End Collisions and Vehicle Rollovers.*

In the regular case, a rigid 1800-kg barrier is used for rear-end collisions in order to examine, at a speed of about 50 km/h, the tightness of the gas tank, the vehicular structure, the occupant stress, the stability of the passenger compartment, and the rescue behavior of the vehicle (see case 3 of Fig. 20.26). Several other rear-end impact tests exist, as revealed by the arrangements of the Porsche safety development shown in Fig. 20.26.

Case	Crash Kind	Configuration	Speed [km/h]	Vehicle Load	Test Reason
1	180°	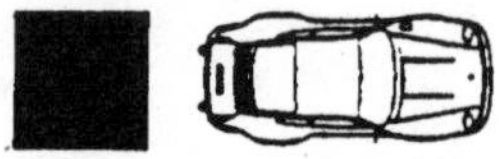	4–8	Curb Weight Test Weight	bumper
2	180° 1100 kg	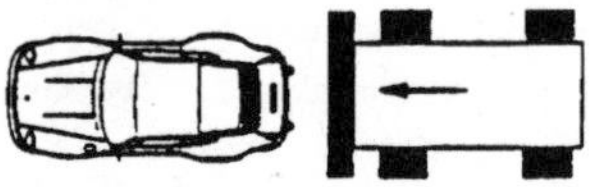	38–52	- Curb Weight - 2 Occupants and Additional Load, - 5 Occupants	gas tank tightness, vehicle structure, occupant stress, passenger compartment, rescue behavior
3	180° 1800 kg	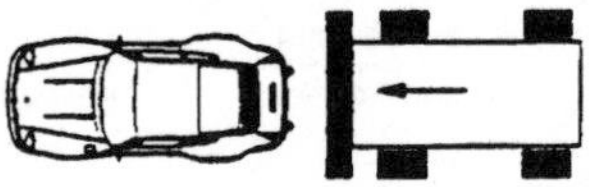	48–52	- 2 Occupants and Additional Load,	gas tank tightness, vehicle structure, occupant stress, passenger compartment, rescue behavior
4	45° right / left 1800 kg		50	2 Occupants and Additional Load	gas tank tightness, vehicle structure, occupant stress, passenger compartment, rescue behavior
5	40 % Overlapping 1100 kg	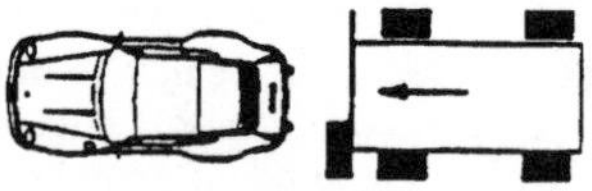	15	1 Occupant	repair costs
6	Vehicle / Vehicle	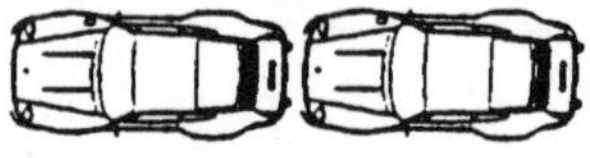	50–80	2 Occupants and Additional Load	gas tank tightness, vehicle structure, occupant stress, passenger compartment, rescue behavior
7	Rollover		48–52	2 Occupants and Additional Load	occupant stress, vehicle structure, gas tank tightness, passenger compartment, rescue behavior

FIGURE 20.26   Selection of rear-end crash arrangements of the Porsche safety development.[633]

With respect to vehicle rollovers, two simulation procedures exist:

1. The tightness of the gas tank is tested following accident simulations (like front, lateral, and rear-end tests) by means of rotating the vehicle in steps of 90° (length of time in each position: 5 min). In this way, the gravity valves, which should prevent gasoline from leaking out of the gas tank, are tested. The gravity valves should shut off the gas feed already at small inclined positions. Otherwise, in case of a rollover accident, gasoline could leak out and cause a fire hazard, environmental pollution, etc.

2. In order to test occupant behavior, the vehicle is subjected to a dynamic rollover test. In this connection, the vehicle is placed on a sledge at an inclination of 23° (Fig. 20.26, step 7). The sledge rides with a speed of 48 km/h and is slowed down by an exactly defined deceleration process. According to this procedure, the vehicle falls from the sledge and performs a rotational movement (Fig. 20.27).

It is interesting to note that in comparison to the front or lateral collision (time span of about 100 ms), the rollover test shows much longer time periods of several seconds (Fig. 20.27) which guarantee a high surviving rate without serious injuries for the buckled-up occupants. In contrast to the front impact, the occupants are not generally endangered by head accelerations. More critical are the forces and torques in the neck area.

20.3.5 Biomechanical Stress Limits

With respect to vehicle safety, *biomechanics* is a measurement for describing the injury mechanisms and an instrument for determining the mechanical endurance of the human body. These are achieved by the so-called dummies which are used during crash tests in order to record biomechanical stresses. These dummies are equipped with sensors for measuring accelerations, forces, and distances.

In order to determine stress limits, biomechanical tests are required on either living individuals who volunteer or theoretical and physical models. As a basis for simulations with physical models, irreversible and/or reversible injuries are experimentally investigated on dead bodies or animals. These basic biomechanical tests lead to relationships between stress severities and stress criteria (for example, with respect to forces and accelerations). From this procedure, we arrive at protection criteria that can be physically measured or whose corresponding limits should not be exceeded.

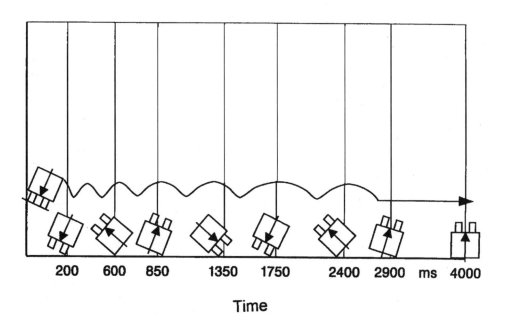

200   600   850   1350   1750   2400 2900   ms   4000

Time

FIGURE 20.27   Rotational movement of a motor vehicle during the rollover test.[642]

Even though the field of biomechanics contains more likely medical aspects, it is used here to derive substantial criteria for defining and measuring stress limits on the human being in vehicle crashes. The knowledge of these limits is important for the automotive industry in order to evaluate reasonable occupant stresses in crash situations.

Among others, the stress limits describe fractures, organ damage, and many other injuries. Classification is done by the abbreviated injury scale (AIS) or the overall abbreviated injury scale (OAIS). AIS and OAIS evaluate individual, as well as overall injuries, in steps from 0 to 6. The numbers correspond to the increasing severity of an injury from 0 (not injured) to light, to average, and to serious injuries up to 6 (fatally injured).

However, the stress limits are different from one individual to another. For example, they strongly depend on the age, sex, mass, and mass distribution of the human being. In this connection, it is not possible to build a dummy in the form or weight of every body. The configuration of a dummy also differs greatly from the anatomy of the human being. First of all, the mass distribution and the modeling of the internal organs of humans are different from dummies. Dummies are expensive and should be reusable. This assumes a high stress resistance and a certain nondestructive testing, which, unfortunately, the human being does not exhibit. Therefore, tests with post-mortal test objects (PMTO) are still being conducted today.

In the following subsections, several stress values for external and internal injuries are described in detail.

20.3.5.1   External Injuries.   Figure 20.28 shows biomechanical stress limits for important body parts, such as the brain, skull, forehead, cervical vertebra, thorax, pelvis, thigh, and shinbone. The stress limits are expressed by accelerations (a), forces (F), distances (s), etc.

For example, thorax injuries can originate from an impact against the steering wheel or instrument panel. As shown in Fig. 20.28, the impact reaction force in this case should not exceed 8000 N in order to avoid injuries; the deformation distance of the thorax should be less than 6 cm. Accordingly, detailed information on stress limits, often expressed by g's, can be found in Fig. 20.28 for every part of the body.

For instance, the sketch in Fig. 20.28 shows different accelerations, in g's, for different parts of the skull that should not be exceeded in case of a crash.

20.3.5.2   Internal Injuries.   It is still a complicated issue to evaluate a human being's internal injuries. The biggest problem is the stress placed on the brain and on the cervical vertebra. Figure 20.29 shows the time characteristic deceleration curve for a passenger car for an impact against a wall at 50 km/h and the head deceleration curves of a buckled-up and an unbuckled driver. It is evident that the head deceleration of a buckled-up driver (the upper part of Fig. 20.29) is lower and more uniform than that of an unbuckled driver (the lower part of Fig. 20.29). For the unbuckled case, decelerations of about $100g$ are reached, which, depending on the action time, may lead to irreversible brain damage in addition to mechanical bodily injuries.

The deceleration-time function of the "Patrick curve" reveals that the height of the deceleration stress on the brain has to decrease with an increase in action time (see Fig. 20.30) in order to avoid irreversible brain damages.

For example, the high deceleration, x, in Fig. 20.30 can only be endured for the small action time, y. A lower deceleration, z, can be endured for a longer action time, w. The higher the deceleration is, the more the brain is pressed and contused against the scalp. A heavy contusion can only be endured for a short period of time. In contrast, if the deceleration is relatively low and the brain is only slightly pressed against the scalp, this deceleration could be endured for a relatively longer period of time.

The neck is just as critical as the head. Acting as a connecting element between the torso and the head, the neck is more or less exposed to all kinds of stresses in traffic accidents. This primarily concerns the cervix, which becomes highly stressed in case of a relative forward motion by the head to the torso (flexion) and in case of a relative backward motion to the torso (extension). These motions cause traction, pressure, and gravity forces as well as torques. Depending on the state of the muscular system and on the behavior of the occupants, serious injuries could

Body Part	Mechanical Sizes	Stress Limits
Whole Body	$a_{x,max}$ $\bar{a}_x$	40 ... 80 g 40 ... 45 g, 160 ... 220 ms
Brain	$a_{x,max}$; $a_{y,max}$	100 ... 300 g WSU - curve with 60 g, t > 45 ms 1800 ... 7500 rad/sec^2
Skull	$a_{x,max}$; $a_{y,max}$	80 ... 300 g depending on the size of he impact area
Forehead	$a_{x,max}$ F_x	120 ... 200 g 4000 ... 6000 N
Cervical Vertebra	$a_{x,max}$ Thorax $a_{y,max}$ Thorax F_x α_{max} foreward α_{max} backward	30 ... 40 g 15 ... 18 g 1200 ... 2600 N shear stress 80° ... 100° 80° ... 90°
Thorax	$a_{x,max}$ F_x s_x	40 ... 60 g, t > 3 ms 60 g, t < 3 ms 4000 ... 8000 N 5 ... 6 cm
Pelvis-Thigh	F_x $a_{y,max}$	6400 ... 12500 N force application into the knee 50 ... 80 g (pelvis)
Shinbone	F_x E_x M_x	2500 .. 5000 N 150 ... 210 Nm 120 ... 170 Nm

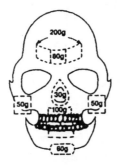

Legend:
```
WSU = Wayne State University Cerebral Tolerance Curve
  g = gravitational acceleration [9.81 m/s²]
  a = acceleration/deceleration [m/s²]
  x = longitudinal direction
  y = lateral direction
  F = force [N] → Newton = [kg m/s²]
  α = angle [°]
  s = distance [cm]
  t = time [ms]
  E = energy [Nm] → Newtonmeter = [kg m²/s²]
  M = torque [Nm]
```

FIGURE 20.28   Biomechanical stress limits for the human being.[642]

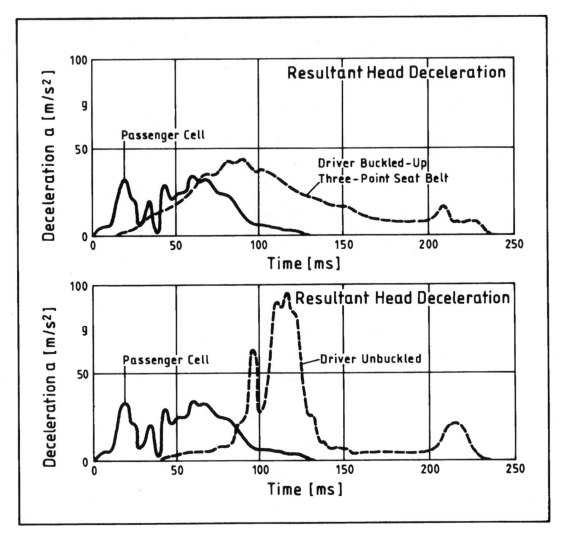

FIGURE 20.29   Time characteristic deceleration curve of a passenger car in the case of an impact against a wall at 50 km/h and head decelerations of a buckled-up/unbuckled driver.[580]

result. Limiting curves for stresses do not exist so far because, by using test persons, tests could only be conducted up to the maximum level of pain.

For evaluating head injuries in case of front impacts, the head injury criterion (HIC) was developed. It represents the allowable stress on the head. HIC is expressed as follows:[642]

$$\text{HIC} = \left(\frac{1}{t_2 - t_1} \int_{t_1}^{t_2} a_{\text{res}}\, dt \right)^{2.5} (t_2 - t_1) < 1000 \tag{20.6}$$

In this equation the resultant head acceleration or deceleration, a_{res}, is expressed as multiples of gravitational acceleration, in g's, and time, in milliseconds; t_1 and t_2 are arbitrary points of time and a_{res} is the resultant head acceleration which is a function of time. As a limiting value for HIC,

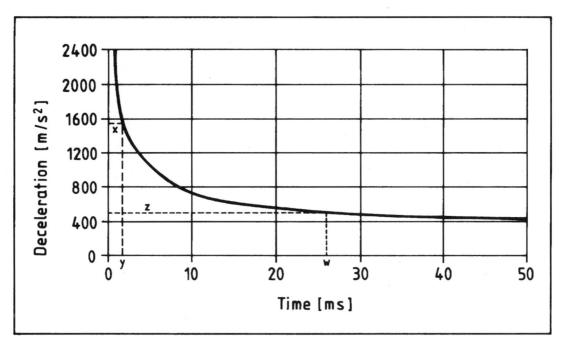

FIGURE 20.30   Patrick curve (a scale for evaluating stresses on the human brain).[642]

the number 1000 was chosen. As can be seen from Fig. 20.31, the curve characteristics of acceleration-time functions have a distinct influence on the size of HIC. Despite using similar maximum accelerations of 1000, the HIC value varies between 246 and 1000 with respect to a triangular, sinusoidal, or rectangular curve characteristic (see examples in Fig. 20.31).

The stress on the brain depends on the resultant deceleration. If, for example, a rectangular curve characteristic is present (Fig. 20.31c), then the maximum deceleration would suddenly have an effect on the brain. That means the brain would be crushed and would not be able to regenerate because of the high deceleration which is in effect during the whole time period.

With respect to the sinusoidal curve characteristic, deceleration gradually increases, reaches a maximum, and then gradually decreases (Fig. 20.31b). The stress on the brain is essentially lower.

The triangular curve characteristic shows the best solution because deceleration increases steadily up to a maximum which is then followed by an abrupt decrease down to zero (Fig. 20.31a).

The head protection criterion (HPC) denotes the European proposal for describing the injury criterion of the head in case of a lateral impact. HPC is once again expressed as a time integral with respect to the resultant head acceleration [see Eq. (20.6)].

Limiting values from different standards for HIC, HPC, and accelerations, as protection criteria with respect to the head area, are given in Table 20.13.

The thoracic trauma index (TTI) was developed as a protection criterion in case of a lateral collision for the chest stress. It is calculated from the following formula:

$$TTI = 0.5 \, (G_R + G_{LS}) \qquad (20.7)$$

where G_R = the maximum measured rib acceleration, in g's
$\quad G_{LS}$ = the maximum value of the acceleration of the lower vertebral column

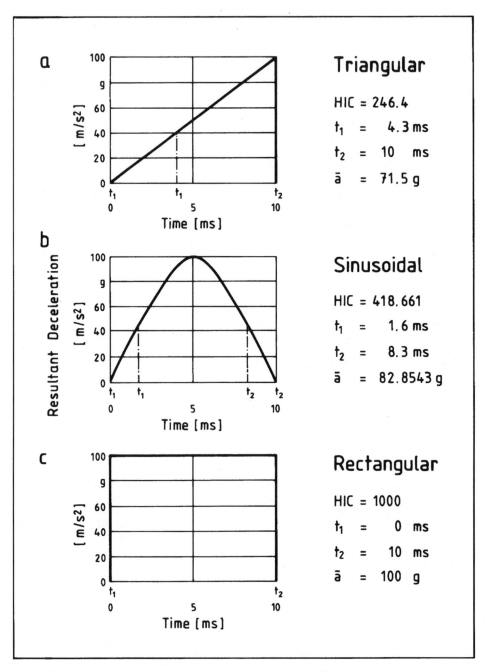

FIGURE 20.31 Examples for calculating the head injury criterion (HIC).[642]

TABLE 20.13   Existing Test Limits for the Head Area[233]

Limiting value	Standard	Description
HIC = 1000	FMVSS 208	Front crash; from the resultant head acceleration of the dummy
HIC = 1000	FMVSS 213	Examination of child restraint systems; from the resultant head acceleration of the child dummy
$a_{3ms} = 80g$	FMVSS 201	Head skull
$a_{3ms} = 80g$	EWG 74/90	Pendulum test; acceleration of the test head, deceleration should not exceed $80g$ in 3 ms
	EWG 78/932 ECE R 12 ECE R 21	
$a_{5ms} = 150g$ $a_{max} = 300g$	ECE R 22	Impact absorber test for helmets
HPC = 1000	EEVC proposal, lateral impact	Integration over time of the resultant head acceleration

Note: HIC = head injury criterion; a_{3ms}/a_{5ms} = acceleration/deceleration in 3/5 ms; HPC = head protection criterion; FMVSS = Federal Motor Vehicle Safety Standard; EWG = European Economic Community (EEC); ECE R = Economic Commission of Europe, Regulation; and EEVC = European Experimental Vehicle Committee.

TTI is expressed in g's; limiting values are given in Table 20.14

As indicated earlier, the assessment of protection criteria for the human body is normally derived from stress limits which are then applied in measurement systems and mathematical procedures. A summary of existing legal test limits, with respect to chest, pelvis, belly, and abdomen areas, are given in Table 20.14.

20.3.6 Body Design

The superstructures of all kinds of vehicles are known as bodies. With respect to vehicle safety, the body design has, first of all, to guarantee a high stability for the passenger compartment combined with a sufficient energy absorption of the vehicle's forward and rearward structure as well as doors. The exterior paneling parts of the body should normally provide a good surface quality with high rigidity against dents and provide impact stability as well as a nonaggressive crash behavior. That means, in case of an accident, the paneling should not shatter as this would drastically increase the injury severity of accident participants.[229] In Table 20.15, demands on different vehicle areas relating to stability, rigidity, and energy absorption are listed.

20.3.6.1 Influence of Body Design on Safety in the Case of Front Impact.   The forward structure has to be designed in such a way that both pedestrian protection and protection at low-impact speed are achieved. Today's bumpers are made in such a way that up to an impact speed of 4 km/h, absolutely *no* damages are sustained, and up to an impact speed of 16 km/h (10 mi/h), no structural damages occur. Figure 20.32 shows the "reversible" deformation of a crush zone in the front area.

For a front collision against a rigid obstacle (90° in the longitudinal direction of the vehicle), Figure 20.33 qualitatively shows the characteristic curves for deceleration, speed, and deformation distance as a function of time.

According to Fig. 20.33, the vehicle impacts against a wall. The safety structure absorbs the energy and crumbles. Depending on the structural elements which are hit in sequence, inconsistencies would result with respect to the deceleration curve (see Fig. 20.33). Some parts would crumble easier (offer hardly any resistance) than others which would crumble at higher forces.

TABLE 20.14 Existing Test Limits for the Chest, Pelvis, Belly, and Abdomen Areas[233]

Limiting value	Standard	Description
SI = 1000		Front crash; from the resultant chest acceleration of the dummy
s = 50.8 mm s = 76.2 mm	FMVSS 208	Front crash; chest compression relative to the vertebral column, corresponding value by using air-bag systems
a_{3ms} = 60g	FMVSS 208	Front crash; from the resultant chest acceleration of the dummy
a_{3ms} = 60g	FMVSS 213	Examination of child restraint systems; from the resultant chest acceleration of the child dummy
a_{3ms} = 50g	ECE R 44	Examination of child restraint systems; from the resultant chest acceleration of the child dummy
delta a_{3ms} = 30g	ECE R 44	Examination of child restraint systems; difference in the vertical acceleration measured between belly and head of the child dummy
F = 11,100 N	FMVSS 203 EWG 74/297 ECE R 21	Examination of the steering system for accident impacts; the force from the steering system on the test body should not exceed 11,100 N
TTI = 85g (4 doors) TTI = 90g (2 doors)	FMVSS 214	Lateral crash; calculation from rib and vertebral column accelerations
Pelvis < 130g	FMVSS 214	Lateral crash; the lateral acceleration of the pelvis should not exceed 130g
RDC = 42 mm	EEVC proposal, lateral impact	Maximum rib compression
VC = 1 m/s	EEVC proposal, lateral impact	Calculation from the product of related rib compression distance and speed (pelvis)
PSPF = 6 kN	EEVC proposal, lateral impact	Peak force in the public symphysis (pelvis)
APF = 2.5 kN	EEVC proposal, lateral impact	Sum of the three abdomen forces

Note: See also Table 20.13 and Fig. 20.28; SI = safety index; TTI = thoracic trauma index, in g's; RDC = rib deflection criterion, mm; VC = viscous criterion, m/s; PSPF = public symphysis peak force, kN; and APF = abdomen peak force, kN

TABLE 20.15 Demands on Stability, Rigidity, and Energy Absorption for Different Vehicle Areas[580]

Demand area	Stability	Rigidity	Energy absorption
Body platform	++	++	—
Crash area (front + rear)	+	++	++
Doors	++	+	++
Connection (chassis + engine)	++	++	—

Note: — = low demand; + = high demand; ++ = very high demand.

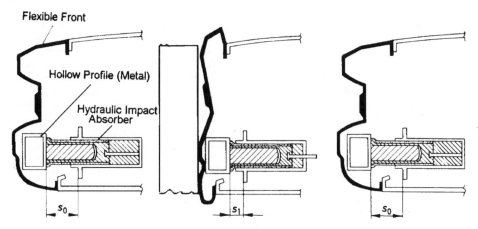

FIGURE 20.32   "Reversible" deformation of a crush zone in the front area.[580]

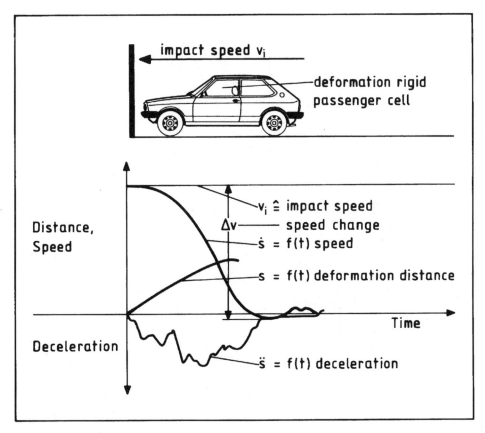

FIGURE 20.33   Characteristic curves for deceleration, speed, and deformation distance in case of a front impact.[642]

The speed of the vehicle decreases more and more until it comes to a standstill. A major part of energy, which is absorbed by the vehicle during the crumbling process, is lost as heat (plastic portion). The remainder of energy is still accumulated in the crumbled parts of the vehicle and becomes effective like a spring (elastic or flexible portion). The vehicle then rebounds from the wall, and this is characterized by the negative speed in Fig. 20.33. Thus, time is still going by until the vehicle finally comes to a complete stop. The flexible portion of the vehicle is estimated to be about 10 percent. That means, with respect to an impact against a rigid wall at a speed of 50 km/h, the overall speed change is about 55 km/h. Consequently, the action time during the deceleration process is longer for 55 km/h than for 50 km/h, and thus the probability of more serious accident consequences, such as critical brain damage, is higher.

However, the front structure of a vehicle is not only engineered for the rare case of a single accident against a rigid wall (obstacle) but also for the front collision, for example, of two vehicles with different masses. For the latter case, compatibility has to be regarded in addition, which means care must be taken that a favorable energy distribution would take place on both impact zones. Figure 20.34 shows the ideal characteristics of a forward structure relative to both partner and self-protection.

One important safety problem is represented by the engine transmission block which, in most cases, is built into the forward structure of the vehicle and shifts downward in case of front

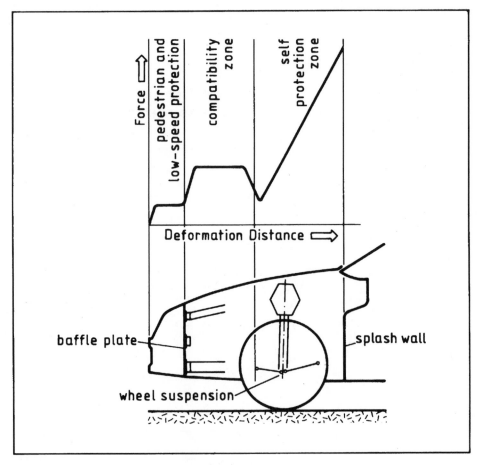

FIGURE 20.34   Ideal characteristics of a forward structure relative to both partner and self-protection.[580]

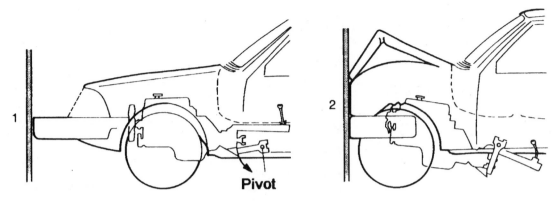

FIGURE 20.35 Structural measures for shifting the engine block downward in case of a crash (Volvo).[580]

impacts by appropriate measures so that it does not push into the passenger cell. Figure 20.35 provides a good solution.

This technique is especially important in small, light vehicles. With its "Vision A 93," Mercedez-Benz presented a passenger car which, despite its small length of 3350 mm, has a relatively large deformation distance with a low intrusion risk. This is also achieved by pushing the engine downward in case of a front impact. Therefore, in the area of the deformation zone, block-forming aggregates no longer exist. Also, the stresses placed on the occupants can be minimized in case of a lateral impact by using this underfloor concept, which is supported by high seat positions and a rigid underfloor structure.

The energy of a front impact is absorbed by the longitudinal frame rail's structure, which must be deformable. Two design modes exist:

1. The frame rails can be prebent. In doing so, an exact definition of deformation originates from the bending moments in case of a front crash. The deformation locations are determined by the selected cross-sectional areas (inertia moments).

2. By stiffening the sheet-metal fields, deformations in the longitudinal direction can occur at a certain location under normal forces. As long as feasible, the force-time characteristic should be held constant.

In order to design vehicles for front collisions with only a partial overlapping, it is important to redirect the offset forces to other, not directly impacted, vehicle areas. By doing so, any over-stressing at one location and, consequently, high local deformations could be avoided. For instance, Mercedez-Benz is building three-armed, reinforced, longitudinal rails into the vehicle bodies. The so-called fork rails redirect the deformation forces to the floor assembly, the transmission tunnel, and the lateral pillars in order to provide good protection for the occupant. The fork rails are especially able to reduce the expected high local deformations in the front face area and the front leg room of the occupants.

The oblique forward impact (see Fig. 20.22) which often occurs or the pole impact (see Fig. 20.24, case 9) can only be absorbed by a rigidly dimensioned cross-rail or better still by an overall cross-front. The impact forces can be directed either to the floor structure or to other structural areas in addition. In general, for front oblique impacts, the forward composite body should be sufficiently rigid to protect the front leg room of the driver and of the passenger on the right side in the front by preventing, for example, the intrusion of the front wheels into the safety cage.

20.3.6.2 *Influence of Body Design on Safety in the Case of a Lateral Impact.*   Lateral impact is another very frequent collision type after front impact (see Figs. 20.22 and 20.23). The most serious injuries generally occur when the passenger cell of the vehicle receives a direct hit from another vehicle or an obstacle.

The separation of the passenger car structure into a rigid passenger cell and a deformation area, like for the front impact, is not possible in case of a lateral impact because of the limited width of the vehicle. For an increase in lateral protection, the vital point is an adjusted rigidity of the side of the vehicle's structure and a uniform force application at the beginning of the deformation. The more rigid the side structure is, the lower the absolute intrusion speed of the colliding vehicle is. That means the speed change which takes place leads to a lower stress level for the occupants. Double steel tube construction in the doors and noninterrupted reinforcements of the B pillar (the center pillar) can lead to a stabilization of the passenger cell. It is not the strength of one individual pillar which dictates how effective the forces can be. This is more likely related to the type or manner according to which the different pillars are connected to the other safety network, joined by means of pillars and sheet metals.

In modern vehicle design, a safety cross-rail below the dashboard and between the two A pillars (front pillars) offers increased lateral rigidity in the forward vehicle area, and is used by VW. The safety cross-rail is additionally attached to the center tunnel and to the cross-bar inside the dashboard in order to reduce the free buckling length in the lateral direction of the vehicle. The door sills are stiffened and guarantee increased safety not only in case of a lateral impact but also in case of a front impact (100 or 50 percent overlapping). The removal of compact components, such as the electric motor for lifting windows, from the direct impact area on doors diminishes the stress on occupants. An improved lateral rigidity of the seats, an appropriate design of the door covering, including padding measures, and a reinforcement of the body shell also lead to a considerable decrease in occupant stress in case of a lateral impact. Nonsplintering plastics used for components in the compartment also result in reduced injury potential for occupants.

A cross-rail system at the B pillar (center pillar) would lead to a strong increase in lateral rigidity. However, such a measure is not possible in today's passenger cars with seats for longitudinal displacements. If the force application occurs at the B pillar, only deformations of the B pillar itself, as well as of the side sill-floor area, are available for energy absorption. However, an increase in lateral rigidity is realized without great difficulties at the A and C pillars. Therefore, between the pillars, the forces have to be transmitted by the door sills and by bars installed into the doors.

20.3.6.3 *Influence of Exterior Paneling Design on Safety.*

The design of exterior paneling parts can have serious consequences for traffic safety. Thus, with respect to all structural considerations, they should ensure that an endangerment by external projections should be excluded for pedestrians and two-wheelers.[229] Furthermore, very low-level bumpers, reversibly deformable forward structures, padded engine hoods as well as paddings at the windshield frame and at the roof edge (for example, polymer materials) are helpful for protecting pedestrians.

In general, with respect to pedestrian accidents, the aim should be to ensure that the human body is smoothly "shoveled up" (without a direct rebound on the street) and that critical body parts, especially knee joints, should not be hit immediately. The head impact on the vehicle surface should take place at a speed which is as low as possible and at an angle which is as flat as possible. For pedestrian protection, such measures have often been realized, based on aerodynamic results.

In case of a front collision between a bicyclist and a passenger car, the head impact spot is not located—not even at low speeds—on the mostly smooth and flexible front hood, as in the case of pedestrians, but farther backward at the end of the hood on the windshield or the A pillar. A bicyclist bouncing laterally against the vehicle can hit the lateral roof frame with his head. Therefore, this area also has to be designed in a more flexible manner.

Meanwhile, a number of other measures can help to reduce the injury risk for other traffic participants. In addition to the protection devices which have been discussed, other important features exist such as rounded body structures, a flexible hood, bumpers with an elastic plastic covering, easily fold-away outside mirrors, covered wiper axes, and concealed door handles.

Future safety devices which can be expected are:

- Yielding headlights in case of an impact
- Weakening of the hood front end by performation
- Weakening of the fenders by performation
- Flexible hood mask by a very thin sheet metal
- Multilayer A pillar and roof frame with an outside skin of a thin sheet metal, which is easily deformed as a result of a head impact

20.3.7 Preliminary Conclusion

The crash behavior of a motor vehicle reveals something about its passive safety. In this connection, it is possible to design motor vehicles which are strong and rigid enough so that any damage to their body in case of an accident is relatively small. A vehicle equipped with such a rigid forward structure would suddenly stop when it impacts a rigid obstacle. This crash type places a high risk on the occupants because of the considerable stresses imposed on the human being. Thus, the whole forward structure has to be designed by means of a preprogrammed deformation in such a way as to be able to protect the occupants in case of a front impact against a rigid obstacle.

However, the passenger cell has to be designed in a way that offers the highest possible deformation strength. If this surviving space is not guaranteed, then restraint systems and air bags will not help in this case. Otherwise, the safest passenger cell would be inefficient when unbuckled occupants hit the steering wheel, the interior covering, or the windshield.

Many years have passed since these safety considerations were considered in the design of motor vehicles. Key words relevant to the history of development are: deformation rigidity of the passenger cell; behavior of forward, rearward, and lateral structure in case of an accident; door reinforcements and safety locks; safety glazing; soft design of all critical parts in the passenger cell and on the outside of the body; safe anchorage of seats; seat belts; impact cushions; etc.

Much has been achieved with respect to active and passive safety. Active safety helps to prevent accidents. Passive safety of a motor vehicle takes action when active safety fails because the limits of physics cannot be removed not even through the support of ABS systems, etc., either.

During the last few years, the demand for exterior passive safety has stood in the forefront. Motor vehicles should be able to deal gently with other traffic participants, including pedestrians, two-wheel riders, occupants of smaller, lighter-weight vehicles, and occupants of passenger cars in case of accidents involving trucks and buses. In this connection, the term "compatibility" summarizes the mutual behavior of motor vehicles in case of an accident with respect to partner's protection.

CHAPTER 21
SUMMARY OF PART 2 "ALIGNMENT"

21.1 GENERAL

Part 2, "Alignment," represents the core of information presented in this handbook and with Part 1, "Network," and Part 3, "Cross Sections" fills in an existing gap on the international bookshelf. As elsewhere in the book, most chapters include two subchapters. Subchapter 1 gives a concise presentation of recommendations for practical design task. Subchapter 2 discusses the recommendations of Subchapter 1 in-depth and associates them with existing research results and/or practical experience from all over the world.

Until now, "safety" has been considered in highway design more or less qualitatively, if at all. Therefore, this handbook is based on the research of the authors and emphasizes highway geometric design and traffic safety which has led to the development of three quantitative safety criteria for distinguishing good, fair, and poor design practices on both planned and existing roadway sections and a safety module for evaluating road networks in rural and suburban areas. The safety criteria are directed toward the achievement of:

- Design consistency (safety criterion I)
- Operating speed consistency (safety criterion II)
- Driving dynamic consistency (safety criterion III)

in highway design.

Part 2, "Alignment," refers to the principles and methods of the geometric design of nonbuilt-up roads in rural and suburban areas and to the limiting and standard design values associated with them. This information is equally applicable to new design as well as to redesigns and/or restoration strategies of the existing road network. All relevant aspects presented and discussed in the various chapters always fall within the context of safety in order to substantially comply with the worldwide effort of reducing the number and the severity of road accidents.

21.2 DETAILS

The discussion of alignment design starts with an historical overview from ancient to modern times and highlights the various important period milestones through about 70 centuries of highway engineering up to the 1970s. The historical overview is followed by sorting Part 2, "Alignment," within the whole framework of highway design and its various components.

Highway design is strongly influenced today by sometimes stringent environmental requirements. Therefore, the relevant steps for conducting an environmental compatibility study are provided as a starting point for the basic procedure in road planning.

The next chapter is devoted to the most important impact on alignment design, that is, speed. In this connection, two speed concepts are of importance: the design speed concept and the operating speed, expressed by the 85th-percentile speed of passenger cars under free-flow conditions. A new design parameter, called curvature change rate of the single curve, was developed to best describe the functional relationship between road characteristics and operating speed. Based on this knowledge, operating speed backgrounds for different countries were developed on an international basis to describe the driving behavior of motorists at curved sites and on tangent segments. The exact definition of design speeds and the possibility of describing operating speeds with respect to road characteristics represent fundamental assumptions for the establishment of safety evaluation processes, the solution of driving dynamic requirements, the examination of accident characteristics, and the sound assessment of design elements in plan, profile, and cross section.

Introduction of the design and operating speed concepts to address safety problems of a highway has led to the development of two quantitative safety criteria for the application, in particular, of two-lane rural roads—the most dangerous road type. The two criteria can be implemented on every individual curve or tangent segment of a route and rank the site to three design quality classes, that is, good, fair, and poor design. For new alignment design, only good designs are allowed, while for redesigns and restoration strategies, fair designs can also be tolerated, in particular, for economical and environmental reasons. The first criterion warrants design consistency by letting the design speed comply with the actual operating speed chosen by drivers. The second developed safety criterion warrants operating speed consistency by maintaining speeds, selected by drivers, on subsequent curved or tangent sites so they do not vary beyond justified operating speed differences. Satisfaction of both safety criteria leads to sound alignment design, nonsatisfaction reveals fair or even poor design practices, as corresponding research and accident experience verified. One important goal of the book is that it is not solely directed to new designs—the basic concept of most national and international contributions—but it is also strongly related to the safety evaluation of existing (old) alignments in order to give the responsible authorities accurate information about appropriate countermeasures. Furthermore, relation design backgrounds for different countries were developed on an international basis in order to inform the highway engineer about sensible or nonsensible alignment relations between successive design elements.

Driving dynamic considerations are considered in most highway geometric design guidelines at least qualitatively. To come to a quantitative evaluation, all driving dynamic fundamentals are discussed, analyzed, and evaluated in detail such as vehicle dynamics, skid resistance, friction, and superelevation rate with respect to different topography classes. Arrangements of maximum permissible tangential and side friction factors were established for design purposes (horizontal/vertical alignment and sight distance), based on our own research, international experience, and actual skid resistance backgrounds on two continents. For a quantitative evaluation of driving dynamic safety at curved sites, safety criterion III (achieving driving dynamic consistency) was developed, which compares side friction assumed in terms of the design speed with side friction demanded in terms of the actual operating speed. The assessed design levels for safety criterion III are again in harmony with the road characteristic, accident rates, and operating speeds. As the corresponding safety analysis indicates, the satisfaction of this safety criterion significantly reduces the risk potential when driving over a specified curve site.

After the discussion of speed and safety considerations, the methodology of a consistent alignment design flow from the horizontal alignment → vertical alignment → cross-sectional issues → sight distance requirements → to the three-dimensional alignment is presented, including the respective safety impacts.

On the ground of the preceding discussions, the governing aspects of individual design elements and interrelationships are described for the horizontal and vertical alignment, treated analytically, and evaluated with respect to their application range, limiting and standard values, geometry, types, and safety considerations, in particular. The same is true for the design elements of the cross section in the course of the alignment as well as stopping and passing sight distances.

The alignment design was presented in the conventional way of consequent two-dimensional design for the horizontal and vertical plane, as well as for the cross section, resulting in the three-dimensional alignment features. Beyond the aspects of an esthetically pleasing roadway, the aspects of safety, consequently, are also considered.

In connection with the chapters dealing with highway design, numerous specific or new fields of interest are incorporated to give the reader a broad overview and, at the same time, exact information about the newest developments of today's knowledge.

After the development of quantitative safety evaluation procedures and the treatment of all relevant fields in highway design, the establishment of a safety module for examining entire road networks and individual roadway sections within these networks on a superior level could be achieved. Furthermore, a generally valid methodology is presented to evaluate quantitatively alignment segments and sequences with respect to safety processes. Numerous international case studies provide major insights into the application mode and range for examining new designs and existing alignments with respect to redesigns and appropriate countermeasures for the improvement of safety standards.

Besides the road, the human being and the motor vehicle play important roles when evaluating the overall term "traffic safety." Therefore, two additional chapters about human factors and vehicular safety are provided to inform the civil engineer, in particular, about those important fields which should be considered in the future more carefully when designing or improving roadways.

CROSS SECTIONS
OF
NONBUILT-UP ROADS (CS)

CHAPTER 22
METHODICAL PROCEDURE

In contrast to Part 2, "Alignment (AL)," which is mainly based on scientific backgrounds and research assumptions for the mostly quantitative assessment and evaluation of the individual alignment design parameters, Part 3, "Cross Sections of Nonbuilt-Up Roads (CS)," is primarily related to practical experience. Therefore, it should be understood that when attempting to explain the corresponding assessments and arrangements for the individual cross-sectional design features, some related research may be missing. The information presented here is related to the official guidelines of a number of countries, national and international research with special emphasis on traffic safety, and practical experience.

Even though the present work is based on available research and experience in this specific area of interest, no one should expect that all the background questions related to the final elaboration of Part 3, "Cross Sections," are answered here because solutions are often based on practical experience and judgment, on economic and environmental assumptions, and on administrative requests, which cannot always be explained on a purely scientific basis.

22.1 CROSS-SECTIONAL DESIGN

Overall, Part 3, "Cross Sections," is a combination of a number of components or elements, such as the width of the design vehicle, the width of lateral moving and safety spaces, and so forth, that have to be properly assembled.

One of the most important differences between Parts 3, "Cross Sections," and 2, "Alignment," is that the latter is based on the more or less quantitative selection and proper arrangement of individual design elements and element sequences, while the first is based on the assembly of single cross-sectional components or elements in a way similar to the building kit principle.

In spite of this, an attempt was made here to discuss and analyze all available scientific and practical background information regarding Part 3, "Cross Sections (CS)," and to arrange it in an applicable orderly manner.

Important topics covered here include the following issues:

- Discussion of the most important cross-sectional elements
- Identification of representative parameters and procedures for evaluating these elements based on an international survey
- Comparison of cross-sectional parameters of different guidelines
- Review of cross-sectional design with special emphasis on safety

Again, as in Part 2, "Alignment (AL)," Part 3, "Cross Sections (CS)," is divided, to a major extent, into the two subchapters:

- *Subchapter 1:* "Recommendations for Practical Design Tasks"
- *Subchapter 2:* "General Considerations, Research Evaluations, Guideline Comparisons, and New Developments"

Detailed information on the contents of the two subchapters and their interrelationships is presented in the "Preface" of this book.

22.2 TRAFFIC QUALITY/CAPACITY

Techniques and procedures for adjusting operational and highway design impacts to compensate for less than ideal conditions are found for many countries in guidelines and/or standards such as the American Highway Capacity Manual (HCM)[711] or the German "Guidelines for the Design of Roads, Part: Cross Sections."[77,247]

The Americans always use the term "capacity" to express the maximum hourly rate at which vehicles can reasonably be expected to traverse a point or a uniform section of a "lane" during a given time period under prevailing roadway and traffic conditions. Furthermore, capacity is used as a design control in such a way that the number of lanes should be sufficient enough to accommodate the design volumes for the selected level of service. Thus, the determination of the number of lanes is the most important issue, using concepts such as those in the *Highway Capacity Manual.*[711]

In contrast, "standard cross sections" were defined in Germany in Ref. 247 for all operational ranges in order to achieve uniformity and consistency in the design, construction, and operation of roads while, at the same time, considering economics.

Thus, the German method provides standard cross sections, including the number of lanes and other exactly defined and important cross-sectional elements. Through a thorough examination procedure, which is based on design traffic volumes, alignment characteristics, speeds, and so forth, the selected standard cross section is evaluated as to whether or not it is able to fulfill the assumed requirements, especially with respect to guaranteeing a certain traffic quality and safety in addition to economic issues.

Both design methods lead to similar results in defining reliable cross sections. Both methods are complex because they include numerous influencing parameters. Capacity, combined with the level of service concept, is the main issue in the United States—traffic quality is indirectly included. In Germany, however, traffic quality is the main issue, and capacity is indirectly included.

The meaning of cross-sectional design in most European countries is *standard cross-sectional design.* The German approach is considered to be the most advanced one for assembling standard cross sections through an exact definition of basic design elements. On the other hand, the approach in the *Highway Capacity Manual,* which is widely used, is considered to be the best in terms of assessing and examining traffic quality and/or capacity on the basis of the principle of level of service.

The advantages of both approaches are combined in the new Greek "Guidelines for the Design of Highway Facilities, Part: Cross Sections."[354,454] The establishment of standard cross sections mainly follows the German approach, with some additions from the Austrian guidelines.[36] On the other hand, for the evaluation of traffic quality and the control of capacity, the recommendation is to follow the *Highway Capacity Manual* approach. However, it was decided to exclude detailed traffic quality/capacity issues from Part 3, "Cross Sections," since this handbook concentrates, first of all, on highway geometric design and safety.

CHAPTER 23
OVERVIEW*

23.1 RECOMMENDATIONS FOR PRACTICAL DESIGN TASKS

Subchapter 1, to a major extent, is based on the German "Guidelines for the Design of Roads,"[247] since the authors are convinced that the German cross-sectional framework, as well as its practical and theoretical background, represent one of the soundest solutions in this field.[430]

The fundamental issues of Subchapter 2 will be discussed, examined, and completed in detail in Subchapter 2 with respect to safety evaluations, guideline comparisons, and so forth.

The *cross section* of a roadway is the view obtained in a section between the right-of-way lines cut perpendicular to the direction of travel along the road. It includes features on the traveled portion of the road used by vehicular traffic as well as on the roadside. There is strong consensus in the highway engineering community that the design of cross-sectional elements influences a roadway's cost, operation, and safety.[257]

23.1.1 Contents

Part 3, "Cross Sections (CS)," contains principles and methods, as well as limiting and standard values, for new designs, redesigns, and restoration strategies of nonbuilt-up roads outside built-up areas.

Subchapters 1 and 2 deal together with the geometric design of the cross sections, the corresponding cross-sectional elements, the roadside design, and the clear recovery area.

23.1.2 Range of Validity

Roads for public travel can be categorized, according to Part 1, "Networks" (Fig. 3.1), by

- Location (outside or in built-up areas)
- Adjacent area (without buildings or with buildings)
- Relevant design function:

    Mobility (connector)

    Access (collector)

    Local or pedestrian

for the five road category groups A to E in Table 6.1 of Part 2, "Alignment (AL)," and the resulting 14 road categories according to Table 6.2.

Part 3, "Cross Sections (CS)," is valid for new designs, redesigns, and rehabilitation strategies for roads of category group A and road categories B II and B III.

---

*Elaborated based on Refs. 77, 247, and 454.

Further information about the range of validity is included in the corresponding Sec. 6.1.2. Interrelationships between "CS" and other design parts are presented in Fig. 6.1, while the desirable structure of modern recommendations for highway geometric design is presented in Table 6.4.

23.1.3 Fundamental Issues

Each road section in the road network has to fulfill a certain task. The traffic assumptions given for the cross sections must consider the regional planning targets in rural areas. In urban areas, the assumptions must be directed toward the master traffic plans of municipalities.

A standard cross section is the cross section of a road perpendicular to the road axis. A standard cross section defines the typical technical layout for a specific stretch of roadway for the existing or planned traffic and alignment conditions.

Safety, capacity, and economy are essential considerations in the selection of cross-sectional elements. The effects of the road on the environment and surrounding topography must be taken into consideration when selecting a suitable standard cross section. A favorable cost effectiveness of the total investment, however, should always be sought.[686]

Fundamentally, an examination must be made as to whether continuing a standard cross section leads to disproportionately strong interferences with the vicinity of the road (built-up areas, landscape, and so forth).

When other than standard dimensions are used, it should be kept in mind that shorter dimensions reduce the traffic-related technical standard of the road, while wide cross sections provide a more comfortable situation for all vehicle types. The task of the design engineer must be to find a relationship between traffic engineering demands, with respect to safety, and ease for all traffic participants and the requirements for an environmentally justified project while considering, at the same time, economic conditions.

For interstates, a structural separation between opposite traffic directions must be provided. Other main traffic routes with four or more lanes also have to be provided with a structural separation.

For traffic safety reasons, separate travel paths for motor vehicles, bicyclists, and pedestrians should be provided wherever possible.

For parking on multilane median-separated roads of category group A and road categories B II and B III, dedicated areas have to be assigned outside the traveled way. Such areas have to be planned for interstates, expressways, and normally also for other major roads outside the through lanes and have to be structurally separated from the roadway.

It should be noted that the "Recommendations for Practical Design Tasks" presented here should not be regarded as fixed rules.

23.2 GENERAL CONSIDERATIONS, RESEARCH EVALUATIONS, GUIDELINE COMPARISONS, AND NEW DEVELOPMENTS

Part 3, "Cross Sections," is applicable to nonbuilt-up roads outside built-up areas.

Bicycle and/or pedestrian routes, designed independently of the roadway, are only partially addressed here.

Lane widths, shoulder widths, medians, etc. should be adjusted to traffic requirements and characteristics of the terrain. This means that the cross section may vary over a particular route because the controlling factors vary. However, the basic requirements are that changes in cross section standards should not be made unnecessarily, the cross section standards should be uniform within each subsection of the route, and any change of the cross section should be gradually and logically effected over a sufficient transition length.

In exceptional cases, it may be necessary to accept isolated reductions in cross section standards, for example, when an existing narrow structure has to be retained because it is not economically feasible to replace it. In such cases, a proper application of traffic signs and road markings is required to warn motorists of the discontinuity in the road.[663]

In an overview of cross-sectional elements, Hall, et al. report:[257]

> The most obvious element of a roadway is the travel lane, its width constitutes a basic cross section characteristic. A related parameter, the number of lanes, may be increased to accommodate the travel demand for the roadway. Even on the simplest highway, with one lane of travel in each direction, a minimal clearance separating the opposing movements is often provided. The cross section of a highway also includes the superelevation across the travel lane; small superelevations or cross slopes are employed in tangent sections to facilitate drainage, while higher superelevation rates are employed on horizontal curves to help to counteract centrifugal acceleration.
>
> Shoulders are often utilized for design and operational reasons. The width and superelevation of shoulders are both cross section characteristics. On some facilities, curbs are used to restrict traffic movements or to facilitate drainage. Medians may be installed to separate the opposing directions of traffic on multilane highways; the presence or absence of a median and the median width are cross section design elements. Pedestrian sidewalks are also common cross section features. Some countries, such as Germany and Japan, include bicycle lanes among roadway cross section elements.
>
> The portion of the cross section outside the area normally used for vehicular, bicycle and pedestrian travel may serve multiple purposes, including future expansion and recovery room for errant vehicles. The clearance distance to rigid fixed objects is a very important cross section parameter. The roadside slope and ditch design affect maintenance and the potential for vehicle recovery in this area. The application of breakaway object designs, roadside barriers, and crash attenuators is often considered in cross section plans.[257]

Of the many cross-sectional roadway elements discussed here, an illustration is given in Figure 23.1 for those elements typically found on two-lane rural roads. Illustrations of cross-sectional features for multilane roads are presented in the next chapter.

In Subchapter 1, recommendations are given for the design of reliable standard cross sections and cross-sectional elements. They are discussed, compared, and evaluated in Subchapter 2 with respect to other guidelines not only for new highway and road construction but also for the increasingly more common roadway reconstruction.[710]

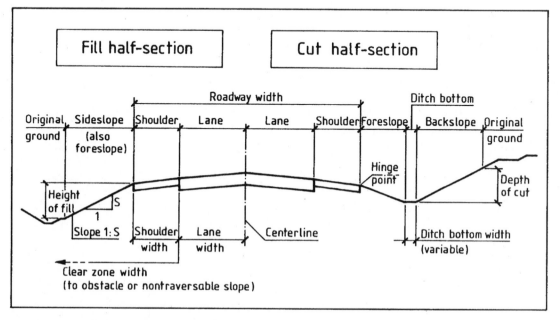

FIGURE 23.1   Cross-sectional design features and terms.[710,729]

CHAPTER 24
FUNDAMENTALS FOR THE DIMENSIONS OF THE CROSS-SECTIONAL DESIGN ELEMENTS*

24.1 RECOMMENDATIONS FOR PRACTICAL DESIGN TASKS

Because the standard cross section is of great importance in the design process, a thorough description and definition of the corresponding elements will be made in the following.

Influencing factors that have to be considered in the design of the cross section of a road are:

- Network function
- Traffic function

  Speed

  Traffic volume and composition

- Environmental factors
- Vicinity and design function
- Operational demands

Cross sections are classified by groups designated by the letters a to f. The main feature of the cross section group is the basic lane width. This width consists of the width of the design vehicle and the width of the lateral moving space (see Tables 24.1 and 24.2). Furthermore, lanes for opposing traffic that lie immediately adjacent to each other should be wider than would otherwise be required.[247]

24.1.1 Basic Dimensions

The principal elements of standard cross sections can be seen in Figs. 24.1 to 24.3 for

- Two-lane rural roads
- Four-lane median-separated rural roads

The design vehicle for motorized traffic is 2.50 m wide and 4.00 m high (see Tables 24.1 and 24.2).

---

*Elaborated based on Refs. 36, 77, 247, 430, and 454.

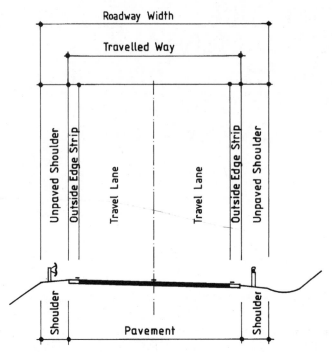

FIGURE 24.1 Example of cross-sectional elements for a two-lane rural road without areas for pedestrian and bicycle traffic. (Elaborated and modified based on Ref. 36.)

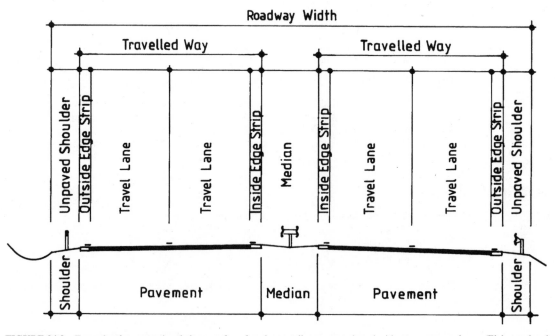

FIGURE 24.2 Example of cross-sectional elements for a four-lane median-separated road without emergency lanes. (Elaborated and modified based on Ref. 36.)

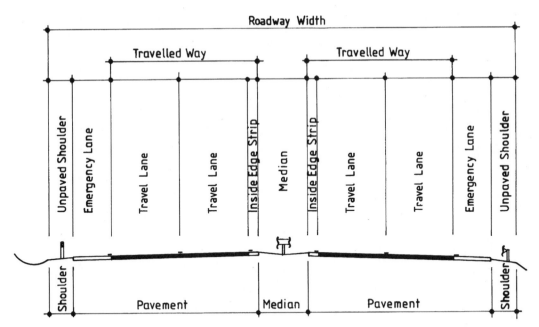

FIGURE 24.3 Example of cross-sectional elements for a four-lane median-separated road with emergency lanes. (Elaborated and modified based on Ref. 36.)

For bicycles and pedestrians, standard design dimensions of 0.80 and 0.75 m wide and 2.00 m high are used (see Table 24.2).

Lateral Moving Space. The lateral moving space is the space needed by a nontrack-guided vehicle to compensate for driving and steering uncertainties as well as for safety distances for lateral projecting parts (like mirrors) or lateral loading overhangs.

The width of the lateral moving space depends on the expected traffic speeds, traffic volumes (frequency of opposing traffic and/or passing maneuvers), and traffic composition (percentage of trucks).

On the basis of these requirements, the width of the lateral moving space is 1.25 m for roads in cross section group a, and then decreases by increments of 0.25 to 0.00 m for roads in cross section group f (see Table 24.1, col. 4).

For bicycle traffic, the width of the lateral moving space is 0.10 m on each side (Table 24.2, col. 4).

For pedestrian traffic, no lateral moving space is necessary.

The vertical moving space for motor vehicle traffic is the space needed to compensate for load constraints and the resulting vehicle vibrations on uneven pavements. It is generally 0.25 m (Table 24.2, col. 7).

For pedestrian and bicycle traffic, the upper moving space is 0.25 m (Table 24.2, col. 7).

Basic Lane Width. For the individual cross section groups, the basic lane widths shown in col. 5 of Table 24.1 are obtained by adding the width of the lateral moving space (col. 4) to the width of the design vehicle (col. 3).

Opposing Traffic Width Increase. Even on the simplest highway that has one lane of travel in each direction, an increase in the opposing traffic width separating the opposing movements has to be provided. The increase in the opposing traffic width increase should be 0.25 m in each direction of travel (see Table 24.1, col. 6).

TABLE 24.1 Proposed Widths of Cross-Sectional Elements*

Cross section group (1)	Number of lanes, s (2)	Design vehicle, m (3)	Lateral moving space, m (4)	Basic lane width, m (5)	Opposing traffic width increase, m (6)	Travel lane width without opposing traffic, m (7a)	Travel lane width with opposing traffic, m (7b)	Edge strip, m (8)	Median left turn, m — No (9a)	Median left turn, m — Yes (9b)	Paved shoulder or emergency/multiple purpose lane, m (10)	Unpaved shoulder, m — If col. 10 is present (11a)	Unpaved shoulder, m — If col. 10 is not present (11b)	Unpaved shoulder, m — Besides sidewalk/bicycle lane (11c)	Outer separation, m (12)	Sidewalk/bicycle lane, m (13)
a	6 / 4	2.50	Outside 1.25, Others 1.25 (1.00)†	Outside 3.75, Others 3.75 (3.50)†	—	Outside 3.75, Others 3.75 (3.50)†	—	Outside 0.50, Inside 1.00 (0.75)†	≥4.00 (3.50)†	—	2.50	1.50	—	—	3.00	—
b	6	2.50	1.00	3.50	—	3.50	3.75	0.50	≥3.00	—	2.00	1.50	—	—	3.00	—
	4				—			0.50	≥3.00	—	—		—	—	3.00	—
	2+1				0.25			0.25	—	—	—		2.50/1.50	—	—	—
	2				0.25			0.25	—	—	1.50		2.00	—	1.75	≥2.00 to ≤2.50
c	4	2.50	0.75	3.25	0.25	3.25	—	0.50	≥2.00	5.25	2.00	1.50	1.50	≥0.50	1.75	≥2.00 to ≤2.50
	2					—	3.50	0.25	—	—	—					
d	2	2.50	0.50	3.00	0.25	3.00	3.25	0.25	—	—	—	—	1.50	≥0.50	1.75	≥2.00 to ≤2.50
e	2	2.50	0.25	2.75	0.25	—	3.00	0.25	—	—	—	—	1.50	≥0.50	1.75	≥2.00 to ≤2.50
f	(2)†	(2.50)†	(0.00)†	(2.50)†	(0.25)†	—	(2.75)†	—	—	—	—	—	(1.00)†	(≥0.50)†	(1.25)	≥2.00 to ≤2.50

*Elaborated and modified based on Refs. 247 and 454.

†() = exceptional cases.

NOTE: "Outside" means "outside lane"; "Others" means "other lanes", etc.; see Figs. 24.1, 24.2, and 25.1.

24.1.2 Clearance[247]

Clearance is that space of the road cross section that must be kept free of rigid obstacles. It is made up of traffic spaces and vertical and lateral safety spaces (see Fig. 24.4). Clearance dimensions are given in Table 24.2.

Traffic Space.   The traffic space for motor vehicle traffic is made up of the space of the design vehicle, lateral and vertical moving space, opposite width increase, spaces above the edge strips, and paved shoulders. Its height is 4.25 m (Fig. 24.4).

The traffic space for bicycle traffic is 1.00 m wide and 2.25 m high for each driving strip.

The traffic space for pedestrian traffic is 0.75 m wide and 2.25 m high for each walking strip.

The overall dimensions of traffic spaces can be derived from Tables 24.1 and 24.2.

Vertical Safety Space S_V.   The height of the vertical safety space is 0.25 m for motor vehicle traffic. The necessary clearance height is 4.50 m (Table 24.2). An extension of the clearance height to 4.70 m is normally recommended to accommodate a pavement overlay in the future. For interstates, a clearance height of 5.00 m is reasonable.

For sidewalks and bicycle lanes, the vertical safety space is 0.25 m and the clearance height is normally 2.50 m. Traffic signs are allowed to penetrate the clearance area up to the limit of the traffic space (Fig. 24.4).

Lateral Safety Space S_l

Motor Vehicle Traffic S_{IV}.   The lateral safety space is based on the fact that motor vehicles cannot pass lateral obstacles such as utility poles, walls, and so forth with a distance of zero.

Another function of the lateral safety space is to offer a chance to recover control for those motorists who run off the pavement due to driving errors or accident impacts. In any case collisions of those vehicles with rigid obstacles must be avoided. Therefore, the lateral safety space has to be kept free from any kind of rigid objects. It is obvious that greater lateral distances to obstacles are needed as speeds increase. As a consequence, the lateral safety space is speed-dependent. Exceptions are longitudinal appurtenances for protection purposes such as guardrails or barriers.[78]

The width of the lateral safety space is measured from the edge of the traffic space to the outside of the roadway (Fig. 24.4). The necessary width of the lateral safety space, S_{IV}, depends on the maximum permissible speed (see Table 24.2):

$$\max V_{perm} > 70 \text{ km/h} \qquad S_{IV} \geq 1.25 \text{ m}$$

$$\max V_{perm} \leq 70 \text{ km/h} \qquad S_{IV} \geq 1.00 \text{ m}$$

$$\max V_{perm} \leq 50 \text{ km/h} \qquad S_{IV} \geq 0.75 \text{ m}$$

TABLE 24.2   Standard Values for Clearance*

Kind of traffic	Maximum permissible speed, V_{perm}, km/h	Basic dimensions, km/h	Width of lateral moving space, m	Width of lateral safety space, S_l, m	Design height, m	Height of vertical moving space, m	Height of vertical safety space, S_v, m	Height of clearance, m
(1)	(2)	(3)	(4)	(5)	(6)	(7)	(8)	(9)
Motorized traffic	>70	2.50	According to the cross section group 0.0 to 1.25 (see Table 24.1)	≥1.25	4.00	0.25	0.25	4.50
	≤70	2.50		≥1.00	4.00	0.25	0.25	4.50
	≤50	2.50		≥0.75	4.00	0.25	0.25	4.50
Bicycle traffic		0.80	0.10	0.25	2.00	0.25	0.25	2.50
Pedestrian traffic		0.75	—	—	2.00	0.25	0.25	2.50

*Elaborated based on Refs. 247 and 454.

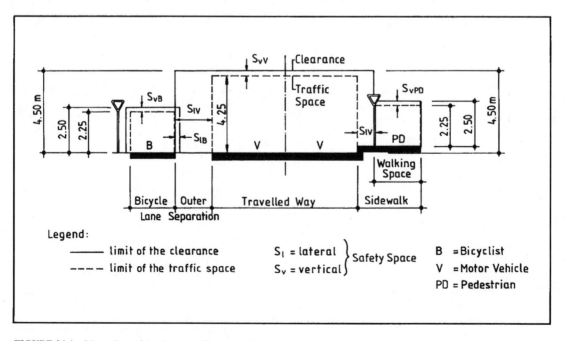

FIGURE 24.4   Dimensions of the clearance (Germany). (Elaborated based on Ref. 247.)

Bicycle Traffic S_{1B}. The width of the lateral safety space is 0.25 m (Table 24.2).

Traffic signs and traffic installations are allowed to penetrate the clearance area up to the border line of the traffic space (Fig. 24.4).

Pedestrian Traffic S_{1PD}. For pedestrians, no lateral safety space is needed. Sidewalks that are directly adjacent to other traffic spaces are made up of the traffic space for pedestrians (walking space) and the corresponding safety space of the adjacent traffic space.

Combined Cross-Sectional Parts. If portions of different traffic modes are added to an overall cross section, the lateral safety spaces for the individual traffic modes are allowed to overlap. With respect to the distance between two traffic spaces, the wider of the lateral safety spaces is then the relevant one (see Fig. 24.4).

24.1.3 Cross-Sectional Elements[247]

The elements of the road cross sections are shown in Table 24.1.

Pavement. The pavement width consists of the lane widths, the widths of the edge strips, and the widths of paved shoulders (for example, emergency lanes), if present (see Figs. 24.1 to 24.3).

Travel Lane (Col. 7, Table 24.1). The most obvious element of a roadway is the travel lane; its width constitutes a basic cross-sectional characteristic. As a related parameter, the number of lanes to accommodate the travel demand for the roadway also represents important characteristic issues.[257]

The lane widths of the individual cross section groups are developed from the basic lane widths and the increase in the opposing traffic width.

For multilane highways and freeways separated by medians, the lane widths correspond to the basic lane widths (compare col. 5 and col. 7a of Table 24.1).

For two-lane roads, the lane width (col. 7b) results from the basic lane widths (col. 5) plus the increase in the opposing traffic width (col. 6).

Edge Strip (Col. 8, Table 24.1). Edge strips belong to the pavement, and denote the delineation. They are assigned to the cross section groups a to e.

The edge strip width is 0.25 m for two-lane cross sections, as well as for the intermediate cross section of type b2 + 1. For multilane standard cross sections with medians, edge strips of 0.50 m wide must be used, except for the cross section group a, where a width of 1.00 m (exceptional case = 0.75 m) must be provided for safety reasons for the inside edge strip adjacent to the median (see Table 24.1 and Fig. 25.1).

Median (Col. 9, Table 24.1). Medians serve as the structural separation of lanes for traffic in opposite directions. The standard widths for different cross section groups are

4.00 m for the cross section group a

3.00 m for the cross section group b

2.00 m for the cross section group c

where land acquisition is limited.

Wider medians can be considered because of safety demands where land acquisition is not a major problem.

Using New Jersey–type concrete barriers for roads of cross section group c, the width of the median can be reduced to 1.50 m (Fig. 25.3).

The necessary stopping sight distance must always be provided. If this is not possible for the inside lane of multilane median-separated roads on left-handed curves, and if the selection of a larger radius of curve is impossible, the necessary increase in width of the median can then be achieved by

- A separate design of the two directional parts of the cross section
- An eccentric placement of the guardrail in the median.

In cases where these measures are not technically and/or economically feasible, the introduction of a local speed limit is necessary, at least under wet pavement conditions. Normally, a median should be provided in such a way that the stopping sight distance is not impaired.

For roads with numerous intersections and left turns, the median for the cross section group c is 5.25 m (Table 24.1, col. 9b).

Outer Separation (Col. 12, Table 24.1).   The outer separation is the structural separation between the pavement for through traffic lanes and ramps, frontage roads, bicycle lanes, and sidewalks.

The following minimum widths are recommended for the outer separation of different cross section groups:

a and b: 3.00 m (separation of two traveled ways for motor vehicles)

b, c, d, and e: 1.75 m (separation of different traffic modes)

f: 1.25 m (separation of different traffic modes)

Wider outer separations should be considered where land acquisition is not a major problem.

Emergency Lane (Col. 10, Table 24.1).   Emergency lanes offer vehicles the possibility of escaping to the outside or of stopping. In the case of accidents or during construction periods, they make it easier to detour traffic. In winter time, they may be used to store snow after removal from the road. They also make maintenance operations easier.

The widths of emergency lanes are 2.50 m for roads of cross section group a and 2.00 m for multilane roads of cross section groups b and c.

Multiple Purpose Lane (Col. 10, Table 24.1).   Multiple purpose lanes can only be assigned to two-lane, nonbuilt-up roads of the cross section group b. They are used for slower traffic, maintenance, and stopping in emergencies. The width of the multiple purpose lane is 1.50 m.

Multiple purpose lanes help traffic to become disentangled. As described in Sec. 25.2.2.1, this cross-sectional type is relatively unsafe. For this reason, the corresponding cross section b2s in Fig. 25.1 is shown in parentheses and should only be considered in exceptional cases such as a high proportion of agricultural vehicles and no existence of parallel farm roads.

Unpaved Shoulder (Col. 11, Table 24.1).   *Shoulders* are that portion of the highway immediately adjacent to the outside edge of the pavement. They are the areas where guardrails and traffic signs may be placed and provide a working space for maintenance crews and space for snow storage after removal from the road.

The following shoulder widths are recommended for different cross section groups (Col. 11b):

b: 2.00 m

c, d, and e: 1.50 m

f: 1.00 m

If the shoulder is adjacent to an emergency/multiple purpose lane, then its width is normally 1.50 m (col. 11a).

The intermediate cross section type b2 + 1 must be considered as a special case (see Secs. 25.1.2 and 25.2.2). In this case, a shoulder width of 2.50 m must be provided next to the one-lane portion of the cross section and 1.50 m next to the two-lane portion of the cross section (Fig. 25.3).

Bicycle Lane (Col. 13, Table 24.1).   A single bicycle lane is normally 1.00 m wide and a double bicycle lane is normally 2.00 m wide. For operational and maintenance services, a width of 2.25 m is recommended.

Bicycle lanes should normally consist of two strips. The location of the bicycle lane within the cross section must be at least far enough from the travel lane to avoid a wide-open car door.

Sidewalk (Col. 13, Table 24.1).   The width of a sidewalk, which is separated from traffic lanes by a curb, is made up of the walking space and the width of the lateral safety space for the motorized traffic (Fig. 24.4). The minimum width of a sidewalk consisting of two strips adjacent to the curb should be 2.25 m.

Sidewalks which are divided by outer separations from the traffic lanes should be at least 2.00 m wide for maintenance reasons as long as they do not lie directly beside bicycle lanes.

Combined Sidewalks and Bicycle Lanes (Col. 13, Table 24.1).   On rural roads, combined sidewalks and bicycle lanes on one side of the road can be considered a standard solution. They should be 2.50 m wide. Again, for operational and maintenance services, a minimum width of at least 2.25 m is recommended. However, widths of more than 2.50 m should not be used, since then the combined sidewalk and bicycle lane could be used as a path for motor vehicles.

Curbs and Side Gutters.   On nonbuilt-up roads, curbs should not be provided, since those roads basically should be drained openly for environmental, safety, and economic reasons. Furthermore, sidewalks and bicycle lanes are safer if they are located beyond an outer separation rather than next to a curb.

The accident situation with respect to cross-sectional elements and their relationships is dealt with in detail in Sec. 24.2.4.

24.2 GENERAL CONSIDERATIONS, RESEARCH EVALUATIONS, GUIDELINE COMPARISONS, AND NEW DEVELOPMENTS

As previously stated, the makeup of the cross section is similar to the building kit principle. Since most of the definitions and assessments are related to practical experiences and technical rules, no relevant research studies are discussed further. In this connection, Brilon, the chairman of the Committee for the development of the new "German Guidelines for the Design of Roads, Part Cross Sections," explains.[78] Cross-sectional design is, in the countries studied so far, to a large extent arbitrary and is based on assessments in a sense of a generally accepted convention. However, such an assessment based on a convention can only be accepted if it can be verified by examination and application and if it finds broad agreement. In this way the arbitrariness is limited.

Many theories for justifying cross-sectional design are still in the development stage. Of course, numerous isolated theories do exist with respect to individual cross-sectional parts and also several empirical theories. Nevertheless, an overall theoretical framework (comparable to the driving dynamic basic concept or to the developed safety criteria in Part 2, "Alignment"), through which individual cross-sectional features could be derived, does not exist.

Brilon calls the design of cross sections based on the makeup of individual elements a deterministic approach. He supports the concept of a holistic approach, that is, the assessment of the cross section as a whole. As an example, he cited the development of intermediate cross sections, as presented in Secs. 25.1.2 and 25.2.2. However, he concludes that, at least in the near future, the holistic approach cannot be realized because of the enormous research costs, and he expects that the resulting cross sections would not deviate much from the existing ones.[78]

24.2.1 Basic Dimensions*

Many guidelines use the deterministic approach for cross-sectional design. The cross section of a roadway is composed of many individual elements with certain widths. Each of these elements is considered by itself and its dimensions are established individually. Interrelationships between these individual elements are considered only to a limited extent in the overall design of the cross section.

---

*Elaborated based on Ref. 78.

The individual elements of the design are[78]

- Design vehicle
- Moving space
- Opposing traffic width increase
- Lateral safety space
- Travel lane
- Edge strip
- Shoulder
- Median

An important basic assumption for coordinating cross sections is a uniform design vehicle. In this connection, the German and the Dutch guidelines[141,247] explicitly describe a design vehicle. On the other hand, most of the investigated national guidelines seem to use similar dimensions. For this reason, and taking into account the German and Dutch guidelines, a design vehicle of 2.50 m wide and 4.00 m high is recommended at least throughout Europe and is confirmed by Ref. 153. However, it should be mentioned here that Canada and the United States use other design vehicles. The width of the largest design vehicle in those two countries is 2.60 m and the height corresponds to 4.10 m.

Motor vehicles need greater widths on the travel lanes than the dimensions of the regular design vehicle require. This additional width is designated as *lateral moving space.* Even on straight, level roads, steering corrections are necessary; they increase in the lateral direction with increasing speed of the motor vehicle. Another unavoidable width increase is related to the widening of the vehicle track on curves (see Secs. 14.1.4 and 14.2.4). Furthermore, the lateral moving space should guarantee a certain safety level in order to compensate for lateral loading overhangs. Other factors may also influence the lateral moving space, for example, side wind forces, pavement irregularities, driving dynamic forces, and so forth.

In conclusion, a motor vehicle needs more space on the road than its regular width actually dictates. In any case, the lateral moving space should depend on the speed. This aspect is fundamentally considered by the relationship in Fig. 24.5, which led to the dimensions shown in col. 4 of Table 24.1.

A further specific element of cross-sectional design on two-lane roads is the *opposing traffic width increase.* Many experts are afraid that the collision risk between vehicles traveling in opposite directions is especially high on such roads. Therefore, for all two-lane roads with opposing traffic, a uniform increase of 0.25 m is provided to the width of each lane (see Col. 6 of Table 24.1). This assessment is based on the research of Brilon and Doehler,[75] who recognized that drivers tend to shy away from the centerline of an undivided highway.

24.2.2 Clearance

In studying different European and other international design guidelines, a presentation of the dimensions of the traffic space and the clearance according to Fig. 24.4 was found only in the Austrian guidelines for the design of roads.[36] For comparative reasons, the Austrian approach will be briefly discussed based on Fig. 24.6.

Traffic Space (Fig. 24.6).   Traffic space serves traffic operations. The width of the traffic space for motor vehicle traffic is equal to the pavement width.

The height of the traffic space is equal to 4.20 m.

This space has to be kept obstacle-free. Traffic signs can, for a short distance, intrude into the traffic space, but only in substantiated exceptional cases or for temporary arrangements. However, under no circumstances should they intrude into the travel lanes.

Traffic Space for Pedestrian Traffic (Fig. 24.7).   The width of traffic space for pedestrian traffic is equal to the width of the sidewalk.

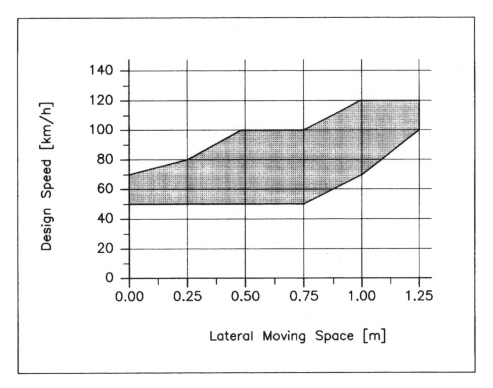

FIGURE 24.5   Relationship between lateral moving space and design speed.[78]

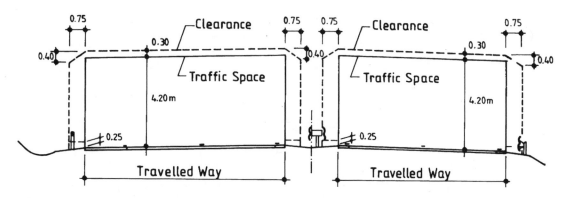

FIGURE 24.6   Traffic space and clearance for the motor vehicle traffic (Austria). (Elaborated based on Ref. 36.)

The height of the traffic space is equal to 2.20 m.

Single traffic sign poles or other vertical guidance equipment, such as street lights and so forth, may be placed in the pedestrian traffic space area. A minimum obstacle-free width of 1.00 m should be provided. By comparing Figs. 24.6 and 24.7 (Austria) with Fig. 24.4 (Germany), it can be concluded that similar results could be observed.

Clearance.   Clearance is that space that must be kept obstacle-free for traffic operations.

Branches of trees and bushes, traffic signs, other vertical guidance equipment, light poles, and so forth are allowed to extend into the clearance area.

Clearance for Motor Vehicle Traffic (Fig. 24.6).   The width of the clearance is equal to the traffic space plus the lateral safety spaces of 0.75 m on both sides.

The height of the clearance is equal to 0.30 m plus the height of the traffic space. Elevated paved shoulders, such as sidewalks up to a height of 0.25 m, are not given special consideration. A sloped area, as shown in Fig. 24.6, is allowed as a part of the height limitation.

Clearance for Pedestrian Traffic (Fig. 24.7).   The width of the clearance is equal to the traffic space.

The height of the clearance is equal to 0.30 m plus the height of the traffic space.

A comparison of the German approach (Fig. 24.4) and the Austrian approach (Figs. 24.6 and 24.7) reveals great similarities between the two guidelines. With respect to the lateral safety space, however, one important difference exists. For instance, in Germany the lateral safety space is dependent on the permissible speed and varies between 0.75 and 1.25 m (Table 24.2, col. 5). On the other hand, in Austria this dimension is kept constant at 0.75 m.

Since the lateral deviations of motor vehicles also increase with increasing speeds, the authors decided to consider the speed-dependent dimensions for the lateral safety space according to the German assumptions, in Sec. 24.1.2.

With respect to the lateral safety space, the literature review revealed, besides Austria and Germany, other explicitly defined values only for Sweden.[686] These values are also speed dependent and range from 0.05 m ($V_d \leq 30$ km/h) to 1.75 m ($V_d = 110$ km/h). The values of various other guidelines about lateral clearances are not directly comparable with the explicitly defined values of Austria, Germany, and Sweden.

With respect to the vertical clearances, the information given in Table 24.3 was found.

A comparison of the vertical clearances given in Table 24.3 indicates that the value of 4.70 m, which was recommended in Sec. 24.1.2, may be considered reasonable.

In conclusion, the recommendations given in Subchapter 24.1 for the dimensions of the traffic spaces and clearances offer a sound, economic solution.

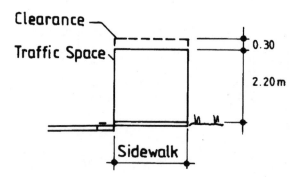

FIGURE 24.7   Traffic space and clearance for the pedestrian traffic (Austria). (Elaborated based on Ref. 36.)

TABLE 24.3 Vertical Clearances in Various Countries and "CS"

Country	Vertical clearance, m	Remarks
AGR*	4.50	Minimum
Austria[36]	4.50	—
Canada[608]	4.65/5.00	4.50 (restoration and rehabilitation projects)
France[700]	4.50	4.30 (exception)
Germany[247]	4.50	4.70 (recommended)
Greece[454]	4.50	4.70 (recommended)
Sweden[686]	4.70	—
Switzerland[687]	4.50	4.20 (minimum)
TEM[707]	4.80	5.00 (recommended)
United States[5]	5.00	4.40 (minimum)
Part 3, "Cross Sections (CS)"	4.50	4.70 (recommended) 5.00 (recommended for interstates)

*ARG = European Agreement on Major Roads for International Traffic.

24.2.3 Cross-Sectional Elements*

This section provides design rules for the different elements of a cross section, that is, travel lane, edge strip, emergency/multiple purpose lane, unpaved shoulder, median, etc. Furthermore, guideline comparisons with respect to the previously mentioned elements are presented. A major part of this section deals with the relationships and interrelationships between the accident situation and the cross-sectional elements.

24.2.3.1 Travel Lane.   *Travel lanes* are that portion of the highway intended for use by general traffic. The travel lane width of a two-lane rural road is measured from the centerline of the highway to the edge line or to the joint separating the lane from the shoulder.[729]

As reported in Ref. 257, a fundamental feature of the roadway cross section is the width of a travel lane, which must be sufficient to accommodate the design vehicle, allow for imprecise steering maneuvers, and provide clearance for traffic traveling in opposite directions in adjacent lanes. Truck widths (2.6 m in the United States, 2.50 m in Europe) define the absolute minimum lane width on those roadways where trucks are expected. Additional space to accommodate rearview mirrors and the lateral movement of vehicles within the lane requires a lane width of at least 3.0 m. Table 24.4 shows typical lane width design values for various countries. An international survey found that lane widths vary from nation to nation but within reasonably narrow ranges: typically 3.50 to 3.75 m for freeways (interstates), 3.00 to 3.75 m for major roads (arterials), and normally 2.75 to 3.30 m for minor or local roads.[257]

The design values of "CS" selected in Figure 25.1 represent very reasonable assumptions in comparison to most of the studied countries listed in Table 24.4.

24.2.3.2 Edge Strip.   The *edge strips* serve both to compensate for possible inaccuracies in the pavement construction process and to carry the edge pavement markings. On interstates the wider edge strip at the left side of the inside lane may have an additional safety effect in cases of emergency stops (see Fig. 25.1).

24.2.3.3 Median.   A *median* is highly desirable on highways with four or more lanes. A *median* is defined as the portion of a divided highway separating the travel paths for traffic in opposite

---

*Elaborated based on Ref. 257.

TABLE 24.4   Typical Lane-Width Design Values in Various Countries and "CS"[257]

Country	Roadway classification		
	Freeway (interstate)	Motor road (arterial)	Minor or local road
Brazil	3.75 m	3.75 m	3.0 m
Canada	—	3.0 to 3.7 m (rural connector)	3.0 to 3.3 m (rural local)
China	3.5 to 3.75 m	3.75 m	3.5 m
Czech Republic	3.5 to 3.75 m	3.0 to 3.5 m	3.0 m
Denmark	3.5 m	3.0 m	3.0 to 3.25 m
France	3.5 m	3.5 m	3.5 m
Germany	3.5 to 3.75 m	3.25 to 3.5 m	2.75 to 3.25 m
Hungary	3.75 m	3.5 m	3.0 to 3.5 m
Indonesia	3.5 to 3.75 m	3.25 to 3.5 m	2.75 to 3.0 m
Israel	3.75 m	3.6 m	3.0 to 3.3 m
Japan	3.5 to 3.75 m	3.25 to 3.5 m	3.0 to 3.25 m
The Netherlands	3.50 m	3.10 to 3.25 m	2.75 to 3.25 m
Poland	3.5 to 3.75 m	3.0 to 3.5 m	2.5 to 3.0 m
Portugal	3.75 m	3.75 m	3.0 m
South Africa	3.70 m	3.1 to 3.7 m (rural) 3.0 to 3.7 m (urban)	2.25 to 3.0 m
Spain	3.5 to 3.75 m	3.0 to 3.5 m	3.0 to 3.25 m
Sweden	—	3.75 m (rural undivided)	—
Switzerland	3.75 to 4.0 m	3.45 to 3.75 m	3.15 to 3.65 m
United States	3.6 m	3.3 to 3.6 m	2.7 to 3.6 m
Venezuela	3.6 m	3.6 m	2.6 m
Yugoslavia	3.5 to 3.75 m	3.0 to 3.25 m	2.75 to 3.0 m
"CS" and Greece	3.5 to 3.75 m	3.25 to 3.5 m	3.0 to 3.25 m (2.75 m exception)

directions. The median width is expressed as the dimension between the inside pavement edges. Medians allow the storage of snow after removal from the road and minimize headlight glare. Wider medians than those usual in Europe, as, for example, in Canada and the United States, provide a recovery area for out-of-control vehicles and a stopping area in case of emergencies.[5]

Because of their relatively poor safety record, undivided four-lane highways should no longer be built and medians should be incorporated in all new designs.

Table 24.5 shows standard median widths reported by an international survey[257] from various countries. Less is known about the design of median widths for safety and operations than about many other cross-sectional elements, and this is reflected in the variations shown in the table. For example, minimum freeway median width ranges from 1.5 to over 4.5 m. In general, consensus values have not been reached for median widths for any category of roadway.[257]

24.2.3.4   Shoulder.   *Shoulders* are typically designed and intended to accommodate occasional use by vehicles but not continual travel. Part or all of the shoulder may be paved. Paved shoulders are often called *emergency lanes* or *multiple purpose lanes.*

According to Refs. 5 and 257, shoulders are used for emergency stopping; for parking of nonfunctional vehicles; and for lateral support of the subbase, base, and surface courses of the travel

TABLE 24.5 Typical Median-Width Design Values in Various Countries and "CS"[257]

Country	Roadway classification		
	Freeway (interstate)	Motor road (arterial)	Minor or local road
Brazil	2.0 to 6.0 m	2.0 to 6.0 m	2.0 to 6.0 m
China	1.5 to 3.0 m	1.5 to 3.0 m	—
Denmark	3.0 m	2.0 m	—
France	12 m; 3 m curbed	12 m; 3 m curbed	12 m; 3 m curbed
Germany	3.0 to 3.5 m	2.0 to 3.0 m	—
Greece	3.0 to 3.5 m	2.0 to 3.0 m	—
Hungary	3.0 m	1.5 to 3.0 m; 2.5 m curbed	—
Indonesia	2.0 to 2.5 m	1.5 to 2.0 m	1.0 m
Israel	3.0 m	2.5 m for 4-lane roadways	—
Japan	>4.5 m	>1.75 m	>1.0 m
The Netherlands	12.0 m	3.0 to 4.5 m	—
Poland	3.5 to 5.0 m	3.0 to 5.0 m	—
Portugal	2.0 to 6.0 m, curbed	2.0 to 6.0 m, curbed	2.0 to 6.0 m, curbed
South Africa	—	9.2 m rural, 1.5 m urban, curbed	—
Spain	10 to 12 m; 3.0 m curbed	—	—
Switzerland	3.5 m; 2.0 m curbed	—	—
United Kingdom	4.5 m	4.0 m rural, 1.8 to 3.0 m urban	—
United States	Minimum 3.0 m	1.2 to over 20 m	—
Yugoslavia	4.0 m	4.0 m	3.0 m
"CS"	3.5 to 4.0 m	2.0 to 3.0 m	—

lanes. On some roadways, shoulders may be used for pedestrian and bicycle traffic where no separate paths are provided for those functions. On divided highways, shoulders are generally provided on both the median (known as inside edge strip in several countries) and the outside of the traveled way. Shoulders should be wide enough to adequately fulfill their purpose, but excessive widths encourage drivers to use them as additional travel lanes,[257] for example, in countries like Greece.

The international survey responses, displayed in Table 24.6, demonstrate that there is no international consensus on appropriate shoulder widths. For freeways and certain types of divided highways, some nations use different widths for inside and outside shoulders, while others do not. Almost all countries use narrower shoulders for classifications of roadways with lower design speeds. Several studies, summarized in Ref. 778, have found that wider shoulders result in significant safety benefits.[257]

Meanwhile, many countries recommend that shoulders be differentiated from the travel lanes by the use of color or texture to discourage their use as travel lanes and to alert drivers when they depart from the travel lane. Visual contrast can be accomplished by using different colors of pavement for shoulders and through lanes or by striping and pavement markings; texture contrast can be attained by varying the aggregate content of the pavement.[257] Many jurisdictions have experienced a safety benefit from the placement of corrugated depressions (rumble strips) on the shoulder.[266]

TABLE 24.6   Typical Shoulder-Width Design Values in Various Countries and "CS"[257]

Country	Roadway classification		
	Freeway (interstate)	Major road (arterial)	Minor or local road
Brazil	3.0 m left; 1.0 m right	2.5 m	1.5 to 2.5 m
Canada	—	1.5 m to 3.0 m (rural connector)	1.0 m (rural local)
China	2.0 to 3.25 m	0.75 to 2.5 m	0.5 to 1.5 m
Czech Republic	1.5 to 2.5 m	0.25 to 1.5 m	—
Denmark	3.5 m	2.5 m	1.0 m
France	3.0 m +0.75 m unpaved	2.5 m +0.75 m unpaved	2.5 m +0.75 m unpaved
Germany	2.0 to 2.5 m paved; 1.5 unpaved	2.0 m paved; 1.5 unpaved	1.0 to 1.5 m
Greece	2.0 to 2.5 m paved; 1.5 unpaved	2.0 m paved; 1.5 unpaved	1.0 to 1.5 m
Hungary	4.0 m	2.0 to 2.5 m	0.75 to 1.5 m
Indonesia	1.5 to 4.0 m	1.0 to 3.0 m	0.25 to 2.5 m
Israel	3.0 m	3.0 m	2.0 to 2.5 m
Japan	>2.5 m	>1.75 m	>0.5 m
The Netherlands	1.25 m	0.20 to 0.45 m	0.15 to 0.45 m
Poland	2.5 to 3.0 m	2.0 to 2.75 m	1.0 to 1.5 m
Portugal	3.0 m left; 1.0 m right	2.5 m	1.5 to 2.5 m
South Africa	>2.0 m	Rural 1.0 to 3.0 m; urban not essential	—
Spain	0.5 to 1.0 m left; 2.5 to 3.0 m right	1.5 to 2.5 m	0.5 to 2.0 m
Sweden	—	0.75 m (rural undivided)	—
Switzerland	1.0 to 2.5 m	0.5 to 1.5 m	No shoulders
United Kingdom	—	Rural: 2.75 m left; 3.3 m right Urban: 2.75 m	—
United States	3.0 to 3.6 m right; 1.2 to 3.6 m median	1.2 to 2.4 m	0.6 to 2.4 m
Venezuela	2.4 m left; 1.2 m right	0.6 m left; 1.2 m right	—
Yugoslavia	1.5 m	1.35 to 1.5 m	1.2 to 1.35 m
"CS"	2.0 to 2.5 m paved; 1.5 m unpaved	1.5 m to 2.0 m paved; 1.5 m to 2.0 m unpaved	1.5 m (1.0 m exception)

24.2.3.5 Sidewalk and Bicycle Lane.   Sidewalks are mainly provided in urban and suburban areas. AASHTO standards[5] specify widths of 1.2 to 2.4 m for both residential and commercial areas, assuming a 0.6-m planting strip is located between the sidewalk and the curb at the edge of the traveled way. In the absence of such a planting strip, the sidewalk width should be increased by 0.6 m. When sidewalks are constructed along rural roads, they should be well removed from the traffic lanes. In some installations, pedestrians and bicyclists may share a lane separate from the vehicular lanes of a roadway; if volumes are sufficient, however, both sidewalks and bicycle lanes may be built[257] (see Sec. 24.1.3).

24.2.4 Safety Considerations

With regard to the cross-sectional elements, two important questions are frequently discussed in the literature:

1. How do they influence speed, traffic quality, and/or capacity? (As already mentioned in Sec. 22.2, this question will not be discussed in "CS").
2. How do they influence the accident situation? (This question is discussed in the following).

Past studies have revealed that of more than 50 roadway-related features which can significantly affect crash experience, cross-sectional elements are among the most important.[317,778] Such elements include lane width, shoulder width, shoulder type, roadside features (for example, side-slope, clear zone, and placement and types of roadside obstacles), bridge width, and median width, among others.

In addition to these elements, multilane design alternatives may also be considered where basic two-lane roads are not adequate. Such alternatives include the addition of through lanes, passing lanes, various median designs, intermediate lanes (two-way, alternating), and others. Such design alternatives can affect traffic operations, as well as safety, along a highway section.[729]

By an in-depth literature review[109,440] for two-lane rural roads, it was found that, to date, the cross-sectional elements pavement (lane)-width and shoulder width have been investigated the most with regard to traffic safety aspects. However, the combination of lane and shoulder widths plus median, if any, comprises the roadway width. Total roadway width is among the most important cross-sectional considerations in the safety performance of a two-lane highway.[710,729]

Of the many cross-sectional roadway elements that are discussed in the following chapters, an illustration is given in Fig. 23.1 for those typically found on two-lane rural roads.

24.2.4.1 Pavement/Lane Width.   Over the years considerable research has attempted to quantify how accidents change with different lane widths for different types of roads. As with most accident studies, it has been difficult to isolate the effect of one factor, in this case lane width, from other influencing factors, including shoulder width, number of lanes, and volume.[503]

Research studies have generally shown that adequate pavement widths are necessary for safe driving operations. The necessary widths are generally the sum of the dimension of design vehicles and lateral clearances for transportation and safety maneuvers. If these widths are not selected correctly, impairment of traffic safety can occur. Therefore, a certain correlation between pavement width and traffic safety can be expected.

An in-depth literature review conducted by the authors in Ref. 109 has generally shown that accident rates or accident frequencies decrease with increasing pavement/lane width between 5.50/2.75 m and 7.50/3.75 m on two-lane rural roads. These statements are confirmed by the following authors from different countries between the 1940s and the 1980s:

- Baldwin, United States, 1946[43] (see Fig. 24.8)
- Cope, United States, 1995:[128] Increase in pavement widths from 5.5 to 6.7 m → decrease of accident rate from 1.4 to 0.9 (accidents per 10^6 vehicle kilometers)
- Bitzl, Germany, 1961/1967[53,57] (see Fig. 24.8)

- Winch, Canada, 1963[762]
- Balogh, Hungary, 1967[44]
- Silyanov, former U.S.S.R., 1973[649] (see Fig. 24.8): AR = 1/(0.173 w − 0.21) (accidents per 10^6 vehicle kilometers), where w = pavement width for 4 to 9 m
- Pignataro, United States, 1973:[578] Increase in pavement width from 5.0 to 7.5 m → decrease of accident rate from 3.4 to 1.5 (accidents per 10^6 vehicle kilometers)
- Kunze, Germany, 1976[378] (see Fig. 24.8)
- Krebs and Kloeckner, 1977[368] (see Fig. 24.8): Establishment of a negative linear relationship between pavement width and accident risk
- Zegeer, Deen, Mayes, United States, 1981:[774] Accident rate decreases as pavement width increases up to about 7.25 m
- McCarthy, Scruggs, and Brown, United States, 1981:[500] Widening lane widths from 2.7 and 3.0 m to 3.4 and 3.7 m → reduction of injury-fatality accident rate of 22 percent

Figure 24.8 shows some of the relationships between accident rate and pavement width as derived from the results of several of the previous studies for two-lane rural roads.

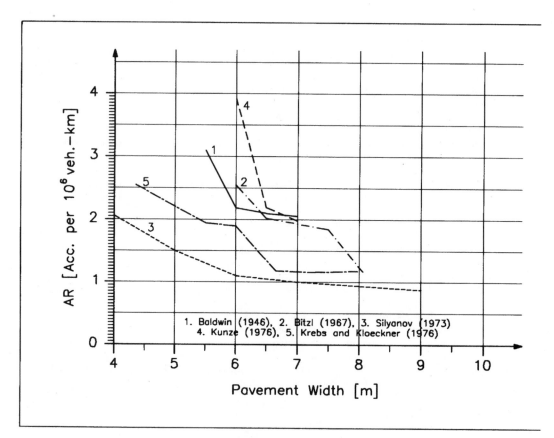

FIGURE 24.8   Relationship between accident rate and pavement width.[109]

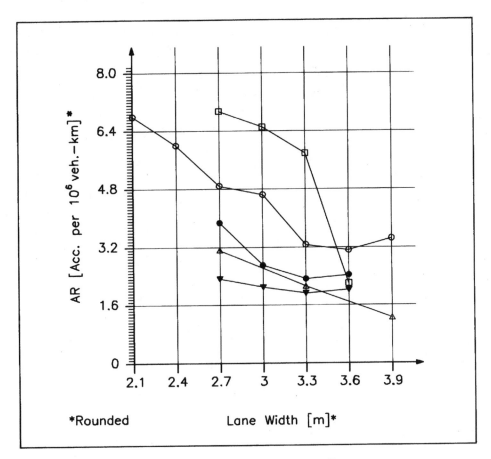

FIGURE 24.9   Accident rates based on lane width for two-lane rural roads.[52]

There is enough empirical evidence to show that accident rates increase with decreasing lane width, and it is also presented in Fig. 24.9.[52] The figure shows the results of five studies between lane widths and accident rates in different states of the United States during the 1970s.

These results are also confirmed by O'Cinneide,[553] whose review contained the following references in addition to others. He reported:

- Hedman, Sweden, 1990[273] noted that some results indicated a rather steep decrease in accidents with increased width of carriageway from 4 to 7 m but little additional benefit was gained by widening the carriageway beyond 7 m.
- As reported for Denmark,[704] if the lane width increases, the relative accident frequency decreases: for road widths less than 6 m, there was a significant increase in the risk of both injury accidents and severe injury accidents.
- Srinivasan[669] reported that the accident rate of a 5-m road was about 1.7 times that of a 7.5-m road.
- National Cooperative Highway Research Program (NCHRP) Report 197[709] suggested that widening lanes from 2.7 to 3.7 m would reduce accidents by 32 percent.

An extensive "before" and "after" investigation by Hiersche, Lamm, et al.,[286] covering 3 y of accident experience (1428 accidents) on 28 sections with a section length of about 90 km that was redesigned according to the German design guidelines,[241] established the results given in Fig. 24.10 between accident rate/accident cost rate and the pavement width. The regression curves in Fig. 24.10 are based on the vehicle kilometers traveled and are calculated for every pavement width class. The relationship between accident rate and pavement width is in agreement with prior research; that is, the risk of being involved in an accident decreases as pavement width increases.

Contrary to the opinions of many experts, Fig. 24.10 shows that the accident cost rate, an indicator of accident severity, increases as pavement width increases, even though the investigated road sections were designed according to the—at that time—new German design guidelines.[241] This result may be explained by the fact that wide pavements are usually assigned to high design speed levels. With high design speeds, operating speeds are also usually high. Consequently, as operating speeds increase, the severity of an accident, as measured by the accident cost rate, is likely to increase, too.

The influence of pavement width on accident rate and accident cost rate has also been analyzed by Leutzbach and Zoellmer[476] in Germany. About 1500 km of two-lane rural roads with corresponding accident data from 1978 to 1985 were analyzed. Pavement widths were arranged in 10 classes with increments of 50 cm. The relationships between accident rate/accident cost rate and pavement width are shown in Figs. 24.11 and 24.12.

These figures indicate that pavement width influences accident rate (risk) and accident cost rate (severity) in different ways. For instance, an increase in pavement width up to 8.50 m resulted in a decrease in the accident rate. For pavements greater than 8.50 m, the accident rate experienced a slight increase (see Fig. 24.11).

The pavement width has distinct effects on the traffic flow and the accident situation. Of special interest are run-off-the-road accidents and passing/rear-end accidents. Therefore, an additional analysis was conducted for these accident types, which are responsible for about 60 percent of all accidents in this study.[476] The results revealed trends similar to those presented in Fig. 24.11 for the accident rate and Fig. 24.12 for the accident cost rate.

Because of contradictions in the results concerning the effect of pavement width on accident cost rate (see Figs. 24.10[286] and 24.12[476]), further research is needed.

Finally, statistical analyses by Zegeer, et al.[782] of accident relationships based on an analysis of 10,900 horizontal curves on two-lane rural highways in Washington State with corresponding accidents (12,123 accidents), geometric, traffic, and roadway data variables determined a 21 percent accident rate reduction for 1.2 m of lane widening.

Krebs and Kloeckner[368] superimposed the pavement width on the radius of curve, R, the longitudinal grade, G, and the average annual daily traffic, AADT. The results of these superimpositions are shown in Figs. 24.13 to 24.15.

In general, it can be concluded that the accident rate decreases in all three classes of radii of curve with increasing pavement width (Fig. 24.13). For radii of curve less than 400 m, the highest accident risk could be observed. In all radii classes, the decline of the accident rate is nearly parallel.

The accident rate also decreases for the three classes of longitudinal grades (Fig. 24.14). Grades of $G \geq 6$ percent show the highest accident rates.

For increasing pavement widths, the accident rate also decreases with respect to the three classes of AADT; however, the differences between the classes are minor (Fig. 24.15). In general, AADT values less than 5000 vehicles/day reveal higher accident rates than those greater than 10,000 vehicles/day. The highest accident rates could be observed for narrow pavement widths for all three AADT classes.[368]

Furthermore, it should be kept in mind that in Fig. 9.32 a nomogram to determine the accident rate by superimposing pavement width, radius of curve, and longitudinal grade for two-lane rural roads was developed. As could be seen from this figure, pavement width played an important role, in addition to radius of curve and longitudinal grade, for evaluating expected accident rates.[390]

Research studies reported here have generally shown that accident rates decrease with an increase in pavement width up to 7.0/7.5 m on two-lane rural roads. This increase in traffic safety was evident for all classes of radii of curve, gradients, and traffic volumes.

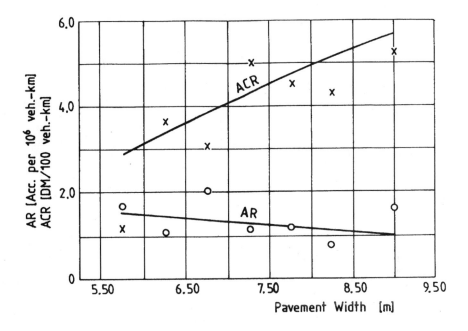

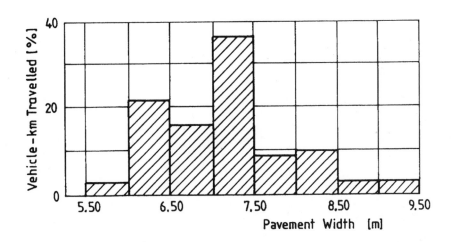

Legend: DM = German marks

FIGURE 24.10 Accident rate and accident cost rate versus pavement width.[286]

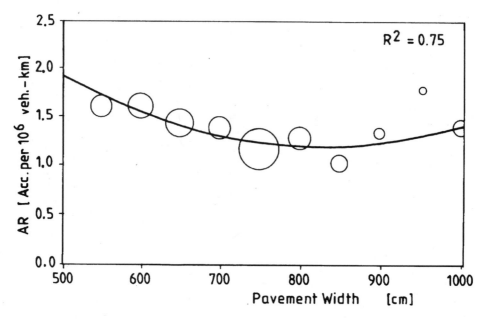

FIGURE 24.11   Accident rate with regard to the pavement width for all accident types.[476]

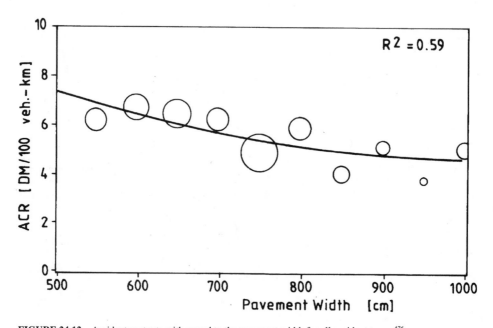

FIGURE 24.12   Accident cost rate with regard to the pavement width for all accident types.[476]

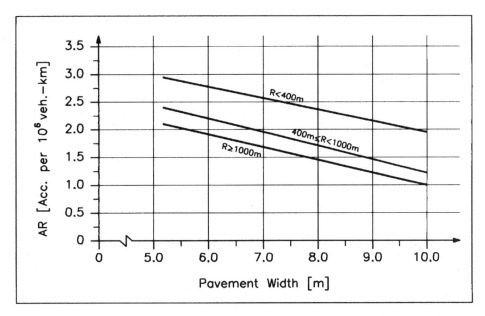

FIGURE 24.13   Accident Rate as function of the pavement width for different classes of radii of curve.[368]

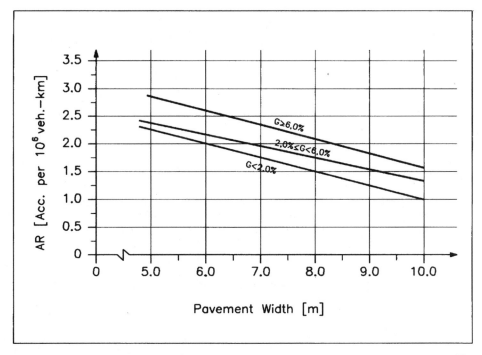

FIGURE 24.14   Accident rate as function of the pavement width for different classes of longitudinal grades.[368]

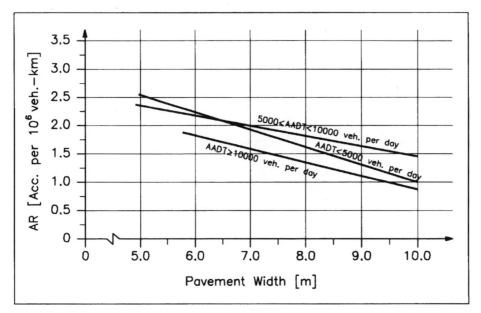

FIGURE 24.15   Accident rate as function of the pavement width for different classes of average annual daily traffic.[368]

A similar statement was given by O'Cinneide:[553] As the lane width increases above the minimum, the accident rate decreases. However, the marginal rate diminishes with increased lane width.

With respect to U.S. research, it can be concluded:[503]

- Lane widths of 3.0 m are considered the minimum appropriate lane width for rural highways.

- Lane widths of 3.3 m have substantially lower accident rates than lane widths of 3.0 m, particularly when narrow shoulders exist.

- For higher-speed, higher-class highways, lane widths of 3.6 m are considered appropriate, regardless of the lack of a quantitative safety benefit of 3.6-m versus 3.3-m lanes.

- On multilane roads, the more lanes that are provided in the traveled way, the lower the accident rate.[553]

These statements confirm the previously discussed research results for Europe and justify the lane-width values considered as appropriate in Table 24.1, col. 7 as well as in Fig. 25.1.

24.2.4.2   Other Cross-Sectional Elements and Interrelationships.

Shoulder.   Wide lanes and shoulders provide motorists with the opportunity for safe recovery when their vehicles run off the road (an important factor in single-vehicle accidents) and increased lateral separation between overtaking and meeting vehicles (an important factor in sideswipe and head-on accidents). Additional safety benefits include reduced interruption from both emergency stopping and road maintenance activities, less wear at the lane's edge, improved sight distance at critical horizontal curves, and improved roadway surface drainage.[710]

Cirillo and Council[112] reported about shoulder widths in the United States. Most studies agree that shoulders up to 1.8 m wide on facilities with greater than 1000 Average Annual Daily Traffic (AADT) provide a safety benefit. The effect of shoulders wider than 1.8 m is so far not clear.

In this connection, O'Cinneide[553] reports the following. There have been a number of studies of the relationship between the shoulder width and the accident rate. As noted by Hedman,[273] more recent studies show a decrease in accidents with increases in shoulder width from 0 to 2 m, and little additional benefits are obtained above 2.5 m.

On tangents, as the shoulder width increases beyond the minimum, the safety benefit becomes insignificant; on curves, as the shoulder width increases, the accident rate decreases; and on multilane divided highways, as the median shoulder width increases, accidents increase. For this reason, median shoulders are not included in the design standards of some European countries.[552] Therefore, in Part 3, "Cross Sections (CS)," the median shoulder is reduced to an edge strip of between 0.50 and 1.00 m wide (also see Fig. 25.1).

As Transportation Research Board (TRB) Special Report 214[710] noted, the literature does not provide an entirely consistent model of the simultaneous effects of lane width and shoulder type on accidents. It also noted that accident rates decrease with increases in lane and shoulder width and that widening lanes has a greater safety benefit than widening shoulders.[553]

An interesting interpretation of the relationships between cross-sectional elements is given by McLean (Australia)[507] in Table 24.7 for accident rate adjustment factors. This table was developed in reference to a base accident rate of 1.8 total accidents/million vehicle kilometers for a cross section of 3.25-m lanes and 1.8-m unpaved shoulders. The accident rate adjustment factors given in Table 24.7 were developed from a combination of U.S. accident-width relations[317,779] and results of Australian studies[28,101] and were employed to estimate accident rates for other cross sections.

As can be seen from Table 24.7, lane widths from 3.50 to 3.70 m have the best accident rate adjustment factors. Those of unpaved shoulders are basically higher than those of paved shoulders, which means a lower safety gain. Both types show the best results for shoulder widths from 2.00 to 2.50 m. However, the safety gain between a 2.00- and 2.50-m shoulder width is insignificant.

These findings confirm the previous statements and, to some degree, those given in Table 24.1 and Fig. 25.1 with respect to the proposed lane and shoulder widths in this handbook.

Furthermore, Armour (Australia)[28] derived relative accident rates for roads with paved and unpaved shoulders (Table 24.8). Although the author himself evaluates the values of Table 24.8 only as indicative, the differences in accident rates between straight and curved roadway sections with unpaved or paved shoulders yield the following interesting trends:

TABLE 24.7   Accident Rate Adjustment Factors[507]

Element	Width, m	Adjustment factor
Traffic lane	2.75	1.00
	3.00	0.90
	3.25	0.81
	3.50	0.75
	3.70	0.73
Paved shoulder	0.00	1.00
	0.50	0.93
	1.00	0.82
	1.50	0.70
	2.00	0.58
	2.50	0.52
Unpaved shoulder	0.00	1.00
	0.50	0.96
	1.00	0.89
	1.50	0.82
	2.00	0.75
	2.50	0.71

TABLE 24.8   Relative Accident Rates[28]

Curvature	Shoulder type		
	Unpaved	Paved	All
Straight	1.0	0.3	0.7
Curve	5.9	1.5	3.5
All	1.8	0.6	

Note: Accident rates are relative to the accident rate for a straight road section with unpaved shoulders.

1. Curved roadway sections are significantly more dangerous than tangent sections. Also compare Tables 9.10, 10.12, and 18.14.

2. The influence of paved shoulders on the accident situation appears to be significantly more favorable than that of unpaved shoulders.

In this connection, a German research report[68] concluded the following. Compared to unpaved shoulders, paved shoulders yield, on average, about 25 percent lower accident rates and about 10 percent lower accident cost rates.

U.S. research results concerning the safety effectiveness of lanes and shoulders are given in Ref. 729, where the following statement has been made:

> Numerous studies have been conducted in recent years to determine the effects of lane width, shoulder width, and shoulder type on accident experience. However, few of them were able to control for roadside conditions (e.g., clear zone, sideslope), roadway alignment and other factors which, together with lane and shoulder width, influence accident experience. Also, since lane and shoulder width logically affect some accident types (e.g. run off the road, head-on), but not necessarily other accident types (e.g., angle, rear-end), there is a need to express accident effects as a function of those related accident types.

A 1987 Federal Highway Administration (FHWA) study by Zegeer, et al. quantified the effects of lane width, shoulder width, and shoulder type on highway crash experience based on an analysis of data for nearly 8000 km of two-lane highway from seven states.[779] Accident types found to be related to lane and shoulder width, shoulder type, and roadside condition include run-off-the-road, head-on, and opposite and same direction sideswipe accidents, which together were termed as *related accidents*. An accident prediction model was developed and used to determine the expected effects of lane and shoulder widening improvements on related accidents.

As shown in Table 24.9, lane widening of 0.30 m (for example, from 3.00- to 3.30-m lanes) would be expected to reduce related accidents by 12 percent and widening of 1.20 m (for example, from 2.50- to 3.70-m lanes) should result in a 40 percent reduction in related accident types.

Reductions in related accidents due to widening paved or unpaved shoulders are given in Table 24.10. For example, widening 0.60-m unpaved shoulders to 2.40 m will reduce related accidents by 35 percent (that is, for a 1.80-m increase in unpaved shoulders). Adding 2.40-m paved shoulders to a road with no shoulders will reduce related accidents by approximately 49 percent.[779] It should be noted that the predicted accident reductions given in Tables 24.9 and 24.10 are only valid when the roadside characteristics (side slope and clear zone) are reestablished as they were before the lane or shoulder widening. (The dimensions in Tables 24.9 and 24.10 were converted from the imperial system to the international system, and have to be regarded as rounded values.)

When two or more roadway improvements are proposed simultaneously, the accident effects are not cumulative. For example, implementing two different improvements having accident reductions of 20 and 30 percent will not result in a combined 50 percent accident reduction.[729]

TABLE 24.9 Percentage of Accident Reduction of Related Accident Types for Lane Widening Only[729,779]

Amount of lane widening, m*	Percent reduction in related accident types
0.30	12
0.60	23
0.90	32
1.20	40

*Rounded.

Note: These values are only valid for two-lane rural roads.

TABLE 24.10 Percentage of Accident Reduction of Related Accident Types for Shoulder Widening Only[729,779]

Shoulder widening per side, m*	Percent reduction in related accident types	
	Paved	Unpaved
0.60	16	13
1.20	29	25
1.80	40	35
2.40	49	43

*Rounded

Note: These values are only valid for two-lane rural roads.

With respect to U.S. research, supported by Australian and European experience, it can be concluded, according to Ref. 503:

- The presence of a shoulder is associated with a significant accident reduction for various lane-width categories, particularly for shoulder widths of at least 1.0 to 1.2 m.
- For 3.0-m lanes, a shoulder of 1.5 m or wider appears to be needed to significantly affect accident rate.
- For 3.30- and 3.60-m lane widths, shoulders of 1.0 m or wider have significantly beneficial effects.[503]
- Roads with stabilized shoulders, such as asphalt or portland cement concrete, have lower accident rates than nearly identical roads with unstabilized earth, turf, or gravel shoulders.[710]
- Shoulders wider than 2.5 m provide little additional safety benefit. As the median shoulder width increases, accidents increase.[553]

These statements confirm the previously discussed research results and justify the shoulder-width values considered as appropriate in Table 24.1 and Fig. 25.1.

Median.[729]   The median is another important element of the cross section. In this connection, only a few accident-related studies exist, since in most countries the median width is standardized with dimensions up to 4 m because of land constraints (Table 24.5), and the opposing traffic streams are protected against each other by safety guardrails (fences).

Srinivasan[669] found that, on high-speed roads with two or more lanes in each direction, medians improve safety in a number of ways, for example, by reducing the number of head-on collisions. The Danish design standards[704] include a table showing the relationship between the median width, the accident frequency of the through section, and a severity index for medians with and without a crash barrier. Medians, particularly with barriers, reduce the severity of accidents; however, medians wider than 3.0 to 4.0 m show little additional benefit.[553]

However, in countries like Canada and the United States, land constraints often do not stand in the forefront and much wider medians are designed.

In this respect, the following was reported in Ref. 729. Elements of median design which may influence accident frequency or severity include median width, median slope, median type (raised or depressed), and presence or absence of a median barrier. Wide medians are considered desirable because they reduce the likelihood of head-on crashes between vehicles traveling in opposite directions. Median slope and design can affect rollover accidents and also other single-vehicle crashes (fixed object) and head-on crashes with traffic in opposite directions. The installation of median barriers typically increases overall accident frequency due to the increased number of crashes into the barrier but reduces crash severity, as a result of a reduction or elimination of head-on impacts with traffic in the opposite direction. A controlling factor in median width is often the limited amount of highway right-of-way available.

A comparison was made of the safety of raised (mound) medians versus depressed (swale) medians in a 1974 Ohio study.[199] Using a sample of rural interstates, all having 25-m-wide medians and other similar geometrics, the accident experience was compared between the two median designs. The typical median cross sections for the sample mound and swale medians used in the study are shown in Fig. 24.16.

No differences were found in the number of injury accidents, rollover accident occurrence, or overall accident severity between the raised and depressed median designs. However, a significantly lower number of single-vehicle median-involved crashes was found on sections with depressed medians compared to raised medians. The authors concluded that this may indicate that mildly depressed medians provide more opportunity for encroaching vehicles to return safely to the roadway.[729]

A 1973 study[203] compared the crash experience of various median widths, median types (raised versus depressed), and slopes on interstate and turnpike roads in Kentucky. As shown in Fig. 24.17, highways with at least 9-m-wide medians had lower accident rates than those with narrower median widths. For wider medians, a significant reduction was also found in the percentage of accidents involving a vehicle crossing the median.

The authors[203] recommended minimum median widths of 9 to 12 m, slopes of 1:6 or flatter and, on roadway sections where guardrail is installed, 3.5-m paved medians. Raised medians were found to be undesirable based both on accident experience and on less-than-ideal surface drainage.[729]

Taken together, the two median studies indicate that where a wide median width can be provided (for example, 25 m), a mildly depressed median (depressed by 1.2 m with 1:8 downslopes) and a mound median (1:3 upslope) provide about the same crash experience. However, in cases with narrower medians (for example, 6 to 9 m), slopes of 1:6 or flatter are particularly important. Deeply depressed medians with slopes of 1:4 or steeper are clearly associated with a greater occurrence of overturned crashes. While accident relationships are unclear for median widths of

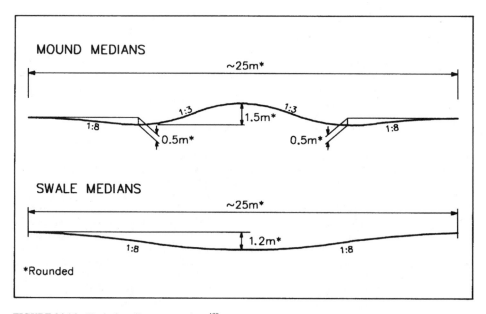

FIGURE 24.16   Typical median cross sections.[199]

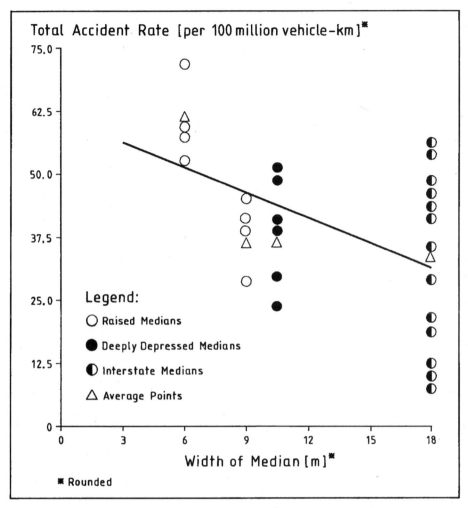

FIGURE 24.17 Total accident rate versus median width.[203]

Note: **This figure is based on multilane Interstate and turnpike roads in rural areas**

less than 6 m, wider medians, in general, are better, and median widths in the range of 18 to 25 m or more with flat slopes appear to be desirable, where feasible.[729]

However, it should be kept in mind that those wide medians can only be applied in countries with no land constraints. Therefore, for most countries, medians of up to 4.0 m with safety guardrails or barriers are recommended according to Table 24.5.

Edge Strip. *Edge strips* are provided to carry the edgeline and to separate the traveled way from adjacent cross-sectional elements. Driving down a dark road on a misty night is never pleasant. The only comfort comes from centerlines and edgelines. These pavement markings, along with lane lines, are important driving aids. The driver's manual advises watching the edgeline when blinded by oncoming headlights. Lane lines organize vehicles into efficient lanes on multilane roads. Centerlines help oncoming vehicles to avoid collisions. Even in daylight, pavement markings make it possible for vehicles to travel more safely.[514]

Published literature suggests that existing longitudinal pavement markings reduce crashes by 21 percent, and edgelines on rural two-lane highways reduce crashes by 8 percent.[514]

As reported by Lum and Hughes,[488] the use of edgelines has become an accepted practice in most countries. The normal edgeline width is between 10 and 20 cm in the United States. In the early 1980s, two research studies were conducted to evaluate drivers' traffic performance in response to different edgeline widths. A 1980 study evaluated the effects of 10-, 15-, and 20-cm-wide edgelines on the performance of 16 male subjects ages 21 to 25.[541,542] The subjects drove on an isolated and controlled 7.3-m two-lane road used as a test course between midnight and 3 a.m. *Good driving* was defined as the ability to remain in the center of the lane. Overall, the subjects performed better—weaving less—on 10 curved sections of the test course using 20-cm edgelines than with 10-cm edgelines. Further analysis showed that subjects dosed to a 0.05 or 0.08 blood alcohol content (BAC) level exhibited more "good" driving traits on the wider edgeline course.

The second study, conducted in 1982 in Australia, evaluated the effect of different delineation treatments on the performance of 36 male subjects driving on a closed test route with horizontal curves and tangents.[314] For the 12 curves analyzed, edgelines added little improvement to drivers' lateral positioning of their vehicles over long-range delineations (for example, chevrons, post-mounted delineators). Subject drivers, however, with a blood alcohol (BAC) of 0.05 showed significantly better lateral positioning of their vehicles at the critical midcurve point.[488]

Specific study findings[488] suggest that 20-cm edgelines could potentially be cost-effective in reducing run-off-the-road accidents on two-lane rural roads with pavement widths of at least 7.3 m, unpaved shoulders, and an average ADT of 2000 to 5000 vehicles/day.

The 20-cm edgelines may be appropriate as a safety improvement when applied at spot locations such as isolated horizontal curves and approaches to narrow bridges.[488]

Overall, the authors of this handbook are convinced that wide edgelines (25 cm) on two-lane rural roads reduce accidents on curves and have to be considered as an additional reliable countermeasure, for example, with respect to the fair design ranges of safety criteria I to III developed in Part 2, "Alignment."

24.2.4.3   Preliminary Conclusion.   Generally, wider lanes and/or shoulders will result in fewer accidents and lower accident rates (risks).[109,553,729] The relationship between lane width and accident cost rate has to be further investigated.[286]

From a traffic safety point of view, it can be stated that by introducing lane widths of 3.50 or 3.75 m and paved shoulder widths of 2.0 to 2.5 m, very favorable results can be expected. These recommendations apply mainly to high-speed and/or high-volume roads with high/medium truck traffic because these dimensions contain generous ranges for:

- Lateral safety spaces (multilane and two-lane)
- Lateral moving spaces (multilane and two-lane)
- Opposing traffic width increases (two-lane) (see Tables 24.1 and 24.2)

On the basis of the research studies discussed here, it can be stated that all proposed standard cross sections with lane widths of 3.50 to 3.75 m are very favorable from a traffic safety viewpoint (see Fig. 25.1). This is true for all road categories, with certain exceptions such as the standard cross sections b2s and b2 (see Sec. 25.2.2.1).

With respect to the recommended shoulder widths in Fig. 25.1 (paved or unpaved or a combination of both), the in-depth literature review has indicated that the recommended dimensions will lead to adequate lateral roadway spaces. For traffic safety reasons, an unpaved shoulder width of 1.50 m for all two-lane roadway types seems to be appropriate (see Fig. 25.1) on the basis of the previously discussed research and is, at the same time, economically and environmentally feasible.

For low-volume roads in particular, the research conducted by Zegeer, et al.[783] in the United States revealed that, on roads with lane widths of at least 3.00 m, the accident rates were lower when wide shoulders were present than when narrow shoulders were. The wide shoulder roadway type corresponds to the standard cross sections d2 and e2 in Fig. 25.1. These road types represent acceptable cross sections in the range of medium- and/or low-volume roads. But further research is urgently needed in this field. For additional information about traffic safety issues of cross-sectional elements and standard and/or intermediate cross sections, see Chap. 25.

In conclusion, the following statements can be made:

- A distinct tendency for accident rates to decrease with increasing lane width (up to 3.50 to 3.75 m) and with increasing paved shoulder width (up to 2.00 to 2.50 m) for high-speed and/or high-volume roads was established.
- For low-volume roads, a lane width of at least 3.00 m and an unpaved shoulder width of 1.50 m are recommended.
- Highways with four or more lanes have to be separated by a median.
- Edgelines increase safety.

However, cross-sectional elements alone cannot describe the accident situation satisfactorily and other influencing parameters, such as alignment, traffic volume, and so forth, also have to be considered.

CHAPTER 25
CROSS SECTION DESIGN

25.1 RECOMMENDATIONS FOR PRACTICAL DESIGN TASKS

When selecting the cross section of a road, attention must be given to the fact that traffic quality (capacity) and safety depend on

- Cross section design
- Intersection/interchange design
- Alignment
- Operational mode of the road, for example, as exclusive motor vehicle road or as road for the general traffic

25.1.1 Standard Cross Sections*

In order to achieve uniformity and consistency in the design, construction, and operation of roads, standard cross sections (SCSs) will be presented in the following for all operational ranges. These standard cross sections should remain uniform along relatively long roadway sections.

All standard cross sections are composed of cross-sectional elements, as shown in Table 24.1. The standard cross sections for nonbuilt-up roads are schematically shown in Fig. 25.1 for:

- Multilane median-separated roads
- Two-lane roads

Furthermore, examples for intermediate cross sections (ICSs) are given in Fig. 25.3.

The abbreviations in parentheses in Fig. 25.1 refer to the roadway width of the cross section, in meters. For example, SCS a6ms means that the overall width of this standard cross section is 37.5 m.

The other designations of the standard cross sections are based on the cross-sectional elements presented in Table 24.1. For instance, standard cross section b6ms in Fig. 25.1 indicates (compare Table 24.1):

- Standard cross section is 33.0 m wide
- b is a cross section group with a lane width of 3.50 m (see col. 7a of Table 24.1)
- 6 is the number of lanes for both directions of travel
- m is the median
- s means a paved shoulder was used, for example, as emergency lane/multiple purpose lane.

---

*Elaborated based on Refs. 247 and 454.

Multilane

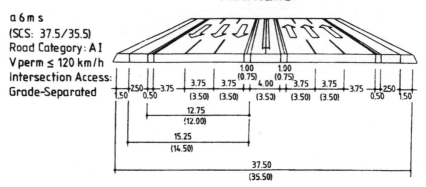

a 6 m s
(SCS: 37.5/35.5)
Road Category: A I
V perm ≤ 120 km/h
Intersection Access:
Grade-Separated

a 4 m s
(SCS: 30/28.5)
Road Category: A I
V perm ≤ 120 km/h
Grade-Separated

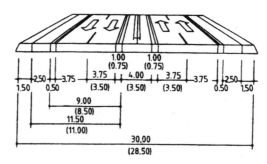

( ) = Exceptional Values

b 6 m s
(SCS: 33.0)
Road Category: A II
V perm ≤ 110 km/h
Grade-Separated

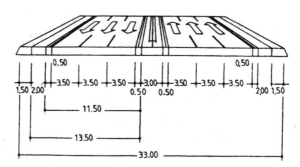

(*Note:* Road categories A I to A V and B II to B IV are defined in Fig. 3.1 and Table 6.2.)
All dimensions are in meters:
a to e = cross section groups according to Tables 24.1 and 25.1
    m = median
     s = paved shoulder (emergency/multiple purpose lane)
  SCS = standard cross section
  V_{perm} = permissible speed limit

FIGURE 25.1 Examples of standard cross sections for nonbuilt-up roads. (Elaborated, modified, and extended based on Refs. 247 and 454.)

b 4ms
(SCS : 26.0)
Road Category : AII, BII
AII : V perm ≤ 110 km/h
BII : V perm ≤ 90 km/h
Grade-Separated

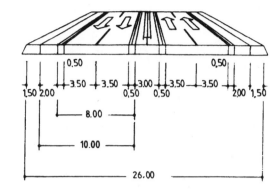

c 4ms
(SCS : 24.0)
Road Category : AII, BII
AII : Vperm ≤ 110 km/h
Grade-Separated / AT-Grade
BII : Vperm ≤ 90 km/h
Grade-Separated / AT-Grade

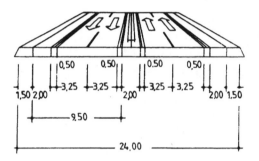

Two-Lane

(b2s)*
(SCS : 14.0)
Road Category : AII , AIII
AII , AIII : Vperm ≤ 90 km/h
AT-Grade
Number of Slow (Farm) Vehicles
> 10 per hour (AII)
> 20 per hour (AIII)

*Parantheses mean relatively
 unsafe cross section, compare
 Sec. 25.2.2.1

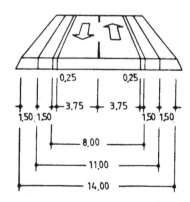

FIGURE 25.1   (*Continued*) Examples of standard cross sections for nonbuilt-up roads.

(b2)＊
(SCS: 12.0)
Road Category : AII, AIII
AII, AIII : Vperm ≤ 90 km/h
AT-Grade
High Truck Traffic

＊Parantheses mean relatively
unsafe cross section, compare
Sec. 25.2.2.1

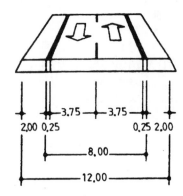

c 2
(SCS: 10.5)
Road Category : AII
AII: Vperm ≤ 90km/h
AT-Grade

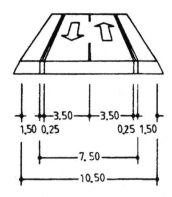

d 2
(SCS: 10.0)
Road Category: AIII, AIV, BIII, BIV
AIII: $V_{perm} \leq 90\,km/h$
AIV: $V_{perm} \leq 80\,km/h$
BIII: $V_{perm} \leq 70\,km/h$
BIV: $V_{perm} \leq 60\,km/h$
AT-Grade

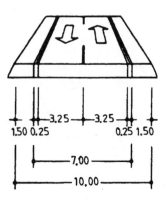

FIGURE 25.1   (*Continued*) Examples of standard cross sections for nonbuilt-up roads.

e 2
(SCS: 9,5)
Road Category: AIV, AV
AIV: V_{perm} ≦ 80 km/h
AV : V_{perm} ≦ 70 km/h
AT-Grade

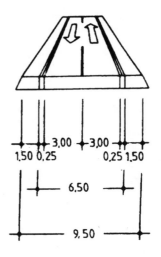

f 2 (Exception)
(SCS : 7,5)
Road Category: AV
V_{perm} ≦ 50 km/h
AT-Grade

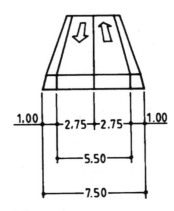

FIGURE 25.1 (*Continued*) Examples of standard cross sections for nonbuilt-up roads.

For example, adding one basic lane in each direction to an a6ms would result in an a8ms, and so forth. Or, for a typical two-lane standard cross section e2, the meaning, according to Fig. 25.1, is

- Standard cross section is 9.5 m wide
- e is a cross section group with a lane width of 3.00 m (col. 7b of Table 24.1)
- 2 is the number of lanes, one lane for each direction of travel.

The other information about the standard cross sections in Fig. 25.1, about road categories A I to A V and B II to B IV, or the statements about permissible speed limits V_{perm} and intersection access are based on Table 6.2.

TABLE 25.1   Estimations of Traffic Volume Ranges for Standard Cross Sections*

Typical road category† (1)	Traffic volume range, vehicles/h (2)	Typical standard cross section† (3)	Operational mode (4)	Permissible speed limit, km/h (5)	Intersection access (6)	Design speed V_d, km/h (7)
A I	3100–5200	a6ms	EMVR	≤120	GS	(130) 120 (110)
	2100–3450	a4ms	EMVR	≤120	GS	(130) 120 (110)
	1050–2300	b2 + 1‡	EMVR	≤100	GS	100 90 (80)
A II	3100–5200	b6ms	EMVR	≤110	GS	(120) 110 100 90 (80)
	2100–3450	b4ms	EMVR	≤110	GS	110 100 90 (80)
	2000–3300	c4ms	EMVR	≤110	GS (AG)	110 100 90 (80)
	1050–2300	b2 + 1‡	EMVR	≤ 90	GS (AG)	(100) 90 80 (70)
	950–2100	b2s	RGT	≤ 90	AG	(100) 90 80 (70)
	800–2000	b2	RGT	≤ 90	AG	90 80 (70)
	700–1950	c2	RGT	≤ 90	AG	90 80 70
A III	900–2000	b2s	RGT	≤ 90	AG	90 80 70 (60)
	750–1850	b2	RGT	≤ 90	AG	90 80 70 (60)
	500–1950	d2	RGT	≤ 90	AG	90 80 70 60
A IV	500–1850	d2	RGT	≤ 80	AG	80 70 60 (50)
	350–2050	e2	RGT	≤ 80	AG	80 70 60 (50)
B II	2100–3500	b4ms	EMVR	≤ 90	GS	(100) 90 80 70 (60)
	1950–3250	c4ms	EMVR	≤ 90	GS (AG)	(100) 90 80 70 (60)
B III	500–1950	d2	RGT	≤ 70	AG	70 60 (50)
B IV	500–1950	d2	RGT	≤ 60	AG	60 50

*Elaborated based on the Greek guidelines.[354,454]

†Other combinations of road categories (col. 1) and standard cross sections (col. 3) are also possible.

‡Intermediate cross section (see Fig. 25.3).

Note: EMVR = exclusive motor vehicle road; RGT = road for general traffic; GS = intersection access, grade-separated (controlled); and AG = intersection access, at-grade (free).

Estimates on traffic volume ranges for the standard cross sections according to the Greek guidelines[354,454] are given in Table 25.1. However, because of the numerous influencing factors and the complex relationships, these estimations are rough in nature and should only be considered as approximate guiding principles. Also, they may differ from country to country.

The estimates of traffic volume ranges given in Table 25.1 are estimated for *one* direction of traffic in the case of multilane roads and for *both* directions of traffic in the case of two-lane roads. Generally, they are related to road sections without intersections and/or interchanges, a truck percentage of 0 percent, a flat gradient, and a curvature change rate of the single curve $CCR_S = 0$ gon/km. Furthermore, Table 25.1 provides valuable insights about the standard cross sections with respect to operational modes (exclusive motor vehicle road or road for the general traffic), permissible speed limits, access control, and recommended design speeds.

When selecting standard cross sections for two-lane roads for which traffic volumes reach the upper limit of the traffic volume ranges in col. 2 of Table 25.1, one should keep in mind that a roadway section with an additional lane, such as the intermediate type b2 + 1, or even a four-lane divided road, can often be adapted favorably to critical sites and to the landscape with little impact on the environment. This is due to the fact that the application of a two-lane road for a high-volume traffic would lead to greater requirements for sufficient passing sight distances and for selecting generous alignment design parameters. Generally, the quality of traffic flow is better on a four-lane divided road, at least with respect to passing and weaving maneuvers. Furthermore, structural separation by a median is especially favorable for traffic safety.

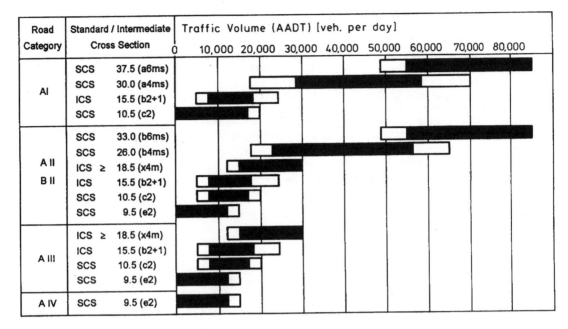

Road Category	Standard / Intermediate Cross Section		Traffic Volume (AADT) [veh. per day]

Note: ICS = intermediate cross section (Fig. 25.3).

SCS e2 and SCS d2 represent similar traffic volume ranges; therefore, SCS d2 is not listed.

FIGURE 25.2 Examples for the preselection of standard/intermediate cross sections. (Elaborated and modified based on Refs. 79 and 247.)

An interesting new approach for preselecting standard/intermediate cross sections was developed for the new German guidelines, 1996 edition.[79,247] Figure 25.2 illustrates the typical operational ranges for the possible preselection of standard/intermediate cross sections. The black sections of the bars correspond to the traffic volume ranges for which the standard/intermediate cross section is appropriate. Within this range the road is adequate for the corresponding traffic volume of the cross section, independent of restrictions caused by heavy truck traffic, longitudinal grade, and curvature change rate of the single curve. In the white sections of the bars, those restrictions may play a restrictive role. Note that Fig. 25.2 is related specifically to German traffic conditions; however, estimates about traffic volume ranges with respect to certain cross section types may be interesting for other countries, too.

After the preselection of the "Standard/Intermediate Cross Sections", as based on Table 25.1 or Fig. 25.2, a safety audit should be conducted to determine whether or not the selected cross section can be regarded as safe and sound for the given traffic and alignment conditions. In this connection, the safety information about the individual and combined design parameters given so far in the Parts "Alignment" and "Cross Sections," as well as the developed Safety Criteria can be taken as a basis. With respect to traffic quality issues the Highway Capacity Manual approach should be followed.[711]

25.1.2 Intermediate Cross Sections*

Recent investigations about the standard cross sections and their operational ranges, defined in Sec. 25.1.1, have revealed that it would be desirable to complete them by specific operational

---

*Elaborated based on Refs. 30, 68, 77, and 509.

ranges of so-called intermediate cross sections. The results of the investigations furthermore reveal that for the evaluation of roads, the guidance at intersections also has to be considered more seriously in the future (for example, at-grade with right-of-way, traffic lights, or grade-separated guidance). The same is true with respect to the operational mode, which means whether the road should be designated as exclusive motor vehicle road or if all traffic modes should be permitted as road for general traffic. Exclusive motor vehicle roads mean that only motor vehicles with speeds greater than 60 km/h (or, better, 80 km/h) are allowed to use the road.

Interstates will not be discussed in this chapter because they must be planned and designed not on a local but on a statewide level and because their cross sections should correspond, without exception, to the multilane standard cross section groups a or b, as shown in Fig. 25.1.

The following is a discussion of the recent knowledge for the sound design of nonbuilt-up roads that requires a somewhat different outlook as compared to the planning and design of the standard cross sections in Fig. 25.1.

This new knowledge could lead to considerable safety advantages, lower demand for space, smaller capital investment and maintenance costs, and less interference with nature and landscape. Figure 25.3 shows the recommended intermediate cross sections for future use.

The cross section types in Fig. 25.3 were evaluated on the basis of safety and traffic quality investigations, which are discussed in detail in Sec. 25.2.2.

In this respect, the three-lane b2 + 1 cross section and the four-lane divided x4m cross section should especially be noted, and they have been considered already in Fig. 25.2.

Intermediate cross sections bridge the gap between the normal capacity of a "regular" two-lane standard cross section (about 12,000 vehicles/day) and the minimum traffic volume of a four-lane divided standard cross section (about 25,000 vehicles/day).[30,68]

25.1.2.1   *Traffic Volume Ranges.*

For the appropriate selection of intermediate cross sections, the following recommendations, based on Fig. 25.4, are given:

- If the expected traffic volume is greater than 25,000 vehicles/day, then four-lane median-separated standard cross sections should be provided. Guidance at intersections should be "grade separated." The road has to be operated as exclusive motor vehicle road.

- For traffic volumes between 15,000 and 25,000 vehicles/day, either the intermediate cross-sectional type b2 + 1 (up to about 20,000 vehicles/day) or x4m is appropriate. The b2 + 1 cross section requires operation as an exclusive motor vehicle road. For roads with cross section b2 + 1, guidance at intersections should normally be grade-separated, in order to achieve an adequate quality of traffic flow, especially with respect to travel times.

- If the traffic volume is between 10,000 and 15,000 vehicles/day, the cross section type b2 + 1 may be adequate. However, note that this road type always has to be operated as an exclusive motor vehicle road, which often becomes difficult in this volume range. A grade-separated guidance at intersections increases traffic safety and guarantees a good connector quality (high travel speeds). At-grade design is also possible in this case.[68]

- For traffic volumes between 5000 and 10,000 vehicles/day, the application of intermediate cross sections is normally not desirable. An efficient solution in this range is the two-lane cross section type c2r with separate sidewalks and bicycle lanes (Figs. 25.3 and 25.4) or the standard cross sections c2, d2, and e2 (Figs. 25.1 and 25.2).

25.1.2.2   *Design and Traffic Guidance*

b2 + 1 Cross Section.   The b2 + 1 cross section has three lanes, with the center lane being assigned alternatively to both driving directions, allowing passing possibilities in the direction with two lanes (Figs. 25.3 and 25.5).

The cross section b2 + 1 is not only appropriate for new designs but also for redesigns because of its high safety level. This cross section can also be laid out through the rearrangement of road markings on existing roads.

c2r
(SCS: 14.0)
Road Category: A II, A III
V$_{perm}$ ≤ 90 km/h
AT Grade

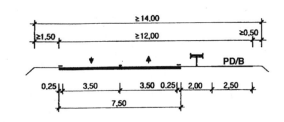

b2+1 (ICS: 15.5 / 14.0)
Road Category: A I, A II
A I: V$_{perm}$ ≤ 100 km/h
Grade Separated
A II: V$_{perm}$ ≤ 90 km/h
Grade Separated (AT Grade)

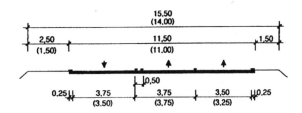

x4m (ICS: ≥ 18.5 / 17.0)
with Barrier (New-Jersey-
Type)
Road Category: A III, B II,
B III
A III, B II : V$_{perm}$ ≤ 90 km/h
B III: V$_{perm}$ ≤ 70 km/h
Grade Separated (AT Grade)

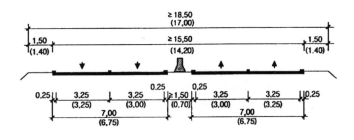

x4m with
Guardrail

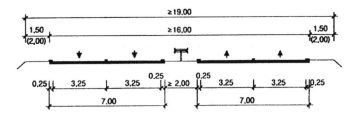

( ) exceptional values

All dimensions are in meters:
  ICS = intermediate cross section
V_{perm} = permissible speed limit, km/h
    b = cross section group with a basic lane width of 3.50 m without opposing traffic and of 3.75 m with opposing traffic (Table 24.1)
    r = additional sidewalk and bicycle lane
    x = designation for cross section not fitting directly to the previous systematics
    m = median
   PD = pedestrian
    B = bicyclist

FIGURE 25.3   Intermediate cross sections for nonbuilt-up roads. (Elaborated and completed based on Refs. 30, 68, and 77.)

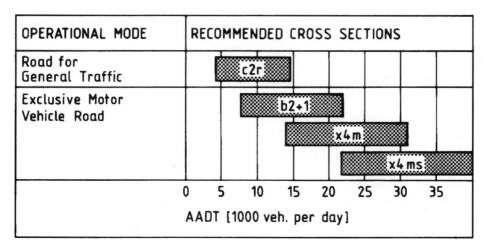

FIGURE 25.4   Traffic volume ranges for intermediate cross sections. (Elaborated based on Ref. 68.)

Important additional design features for the design of roads with b2 + 1 cross sections are described as follows:[30,68]

- Length of one- or two-lane sections
- Location of critical changes in cross section (merging two lanes into one lane)
- Characteristics at intersections and in narrow curves
- Design of tapers
- Road marking and signing

With respect to traffic flow, long two-lane sections (to break up vehicle platoons) and short one-lane sections (to decrease the chance of building up new groups of vehicles, and to minimize the chance of disregarding passing restrictions) are very favorable.

Section lengths.   Fundamentally, section lengths between 1000 and 1400 m have proven to be favorable. Section lengths of less than 800 m should not be used because of the required passing lengths. Section lengths of more than 2000 m should also be avoided because groups of vehicles build up on the one-lane section (Fig. 25.5). Generally, the individual section lengths should be designed in such a way that a vehicle group build-up on one-lane sections can be broken up on two-lane sections.[30,68]

Location of critical transition areas.[30,50,509]   With respect to traffic safety on roads with a b2 + 1 cross section, the design of road markings and signing in areas where the lanes are changing is especially important. Recommendations for marking and signing transition areas are shown in Fig. 25.5. In this figure, the transition from a two-lane section to a one-lane section is considered critical (merge area), but the transition from a one-lane section to a two-lane section is considered uncritical (diverge area). Transitions in the cross section should be clearly visible for perception and psychological reasons. Therefore, the taper inclination length should be short, that is, between 5 and 10 m. Critical tapers should be approximately 180 m long; if they are easily seen, they could be reduced to 160 m. Uncritical tapers should at least be 30 m long.

Another interesting approach for the layout of intermediate cross sections, which can also be recommended, is related to the "French Highway Design Guide"[700] (see Figs. 13.1 and 25.17). However, this layout is more expensive because of the considerably longer merging and diverging areas but may be safer, especially in hilly and mountainous topography.

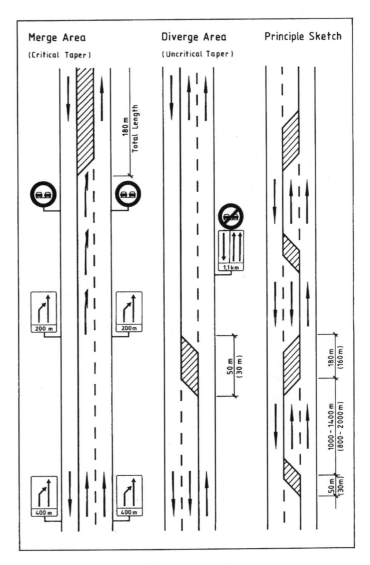

() exceptional values

FIGURE 25.5 Sketch of design and traffic guidance for b2 + 1 cross sections. (Elaborated based on Refs. 30, 50, 201, and 509.)

Route guidance signs and passing zone signs are especially important for driving at night or on wet and icy road surfaces. To reduce passing pressure, it has been proven successful to indicate at regular intervals on the one-lane section the start of upcoming two-lane sections.

If the two-lane sections are located on upgrade sections, a better separation of fast- and slow-moving traffic can be achieved by means of passing restrictions for trucks.

Where trucks make up 15 percent or less of traffic, section lengths of between 1000 and 1400 m are acceptable (Fig. 25.5); for higher truck percentages, shorter section lengths are recommended. However, often local limiting conditions, like the locations of upgrade sections as well as intersections, play a more important role for the allocation of areas than traffic-related requirements.

Therefore, the following conditions should be considered when determining transition allocations:

- Transition areas should be assigned in those places where a driver can oversee his or her own lane-changing action as well as those of other motorists, that is, in easily visible areas with generous alignment. Critical transition areas (changing from a two-lane to a one-lane section) have to be avoided in complex sections or sections that are hard to see.

- Transition areas should not be placed on slippery road sections, for example, bridges.

- In narrow curves, two-lane sections should be placed on the outer section of the curve. Otherwise, it would be safer to interrupt the three-lane section before the curve and to mark the center lane as a prohibited area. Also, a structural separation in the direction of travel could be considered in this case.

- On upgrade sections, the two-lane section should be assigned to the upgrade direction of travel. For determining the beginning and the end of the two-lane section, see Sec. 13.1.2. In areas of crest vertical curves, four-lane sections can offer good solutions. Structural separation of the directions of travel are recommended here.

- With regard to traffic safety, one-lane sections are recommended when entering a village, thus ensuring that at least the motor vehicles driving in groups enter the village at low speed levels.[30,50,509]

Intersections and curves.[30,68]

Grade-separated intersections.   A grade-separated guidance at intersections has proven to be efficient. This could be advantageous especially where long two-lane sections start with a lane addition in the intersection area (Fig. 25.6a). When an intersection has to be placed in an area of critical transition, a lane drop should be established prior to the intersection in order to avoid critical lane changing ahead of the exit area (Fig. 25.6b).

At-grade intersections.   Traffic should enter at-grade intersections on one lane (Fig. 25.6c).

In order to avoid a critical transition in an intersection area, a lane drop should start prior to the at-grade intersection (Fig. 25.6d). The passing lane of the b2 + 1 cross section is not allowed to run directly into the left-hand turn lane in the intersection area.

If sufficient capacity cannot be achieved by means of at-grade guidance and right-of-way rules indicated by traffic signs or if traffic safety would be impaired because of high approach speeds, then a traffic regulation scheme using traffic signals should be introduced. Of course, it would be better if grade separation could be provided.

Road signs and markings.[30,68]   Figure 25.7 shows the recommended design for lane drop and lane addition. Road markings between the driving lane of a one-lane section and the passing lane of the opposite direction always have to be in the form of double lines (also see Fig. 25.3). The end of a two-lane section has to be indicated by means of arrows on the final 160 m and by means of guidance signs at a distance of 200 m (if possible, also at 400 m) prior to a lane drop (see Fig. 25.7).

x4m Cross Section.   Guardrails and New Jersey–type concrete barriers would lead to a reduction in the median width needed on multilane roads.

Therefore, they could be used for traffic separation on existing roads or in the case of constraints (Fig. 25.3).

The intermediate cross section x4m is also recommended for supplementary redesign or rearrangement of road markings, based on its beneficial directional separation and the resulting improvement of traffic safety. This also applies if no more than the indicated exceptional values of the cross-sectional elements in Fig. 25.3 can be applied because of existing constraints.

Based on a lane width of 3.25 m with no provision for an emergency lane, a speed limit is necessary for this cross section type. Emergency pull-off areas have to be provided at appropriate intervals.[30,68]

25.1.2.3 Preliminary Conclusion.   Based on Refs. 30, 68, 69, and 259, the following conclusions could be drawn:

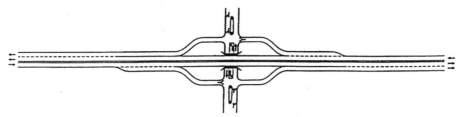

a) Grade-Separated Intersection for a Road with "b2+1" Cross
   Section, Combined with an "Uncritical" Transition

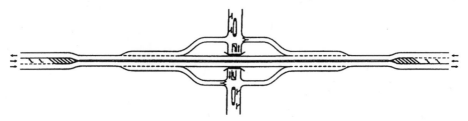

b) Grade-Separated Intersection for a Road with "b2+1" Cross
   Section, Outside of Transition Locations or in the Area of a
   "Critical" Transition

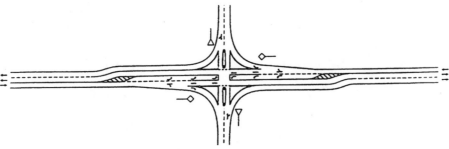

c) At-Grade Intersection for a Road with "b2+1" Cross Section,
   Combined with an "Uncritical" Transition

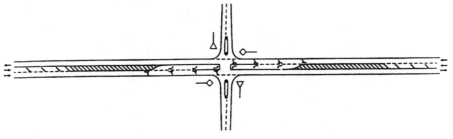

d) At-Grade Intersection for a Road with "b2+1" Cross Section,
   Outside of Transition Locations or in the Area of a "Critical"
   Transition

FIGURE 25.6 Design of b2 + 1 cross sections in intersection areas. (Elaborated based on Refs. 30 and 68.)

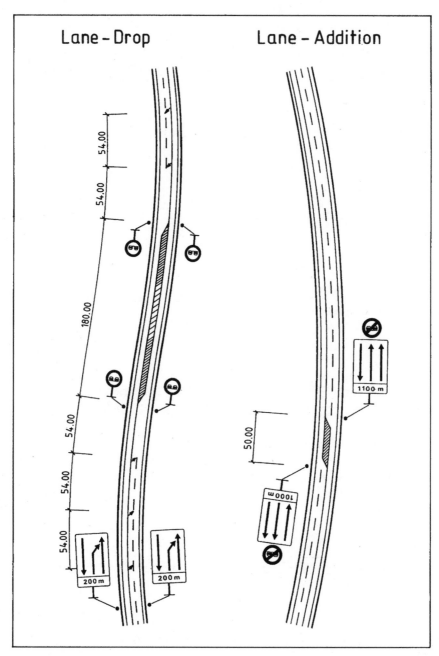

FIGURE 25.7 Example for the design of areas for lane drop and lane addition (cross section b2 + 1). (Elaborated based on Refs. 30 and 509.)

Intermediate cross sections bridge the gap between the capacity of regular two-lane rural roads (about 12,000 vehicles/day) and the minimum traffic volume on four-lane divided highways (about 20,000 vehicles/day). However, for future applications only the b2 + 1 and x4m cross sections are recommended (Fig. 25.3).

Note that both cross sections can only be used as exclusive motor vehicle roads. That means these road types cannot be used by pedestrians, bicyclists, agricultural vehicles, etc. For safety reasons, grade separations are recommended.

However, it should be pointed out that the capacity of a highway is affected not only by its cross section and alignment but also by the design of intersections (grade-separated, at-grade, signal controlled or not) and the type of highway operation (road for general traffic or exclusive motor vehicle road).

In conclusion, it can be stated that the intermediate cross sections b2 + 1 and x4m should be recommended for new designs, redesigns and restoration, rehabilitation, and resurfacing projects in order to close the gap with respect to the traffic quality and safety between regular two-lane standard cross sections and regular four-lane median-separated standard cross sections.

For this reason, these two intermediate cross section types were integrated into the framework of this book (Fig. 25.3).

Finally, based on the highway operation experience of road maintenance crews, no reason exists to question the application of the b2 + 1 and the x4m cross sections.[30,68,69,259]

25.1.3 Specific Cross-Sectional Issues*

25.1.3.1 Roadside Conditions.   Roadside conditions are another part of cross-sectional elements that affect crash frequency and severity. This is due to the high percentage of crashes, particularly on rural two-lane roads, which involve a run-off-the-road vehicle. Providing a more "forgiving" roadside that is relatively free of steep slopes and rigid objects would allow many of these vehicles to recover without having a serious crash.[729]

In this subsection, the design of standard side slopes is discussed. Accident experience related to roadside conditions in general is dealt with in Sec. 25.2.3.1.

Side slopes should be designed to ensure stability of the road and to provide an out-of-control vehicle with a reasonable opportunity to recover. Embankments and cuts should have a uniform slope for heights greater than or equal to 2.00 m (see Fig. 25.8).

The standard side slope is

$$1{:}n = 1{:}1.5 \tag{25.1}$$

For heights of less than 2.00 m, a constant side-slope width of w = 3.00 m should be applied instead of the standard side slope. In this way, the slope becomes more level as its height decreases (Fig. 25.8).

An alternative side-slope design can be considered with regard to:

- Soil mechanics
- Adaptation of the road to the topography
- Emission protection
- Avoidance of snowdrifts

If other side slopes $1{:}n$ are used, then the following equation for determining the width of the side slope, w, applies for side-slope heights of less than 2.00 m:[247]

$$w = 2\,n \tag{25.2}$$

For heights of side slopes greater than or equal to 2.50 m, guardrails should be installed for safety reasons. This is not necessary for side slopes of 1:5 or less, since the guardrail itself pre-

---

*Elaborated based on Refs. 247 and 454.

Height of Sideslope h	h ≧ 2.0 m	h < 2.0 m
Embankment	Foreslope 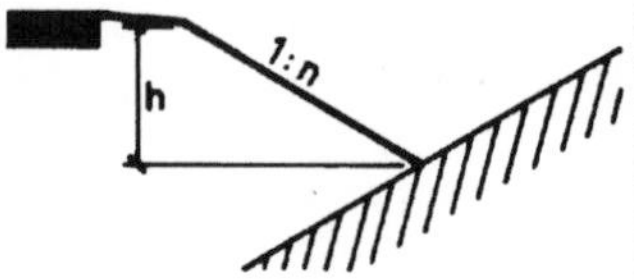 	Foreslope 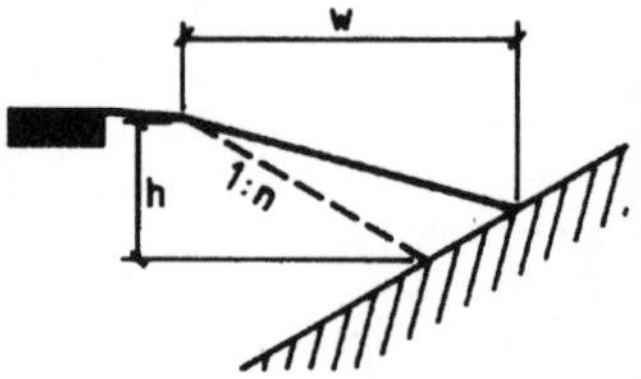
Cut	Backslope 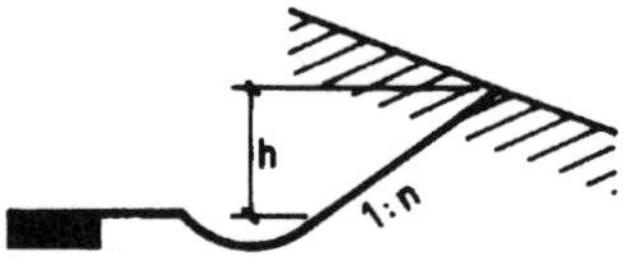 	Backslope 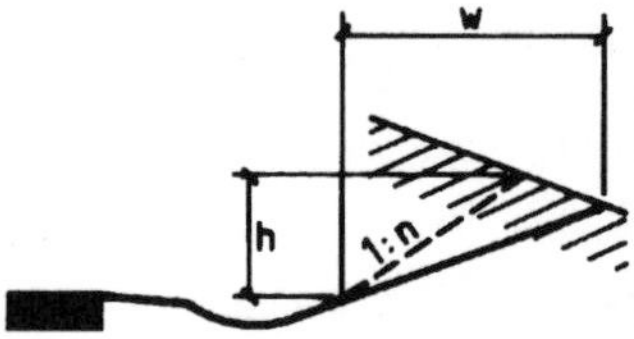
Standard Sideslope	1 : 1.5	w = 3.0 m
General Dimensions of Sideslope	1 : n	w = 2 n
Tangent Length of Rounding	3.0 m	1.5 h

* Quadratic parabolas are used for rounding.

FIGURE 25.8   Design of standard side slopes. (Elaborated based on Ref. 247.)

sents an obstacle which may be more dangerous than the side slope itself. The transition zone between the side slope and the original ground should be rounded by a quadratic parabola. The side-slope design according to Fig. 25.8 is more or less for countries with limited possibilities for land acquisition. As will be shown in Sec. 25.2.3.1, countries like Australia, Canada, and the United States use embankment slopes of up to 1:6 (see Table 25.5).

25.1.3.2 Superelevation Rates of the Cross-Sectional Elements.   On tangents, pavements should be provided with superelevation rates (cross slopes) of 2.5 percent for the drainage of surface water. Detailed information about superelevations on tangents and curves is discussed in detail in Secs. 14.1.2 and 14.2.2. Furthermore, minimum and maximum superelevation rates are presented for various countries worldwide in Table 14.4.

Emergency lanes, multiple purpose lanes, and paved shoulders should be provided with the same superelevation rates as the travel lanes. Unpaved shoulders where the pavement is drained should be sloped by 12 percent, while in other areas it should be 6 percent. Stabilized shoulders and outer separations should be sloped by 4 percent.

Sidewalks and bicycle lanes should be provided with superelevation rates (cross slopes) of 2.5 percent.[247]

25.1.3.3 *Facilities in the Cross Section.*

Traffic and protection measures, such as signing and traffic lights, guardrails, antiglare screens, lighting, and noise barriers, have to be located in such a way that clearance dimensions are observed (see Secs. 24.1.2 and 24.2.2).

Utility lines should be located at a considerable distance from the traveled way, paved shoulders, and drainage facilities. Utility lines in unpaved shoulders can be affected by measures to set up traffic signs or guardrails. Therefore, a minimum distance of 2.00 m has to be provided from the edge of the traffic space.

Trees should be planted at a minimum distance of 3.00 m from the edge of the traffic space. For roadway sections where the likelihood of run-off-the-road accidents is high, for example, on narrow curves of roads of category group A, the minimum distance should be at least 4.50 m. On existing roads, the danger from trees can be taken care of by using guardrails.

The inside areas of curves must be kept free from plantings to allow sufficient sight distances.[247]

25.1.3.4 *Drainage.*

It is reported in the French "Highway Design Guide":[700] The road must include storm-water collection and drainage systems. These systems must include water treatment in the more sensitive areas (around sources of drinking water, swimming areas, or fish breeding areas, etc.). Estimating the size of these systems is not always easy at the outset of a project, thus, it is advisable to reserve the necessary space in areas which seem most appropriate.

The choice and the design of these systems must be based on the rainfall data, the geometric and physical characteristics of the road, and the safety constraints.

In general the following issues should be considered:

- On embankments where side-slope erosion might occur, it is advisable to construct a longitudinal water collecting element on the top of the side slope which will channel the water down the side slope at suitable points.

- In cuts, water is collected and evacuated laterally through surface structures (channels, ditches, and gutters) along with, if required, subsurface drains. Techniques such as *catch drains* can also be used.

The safety considerations to be given priority are the following:

- The best solutions, from the safety point of view, are flat gutters or very shallow gutters, covered ditches or shallow ditches (20 cm deep or less), and channels (may be out of concrete when there is a frontage access), which may be linked to a subsurface internal drainage system.

- Ditches of moderate depth (less than 50 cm) or ditches with gentle slopes (roughly less than 25 percent) are acceptable.

- Deep ditches (more than 50 cm), except those with gentle slopes, and ditches at the foot of embankment slopes are to be avoided because of their detrimental effects on safety; if they cannot be avoided, plan to protect them with safety guardrails taking into account the conditions concerning lateral clearance.[700]

25.1.4 Additional Areas along the Roadside

25.1.4.1 *Pedestrian and Bicyclist Traffic.**

On roads for general traffic outside built-up areas, it is normal for pedestrian and/or bicycle traffic to use combined paths parallel to the traveled way without a discrete separation. However, safer solutions are provided if sidewalks and bicycle lanes are designed:

- As independent side paths
- Parallel to the traffic lanes separated by drainage areas or outer separations (Fig. 25.9).

As independent paths, sidewalks and bicycle lanes offer maximum traffic safety.

---

*Elaborated based on Refs. 247 and 454.

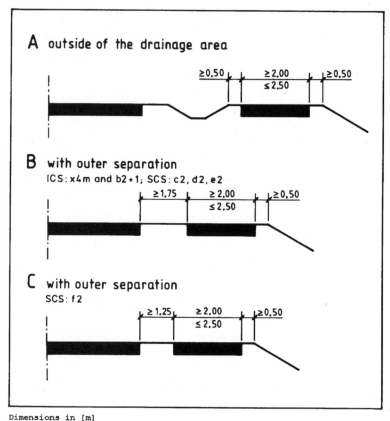

Dimensions in [m]

FIGURE 25.9   Combined sidewalk and bicycle lane.[247]

In addition, the recommendations for the design of bicycle lanes and sidewalks presented in Fig. 24.4 are valid.

Combined sidewalks and bicycle lanes have to be designed according to Fig. 25.9. On nonbuilt-up roads, the establishment of sidewalks and bicycle lanes beyond the drainage area offers significant advantages for pedestrian/bicyclist traffic (separation of traffic modes, no undesirable parking possibility, independent alignment, better adaptation to topography, preservation of the drainage area as a natural planting space, facilitation of winter service, and a lessening of glare danger).

For combined sidewalks and bicycle lanes, the operational ranges in Table 25.2 are valid. If the expected number of pedestrians (or bicyclists) exceeds the values in Table 25.2, then sidewalks or bicycle lanes or a combination of both should be provided. For lower operational ranges, more dangerous solutions may be considered, for example, the use of paved shoulders or of multiple purpose lanes (see standard cross section b2s in Fig. 25.1).

Combined sidewalks and bicycle lanes can be provided on both sides of the road. However, a combined sidewalk and bicycle lane on one side only is recommended as an economic solution.[247] See, for example, the intermediate cross section type c2r in Fig. 25.3.

25.1.4.2   Farm Traffic.   Farm traffic can be allowed on regular travel lanes, on multiple purpose lanes (standard cross section b2s, see Fig. 25.1), or on parallel paths. Guidance on parallel paths is necessary for exclusive motor vehicle roads, and is desirable, for safety reasons, on roads for the general traffic.

TABLE 25.2   Operational Ranges for Sidewalks and Bicycle Lanes[247]

Motor vehicle traffic, vehicles/day	Bicycle lane, bicycle/moped traffic (rush hour)	Sidewalk, pedestrian traffic (rush hour)	Combined sidewalk and bicycle lane, pedestrian and bicyclist traffic (rush hour)
<2500	90	60	75
2500–5000	30	20	25
5000–10000	15	10	15
>10000	10	5	10

Note: Rush hour is estimated to be 20 percent of the daily pedestrian and bicycle traffic.

If combined pedestrian and bicycle traffic is to be considered, and if farm and forest paths have to be developed beyond the side slopes, it is advisable to combine farm and forest traffic with pedestrian and bicycle traffic on separated parallel paths. The combined lateral paths can offer the maximum possible safety and capacity for all traffic participants and can be well adapted to road surroundings.[247]

25.2 GENERAL CONSIDERATIONS, RESEARCH EVALUATIONS, GUIDELINE COMPARISONS, AND NEW DEVELOPMENTS

Road cross sections, alignments, and intersections are essential elements of road characteristics, and together they influence traffic safety. Therefore, they have to be in harmony with each other.

Traffic flow safety depends on a number of influencing factors. Besides traffic volume and composition, the cross section design of the road is of special significance. Since with decreasing lane widths the moving spaces also decrease, the risk with regard to opposing and passing maneuvers increases if the speed is not reduced accordingly.

Therefore, cross-sectional features are very important for road characteristics. One of the strongest impacts on traffic safety is whether or not, for example, a road is separated by a median. A median separation produces an enormous positive effect on traffic flow and safety. For multilane median separated highways combined with grade-separated intersections, it was found that

- The personal injury accident rate was less than half
- The fatality rate was only a quarter

of the values corresponding to two-lane highways with at-grade intersections. Furthermore, with respect to safety, the b2 + 1 intermediate cross section also has significant safety advantages over the traditional two-lane rural standard cross sections, as will be shown in Sec. 25.2.2.1. Since with the b2 + 1 cross section a good traffic quality can also be achieved, this cross section should always be considered for the traffic volume ranges shown in Figs. 25.2 and 25.4 when selecting appropriate cross section types. These factors are of special interest when a decision has to be made, depending on traffic volume and economic design considerations, to select either a two-lane, three-lane, or four-lane divided standard (intermediate) cross section.

25.2.1 Standard Cross Sections*

In addition to the design elements of the horizontal and vertical alignment, which essentially result from the demands of operating speed consistency, driving dynamics, driving psychology, and sight

---

*Elaborated based on Ref. 153.

distance, the selected cross section also decisively influences traffic quality and traffic safety. For this reason, several of the investigated design guidelines apply to certain standard cross sections in individual road categories often based on road network functions. In addition, environmental demands and landscape protection should be considered in the selection and design of the cross section.

The lane width is determined by the maximum assumed design vehicle width (normally 2.50 m), the lateral moving or safety space and the width increase for opposing traffic, as well as consideration of lateral obstacles. Normally, these conditions lead to travel lane widths of between 3.0 and 3.75 m, depending on the cross section type. The width of the traveled way depends on the number of travel lanes and the edge strips (see Figs. 24.1 to 24.3).

In addition, paved shoulders are often provided for stopping in case of emergencies, as well as on two-lane rural roads for pedestrian bicycle traffic; if not, solutions according to Fig. 25.9 are provided. At all cross sections unpaved shoulders are provided, on which, among other items, marker posts and guardrails can be positioned. The traveled way and shoulders, as well as any existing medians, form the roadway width (roadbed) (see Fig. 25.10). Furthermore, the drainage area and the side slopes are also part of the road structure.

25.2.1.1   Types of Cross Section in Various Guidelines.

The guidelines investigated did not always give information on the design vehicles, lateral moving spaces, and opposing traffic width increases on which the geometric design of the cross section is based.

For a comparison of guidelines in different countries, the following four basic types of cross sections on rural roads could be differentiated (see Fig. 25.10):

- A four-lane cross section with structural separation
- A four-lane cross section without structural separation
- A two-lane cross section with paved shoulders
- A two-lane cross section without paved shoulders[153]

For most countries, cross section types were found with features corresponding qualitatively to the ones shown in Fig. 25.10, and most guidelines consider travel lanes, including edge strips as traveled way or traffic area (see also Figs. 24.1 to 24.3).

25.2.1.2   Comparison of Cross Section Types.

In all guidelines investigated, lane width is considered as the most important design element of the road cross section. Lane widths normally vary—disregarding overwide cross sections—between 2.75 and 3.25 m for roads with low traffic volumes and between 3.50 and 3.75 m for roads with high traffic volumes.

Opposing traffic width increases are considered only in certain countries, where they normally amount to 0.25 m for each relevant traffic lane.

In some countries, only cross sections with paved shoulders are used, whereas in other countries they are absent, at least in connection with standard cross sections. However, paved or unpaved shoulders are not, in all cases, adequate for safe stopping of motor vehicles or for emergency situations.

In contrast to the already far-reaching uniformity in traffic lane widths, roadway widths in individual countries still show considerable differences.

Larger roadway widths are more likely in countries featuring a predominantly flat topography. Countries with more mountainous topography or long travel distances seem to have narrower roadway widths. Standardization in this area would be desirable at least in Europe or within a specific country. Even though the roadway width does not have an influence on the purely technical driving maneuvers on the road, it is of significance for the perception of the road by the motorist.

With a few exceptions, four-lane cross sections without structural separations (Fig. 25.10) are no longer recommended in most guidelines because of the extremely low safety level of this road type due to head-on accidents as a consequence of passing maneuvers.[153]

With regard to three-lane cross sections, it should be noted that this type is being used more and more in several countries.

However, three-lane roads should be regarded as an independent cross section type. For example, in France there are already more than 3000 km of this road type. Recent research results and practical experiences regarding three-lane cross sections are reported in Secs. 25.1.2 and 25.2.2.

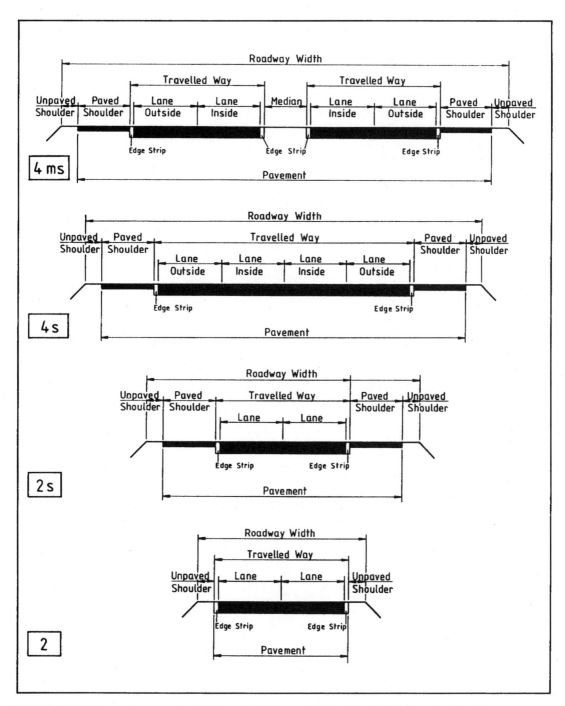

FIGURE 25.10 Cross section types and their characteristic elements. (Elaborated and modified based on Ref. 153.)

Based on the previous discussions, analyses, and evaluations, the standard cross sections (SCS) given in Fig. 25.1 were selected as "Recommendations for Practical Design Tasks." Based on our own experience and on Refs. 247 and 454, the selected standard cross sections are considered as sound. Note that other combinations between road categories and standard cross sections may also be possible to a certain extent, based on the kit-building principle of Part 3, "Cross Sections."

With regard to SCS, b2s and SCS b2 (Fig. 25.1), it should be considered that the favorable operational advantages (such as suitability for slow-moving vehicles or for a high volume of truck traffic) may be affected by safety disadvantages according to Sec. 25.2.2.1.

25.2.2   Intermediate Cross Sections*

With regard to capacity and safety of rural roads, most experts agree that a ratio of 3 to 1 between four-lane and two-lane roads is valid. In addition, the level of service on four-lane roads is higher. They allow passing maneuvers, thus reducing the risk of head-on collisions (in cases of structural separation) and make it easier to attain a desired travel speed. Both road types differ considerably with regard to the required right-of-way and construction costs.

The following considerations are aimed at testing cross sections that require less right-of-way and lower construction costs than regular four-lane roads but provide a greater degree of safety and a higher level of service than conventional two-lane roads.

Within the scope of the study "Application of Intermediate Cross Sections," conducted by Brannolte, et al.,[68] which started with an analysis of the available international experiences and relevant guidelines, a survey was made of three-lane type road sections with a length of 3 km or more in Germany (West). The three-lane roads (intermediate cross sections) studied had roadway widths between 14 and 22 m. Accident and traffic volume (AADT) data were considered.

At a later stage, standard cross sections of two-lane and four-lane roads were also included in the survey. This was necessary to allow a more comprehensive consideration of the safety level of three-lane road sections in order to compare their respective safety characteristics with those of other cross sections.

In addition, test roads were selected to allow a "before" and "after" comparison of accidents and traffic flow characteristics of the various types of cross sections. The key question of the study was whether or not, and under what conditions, the so-called b2 + 1 cross section should be considered in the future (Fig. 25.3). The b2 + 1 cross sections have three lanes with the center lane assigned alternately to both driving directions.[68]

25.2.2.1   Traffic Safety and Traffic Flow.   In order to evaluate traffic safety on different cross section types (Fig. 25.11), an analysis of the severity and number of accidents was conducted. In respect to this, adjusted accident cost rates (ACRa) were calculated. For each group of cross section types, the number of fatalities, as well as serious and light injuries, were noted. The results are shown in Fig. 25.12:[30,69]

- The adjusted accident cost rates (ACRa) for the individual cross section groups shown in the top part of Fig. 25.12 describe the safety level of road cross sections with regard to the severity and number of accidents. Also noted in Fig. 25.12 are left turn (LT) and right turn/crossing (RT/C) accidents. These reflect the accident types on at-grade intersections.

- The accident rates (AR) for accidents involving personal injury and heavy property damage (middle part) denote the frequency of accidents for the individual road groups for the investigated categories, based on the same exposure (vehicle kilometers traveled) without considering accident severity.

- The relevant average accident severity (AS) of accidents involving personal injury is shown in the bottom part of Fig. 25.12, as an adjusted cost estimate per accident (ASa).[168]

---

*Elaborated based on Refs. 30, 50, 68, 69, and 259.

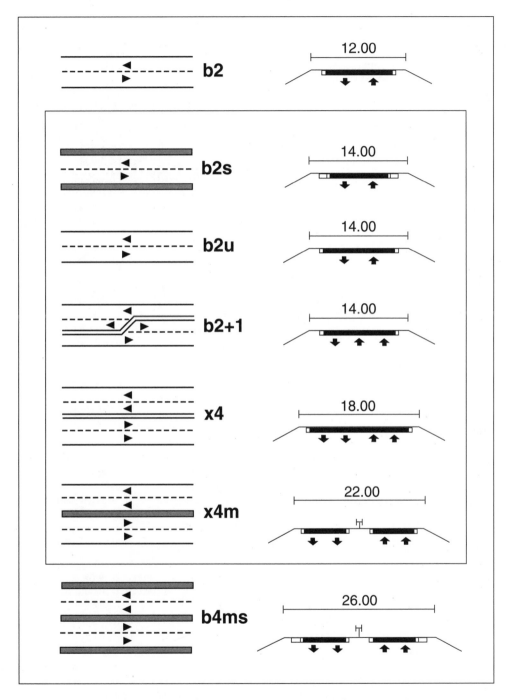

FIGURE 25.11   Intermediate cross sections (roadway width = 14 to 22 m).[68] (*Note:* For a definition of symbols, see Figs. 25.1 and 25.3. The letter *x* means that cross sections with varying lane width are included.)

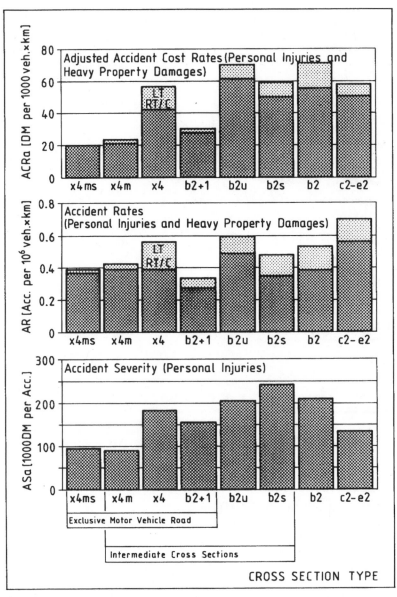

Legend:

    a = adjusted
    LT = Left Turn Accident
    RT = Right Turn Accident
    C = Crossing Accident
    DM = German Marks

FIGURE 25.12 Evaluation of traffic safety for intermediate cross sections compared to standard cross sections. (Elaborated based on Refs. 68 and 69.) (*Note:* For a definition of symbols, see Figs. 25.1, 25.3, and 25.11.)

Figure 25.12 leads to the following conclusions:

- Four-lane cross sections with structural directional separation (x4ms and x4m) offer the highest safety level. The four-lane cross section x4ms with paved shoulders is slightly safer than the four-lane cross section x4m without paved shoulders. On the other hand, the four-lane cross section x4m with structural separation is much safer than the four-lane cross section x4 without structural separation.

  Without considering intersection-related accidents LT and RT/C, the accident rates are the same for both cross section types (x4m and x4). On the contrary, the safety difference results exclusively from the nearly double-accident severity of accidents involving personal injury (bottom part of Fig. 25.12). For instance, the number of fatalities per 100 accidents with personal injury is, on roads with four-lanes of the type x4 without structural separation, more than twice as high as that of the type x4m with structural separation protected by guardrails or concrete barriers (Fig. 25.3).

- The cross sections b2s and b2u,* which have the same pavement width (Fig. 25.11), are about twice as unsafe as the cross section type b2 + 1. This is related to the larger accident frequency (higher accident rates) as well as to the increase in severe accidents involving personal injuries. For this reason, the cross section type b2s is shown in parentheses in Fig. 25.1. This means that the highway engineer should be aware of the fact that by using this cross section type, for example, in the case of slow-moving (farm) vehicles, a low safety level should be expected. Furthermore, the cross section type b2u* should no longer be applied.

- With respect to lack of traffic safety, the standard cross section b2 (see Fig. 25.12) corresponds to the wider cross sections b2s and b2u. This cross section is also shown in parentheses in Fig. 25.1, since research investigations have revealed that two-lane rural roads with a 7.0- to 7.5-m pavement width and a 10.0- to 10.5-m roadway width, such as the standard cross sections c2 and d2 in Fig. 25.1, perform better in terms of adjusted accident cost rate and adjusted accident severity than the standard cross section b2. Furthermore, in Fig. 25.12 all roads with cross sections narrower than b2 are combined in the group c2–e2. It is evident that this group can be used without affecting the safety level negatively in comparison to the cross section b2 (8.0-m pavement width).

- The favorable results of the narrower cross sections result first of all from the lower accident severity. A possible reason for the higher accident severity of the b2 cross section could be related to the wider cross section width, which in turn encourages higher speeds and more passing maneuvers that may not always be completed in time. However, it should be noted that considering the accident rate, the b2 cross section performs better than the cross section group c2–e2.

In any case, cross section types b2s and b2 may lead to an increased accident severity. Therefore, the highway engineer should consider this fact when using these types of cross sections.[30,68,69]

Note that the intermediate cross section b2 + 1 (Fig. 25.12) reveals significantly better safety results than all the other two-lane cross sections in terms of the adjusted accident cost rate and the accident rate.

Further advantages of the cross section b2 + 1 are

- Based on the possibility of systematic passing maneuvers, not only could improvements in traffic safety be noted but also improvements in the quality of traffic flow and acceptance by road users.[68,509]

- The need for sufficient passing sight distance, as required for all other two-lane cross sections, is not needed here, since passing maneuvers are possible at low risk.[30,68,69]

---

*b2u refers to two-lane cross sections for which no delineations exist between traffic lanes and the paved shoulders, so that the width corresponds to the cross section type b2s (see Fig. 25.11).

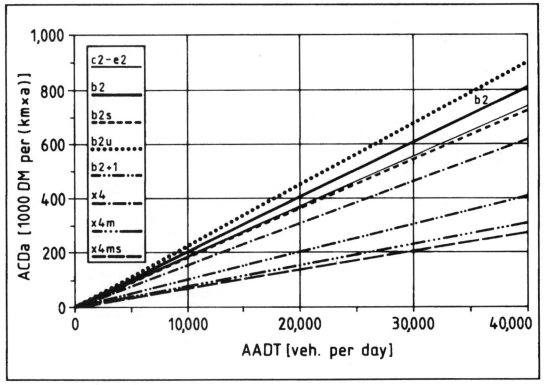

Legend: a = adjusted

FIGURE 25.13   Adjusted accident cost densities (ACDa) for roads with different cross sections (without the left-turn and right-turn/crossing intersection types).[68]

The comparison based on Fig. 25.13 also reveals that the intermediate cross section b2 + 1 is significantly more favorable than all other two-lane intermediate and standard cross sections and the four-lane undivided cross section type x4. The best results are achieved by the four-lane intermediate cross section types x4m and x4ms. In this connection, it can be stated that the presence of paved shoulders leads to an accident rate approximately 25 percent lower and to an accident cost rate approximately 10 percent lower.

With respect to traffic flow, it can be stated:

- In comparison to the b2 + 1 cross section, the speed behavior on the other two-lane intermediate cross sections (b2s and b2u) shows hardly any difference.
- Paved shoulders or overwide lanes have a favorable effect on breaking up groups of vehicles. However, critical passing maneuvers that interfere with or endanger traffic in the opposite direction are frequently observed. The b2 + 1 cross section is also very suitable for breaking up groups of vehicles, and in addition for avoiding critical driving situations.
- Based on existing speed-volume relationships, no exact information about the capacity of b2 + 1 cross sections could be obtained. The traffic volumes on the investigated sections are limited by the capacity of the adjacent sections with standard cross sections.

A survey taken among drivers on several b2 + 1 cross sections revealed a high acceptance level of this cross section type.

25.2.2.2 Design Alternatives. In connection with the present section, the interested reader may also consult Secs. 13.1.2 and 13.2.2 "Auxiliary Lanes on Upgrade Sections."

The design and operation of intermediate cross sections in selected countries is dealt with in the following.

Canada.† Passing lanes are also known as *three-lane highways* or *intermediate cross sections* of the type b2 + 1.

As reported by Frost and Morrall, a *passing lane* in Canada is defined as an auxiliary lane provided on a two-lane highway to enhance passing opportunities. Passing lanes are distinct from climbing lanes which are used in hilly or mountainous terrain. The distinction is that climbing lanes are provided to allow faster vehicles to pass slower vehicles on particular upgrades, whereas passing lanes are provided to increase passing opportunities on extended sections of highway. Passing lanes offer a low-cost, environmentally friendly alternative to major reconstruction of two-lane roads or "twinning" to a four-lane divided standard. Passing lanes are used as an intermediate level of road improvement on highway sections which may not warrant four lanes but exhibit deteriorating levels of service and periods of reduced speeds, increased time spent traveling in platoons, and passing demands which exceed passing opportunities. Depending on terrain and traffic composition, passing lanes may offer a low-cost alternative to major reconstruction or "twinning" on roads with traffic volumes of 2500 AADT to a maximum of 10,000 AADT.[201]

A number of sections of two-lane highways with wide paved shoulders in western Canada have been retrofitted with passing lanes. Two typical design examples including the operational characteristics are shown in Fig. 25.14. The retrofitting, which has resulted in a system of passing lanes on the Trans-Canada Highway in the Mountain National Parks, was accomplished by shifting the center line 1.8 m. The resulting lane configuration provides two 3.7-m lanes, one 3.6-m lane, and 1.2-m paved shoulders. If the highways were to be resurfaced, in mountainous terrain a passing lane section would be expanded to provide a minimum desirable shoulder width resulting in a cross section of 15.6 m, including concrete barriers (Fig. 25.14).

With respect to climbing and passing lanes, it is reported[334] that eight of 10 highway systems in Canada have climbing and passing lanes for a total of approximately 1700 installations of climbing and passing lanes reported in Canada. For example, on the Trans-Canada Highway, this highway type represents approximately 52 percent of the system length.

The passing lane system used on the Trans-Canada Highway is shown in Fig. 25.15. One interesting difference between Canada and Germany is that in Germany the high-speed lane must merge into the slower-speed outside lane (Fig. 25.5), while Fig. 25.15, which is for Canada, indicates that drivers must show a degree of common sense and cooperation when merging. Many Canadian highway systems place the onus on the driver in the outside (slow) lane to give way to drivers in the passing lane. Properly designed and signed merge areas have not been found to be high accident locations in either Canada or Germany.[201]

Figure 25.16 shows various configurations of passing lanes used in Canada.

According to the "Design Guidelines for Passing Lanes" developed in Alberta, Canada, the following criteria for location and spacing of passing lanes are to be considered:[11,526]

1. Avoid highway sections with intersections to prevent confusion with turning bays. Otherwise an intersection treatment is required.

2. Avoid costly physical constraints such as bridges and culverts.

3. Improve areas of traffic congestion. Passing lanes in close proximity to four-lane sections or downstream from climbing lanes are not particularly effective in improving the overall level of service.

4. Space lanes at appropriate intervals to relieve driver's frustration.

5. If possible, locate solid barrier line segments to achieve an increase in net passing opportunities.

---

*Elaborated with the support of C. Schulze based on Ref. 637.

†Elaborated based on Refs. 11, 201, and 526.

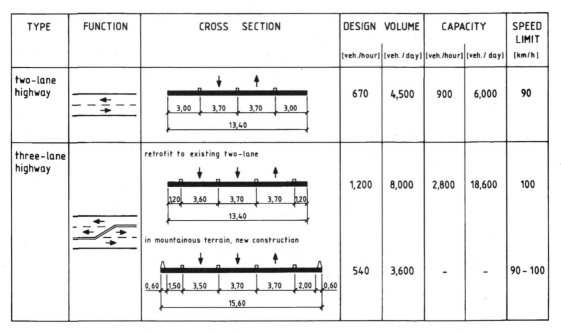

TYPE	FUNCTION	CROSS SECTION	DESIGN VOLUME		CAPACITY		SPEED LIMIT
			[veh./hour]	[veh./day]	[veh./hour]	[veh./day]	[km/h]
two-lane highway		3,00 3,70 3,70 3,00 13,40	670	4,500	900	6,000	90
three-lane highway		retrofit to existing two-lane 1,20 3,60 3,70 3,70 1,20 13,40	1,200	8,000	2,800	18,600	100
		in mountainous terrain, new construction 0,60 1,50 3,50 3,70 3,70 2,00 0,60 15,60	540	3,600	–	–	90 – 100

FIGURE 25.14   Retrofitting of a two-lane highway into a three-lane highway and operational characteristics, Canada.[201]

6. Generally, passing lanes should not exceed about 25 percent of the highway section length for each direction of travel.[11,526]

Based on the safety analysis presented in Ref. 201, passing lanes in Canada are considered to provide safety benefits both upstream and downstream in addition to along their length. Available accident data in Canada suggest that passing lanes change the distribution of accident severity by reducing the number of collisions within its effective length.

France.[700]   Special consideration is devoted in France to roadway sections with passing capabilities when traffic volumes are near congestion levels and especially when sight distances are inadequate. Passing capabilities can be provided in France either by adding one extra lane, thus upgrading the two-lane highway to a "2 + 1" type of cross section or by adding two extra lanes, one in each direction of travel, separated by a median (a kind of "2 + 2" type of cross section). The layout of both types of cross section common in France is illustrated in Fig. 25.17.

The optimal length of passing lane segments is normally considered to vary between less than 1000 and 1250 m in flat terrain. This corresponds well to the design of Fig. 25.5. Depending on the traffic volume and the existing alignment, the spacing of passing lanes should not be less than 4 to 5 km. This applies especially in cases e and f of Fig. 25.17, since more dense passing lanes supplied with a median may give the drivers the false impression of moving on a freeway.

Case a of Fig. 25.17, in which the passing lane can be used simultaneously in both directions of travel, should be the exception rather than the rule.

As indicated by other countries' experience, French studies show that road sections with "2 + 1" type cross sections, based on cases b to d of Fig. 25.17, are not more dangerous than two-lane roads, assuming that each lane is not narrowed down to 3 m, a visibility of over 500 m can be guaranteed and certain precautions are being taken in the crossing layout and the treatment of shoulders.[571] On average they are less dangerous!

However, it is important:[700]

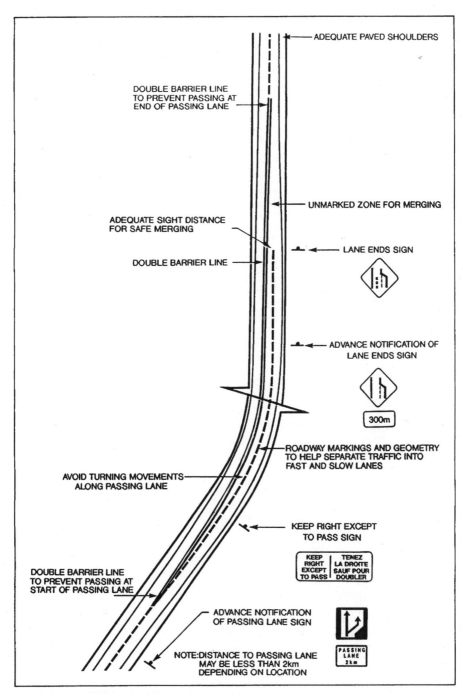

FIGURE 25.15   Signing and marking for the Trans-Canada Highway Mountain National Parks passing lane system (AADT > 4000 vehicles/day).[11,201]

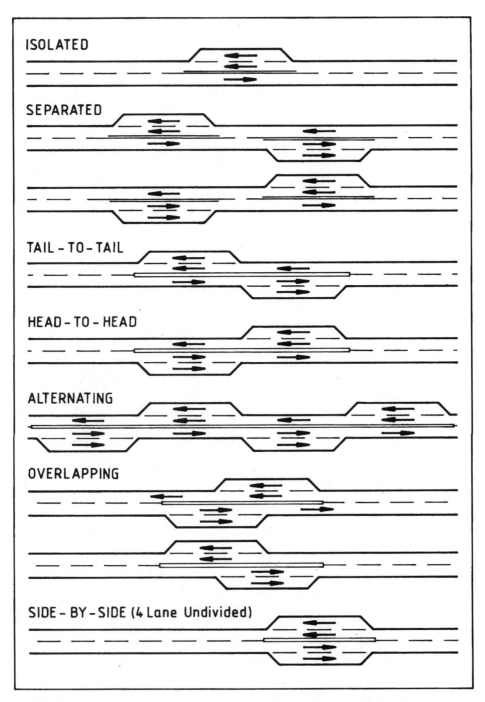

FIGURE 25.16   Alternative configurations for passing lanes, Canada (AADT > 4000 vehicles/day).[11,201]

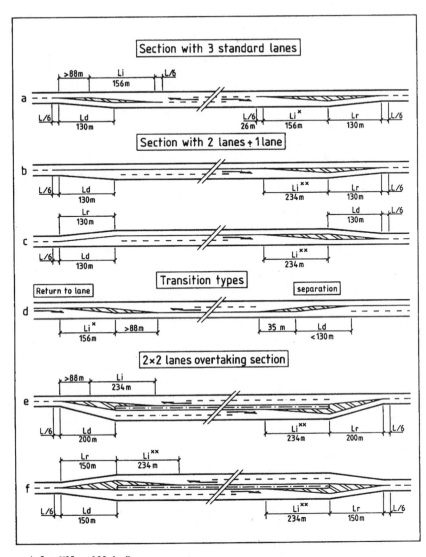

* for V85 = 100 km/h,
** for V85 = 120 km/h.

Legend:

Li   = merge taper length (lane drop) [m],
Ld   = diverge taper length (lane addition) [m],
Lr   = lateral displacement taper length [m],
Ld, Lr = $(4,000 + 3,600 \cdot d)^{1/2}$,
d   = lateral displacement [m].

Note: The lengths shown in the sketches above correspond
       to highway displacements of 3.50 m wide.

FIGURE 25.17 Typical layouts for intermediate cross sections, France.[700] (*Note:* See also Fig. 13.1.)

- To respect the precautions concerning intersections (Fig. 25.6) or frontage access
- To respect the conditions of safety in terms of the shoulders and rigid obstacles
- Not to reduce the lane width to 3.0 m (in the normal case, the lane width is 3.5 m)
- To assign the extra lane as soon as the overtaking visibility is less than 500 m (it can also be planned along the route in systematic distances)
- Not to make the allocation of the extra lane to one direction of traffic longer than 1200 m, excluding the merge taper section (a distance less than 1 km is generally recommended)

Typical layouts for intermediate cross sections in France are shown in Fig. 25.17. At the beginning or end of the hatched markings, a continuous line with a minimum length of $L/6$ is always provided.[700]

Norway.   In Norway it is proposed[653] to coordinate overtaking requirements with traffic volume, allowing overtaking possibilities every 5 km, as follows:

AADT ≤ 1,500 vehicles/day: 1 overtaking possibility every 5 km

1500 < AADT ≤ 5000 vehicles/day: 2 overtaking possibilities every 5 km

AADT > 5000 vehicles/day: 3 overtaking possibilities every 5 km

Passing lanes are considered favorable when AADT is 10,000 vehicles/day or more and passing sight distances between 300 and 450 m are not available.

In many cases, a passing lane is a better solution than sight distance improvements in terms of both cost and traffic considerations. The road is then built as a three-lane road, with the center lane marked for passing (Fig. 25.18). The driving direction on the center lane should alternate between the different stretches of the road. Passing lanes should have the same width as the other lanes.

Passing lanes improve traffic flow and traffic safety. Nonetheless, a passing lane should not be used in the place of a four-lane road. Undesirable situations can easily arise when traffic is heavy, especially at the beginning and the end of the passing lane segments.[653]

Typical layouts with traffic signs and road markings for passing lanes in Norway are shown in Fig. 25.18.

United States.*   In the United States, intermediate cross sections in the sense described in Sec. 25.1.2 are not directly defined. Instead, multilane design alternatives have been used and practiced to address the considerable safety and operational problems existing on some higher-volume two-lane highways.

On the topic of cross sections with three lanes, AASHTO[5] reports the following.

An added lane can be provided in one or both directions of travel. They are provided to improve traffic operations at bottlenecks and to reduce delays caused by inadequate passing opportunities over significant lengths of two-lane highways. The location of the added lane needs to appear logical to the driver. The value of it is more obvious at locations where passing sight distance is restricted than on long tangents which provide passing opportunities. On the other hand, the selection of a site should recognize the need for adequate sight distance at both the lane addition (diverge) and lane drop (merge) tapers. A minimum sight distance of 300 m (1000 ft) on the approach to each taper is recommended.

A minimum length of 300 m (1000 ft), excluding tapers, is needed to ensure that delayed vehicles have an opportunity to complete at least one pass in the added lane. The optimal length is usually 800 to 1600 m (0.5 to 1.0 mi). Added lanes have to be designed systematically at regular intervals and alternatively for each direction of travel. A hard shoulder at least 1.2 m (4 ft) wide is necessary on the passing lane section and, whenever possible, the shoulder width should match that of the adjoining two-lane highway. The transition tapers at each end of the section should be designed to encourage safe, efficient operation.

The lane-drop taper length should be computed from the formula

---

*Elaborated based on Refs. 5 and 729.

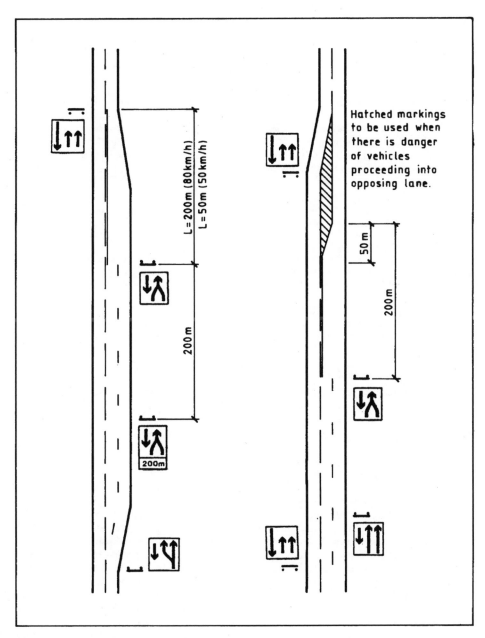

FIGURE 25.18 Typical layouts for passing lanes, Norway.[653]

$$L = \frac{W \cdot V}{1.6} \tag{25.3}$$

where L = length, m
   W = width, m
   V = speed, km/h

The recommended length for the lane addition taper is one-half to two-thirds of the lane-drop length.[5]

In the state of California, the length of a passing lane is a function of volume, with the shortest recommended length being 800 m and the maximum length being 2400 m. Diverge tapers are 76 m and merge tapers 183 m in length. Hard shoulders vary between 0.60 and 2.40 m.[528]

In addition, the FHWA study in 1992[729] comments upon multilane design alternatives in the United States:

> A majority of two-lane highways carry relatively low traffic volumes and experience few operational problems. However, considerable safety and operational problems exist on some higher volume two-lane highways. Such problems are often due to inadequate geometry (steep grades, poor sight distance), the lack of passing opportunities (due to heavy oncoming traffic and/or poor sight distance), or turns at intersections and driveways. While a major reconstruction project may be used to reduce the problem (e.g., widening to a four-lane facility or major alignment changes), other lower-cost alternatives have been used successfully to reduce accident operational problems.[265]

As illustrated in Figure 25.19, a 1985 study by Harwood and St. John evaluated the following five different operational and safety treatments as alternatives to basic two-lane highways:[262]

1. Passing lanes

2. Short four-lane sections

3. Shoulder use sections (that is, shoulders used as driving lanes)

4. Turnout lanes (a widened, unobstructed area on a two-lane highway allowing slow vehicles to pull off through lane to allow other vehicles to pass)

5. Two-way, left-turn lanes (TWLTLs)[729]

The nearest case with respect to intermediate cross sections ("2 + 1" type), as understood here, is represented by Fig. 25.19 (alternative 1).

Alternatives 2 to 5 are discussed by AASHTO, as follows.

Short four-lane sections have to be regarded as a special case of passing lanes, since they are developed by providing a passing lane simultaneously in both directions of travel. They are designed to meet the desired frequency of safe passing zones or to eliminate interference from low-speed heavy vehicles, or both. Short four-lane sections are particularly advantageous in rolling terrain, especially where alignment is winding or the profile includes critical lengths of grade.[5] The desired length of short four-lane sections is 1600 to 2400 m, since this length is sufficiently long to dissipate most queues formed, depending on volume and terrain conditions.[711] Sections longer than 3.2 km may mislead drivers to perceive the section as a four-lane highway.

Shoulder use sections, when applied, are covered by legislation and are indicated by special signing. On such sections slow-moving vehicles are permitted to use the paved shoulder only long enough for the following vehicles to pass and then return to the through lane. Shoulder use sections should be 300 to about 5 km long and should be provided with shoulders of at least 3.0 m wide.[5]

Turnouts are constructed as widened, unobstructed shoulders where a driver of a slow-moving vehicle is expected to pull out of the through lane and remain in it only long enough for the following vehicles to pass before returning to the through lane. Turnouts are suitable on low-volume roads, in mountainous terrain, or in coastal and scenic areas where large trucks and recreational vehicles exceed 10 percent of the vehicle volume. Prerequisites for the provision of a turnout are

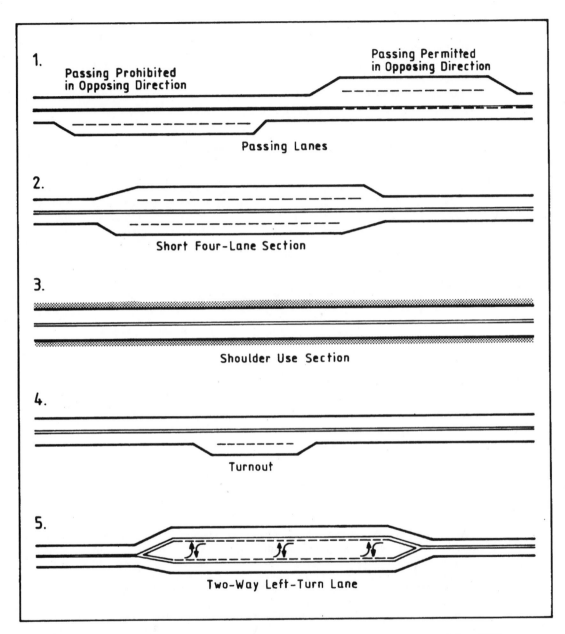

FIGURE 25.19 Typical multilane design alternatives compared to a basic two-lane rural highway, United States.[262,729]

1. A minimum sight distance of 300 m in each direction
2. An available width from the edge of the traveled way of 4.8 m
3. A firm, smooth surface
4. A minimum length, including taper, ranging from 60 m at a 40-km/h approach speed to 165 m at a 95-km/h approach speed
5. A maximum length of 180 m to avoid use of the turnout as a passing lane
6. A minimum width of 3.6 m with a desirable width of 4.8 m[5]

Two-way, left-turn lanes (*TWLTL*) are found in the United States on two-lane highway sections in urban and suburban areas with high left-turn volumes.[543] All vehicles turning left from both directions of travel can use the TWLTL. Where TWLTL are constructed, passing is prohibited and the design speeds of the section range from 40 to 80 km/h.

With respect to the typical operational treatments used on two-lane rural highways in Fig. 25.19, accident reductions due to making such design improvements are given in Table 25.3. According to this table, the provision of passing lanes, short four-lane sections, and turnout lanes led to accident reductions of 25 to 40 percent. Two-way, left-turn lanes were found to reduce accidents by approximately 35 percent in suburban areas and from 70 to 85 percent in rural areas. No known accident effects were found for shoulder use sections.[262,264]

TABLE 25.3 Accident Reductions due to Two-Lane Highway Improvements in the United States[262,264,729]

Multilane design alternative	Type of area	Percent reduction in accidents	
		Total accidents	F + I accidents
Passing lanes	Rural	25	30
Short four-lane section	Rural	35	40
Turnout lanes	Rural	30	40
Two-way, left-turn lane	Suburban	35	35
Two-way, left-turn lane	Rural	70–85	70–85
Shoulder use section	Rural	No known significant effect	

Note: F + I = fatal plus injury accidents. These values are only for two-lane roads in rural or suburban areas.

As noted in Ref. 729, the reader should use caution regarding the accident effects of the design alternatives in Fig. 25.19, since accident experience may vary widely depending on the specific traffic and site characteristics. Also, while such alternatives may reduce some safety and operational problems, other problems may be created in some cases. For example, at rural locations where passing zones exist, using TWLTLs can create operational problems with respect to same-direction passing maneuvers.[729]

Other Countries. Specific designs of intermediate cross sections or passing lanes have a tradition in a number of countries, for example, Australia,[37,525] Switzerland,[142,570,693] and the United Kingdom.[139]

In Australia many passing lanes have been built since the 1970s as a result of increased efforts to improve operation and safety on two-lane rural highways, especially high-volume roads, by low-cost measures. An example of a typical passing lane in Australia is depicted in Fig. 25.20. According to Morrall and Hoban,[525] the passing lane has been divided into five zones. Proceeding from upstream to downstream in the direction of traffic flow on the lane, the five zones are as follows:

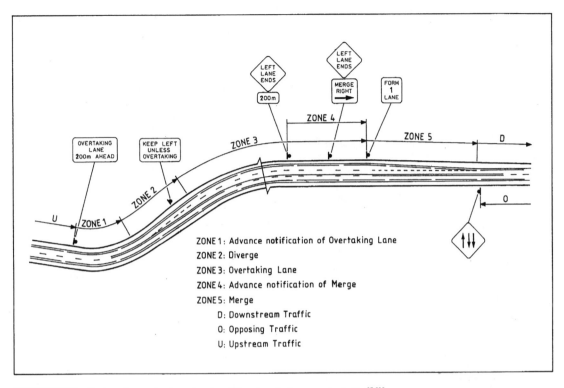

FIGURE 25.20   Design of a passing lane showing different analysis zones, Australia.[37,525]

- *Zone 1:* Advance notification of the overtaking lane
- *Zone 2:* Diverge area
- *Zone 3:* Overtaking section
- *Zone 4:* Advance notification of the merge (note that this area overlaps with the overtaking section)
- *Zone 5:* Merge area

This layout is typical for most of the investigated countries (see also Figs. 25.5, 25.7, 25.15, and 25.18). Note, in Australia there is left-handed traffic.

There are various proposals for intermediate cross section types in Switzerland, the United Kingdom, and several other countries.

25.2.2.3   Preliminary Conclusion.   Passing lanes, also called *intermediate cross sections,* are additional parallel auxiliary lanes provided on two-lane undivided highways for the exclusive purpose of improving passing opportunities. Passing lanes should be considered as a cost-effective geometric improvement on two-lane roads where the length and location of passing zones on the existing highway are less than desirable, and the traffic volume is high enough that the level of service is noticeably low. Passing lanes should also be considered on new construction or major realignment projects to achieve the desired level of service. Passing lanes (intermediate cross sections) may also be a cost-effective solution where:

1. Volumes on a two-lane highway are increasing and will soon warrant twinning
2. Where the provision of passing lanes may postpone the construction of a divided facility[11]

According to Ref. 571, there is room for intermediate types of road between the ordinary two-lane rural road on which the traffic volume does not exceed 12,000 vehicles/day (problems appear from 5000 onwards) and the four-lane divided motorway of normal characteristics, well suited for a traffic volume of 20,000 to 35,000 vehicles/day and even more. Intermediate types are (Fig. 25.3):

- Four-lane divided motorways with reduced characteristics for reasonable traffic volumes
- Highways with limited access if a motorway is not justified in the medium term.

As far as two-lane roads are concerned, German studies show that in safety and traffic flow terms the best performance type is the three-lane road with the central lane alternately assigned to one direction of travel and then to the other one.[571]

With respect to the previously discussed experience and research, it can be stated that intermediate cross sections or passing lanes have been proven to improve operational and safety performance of high-volume two-lane rural highways as long as type, location, and layout are used appropriately. Therefore, these road types represent sound low-cost solutions as long as upgrading to four-lane divided motorways is not economically feasible.

The recommendations in Sec. 25.1.2 represent a reliable basis for the implementation of different types of intermediate cross sections (see Figs. 25.3 to 25.7).

25.2.3 Specific Cross-Sectional Issues

A substantial amount of research has been conducted on accidents and roadside design elements, fixed objects, roadside features, and so forth. On the basis of the available literature, the following represents a summary of the key findings related to roadside design and traffic safety.

With respect to roadside elements it is reported by Hall, et al.[257] Highway geometric design with respect to cross sections deals almost exclusively with the geometric features of the roadway—most commonly lane width, lane configuration, and shoulder width and type. Alignment and curvature are also addressed. In a "perfect" world, the highway engineer would need to look no further than this because motorists would never "accidentally" leave the paved portion of the roadway. However, in spite of engineers' best efforts to design fail-safe facilities, accident data demonstrate that motorists leave the roadway for numerous reasons—with errors in judgment leading the list. In this connection, many engineers today recognize that they can almost always lessen the severity of an accident when a driver does run off the roadway. The concept applied to decreasing the severity of run-off-the-road accidents has been called *the forgiving roadside*. The United States, where this concept originated, has more extensive experience than other countries; therefore, the following discussion is based largely on U.S. experience.[257]

Roadside encroachments begin when the vehicle inadvertently leaves the travel lanes, veering toward the roadside. Most encroachments are quite harmless: the driver is able to regain control of the vehicle on the shoulder and safely returns to the travel lanes. When coupled with nearby roadside hazards, however, encroachments can result in roadside accidents. Such accidents comprise a significant number of the accidents that occur on two-lane rural roads. More than 30 percent of all accidents involve single vehicles running off the road.[232,710]

Furthermore, there is a higher rate of run-off-the-road accidents at or near horizontal curves. A 1976 study[767] of 300 fatal accidents that involved roadside objects in Georgia found that over one-half of the collisions occurred at or near horizontal curves of greater than 220 gon/km. This fact is often confirmed in Part 2, "Alignment." For example, see Tables 9.10, 10.12, and 18.14.

In general, it can be stated that on horizontal curves of two-lane rural highways, ROR accidents on left-handed curves appear to occur more frequently than on right-handed curves (see Fig. 25.24). The results of a 1978 study[80,569] showed that the proportion of ROR accidents to the outside of curves increases with increasing curvature change rate of the single curve. The same study also found a higher incidence of run-off-the-road *right* accidents on tangents than of run-off-the-road *left* accidents on tangents on undivided roads.[503]

25.2.3.1 Roadside Conditions.* The concept of a roadside "clear zone" that is relatively flat and free of obstructions was developed in the United States as early as 1967.[3] According to an international overview,[257] it has become generally accepted that the width of the clear zone should depend on speed and that it should be wider at those locations (like the outside of horizontal curves) where vehicles are more likely to leave the traveled portion of the roadway. Although the idea of a clear zone originated in the United States and much of the early research was done there, over the past 25 y, the concept has been accepted as an important cross section design element in other countries.[257]

On the topic of roadside conditions, the following is reported in Refs. 710 and 729. Past research on the safety of the roadside environment has produced important improvements in roadside hardware, including, for example, the development of barriers that better contain and more safely redirect errant vehicles and sign and light supports that break away on impact, causing little damage to the striking vehicle and its occupants. In addition, several design standards already require that for clear recovery areas, the border should begin at the edge of the travel lanes, the slopes should be traversable, and the area should be free of hazardous obstacles. Improved designs for drainage structures, such as culvert headwalls, reduce the hazard posed by unforgiving roadside obstacles. Also, specifications for side-slope and ditch configuration now recognize the safety benefits as well as the more conventional objectives of construction economy, maintainability, and slope stability.

Entry of an errant vehicle into the roadside border does not in itself mean that an accident is inevitable. Although some danger always exists, the chances of a safe recovery are excellent if the border is reasonably smooth, flat, and clear of fixed objects and other nontraversable hazards. The chances of successful recovery diminish as the ground slope within the border increases and the width decreases. Although there are no clear breakpoints, safety researchers in the United States generally agree that at speeds of approximately 90 km/h, "safe" clear zones should have side slopes no steeper than about 1:6 and should extend outward at least 10 m from the edge of the travel lanes. When the border is flat, unintended encroachments on tangent alignments seldom extend beyond the 10-m range.

Much of what is known about roadside safety relationships remains qualitative in nature, and only tentative steps have been taken to develop comprehensive accident models. Previous studies have found significant relationships between accident rates and composite measures of roadside condition.[113,200,766,776,777]

The relative hazard of the roadside may be described in terms of several characteristics, including:

- Roadside recovery distance (or roadside clear zone)
- Side slope (foreslope)
- Presence of specific roadside obstacles (for example, trees, culverts, utility poles, and guardrails)

Both the severity of crashes and the crash frequency are affected by such roadside features. Following is a discussion of these roadside characteristics.[710,729]

Roadside Recovery Distance/Clear Zone. The roadside recovery distance is a relatively flat, unobstructed area adjacent to the travel lane (that is, edge strip or edge line), where there is a reasonable chance for an off-the-road vehicle to recover safely.[779] Therefore, it is the distance from the outside edge of the travel lane to the nearest rigid obstacle.

A 1982 study[232] determined the effect of clear zone policy on the single-vehicle accident rate. As shown in Fig. 25.21, single-vehicle accidents per kilometer per year are highest for roads with

---

*Elaborated based on Refs. 710 and 729.

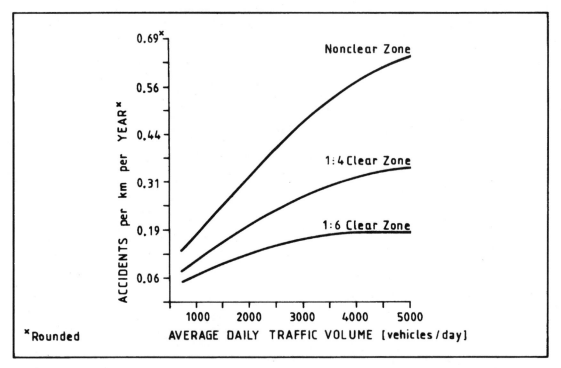

FIGURE 25.21   Relationship between single-vehicle, run-off-the-road accidents per kilometer per year and ADT for two-lane high-ways.[779]

a nonclear zone, next highest for a 1:4* clear zone policy (that is, same clear area with a 1:4 side slope), and lowest for a 1:6 clear zone. This study indicates the high potential for safety benefit resulting from increased roadside clear zones.

Accident reduction factors due to increasing roadside clear recovery distance are given in Table 25.4.[779]

Similar results were found in Ref. 710, as shown in Fig. 25.22. This figure reveals a significant relationship between the roadside recovery distance and the accident rates on two-lane rural roads. According to Fig. 25.22, an increase in the clear recovery area from 1.50 to 6.25 m, as an example, reduces the relative accident rate of single-vehicle, head-on, and sideswipe accidents by approximately 35 percent. (Note that the clear recovery distance in Fig. 25.21 and Table 25.4 is measured from the outside travel lane edge, whereas in Fig. 25.22 it is measured from the outside shoulder edge.)

This clear zone policy often cannot be applied, for example, in Europe because of land acquisition constraints and costs. Therefore, instead of increasing roadside clear zones, guardrails, which are themselves obstacles that can be struck by vehicles, are often installed in Europe.

Side slope.   According to Ref. 5, three regions of the roadside are important when evaluating the safety aspects: the top of the slope (hinge point), the foreslope, and the toe of the slope (intersection of the foreslope with level ground or with a backslope, forming a ditch).

The hinge point contributes to loss of steering control because the vehicle tends to become airborne in crossing this point. The foreslope region is important in the design of high slopes where

---

*U.S. terminology: 4:1.

TABLE 25.4   Accident Reduction Factors due to
Increasing Roadside Clear Recovery Distance[779]

Amount of increased roadside recovery distance, m*	Percent reduction related accident types
1.50	13
2.50	21
3.00	25
3.75	29
4.50	35
6.25	44

*Rounded.

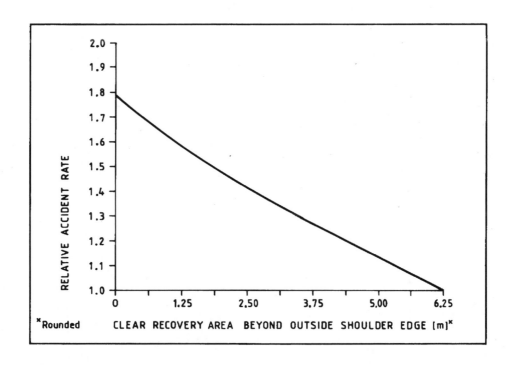

Notes: Accident relationship covers single-vehicle, sideswipe, and opposite-
direction accident on two-lane rural highways. Clear recovery area is
measured from the outside shoulder edge to the nearest roadside hazard.
Relative accident rate is defined as a multiple of the accidents per
million vehicle-km for a clear recovery area of 6.25 m.

FIGURE 25.22   Normalized relationship between accidents and width of clear recovery area.[779]

a driver could attempt a recovery maneuver or reduce speed before impacting the ditch area. In many situations, the toe of the slope is within the clear zone and the probability of reaching the ditch is high, in which case safe transition between fore- and backslopes should be provided (see Fig. 23.1).[5]

Research[535] in these three regions of the roadside has found that rounding at the hinge point, though not necessarily from a vehicle rollover standpoint, can significantly reduce its hazard potential. Rounded slopes reduce the chances of an errant vehicle becoming airborne, thereby reducing the hazard of encroachment and affording the driver more control over the vehicle. Foreslopes steeper than 1:4 are not desirable because their use severely limits the choice of backslopes. Slopes 1:3 or steeper are recommended only where site conditions do not permit use of flatter slopes. When slopes steeper than 1:3 must be used, a roadside barrier has to be provided,[5] which is the typical case in Europe.

As reported in Ref. 729, the steepness of the roadside slope or side slope, also called *foreslope,* is a cross-sectional feature which affects the likelihood that an off-the-road vehicle will roll over or recover back into the travel lane. Until recently, little was known about the relationship between accidents and side slopes. As part of their 1987 study, Zegeer, et al. developed relationships between single-vehicle crashes and field-measured side slopes from 1:2 to 1:7 or flatter for 2840 km of roadway in the United States.[779] As shown in Fig. 25.23, single-vehicle accidents (as a ratio of accidents on a 1:7 slope) are highest for slopes of 1:2 or steeper and drop only slightly for 1:3 slopes. Single-vehicle accidents then drop linearly (and significantly) for flatter slopes. This plot represents the effect of side slope after controlling for ADT and roadway features.[779]

The use of flatter slopes not only reduces the accident rate but it may also reduce rollover accidents, which are typically quite severe. In fact, injury data revealed that 55 percent of run-off-the-road rollover accidents result in occupant injury and 1 to 3 percent end in death. Of all other accident types, only pedestrian accidents and head-on crashes result in higher injury percentages.[729,779]

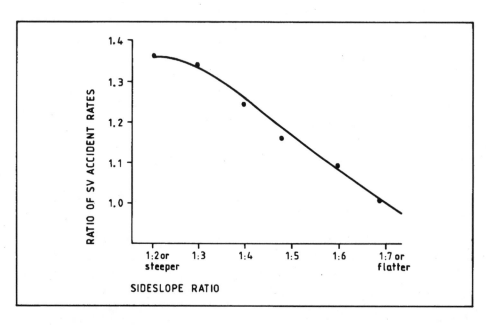

Note: These values are only valid for two-lane rural roads, and include adjust-
        ments for ADT, lane width, shoulder width, and recovery distance.

FIGURE 25.23   Plot of single-vehicle (SV) accident rate for a given side slope versus single-vehicle accident rate for a side slope of 1:7 or flatter.[779]

In this connection, Ajluni[9] reported the following about roadside terrain slopes, which frequently initiate rollover accidents:

- The likelihood of a rollover increases with embankment steepness, height, and ditch depth
- About 25 percent of all off-roadway accidents result in a rollover
- Ejection is the leading cause of serious and fatal injuries, accounting for more than half of the fatalities incurred in rollover accidents
- The fatality rate for occupants of rollover vehicles is approximately twice that for occupants of vehicles in nonrollover impacts
- The vast majority of rollovers occurs within approximately 10 m (30 ft) of the roadway. Relatively few rollover occur or are initiated on the shoulder.

Consequently, it has been concluded that

- All roadside terrain slope breaks should be rounded as much as possible to reduce the potential of vehicle rollover due to tripping on sag vertical curves. The need for adequate rounding of crest vertical curves, such as the break line of shoulder and side slopes, also cannot be overemphasized. Such rounding not only affords drivers greater opportunity to maintain or regain control of their vehicles, but also decreases the likelihood of rollover by preventing the vehicle from reaching large values of angular momentum about the roll axis.
- Ditches with front slopes no steeper than 1:3 appear relatively safe with respect to vehicle rollover potential.[9]

Based on the literature review, a slope steepness of 1:3 seems to be the turning point for side-slope safety, expressed as rollover potential. This statement is confirmed by Hall, et al.,[257] who reported the following.

To achieve maximum safety, then, slopes should be as flat as practical and liberally rounded at the top and bottom. A slope of 1:6 or flatter is required for recovery, but slopes up to 1:3 can be traversed if they are smooth and unobstructed. A runout area at the toe of the slope is essential if crash severities are to be kept to a minimum.

Typical values of embankment slope standards from an international overview are summarized in Table 25.5.[257]

Even though flatter side slopes are very favorable from the point of view of traffic safety, this cross-sectional feature is rarely used in Europe because of the previously mentioned constraints, and is normally replaced by a combination of guardrail and steeper slopes (see Fig. 25.8).

25.2.3.2 *Superelevation Rates of the Cross-Sectional Elements.* All relevant information about superelevation rates is discussed in Sec. 25.1.3.2 and especially in Secs. 14.1.2 and 14.2.2, "Superelevation." Special attention should be given to Sec. 14.2.2.2, "Safety Considerations."

It is interesting to note that an international overview[257] found a relatively high level of agreement about minimum superelevation rates. It appears that a superelevation rate (cross slope of 2.0 to 2.5 percent) is the most widely accepted value for design and that there is little or no variation by roadway class.

25.2.3.3 *Facilities in the Cross Section.* * This subsection may only partially apply to western European conditions and standards, but it does apply to a number of other countries for which research experiences, gained in the United States, may be of interest.

While previous discussions have addressed general roadside improvements with respect to clear zones and side slopes, recent studies have also quantified the effects of more specific roadside obstacle improvements.

According to Ref. 257, man-made roadside hardware is an important category of obstacle that can be struck by an errant motorist. The most common appurtenances include sign and luminaire

---

*Elaborated based on Refs. 502 and 729.

TABLE 25.5   Typical Values of Embankment Slopes in Various Countries and "CS"*

Country	Roadway classification		
	Freeway (interstate)	Major road (arterial)	Minor or local road
Australia	—	≤1:6 up to 10 m high; then 1:1.5 to 1:2	—
Canada	—	Rural collector 1:2 to 1:6	Rural local 1:2 to 1:3
China	1:1 to 1:1.75	1:1 to 1:1.75	1:1 to 1:1.75
Czech Republic	1:2.5 if <3.0 m 1:1.5 if ≥3.0 m	1:2.5 if <3.0 m 1:1.5 if ≥3.0 m	1:2.5 if <3.0 m 1:1.5 if ≥3.0 m
Denmark	1:2	1:2	1:2
France	Based on geotech design	Based on geotech design	Based on geotech design
Germany	1:1.5	1:1.5	1:1.5
Greece	1:1.5	1:1.5	1:1.5
Hungary	1:2 to 1:2.5	1:1.5 to 1:2.5	1:1.5
Japan	1:1.5 to 1:2	1:1.5 to 1:2	1:1.5 to 1:2
The Netherlands	1:2 to 1:3	1:2 to 1:3	1:2 to 1:3
Poland	1:5 to 1:3	1:5 to 1:3	1:5 to 1:3
Portugal	Based on geotech design; maximum = 1:1.5 without barrier	Based on geotech design; maximum = 1:1.5 without barrier	Based on geotech design; maximum = 1:1.5 without barrier
South Africa	—	Rural 1:4	—
Spain	1:6 to 1:2	1:4 to 2:3	2:3 to 1:2
Sweden	—	Rural undivided 1:3	—
Switzerland	1:2 to 2:3 1:3 and Δh > 4 m	1:2 to 2:3	1:2 to 2:3
United States	1:6 for low fill height; 1:3 for moderate height	1:6 for low fill height; 1:3 for moderate height	1:2 maximum
Venezuela	1:6 for <1.1 m height 1:2 for >4.5 m height	—	—
Yugoslavia	Varies with barrier and fill height	Varies with barrier and fill height	Varies with barrier and fill height
CS	1:1.5 to 1:6	1:1.5 to 1:6	1:1.5 to 1:6

*Elaborated based on Ref. 257.

supports.[720] In the United States and in many other countries, virtually all such supports are designed to yield on impact, thereby preventing sudden vehicle deceleration and occupant injuries. Small signs typically yield by bending or fracturing, while larger ones give way through a slip-base and hinge combination. Cantilevered and overhead signs, which cannot be redesigned to enhance safety, are usually shielded.[257] Other roadside obstacles include trees, mailboxes, culverts, guardrails, and fences.

An interesting phenomenon regarding the hazardous effects of highway features and roadside objects is the effect of roadway geometry. According to Pilkington,[579] both horizontal curvature and grade have an effect on run-off-the-road and fixed-object accidents, with horizontal curvature being the most frequently mentioned geometric characteristic. The run-off-the-curve accident occurs more often at the outside of a left-hand curve, as opposed to a right-hand curve or the inside of either curve. Departure percentages are presented in Fig. 25.24.[569]

However, the figure is biased with respect to the comparison between tangent and curved sections because the percentages are not length-related. Despite this fact, it can be concluded that ROR accidents encroach 3 times more to the outside than to the inside of the curve.

Utility Poles.[729]   The utility pole accident is the most frequent and severe roadside accident involving a man-made object; thus, the development and use of countermeasures to reduce this threat to highway safety will be cost-effective. However, any discussion of utility pole accidents and ways of reducing them should be put into context of the larger highway safety problem and the roadside safety problem. Utility pole accidents are a subset of this latter safety problem. Their reduction alone may have little effect upon the larger highway safety problem and in the current trend of annual reductions in the highway fatality rate.[579]

It is reported in Ref. 729 that improvements which should reduce the frequency of utility pole crashes include relocating the poles further from the roadway, increasing pole spacing, removing the poles and burying the utility lines, and multiple pole use (that is, removing poles on one side of the road and using poles on the other side to carry multiple electric and/or utility lines). However, on rural roads with relatively low traffic volumes, burial of lines is seldom practical. For reducing crash severity, breakaway utility poles should be used on a more widespread basis.

Reductions in utility pole crashes, due to such utility pole treatments, were defined in a 1983 study by Zegeer and Parker.[775] The study analyzed traffic, accident, roadway, and utility pole data

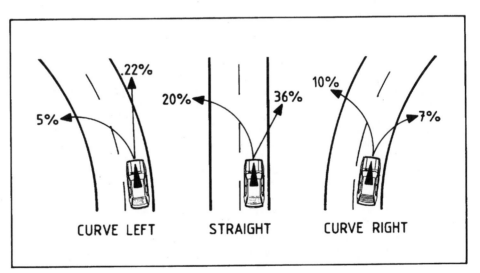

Database: 6313 ROR accidents, not length-related.

FIGURE 25.24   Departure percentages by horizontal alignment. (Elaborated and modified based on Ref. 569.)

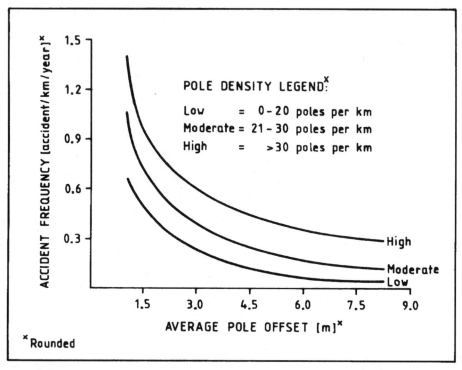

FIGURE 25.25   Relationship between frequency of utility pole accidents and pole offset for three levels of pole density.[775]

for 4000 km of two-lane and multilane roads in urban and rural areas in four states in the United States. The resulting accident relationships with pole offset and poles per kilometer are given in Fig. 25.25.

The nomograph shown in Fig. 25.26 was developed for utility pole accidents (per kilometer per year) as a function of ADT, pole density (number of poles per kilometer), and pole offset (average distance of the utility poles from the edgeline). This nomograph shows, for example, that a road with an ADT of 10,000 vehicles/day and about 40 poles/kilometer, offset 1.5 m from the edgeline, can be expected to have approximately 0.75 utility pole accidents/km/y. If a countermeasure was implemented to offset the poles to 4.50 m, the nomograph shows that the expected accidents afterwards would be about 0.375, or about a 50 percent reduction in utility pole accidents.[729,775]

Other Obstacle Types.[729]   In a 1987 FHWA study by Zegeer, et al., a model was developed for accidents involving trees, mailboxes, guardrails, and fences. Accident reductions were determined for clearing or relocating such obstacles further from the roadway. The model only applies to obstacle distances between 0 and 9 m from the outside edge of the travel lane (that is, edgeline).[780]

Table 25.6 gives the estimated percent reductions, as a function of increased lateral offset from the edge of the travel way to the obstacle.

Accidents involving trees can be reduced as shown in Table 25.6. For example, clearing trees by 3.0 m (for example, from 2.5 to 5.5 m) will reduce accidents involving trees by an expected 57 percent. These values indicate that if trees were cleared back from the roadway, run-off-the-road vehicles would have additional roadside areas to recover, assuming the trees were not on a steep

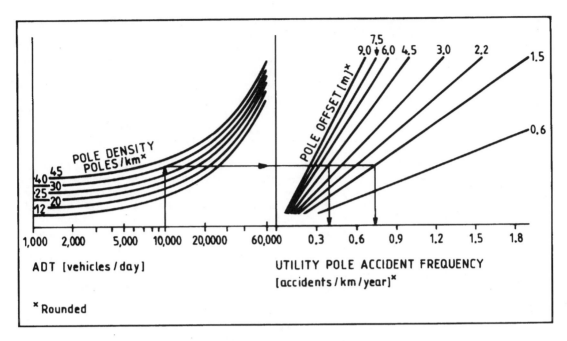

FIGURE 25.26   Nomograph for predicting utility pole accident frequency.[775]

TABLE 25.6   Percent Reduction in Specific Types of Obstacle Accidents due to Clearing/Relocating Obstacles Further from the Roadway[503,780]

Amount of increase in the lateral offset to obstacle, m*	Percent reduction in obstacle accidents by obstacle type			
	Trees	Mailboxes, culverts, and signs	Guardrails	Fences and/or gates
0.9	22	14	36	20
1.5	34	23	53	30
2.5	49	34	70	44
3.0	57	40	78	52
4.0	66	—	—	—
4.5	71	—	—	—

*Rounded.

Note: Relocation of obstacle to specified distance is generally not feasible.

side slope. Since trees are the fixed objects which are most often struck on many rural roads, clearing trees back from the road (particularly on roads with severe alignment) can be an effective roadside safety treatment.[780]

It should be noted, however, that most of the western European countries do not support the removal of trees for esthetic and environmental reasons. Trees should be protected by guardrails in the future to lessen the degree of high accident risk and severity.

Culvert headwalls can result in serious injury or death when struck at moderate or high speeds on rural roadways. While relocating such culverts further from the roadway may be feasible under certain conditions, the ideal solution would be to reconstruct the drainage facilities so that they are flush with the roadside terrain and present no obstacle to motor vehicles. Such designs would essentially eliminate culvert accidents, although run-off-the-road vehicles could still strike other obstacles (for example, trees) beyond the culverts or roll over on a steep side slope. Accident reductions, shown in Table 25.6, correspond to placement of culvert headwalls further from the roadway. For example, a 40-percent reduction in culvert hits is expected for culverts located 4.5 m from the road compared to 1.5 m (that is, a 3.0-m difference in distance).[729,780]

Further percentage reductions in accidents, such as for signs, guardrails, and fences, are also shown in Table 25.6.

Crash Severity of Obstacles.[729]   In addition to crash frequency, the severity of crashes involving specific roadside obstacles is also important. A 1978 FHWA study by Perchonok, et al. analyzed accident characteristics of single-vehicle crashes, including crash severity related to types of objects struck.[569] For nonrollover fixed-object crashes, the obstacles associated with the highest percent of injury occurrences are, in order, bridge or overpass entrances, trees, field approaches (that is, ditches created by driveways), culverts, embankments, and wooden utility poles. The actual percent injuries and fatalities of these crashes are shown in Table 25.7. Obstacle types with the lowest crash severity include small sign posts, fences, and guardrails.[569,729]

Guardrails/Traffic Barriers.[257,729]   Guardrails are installed along roadways to keep a vehicle from striking a more rigid obstacle or from rolling down a steep embankment. When installed, a guardrail is generally positioned at the greatest practical distance from the roadway to reduce the incidence of guardrail impact. The presence of these devices is the highway engineer's admission that it was practically or economically impossible to eliminate a hazard and that it was necessary to shield traffic from an object. The high number of fixed-object fatalities in which collision with a traffic barrier is identified as the most harmful event demonstrates that the shielding is not a panacea.[720] Traffic barriers are expensive to design and install; once installed, the highway agency has the responsibility (and expense) of maintaining the barriers.[257]

It is not often feasible to relocate guardrails further from the roadway along a section unless some flattening of the roadside occurs. However, when it is feasible to flatten roadsides to a relatively mild slope (for example, 1:5 or flatter) with appropriate removal of obstacles, then

TABLE 25.7   Severest Injury by Object Struck in Nonrollover Accidents[569]

Object	Accident sample size	Percent injured	Percent killed
Bridge or overpass entrance	88	75.0	15.9
Tree	667	67.9	7.2
Field approach	75	66.7	1.3
Culvert	231	62.3	6.1
Embankment	406	57.6	4.4
Wooden utility pole	598	51.2	2.3
Bridge or overpass siderail	82	51.2	2.4
Rock(s)	73	49.3	1.4
Ditch	368	48.9	1.1
Ground	153	48.4	3.3
Trees/brush	255	38.4	2.0
Guardrail	284	31.7	1.8
Fence	325	24.3	0.3
Small sign post	76	22.4	1.3
Total	3681	50.8	3.6

guardrails should be removed, since the guardrail itself presents an obstacle which vehicles can strike. The accident reductions in Table 25.6 for guardrail placement illustrate the crash benefits of relocating guardrails.[780]

With respect to the clear zone concept, to date sufficient research work in Europe does not exist, at least not in Germany. The guardrail solution is far too simple to be acknowledged in general as a final remedy or as an appropriate excuse for space constraints.

For example, during a meeting of Committee 3.8 "Traffic Accidents" of the German Road and Transportation Research Association in 1997, the main author used the term *recovery area*. As he was asked about the meaning of this term, he really had difficulties in finding an appropriate expression in the German language. This simple example shows that important topics of roadside safety conditions have not yet been adequately considered in European highway engineering; for example, the guardrail concept cannot be accepted as a final solution for everything with respect to roadside safety. Therefore, research is urgently needed.

25.2.3.4 *Drainage.*

Drainage issues are discussed in Sec. 25.1.3.4. However, with respect to the design of standard slopes of ditches, the following supplementary information is of interest.

According to Ref. 569, ditches represent another common roadside feature that can be a great safety concern. Injury rates, which can be affected by ditch depth and slope, revealed that deep ditches had a 20-percent higher injury rate than shallow ones. For example, it was found that ditches which were 0.9 m or deeper were associated with a higher percent of injury accidents (61 percent) than ditches 0.3 to 0.6 m deep (54 percent injury). The percent of fatal accidents was about the same for each depth category (that is, about 5 percent for both the 0.3- to 0.6-m and the 0.9-m and deeper groups).[569]

Therefore, the side slopes of ditches should be flat enough to minimize the possibility of errant vehicles overturning. A maximum slope of 1:4 is recommended; however, if possible, a slope of 1:6 is desirable.[37]

If there is a high probability of encroachment onto a steep foreslope and right-of-way is limited, an enclosed drainage system is recommended.[257]

25.2.4 Additional Areas along the Roadside and Distance between Railroad and Roadway*

In this book, the "roadway" is most important. However, this section is mainly focused on the railroad point of view.[207] In spite of this fact, the safety assessments (expressed by distances) are, to a large extent, sufficient for both the railroad and the roadway.

In the following, important considerations about sections on railroads and roads sharing the same corridor are discussed.

When planning railroad lines in conjunction with roads or vice versa, sharing the same corridor can become useful or even necessary, based on the following reasons:

- Alignment design
- Economic
- Environmental
- Traffic political

Figure 25.27 shows a good example of parallel routing of a high-speed railroad line and State Road 3 in southwestern Germany.

Fundamentally, railroad and road alignments can be located as near as possible to each other, as long as it is

---

*Elaborated with the support of R. Priess based on Refs. 207 and 584.

FIGURE 25.27   Good example of parallel routing of a high-speed railroad line and State Road 3 in southwestern Germany. (*Source: Mailaender Ingenieur Consult.*)

- Technically possible and economically sensible
- Possible based on the operational safety of the railroad (for example, signal sight distance)
- Permissible for the safety of road traffic participants (for example, misinterpretation by the motorist with respect to opposing trains, snowploughs, etc.)
- Possible considering subsequent expansions of both transportation elements
- Possible in terms of the requirements of the construction and maintenance of the traffic routes.

Further issues for the assessment of the distance between railroad and roadway can result from

- Landscape cultivation requirements (location of biotypes)
- Agricultural requirements for the economical use of fields under cultivation
- Planning laws and laws on right-of-way acquisition

The distance between the alignments of the railroad and the road has to consider the dimensions of the railroad and road cross sections, including required installations like poles, cable housings, retaining walls, drainage ditches, side slopes, etc., as well as possibly required expansions to both the railroad and the road.

The required distance, *E,* between railroad and roadway can be approximately determined with respect to the mutual height difference, *H,* according to the diagram in Fig. 25.28. The goal is the protection of railroad sections from motor vehicles, which may run off the road (ROR), as well as the protection of roads from derailed trains.

Parallel routing in the sense of the diagram in Fig. 25.28 includes a tangential (or approximately tangential) mutual course of the alignments. Selection and location of the outer separation types depend on the mutual height difference, *H,* and the horizontal distance, *E,* between the railroad and the road.

The diagram contains nine cross-sectional types with the relevant distance elements considering subgrade width, overhead power system, drainage ditches, cable housings, retaining walls, and slopes. Furthermore, space for expansion areas, as well as for protective barriers or fences, also have to be considered, if necessary.

The lateral distances, *E,* and the vertical distances, *H,* are defined as follows:

E = distance between the center of the outside rails and the edge of the pavement of the adjacent road (driving lane or, if present, emergency lane or stabilized shoulder)

H = distance between the top of the rail and the upper edge of the road surface

The areas of the diagram in Fig. 25.28 are subdivided as follows:

- *Area—only permissible with specific precautions:* Distances *E* of less than 8.90 m normally require specific precautions for the safety of the railroad operation, since the railroad section has to be protected from ROR motor vehicles and their loads or the road from derailed trains.

- *Area—to be avoided if possible* (*type 2*): This area represents distances between 9.00 and 14.35 m. An endangerment could develop here by ROR motor vehicles and their loads for the railroad section and the poles of the overhead power system. This is especially true for convex curvatures of the road alignment compared to the observed railroad line. Therefore, it is necessary in this case to connect retaining walls and crash barriers that are tied into each other (if possible, monolithic). The barrier should be designed for headlong fall safety and should be able to absorb the impact loads resulting from heavy vehicles. Roadside safety barriers have to be provided.[210]

- *Area—with retaining wall for the road* (*type 1*): Two possibilities exist for this type: a high retaining wall with level connections at the bottom or a lower retaining wall combined with side slopes. Normally, the same safety measures have to be provided here as for type 2. (Compare types 8 and 9 concerning retaining walls for the railroad.)

- *Area—sloped areas* (*types 3 and 6*): In these areas, one part can be created as a side slope and the rest as level connection. In the upper part (case 3), guardrails are necessary for roads with distances (*E* ≤ 14.35 m).[210]

- *Area—pure side slopes* (*types 4 and 7*): In this important limiting case, the design of the terrain between both traffic routes can only be a continuous side slope (fat line for slope 1:1.5; dotted line for slope 1:2).

- *Area—road and railroad on the same level* (*type 5*): In this case, the distance, *E,* should not be less than 14.35 m. Guardrails should be provided at the roadside edge in areas with *E* ≤ 14.35 m.[210]

- *Area—with retaining wall for the railroad* (*types 8 and 9*): The same statements discussed for type 1 are valid here. However, in these cases, it is possible that a derailed train might not be stopped by retaining walls from hitting the road because of the great mass. To build retaining walls to withstand those large impact loads would be very costly, considering the relatively rare event of train derailments.

For new designs, redesigns, and restoration strategies, the railroad or highway engineer can, at once, determine roughly the lateral distance, *E,* between the railroad and the road based on the

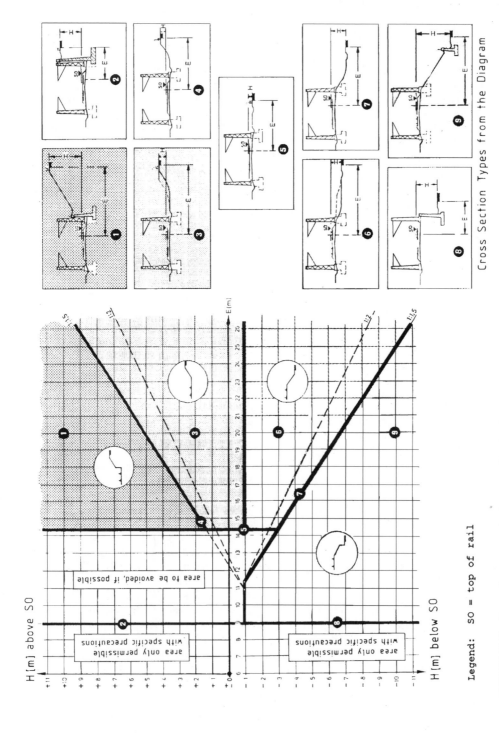

FIGURE 25.28 Types of outer separation between railroad and roadway as function of the distance, *E*, and the mutual height difference, *H*. (Elaborated based on **Ref. 207**.)

mutual height difference, *H,* depending on the terrain and the corresponding outer separation type (Fig. 25.28). Normally, these distances are adequate to avoid conflicts between road and railroad vehicles.

The procedure[207] is being used with great success during the planning stages for the new high-speed railroad system in Germany. Especially in densely populated, industrialized countries, the frequency of encounters between roads and railroads is constantly increasing and, consequently, the parallel routing of both transportation elements becomes an important issue in modern traffic route engineering. One important goal is to use, wherever possible, only one corridor for both traffic modes to reduce detrimental impacts on the environment.[357]

Figure 25.29 shows a typical example for the parallel guidance of a German high-speed railroad section and farm roads on both sides. The minimum lateral distance, *E,* depends on the space needed for cable housings, poles for overhead power systems, as well as railroad and roadway drainage. According to Fig. 25.28, a lateral distance of $E < 8.90$ m is permissible only if specific precautions are provided for securing the railroad operation. In this connection, it was decided by the responsible officials that guardrails were sufficient protective measures but only for farm roads (not for regular roads!). This decision is very important, since very often farm roads parallel railroad sections.

Finally, antiglare screens have to be provided as necessary, depending on the grade, the curvature, and the relative heights of the road.

When planning antiglare facilities, the signal distance should not be impaired and every possible confusion over signals for the railroad and traffic lights on the road has to be eliminated.

In conclusion, the procedure presented for determining adequate distances between railroad and road[207] has proven to be reliable in practice. However, the results of Fig. 25.28 are to be considered as recommendations for the design engineer and not as fixed rules.

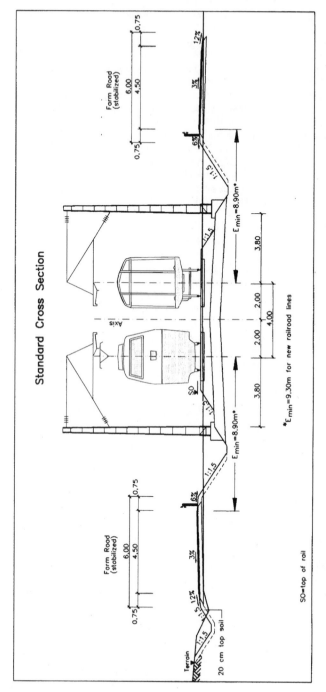

FIGURE 25.29 Parallel guidance of a high-speed railroad section and farmroads on both sides, Germany. (*Source:* Mailaender Ingenieur Consult.)

CHAPTER 26
SUMMARY OF PART 3 "CROSS SECTIONS"

As Part 2, "Alignment (AL)," Part 3, "Cross Sections (CS)" was also separated, to a major extent, into two subchapters:

1. Recommendations for practical design tasks
2. General considerations, research evaluations, guideline comparisons, and new developments

In this way the reader is provided with all the necessary experience and background information for selecting a sound cross section for new alignments or evaluating an existing one in terms of geometric design, operation, environmental compatibility, economic feasibility, and—above all—safety.

The material presented herein mainly refers to cross sections of nonbuilt-up roads and focuses on:

- Discussing major cross-sectional elements
- Providing the assumptions and procedures for assessing each individual cross-sectional element
- Comparing various cross-sectional parameters from national guidelines and practices worldwide
- Reviewing the safety aspects associated with each specific design element

The development of Part 3, "Cross Sections," differs from Part 2, "Alignment" in that it provides information based on national or international experience, engineering judgment, and, to some extent, on accident history rather than on rigorous scientific, or physical laws, and/or quantitative safety evaluation processes. Despite this fact, the highway community, including agencies, engineers, and academia, will find all the necessary information about the design elements of cross sections such as pavement/lane, shoulder, median, edgeline width, etc. Overall, the main cross-sectional elements and, therefore, the cross sections themselves are based on considerations about clearance, traffic, and safety spaces as well as on opposing traffic width increases with respect to speed.

Furthermore, a presentation of the various cross-sectional elements is given, including their dimensions, as they are categorized by different driving behavioral characteristics (for example, speed). Since there is no one-to-one correlation between the dimensions of each design element and its operational features, a thorough discussion is given about the identified impacts of the various possible implementation schemes of each design element individually and in combination with other elements. In this connection, special emphasis is placed on traffic safety considerations in order to evaluate the impact of the main cross-sectional elements (like pavement/lane, paved/unpaved shoulder, median, edge strip width, etc.) on the accident situation.

Throughout this part of the book, the adoption of the concept of the *standard cross section* has to be considered as a key issue. This concept has been successfully practiced for years, in Europe, especially in Germany. Adoption of this concept enabled the authors to present standard cross sections in a concise, systematic, methodical, and sequentially consistent manner.

Following this discussion, the issue of cross section design is addressed, which represents a critical design step for both the highway agencies and the designer, since the operational and economic feasibility of the highway are strongly influenced by the cross section selected. In a separate chapter, various compatible standard cross sections for each road category are given, and they are grouped into multilane and two-lane roads. For every standard cross section, the corresponding operational characteristics of design speed range, type of intersection, traffic volume ranges, permissible speed limits, and associated vehicle-type usage were identified and recommended for practice.

To close the gap between two-lane and multilane median separated standard cross sections, the subject of the so-called intermediate cross sections was thoroughly discussed with respect to traffic volume and operational and safety issues. Intermediate cross sections represent the alternative solution and accepted practice between high-volume two-lane roads and divided highways when the latter are not economically feasible. All design impacts associated with the implementation of these types of cross sections and especially those of the b2 + 1 type, such as location, design, and traffic guidance of critical and uncritical areas, approach to intersections, etc., are completely discussed. In-depth investigations were conducted to assign safety levels to the various standard and intermediate cross sections in order to provide the highway engineer with sound solutions for individual design cases.

Finally, recommendations for providing separate bicycle and/or pedestrian lanes parallel to the traveled way are given.

The last chapter is devoted to specific cross-sectional issues, again directed mainly at safety considerations. In this connection, roadside conditions with respect to recovery distances/clear zone concepts, side slopes, and the crash severity of obstacles are of major importance. In addition, questions about superelevation rates, drainage requirements, and the distances between railroad and highway alignments are addressed.

Finally, it can be concluded, as in Ref. 257, that a great number of issues to enhance traffic safety have been considered with respect to standard and intermediate cross sections and cross-sectional elements. In addition, many efforts have been made to improve roadside design and to alleviate, in particular, the fixed-object problem.

The research investigations and cross-sectional guidelines studied revealed a more or less general consensus on lane and shoulder width, as well as for road surface superelevation (cross slope). Other subject areas, especially with respect to roadside features, like embankment slopes, ditch design, or use of guardrails, were found to be less consistent between the investigated countries.

GENERAL CONCLUSION OF PARTS 1 TO 3

The goal of this book is to provide a comprehensive presentation on the relationships and interrelationships between highway design, driving behavior, driving dynamics, and accident characteristics. Based on this knowledge, sound recommendations were developed regarding the design rules of today, while always attempting to consider traffic safety quantitatively. Another important goal was the summarization of background research and practical acknowledged experience.

The scope of this book encompasses all relevant fields in network, alignment, and cross section design with special emphasis on traffic safety.

The book is, to a large extent, based on extensive research conducted during the last three decades by the main author, analyzing and evaluating highway geometric design and traffic safety. The authors have developed three safety criteria for distinguishing good, fair (tolerable), and poor design practices on newly designed or existing roadway sections and a safety module for evaluating road networks. The work takes into account numerous research impacts and practical experiences from colleagues and organizations on an international level.

Additional information dealing with traffic safety issues worldwide, environmental protection, human factors, and vehicular safety have also been included.

In conclusion, sound design rules in agreement with quantitative safety evaluation processes and, at the same time, coordinated with ecological and economical demands should constitute the framework for modern highway design solutions.

Overall, the book is an invaluable source of information for educators, students, scientists, highway agencies, and consultants in the field of highway design and traffic safety engineering.

REFERENCES

The authors apologize for any inaccuracies that may have resulted as a consequence of the process of translation.

1. AASHO, "A Policy on Highway Types (Geometric)," Special Committee on Design Policies, American Association of State Highway Officials, Washington, D.C., U.S.A., 1940.

2. AASHO, "A Policy on the Geometric Design of Rural Highways," American Association of State Highway Officials, Washington, D.C., U.S.A., 1965.

3. AASHTO, "Highway Design and Operational Practices Related to Highway Safety," American Association of State Highway Officials, Washington, D.C., U.S.A., 1967.

4. AASHO, "Guidelines for Skid Resistant Pavements," American Association of State Highway and Transportation Officials, Washington, D.C., U.S.A., 1976.

5. AASHTO, "A Policy on Geometric Design of Highways and Streets," American Association of State Highway and Transportation Officials, Washington, D.C., U.S.A., 1984, 1990, 1994.

6. Adam, S., "Road Safety-Concrete Barriers and Arrestor Beds," *Revue Générale des Routes et des Aérodromes,* vol. 598, France, June 1983, pp. 81–92 (in French-English translation available from Australian Road Research Board, ARRB).

7. Adams, J. L., Chap. 2 in *Risk and Freedom; the Record of Road Safety Regulation,* Transportation Publishing Projects, Cardiff, United Kingdom, 1985, pp. 17–28.

8. Administration of the German Reichsbahn, *One Hundred Years of German Railroads,* Ullsteinhaus, Berlin, 1935.

9. Ajluni, K. K., "Rollover Potential of Vehicles on Embankments, Sideslopes, and Other Roadside Features," *Public Roads,* vol. 52, no. 4, U.S.A., 1989, pp. 107–113.

10. A.K.G. Software, *Manual for Highway Geometric Design, Surveying, Graphic,* Ballrechten-Dottingen, Germany, 1992, 1995.

11. Alberta Transportation, *Highway Geometric Design Guide, Alignment Elements,* Alberta, Canada, 1995.

12. Al-Kassar, B., G. Hoffmann, and D. Zmeck, "The Influence of Road Sectional Features on the Momentary Speed of Free-Moving Passenger Cars," Research Road Construction and Road Traffic Technique (Forschung Strassenbau und Strassenverkehrstechnik), vol. 323, published by the Minister of Transportation, Bonn, Germany, 1981.

13. Allen, R. W., et al., "Analytical Modeling of Driver Response in Crash Avoidance Maneuvering," vol. 1, "Technical Background," DOT HS 807 270, National Highway Traffic Safety Administration, Washington, D.C., U.S.A., 1988.

14. Allen, R. W., et al., "Vehicle Stability and Rollover," DOT HS 807 956, *National Highway Traffic Safety Administration,* Washington, D.C., U.S.A., 1992.

15. Allen, R. W., and T. J. Rosenthal, "Requirements for Vehicle Dynamics Simulation Models," SAE Paper 940175, Society of Automotive Engineers, Warrendale, Pennsylvania., U.S.A., 1994.

16. Allen, R. W., T. J. Rosenthal, and D. H. Klyde, "Application of Vehicle Dynamic Modeling to Interactive Highway Safety Design," paper no. 951050, presented at the 74th Annual Meeting of the Transportation Research Board, Washington, D.C., U.S.A., 1995.

17. Allsop, R., "Summary and Conclusions of (European Transport Safety Council) ETSC-Symposium Reducing Speed-Related Casualties, The Role of the European Union," ETSC, Brussels, Belgium, May 3, 1995.

18. Al-Masaeid, H. R., M. Hamed, M. Abou-Ela, and A. Ghannam, "Consistency of Horizontal Alignment for Different Vehicle Classes," *Transportation Research Record,* vol. 1500, U.S.A., 1995, pp. 178–183.

19. Alrutz, D., and V. Meewes, "Investigations of the Bicycle Traffic in Cologne," Association of Car-Insurances (HUK), Cologne, Germany, 1980.

20. Alrutz, D., H. W. Fechtel, and J. Krause, "Documentation for Securing the Bicycle Traffic," *Accident and Safety Research,* vol. 74, Federal Highway Research Institute (BASt), Bergisch, Gladbach, Germany, 1989.

21. American Society of Civil Engineers, *Practical Highway Esthetics,* New York, U.S.A., 1977.

22. Anderson, L. D., "Applying Geographic Information Systems to Transportation Planning," Federal Highway Administration, paper prepared for Transportation Research Board, 70th Annual Meeting, Washington, D.C., U.S.A., 1991.

23. Andersson, K., and G. Nilsson, "Models for a Description of Relationships between Accidents and the Road's Geometric Design," VTI Report 153, Swedish Road and Traffic Research Institute, Linkoeping, Sweden, 1978.

24. Andrews, M., S. Jacobi, U. Reuter, and V. Salzer, "Porsche's Contribution to Safer and More Comfortable Driving in the Future," Dr.-Ing. h.c. F. Porsche AG, Porschestr., Weissach, Germany, 1994.

25. Angenendt, W., H. Erke, G. Hoffmann, E. A. Marburger, W. Molt, and G. Zimmermann, "Situation-Related Safety Criteria in Road Traffic," Project Group, Research Report of the Federal Highway Research Institute (BASt), Bergisch Gladbach, Germany, 1987.

26. Angerer, F., et al., "Road and Environment," German Road and Transportation Research Association, Cologne, Germany, 1980.

27. Appel, H., et al., "Driver-Vehicle-Behavior in Critical Situations," Project Group of the Federal Highway Research Institute (BASt), Cologne, Germany, 1979.

28. Armour, M., "The Relationship between Shoulder Design and Accident Rates on Rural Highways," *Proc. 12th Australian Road Research Board Conference,* vol. 12, no. 5, Australia, 1984, pp. 49–62.

29. Ashton, S. J., "Vehicle Design and Pedestrian Injuries," in A. J. Chapman, F. M. Wade, and H. C. Foot, eds., *Pedestrian Accidents,* John Wiley and Sons, New York, U.S.A., 1982.

30. Association of Car-Insurances (HUK), "Planning and Design of Rural Roads—PELa," recommendation no. 9, Cologne, Germany, 1993.

31. Association of the German Automobile Industry, "Facts and Numbers of the Motor Vehicle Economy," *52/56/59 Proceedings,* Frankfurt, Germany, 1988, 1992, 1995.

32. Auberlen, R., "Controlled Force-Interactions or Pattern," *Road and Interstate (Strasse und Autobahn),* vols. 4, 5, 6, Germany, 1969.

33. Auffenberg, A., and D. Schwell, "Federal Autobahnen in the Years 1991–1994," *Road and Interstate (Strasse und Autobahn),* vol. 4, Germany, 1995, pp. 197–209.

34. Australian Transport Advisory Council, "The National Road Safety Strategy," Canberra, Australia, 1992.

35. Austrian Research Association for Traffic and Road Engineering, "The New Austrian Guidelines for the Alignment of Roads, Principles and Explanations, RVS 3.23," vol. 76, Vienna, Austria, 1981.

36. Austrian Research Association for Traffic and Road Engineering, "Guidelines and Standards for Road Construction, RVS 3.31, Cross Section Elements; Traffic Space and Clearance," Vienna, Austria, 1995.

37. AUSTROADS, *Rural Road Design: Guide to the Geometric Design of Rural Roads,* National Office, Haymarket, New South Wales, Australia, 1989.

38. AUSTROADS, "Road Safety Audit," Publication No. AP-30/94, Sydney, Australia, 1994.

39. Babkov, V. F., "Road Design and Traffic Safety," *Traffic Engineering and Control,* vol. 9, United Kingdom, 1968, pp. 236–239.

40. Babkov, V. F., *Road Conditions and Traffic Safety,* Mir Publishers, Moscow, 1975.

41. Baker, S. P., B. O'Neill, and R. S. Karpf, *The Injury Fact Book,* D.C. Heath and Company, Boston, Massachusetts, U.S.A., 1984.

42. Bald (1987), cited by K. Pfundt, "Accident Situation on Two-Lane Rural Roads—A Problem Outline and Comments to Open Questions for Preparing Expert Discussions," Office for Traffic Safety, Cologne, Germany, 1992.

43. Baldwin, D. M., "The Relation of Highway Design to Traffic Accident Experience," Convention Group Meetings, AASHO, Washington, D.C., U.S.A., 1946, pp. 103–109.

44. Balogh, T., "Effect of Design Parameters of Public Highways on Traffic Safety," *Kozlekedestudomanyi,* vol. 17, Hungary, 1967, pp. 394–403.

45. Baumann, N., "Relationship between Traffic Accidents and Traffic Conditions on Two-Lane Rural Roads," German Road Congress, German Road and Transportation Research Association, Cologne, Germany, 1984.

46. Behrens, Gotzen, Richter, Stuertz, Suren, Wanderer, and Weber, "Local Accident Inquiries," Research Report of the Federal Highway Research Institute (BASt), Bergisch Gladbach, Germany, 1978.

47. Bergh, T., and A. Carlsson, "Design Criteria and Traffic Performance Research in New Swedish Guidelines on Rural Highways," International Symposium on Highway Geometric Design Practices, Boston, Massachusetts, U.S.A., August 1995.

48. Berkowitz, A., "Motorcycle Fatality Experience Based on FARS Data 1976–1979," U.S. Department of Transportation, National Highway Traffic Safety Administration, Washington, D.C., U.S.A., 1981.

49. Berkowitz, A., "The Effect of Motorcycle Helmet Usage Laws on Head Injuries, and the Effect of Usage Laws on Helmet Wearing Rates," U.S. Department of Transportation, National Highway Traffic Safety Administration, Washington, D.C., U.S.A., 1981.

50. Bickelhaupt, R., "Evaluation of b2 + 1 Cross Sections," Publications of the Institute for Highway and Railroad Engineering, University of Karlsruhe (TH), vol. 49, Germany, 1996, pp. 57–78.

51. *Bild-Zeitung,* "50,000,000 Traffic Fatalities up to the Year 2030," March 15, 1995, p. 2.

52. Bissell, H. H., G. B. Pilkington, J. M. Mason, and D. L. Woods, "Roadway Cross Section and Alignment," *Synthesis of Safety Research Related to Traffic Control and Roadway Elements,* vol. 1, Report No. FHWA-TS-82-232, Federal Highway Administration, Washington, D.C., U.S.A., 1982.

53. Bitzl, F., "Investigations about Accident Causes—Influence of Different Factors," XI. International Road Congress 1959, Rio de Janeiro, *Research Road Construction and Road Traffic Technique (Forschung Strassenbau und Strassenverkehrstechnik),* Minister of Transportation, vol. 15, Bonn, Germany, 1961, pp. 106–108.

54. Bitzl, F., "The Safety Level of Roads," *Research Road Construction and Road Traffic Technique (Forschung Strassenbau und Strassenverkehrstechnik),* vol. 28, Minister of Transportation, Bonn, Germany, 1964.

55. Bitzl, F., "Influence of Road Features on Traffic Safety," German Road and Transportation Research Association, planning meeting, July 1, 1965, Kirschbaum Publishers, Bad Godesberg, Germany.

56. Bitzl, F., and J. Stenzel, "The Influence of Non-Sufficient Sight Distances on Traffic Safety," *Journal for Traffic Laws (Zeitschrift fuer Verkehrsrecht),* vol. 11, Germany, 1966, pp. 233–238.

57. Bitzl, F., "The Influence of Road-Characteristics on Traffic Safety," *Research Road Construction and Road Traffic Technique (Forschung Strassenbau und Strassenverkehrstechnik),* vol. 55, Minister of Transportation, Bonn, Germany, 1967.

58. Bitzl, F., "The Term of the Safety Level in Road Traffice Technique" *Road and Construction (Strassen- und Tiefbau),* vol. 12, Germany, 1960, pp 943–956.

59. Bjornskau, T., "Can Road Traffic Law Enforcement Permanently Reduce the Number of Accidents?" *Proceedings, Road Safety and Traffic Environment in Europe, Gothenburg, Sweden,* VTIrapport 365A, Swedish Road and Traffic Research Institute, Linkoeping, Sweden, 1990, pp. 121–145.

60. Boelcke, W. A., "The Importance of the Road in the 18th and 19th Century," *The Road in the Change of the Centuries, Strabag Proceedings,* vol. 50, Cologne, Germany, 1995, pp. 46–49.

61. Bopp, D., "Program Description of HIDES, Program System CEP," IBM, Frankfurt, 1970.

62. Bosch, R., GmbH, *Automotive Handbook,* 3d ed., Society of Automotive Engineers, Warrendale Pennsylvania, U.S.A., 1993.

63. Boston University, "Impact Calculated in Advance," *Focus, the Modern Newsmagazine,* vol. 10, March 3, Germany, 1997.

64. Braess, H., "Vehicle Engineering Methods for Examining the Side Wind Behavior of the System Driver-Vehicle," *German Motor Vehicle Research and Road Traffic Technique,* vol. 193, Germany, 1968.

65. Braess, H.-H., "Active and Passive Safety in Road Traffic," *Journal for Traffic Safety (Zeitschrift fuer Verkehrssicherheit)* vol. 2, pp. 50–52, 1996.

66. Brannolte, U., "Traffic Flow at Upgrade Sections of Directional Lanes," *Research Road Construction and Road Traffic Technique (Forschung Strassenbau und Strassenverkehrstechnik),* vol. 318, Minister of Transportation, Bonn, Germany, 1980, pp. 1–99.

67. Brannolte, U., and S. Holz, "Simulation of Traffic Flow on Rural Roads, Model Extension," *Research Road Construction and Road Traffic Technique (Forschung Strassenbau und Strassenverkehrstechnik),* vol. 402, Minister of Transportation, Bonn, Germany, 1983.

68. Brannolte, U., J. Dilling, W. Durth, G. Hartkopf, V. Meewes, W. Reichelt, H. Schliesing, and P. Stievermann, "Application of Intermediate Cross Sections," Research Reports of the Federal Highway Research Institute (BASt), Division for Accident Research, vol. 265, Bergisch Gladbach, Germany, 1993.

69. Brannolte, U., et al., "Accident Evaluation for Selected Road Sections with Intermediate Cross Sections," Research Project 8527/6, German Federal Highway Research Institute (BASt), Bergisch Gladbach, Germany, 1991; corresponds to: "Safety Evaluation of Cross Sections of Non Built-Up Roads," Reports of the Federal Highway Research Institute (BASt), Traffic Technique, vol. V 5, Bergisch Gladbach, Germany, 1993.

70. Braun, H., P. Kupke, J. Ihme, D. Maus, and K.-D. Schlichting, "Definition of Critical Situations in Motorized Traffic," Research Report of the Federal Highway Research Institute (BASt), Bergisch Gladbach, Germany, 1982.

71. Brenac, T., "Speed, Safety and Highway Design," *Recherche Transports Sécurité,* English Issue, no. 5, 1990.

72. Brenac, T., *Annex IX (E) Curves on Two-Lane Roads, Safety Effects of Road Design Standards,* SWOV, Leidschendam, The Netherlands, 1994.

73. Brieter, K., "Crash in the Bee-House," General German Automobile Club (ADAC), *Motorworld,* no. 9, Munich, Germany, 1995, pp. 6–10.

74. Brilon, W., "Accident Situation and Traffic Flow," *Research Road Construction and Road Traffic Technique (Forschung Strassenbau und Strassenverkehrstechnik),* vol. 201, Minister of Transportation, Bonn, Germany, 1976.

75. Brilon, W., and M. Doehler, "Track Control on Two-Lane Rural Roads with Opposing Traffic," *Road Traffic Technique (Strassenverkehrstechnik),* vol. 3, Germany, Germany, 1978, pp. 79–82.

76. Brilon, W., and U. Brannolte, "Simulation Model for the Traffic Flow on Two-Lane Roads with Opposing Traffic," *Research Road Construction and Road Traffic Technique (Forschung Strassenbau und Strassenverkehrstechnik),* vol. 239, Minister of Transportation, Bonn, Germany, 1977.

77. Brilon, W., and F. Weiser, "Actualization of the Guidelines for the Design of Roads, Part: Cross Sections (RAS-Q, 1982)," Final Report of the Research Contract, FE-No. 02.151 R 92 F of the Minister of Transportation, Ruhr-University, Institute for Traffic Engineering, Bochum, Germany, 1993.

78. Brilon, W., "Design Fundamentals for Dimensioning Cross Sections Outside Built-Up Areas," Commemorative Volume to the 60th Birthday of Univ.-Prof. Dr.-Ing. Walter Durth, Technical University of Darmstadt, Department: Road Design and Road Operation, Darmstadt, Germany, 1995.

79. Brilon, W., "The New German Guidelines, Part: 'Cross Sections (RAS-Q 96)': What is New?" *Road Traffic Technique (Strassenverkehrstechnik),* vol. 11, Germany, 1996, pp. 529–532.

80. Brinkman, C. P., and K. Perchonok, "Hazardous Effects of Highway Features and Roadside Objects—Highlights," *Public Roads,* vol. 43, no. 1, U.S.A., 1979, pp. 8–14.

81. Brinkman, C. P., "Safety Studies Related to RRR Projects," *Transportation Engineering Journal of ASCE,* vol. 108, 1982, pp. 307–312.

82. Brinkman, C. P., and S. A. Smith, "Two-Lane Rural Highway Safety," *Public Roads,* vol. 8, no. 2, U.S.A., 1984, pp. 48–53.

83. Brown, R. G., "Report on Arrestor Bed Tests Carried out by the NSW Department of Main Roads in March 1982," *Proceedings, 12th ARRB Conference,* vol. 12, no. 7, Australia, 1984, pp. 110–125.

84. Brownlee, K. A., *Statistical Theory and Methodology in Science and Engineering,* John Wiley & Sons, New York, U.S.A., London, United Kingdom, 1960.

85. Brude, U., J. Lassen, and H. Thulin, "Influence of Road Alignment on Traffic Accidents," *VTI Meddelande 235,* Linkoeping, Sweden, 1980.

86. Bruederle, R., "Stronger Penalties for the Non-Usage of Seat Belts," *World on Sunday (Welt am Sonntag)*, no. 5, Germany, February 2, 1997, p. 3.

87. Bruehning, E., and R. Voelker, "Accident Risk of Children and Juveniles as Pedestrians and Drivers of Motorvehicles in Road Traffic," *Journal for Traffic Safety (Zeitschrift fuer Verkehrssicherheit)*, vol. 2, Germany, 1990, pp. 51–60.

88. Bruehning, E., and S. Berns, "Traffic Safety in Eastern and Western Europe at the Beginning of the Nineties," *Proceedings of the Conference Road Safety in Europe, Berlin, Germany,* September 30–October 2, 1992, VTIrapport 380 A, Swedish Road and Traffic Research Institute, Linkoeping, Sweden, 1992.

89. Brummelaar, T., "Where are the Kinks in the Alignment?," *Transportation Research Record,* vol. 556, U.S.A., 1975, pp. 35–50.

90. Bubb, H., "Analysis of Speed Perception in Vehicles," *Journal for Work Sciences (Zeitschrift fuer Arbeitswissenschaften)*, vol. 31, no. 2, Germany, 1977, pp. 103–111.

91. Buehlmann, F., H. P. Lindenmann, and P. Spacek, "Sight Distances, Examination of the Fundamentals to the Swiss Norm SN 640 090, Design Principles, Sight Distances," Institute for Traffic Planning, Transport Technique, Highway- and Railroad Construction, ETH-Zuerich, Publication of the IVT, no. 89, Zuerich, Switzerland, 1991.

92. "Building Instructions for National Autobahnen (BAURAB TG)," *Proceedings: The Road,* vol. 28, Volk and Reich Publishing House, Berlin, Germany, 1943.

93. Burg, H., and H. Rau, *Handbook for the Reconstruction of Traffic Accidents,* Publishers Information Ambs Company, Kippenheim, Germany, 1981.

94. Burkhard, T. W., "Project Study: State Route L 509 (Landau-Eschbach, Germany)," Graduate Study Work, Institute for Highway and Railroad Engineering, University of Karlsruhe (TH), Germany, 1979.

95. Cable News Network (CNN), "Television Report on Helmet Use," Atlanta, Georgia, U.S.A., November 14, 1990.

96. Cameron, M., S. Newstead, and P. Vulcan, "Analysis of Reductions in Victorian Road Casualties, 1989 to 1992," *Proceedings 17th ARRB Conference,* Australia, part 5, 1994, pp. 165–182.

97. Campbell, R. C., *Statistics for Biologists,* Cambridge University Press, Cambridge, United Kingdom, 1974.

98. Carlquist, J. C. A., "Safe or Dangerous? An International Comparison of Traffic Accident Figures," *Highways and Traffic Engineering,* vol. 30, no. 11, 1966.

99. Carré, Lassarre, and M. Liger, "Fluidité—sécurité: faut-il choisir?" *Transport-Environment-Circulation,* vol. 66, 1984, pp. 14–19.

100. Cassione, "The Forth International Road Congress," *Traffic Technique (Verkehrstechnik)*, vol. 48–50, Germany, 1923, pp. 421–423, 430–432, 440–441.

101. Catchpole, J., "Review of the Relationship between Road Shoulder Treatments and Accidents," Appendix C in *Optimum Traffic Lane, Seal and Pavement Widths for Non-Urban Roads,* Contract Report TE 04 0595/1, Australian Road Research Board, Melbourne, Australia, 1990.

102. Ceder, A., and M. Livneh, "Relationship between Road Accidents and Hourly Traffic Flow 1.: Analysis and Interpretation," *Accident Analysis and Prevention,* vol. 14, no. 1, U.S.A., 1982, pp. 19–34.

103. CETUR, "Sécurité des Routes et des Rues," Centre des Etudes des Transports Urbains, Bagneux, France, 1992.

104. Choueiri, E. M., "Statistical Analysis of Operating Speeds and Accident Rates on Two-Lane Rural Highways," dissertation submitted to the Department of Civil Engineering, Clarkson University, Potsdam, N.Y., in partial fulfillment of the requirements for the degree of Doctor of Philosophy, U.S.A., 1987.

105. Choueiri, E. M., and R. Lamm, "A Comparative Analysis of Motorcycle Accident Statistics in Western Europe and the United States, 1970–1987," *Proceedings of the 1991 International Motorcycle Conference, Safety—Environment—Future,* Institute for Two-Wheel-Safety, Bochum, Germany, vol. 7, 1991, pp. 7–29.

106. Choueiri, E. M., "Analysis of Accident Experiences in the U.S.A. and Western European Countries from 1970 through 1980," thesis submitted to the Department of Civil and Environmental Engineering, Clarkson University, Potsdam, N.Y., in partial fulfillment of the requirements for the degree of Master of Science, U.S.A., 1984.

107. Choueiri, E. M., R. Lamm, B. M. Choueiri, and G. M. Choueiri, "An Analysis of Bicycle Accidents in Western Europe and the United States (1975–1989)," *Proceedings of Road Safety in Europe, Berlin, Germany, September 30—October 2, 1992,* VTIrapport 380A, Swedish Road and Traffic Research Institute, Linkoeping, Sweden, 1992, pp. 87–106.

108. Choueiri, E. M., R. Lamm, B. M. Choueiri, and G. M. Choueiri, "The Relationship between Highway Safety and Geometric Design Consistency: A Case Study," *Proceedings of the Conference Road Safety in Europe and Strategic Highway Research Program (SHRP), Lille, France, September 26–28, 1994,* VTI konferens, no. 2A, part 2, 1995, pp. 133–151.

109. Choueiri, E. M., R. Lamm, J. H. Kloeckner, and T. Mailaender, "Safety Aspects of Individual Design Elements and Their Interactions on Two-Lane Highways: International Perspective," *Transportation Research Record,* vol. 1445, U.S.A., 1994, pp. 34–46.

110. Choueiri, E. M., R. Lamm, G. M. Choueiri, and E. M. Choueiri, "An International Investigation of Road Traffic Accidents," *Proceedings of the Conference Road Safety in Europe and Strategic Highway Research Program (SHRP), Lille, France, September 26–28, 1994,* VTI konferens, no. 2A, part 1, 1995, pp. 67–98.

111. Cirillo, J. A., "Interstate System Accident Research Study II, Interim Report II," *Public Roads,* vol. 35, no. 3, pp. 71–75, U.S.A., 1968.

112. Cirillo, J. A., and F. M. Council, "Highway Safety: Twenty Years Later," Transportation Research Circular 1068, Transportation Research Board, Washington, D.C., U.S.A., 1986, pp. 90–95.

113. Cleveland, D. E., and R. Kitamura, "Macroscopic Modeling of Two-Lane Rural Roadside Accidents," *Transportation Research Record,* vol. 681, U.S.A., 1978, pp. 53–62.

114. Cleveland, D. E., L. P. Kostyniak, and K. L. Ting, "Geometric Design Element Groups and High-Volume Two-Lane Rural Highway Safety," *Transportation Research Record,* vol. 960, U.S.A., 1994, pp. 1–13.

115. Coburn, T. M., "Accident Speed and Layout Data on Rural Roads in Buckinghamshire," Road Research Laboratory, Crowthorne, United Kingdom, 1952.

116. Coburn, T. M., "The Relation between Accidents and Layout on Rural Roads," *International Road Safety and Traffic Review,* vol. 10, 1962, pp. 15–20.

117. Cocks, G. C., and L. W. Goodram, "The Design of Vehicle Arrestor Beds," *Proceedings, 11th ARRB Conference,* vol. 11, no. 3, Australia, 1982, pp. 24–34.

118. Cohen, A. S., "Influencing Factors on the Usuable Field of Vision," Research Reports of the Federal Highway Research Institute (BASt), vol. 100, Division for Accident Research, Bergisch Gladbach, Germany, 1984.

119. Cohen, A. S., "Information Deficits during Driving at Night," in *Possibilities and Limitations of Visual Perception in Road Traffic, Proceedings—Accident and Safety Research in Road Traffic,* vol. 57, pp. 45–78, Federal Highway Research Institute (BASt), Bergisch Gladbach, Germany, 1986.

120. Cohen, A. S., "Perception and Estimation of Speeds," in *Possibilities and Limitations of Visual Perception in Road Traffic, Proceedings—Accident and Safety Research in Road Traffic,* Federal Highway Research Institute (BASt), vol. 57, Bergisch Gladbach, Germany, 1986.

121. Cohen, A. S., "Endangerment of Novice Drivers as well as Acquiring a Traffic Justified Orientation and Their Deficits at Night," *Journal for Traffic Safety (Zeitschrift fuer Verkehrssicherheit),* vol. 4, Germany, 1994, pp. 156–161.

122. Cohen, A. S., "Traffic Signs as Disturbed Communication System," *Journal for Traffic Safety (Zeitschrift fuer Verkehrssicherheit),* vol. 2, Germany, 1995, pp. 73–77.

123. Cohen, A. S., "Traffic Justified Visual Information Input," *Journal for Traffic Safety (Zeitschrift fuer Verkehrssicherheit),* vol. 2, Germany, 1997, pp. 83–85.

124. Commentary to the Guidelines for the Design of Rural Roads, Part: Alignment, Section: "Parameters of the Alignment (RAL-L-1, 1973)," German Road and Transportation Research Association, Cologne, Germany, 1979.

125. Committee of State Road Authorities, "Geometric Design of Rural Roads," Draft TRH 17, Pretoria, South Africa, 1984.

126. Consiglio Nazionale Delle Ricerche, "Norme sulle Caratteristiche Geometriche delle Strade Extraurbane," Rome, Italy, 1980.

127. Consumer Union, "Bike Helmets," *1992 Buying Guide Issue,* U.S.A., December 1991.

128. Cope, A. J., "Traffic Accident Experience—Before and After Pavement Widening," *Traffic Engineering,* U.S.A., 1955, pp. 114–115.

129. Cron, F., "The Art of Fitting the Highway to the Landscape," in B. Snow, *The Highway and The Landscape,* Rutgers University Press, New Brunswick, N.J., U.S.A., 1959, pp. 78–109.

130. Dames, J., R. Merckens, and J. Bergmann, "New Determination of the Evaluation Background for the Results of Skid Resistance Measurements," *Research Road Construction and Road Traffic Technique (Forschung Strassenbau und Strassenverkehrstechnik),* vol. 413, Minister of Transportation, Bonn, Germany, 1984.

131. Dames, J., "Data Collection and Evaluation of the Skid Resistance of Roads," *Road and Interstate (Strasse und Autobahn),* vol. 2, Germany, 1992, pp. 88–93.

132. Dart, O. K., and L. Mann, "Relationship of Rural Highway Geometric to Accident Rates in Louisiana," *Highway Research Record,* vol. 312, U.S.A., 1970, pp. 1–16.

133. Dart, O. K., and W. W. Hunter, "An Evaluation of the Halo Effect in Speed Detection and Enforcement," *Transportation Research Record,* vol. 609, U.S.A., 1976.

134. Datta, T., D. Perkins, J. Taylor, and H. Thompson, "Accident Surrogates for Use in Analyzing Highway Safety Hazards," Report no. FHWA/RD-82/104, Federal Highway Administration, Washington, D.C., U.S.A., 1983.

135. Deacon, J., "Relationship between Accidents and Horizontal Curvature," Appendix D, "Designing Safer Roads," in *Practices for Resurfacing, Restoration, and Rehabilitation,* Transportation Research Board, SR 214, Washington, D.C., U.S.A., 1986.

136. Delmarcelle, A., "Problem of Accident Black Spots Development," Report to World Congress Montreal, PIARC Committee C13, Road Safety (Draft), Canada, September 4–8, 1995.

137. Department of Transport, "Road Safety Audits," Advice Note HA 42/90, Department of Transport: Highway, Safety and Traffic, United Kingdom, 1990.

138. Department of Transport, "Road Safety Audits," Department Standard HD 19/90, Department of Transport: Highway, Safety and Traffic, United Kingdom, 1990.

139. Department of Transport, "Road Layout and Geometry—Highway Link Design," Departmental Standard TD 9/81 and TD 9/93, Her Majesty's Stationery Office, London, 1981, 1993 and, Department of Transport, Departmental Advice Note TA 43/84 and TA 43/94, Subject Area, "Highway Link Design, Geometric Alignment Standards," United Kingdom, 1984, 1993.

140. Department of Transport, Scottish Office & Welsh Office, "Road Safety Report," United Kingdom, 1995.

141. Dienst Verkeerskunde, "Guidelines for the Design of Niet Autosnelwegen," Commissie Rona, Rijkswaterstaat, The Netherlands, 1980–1986.

142. Dietrich, K., E. Boppart, H.-P. Lindenmann, and P. Spacek, "Intermediate Types—An Investigation on Possible Operational Forms of Major Highways," Federal Technical University of Zuerich, Institute for Traffic Planning and Transport Technique, ETH Zuerich, Publication of the IVT, vol. 83, no. 2, Zuerich, Switzerland, 1983.

143. Dilling, J., "Driving Behavior of Motor Vehicles at Curved Roadway Sections," *Research Road Construction and Road Traffic Technique (Forschung Strassenbau und Strassenverkehrstechnik),* vol. 151, Minister of Transportation, Bonn, 1973.

144. Donges, E., "A Cybernetic Two-Level Model of the Human Steering Behavior of the Motor Vehicle," *Journal for Traffic Safety (Zeitschrift fuer Verkehrssicherheit),* vol. 3, Germany, 1978, pp. 98–112.

145. Donges, E., "Aspects of Active Safety when Steering Passenger Cars," *Automobile-Industry (Automobil-Industrie),* vol. 2, Germany, 1982, pp. 183–190.

146. Dorsch, M. M., A. J. Woodward, and R. L. Somers, "Do Bicycle Safety Helmets Reduce Severity of Head Injury in Real Crashes?" *Accident Analysis & Prevention,* vol. 19, no.3, pp.183–190, U.S.A., 1987.

147. Draaisma, J. M. T., "Evaluation of Trauma Care—with Emphasis on Hospital Trauma Care," Nijmegen, The Netherlands, 1987.

148. Dunlap, D. F., P. S. Fancher, R. E. Scott, C. C. MacAdam, and L. Segel, "Influence of Combined Highway Grade and Horizontal Alignment on Skidding," NCHRP Report 184, Transportation Research Board, National Research Council, Washington, D.C., U.S.A., 1978.

149. Durth, W., "A Contribution for the Extension of the Model for Driver, Vehicle and Road in Highway Planning," *Research Road Construction and Road Traffic Technique* (*Forschung Strassenbau und Strassenverkehrstechnik*), vol. 163, Minister of Transportation, Bonn, Germany, 1974.

150. Durth, W., G. Koerner, and K. Manns, "Examination of Driving Specific Basic Values of the German Guidelines for the Design of Rural Roads (RAL-L)," *Research Road Construction and Road Traffic Technique* (*Forschung Strassenbau und Strassenverkehrstechnik*), vol. 365, Minister of Transportation, Bonn, Germany, 1982.

151. Durth, W., B. Biedermann, and B. Vieth, "Influences of the Increase of Speeds and Accelerations of Motor Vehicles on the Design Speed," *Research Road Construction and Road Traffic Technique* (*Forschung Strassenbau und Strassenverkehrstechnik*), vol. 385, Minister of Transportation, Bonn, Germany, 1983.

152. Durth, W., and K. Habermehl, "Passing Maneuvers on Two-Lane Rural Roads," *Research Road Construction and Road Traffic Technique* (*Forschung Strassenbau und Strassenverkehrstechnik*), vol. 489, Minister of Transportation, Bonn, Germany, 1986.

153. Durth, W., and G. Beys-Kammarokos, "Comparison of the Guidelines for Road Design in the Countries of the European Community," Research Contract "Project Road Safety Year-Grant no. VII-B-336" of the Commission of the European Community, Technical University of Darmstadt, Department: Road Design and Road Operation, Darmstadt, Germany, 1987.

154. Durth, W., and N. Wolff, Part I: "Examination of the Effectiveness of Relation Design for Roads of Category Group B," Part II: "Examination of the Effectiveness of Relation Design for Roads of Category Group C (Local Through Roads)," *Research Road Construction and Road Traffic Technique* (*Forschung Strassenbau und Strassenverkehrstechnik*), vol. 540, Minister of Transportation, Bonn, Germany, 1988.

155. Durth, W., and J. S. Bald, "Risk Analysis in Road Engineering," *Journal for Traffic Safety* (*Zeitschrift fuer Verkehrssicherheit*), vol. 1, Germany, 1989, pp. 17–24.

156. Durth, W., and C. Levin, "Long Term Development of Eye and Vehicle Heights," *Road and Interstate* (*Strasse und Autobahn*), vol. 12, Germany, 1991, pp. 684–686.

157. Durth, W., and C. Levin, "Specified Design of Crest Vertical Curves," *Research Road Construction and Road Traffic Technique* (*Forschung Strassenbau und Strassenverkehrstechnik*), vol. 600, Minister of Transportation, Bonn, Germany, 1991.

158. Durth, W., E. Dengler, and T. Ferrero, "Literature Review: Alignment and Cross Section," Contract of the German Federal Research Institute (BASt), Technical University of Darmstadt, Department: Road Design and Road Operation, Darmstadt, Germany, 1991.

159. Durth, W., and C. Lippold, "Adjustment of the German Design Guidelines for the Alignment (RAS-L-1, 1984) to Newer Design Guidelines," Research Contract FE-no. 6.2.2/91 of the Federal Minister of Transportation, Technical University of Darmstadt, Department: Road Design and Road Operation, Darmstadt, Germany, 1993.

160. Durth, W., and C. Lippold, "Fundamentals of Road Design in the Federal Republic of Germany," Technical University of Darmstadt, Department: Road Design and Road Operation, Darmstadt, Germany, 1993.

161. Durth, W., E.-U. Hiershe, R. Lamm, et al., "Winter Service on Bicycle Lanes," Research Contract FE 77329/90 of the Federal Minister of Transportation, Final Report, Darmstadt, Karlsruhe, Germany, 1995.

162. Durth, W., and C. Lippold, "Proposal for the New Guidelines for the Design of Roads, Part: Alignment (RAS-L, 1994)," *Road and Interstate* (*Strasse und Autobahn*), vol. 2, Germany, 1995, pp. 80–86.

163. Durth, W., G. Weise, A. Bark, C. Lippold, and A. Sòssoumihen, "Examination of the Relation Design of Roads of Category Group A," Interim Report, Research Contract 02.153 R 93 E of the German Minister of Transportation, Darmstadt, Germany, 1995.

164. Eberhard, O., "Development of an Operating Speed Background for Roadway Sections with Grades ≥ 6 Percent, as well as Analysis and Evaluation of Selected Road Sections, Based on Three Safety Criteria," Masters Thesis, Institute for Highway and Railroad Engineering, University of Karlsruhe (TH), Germany, 1997.

165. Economic Commission for Europe, "Statistics of Road Traffic Accidents in Europe and North America," United Nations, New York, and Geneva, Switzerland, 1995.

166. EJPD Expert Group Traffic Safety of BAP, "Safety in Road Traffic: Strategies and Measures for the 1990s," EJPD, Bern, Switzerland, 1993.

167. Emde, W., et. al. "Uniform Costs for the Economic Evaluation of Road Traffic Accidents," *Road and Interstate (Strasse und Autobahn),* vol. 9, Germany, 1979, pp. 397–398.

168. Emde, W., et. al. "Costs for the Economic Evaluation of Road Traffic Accidents, Price Status 1985," *Road and Interstate (Strasse und Autobahn),* vol. 4, Germany, 1985, pp. 159–162.

169. Emmerson, J., "A Note on Speed—Road Curvature Relationships," *Traffic Engineering & Control,* vol. 12, no. 7, United Kingdom, 1970, p. 369.

170. Engineering Office Dr. Burg and Partner, "CARAT, Computer Aided Reconstruction of Accidents in Traffic," Personal Informations of Dr. Burg and Handbook *Vehicle Dynamics and Collision Analysis CARAT 1.0,* IbB, Wiesbaden, 1995, Germany.

171. Ervin, R. D., C. C. Mac Adam, and M. Barnes, "Influence of the Geometric Design of Highway Ramps on the Stability and Control of Heavy-Duty Trucks," *Transportation Research Record 1052,* Washington, D.C., U.S.A., 1985.

172. Ervin, R. D., R. L. Nisonger, C. C. Mac Adam, and P. S. Fancher, "Influence of Size and Weight Variables on the Stability and Control Properties of Heavy Trucks," Report no. FHWA/RD-83/029, Federal Highway Administration, Washington, D.C., U.S.A., 1986.

173. "European News about Traffic Safety," European Traffic Safety Federation and Supported by the European Commission, no. 6, 1996, pp. 7, 10, and 11.

174. European Transport Safety Council, "Reducing Traffic Injuries through Vehicle Safety Improvements; The Role of Car Design," ETSC, Brussels, Belgium, 1993.

175. European Transport Safety Council, "Reducing Traffic Injuries Resulting from Excess and Inappropriate Speed," ETSC, Brussels, Belgium, 1995.

176. Euting, *The Construction of Rural Roads,* Teubner-Publishing House, Leipzig and Berlin, Germany, 1920.

177. Evans, L., and P. Wasielewski, "Do Accident-Involved Drivers Exhibit Riskier Everyday Driving Behavior?" *Accident Analysis and Prevention,* vol. 14, U.S.A., 1982, pp. 57–64.

178. Evans, L., and M. C. Frick, "Helmet Effectiveness in Preventing Motorcycle Driver and Passenger Fatalities," *Accident Analysis and Prevention,* vol. 20, no. 6, U.S.A., 1988, pp. 447–458.

179. Evans, L., *Traffic Safety and the Driver,* Van Nostrand Reinhold, New York, U.S.A., 1991.

180. Farah, E., "La Route Sanglante: Les Chiffres, Les Causes," le Commerce du levant, Beirut, Lebanon, 1995.

181. Farm, G., "Opening Speech," *Proceedings, Road Safety and Traffic Environment in Europe, Gothenburg, Sweden,* VTIrapport 362A, Swedish Road and Traffic Research Institute, Linkoeping, Sweden, 1990, pp. 1–5.

182. "Fatal Accident Reporting System," NHTSA, U.S. Department of Transportation, National Center for Statistics and Analysis, Washington, D.C., U.S.A., editions up to 1985.

183. Federal Highway Research Institute (BASt), "Guidelines for the Design of Auxiliary Lanes on Upgrade Sections," Bergisch Gladbach, Germany, 1985.

184. Federal Highway Research Institute (BASt), "International Road Traffic and Accident Database," Bergisch, Gladbach, Germany, 1995.

185. Federal Highway Research Institute (BASt), "Belts, Child Seats, Helmets, and Protective Clothing," Information 8, Bergisch Gladbach, Germany, 1996.

186. Federal Ministry of Transportation, "Safety in Road Traffic; Report on Accident Prevention in Road Traffic and Rescue Operations in the Years 1992 and 1993," Federal Printed Matter 12/8335, Bonn, Germany, 1994.

187. Fee, J. A., "European Experience in Pedestrian and Bicycle Facilities," U.S. Department of Transportation, Federal Highway Administration, Washington, D.C., U.S.A., 1975.

188. Felke, D., "The Effectiveness of Speed Limits by Police Surveillance," in *Police Information of the State of Hessen, Germany,* vol. 2, 1980, pp. 10–15.

189. Felke, D., "Activities by Police in Traffic—A Contribution to Traffic Safety," *Journal for Traffic Safety (Zeitschrift fuer Verkehrssicherheit),* vol. 2, Germany, 1983, pp. 63–66.

190. Felke, D., "15 Years of Traffic Surveillance at the Elzer Mountain," in *Police Information of the State of Hessen, Germany,* vol. 11, 1988, pp. 19–22.

191. Feuchtinger, M. E., and C. Christoffers, "Driving Dynamic Investigations as Measure of Road-Traffic-Safety," *Journal for Traffic Safety (Zeitschrift fuer Verkehrssicherheit),* vol. 1, Germany, 1953.

192. Fieldwick, P. T., and R. J. Brown, "The Effect of Speed Limits on Road Casualties," *Traffic Engineering and Control,* vol. 28, no. 12, 1987.

193. Fife, D., J. Davis, L. Tate, J. K. Wells, D. Mohan, and A. Williams, "Fatal Injuries to Bicyclists: The Experience of Dade County, Florida," *The Journal of Trauma,* vol. 23, U.S.A., 1983.

194. Finch, D. J., P. Kompfner, C. R. Lockwood, and G. Maycock, "Speed, Speed Limits and Accidents," Project Report 58, Transport Research Center, Crowthorne, United Kingdom, 1994.

195. Fink, K. L., and R. A. Krammes, "Tangent Length and Sight Distance Effects of Accident Rates at Horizontal Curves on Rural Two-Lane Highways," paper presented at Transportation Research Board, 74th Annual Meeting, Washington, D.C., U.S.A., 1995.

196. Fiolic, R., "Crawling Lanes on Upgrades," Federal Ministry for Buildings and Technique, *Road Research,* vol. 57, Vienna, Austria, 1976.

197. Fites, L. A., and M. M. Jacobs, "Fundamentals of Geometric Design," Institute of Transportation and Traffic Engineering, University of California, Berkeley, California, U.S.A., 1971.

198. Flagstad, G., "Trafikksikkerhet og Vegutforming," *Norsk Vegtidsskrift,* vol. 43, Norge, 1967, pp. 81–87.

199. Foody, T. J., and T. B. Culp, "A Comparison of the Safety Potential of the Raised vs. Depressed Median Design," paper presented at the 1974 Annual Meeting, Transportation Research Board, Washington, D.C., U.S.A., 1974.

200. Foody, T. J., and M. D. Long, "The Identification of Relationships between Safety and Roadway Obstructions," Columbus Bureau of Traffic, Ohio Department of Transportation, 1974.

201. Frost, U., and J. Morrall, "A Comparison and Evaluation of the Geometric Design Practices with Passing Lanes, Wide-Paved Shoulders and Extra-Wide Two-Lane Highways in Canada and Germany," paper presented at the International Symposium on Highway Geometric Design Practices, Transportation Research Board, Boston, Massachusetts, U.S.A., 1995.

202. Gambard, J.-M., and G. Louah, "Vitesses Pratiquées et Géométrie de la Route," SETRA, Paris, France, 1986.

203. Garner, G. R., and R. C. Deen, "Elements of Median Design in Relation to Accident Occurrence," paper prepared for Transportation Research Board Annual Meeting, Washington, D.C., U.S.A., 1973.

204. Gartner, W. B., and M. R. Murphy, "Pilot Workload and Fatigue: a Critical Survey of Concepts and Assessment Techniques," NASA TN-D-8365, National Aeronautics and Space Administration, Washington, D.C., U.S.A., 1976.

205. Geissler, E. H., "A Three Dimensional Approach to Highway Alignment Design," *Highway Research Record 232,* U.S.A., 1968, pp. 16–28.

206. Geissler, E. H., and A. Aziz, "Evaluation of Complex Interchange Designs by Three Dimension Models," *Highway Research Record,* Washington, D.C., U.S.A., vol. 270, 1969.

207. German Railroad Inc., "Design of Railroad Facilities—General Design Fundamentals," Official Regulations (DS 800 01), Business Sector: Network, Munich, Germany, 1993.

208. German Road and Transportation Research Association, "German Standard for Road Skid Resistance and Traffic Safety for Wet Conditions," Cologne, Germany, 1968.

209. German Road and Transportation Research Association, *German Road Congress,* Berlin, 1980, Kirschbaum Publishers, Bonn, Germany, 1981.

210. German Road and Transportation Research Association, *Guidelines for Passive Protection Facilities (RPS),* Cologne, Germany, 1989 ed.

211. German Road and Transportation Research Association, "Methodology for the Investigation of Road Traffic Accidents," Cologne, Germany, 1991.

212. Gerondeau, C. (chmn.), "Report of the High Level Expert Group for a European Policy for Road Safety," OECD, Paris, France, 1991.

213. Gibson, J. J., *The Perception of the Visual World,* Houghton Mifflin, Boston, Massachusetts, U.S.A., 1950.

214. Gimotty, P. A., et al., "Statistical Analysis of the National Crash Severity Study Data," University of Michigan, Highway Safety Research Institute, Ann Arbor, Michigan, U.S.A., 1980.

215. Glaeser, H. H., *Novel of the Road,* Udo Pfriemer Book Publishing Company, Wiesbaden, Germany, 1987.

216. Glennon, J. C., "An Evaluation of Design Criteria for Operating Trucks Safety on Grades," Texas Transportation Institute, *Highway Research Record 312,* Washington, D.C., U.S.A., 1970, pp. 93–112.

217. Glennon, J. C., T. R. Neuman, and J. E. Leisch, "Safety and Operational Considerations for Design of Rural Highway Curves," Report no. FHWA/RD-86/035, Federal Highway Administration, Department of Transportation, Washington, D.C., U.S.A., 1985.

218. Glennon, J. C., "Effect of Alignment on Highway Safety: A Synthesis of Prior Research," TRB State-of-the-Art Report 6, Transportation Research Board, Washington, D.C., U.S.A., 1987.

219. Glennon, J. C., "Effect of Sight Distance on Highway Safety, Relationship between Safety and Key Highway Features," State-of-the-Art Report 6, Transportation Research Board, Washington, D.C., U.S.A., 1987, pp. 64–77.

220. Godthelp, H., "Vehicle Control During Curve Driving," *Human Factors,* vol. 28, no. 2, U.S.A., 1986, pp. 211–221.

221. Godthelp, H., G. J. Blaauw, and J. Moraal, "Studies on Vehicle Guidance and Control," Transportation Research Record, vol. 1047, U.S.A., 1986, pp. 21–29.

222. Goerich, H.-J., "System for the Establishment of the Actual Friction Potential of a Passenger Car in Operation," Association of German Engineers (VDI), Progress-Report 12, no. 181, Duesseldorf, Germany, 1993.

223. Goldberg, S., "Detailed Investigation of Accidents on National Roads in France," *International Road Safety and Traffic Review,* vol. 10, 1962, pp. 23–31.

224. Goldenbeld, C., "Police Enforcement: Theory and Practice," paper presented at PTRC Summer Annual Meeting 1995, SWOV, Leidschendam (Draft), The Netherlands, 1995.

225. Goodell-Grivas, Inc., "Highway Safety Evaluation, Procedural and Instructors Guide," Prepared for: U.S. Department of Transportation, Federal Highway Administration, Washington, D.C., U.S.A., 1981.

226. Gopher, D., and R. Braune, "On the Psychophysics of Workload: Why Bother with Subjective Measures?" *Human Factors,* vol. 26, no. 5, U.S.A., 1984, pp. 519–532.

227. Gordon, D. A., "Static and Dynamic Visual Fields in Human Space Perception," *Journal Optical Society of America,* vol. 55, U.S.A., 1965, pp. 1296–1302.

228. Goyal, P. B., "Friction Factors for Highway Design Regarding Driving Dynamic Safety Concerns in the State of New York," Masters Thesis, Clarkson University, Potsdam, N.Y., U.S.A., 1987.

229. Grabner, J., and R. Nothhaft, *Design of Passenger Car Bodies,* Springer Verlag, Berlin, Heidelberg, Germany, 1991.

230. Graevell, "Superelevation and Widening in Chaussee-Curves," *Traffic Technique (Verkehrstechnik),* vols. 5 and 9, Germany, 1920.

231. Graham, J. D., and Y. Lee, "Behavioral Response for Safety Regulation: The Case of Motorcycle Helmet-Wearing Legislation," *Policy Sciences,* vol. 19, U.S.A., 1986.

232. Graham, J. L., and D. W. Harwood, "Effectiveness of Clear Recovery Zones," NCHRP Report no. 247, Transportation Research Board, Washington, D.C., U.S.A., 1982.

233. Grandel, J., and C.-F. Mueller, "Vehicle Safety and Crash-Tests," DEKRA, *Technical Journal 49,* U.S.A., 1995.

234. Greek Ministry for Public Works, "Cross Section Design of Greek Roads," Bulletin no. 103/1E, 60-62, Direction for the Design of Roads and Bridges, Athens, Greece, 1962.

235. Gregory, R., *Eye and Brain: Psychophysiology of Vision,* Fischer Publishing House, Stuttgart, Germany, 1972.

236. Guenther, A. K., and R. Lamm, "New Developments in the CAD-Field for Highway and Railroad Engineering," Engineering Qualification for the Year 2000, 1st Symposium for Computer Operation in Technical Education, Karlsruhe, Germany, 1989, *Computer and Education,* vol. 2, Leuchtturm Publishers, Alsbach, Germany, pp. 91–105.

237. Guenther, A. K., and Lamm R., "Road Design and Construction with the Data-Processing Program System: Combination," *Road and Construction (Strassen und Tiefbau),* vol. 3, Germany, 1990, pp. 14–20.

238. Guenther, A. K., and R. Lamm, "A VERBUND ut—es vasutepitesi szamitogepes tervező programrend-szer (Data Processing System for the Design of Roads and Railroads: Combination)," *Közlekedesepites—es melyepitestudomanyi szemle (Journal for Traffic and Construction Sciences),* vol. 11, Germany, 1991, pp. 443–452 (in Hungarian).

239. Guenther, F., "The Behavior of Passenger Car Tires under Longitudinal and Side Forces," Institute for Mechanical-Engineering, University of Karlsruhe (TH), Germany, 1992.

240. Guidelines for the Design of Rural Roads (RAL), Part II: "Alignment (RAL-L)," Section 2: "Three Dimensional Alignment," German Road and Transportation Research Association, Cologne, Germany, 1970.

241. Guidelines for the Design of Rural Roads (RAL), Part II: "Alignment (RAL-L)," Section 1: "Elements of the Alignment (RAS-L-1)," German Road and Transportation Research Association, Cologne, Germany, 1973.

242. Guidelines for the Design of Roads (RAL), Part: "Road Network (RAL-N)," German Road and Transportation Research Association, Cologne, Germany, 1977.

243. Guidelines for the Design of Roads (RAS), Part: "Alignment (RAS-L)," Section 1: "Elements of the Alignment (RAS-L-1)," German Road and Transportation Research Association, Cologne, Germany, 1984.

244. Guidelines for the Design of Roads (RAS), Part: "Guide for the Functional Classification of the Road Network (RAS-N)," German Road and Transportation Research Association, Cologne, Germany, 1988.

245. Guidelines for the Design of Roads, Part: "Landscape Cultivation (RAS-LP): Section 1: Landscape Justified Planning, Section 2: Realization of Landscape Cultivation," German Road and Transportation Research Association, Cologne, Germany, 1993.

246. Guidelines for the Design of Roads (RAS), Part: "Alignment (RAS-L)," proposals, 1993 and 1995, elaborated in connection with Ref. 159; New Edition of RAS-L, German Road and Transportation Research Association, Cologne, Germany, 1995.

247. Guidelines for the Design of Roads (RAS), Part: "Cross Sections (RAS-Q)," German Road and Transportation Research Association, 1982; proposal 1993, Elaborated in Connection with Ref. 77; New Edition of RAS-Q, Cologne, Germany, 1996.

248. Gwynn, D. W., "Relationship of Accident Rates and Accident Involvements with Hourly Volumes," *Traffic Quarterly,* vol. 29, 1967, pp. 407–418.

249. Habermehl, K., "A Contribution for the Design of Passing Sections," Ph.D. Dissertation, Technical University of Darmstadt, Department: Road Design and Road Operation, Darmstadt, Germany, 1987.

250. Habermehl, K., *Sight Distances in Highway Design,* Commemorative Volume to the 60th Birthday of Univ.-Prof. Dr.-Ing. Walter Durth, Technical University of Darmstadt, Department: Road Design and Road Operation, Darmstadt, Germany, 1995.

251. Hacker, W., and P. Richer, *Physical Stress,* Springer Verlag, Berlin, Heidelberg, Germany, 1984.

252. Hagenzieker, M. P., "Strategies to Increase the Use of Restraint Systems," *Proceedings of a Workshop Organized by SWOV and VTI,* Gothenburg, Sweden, September 18–20, 1991, R-91-60, SWOV, Leidschendam, The Netherlands, 1991.

253. Hajela, G. P., "Compiler, Resume of Tests on Passenger Cars on Winter Driving Surfaces, 1939–1966," National Safety Council, Committee on Winter Driving Hazards, Chicago, 1968, 165 pp.

254. Hakkert, A. S., R. E. Allsop, and W. A. Leutzbach, "Comparison of Road Safety in the Federal Republic of Germany and Great Britain," Federal Highway Research Institute (BASt), Bergisch, Gladbach, Germany, 1987.

255. Halcrow Fox and Associates, "The Effect on Safety of Marginal Design Elements," A Report for the Department of Transport, United Kingdom, 1981.

256. Hale, A. R., and M. Hale, "Accidents in Perspective," *Occupational Psychology,* vol. 44, U.S.A., 1970, pp. 115–122.

257. Hall, L. E., R. D. Powers, D. S. Turner, W. Brilon, and S. W. Hall, "Overview of Cross Section Design Elements," International Symposium on Highway Geometric Design Practices, Transportation Research Board, Boston, Massachusetts, U.S.A., August 1995.

258. Hanke, H., "Effects of Alignment Design on Traffic Safety," *VSVI Journal,* vol. 1, Wiesbaden, Germany, 1995.

259. Hartkopf, G., "Evaluation of Intermediate Cross Sections of Roads," *Road and Interstate (Strasse und Autobahn),* vol. 11, Germany, 1987.

260. Hartkopf, G., "The Functional Categorization of Highways and its Importance for Highway Design," *Road and Interstate (Strasse und Autobahn),* vol. 6, Germany, 1988, pp. 210–222.

261. Hartunian, N. S., C. N. Smart, T. R. Willemain, and P. Zador, "The Economics of Deregulation: Lives and Dollars Lost Due to Repeal of Motorcycle Helmet Laws," *Journal of Health Politics, Policy and Law,* vol. 8, U.S.A., 1983.

262. Harwood, D. W., and A. D. St. John, "Passing Lanes and Other Operational Improvements on Two-Lane Highways," Report no. FHWA/RD-85/028, Federal Highway Administration, Washington, D.C., U.S.A., 1985.

263. Harwood, D. W., "Multilane Design Alternatives for Improving Suburban Highways," NCHRP Report no. 282, Transportation Research Board, Washington, D.C., U.S.A., 1986.

264. Harwood, D. W., and C. J. Hoban, "Low-Cost Methods for Improving Traffic Operations on Two-Lane Roads: Informational Guide," Report no. FHWA-IP-87-2, Federal Highway Administration, Washington, D.C., U.S.A., 1987.

265. Harwood, D. W., J. M. Mason, W. D. Glauz, B. T. Kulakow, and K. Fitzpatric, "Truck Characteristics for Use in Highway Design and Operation," Report nos. FHWA-RD-89-226 and 227, Federal Highway Administration, Washington, D.C., U.S.A., 1990.

266. Harwood, D. W., "Synthesis of Highway Practice 191: Use of Rumble Strips to Enhance Safety," Transportation Research Board, National Research Council, Washington, D.C., U.S.A., 1993.

267. Harwood, D. W., and J. M. Mason, "Horizontal Curve Design for Passenger Cars and Trucks," paper presented at the 72nd Annual Meeting of the Transportation Research Board, Washington, D.C., U.S.A., 1993.

268. Harwood, D. W., D. B. Fambro, B. Fishburn, H. Joubert, R. Lamm, and B. Psarianos, "International Sight Distance Design Practices," International Symposium on Highway Geometric Design Practices, Transportation Research Board, Boston, Massachusetts, U.S.A., August 1995.

269. Hauer, E., "International Symposium on the Effects of Speed Limits on Traffic Accidents and Transport Energy Use: Speed Enforcement and Speed Choice," Organization for Economic Cooperation and Development, Paris, France, 1981.

270. Hayward, J. C., "Highway Alignment and Superelevation: Some Design—Speed Misconceptions," *Transportation Research Record,* vol. 757, U.S.A., 1980, pp. 22–25.

271. Hayward, J. C., R. Lamm, and A. Lyng, "Geometric Design," *Survey of Current Geometric and Pavement Design Practices in Europe,* International Road Federation, Washington, D.C., U.S.A., 1985.

272. Head, J. A., "Predicting Traffic Accidents from Roadway Elements on Urban Extensions of State Highways," *Highway Research Board Bulletin,* vol. 208, U.S.A., 1959, pp. 45–63.

273. Hedman, K. O., "Road Design and Safety," *Proceedings of Strategic Highway Research Program and Traffic Safety on Two Continents,* Gothenburg, 1989, VTI Report 315A, Sweden, 1990.

274. Heger, R., "Driving Behavior and Driver Mental Workload as Criteria for Highway Geometric Design Quality," paper presented at the International Symposium on Highway Geometric Design Practices, Boston, Massachusetts, U.S.A., 1995.

275. Heger, R., "Driving Behavior and Mental Workload," unpublished manuscript, Dresden, Germany, 1996.

276. Heger, R., and G. Weise, "Driving Behavior and Activation of Motorists as Criteria for Evaluating the Alignment of Two-Lane Rural Roads," in B. Schlag, *Progresses of the Traffic Psychology 1996,* Congress Report, German Psychologists Publishing House, Bonn, Germany, 1997.

277. Heinemann, R., "Swept Curves, Functional Determination and Application on the Geometry of the Travelled Way," Publications of the Institute for Highway and Railroad Engineering, vol. 9, University of Karlsruhe (TH), Germany, 1972.

278. Heinrich, H. C., "Aspects of Visual Perception in Road Traffic," *Journal for Traffic Safety* (*Zeitschrift fuer Verkehrssicherheit*), vol. 1, Germany, 1987, pp. 42–45.

279. Heinrich, H. C., and K. M. Porschen, "The Importance of Interactive Accident Models for Traffic Safety Research," *Journal for Traffic Safety* (*Zeitschrift fuer Verkehrssicherheit*), vol. 1, Germany, 1989, pp. 8–16.

280. Helander, M., "Drivers' Physiological Reactions and Control Operations as Influenced by Traffic Events," *Journal for Traffic Safety* (*Zeitschrift fuer Verkehrssicherheit*), vol. 3, Germany, 1974, pp. 174–187.

281. Helms, "Following Costs of Road Traffic Accidents 1968 by Damage Modes," *Journal for Traffic Safety* (*Zeitschrift fuer Verkehrssicherheit*), vol. 4, Germany, 1971, pp. 230–240.

282. Herrin, G. D., and J. B. Neuthardt, "An Empirical Model for Automobile Driver Curve Negotiation," *Human Factors,* vol. 16, no. 2, U.S.A., 1974, pp. 129–133.

283. Herring, H. E., "Simultaneous Power Transmission between Motor Vehicle and Road Surface," Ph.D. Dissertation, University of Karlsruhe (TH), Germany, Faculty for Civil and Surveying Engineering, 1977.

284. Hicks, T. G., and W. W. Wierwille, "Comparison of Five Mental Workload Assessment Procedures in a Moving-Base Driving Simulator," *Human Factors,* vol. 21, no. 2, U.S.A., 1979, pp. 129–143.

285. Hiersche, E.-U., "The Significance and Determination of Sight Distances of Roads," *Research Road Construction and Road Traffic Technique (Forschung Strassenbau und Strassenverkehrstechnik),* vol. 67, Minister of Transportation, Bonn, Germany, 1968.

286. Hiersche, E.-U., R. Lamm, K. Dieterle, and A. Nikpour, "Effects of Highway Improvements Designed in Conformity with the RAL-L on Traffic Safety of Two-Lane Rural Highways," *Research Road Construction and Road Traffic Technique (Forschung Strassenbau und Strassenverkehrstechnik),* vol. 431, Minister of Transportation, Bonn, Germany, 1984.

287. Hiersche, E.-U., S. Knepper, H. Messmer, and A. Nikpour, "Evaluation of the Driving Behavior on Wet Road Surfaces Using the Conductometric Detector," *Research Road Construction and Road Traffic Technique (Forschung Strassenbau und Strassenverkehrstechnik),* vol. 575, Minister of Transportation, Bonn, Germany, 1989.

288. Hiersche, E.-U., and S. Knepper, "Comparative Measurements between the Stuttgarter Tribometer and the Sideway-Force Coefficient Routine Investigation Machine in Respect of their Possibilities to Use as Part of Road Maintenance Management," *Research Road Construction and Road Traffic Technique (Forschung Strassenbau und Strassenverkehrstechnik),* vol. 582, Minister of Transportation, Bonn, Germany, 1990.

289. Hiersche, E.-U., H. E. Herring, and V. Redzovic, "Test Rides for Evaluating the Influence of Negative Superelevation Rate on Driving Dynamic Safety with Regard to Side Friction Demand when Changing Lanes and During Braking Maneuvers," Research Contract FE-04.154 G 91 D of the Federal Minister of Transportation, University of Karlsruhe (TH), Institute for Highway and Railroad Engineering, Karlsruhe, Germany, 1994.

290. Highway Research Board, "Report of the Committee on Roadside Development," Washington, D.C., U.S.A., 1949.

291. Hilger, S., "Application Aspects of GIS in the Environmental Compatibility Examination (ECE)—The Use of Digital Surface Models for the Sight-Distance-Analysis," *Salzburg Geographical Materials,* vol. 15, Salzburg, Austria, 1990, pp. 85–95.

292. Hocking, R. R., "The Analysis and Selection of Variables in Linear Regression," *Biometrics,* vol. 32, 1976.

293. Hoffmann, H. G., "The Effect of Up-Grade Sections on the Traffic Flow of Freeways (Autobahnen)," *Research Road Construction and Road Traffic Technique (Forschung Strassenbau und Strassenverkehrstechnik),* vol. 63, Minister of Transportation, Bonn, Germany, 1967.

294. Hoffmann, R., "Lessons of Traffic Accident Statistics for the Design of the Reichsautobahn," *Alignment Principles of the Reichsautobahnen, Proceedings: The Road (Die Strasse),* vol. 28, Volk and Reich Publishing House, Berlin, Germany, 1943.

295. Holzinger, J., and A. Weinzerl, "Electronic Driver Information System," *Automobile Technical Journal (ATZ),* Special Issue: Motor and Environment, Germany, 1992, pp. 54–55.

296. Hughes, W. E., "Safety and Human Factors: Worldwide Review," International Symposium on Highway Geometric Design Practices, Transportation Research Board, Boston, Massachusetts, U.S.A., August, 1995.

297. Hurt, H. H., J. V. Ouellet, and D. R. Thom, "Motorcycle Accident Cause Factors and Identification of Countermeasures," Technical Report: DOT-HS-805-862, U.S. Department of Transportation, National Highway Traffic Safety Administration, Washington, D.C., U.S.A., 1981.

298. IATSS (International Association of Traffic and Safety Service), "Statistics '93 Road Accidents Japan," Traffic Bureau, National Police Agency, Tokyo, Japan, 1994.

299. IATSS, "White Paper on Traffic Safety '94," abridged edition, Tokyo, Japan, 1994.

300. IBM Corporation, "CEP-HIDES (Highway Design System)," Program Information Department (PID), Hawthorne, New York, U.S.A., 1970.

301. IBRD (International Bank for Reconstruction and Development) & World Bank, "Road Safety for Central & Eastern Europe; A Policy Seminar," Budapest, Hungary, October 17–21, 1994.

302. JHT (Institution of Highways and Transportation), "Guidelines for the Safety Audit of Highways," London, United Kingdom, 1990.

303. Infrastructure Development Institute of Japan, "Michi-Roads in Japan," Tokyo, 1995.

304. Institution of Highways and Transportation and Department of Transport, "Roads and Traffic in Urban Areas," Her Majesty's Stationery Office, London, 1987.

305. "Instructural Guide for the Environmental Compatibility Study in Highway Planning (ECS)," German Road and Transportation Research Association, Committee: Highway Geometric Design, Cologne, Germany, 1990 ed.

306. International Commission IV/ATR-FG-VSS, "Road-Projects, Fundamentals," Final Report 2, *Road and Traffic,* vol. 2, Switzerland, 1974.

307. Ippen, H., "Vehicle-Electronics: Rapid Controlling," *Motor Vehicle Technique (Kraftfahrzeugtechnik),* vol. 7, p. 40, Germany, 1993.

308. (Institute of Transportation Engineers), "The Traffic Safety Toolbox, A Primer on Traffic Safety," Washington, D.C., U.S.A., 1993.

309. ITE Technical Council Committee 4S-7, "Road Safety Audit: A New Tool for Accident Prevention," *ITE Journal,* February U.S.A., 1995, pp. 15–19.

310. IVM Technical Consultants Wolfsburg GmbH, "Accident Reconstruction Programs HVOSM and SMAC Validated," *Automobile Technical Journal (ATZ),* vol. 96, no. 2, Germany, 1994, p. 125.

311. Jacobs, G. D., and S. Kirk, "Road Safety in Developing Countries," *Proceedings, International Road Federation XIIIth World Meeting, Toronto, Ontario, Canada,* June 1997, Preprint Paper no. 302-E, Topic 12, and CD-ROM, Canada, 1997.

312. Janssen, E. G., J. P. Pauwelussen, J. S. H. M. Wismans, L. T. B. van Kampen, and C. C. Schoon, "Developments Towards Sustainable Vehicular Safety," Final Report, R-95-76, SWOV, Leidschendam, The Netherlands, 1995.

313. Jex, H. R., "Measuring Mental Workload: Problems, Progress, and Promises," *Human Mental Workload,* North Holland, Amsterdam, The Netherlands, 1988, pp. 5–39.

314. Johnston, I. R., "The Effects of Roadway Delineation on Curve Negotiation by Both Sober and Drinking Drivers," AAR no. 128, Australian Road Research Board, Victoria, Australia, 1983.

315. Jorgensen, N., "European Trends in Road Safety and Road Safety Research," *Proceedings, Roads and Traffic Safety on Two Continents, Gothenburg, Sweden,* VTIrapport 328 A, Swedish Road and Traffic Research Institute, Linkoeping, Sweden, 1988, pp. 25–35.

316. Jorgensen, R. E., "Evaluation of Criteria for Safety Improvements on the Highway," U.S. Department of Commerce, Bureau of Public Roads, Washington, D.C., U.S.A., 1966.

317. Jorgensen, R. E. and Associates, "Cost and Safety Effectiveness of Highway Design Elements," NCHRP Report 197, Transportation Research Board, National Research Council, Washington, D.C., U.S.A., 1978.

318. Kahl, K. B., and D. B. Fambro, "Investigation of Objects Affecting Stopping Sight Distances," Part of the National Cooperative Highway Research Program, Project 3-42, "Determination of Stopping Sight Distances," Texas A&M University, College Station, Texas, U.S.A., 1997.

319. Kaiser, A., C.-S. Boettcher, W. Faisst, and T. Kiefer, "Design of the Vehicle Structure for a Frontal Offset Crash-Impact-Simulation and Test, *Developments in Body Manufacturing,* VDI-Reports 968, Meeting Hamburg, May 1992, VDI-Publishing House (VDI-Verlag), Duesseldorf, Germany, 1992, pp. 93–108.

320. Kakavoutis, J., "Relationship between Driving Behavior and Curvature Change Rate on Multilane Roads," Graduate Study Work, Institute for Highway and Railroad Engineering, University of Karlsruhe (TH), Germany, 1975 (speed measurements were repeated by R. Lamm in 1995).

321. Kalender, U., "Superelevation and Driving Safety, Possible Influences of Negative Superelevation," *Research Road Construction and Traffic Technique (Forschung Strassenbau und Strassenverkehrstechnik),* vol. 173, Minister of Transportation, Bonn, Germany, 1974.

322. Kamplade, J., "Requirements on Skid Resistance of Pavements—Skid Resistance and Traffic Safety," *Road and Interstate (Strasse und Autobahn),* vol. 3, Germany, 1995, pp. 149–158.

323. Kanellaidis, G., J. Golias, and S. Efstathiadis, "Driver's Speed Behavior on Rural Road Curves," *Traffic Engineering and Control,* vol. 31, no. 7/8, United Kingdom, 1990, pp. 414–415.

324. Kanellaidis, G., A. Loizos, and B. Papavasiliou, *Contribution to the Preparation of the Guidelines for the Geometric Design of Rural Highways: Skid Resistance Measurements,* vol. 1, Research Report, National Technical University of Athens, Greece, 1993.

325. Kanellaidis, G., B. Psarianos, J. Golias, G. Glaros, B. Papavasiliou, S. Efstathiadis, and S. Vardaki, *Contribution to the Preparation of the Geometric Design of Rural Highways,* vol. 2, Research Report, National Technical University of Athens, Greece, 1993.

326. Kasper, H., W. Schuerba, and H. Lorenz, *The Clothoid as an Element of Horizontal Alignment,* F. Dümmlers, Publishing House, Bonn, Germany, 1954.

327. Kaule, G., et al., "Road Network Design with Regard to the Decreasing Pollution of Soils," *Road and Interstate (Strasse und Autobahn),* vol. 9, Germany, 1992, pp. 497–500.

328. Kawczynski, M., "Design Practice in Highway Aesthetics," Report by MAK Engineering, Prepared for the Ministry of Transportation and Highways, Highway Engineering Branch, Victoria, British Columbia, Canada, 1994.

329. Kayser, H. J., N. Otten, and S. Hahn, "Simulation Tests for Passing Maneuvers of Motorists on Two-Lane Rural Roads," Publications of the Institute for Highway Engineering, Soil and Tunnel Construction, vol. 20, Technical University of Rheinland-Westfalen (RWTH), Aachen, Rheinland-Westfalen, Germany, 1986.

330. Kayser, H. J., W. Durth, N. Otten, and K. Habermehl, "Comparison of the Results of Field- and Simulation Experiments for Passing Maneuvers of Motorists," Research Reports of the Federal Highway Research Institute (BASt), Division for Accident Research, Bergisch Gladbach, Germany, 1989.

331. Kellermann, G., "Speed Behavior on the German Interstate (Autobahn) Network," *Road and Interstate (Strasse und Autobahn),* vol. 5, Germany, 1995, pp. 283–287.

332. Kendrick, M., "How Can Technical Means Influence Human Behavior," ECMT (European Conference of Ministers of Transportation) Conference, Hamburg, Germany, 1988.

333. Kentner, W., "The Traffic Safety as Economic Planning Value," *Road and Interstate (Strasse und Autobahn),* vol. 23, 1972, pp. 642–647.

334. Kilburn, P., "A Study of Vehicle Speeds at Climbing and Passing Lane Locations in Alberta," International Road Federation Conference, vol. 5, 1994, pp. B45–B64.

335. Klebelsberg, D., *Traffic Psychology,* Springer Verlag, Berlin, Heidelberg, Germany, 1982.

336. Klein, R., D. Lehnert, and A. Holderbaum, "Model Tests for the Drainage of Water on Distortion Sections," *Research Road Construction and Road Traffic Technique (Forschung Strassenbau und Strassenverkehrstechnik),* Minister of Transportation, Bonn, Germany, vol. 250, part I, 1978.

337. Kloeckner, J. H., "The Accident Situation on Roads with Three Directional Lanes," Publications of the Institute for Highway and Railroad Engineering, vol. 14, University of Karlsruhe (TH), Germany, 1975, pp. 58–75.

338. Kloeckner, J. H., "Property Damages of Road Traffic Accidents as Evaluation Value in Accident Statistics," Institute for Highway and Railroad Engineering, vol. 15, University of Karlsruhe (TH), Germany, 1976.

339. Kloeckner, J. H., and G. Maier-Strassburg, "Accident Characteristics of the Standard Cross Section RQ 12.50," *Research Road Construction and Road Traffic Technique (Forschung Strassenbau und Strassenverkehrstechnik),* vol. 331, pp. 1-21, Minister of Transportation, Bonn, Germany, 1981.

340. Kloeckner, J. H., R. Lamm, E. M. Choueiri, and T. Mailaender, "Traffic Accidents Involving Accompanied and Unaccompanied Children in the Federal Republic of Germany," *Transportation Research Record,* vol. 1210, U.S.A., 1989, pp. 12–18.

341. Kloeckner, J. H., "Aspects for the Development of the Bicycle Traffic," unpublished manuscript, Federal Highway Research Institute (BASt), Bergisch, Gladbach, Germany, 1990.

342. Knoche, G., "Influence of Bicycle Lanes on Traffic Safety," Research Reports of the Federal Highway Research Institute (BASt), Division for Accident Research, vol. 62, Bergisch, Gladbach, Germany, 1981.

343. Knoflacher, H., "Results and Experiences of Accident Analyses," *Road-Skid-Resistance, Traffic Safety on Wet Pavements,* vol. 2, Institute of Road- and Traffic-Engineering, Technical University, Berlin, Germany, 1968, pp. 151–156.

344. Knoflacher, H., "Interrelation Between Road Construction and Traffic Safety," *Road and Traffic (Strasse und Verkehr),* vol. 61, Switzerland, 1975, pp. 414–420.

345. Kockelke, W., and J. Steinbrecher, "Driving Behavior Investigations with Respect to Traffic Safety in the Area of Community Entrances," Report to the Research Project 8363 of the German Federal Research Institute (BASt), no. 153, Bergisch Gladbach, Germany, 1987.

346. Koeppel, G., "The Permissible Radii of Curve Relation for Reversing Clothoids," *Road and Interstate (Strasse und Autobahn)*, vol. 9, Germany, 1968, pp. 319–321.

347. Koeppel, G., and H. Bock, "Curvature Change Rate, Consistency, and Driving Speed," *Road and Interstate (Strasse und Autobahn)*, vol. 8, Germany, 1970, pp. 304–308

348. Koeppel, G., and H. Bock, "Operating Speed as a Function of Curvature Change Rate," *Research Road Construction and Road Traffic Technique (Forschung Strassenbau und Strassenverkehrstechnik)*, vol, 269, Minister of Transportation, Bonn, Germany, 1979.

349. Koeppel, G., "Development of the Design of Radius of Curve, Superelevation Rate and Stopping Sight Distance with Regard to Road Geometry," *Research Road Construction and Road Traffic Technique (Forschung Strassenbau und Strassenverkehrstechnik)*, vol. 429, Minister of Transportation, Bonn, Germany, 1984.

350. Koeppel, G., "Highway Geometric Design—Review and Outlook," *Road and Interstate (Strasse und Autobahn)*, vol. 9, Germany, 1986, pp. 395–402.

351. Koerner, G., "To the Problem of Sight-Distance Design for Alignment Fundamentals of Road Design," Ph.D. Dissertation, Technical University of Darmstadt, Department: Water and Traffic, Darmstadt, Germany, 1982.

352. Kollartis, S., "SPANS—Concept and Function of an Innovative GIS," University of Salzburg, Institute for Geography, AKG Software Consulting, Ballrechten-Dottingen, Germany, 1990.

353. Kontaratos, M., B. Psarianos, and A. Yotis, "Minimum Horizontal Curve Radius as Function of Grade Incurred by Vehicle Motion in Driving Mode," *Transportation Research Record,* vol. 1445, U.S.A., 1994, pp. 86–93.

354. Kontaratos, M., R. Lamm, B. Psarianos, and E. Kassapi, "Capacity Analysis of Standard Cross-Sections: A Case Study of Greece," XXth World Road Congress Montreal, September 3–9, 1995, Individual Papers, Committees and Working Groups 20.52.E, Canada, pp. 129–131.

355. Koornstra, M. J., M. P. M. Mathijssen, J. A. G. Mulder, R. Roszbach, and F. C. M. Wegman, "Naar een duurzaam veilig wegverkeer; Nationale Verkeersveiligheidsverkenning voor de jaren 1990/2010," SWOV, Leidschendam, The Netherlands, 1992; revised in 1995/1996.

356. Koshi, M., "Road Safety—Success and Failure in Japan," The Opening Address to the 13th ARRB (Australian Road Research Board) /5th REAAA (Road Engineering Association of Asia and Australasia) Combined Conference, 1986.

357. Kracke, R., H. Engelmann, B. Keppeler, K. Prause, F. Fehsenfeld, and W. Kuehn, "Alignment Design and Traffic Quality of an Interstate-Parallel Railroad Section for High-Speed Traffic," Institute for Traffic Engineering, Railroad Construction and Operation, University of Hannover, Hannover, Germany, 1985.

358. Kramer, U., D. Mary, R. Povel, and W. Zimdahl, "Technical Problems and Solution Attempts for the Research Program: PROMETHEUS of the European Automobile Industry," *Automobile Technical Journal (ATZ)*, vol. 89, no. 3, Germany, 1987, pp. 109–114.

359. Krammes, R. A., et al., "State of the Practice Geometric Design Consistency," Final Report, Contract no. DTFH61-91-C-00050, Federal Highway Administration, U.S. Department of Transportation, McLean, Virginia, U.S.A., 1993.

360. Krammes, R. A., R. Q. Brackett, M. A. Shafer, J. L. Ottesen, I. B. Anderson, K. L. Fink, K. M. Collins, O. J. Pendleton, and C. J. Messer, "Horizontal Alignment Design Consistency for Rural Two-Lane Highways," Report FHWA-RD-94-034, Federal Highway Administration, U.S. Department of Transportation, 1995.

361. Krammes, R. A., and M. A. Garnham, "Review of Alignment Design Policies Worldwide," International Symposium on Highway Geometric Design Practices, Transportation Research Board, Boston, Massachusetts, U.S.A., August 1995.

362. Krammes, R. A., O. Hoon, and K. S. Rao, "Highway Geometric Design Consistency Evaluation Software," *Transportation Research Record,* vol. 1500, U.S.A., 1995, pp. 19–24.

363. Krause, J., and H. P. Reichwein, "Comparison of Limiting Values of Relevant Design Elements for the Design of Roads in the Countries of the European Community," Federal Highway Research Institute (BASt), Bergisch Gladbach, Germany, 1990.

364. Krebs, H. G., "Driving Dynamics and Safety," *Road and Interstate (Strasse und Autobahn)*, vol. 2, Germany, 1970.

365. Krebs, H. G., R. Lamm, and R. Leutner, "Evaluation Criteria with Respect to Recommended Speeds," *Research Road Construction and Road Traffic Technique (Forschung Strassenbau und Strassenverkehrstechnik)*, vol. 126, Minister of Transportation, Bonn, Germany, 1972.

366. Krebs, H. G., "Road Construction—Synthesis of Tradition, Handicraft and Science," *Research between Driving Dynamics and Asphalt Road Construction*, vol. 10, Publications of the Institute for Highway and Railroad Engineering, University of Karlsruhe (TH), Germany, 1973, pp. 61–85.

367. Krebs, H. G., R. Lamm, P. Kupke, and M. Blumhofer, "Introduction of Automatic Radar Devices—Accident and Speed Analysis on Interstate (Autobahn A15)," Speed Limits Outside Built-Up Areas, Research Contract and Publication, Minister of Economy and Technique, State of Hessen, Wiesbaden, Germany, 1974, pp. 23–34.

368. Krebs, H. G., and J. H. Kloeckner, "Investigations of the Effect of Highway and Traffic Conditions Outside Built-Up Areas on Accident Rates," *Research Road Construction and Road Traffic Technique (Forschung Strassenbau und Strassenverkehrstechnik)*, Minister of Transportation, Bonn, Germany, 1977. vol. 223, pp. 1–63.

369. Krebs, H. G., R. Lamm, and J. H. Kloeckner, "Mathematical Investigation of the Three Dimensional Impression of the Perspective Image on the Motorist: Part I, Accident Analysis," Contract no. 2.067 G 80 E, Research Contract and Report for the Minister of Transportation, Germany, 1981.

370. Krebs, H. G., and K. Dieterle, "Behavior of Motor Vehicles on Wet Road Surface," *Research Road Construction and Road Traffic Technique (Forschung Strassenbau und Strassenverkehrstechnik)*, vol. 355, Minister of Transportation, Bonn, Germany, 1982.

371. Krebs, H. G., R. Lamm, and M. Blumhofer, "Evaluation of the Accident Situation on Superelevation Runoff Sections with Insufficient Longitudinal Grade in Comparison to Road Sections with Negative Superelevation," *Research Road Construction and Road Traffic Technique (Forschung Strassenbau und Strassenverkehrstechnik)*, vol. 366, Minister of Transportation, Bonn, Germany, 1982. pp 100–140.

372. Krebs, H. G., and N. Damianoff, "Speed Behavior at Traffic Warning Signs in Curves and at Locations with Speed Limits," *Research Road Construction and Road Traffic Technique (Forschung Strassenbau und Strassenverkehrstechnik)*, vol. 380, Minister of Transportation, Bonn, Germany, 1983.

373. Krebs, H. G., "Commemorative Publication `KREBS'—Contribution to Environmental Protection and Traffic Safety," Publications of the Institute for Highway and Railroad Engineering, University of Karlsruhe (TH), vol. 33, 1986, pp. 115–151.

374. Krempel, G. "Experimental Contribution for Investigations About Vehicle Tires," Ph.D. Dissertation, University of Karlsruhe, Germany, 1965.

375. Kronrumpf, M., *HAFRABA—German Autobahn—Planning 1926–1934*, vol. 7, Archives for the History of Road Engineering, German Road and Transportation Association, Kirschbaum Publishers, Bonn, Germany, 1990.

376. Krupp, R., and G. Hundhausen, "Economic Evaluation of Personal Damages in Road Traffic," Federal Highway Research Institute (BASt), Bergisch Gladbach, Germany, 1984.

377. Kummer, H. W., and W. E. Meyer, "Tentative Skid-Resistance Requirements for Main Rural Highways," National Cooperative Highway Research Program, Report no. 37, Highway Research Board, Washington, D.C., U.S.A., 1967.

378. Kunze, U., "Sight Distance and Cross Section as Influencing Parameters on the Accident Situation," Graduate Study Work, Institute for Highway and Railroad Engineering, University of Karlsruhe (TH), Germany, 1976.

379. Kupke, P., "Simulator Experiments to Alignment-Dependent Driving Behavior and Examination of the Alignment," Publications of the Institute for Highway and Railroad Engineering, University of Karlsruhe (TH), Germany, vol. 16, 1977.

380. Kutscher, J., B. Hartmann, and P. Brieler, "Accident Risk Causes—Empirical Analysis of Passenger Car—Passenger Car Accidents in Berlin (East) in the First Half-Year 1990," *Journal for Traffic Safety (Zeitschrift fuer Verkehrssicherheit)*, vol. 2, Germany, 1991, pp. 54–58.

381. Lachenmayr, B., "Peripheral Vision and Reaction Time in Road Traffic—The Influence of Driver Workload," *Journal for Traffic Safety (Zeitschrift fuer Verkehrssicherheit)*, vol. 4, Germany, 1987, pp. 151–156.

382. Lamm, R., and H. E. Herring, "The Side-Friction Factor in Relation to Speed," *Road and Interstate (Strasse und Autobahn)*, vol. 11, Germany, 1970, pp. 435–443.

383. Lamm, R., and H. E. Herring, "A New Proposal for the Design of Radii of Curve and Superelevation Rates on Rural Highways," *Road and Interstate (Strasse und Autobahn)*, vol. 2, Germany, 1971, pp. 62–65.

384. Lamm, R., and H. G. Schlichter, "Driving Behavior under Different Weather and Daylight Conditions," *Road and Construction (Strassen und Tiefbau)*, vol. 12, Germany, 1971, pp. 873–877.

385. Lamm, R., "Driving Dynamics and Road Characteristics—A Contribution for Highway Design with Special Consideration of Operating Speeds," Thesis for Appointment as University Lecturer, Publications of the Institute for Highway and Railroad Engineering, vol. 11, University of Karlsruhe (TH), Germany, 1973.

386. Lamm, R., "Highway Geometric Design and Speed: Part I: Historic Development, Part II: Speed Concepts in German Design Guidelines, Part III: Road Characteristics and Design Speed," *Road and Construction (Strassen und Tiefbau)* vols. 1, 2, and 3, Germany, 1975, pp. 8–11, pp. 25–29, pp. 26–32.

387. Lamm, R., and J. H. Kloeckner, "Influences of the Motor Vehicle on the Accident Situation," *Police, Technique and Traffic (Polizei, Technik und Verkehr)*, vol. 2, Germany, 1975, pp 44–48.

388. Lamm, R., and J. H. Kloeckner, "Identification and Evaluation of Dangerous Road Sections," *Road and Construction (Strassen und Tiefbau)*, vol. 8, Germany, 1976, pp. 31–36.

389. Lamm, R., and J. H. Kloeckner, "Attempts for Reducing Traffic Accidents in a County," *The County (Der Landkreis)*, vol. 1, Germany, 1979, pp. 33–35.

390. Lamm, R., "Safety Evaluation of Highway Design Parameters," *Road and Construction (Strassen und Tiefbau)*, vol. 10, Germany, 1980, pp. 14–22.

391. Lamm, R., and J. Treiterer, "The Accident Situation in the United States of America and the Federal Republic of Germany: Part I: Comparisons, Trends and Influences of Human Behavior, Site and Time; Part II: Nature of Collision and Kind of Vehicles as Factors in Crash Analysis; Part III: Accident Causes and Out-Look," *Road and Construction (Strassen und Tiefbau)*, vols. 11, 12 (1980), and 1, (1981), pp. 6–16, 10–18, 6–13.

392. Lamm, R., "What Can We Learn from a Comparison of German and American Accident Statistics?" *Proceedings of the Thirty-Fifth Annual Ohio Transportation Engineering Conference,* Conducted by the Department of Civil Engineering, The Ohio State University, in Cooperation with the Ohio Department of Transportation, Columbus, Ohio, U.S.A., 1981, pp. 101–121.

393. Lamm, R., "Is There any Influence on the Accident Situation by the Superimposition of Horizontal and Vertical Curvature in the Alignment?," *Road and Construction (Strassen und Tiefbau)*, vols. 1, 2, and 3, Germany, 1982, pp. 16–19, 21–25, pp. 20–23.

394. Lamm, R., "New Developments in Highway Design with Special Consideration of Traffic Safety," *Proceedings of the Thirty-Sixth Annual Ohio Transportation Engineering Conference,* Conducted by the Department of Civil Engineering, The Ohio State University, Transplex/OSU in Cooperation with the Ohio Department of Transportation, Columbus, Ohio, U.S.A., 1982, pp. 107–119.

395. Lamm, R., and J. H. Kloeckner, "Effective Speed Control on the German Autobahn by Radar Activated Cameras," *Proceedings of the Thirty-Seventh Annual Ohio Transportation Engineering Conference,* Conducted by the Department of Civil Engineering, The Ohio State University, Transplex/OSU in Cooperation with the Ohio Department of Transportation, Columbus, Ohio, U.S.A., 1983, pp. 96–110.

396. Lamm, R., "Driving Dynamic Considerations: A Comparison of German and American Friction Coefficients for Highway Design," *Transportation Research Record,* vol. 960, U.S.A., 1984, pp. 13–20.

397. Lamm, R., and J. H. Kloeckner, "Increase of Traffic Safety by Surveillance of Speed Limits with Automatic Radar Devices: A Long-Term Investigation," *Transportation Research Record,* vol. 974, U.S.A., 1984, pp. 8–16.

398. Lamm, R., F. B. Lin, E. M. Choueiri, and J. H. Kloeckner, "Comparative Analysis of Traffic Accident Characteristics in the United States, Federal Republic of Germany and Other European Countries: 1970–1980," Research Contract and Report for the Alfried Krupp von Bohlen and Halbach Foundation, Essen, Germany, Conducted at Clarkson University, Potsdam, N.Y., U.S.A., 1984 (in English and German).

399. Lamm, R., and J. G. Cargin, "Translation of the Guidelines for the Design of Roads (RAS-L-1)," Federal Republic of Germany, 1984 ed., and the Swiss Norm SN 640 080a, "Highway Design, Fundamentals, Speed as a Design Element," 1981, as discussed by K. Dietrich, M. Rotach, and E. Boppart in *Road*

Design, ETH Zuerich, Institute for Traffic Planning and Transport and Technique, 1983 ed., prepared at Clarkson University for the Safety and Design Division, Federal Highway Administration, U.S. Department of Transportation, 1985.

400. Lamm, R., E. M. Choueiri, and J. H. Kloeckner, "Accidents in the U.S. and Europe: 1970–1980," *International Journal: Accident Analysis and Prevention,* vol. 17, no. 6, U.S.A., 1985, pp. 429–438.

401. Lamm, R., A. Taubmann, and J. Zoellmer, "Comprehensive Study on the Term 'Critical Water' Film Thickness," *Research Road Construction and Road Traffic Technique (Forschung Strassenbau und Strassenverkehrstechnik),* vol. 436, Minister of Transportation, Bonn, Germany, 1985.

402. Lamm, R., and E. M. Choueiri, "Relationship between Design, Driving Behavior and Accident Risk on Curves," *Proceedings of the Fortieth Annual Ohio Transportation Engineering Conference,* Conducted by the Department of Civil Engineering, The Ohio State University in Cooperation with The Ohio Department of Transportation, Columbus, Ohio, U.S.A., 1986, pp. 87–100.

403. Lamm, R., E. M. Choueiri, and J. H. Kloeckner, "Comparative Analysis of Traffic Accident Characteristics in the United States, Federal Republic of Germany and Other European Countries— Extension up to 1983 and Elaboration of a Second Edition," Research Contract and Report for the Alfried Krupp von Bohlen and Halbach Foundation, Essen, Germany, Conducted at Clarkson University, Potsdam, N.Y., U.S.A., 1986.

404. Lamm, R., K. Dieterle, J. Zoellmer, H. G. Schlichter, and A. Taubmann, "Comparative Accident Investigations on Interstate Sections with Superelevation Rates from $e = 1.5\%$ to $e = 2.5\%$," *Research Road Construction and Road Traffic Technique (Forschung Strassenbau und Strassenverkehrstechnik),* vol. 484, Minister of Transportation, Bonn, Germany, 1986.

405. Lamm, R., J. C. Hayward, and G. Cargin, "Comparison of Different Procedures for Evaluating Speed Consistency," *Transportation Research Record,* vol. 1100, U.S.A., 1986, pp. 10–20.

406. Lamm, R., and E. M. Choueiri, "A Design Procedure to Determine Critical Dissimilarities in Horizontal Alignment and Enhance Traffic Safety by Appropriate Low-Cost or High-Cost Projects," Report to the National Science Foundation (Grant-no.: ECE-841 4755), Washington, D.C., U.S.A., 1987.

407. Lamm, R., and E. M. Choueiri, "Recommendations for Evaluating Horizontal Design Consistency, Based on Investigations in the State of New York," *Transportation Research Record,* vol. 1122, 1987, pp. 68–78.

408. Lamm, R., and E. M. Choueiri, "Rural Roads Speed Inconsistencies Design Methods," Research Report for the State University of New York, Research Foundation (Contract no.: RF320-PN72350), Parts I and II, Albany, N.Y., U.S.A., 1987.

409. Lamm, R., and E. M. Choueiri, "The Impact of Traffic Warning Devices on Operating Speeds and Accident Rates on Two-Lane Rural Highway Curves," *Proceedings of the Forty-First Annual Ohio Transportation Engineering Conference,* Conducted by The Ohio State University, Department of Civil Engineering in Cooperation with The Ohio Department of Transportation, Columbus, Ohio, U.S.A., 1987, pp. 171–192.

410. Lamm, R., and E. M. Choueiri, "Investigations about Driver Behavior and Accident Experiences at Curved Sites (Including Black Spots) of Two-Lane Rural Highways in the U.S.A.," *Roads and Traffic 2000,* International Road and Traffic Conference, Berlin, Germany, September 6–9, 1988, *Proceedings Theme 4, Traffic Engineering and Safety,* pp. 153–158.

411. Lamm, R., E. M. Choueiri, and J. C. Hayward, "Tangent as an Independent Design Element," *Transportation Research Record,* vol. 1195, U.S.A., 1988, pp. 123–131.

412. Lamm, R., E. M. Choueiri, J. C. Hayward, and A. Paluri, "Possible Design Procedure to Promote Design Consistency in Highway Geometric Design on Two-Lane Rural Roads," *Transportation Research Record,* vol. 1195, U.S.A., 1988, pp. 111–122.

413. Lamm, R., E. M. Choueiri, and J. H. Kloeckner, "Experiences in Fatalities by Age and Road User Groups—U.S.A. vs. Western Europe 1970–1983," *Proceedings of Roads and Traffic Safety on Two Continents,* Gothenburg, Sweden, VTIrapport 331A, Swedish Road and Traffic Research Institute, Linkoeping, Sweden, 1988, pp. 128–144.

414. Lamm, R., E. M. Choueiri, P. B. Goyal, and T. Mailaender, "An Attempt to Develop Reliable Friction Factors vs. Speed for Design Purposes. A Case Study Based on Actual Pavement Friction Inventories," *Proceedings of the Forty-Third Annual Ohio Transportation Engineering Conference,* Conducted by the Department of Civil Engineering, The Ohio State University in Cooperation with The Ohio Department of Transportation, Columbus, Ohio, U.S.A., 1989, pp. 168–189.

415. Lamm, R., E. M. Choueiri, and T. Mailaender, "Accident Rates on Curves as Influences by Highway Design Elements—An International Review and an In-Depth Study," *Proceedings, Road Safety in Europe,* Gothenburg, Swedish Road and Traffic Research Institute, Linkoeping, Sweden, VTIrapport 344A, 1989, pp. 33–54.

416. Lamm, R., E. M. Choueiri, and T. Mailaender, "Are There Differences in Operating Speeds on Dry Pavements as Compared with Wet?" *Proceedings of the Forty-Third Annual Ohio Transportation Engineering Conference,* Conducted by the Department of Civil Engineering, The Ohio State University in Cooperation with The Ohio Department of Transportation, Columbus, Ohio, U.S.A., 1989, pp. 120–135.

417. Lamm, R., E. M. Choueiri, T. Mailaender, and A. Paluri, "A Logical Approach to Geometric Design Consistency of Two-Lane Roads in the U.S.A." 11th International Road Federation World Meeting, Seoul, Korea, April 16–21, 1989, *Proceedings Summary,* pp. 51/52 and *Proceedings Volume II, Recent Advantages in Design and Construction,* pp. 8–11.

418. Lamm, R., T. Mailaender, and E. M. Choueiri, "New Ideas for the Design of Two-Lane Rural Roads in the U.S.A.," *Road and Construction (Strassen und Tiefbau),* vols. 5 and 6, Germany, 1989, pp. 18–25 and pp. 13–18.

419. Lamm, R., and E. M. Choueiri, "Comparison of the Accident Situation in the U.S.A. and Western Europe from 1970 to 1983," *Bulletin of the Greek Association of Professional Rural and Surveying Engineers,* vol. 93, Athens, Greece, May 1990, pp. 10–17 (in Greek).

420. Lamm, R., E. M. Choueiri, P. B. Goyal, and T. Mailaender, "Design Friction Factors of Different Countries Versus Actual Pavement Friction Inventories," *Transportation Research Record,* vol. 1260, U.S.A., 1990, pp. 135–146.

421. Lamm, R., E. M. Choueiri, and T. Mailaender, "A Case Study Evaluating Traffic Warning Devices with Respect to Operating Speeds and Accident Rates," *Proceedings of Road Safety and Traffic Environment in Europe,* Gothenburg, Sweden, VTIrapport 363A, Swedish Road and Traffic Research Institute, Linkoeping, Sweden, 1990, pp. 113–131.

422. Lamm, R., E. M. Choueiri, and T. Mailaender, "Comparison of Operating Speeds on Dry and Wet Pavements of Two-Lane Rural Highways," *Transportation Research Record,* vol. 1280, U.S.A., 1990, pp. 199–207.

423. Lamm, R., A. K. Guenther, and B. Psarianos, "Automation of Design, Construction, Geodetic Surveying, Accounting and Land Acquisition in Road and Railroad Engineering Based on the German CAD-System Verbund," *Bulletin of the Hellenic Association of Professional Rural and Surveying Engineers,* Athens, vol. 97, November 1990, pp. 5–14 (in Greek).

424. Lamm, R., E. M. Choueiri, and T. Mailaender, "Side Friction Demanded Versus Side Friction Assumed for Curve Design on Two-Lane Rural Highways," *Transportation Research Record,* vol. 1303, U.S.A., 1991, pp. 11–21.

425. Lamm, R., E. M. Choueiri, and T. Mailaender, "Traffic Safety on Two-Continents—A Ten Year Analysis of Human and Vehicular Involvements," *Proceedings of Strategic Highway Research Program (SHRP) and Traffic Safety on Two Continents,* Gothenburg, Sweden, September 18–20, 1991, Swedish Road and Traffic Research Institute, Linkoeping, Sweden, VTIrapport 372A, Part 1, 1991, pp. 121–136.

426. Lamm, R., and E. M. Choueiri, "Identification of Target Groups with High Accident Risk for Different Age Groups and Type of Traffic Participation—Western Europe in Comparison to the U.S.A.," Publications of the Institute for Highway and Railroad Engineering, vol. 40, University of Karlsruhe (TH), Germany, 1992, pp. 67–88.

427. Lamm, R., E. M. Choueiri, T. Mailaender, G. M. Choueiri, and B. M. Choueiri, "Fatality Development of Vulnerable Road User Groups in Europe (1980–1989)—Pedestrians, Cyclists, Teenagers and the Elderly," *Proceedings of Road Safety in Europe,* Berlin, Germany, September 30–October 2, 1992, VTIrapport 380A, Swedish Road and Traffic Research Institute, Linkoeping, Sweden, 1992, pp. 121–140.

428. Lamm, R., T. Mailaender, E. M. Choueiri, and H. Steffen, "Side Friction in International Road Design and Possible Impacts on Traffic Safety," *Road and Construction (Strassen und Tiefbau),* vol. 5, Germany, 1992, pp. 6–25.

429. Lamm, R., and T. Mailaender, "Highway Geometric Design and Practice in Germany with Special Regard of Human Factors and Traffic Safety," Research Report I for CTI Engineering Co. Ltd., Tokyo, Japan; Karlsruhe, Germany, 1992.

430. Lamm, R., and T. Mailaender, "Procedure to Determine Reliable Standard Cross Sections in the Planning Process in Germany with Special Regard of Traffic Quality, Economics and Safety," Research Report II for CTI Engineering Co., Ltd., Tokyo, Japan; Karlsruhe, Germany, 1992.

431. Lamm, R., E.-U. Hiersche, and T. Mailaender, "Examination of the Existing Operating Speed Background of the German Guidelines for the Design of Roads," unpublished manuscript, Institute for Highway and Railroad Engineering, University of Karlsruhe (TH), Germany, 1993.

432. Lamm, R., and T. Mailaender, Translation and Discussion of Refs. 159 and 246 upon Request from CTI Engineering Co. Ltd., Tokyo, and from Texas Transportation Institute, Texas A&M University, College Station, Texas, U.S.A.; Karlsruhe, Germany, 1993.

433. Lamm, R., and T. Mailaender, "What Has to Be Considered in Establishing Modern Highway Geometric Design Guidelines?" Research Report IV for CTI Engineering Co. Ltd., Tokyo, and the Texas Transportation Institute, Texas A&M University, College Station, Texas, U.S.A.; Karlsruhe, Germany, 1993.

434. Lamm, R., T. Mailaender, H. Steffen, and E. M. Choueiri, "Safety Evaluation Process for Modern Highway Geometric Design on Two-Lane Rural Roads," Research Report V for CTI Engineering Co. Ltd., Tokyo, Japan; Karlsruhe, Germany, 1993.

435. Lamm, R., "Design of Motorways and Rural Roads with Special Emphasis on Traffic Safety," Infrastructure Design & Road Safety, OECD Workshop B 3 for CEE's and NIS, 15th-18th November 1994, Prague (Czech Republic); Part 2: Lectures D-94-14 II, Leidschendam, SWOV Institute for Road Safety Research, The Netherlands, 1994.

436. Lamm, R., A. K. Guenther, and B. Grunwald, "Environmental Impact on Highway Geometric Design in Western Europe Based on a Geographical Information System," *Transportation Research Record,* vol. 1445, U.S.A., 1994, pp. 54–63.

437. Lamm, R., A. K. Guenther, and B. Psarianos, "Three New Safety Criteria of Modern Highway Design Integrated in a Geographical Information System," *Technika Chronika, Scientific Journal of the Technical Chamber of Greece,* Section A, vol. 14, no. 4, Athens, Greece, 1994, pp. 155–185; ISSN: 0250-9954.

438. Lamm, R., B. Psarianos, and A. K. Guenther, "Interrelationships between Three Safety Criteria, Modern Highway Geometric Design, as well as High Risk Target Locations and Groups," *Proceedings, The Third International Conference on Safety and the Environment in the 21st Century, Lessons from the Past Shaping the Future,* Tel-Aviv, Israel, 1994, pp. 439–458.

439. Lamm, R., B. Psarianos, M. Kontaratos, and G. Soilemezoglou, "Guidelines for the Design of Highway Facilities, Part A (Draft I)," Ministry for Environment, Regional Planning and Public Works, Athens, Greece, 1994.

440. Lamm, R., B. Psarianos, and T. Mailaender, "Cross Sectional and Alignment Design Elements with Special Emphasis on Traffic Safety and on Operating Speeds," Research Report X for CTI Engineering, Co. Ltd., Tokyo, Japan, 1994.

441. Lamm, R., B. Psarianos, and T. Mailaender, "Traffic Calming Methods in Residential Areas of Towns and Local Through Roads of Villages," Research Report IX for CTI Engineering Co. Ltd., Tokyo, Japan; Karlsruhe, Germany, 1994.

442. Lamm, R., B. Psarianos, and G. Soilemezoglou, "Guidelines for the Design of Highway Facilities, vol. 1: Road Network (Draft I)," Ministry for Environment, Regional Planning and Public Works, Athens, Greece, 1994.

443. Lamm, R., G. Schmidt, B. Psarianos, and T. Mailaender, "Important Specific Issues in Interchange/Intersection—Cross Section—and Alignment Design," Research Report VIII for CTI Engineering, Co. Ltd., Tokyo, Japan; Karlsruhe, Germany, 1994.

444. Lamm, R., and B. L. Smith, "Curvilinear Alignment: An Important Issue for More Consistent and Safer Road Characteristic," *Transportation Research Record,* vol. 1445, 1994, pp. 12–21.

445. Lamm, R., H. Steffen, and A. K. Guenther, "Procedure for Detecting Errors in Alignment Design and Consequences for Safer Redesign," *Transportation Research Record,* vol. 1445, U.S.A., 1994, pp. 64–72.

446. Lamm, R., "Development of a Classification System for a Driving Dynamic Safety Criterion in Europe and the U.S.A." unpublished manuscript, Institute for Highway and Railroad Engineering, University of Karlsruhe (TH), Germany, 1995.

447. Lamm, R., "Highway Geometric Design with Special Emphasis on Traffic Safety—Based on International Research and the New Greek Guidelines for the Design of Highway Facilities, Part:

Alignment," *Proceedings of the First Greek Congress of Highway Engineering,* October 4–7, 1995, Larisa, Technical Chamber of Greece, 1995.

448. Lamm, R., E. M. Choueiri, B. Psarianos, and G. Soilemezoglou, "A Practical Safety Approach to Highway Geometric Design, International Case Studies: Germany, Greece, Lebanon, and the U.S.A.," International Symposium on Highway Geometric Design Practices, Transportation Research Board, Boston, Massachusetts, U.S.A., August 1995.

449. Lamm, R., A. K. Guenther, and E. M. Choueiri, "Safety Module for Highway Design," *Transportation Research Record,* vol. 1512, U.S.A., 1995, pp. 7–15.

450. Lamm, R., S. Kakido, R. Heger, B. Psarianos, and T. Mailaender, "German and Japanese Experiences: Traffic Calming Methods in Residential Areas," First Greek Congress of Highway Engineering, October 4–7, 1995, Larisa, Technical Chamber of Greece, vol. II, Greece, 1995, pp. 455–486.

451. Lamm, R., and B. Psarianos, "Contribution to Session Sight Distance (Austrian, German and Greek Approach)," International Symposium on Highway Geometric Design Practices, Transportation Research Board, Boston, unpublished manuscript, Karlsruhe, Germany; Athens, Greece, 1995, International Elaboration by D. W. Harwood in Ref. 268.

452. Lamm, R., B. Psarianos, E. M. Choueiri, and T. Mailaender, "The Tangent as Dynamic Design Element," *Road and Construction (Strassen und Tiefbau),* vols. 7/8, 9, and 11, Germany, 1995, pp. 16–21, 14–19, and 18.

453. Lamm, R., B. Psarianos, D. Drymalitou, and G. Soilemezoglou, *Guidelines for the Design of Highway Facilities,* vol. 3: *Alignment (Draft III),*" Ministry for Environment, Regional Planning and Public Works, Athens, Greece, 1995.

454. Lamm, R., B. Psarianos, M. Kontaratos, D. Drymalitou, and G. Soilemezoglou, *Guidelines for the Design of Highway Facilities,* vol. 2: *Cross Sections* (Draft II), Ministry for Environment, Regional Planning and Public Works, Athens, Greece, 1995.

455. Lamm, R., B. Psarianos, and T. Mailaender, "International Modern Highway Geometric Design with Special Respect to Safety Evaluation Processes, Relation Design and Dynamic Tangent Issues," Research Report XI for CTI Engineering Co. Ltd., Tokyo, Japan; Karlsruhe, Germany, 1995.

456. Lamm, R., B. Psarianos, and G. Soilemezoglou, *Guidelines for the Design of Highway Facilities,* vol. 1: *Road Network (Draft II),* Ministry of Environment, Regional Planning and Public Works, Athens, Greece, 1995.

457. Lamm, R., R. Steyer, S. Kakido, and B. Psarianos, "Traffic Calming Methods on Local Through Roads of Villages and Small Towns in Germany," First Greek Congress of Highway Engineering, October 4–7, 1995, Larisa, Technical Chamber of Greece, vol. II, Greece, 1995, pp. 430–454.

458. Lamm, R., B. Psarianos, G. Soilemezoglou, and G. Kanellaidis, "Driving Dynamic Aspects and Related Safety Issues for Modern Geometric Design of Non Built-Up Roads," *Transportation Research Record,* vol. 1523, U.S.A., 1996, pp. 34–45.

459. Lamm, R., et al., "Truck Speed Measurements on Auxiliary Lanes in Upgrade and Downgrade Direction for the Examination of Existing Evaluation Speed Backgrounds and for the Establishment of a Sound New Background," unpublished manuscript, Institute for Highway and Railroad Engineering, University of Karlsruhe (TH), Germany, 1996 (see Ref. 638).

460. Lamm, R., O. Eberhard, and R. Heger, "Recommendations Relevant to International Design Standards for Improving Existing (Old) Alignments, Based on Speed and Safety Related Research," International Road Federation XIIIth World Meeting, Toronto, Ontario, Canada, June 1997, Abstracts, p. 124 and CD-ROM of the Proceedings, Canada, 1997.

461. Lamm, R., R. Heger, and O. Eberhard, "Operating Speed and Relation Design Backgrounds—Important Issues to be Regarded in Modern Highway Alignment Design," International Road Federation XIIIth World Meeting, Toronto, Ontario, Canada, June 1997, Abstracts, p. 123 and CD-ROM of the Proceedings, Canada, 1997.

462. Lamm, R., et al., "Relation Design in International Comparison," *Road and Construction (Strassen und Tiefbau),* Germany, 1999 (to be published).

463. Launhardt, *Theory of the Alignment,* Schmorl & von Seefeld Publishing House, Germany, 1887.

464. "Law 1650/1986 for Environmental Protection," Greek Government Bulletin no. 160/A/16.10.86, Ministry of Environment Regional Planning and Public Works, Athens, Greece.

465. "Law about the Environmental Compatibility Examination (ECE) for Public and Private Projects (85/337/EWG)," Published in the Federal-Law-Instructions I, Germany, 1990, pp. 205–214.

466. Lay, M. G., *The History of the Road—From the Beaten Track to the Autobahn (Interstate),*" Campus Publishing House, Frankfurt, Germany, 1994, and *Ways of the World—A History of the World's Roads and of the Vehicles That Used Them,* Rutgers University Press, New Brunswick, N.J., U.S.A., 1992.

467. Leasure, W. A., "The NHTSA IVHS Program for Enhancing Safety Through Crash Avoidance Improvement," *Proceedings, 13, International Technical Conference on Experimental Safety Vehicles,* Paris, France, 1991, pp. 429–437.

468. Lee, C. H., "A Study Into Driver-Speed Behavior on a Curve by Using Continuous Speed Measurement Method," *Proceedings, Australian Road Research Board,* vol. 14, part 4, Australia, 1988, pp. 37–46.

469. Leins, W., "Road Surface and Traffic Safety," presented at the House of Technique, Essen, Germany, 1969.

470. Leisch, J. E., and J. P. Leisch, "New Concepts in Design Speed Application," *Transportation Research Record,* vol. 631, U.S.A., 1977, pp. 4–14.

471. Lenz, K. H., "Motorization and Trends in Road Traffic," *Proceedings of the International Conference Road Safety and Traffic Environment in Europe,* Gothenburg, Sweden, VTIrapport 362A, Swedish Road and Traffic Research Institute, Linkoeping, Sweden, 1990.

472. Leutner, R., "Driving Space and Driving Behavior," Publications of the Institute for Highway and Railroad Engineering, vol. 12, University of Karlsruhe (TH), Germany, 1974.

473. Leutzbach, W., R. Wiedemann, and W. Siegener, "About the Relationship between Traffic Accidents and Traffic Volume on a German Interstate Section," *Accident Analysis and Prevention,* vol. 2, U.S.A., 1970, pp. 92–103.

474. Leutzbach, W., et al., "Relationship between Traffic Accidents and Traffic Conditions on Two-Lane Rural Roads," Research Reports FA 3.123, 1983; FA 3.021, and FA 3.607 for the Minister of Transportation, Germany, 1976.

475. Leutzbach, W., and B. Papavasiliou, "Perceiving Conditions and Safe Behavior in Road Traffic," Research Project 8306, Federal Highway Research Institute (BASt), Karlsruhe, Germany, 1985.

476. Leutzbach, W., and J. Zoellmer, "Relationship between Traffic Safety and Design Elements," *Research Road Construction and Road Traffic Technique (Forschung Strassenbau und Strassenverkehrstechnik),* vol. 545, Minister of Transportation, Bonn, Germany, 1989.

477. Levin, C., "New Ways for Crest Vertical Curve Design," Commemorative Volume to the 60th Birthday of Univ.-Prof. Dr.-Ing. Walter Durth, Technical University of Darmstadt, Department: Road Design and Road Operation, Darmstadt, Germany, 1995, pp. 113–120.

478. Liebmann, *The Construction of Rural Roads,* Goeschenen GmbH, Berlin, Leipzeig, Germany, 1912.

479. Lindemann, H. P., and B. Ranft, "Speed in Curves," ETH-Hoenggerberg, Institute of Highway Planning and Transportation Technique, Zuerich, Switzerland, 1988.

480. Lippold, C., "To the Relation Design of Two-Lane Rural Roads," Commemorative Volume to the 60th Birthday of Univ.-Prof. Dr.-Ing. Walter Durth, Technical University of Darmstadt, Department: Road Design and Road Operation, Darmstadt, Germany, 1995, pp. 121–132.

481. Litzka, J., and E. Friedl, "Investigations about the Relevant Friction Factor," Federal Ministry for Economic Matters, Road Research, vol. 376, Vienna, Austria, 1988.

482. Lloyd, B., "Training Wheels for Adults? No, Try Helmets," *The New York Times,* U.S.A., May 25, 1991.

483. Lobanov, E. M., "The Reaction Time of Motorists on Interstates (Autobahnen)," *Research Volume to Traffic Safety,* vol. 3, Germany, 1978.

484. Loewe, F., *Road Construction Knowledge,* C. W. Kreidels Publishing House, Germany, 1906.

485. Lorenz, H., "The Consistent Alignment Flow," *Alignment Principles of the Reichsautobahnen, Proceedings: The Road (Die Strasse),* vol. 28, Volk and Reich Publishing House, Berlin, Germany, 1943.

486. Lorenz, H., "Optical Guidance," *Road and Construction (Strassen und Tiefbau),* vol. 10, Germany, 1951, pp. 276–280.

487. Lorenz, H., "Modern Alignment Design," *Road and Interstate (Strasse und Autobahn),* vol. 9, Germany, 1954, pp. 370–373.

488. Lum, H. S., and W. E. Hughes, "Edgeline Widths and Traffic Accidents," *Public Roads,* vol. 54, no. 1, U.S.A., 1990, pp. 153–159.

489. Lunenfeld, H., and G. J. Alexander, "A Users' Guide to Positive Guidance," 3d ed., Report, no. FHWA-SA-90-017, U.S. Department of Transportation, Federal Highway Administration, Office of Traffic Operations, Washington, D.C., U.S.A., 1990.

490. Lunenfeld, H., "Human Factors Associated with Interchange Design Features," paper presented at the 71st Annual Meeting of Transportation Research Board, Washington, D.C., U.S.A., 1992.

491. Lupien, C., K. Baass, and P. Barber, "Speed Distance Profiles of Trucks on Grades in the Province of Quebec," International Road Federation XIIIth World Meeting, Toronto, Ontario, Canada, June 1997, Paper no. 673-E, Topic 12, 1997.

492. Mallows, C. L., "Some Comments on CP," *Technometrics,* vol. 15, U.S.A., 1973.

493. Mann, J., "Bike Helmet Apathy," *The Washington Post,* U.S.A., August 2, 1991.

494. Maring, W., and I. van Schagen, "Age Dependence of Attitudes and Knowledge in Cyclists," *Accident Analysis & Prevention,* vol. 22, U.S.A., 1990, pp. 127–136.

495. Martin and Voorthees Associates, "Crawler Lane Study: An Economic Evaluation," Department of the Environment, London, United Kingdom, 1978.

496. Mason, J. M., and J. C. Peterson, "Survey of States' RRR Practices and Safety Considerations," *Transportation Research Record,* vol. 960, U.S.A., 1984, pp. 20–27.

497. Matsumura, T., T. Seo, M. Umezawa, and T. Okutani, "Road Structure and Traffic Safety Facilities in Japan," *Proceedings of Conference on Asian Road Safety* (*CARS*), 1993, pp. 8.13–8.30.

498. Maycock, G., "The Effect on Accidents and Injuries of Excess and Inappropriate Speed," Contribution to the ETSC Symposium Reducing Speed-Related Casualties, The Role of the European Union, ETSC (European Transport Safety Council), Brussels, May 3, 1995.

499. Mazie, D., "Keep your Teen-Age Driver Alive," *Reader's Digest,* June 1991.

500. McCarthy, J., J. C. Scruggs, and D. B. Brown, "Estimating the Safety Benefits for Alternative Highway and/or Operational Improvements," Report no. FHWA/RD-81/179, Federal Highway Administration, Washington, D.C., U.S.A., 1981.

501. McDonald, G. L., "The Involvement of Tractor Design in Accidents," Research Report 3/72, Department of Mechanical Engineering, University of Queensland, St. Lucia, Australia, 1972.

502. McGee, H. W., "Synthesis of Large Truck Safety Research," Wagner-McGee Associates, Alexandria, Virginia, U.S.A., 1981.

503. McGee, H. W., W. E. Hughes, and K. Daily, "Effect of Highway Standards on Safety," National Cooperative Highway Research Program, Transportation Research Board, Report 374, Washington, D.C., U.S.A., 1995.

504. McGuinness, P., "Truck Climbing Lanes on Roadways," Report RT 186, An Foras Forbartha, Dublin, Ireland, 1979.

505. McLean, J. R., "Speeds on Curves: Regression Analysis," Internal Report 200-3, ARRB (Australian Road Research Board), Melbourne, Australia, 1978.

506. McLean, J. R., "Driver Speed Behavior and Rural Road Alignment Design," *Traffic Engineering & Control,* vol. 22, no. 4, United Kingdom, 1981, pp. 208–211.

507. McLean J. R., "Cross Section Design Standards—The Australian Research and Experience," paper prepared for Presentation at the Transportation Research Board 72nd Annual Meeting, Washington, D.C., U.S.A., 1993.

508. McLean J. R., and J. F. Morrall, "Changes in Horizontal Alignment Design Standards in Australia and Canada," International Symposium on Highway Geometric Design Practices, Transportation Research Board, Boston, Massachusetts, U.S.A., August 1995.

509. Meewes, V., and R. Maier, "Model-Test B 33: 2 + 1-Lane Rural Roads—Traffic Safety, Traffic Flow, Opinion Pool," Association of Car Insurances (HUK), Information no. 22, Cologne, Germany, 1984.

510. MEL (Ministère de l'Equipement et du Logement), "Instruction sur les Conditions Techniques D'Aménagement des Routes Nationales," France, Editions 1970, 1975; Instruction sur les Conditions Techniques D'Aménagement des Autoroutes de Liaison, Paris, France, 1971.

511. Mellander, H., "Automotive Crash Safety Engineering—Time for a New Approach," *Proceedings, Road Safety and Traffic Environment in Europe,* Gothenburg, Sweden, VTIrapport 362A, Swedish Road and Traffic Research Institute, Linkoeping, Sweden, 1990, pp. 97–110.

512. Messer, C. J., "Methodology for Evaluating Geometric Design Consistency," *Transportation Research Record,* vol. 757, U.S.A., 1980, pp. 7–14.

513. Meyer, E., E. Jacobi, and E. Stiefel, *Typical Accident Causes in Road Traffic,* vol. III, Board of Trustees, We and the Road (Wir und die Strasse), Munich, Germany, 1961.

514. Miller, R., "Benefit-Cost Analysis of Lane Marking," *Public Roads,* vol. 56, no. 4, U.S.A., 1993, pp. 153–163.

515. Milton, U. S., and J. O. Tsokos, *Statistical Methods in the Biological and Health Sciences,* McGraw-Hill, New York, U.S.A., 1983.

516. Ministère Waloon de L'Equipement et des Transports, "Charactéristiques Routières et Autoroutières," Brussels, Belgium, 1991.

517. Ministry of Transport, "Traffic Safety, The Danish Way!," The Road Directorate, Denmark, 1994.

518. Ministry of Transportation of Ontario, "Ontario Road Safety Agenda," (Draft), Ottowa, Canada, 1994.

519. Mintsis, G., "Speed Distributions of Road Curves," *Traffic Engineering & Control,* vol. 29, no. 1, United Kingdom, January 1988, pp. 21–27.

520. Mitschke, M., J. Maretzke, and H. Otto, "Investigation of the Driving Behavior of Passenger Cars when Cornering," *Research Road Construction and Road Traffic Technique* (*Forschung Strassenbau und Strassenverkehrstechnik*), vol. 466, pp. 35–36, Minister of Transportation, Bonn, Germany, 1986.

521. Mitschke, M., and P. Voelsen, "Driving Behavior of Passenger Cars under Wet Conditions," *Research Road Construction and Road Traffic Technique* (*Forschung Strassenbau und Strassenverkehrstechnik*), vol. 466, pp. 1–34, Minister of Transportation, Bonn, Germany, 1986.

522. Mitschke, M., *Dynamics of Motor Vehicles,* vol. C, *Driving Behavior,* Springer-Verlag, Berlin, 1990.

523. Morales, J. M., and J. F. Paniati, "Two-Lane Traffic Simulation—A Field Evaluation of Roadsim," *Public Works,* vol. 49, no. 3, U.S.A., 1985.

524. Moray, N., "Designing for Transportation Safety in the Light of Perception, Attention, and Mental Models," *Ergonomics,* vol. 33, United Kingdom, 1990, pp. 1201–1213.

525. Morrall, J. F., and C. J. Hoban, "Design Guidelines for Overtaking Lanes," *Traffic Engineering and Control,* vol. 26, no. 10, 1984, pp. 476–483.

526. Morrall, J. F., A. Werner, and P. Kilburn, "Planning and Design Guidelines for the Development of a System of Passing Lanes for Alberta Highways," *Proceedings, Australian Road Research Board Conference,* vol. 13, part 7, Traffic, Australia, 1986.

527. Morrall, J. F., and R. S. Talarico, "Side Friction Demanded and Margins of Safety on Horizontal Curves," *Transportation Research Record,* vol. 1435, U.S.A., 1994.

528. Morrall, J. F., E. Miller Jr., G. A. Smith, J. Feuerstein, and F. Yazdan, "Planning and Design of Passing Lanes Using Simulation Models," *Journal of Transportation Engineering,* vol. 121, no. 1, U.S.A., 1995.

529. Moyer, R. A., "Skidding Characteristics of Automobile Tires on Roadway Surfaces and Their Relation to Highway Safety," Bulletin No. 120, Iowa Engineering Experiment Station, Ames, Iowa, U.S.A., 1934, 128 pp.

530. Mueller, A., W. Achenbach, E. Schindler, T. Wohland, and F.-W. Mohn, "The New Drive-Safety-System Electronics Stability Program of Mercedes-Benz," *Automobile Technical Journal* (*ATZ*), vol. 96, no. 11, Germany, 1994, pp. 656–670.

531. Mulinazzi, T. E., and H. L. Michael, "Correlation of Design Characteristics and Operational Controls with Accident Rates on Urban Arterials," Purdue University, Lafayette, Ind., Joint Highway Research Project Report 35, U.S.A., 1967.

532. Musick, J. V., "Effect of Pavement Edge Markings on Two-Lane Rural State Highways in Ohio," *Highway Research Board Bulletin 266,* Washington, D.C., U.S.A., 1960.

533. Naatanen, R., and H. Summala, *Road User Behavior and Traffic Accidents,* North-Holland, Amsterdam, The Netherlands, 1976.

534. Naess, P., *Environmental Databases in Municipalities,* Published by IBM, Sollentura, Sweden, 1990.

535. National Cooperative Highway Research Program, "Selection of Safe Roadside Cross Sections," NCHRP 158, Washington, D.C., U.S.A., 1975.

536. National Highway Traffic Safety Administration, "Fatal Accident Reporting System 1981," Technical Report DOT HS-806-251, U.S. Department of Transportation, Washington, D.C., U.S.A., 1983.

537. National Road Trauma Advisory Council, "Report of the Working Party on Trauma Systems," Commonwealth Department of Human Services and Health, Canberra, Australia, 1993.

538. National Safety Council, *Accident Facts,* Chicago, Illinois, U.S.A., 1985 ed.

539. National Transportation Safety Board, "Fatal Highway Accidents on Wet Pavements—The Magnitude, Location, and Characteristics," Technical Report: NTSB-HSS-80-1, Washington, D.C., U.S.A., 1980.

540. Naumann, H., "Development of a Program System for the Establishment of Computer Aided Perspective Films," Publications of the Institute for Highway and Railroad Engineering, vol. 17, University of Karlsruhe (TH), Germany, 1977.

541. Nedas, N. D., G. P. Balcar, and P. R. Macy, "Engineering the Way Through the Alcohol Haze," *ITE Journal,* vol. 50, no. 11, Institute of Transportation Engineers, Washington, D.C., U.S.A., 1980.

542. Nedas, N. D., G. P. Balcar, and P. R. Macy, "Road Markings as an Alcohol Countermeasure for Highway Safety: Field Study of Standard and Wide Edgelines (Abridgement)," *Transportation Research Record,* vol. 847, pp. 43–46, U.S.A., 1982.

543. Nemeth, Z., and R. Lamm, "Continuous Two-Way Left-Turn Lanes: An American Solution to Mid-Block Left-Turn Problems," *Road and Construction (Strassen und Tiefbau),* Germany, November/December 1982, vols. 11 and 12, pp. 14–21 and pp. 9–17.

544. Netzer, M., "The Passing Maneuver on Rural Roads with Regard to Traffic Safety," *Research Road Construction and Road Traffic Technique (Forschung Strassenbau und Strassenverkehrstechnik),* vol. 50, Minister of Transportation, Bonn, Germany, 1966.

545. Neuhardt, J. B., G. D. Herrin, and T. H. Rockwell, "Demonstration of a Test-Driven Technique to Assess the Effects of Roadway Geometrics and Development on Speed Selection," Project 326B, Systems Research Group, The Ohio State University, Columbus, Ohio, U.S.A., 1971.

546. Neuzil, D., and J. S. Peet, "Flat Embankment Slope Versus Guardrail: Comparative Economy and Safety," *Highway Research Record 306,* U.S.A., 1970, pp. 10–24.

547. Neuzil, D., "Aesthetic Preference and Perceived Safety in Highway Design Treatments: A Pilot Study," *Transportation Research Record,* vol. 518, U.S.A., 1974, pp. 11–28.

548. New York State Department of Motor Vehicles, *Motorcycle Operator's Manual,* Albany, N.Y., U.S.A., 1989.

549. Niklas, "Benefit-Cost-Analyses of Safety Programs in the Field of Road Traffic," *Proceedings of the Automobile Industry Association,* Frankfurt/Main, Germany, 1970.

550. Nikpour, A., "Procedure for Cost Calculation by 2-digit Accident Types on Two-Lane Rural Roads," Ph.D. Dissertation, Institute for Highway and Railroad Engineering, University of Karlsruhe (TH), Germany, 1986.

551. Nilsson, G., "The Effect of Speed Limits on Traffic Accidents in Sweden," *Proceedings of the OECD International Symposium on The Effects of Speed Limits on Traffic Accidents and Transport Energy Use,* Dublin, Ireland, 1981.

552. O'Cinnéide, D., N. McAuliffe, and D. O'Dwyer, "Comparison of Road Design Standards and Operational Regulations in EC and EFTA Countries," Deliverable 8, EU DRIVE II Project V2002, European Community Brussesl, Belgium, 1993.

553. O'Cinnéide, D., "The Relationship between Geometric Design Standards and Safety," International Symposium on Highway Geometric Design Practices, Transportation Research Board, Boston, Massachusetts, U.S.A., August 1995.

554. O'Flaherty, C. A., *Highways,* vol. 1, Edward Arnold publishers Ltd., London, 1974.

555. Organization for Economic and Cultural Development, "Targeted Road Safety Programmes," Paris, France, 1994.

556. Organization for Economic and Cultural Development Symposium: "Methods for Determining Geometric Design Standards," Helsingoer, Denmark, 1976.

557. Office of Technology Assessment, U.S. Congress, "Technology Assessment of Changes in the Future Use and Characteristics of the Automobile Transport System," vol. II, Technical Report, Washington, D.C., U.S.A., 1977.

558. Official Technical Guidelines for the Design of Rural Roads, *Preliminary Guidelines for the Design of Rural Roads RAL 1937*; *Guidelines for the Design of Rural Roads, Part I: Cross Sections, RAL-Q, 1956*; *Proposal for the Guidelines of Rural Roads RAL, Part II: Alignment (RAL-L), Section 1: Elements of the Alignment, 1959/1963,* 5th ed., Erich Schmidt Publishers, Berlin, Germany, 1965.

559. Opiela, K. S., "Relationships between Highway Safety and Geometric Design," paper presented to the International Symposium Traffic Safety: A Global Issue, Kuwait, January 15–17, 1995.

560. Oppe, S., and M. J. Koornstra, "A Mathematical Theory for Related Long Term Developments of Road Traffic and Safety," M. Koshi, ed., *Transportation and Traffic Theory,* Elsevier, New York, U.S.A., 1990, pp. 113–132.

561. Osterloh, H., "The Space Curve from the Driver's View—An Investigation about the Superimposition of Plan and Profile from the Perspective View," Engineering Office of Krenz-Osterloh, Wiesbaden, Germany, 1968.

562. Osterloh, H., "Mathematical Evaluation of the Three Dimensional Impression of the Perspective View for the Driver," *Research Road Construction and Road Traffic Technique (Forschung Strassenbau und Strassenverkehrstechnik),* vol. 394, Minister of Transportation, Bonn, Germany, 1983.

563. Osterloh, H., *Road Planning with Clothoids and Vehicle Swept-Curve Paths,* Building Publishing House (Bau-Verlag), Wiesbaden, Germany, 1991.

564. Otte, D., "Injury Pattern of Passenger Car Occupants in Case of Accidents—Requirements for Injury Reducing Restraint Systems, Based on Accident Analysis," Traffic Accident Research, Medical University of Hannover, Presentation, House of Technique, Essen, Germany, October 1992.

565. Ottesen, J. L., and R. A. Krammes, "Speed Profile Model for a U.S. Operating-Speed-Based Consistency Evaluation Procedure," paper prepared for the 73rd Annual Meeting of Transportation Research Board, Washington, D.C., U.S.A., January 1994.

566. Pack, A. I., A. M. Pack, E. Rodgman, A. Cucchiara, D. F. Dinges, and C. W. Schwab, "Characteristics of Crashes Attributed to the Driver Having Fallen Asleep," *Accident Analysis and Prevention,* vol. 27, U.S.A., 1995, pp. 769–775.

567. Paisley, J. L., discussions of T. D. Wilson, "Road Safety By Design," *The Journal of the Institution of Highway Engineers,* vol. 15, United Kingdom, 1968, p. 36.

568. Peet, J. S., and D. Neuzil, "Independent Versus Narrow-Median Alignment: Comparative Economy, Safety, and Aesthetics," *Highway Research Record 390,* U.S.A., 1972, pp. 1–14.

569. Perchonok, K., T. Ranney, S. Baum, D. Morris, and J. Eppich, "Hazardous Effects of Highway Features and Roadside Objects, Volume I: Literature Review and Methodology and Volume II: Findings," Report no. FHWA-RD-78-201 & 202, Federal Highway Administration, Washington, D.C., U.S.A., 1978.

570. Permanent International Association of Road Congresses (PIARC), "Question IV: Interurban Roads and Motorways, National Report Switzerland," XVIIth World Road Congress, Brussels, Belgium, September 13–19, 1987.

571. Permanent International Association of Road Congresses (PIARC), "Interurban Roads, Report of the Committee," XXth World Congress, Montreal, Canada, September 1995, pp. 3–9.

572. Persson, J. C., "Traffic Safety Facing Year 2000—A Challenge for the Automotive Industry," *Proceedings, Road Safety and Traffic Environment in Europe,* Gothenburg, Sweden, VTIrapport 362A, Swedish Road and Traffic Research Institute, Linkoeping, Sweden, 1990, pp. 29–37.

573. Pfafferott, I., and R. D. Huguenin, "Behavioral Adaptions to Changes in the Road Transport System—Results and Conclusions of an OECD-Study," *Journal for Traffic Safety (Zeitschrift fuer Verkehrssicherheit),* vol. 37, no. 2, Germany, 1991, pp. 71–83.

574. Pfundt, K., "Comparative Accident Investigations on Rural Roads," *Research Road Construction and Road Traffic Technique (Forschung Strassenbau und Strassenverkehrstechnik),* vol. 82, Minister of Transportation, Bonn, Germany, 1969.

575. Pfundt, K., "Accident Situation on Two-Lane Rural Roads," Research Report (FP 0.9101) of the Federal Highway Research Institute (BASt), Bergisch Gladbach, Germany, 1992.

576. Pietzsch, W., *Road Planning,* Werner-Engineering Texts, Duesseldorf, Germany, 1989.

577. Pigman, J. G., K. R. Agent, J. A. Deacon, and R. J. Kryscio, "Evaluation of Unmanned Radar Installations," paper prepared for the 68th Annual Meeting of Transportation Research Board, Washington, D.C., U.S.A., January, 1989.

578. Pignataro, L. J., *Traffic Engineering, Theory and Practice,* Prentice-Hall, Englewood Cliffs, N.J., U.S.A., 1973.

579. Pilkington, G. B., "Utility Poles—A Highway Safety Problem" *Public Roads,* vol. 52, no. 3, U.S.A., 1988, pp. 61–66.

580. Pippert, H., *Body Engineering, Passenger Cars, Commercial Vehicles, Buses; Lightweight Design, Materials, Production Engineering, Construction and Calculation,* 2d ed., Vogel Publishing House, Wuerzburg, Germany, 1993.

581. Pohl, M., "The Roads of the Antique," *The Road in the Change of the Centuries, Strabag Proceedings,* vol. 50, Cologne, Germany, 1995, pp. 40–45.

582. Porter, R., "Models for Highway Design: Some Construction and Photographic Techniques," *Highway Research Record 270,* U.S.A., 1969, pp. 25–35.

583. Press-Republican, "Report: U.S. Driving Habits Wasteful," Plattsburgh, N.Y., U.S.A., June 11, 1992.

584. Priess, R., "Distance between Railroad and Roadway," unpublished manuscript, Mailaender Ingenieur Consult, Karlsruhe, Germany, 1997.

585. Psarianos, B., "A Contribution to the Development of the Three Dimensional Alignment of Traffic Routes with Special Emphasis on Highways," Ph.D. Dissertation, Department of Surveying, Scientific Proceedings, no. 113, University of Hannover, Hannover, Germany, 1982.

586. Psarianos, B., "Establishment of the Greek Operating Speed Backgrounds," unpublished manuscript, Department of Rural and Surveying Engineering, National Technical University of Athens, Greece, 1994.

587. Psarianos, B., "Establishment of a Classification System for a Driving Dynamic Safety Criterion in Greece," unpublished manuscript, Department of Rural and Surveying Engineering, National Technical University of Athens, Greece, 1995.

588. Psarianos, B., M. Kontaratos, and D. Kasios, "Influence of Vehicle Parameters on Horizontal Curve Design of Rural Highways," International Symposium on Highway Geometric Design Practices, Transportation Research Board, Boston, Massachusetts, U.S.A., August 1995.

589. Pucher, R., "Methods to Increase Safety in Traffic," *Research Works from Road Engineering, New Sequence,* vol. 56, Germany, 1963.

590. Pushkarev, B., "The Paved Ribbon: The Esthetics of Freeway Design," in C. Tunnard and B. Pushkarev, *Man-Made America: Chaos or Control?* Yale University Press, New Haven, Connecticut, U.S.A., 1963.

591. Raff, M. S., "Interstate Highway-Accident Study," *Highway Research Board Bulletin 74,* U.S.A., 1953, pp. 18–43.

592. Rasmussen, J., "Models of Mental Strategies in Progress Plant Diagnosis," J. Rasmussen and W. B. Rouse, *Human Detection and Diagnosis of System Failures,* Plenum, New York, U.S.A., 1981.

593. Read, M., "Road Safety Planning and Targeting," Report to World Congress, Montreal, PIARC (Permanent International Association of Road Congresses) Committee C 13, Road Safety (Draft), September 4–8, Canada, 1995.

594. Reagan, J. A., "The Interactive Highway Safety Design Model: Designing for Safety by Analyzing Road Geometrics," *Public Roads,* vol. 58, no. 1, U.S.A., 1994, pp. 37–43.

595. Reagan, J. A., and W. A. Stimpson, R. Lamm, R. Heger, R. Steyer, and M. Schoch, "Influence of Vehicle Dynamics on Road Geometrics," International Symposium on Highway Geometric Design Practices, Transportation Research Board, Boston, Massachusetts, U.S.A., August 1995.

596. Redzovic, V., "Test Rides for Evaluating the Influence of Negative Superelevation Rate on Driving Dynamic Safety with Regard to Side Friction Demand when Changing Lanes and During Braking Maneuvers," Ph.D. Dissertation, Faculty for Civil and Surveying Engineering, University of Karlsruhe (TH), Karlsruhe, Germany, 1995.

597. Reimpell, J., P. Sponagel, *Vehicle-Technique: Tires and Wheels,* Vogel Publishing House, Wuerzburg, Germany, 1988.

598. Reinboth, K., "To the Pre- and Early History of Automobile Road Construction—A Short Historical Evaluation," *Road and Interstate (Strasse und Autobahn),* vol. 10, Germany, 1994, pp. 639–652.

599. Reinfurt, D. W., D. N. Levine, and W. D. Johnson, "Radar as a Speed Deterrent: An Evaluation," Highway Safety Research Center, University of North Carolina, Chapel Hill, N.C., 1973.

600. Reinhold, "The Danger Level of Roads and Places Considering Traffic Accidents," *Road Construction and Road Traffic Technique (Strassenbau und Strassenverkehrstechnik),* vol. 15, Germany, 1938, pp. 367–370.

601. Reprint of "Highway Geometric Design," Institute for Highway and Railroad Engineering, University of Karlsruhe (TH), Germany, 1994.

602. Ricci, L., "National Crash Severity Study Statistics," Highway Safety Research Institute, University of Michigan, Ann Arbor, Mich., Special Report Contract no. DOT HS-8-01944, U.S.A., 1979.

603. Richter, P., and W. Hacker, *Workload and Stress,* Asanger Publishing House, Heidelberg, Germany, 1996.

604. Richter, P., G. Weise, R. Heger, and T. Wagner, "Driving Behavior and Psychological Activation of Motorists as Evaluation Criteria for the Design Quality of Road Traffic Facilities," Research Report of the German Research Association (DFG), Technical University of Dresden, Germany, 1997.

605. Riemersma, J. B. J., K. W. Mess, and J. A. Michon, "Perception and Maintenance of Speed and Course of Moving Vehicle. A Review of the Literature I," National Defence Research Organization TNO, National Organization for Applied Scientific Research in The Netherlands, Report no. IZF, 1972-C7, The Netherlands, 1972.

606. Ritchie, M. L., W. K. McCoy, and W. L. Welde, "A Study of the Relation between Forward Velocity and Lateral Acceleration in Curves Normal Driving," *Human Factors,* vol. 10, no. 3, U.S.A., 1968, pp. 255–258.

607. Road and Transportation Association of Canada (RTAC), *Geometric Design Standards for Canadian Roads and Streets,* Ottawa, Ontario, Canada, 1981.

608. Road and Transportation Association of Canada (RTAC), *Manual of Geometric Design Standards for Canadian Roads,* 1986 metric edition, Ottawa, Ontario, Canada, 1986.

609. Robertson, L. S., "An Instance of Effective Legal Regulation: Motorcyclist Helmet and Daytime Headlamp Laws," *Law & Society Review,* vol. 10, U.S.A., 1976.

610. Rockwell, T. H., and J. N. Snider, "An Investigation of Variability in Driving Performance in the Highway," Project RF 1455 Final Report, The Ohio State University, Systems Research Group, Columbus, Ohio, U.S.A., 1965.

611. Roenitz, R., H.-H. Braess, and A. Zomotor, "Procedures and Criteria for the Evaluation of the Driving Behavior of Passenger Cars," *Automobile Industry (Automobil-Industrie),* vol. 1, 1977, pp. 29–39 and vol. 3, Germany, 1977, pp. 39–48.

612. Roland, H. E., W. W. Hunter, J. R. Stewart, and B. J. Campbell, "Investigation of Motor Vehicle/Bicycle Collision Parameters," Final Report for the U.S. Department of Transportation, National Highway Traffic Safety Administration, Washington, D.C., U.S.A., 1979.

613. Rompe, K., and B. Heissing, *Unbiased Test Procedure for the Driving Qualities of Motor Vehicles— Lateral and Longitudinal Dynamics,* Publishers TUEV, Rheinland, Cologne, Germany, 1994.

614. Roosmark, P., and R. Fraeki, "Studies of Effects Produced by Road Environment and Traffic Characteristics on Traffic Accidents," *Proceedings of the Symposium on the Use of Statistical Methods in the Analysis of Road Accidents,* Organization for Economic and Cultural Development, 1970.

615. Rotach, M. C., "Trucks on Gradients," *Road and Traffic, (Strasse und Verkehr),* vol. 46, Switzerland, 1960, pp. 444–446.

616. Rothengatter, T., "The Scope of Automatic Detection and Enforcement Systems," *Proceedings, Road Safety and Traffic Environment in Europe,* Gothenburg, Sweden, VTIrapport 365A, Swedish Road and Traffic Research Institute, Linkoeping, Sweden, 1990, pp. 165–176.

617. Rowan, N. J., D. L. Woods, V. G. Stover, D. A. Anderson, J. H. Dozier, and J. H. Johnson, "Safety Design and Operational Practices for Streets and Highways," Technology Sharing Report: 80-228, U.S. Department of Transportation, Washington, D.C., U.S.A., 1980.

618. Rumar, K., "Safety Problems and Countermeasure Effects in the Nordic Countries," International Meeting on the Evaluation of Local Traffic Safety Measures, Paris, France, May 1985.

619. Rumar, K., and L. Stenborg, "The Swedish National Road Safety Programme; A New Approach to Road Safety Work," Swedish National Road Administration, Sweden, 1994.

620. Ruwenstroth, G., et al., "Investigations to Determinants of Speed Selection; Report 3, Situation Adapted Speed Selection on Roads Outside Built-Up Areas (without Interstates)," Report to the Research Project 8525 of the Federal Highway Research Institute (BASt), Bergisch Gladbach, Germany, 1989.

621. Ruyters, H. G. J. C. M., M. Slop, and F. C. M. Wegman, "Safety Effects of Road Design Standards," R-94-7, SWOV, Leidschendam, The Netherlands, 1994.

622. Sayers, M. W., and C. Mink, "Integration of Road Design with Vehicle Dynamics Models Through a Simulation Graphical User Interface (SGUI)," Paper No. 951084, presented at the 74th Annual Meeting of the Transportation Research Board, Washington, D.C., U.S.A., 1995.

623. Schaar, A., "Driving Simulation with Innovative Tools," *Automobile Technical Journal (ATZ),* vol. 95, no. 5, 1993, pp. 256–262.

624. Schiller, H., "The Creation and Design of the Road Network as Regional Planning Task, Part II: "The Evaluation of the Road Condition by Characteristical Traffic Speeds," *Research Road Construction and Road Traffic Technique (Forschung Strassenbau und Strassenverkehrs-technik)*, vol. 116, Minister of Transportation, Bonn, Germany, 1971.

625. Schlichter, H. G., "Road Characteristics—An Analytical Evaluation," *Road and Interstate (Strasse und Autobahn)*, vol. 2, Germany, 1976, pp. 55–58.

626. Schlichter, H. G., "Spatial Alignment of Traffic Routes," Publications of the Institute for Highway and Railroad Engineering, University of Karlsruhe (TH), Germany, vol. 30, 1985.

627. Schliesing, H., "Road Characteristics," *Fundamentals of Road Planning,* Minister of Transportation, Bonn, Germany, 1968.

628. Schmidt, G., "Analysis and Evaluation of Road Sections by Three Safety Criteria," master's thesis, Institute for Highway and Railroad Engineering, University of Karlsruhe (TH), Germany, 1995.

629. Schneider, J., "Accident Analytical and Driving Dynamic Evaluation of the Negative Superelevation," Publications of the Institute for Highway and Railroad Engineering, University of Karlsruhe (TH), Germany, vol. 27, 1982.

630. Schneider, W., *Overall is Babylon,* Econ-Publishing House, Duesseldorf, Germany, 1960.

631. Schneider, W., "Psychological Causes and Background Conditions during Accident Occurrence," *Journal for Traffic Safety (Zeitschrift fuer Verkehrssicherheit)*, vol. 4, Germany, 1977, pp. 140–145.

632. Schoch, M., "Analysis, Development and Evaluation of Theoretical and Practical Side Friction Factors for Highway Design," graduate study work, Institute for Highway and Railroad Engineering, University of Karlsruhe (TH), Germany, 1994.

633. Schoch, M., "The Crash-Behavior of Motor Vehicle Bodies with Special Consideration of Material Use and Design Processing," master's thesis, Institute for Reliability and Failure Analysis in Mechanical Engineering, University of Karlsruhe (TH), Karlsruhe, Germany, 1996.

634. Schoppert, D. W., "Predicting Traffic Accidents from Roadway Elements of Rural Two-Lane Highways with Gravel Shoulders," *Highway Research Board Bulletin,* vol. 158, U.S.A., 1957, pp. 4–26.

635. Schreiber, H., *Symphony of the Road,* Econ-Publishing House, Duesseldorf, Germany, 1959.

636. Schriever, T., and P. Alber, "Development of an Energy Absorbing Front Underride Protection for Commercial Vehicles," *Safety in Road Traffic,* VDI-Reports 1046, VDI-Publishing House (VDI-Verlag), Duesseldorf, Germany, 1993.

637. Schulze, C., "Intermediate Cross Sections—An International Literature Review," unpublished manuscript, Institute for Highway and Railroad Engineering, University of Karlsruhe (TH), Germany, 1995.

638. Schulze, C., "Comparison, Analysis and Evaluation of the Speed Behavior on Auxiliary Lanes of Upgrade Sections, Based on International Design Procedures as well as on Speed Measurements," master's thesis, Institute for Highway and Railroad Engineering, University of Karlsruhe (TH), Germany, 1996.

639. Schulze, C., and R. Lamm, "Design of Auxiliary Lanes in Upgrade Direction," *Road and Construction (Strassen und Tiefbau)* Germany, 1999 (to be published).

640. Schulze, K. H., and L. Beckmann, "Friction Properties of Pavements at Different Speeds," American Society for Testing and Materials Special Technical Publication, no. 326, Germany, 1962.

641. Schwant, W., and J. Stocklose, "Use of Plastic Materials for Safety Components for the Occupant Protection, Especially for the Head Impact," *Plastic Materials in Automobile Design, Compound Systems, Procedures, Applications,* VDI-Publishing House (VDI-Verlag), Duesseldorf, Germany, 1995.

642. Seiffert, U., *Motorvehicle Safety,* VDI-Publishing House (VDI-Verlag), Duesseldorf, Germany, 1992.

643. SETRA/DLI, "Vitesses Pratiquées et Géométrie de la Route," Note d'Information B-C 10, Ministère de l'Equipement, du Logement, de l'Aménagement du Territoire et des Transports, Paris, France, 1986.

644. Shermer, M., *Cycling: Endurance and Speed,* Contemporary Books, Inc., Chicago, Illinois, U.S.A., 1987.

645. Shinar, D., "Curve Perception and Accidents on Curves: An Illusive Curve Phenomenon?" *Journal for Traffic Safety (Zeitschrift fuer Verkehrssicherheit)*, vol. 1, 1977, pp. 16–21.

646. Shinar, D., E. D. McDowell, and T. H. Rockwell, "Eye Movements in Curve Negotiation," *Human Factors,* vol. 19, no. 1, U.S.A., 1977, pp. 33–71.

647. Siegel, S., *Nonparametric Statistics for the Behavioral Sciences,* McGraw-Hill, New York, U.S.A., 1956.

648. Sievert, W., "The Influence of Modern Electronic Systems in Motor Vehicles on Accident Statistics," *Journal for Traffic Safety* (*Zeitschrift fuer Verkehrssicherheit*), vol. 2, Germany, 1994, pp. 72–82.

649. Silyanov, V. V., "Comparison of the Pattern of Accident Rates on Roads of Different Countries," *Traffic Engineering and Control,* vol. 14, United Kingdom, 1973, pp. 432–435.

650. Simon, E., and G. Jonscher, "Pavement Widening in the Curve," *Road and Interstate (Strasse und Autobahn),* vol. 5, Germany, 1963, pp. 155–162.

651. Sloane, E., *The Complete Book of Bicycling,* Simon & Schuster, New York, U.S.A., 1988.

652. Slop, M., "Low-Cost Engineering Measures to Improve Road Safety in Central and Eastern European Countries," A-93-25, SWOV, Leidschendam, The Netherlands, 1993.

653. Smeby, T., "Requirements to Overtaking Facilities," XVIIIth World Road Congress, Brussels, September 13–19, 1987, Question IV: Interurban Roads and Motorways, National Report Norway, Brussels, Belgium, 1987.

654. Smeed, R. J., "Some Statistical Aspects of Road Safety Research," *Journal of the Royal Statistical Society,* vol. CXII, part 1, 1949, pp. 1–34.

655. Smeed, R. J., "The Frequency of Road Accidents," *Journal for Traffic Safety* (*Zeitschrift fuer Verkehrssicherheit*) vols. 2 and 3, Germany, 1974, pp. 95–108, and pp. 151–159.

656. Smith, B., and E. Yotter, "Computer Graphics and Visual Highway Design," *Highway Research Record 270,* U.S.A., 1969, pp. 49–64.

657. Smith, B. L., and R. Lamm, "Coordination of Horizontal and Vertical Alignment with Regard to Highway Esthetics," *Transportation Research Record,* vol. 1445, 1994, pp. 73–85.

658. Smith, S., J. Purdy, H. McGee, D. Harwood, A. St. John, and J. Glennon, "Identification, Quantification, and Structuring of Two-Lane Rural Highway Safety Problems and Solutions," vols. I and II, Report Nos. FHWA/RD-83/021 and 83/022, Federal Highway Administration, Washington, D.C., U.S.A., 1983.

659. Snow, B., *The Highway and the Landscape,* Rutgers University Press, New Brunswick, N.J., U.S.A., 1959.

660. Snyder, J. C., "Environmental Determinants of Traffic Accidents—An Alternate Model," *Transportation Research Record,* vol. 486, U.S.A., 1974, pp. 11–18.

661. Solomon, D., "Accidents on Main Rural Highways Related to Speed, Driver and Vehicle," Bureau of Public Roads, U.S.A., 1964.

662. Sorensen, S., "How to Get Your Child to Wear a Helmet?," *Bicycling,* vol. 30, U.S.A., June 1989.

663. Southern Africa Transport and Communications Commission (SATCC), "Recommendations on Road Design Standards Volume I—Introduction, Geometric Design of Rural Roads," South Africa, 1987.

664. Spacek, P., and P. Dueggeli, "Speeds of Trucks in Upgrades and Downgrades," ETH Zuerich, Institute for Traffic Planning and Transport Technique, IVT Report no. 84/5, Zuerich, Switzerland, 1984.

665. Spacek, P., "Superelevation Rates in Tangents and Curves," ETH Zurich, Institute for Traffic Planning, Transport Technique, Highway and Railroad Construction, Research Report 22/79 of the Swiss Association of Road Specialists, Zuerich, Switzerland, 1987.

666. Sparks, J. W., "The Influence of Highway Characteristics on Accident Rates," *Public Works,* vol. 99, no. 3, U.S.A., 1969.

667. Spoerer, E., *Explorative Analyses of Traffic and Accident Occurrence,* Boehm and Schneider Publishers, Germany, 1965.

668. Sporbeck, O., "Basic Demands and Procedure with the ECE in Road Planning," *Environmental Compatibility Examination, Colloquium of the German Road and Transportation Research Association,* Mannheim, Germany, 1990, pp. 35–40.

669. Srinivasan, S., "Effect of Roadway Elements and Environment on Road Safety," Institution of Engineers, vol. 63, United Kingdom, 1982.

670. St. John, A. D., and D. W. Harwood, "Safety Considerations for Truck Climbing Lanes on Rural Highways," *Transportation Research Record,* vol. 1303, 1991, pp. 74–82.

671. Stahl, G., "Meaning of the RAS-N for Downgrading Measures on Through Roads, Examples for the State of Hessen," *Road and Interstate (Strasse und Autobahn),* vol. 6, Germany, 1988, pp. 233–239.

672. Steffen, H., "Development of a Safety Module for the Examination of Highway Geometric Design and its Implementation in Two Complex CAD-Systems," master's thesis, Institute for Highway and Railroad Engineering, University of Karlsruhe (TH), Germany, 1992.

673. Steffen, H., R. Lamm, and A. K. Guenther, "Safety Examination in Highway Geometric Design by Applying Complex Data-Processing-Systems," *Road and Construction (Strassen und Tiefbau)*, vol. 10, Germany, 1992, pp. 12–23.

674. Steierwald, G., and M. Buck, "Speed Behavior on Two-Lane Rural Roads with Regard to Design, Operational and Traffic Related Conditions," *Research Road Construction and Road Traffic Technique (Forschung Strassenbau und Strassenverkehrstechnik)*, vol. 621, Minister of Transportation, Bonn, Germany, 1992.

675. Steierwald, G., and H.-D. Kuenne, eds., *Urban Traffic Planning. Principles, Methods, Purposes,* Springer Verlag, Berlin, Germany, 1993.

676. Steinfurth, L., "Dangers of the Road from a Traffic Psychological Viewpoint," *Journal for Traffic Safety (Zeitschrift fuer Verkehrssicherheit)*, vol. 4, no. 4, Germany, 1968, pp. 255–259.

677. Stewart, D., and C. J. Chudworth, "A Remedy for Accidents at Bends," *Traffic Engineering and Control,* vol. 31, no. 2, United Kingdom, 1990, pp. 88–93.

678. Stewart. D., "Risk on Roadway Curves," letter to *Traffic Engineering and Control,* vol. 35, no. 9, United Kingdom, 1994, p. 528.

679. Steyer, R., "Examination of the Suitability of the Geographical Information Systems 'SPANS' for Environmental Compatibility Studies and Highway Design," master's thesis, Institute of Traffic Route Construction, Technical University of Dresden, Germany, 1993.

680. Steyer, R., "History of Road Design and Construction," unpublished manuscript, Dresden, Germany, 1996.

681. Stoffel, W., "Geometric and Cinematic Problems of a Highway," Ph.D. Dissertation, University of Bonn, Germany, 1969.

682. Stohner, W. R., "Speeds of Passenger Cars on Wet and Dry Pavements," *Highway Research Board,* Bulletin 139, 1956.

683. Summala, H., T. Nieminen, and M. Punto, "Maintaining Lane Position with Peripheral Vision In-Vehicle Tasks," *Human Factors,* vol. 38, no. 3, U.S.A., 1996, pp. 442–451.

684. Supramaniam, V., G. V. Belle, and J. F. C. Sung, "Fatal Motorcycle Accidents and Helmet Laws in Peninsular Malaysia," *Accident Analysis & Prevention,* vol. 16, pp. 157–162, U.S.A., 1984.

685. Sutton, R., "Who Needs a Helmet?" *Reader's Digest,* U.S.A., May 1991.

686. Swedish National Road Administration, "Standard Specifications for Geometric Design of Rural Roads," Borlaenge, Sweden, 1982, 1986.

687. Swiss Association of Road Specialists (VSS), Swiss Norm SN 640 158, "Cross-Sectional Elements, Clearance," Zuerich, Switzerland, 1967.

688. Swiss Association of Road Specialists (VSS), Swiss Norm SN 640 100a, "Elements of Horizontal Alignment," Zuerich, Switzerland, 1983.

689. Swiss Association of Road Specialists (VSS), Swiss Norm SN 640 090/640 090a "Design, Fundamentals, Sight Distances," Zuerich, Switzerland, 1974, 1992.

690. Swiss Association of Road Specialists (VSS), Swiss Norm SN 640 140 "Alignment, Optical Criteria," Zuerich, Switzerland, 1978.

691. Swiss Association of Road Specialists (VSS), Swiss Norm SN 640 080a/b, "Highway Design, Fundamentals, Speed as a Design Element," Zuerich, Switzerland, 1981, 1991.

692. Swiss Association of Road Specialists (VSS), Swiss Norm SN 640 123, "Alignment, Superelevation Rates in Tangents and Curves," 1969, and Swiss Norm SNV 640 123a, Zuerich, Switzerland, 1988.

693. Swiss Association of Road Specialists (VSS), "Swiss Norm SN 640 080b, 100, 105, 110, 123a, 138a, 141, 145, 195a, 198, and 266, Alignment Features," Zuerich, Switzerland, Various Dates.

694. SWOV, "Framework for Consistent Traffic and Accident Statistical Data Bases," Institute for Road Safety Research, Leidschendam, The Netherlands, 1988.

695. SWOV, "Towards a Sustainable Safe Traffic System in the Netherlands, National Road Safety Investigation 1990–2010," Leidschendam, The Netherlands, 1993.

696. Takoa, G. T., "An Analytical Model for Driver Response," *Transportation Research Record,* vol. 1213, U.S.A., 1989, pp. 1–3.

697. Talarico, R. J., and J. F. Morrall, "Side Friction Factors for Horizontal Curve Design," Department of Civil Engineering, The University of Calgary, Calgary, Alta, Canada, 1993.

698. Talarico, R. J., and J. F. Morrall, "The Cost-Effectiveness of Curve Flattening in Alberta," National Research Council of Canada, NRC.CNRC, reprinted from *Canadian Journal of Civil Engineering,* vol. 21, no. 2, U.S.A., 1994, pp. 285–296.

699. Taragin, A., "Driver Performance on Horizontal Curves," *Public Roads,* vol. 28, no. 2, U.S.A., 1954, pp. 21–39.

700. Technical Recommendations for the General Design and Geometry of Roads, "Highway Design Guide (except for motorways)," Technical Guide August 1994 (Translation August 1995), Document Produced and Distributed by: SETRA, le Service des Etudes téchniques des routes et Autoroutes, Centre de la Sécurité et des Techniques Routières (French Administration for the Technical Studies of Roads and Motorways) Bagneux Cedex, France.

701. Terhune, K., and M. Parker, "Evaluation of Accident Surrogates for Safety Analysis of Rural Highways," Report no. FHWA/RD-86/128, Federal Highway Administration, Washington, D.C., U.S.A., 1986.

702. Terlow, J. C., "Transport Safety: European Co-Operation for the 90s," Westminister Lecture on Traffic Safety, London, Great Britain, 1990.

703. Term Determinations in Road Construction, Part: "Highway Planning and Road Traffic Technique," German Road and Transportation Research Association, Cologne, Germany, 1989.

704. The Roads Directorate 4.30.01, Traffic Engineering, "Roads and Path Types, Catalogue of Types for New Roads and Paths in Rural Areas," The Technical Committee on Road Standards, Copenhagen, Denmark, 1981.

705. Thoma, J., "Speed Behavior and Risks for Different Road Conditions, Weekdays and Daytimes," *Technical Journal for Traffic Safety (Zeitschrift fuer Verkehrssicherheit),* vol. 1, Germany, 1994, pp. 7–11.

706. Toyota, "Toyota and Automotive Safety," Information Booklet, Toyota-cho, Toyota City, Japan, 1996.

707. Trans-European North-South Motorway (TEM), "Standards and Recommended Practice, Design Parameters, Cross-Section," 1983.

708. Transportation Association of Canada, *Manual of Geometric Design Standards for Canadian Roads,* Ottawa, Ontario, Canada, 1986.

709. Transportation Research Board, "Cost and Safety Effectiveness of Highway Design Elements," NCHRPT 197, Washington, D.C., U.S.A., 1978.

710. Transportation Research Board, National Research Council, "Designing Safer Roads," *Practices for Resurfacing, Restoration and Rehabilitation,* Special Report 214, Washington, D.C., U.S.A., 1987.

711. Transportation Research Board, *Highway Capacity Manual,* Special Report 209, Washington, D.C., U.S.A., 1985, 1994.

712. Trapp, K. H., "Investigations about the Traffic Flow on Rural Roads," *Research Road Construction and Road Traffic Technique (Forschung Strassenbau und Strassenverkehrstechnik),* vol. 113, Minister of Transportation, Bonn, Germany, 1971.

713. Trapp, K. H., and F.-W. Oellers, "Road Characteristics and Driving Behavior on Two-Lane Rural Roads," *Research Road Construction and Road Traffic Technique (Forschung Strassenbau und Strassenverkehrstechnik),* vol. 176, Minister of Transportation, Bonn, Germany, 1974.

714. Trapp, K. H., and B. Kraus, "Measurement and Examination of the Influences: Longitudinal Grade Classes, Curvature Change Rate, Passing Possibility, and Lane Width in the Design Procedure (RAS-Q)," *Research Road Construction and Road Traffic Technique (Forschung Strassenbau und Strassenverkehrstechnik),* vol. 381, Minister of Transportation, Bonn, Germany, 1983.

715. Trapp, K. H., "The Influence of Auxiliary Lanes on the Traffic Flow at Upgrade Sections of Two-Lane Rural Roads," *Research Road Construction and Road Traffic Technique (Forschung Strassenbau und Strassenverkehrstechnik),* vol. 304, Minister of Transportation, Bonn, Germany, 1980.

716. Treat, I. R., N. S. Tumbas, S. T. McDonald, D. Shinar, R. D. Hume, R. E. Mayer, R. L. Stansiver, and N. J. Castellan, "TRI-Level Study of the Causes of Traffic Accidents, Executive Summary," Institute for Research in Public Safety, Indiana University, Bloomington, Ind., Final Report, Contract no. DAT HS-

034-3-535, prepared for U.S. Department of Transportation, National Highway Traffic Safety Administration, Washington, D.C., U.S.A., 1979.

717. Trinca, G. W., et al., "Reducing Traffic Injury—A Global Challenge," Royal Australasian College of Surgeons, Melbourne, Australia, 1988.

718. TRL (Transport and Road Research Laboratory), *Towards Safer Roads in Developing Countries, A Guide for Planners and Engineers,* 1st ed., Crowthorne, United Kingdom, 1991.

719. Tunnard, C., and B. Pushkarev, *Man-Made America: Chaos or Control?,* Yale University Press, New Haven, Connecticut, U.S.A., 1963.

720. Turner, D. S., and J. W. Hall, "Severity Indices for Roadside Features," NCHRP Synthesis 202, Transportation Research Board, National Research Council, Washington, D.C., U.S.A., 1994.

721. Undeutsch, U., *Results of Psychological Examinations on Accident Location,* West German Publishing House (Westdeutscher Verlag), Weisbaden, Germany, 1962.

722. Unified Motor Publishers, "Truck/Bus—Catalogues 1985 to 1996," Stuttgart, Germany, 1984 to 1995.

723. United Nations, "Statistics of Road Traffic Accidents in Europe," New York, U.S.A., editions up to 1995.

724. UPI, Environmental and Prognosis Institute, "Consequences of a Global Motorization," UPI—Report no. 35, Heidelberg, Germany, 1995.

725. Urbanik, T., W. Hinshaw, and D. Fambro, "Safety Effects of Limited Sight Distance on Crest Vertical Curves," *Transportation Research Record,* vol. 1208, U.S.A., 1989, pp. 23–35.

726. U.S. Department of Transportation, "Goals Beyond 1980," Interagency Task Force of Motor Vehicle, Washington, D.C., U.S.A., 1977.

727. U.S. Department of Transportation, National Highway Traffic Safety Administration, "The Effect of Motorcycle Helmet Use Repeal. A Case for Helmet Use," Technical Report: DOT-HS-805-312, Washington, D.C., U.S.A., 1980.

728. U.S. Department of Transportation, Federal Highway Administration, *Safety Effectiveness of Highway Design Features, Volume II: Alignment,* Publication no. FHWA-RD-91-045, Washington, D.C., U.S.A., 1992.

729. U.S. Department of Transportation, Federal Highway Administration, *Safety Effectiveness of Highway Design Features, Volume III: Cross Sections,* Publication no. FHWA-RD-91-046, Washington, D.C., U.S.A., 1992.

730. Vaegverket, Division Vaeg & Trafik, "Vagutformning 94, Del 6 Linsefoering," Publication 1994: 052, Borlaenge, Sweden, 1994.

731. Van Winsum, W., and H. Godthelp, "Speed Choice and Steering Behavior in Curve Driving," *Human Factors,* vol. 38, U.S.A., 1996, pp. 434–441.

732. Vasilev, A., cited by Babkov (Ref. 40).

733. Vaughan, R. G., K. Pettigrew, and J. Lukin, "Motorcycle Crashes: A Level Two Study," Traffic Accident Research Unit, Department of Motor Transport, New South Wales, Australia, 1977.

734. Venturino, M., "Time-Sharing and Mental Workload," selected readings in *Human Factors,* U.S.A., 1990, pp. 205–206.

735. Versace, J., "Factor Analysis of Roadway and Accident Data," *Highway Research Board Bulletin,* vol. 240, 1960, pp. 24–32.

736. Voigt, A. P., and R. A. Krammes, "Safety and Operational Evaluation of Alternative Curve Design Approaches for Rural Two-Lane Highways," International Symposium on Highway Geometric Design Practices, Transportation Research Board, Boston, Massachusetts, U.S.A., August 1995.

737. Wacker, J., "Accident Development at Elzer Mountain (Autobahn A 3) between Cologne and Frankfurt from 1961 to 1995," unpublished manuscript, Ministry of Economy and Technique, State of Hessen, Wiesbaden, Germany, 1996.

738. Wallentowitz, H., "Driver-Vehicle-Side Wind," Ph.D. Dissertation, Faculty for Mechanical Engineering and Electrotechnics at the Technical University Carolo-Wilhelmina, Braunschweig, Germany, 1979.

739. Waller, P. F., and L. I. Griffin, "The Impact of Motorcycle Lights on Law," *Proceedings of the American Association for Automotive Medicine,* Vancouver, Canada, 1977.

740. Wanderer, U., and H. Weber, "A Close Examination of Traffic Accidents—Accident Research on Accident Location," *VDI-News (VDI-Nachrichten),* vol. 6, Germany, 1976, pp. 4–5.

741. Watson, G. S., P. L. Zador, and A. Wilks, "The Repeal of Helmet Use Laws and Increased Motorcycle Mortality in the United States, 1975–1978," *American Journal of Public Health,* vol. 70, U.S.A., 1980.

742. Wegman, F. C. M., "Legislation, Regulation and Enforcement to Improve Road Safety in Developing Countries," Contribution to the World Bank Seminar on Road Safety, Washington, D.C., U.S.A., 1992.

743. Wegman, F. C. M., "Evolution of Road Accidents," Contribution to the Road Safety Policy Seminar for Central and Eastern Europe, Budapest, Hungary, October 17–21, 1994.

744. Wegman, F. C. M., "The Road Safety Phenomenon," Contribution to the Organization for Economic and Cultural Development Workshop on Infrastructure Design & Road Safety, Prague, Czech Republic, October 12–14, 1994.

745. Wegman, F. C. M., R. Roszbach, J. A. G. Mulder, C. C. Schoon, and F. Poppe, "Road Safety Impact Assessment. A Proposal for Tools and Procedures for a RIA," R-94-20, SWOV, Leidschendam, The Netherlands, 1994.

746. Wegman, F. C. M., "Road Accidents: Worldwide a Problem That Can Be Tackled Successfully!" Contribution to the PIARC Conference, Montreal, Canada, September 4–8, 1995.

747. Wehner, B., "Results of Skid-Resistance Measurements and Traffic Safety," *Road and Interstate* (*Strasse und Autobahn*), vol. 8, 1965, pp. 261–268.

748. Weise, G., and H.-G. Wiehler, *Road Construction,* vol. 1, VEB Publishers for Civil Engineering, Berlin, Germany, 1978.

749. *Welt am Sonntag* (*World on Sunday*), "U.S. Automobile Manufacturers Warn: Fatal Danger by Air Bags," no. 44, Germany, November 3, 1996, p. 2.

750. West, L. B., and J. W. Dunn, "Accidents, Speed Deviation and Speed Limits," *Traffic Engineering,* U.S.A., 1971.

751. Wierwille, W. W., and J. C. Gutmann, "Comparison of Primary and Secondary Task Measures as a Function of Simulated Vehicle Dynamics and Driving Conditions," *Human Factors,* vol. 20, U.S.A., 1978, pp. 233–244.

752. Wierwille, W. W., and L. Tijerina, "Presentation of the Relationship between the Visual Load of the Driver in the Vehicle and the Occurrence of an Accident," *Journal for Traffic Safety* (*Zeitschrift fuer Verkehrssicherheit*), vol. 2, Germany, 1997, pp. 67–74.

753. Wilde, G. J. S., "Risk Homeostasis Theory and Traffic Accidents; Propositions, Deductions and Discussion of Recent Commentaries," *Ergonomics,* vol. 31, Great Britain, 1988, pp. 441–468.

754. Wildervanck, C., G. Mulder, and J. A. Michon, "Mapping Mental Workload in Car Driving," *Ergonomics,* vol. 21, Great Britain, 1978, pp. 225–229.

755. Wiley, C. C., *Principles of Highway Engineering,* McGraw-Hill, New York, U.S.A., 1928.

756. Wilfert, K., "Development Possibilities in Automobile Construction," *Automobile Technical Journal* (*ATZ*), vol. 8, Germany, 1973, pp. 273–278.

757. Willeke, Boegel and Engels, *Possibilities of Economic Calculations in Road Construction with Special Emphasis on Accident Costs,* vol. 11, Research Report, Institute of Traffic Science, University of Cologne, Germany, 1967.

758. Williams, E. C., H. B. Skinner, and J. N. Young, "Emergency Escape Ramps for Runaway Heavy Vehicles," *Public Roads,* vol. 42, no. 4, 1979, pp. 142–147.

759. Wilson, F. R., J. A. Sinclair, and B. G. Bisson, "Evaluation of Driver/Vehicle Accident Reaction Times," Department of Civil Engineering, The Transportation Group, University of New Brunswick, Frederictown, New Brunswick, Canada, 1989.

760. Wilson, G. F., and R. D. O'Donnell, "Measuring of Operator Workload with the Neuropsychological Workload Test Battery," *Human Mental Workload,* North-Holland, Amsterdam, The Netherlands, 1988, pp. 63–100.

761. Wilson, T. D., "Road Safety by Design," *The Journal of the Institute of Highway Engineers,* vol. 15, 1968, pp. 23–33.

762. Winch, D. M., *The Economics of Highway Planning,* Canadian Studies in Economics, Toronto, Canada, 1963.

763. Winterhagen, J., "Friction Surveillance and Distance Regulation," *Automobile Technical Journal* (*ATZ*), vol. 97, no. 1, Germany, 1995, pp. 22–23.

764. Wong, S. Y., *Theory of Ground Vehicles,* 2d ed., John Wiley and Sons, New York, 1993.

765. World Road Congress, "Question III: Interurban Roads and Motorways," XVII World Congress, National Report Hungary, Sidney, Australia, 1983, pp. 152–154.

766. Wright, P. H., and K. K. Mak, "Single Vehicle Accident Relationship," *Traffic Engineering,* vol. 46, U.S.A., 1976, pp. 16–21.

767. Wright, P. H., and L. S. Robertson, "Priorities for Roadside Hazard Modification: A Study of 300 Fatal Roadside Object Crashes," Georgia Institute of Technology, Atlanta, Georgia, U.S.A., 1976.

768. Yager, M., and R. Van Aerde, "Geometric and Environmental Effects on Speeds of 2-Lane Highways," *Transportation Research,* vol. 17A, no. 4, Great Britain, 1983.

769. Yeh, Y., and C. D. Wickens, "Dissociation of Performance and Subjective Measures of Workload," *Human Factors,* vol. 30, U.S.A., 1988, pp. 111–120.

770. Young, J. C., "Can Safety Be Built Into Our Highways?," *Contractors and Engineers Monthly,* 1951.

771. Zaal, D., "Traffic Law Enforcement: A Review of the Literature," Report no. 53, Monash University Accident Research Centre, Melbourne, Australia, 1994.

772. Zador, P., H. Stein, J. Hall, and P. Wright, "Superelevation and Roadway Geometry: Deficiency at Crash Sites and on Grades (Abridgement)," Insurance Institute for Highway Safety, Washington, D.C., U.S.A., 1985.

773. Zador, P., H. Stein, J. Hall, and P. Wright, "Relationships between Vertical and Horizontal Roadway Alignments and the Incidence of Fatal Rollover Crashes in New Mexico and Georgia," *Transportation Research Record,* vol. 1111, Washington, D.C., U.S.A., 1987, pp. 27–41.

774. Zegeer, C. V., R. C. Deen, and J. G. Mayes, "The Effect of Lane and Shoulder Widths on Accident Reductions on Rural Two-Lane Roads," *Transportation Research Circular 806,* Washington, D.C., U.S.A., 1981.

775. Zegeer, C. V., and M. R. Parker Jr., "Cost-Effectiveness of Countermeasures for Utility Pole Accidents," Report no. FHWA/RD-83/063, Federal Highway Administration, Washington, D.C., U.S.A., 1983.

776. Zegeer, C. V., and M. J. Cynecki, "Determination of Cost-Effective Roadway Treatments for Utility Pole Accidents," *Transportation Research Record,* vol. 970, U.S.A., 1984, pp. 52–64.

777. Zegeer, C. V., and M. R. Parker Jr., "Effect of Traffic and Roadway Features on Utility Pole Accidents," *Transportation Research Record,* vol. 970, U.S.A., 1984, pp. 65–76.

778. Zegeer, C. V., and J. A. Deacon, "Effects of Lane Width, Shoulder Width, and Shoulder Type on Highway Safety, State-of-the-Art, Report 6: Relationship between Safety and Key Highway Features— A Synthesis of Prior Research," Transportation Research Board, Washington, D.C., U.S.A., 1987.

779. Zegeer, C. V., J. Hummer, D. Reinfurt, L. Herf, and W. Hunter, *Safety Effects of Cross-Section Design for Two-Lane Roads,* vols. I and II, Report FHWA-RD-87/008 and 009, Federal Highway Administration, U.S. Department of Transportation, Washington, D.C., U.S.A., 1987.

780. Zegeer, C. V., J. Hummer, D. Reinfurt, L. Herf, and W. Hunter, "Safety Cost-Effectiveness of Incremental Changes in Cross-Section Design—Informational Guide," Report no. FHWA/RD-87/094, Federal Highway Administration, Washington, D.C., U.S.A., 1987.

781. Zegeer, C. V., J. R. Stewart, F. M. Council, and D. W. Reinfurt, "Cost-Effective Geometric Improvements for Safety Upgrading of Horizontal Curves," Report no. FHWA-RD-90-021, Federal Highway Administration, Washington, D.C., U.S.A., 1991.

782. Zegeer, C. V., J. R. Stewart, F. M. Council, and D. W. Reinfurt, "Safety Effects of Geometric Improvements on Horizontal Curves," *Transportation Research Record,* vol. 1356, U.S.A., 1992, pp. 11–19.

783. Zegeer, C. V., R. Stewart, and T. R. Neumann, "Accident Relationships of Roadway Width on Low-Volume Roads," *Transportation Research Record,* vol. 1445, U.S.A., 1994, pp. 160–168.

784. Zeitlin, L. R., "Estimates of Driver Mental Workload: A Long-Term Field Trial of Two Subsidiary Tasks," *Human Factors,* vol. 37, U.S.A., 1995, pp. 611–621.

785. Zibuschka, F., "Influence of Truck-Traffic on Road and Environment as well as Methods for the Explicit Consideration in Traffic Planning," Informations of the Institute for Geotechnique and Traffic Engineering, vol. 7, University for Land Culture, Vienna, Austria, 1982.

786. Zuk, W., "Instability Analysis of a Vehicle Negotiating a Curve with Downgrade Superelevation," *Highway Research Record 390,* Highway Research Board, National Research Council, Washington, D.C., U.S.A., 1972, pp. 40–44.

ADDITIONAL REFERENCES

787. Knowles, W. B., "Operator Loading Tasks," *Human Factors,* vol. 5, U.S.A., 1963, pp. 155–161.

788. Senders, J. W., "The Estimation of Operator Workload in Complex Systems," *Systems Psychology,* McGraw-Hill, New York, U.S.A., 1970, pp. 207–216.

789. Wooldridge, M. D., "Design Consistency and Driver Error," *Transportation Research Record,* vol. 1445, U.S.A., 1994, pp. 148–155.

790. McLean, J. R., "Road Geometric Standards: Overseas Research and Practice," Research Report ARR no. 107, Australian Road Research Board, Victoria, Australia, 1980.

INDEX

PERSONAL INFORMATION

Ruediger Lamm (born in 1937)

1963:	Diploma in Civil Engineering from the University of Karlsruhe (Dipl.-Ing.), Germany
1967:	Ph.D. in Civil Engineering from the University of Karlsruhe (Dr.-Ing.), Germany
1973:	Habilitation in Road Engineering from the University of Karlsruhe (Dr.-Ing. habil.), Germany
1963-1977:	Scientific Assistant, Chief Engineer, Scientific Councilor, and Professor at the Institute for Highway and Railroad Transportation and Engineering, University of Karlsruhe, Germany
1978-1979:	Acting Dean (Sarparast), Engineering Faculty, University of Gilan, Rasht, Iran
1980:	Visiting Professor, The Ohio State University, Columbus, Ohio, U.S.A.
1983-1987:	Professor, Faculty of Civil and Environmental Engineering, Clarkson University, Potsdam, NY, U.S.A.
1978-Present:	University-Professor at the Institute for Highway and Railroad Transportation and Engineering, Faculty of Civil and Surveying Engineering, University of Karlsruhe, Germany

He has authored and co-authored more than 170 papers and research reports in the transportation field.

Basil Psarianos (born in 1954)

1976:	Diploma in Rural and Surveying Engineering from the National Technical University of Athens (Dipl.-Ing.), Greece
1981:	Ph.D. in Road Design from the University of Hannover (Dr.-Ing.), Germany
1984-1987:	Freelance Engineer
1988-1992:	Lecturer for Transportation Engineering at the National Technical University of Athens, Department of Rural and Surveying Engineering, Greece
1993-Present:	Assistant and Associate Professor for Transportation Engineering at the National Technical University of Athens, Department of Rural and Surveying Engineering, Greece

Theodor Mailaender (born in 1949)

1976:	Diploma in Civil Engineering from the Technical College of Darmstadt (Dipl.-Ing.), Germany
1976-1978:	Project Engineer at KWU-Siemens, Erlangen, Germany
1978-1986:	Railroad Engineer, Karlsruhe, Stuttgart, Germany
1986-Present:	President of Mailaender Ingenieur Consult, Karlsruhe, Leipzig, Germany

Elias M. Choueiri (born in 1957)

1979-1984:	Undergraduate and Graduate degrees in Mathematics, Mathematics and Computer Science, and Electrical Engineering from several U.S. colleges and universities
1985:	Master of Science in Civil and Environmental Engineering from Clarkson University, Potsdam, NY, U.S.A.
1987:	Ph.D. in Engineering Science from Clarkson University, Potsdam, NY, U.S.A.
Prior to 1994:	Teaching, Research positions at several U.S. universities and companies
1993-Present:	Director-General in the Ministry of Transport, Beirut, Lebanon
1996-Present:	Lecturer at Notre Dame University, Louaize, Lebanon, and at the National Institute for Management and Development, Beirut, Lebanon

Ralf Heger (born in 1966)

1993:	Diploma in Civil Engineering from the Dresden University of Technology (Dipl.-Ing.), Germany
1993:	Research Scholarship, Texas A&M University, College Station, Texas, U.S.A.
1994-Present:	Scientific Assistant at the Institute of Traffic Route Construction, Dresden University of Technology, Germany

Rico Steyer (born in 1966)

1993:	Diploma in Civil Engineering from the Dresden University of Technology (Dipl.-Ing.), Germany
1993:	Research Scholarship, Texas A&M University, College Station, Texas, U.S.A.
1994-Present:	Scientific Assistant at the Institute of Traffic Route Construction, Dresden University of Technology, Germany

John C. Hayward (born in 1947)

1969:	Bachelor of Science in Civil Engineering from Ohio University, Athens, Ohio, U.S.A.
1971:	Master of Science in Civil Engineering from the Pennsylvania State University, University Park, Pennsylvania, U.S.A.
1974:	Ph.D. in Civil Engineering from the Pennsylvania State University, University Park, Pennsylvania, U.S.A.
1995:	Advanced Management Program, Harvard Business School, Harvard University, Cambridge, Massachusetts, U.S.A.
1971-1974:	Research Assistant, Pennsylvania Transportation Institute, The Pennsylvania State University, University Park, Pennsylvania, U.S.A.

1974-Present: Michael Baker Corporation, Pittsburgh, Pennsylvania, U.S.A.

President of Michael Baker Jr., Inc. (Transportation Engineering) and President of Baker Heavy & Highway, Inc. (Transportation Construction)

Jeffrey A. Quay (born in 1951)

1976:   Bachelor of Science Degree in Civil Engineering from the Pennsylvania State University, University Park, Pennsylvania, U.S.A.

1977-1979:   Assistant Engineer, Dewberry & Davis, Fairfax, Virginia, U.S.A.

1979-1983:   Engineer, Michael Baker Jr., Inc., Pittsburgh, Pennsylvania, U.S.A.

1983-1986:   Engineer, Dewberry & Davis, Fairfax, Virginia, U.S.A.

1987-1990:   Senior Engineer, Fairfax Country Department of Public Works, Fairfax, Virginia, U.S.A.

1990-Present: Project Manager, Michael Baker Jr., Inc., Pittsburgh, Pennsylvania, U.S.A.

All Authors, Associate Authors, and Editors are involved, as researchers and practitioners, in the transportation fields: Highway Geometric Design, Traffic Safety, Driving Behavior, and Driving Dynamics.

ABOUT THE AUTHORS

RUEDIGER LAMM is a professor at the Institute for Highway and Railroad Engineering of the University of Karlsruhe (TH) in Germany, where he taught since 1969. He was acting Dean (Sarparast) of the Engineering faculty of the University of Gilan, Rasht, Iran, and full professor at the faculty of Civil and Environmental Engineering, Clarkson University, Potsdam, N.Y., U.S.A. He was a member of many professional societies on a national and international level. Dr.-Ing. habil. Lamm received the 1993 Best of Session Award, given to him by Committee on Geometric Design from Transportation Research Board, National Academy of Science. He has been active in continuing educational programs and has served as a consultant to transportation agencies, for example in Germany, Greece, and Japan. He has authored and co-authored more than 170 papers and research reports in the transportation field.

BASIL PSARIANOS is an associate professor at the National Technical University of Athens, Greece.

THEODOR MAILAENDER is president of Mailaender Ingenieur Consult, Karlsruhe, Leipzig, Germany.